W9-BGK-796

The Law of Occupational Safety and Health

About the Author

Gary Z. Nothstein, a graduate of The Johns Hopkins University, Ohio State University College of Law and Georgetown University Law School, is a partner in the Baltimore, Maryland and Washington, D.C. law firm of Venable, Baetjer and Howard. He has a national practice as a consultant to management on labor, occupational safety and health and environmental issues, and serves as a member of the OSHA Committee of the American Bar Association and the United States Chamber of Commerce OSHA Task Force.

In addition to teaching the required course on Occupational Safety and Health Law at The Johns Hopkins University, School of Hygiene and Public Health, Division of Environmental Health Engineering, Mr. Nothstein has published several articles in journals and periodicals and given numerous professional seminars and speeches on labor relations, environmental and safety and health topics.

The Law of Occupational Safety and Health

Gary Z. Nothstein
Partner
Venable, Baetjer & Howard
Baltimore, Md. and Washington, D.C.

THE FREE PRESS
A Division of Macmillan Publishing Co., Inc.
NEW YORK

Collier Macmillan Publishers
LONDON

THE FREE PRESS
A Division of Macmillan Publishing Co., Inc.
866 Third Avenue, New York, N.Y. 10022

Collier Macmillan Canada, Ltd.

Library of Congress Catalog Card Number: 81-67217

Printed in the United States of America

printing number

1 2 3 4 5 6 7 8 9 10

Library of Congress Cataloging in Publication Data

Nothstein, Gary Z.
The law of occupational safety and health.

Includes index.
1. Industrial safety—Law and legislation—United States. 2. Industrial hygiene—Law and legislation—United States. I. Title.

KF3570.N68 344.73′0465 81-67217
ISBN 0-02-923110-8 347.304465 AACR2

Acknowledgments

General thanks must go to the law firm of Venable, Baetjer, and Howard, which kept me on its payroll during the two years I labored over this book. Specific thanks are due to my colleagues Jeffrey P. Ayres, Bruce P. Martin, and Stephen P. Carney, who assisted me in the preparation of the manuscript and gave me helpful suggestions on its contents, format, and style. Thanks also to my magnificent, greatly overworked secretary, Carmalene Galaski, and to the wonderful wizard of Wang, Joyce Maddox, for their coordination, organization, and typing of the manuscript. Recognition is also due to Kathleen Astroth, the Wang overseer, Donna Schech, and Janice Sitarik, who all played a large role in the completion of the final manuscript through their efforts in getting the draft onto the Wang system, as did David Durfee, Ellen Smith, Cheryl Bilgin, and Janet Ayres who proofread the manuscript and prepared the tables of cases and statutes and the index.

Last but not least, I must acknowledge the contribution of my wife, Pat, and my children without whose help the book would have been finished six months earlier.

G.Z.N.

Summary of Contents

Table of Contents

Foreword

by
Morton G. Corn, Ph.D.

Professor and Director
Environmental Health Engineering
The Johns Hopkins University
School of Hygiene and Public Health

Former Assistant Secretary of Labor
for Occupational Safety and Health

The Occupational Safety and Health Act of 1970 is one of many statutes passed by U.S. Congress during the past fifteen years for the purpose of regulating the environment. In general, the statutes create and assign to administrative organizations the responsibility for generating and promulgating detailed regulations which affect regulatees. It is not commonly understood that regulations are the products of salaried federal employees earnestly attempting to interpret the wishes of Congress. An extensive legislative history is associated with each statute and serves to guide the bureaucracy when the regulatory issues are particularly complex.

It is inevitable that in the process of creating regulation, there will be those who will disagree with the interpretations of the regulators. The history of each and every one of the environmental statutes demonstrates the rapid increase of case law associated with the particular Act. It is against this background that one must view this book. The author has attempted to explain the Occupational Safety and Health Act of 1970 and to collate and interpret the substantial case law stemming therefrom. It is a particularly fitting and opportune time to undertake this effort because the Act has engendered perhaps the most litigious climate of any of the environmental statutes passed in recent years. The law discussed in this book reflects a progressively increasing occupational health and safety case load at virtually every level of the judiciary, including the Supreme Court.

It is the opinion of many professionals in the occupational safety and health field that we have, as a society, assigned too much responsibility to the courts. They would argue that the courts are establishing policy that should be assigned to the technologists and scientists knowledgeable in these fields. In my opinion, the issues are often too important to leave to the specialists. The issues reflect the changing currents of opinion and social policy in American society. One need only mention the issues of compliance officer right of entry to the workplace, protective measures for susceptible persons on the job, and the establishment of safe working levels for a carcinogenic substance, where current scientific specialists cannot recommend a threshold value for effects. These are issues which demand the wisdom of Solomon. Under these circumstances, I for one believe that the courts have performed remarkably well in the field of occupational safety and health.

There have been conservative and relatively active managements of the Occupational Safety and Health Administration. The standards and policies promulgated by the Agency reflect these leanings. The courts have served the purpose of evolving a body of regulation that is tempered by an objective third party, rather than reflecting the open or perhaps subconscious leanings of agency and administration administrators.

During the formative years of my professional education and apprenticeship as a researcher and then engineering practitioner, the law was hardly mentioned, and it was not necessary for me as an individual to be familiar with the law. Today, the young occupational physician, occupational lawyer, industrial hygienist, or safety engineer must be very familiar with the provisions of the Occupational Safety and Health Act and the regulations stemming therefrom. If he/she is to recommend to clients, be they public officials, management or labor, their responsibilities and duties under the Act, he/she must be familiar with agency promulgations and judicial interpretations. Mr. Nothstein's efforts in this book will provide a very welcome tool to both practitioners and students in the field of occupational safety and health. At the Johns Hopkins University School of Hygiene and Public Health, where we educate and train graduate students in occupational medicine, industrial hygiene, and industrial nursing, Mr. Nothstein presents the required course entitled Occupational Safety and Health Law. Thus, we are stating to the students that the subject matter of this book is an essential ingredient of your professional education.

This text will also be a very useful reference for the professional and the practitioner who seek answers to the "whys," "hows," and "wherefores" of occupational safety and health law, but who do not intend to utilize it in a systematic manner as a text. However, in my opinion, its greatest value will be to the student who desires a thorough, well presented understanding of current occupational safety and health law, and who approaches the book in a systematic manner.

The Law of Occupational Safety and Health

CHAPTER I
Inception and Overview

A. ORIGIN AND OBJECTIVE

In 1970, Congress calculated that there were more than 14,000 job-related deaths and an unknown number of illnesses and injuries resulting from or complicated by exposure to safety and health hazards, including toxic substances, physical stress and hazardous agents in the work environment. Of the nation's 79 million workers, 2,200,000 were disabled on the job and another 5,300,000 suffered lesser job-related illnesses. New cases of occupational diseases totaled 390,000. Lost employee work days totaled 250 million and over $1.5 billion was wasted in lost wages annually. When combined with additional costs in the form of medical expenses, insurance claims, production delays and damage to equipment or material, total losses were estimated by the National Safety Council at $9.3 billion, nearly 1 percent of the Gross National Product. Ashford, N.A., *Crisis in the Workplace* (1976). Congress decided something had to be done!

The Occupational Safety and Health Act of 1970, 29 U.S.C. 651 *et seq.* (the "Act")—the result of over twenty years of congressional debate, controversy, and concerted effort to pass job safety and health legislation—was enacted for the stated purpose of assuring safe and healthful working conditions for the estimated 80 million American workers in approximately 4 million places of employment. To accomplish this goal, Congress, in the Act, authorized the adoption, establishment, and enforcement of standards "to protect employees" from a variety of occupational hazards. In addition to recognizing the need for workplace standards, Congress established the Occupational Safety and Health Administration (OSHA) within the Department of Labor and charged it with the task of administration of the Act and enforcement of the standards which were promulgated. The National Institute for Occupational Safety and Health (NIOSH) was also established within the Department of Health, Education, and Welfare (Health and Human Services) to carry out safety- and health-related research activities. As so constituted, the Act has turned out to be one of the most far-reaching pieces of remedial social legislation ever enacted by the United States Congress.

Recent court decisions have emphasized that the Act, called the most revolutionary piece of labor legislation enacted since the National Labor Relations Act, is indeed remedial and preventive in nature, requiring a liberal construction and broad deference to OSHA's statutory authority. *REA Ex-*

press, Inc. v. Brennan, 495 F.2d 822, 1 OSHC 1651 (2d Cir. 1974); *Brennan v. OSHRC (Gerosa, Inc.),* 491 F.2d 1340, 1 OSHC 1523 (2d Cir. 1974). In *Whirlpool Corp. v. Marshall,* 435 U.S. 1, 8 OSHC 1001 (1980), the Supreme Court noted that:

> The Act in its preamble declares that its purpose and policy is "to assure so far as possible every working man and woman in the Nation safe and healthful working conditions and to preserve our human resource . . ." 29 U.S.C. 765(b).
>
> To accomplish this basic purpose, the legislation's remedial orientation is prophylactic in nature. *See Atlas Roofing Co. v. OSHRC,* 430 U.S. 442, 444–445. The Act does not wait for an employee to die or become injured. It authorizes the promulgation of health and safety standards and the issuance of citations in the hope that these will act to prevent deaths or injuries from ever occurring . . . . In the absence of some contrary indication in the legal history, it and regulations issued under the Act's authority are to be liberally construed to effectuate the congressional purpose. *United States v. Bacto-Unidisk,* 394 U.S. 784, 798; *Lilly v. Grand Truck R. Co.,* 317 U.S. 481, 486.

Other decisions have also noted that the remedial nature of the Act warrants a liberal interpretation of the authority to carry out its purpose of requiring employers to eliminate all foreseeable and preventable hazards. *Industrial Union Department, AFL-CIO v. American Petroleum Institute,* 440 U.S. 906, 8 OSHC 1586 (1980); *AFL-CIO v. Marshall,* 617 F.2d 636, 7 OSHC 1775 (D.C. Cir. 1979); *American Iron and Steel Institute v. OSHA,* 577 F.2d 825, 6 OSHC 1451 (2d Cir. 1978); *United Steelworkers of America, AFL-CIO-CLC v. Marshall,* —F.2d—, 8 OSHC 1810 (D.C. Cir. 1980).

B. HISTORICAL BACKGROUND

1. Early Job Safety and Health Programs

OSHA was not the first response to job-related injury and illness. While private concern over occupational safety and health dates back to the 1870s,[1] governmental involvement did not begin for another fifteen years.[2] The early federal legislative effort concentrated only upon job safety, *i.e.,* work-related accidents, injuries, and deaths, rather than upon occupational health and illness. Health problems were approached on the federal level through research and study programs under the auspices of various agencies.[3]

The states were somewhat more active in enacting job safety and health legislation. Between 1890 and 1920 nearly every state had enacted work safety (and sometimes health) laws. These state laws, however, were basically safety oriented, primarily cosmetic, tended to be poorly enforced, and usually did not provide substantive protection for workers, especially those in the agricultural industry. Moreover, the standards that were in existence varied widely from state to state.[4] An employer generally was not penalized for any violation of the laws but merely was asked to undertake abatement. Additionally, state inspection staffs were inadequate with respect to quality, quantity, and resources. Only a small percentage of workplaces was inspected each year; most of those inspections were limited to workplaces involving hazardous occupations, such as mining. Indeed, many states only legislated for safety and health in specific industries, again usually mining. What enforcement existed was frequently ineffective in reducing injuries, illnesses, and deaths. Thus the states, too, failed with respect to occupational safety and health, except that they usually did have some form of worker-compensation law.

The 1920s and 1930s brought a much greater involvement of the federal government in labor relations, but in the area of occupational safety and health state involvement was still largely preeminent.[5] The Labor Management Relations Act (LMRA),[6] which was passed in 1947, did contain a provision in Section 502 permitting employees to walk off a job that was "reasonably believed" to be abnormally dangerous. In 1948, President Truman initiated the first presidential conference on industrial safety, which continued through the Eisenhower administration. No real legislative advancements were seen, however, until the mid-1960s.[7] During this period, the costs of insurance claims and increased worker's compensation payments, together with lost production time, were presumed to be sufficient incentives for most employers to maintain a safe and healthful workplace. Greater productivity and safer working conditions were thought to go hand in hand, since fewer injuries meant more time on the job and more production. Safety programs featured the promotion of safe work habits, and accidents tended to be associated with worker carelessness. Meanwhile, radical changes were occurring in the nation's industrial technology, creating new types of health and safety hazards.

In the 1960s, Congress enacted the Contract Work Hours and Safety Standards Act,[8] the Construction Safety Act,[9] the Service Contract Act,[10] and the National Foundation on the Arts and Humanities Act,[11] all of which contained certain health and safety provisions but which were applicable only to a limited number of employees and industries. The Federal Metal and Nonmetallic Mine Safety Act[12] and the Federal Coal Mine Health and Safety Act [13] provided a further framework for dealing with safety and health problems in the mining industry. This legislation, however, was sporadic, covered only a very limited portion of the occupational safety and health field, and was not applicable to the majority of employees or employers. An occupational safety and health program was proposed by President Johnson in 1968 but was never enacted. In 1970 Congress enacted the Railroad Safety Act,[14] which contained some employee safety provisions even though it was designed primarily for passenger safety.

2. The Need for a New Safety and Health Law

Injury and illness statistics highlighted the concern and need for additional job safety and health legislation. The nation's work injury and illness rates in most industries were increasing throughout the 1960s and early 1970s.[15] According to organized labor, the major proponents of the new legislation, the existing laws did not effectively cover "in-the-plant" hazards.

The "relevant" testimony of these proponents during the congressional hearings which preceded the enactment of the Act is summarized in the following excerpt:

> The on-the-job health and safety crisis is the worst problem confronting American workers, because each year, as a result of their jobs, over 14,500 workers die. In only four years' time, as many people have died because of their employment as have been killed in almost a decade of American involvement in Vietnam. Over two million workers are disabled annually through job-related accidents.
>
> The economic impact of occupational accidents and diseases is overwhelming. Over $1.5 billion is wasted in lost wages, and the annual loss to the Gross National Product is over $8 billion. Ten times as many man-days are lost from job-related disabilities as from strikes, and days lost productivity through accidents and illnesses are ten times greater than the loss from strikes. [H. Rep. No. 1291, 91st Cong., 2d Sess. 14–15 (1970).]

Although it was true that on-the-job injury and death statistics were (and are) significant when compared to Vietnam war dead, work-related deaths are in fact fourth in frequency compared to other causes, and job-related disabling injuries are third, according to a 1975 compilation of the National Safety Council, see Table A in Appendix 1.

3. Legislative History

The 91st Congress was more receptive than prior congresses to occupational safety and health legislation, and early in its first session passed coal mine, construction, and railroad safety acts. On May 16, 1969, Senator Harrison Williams (D. N.J.) introduced a job safety bill in the Senate. A companion bill was also introduced in the House of Representatives by Congressman William Steiger (R. Wisc.)—hence, the sometime designation as the "Williams-Steiger Bill." Administration bills were also introduced. The early bills, although considerably less detailed than the subsequent bill that was actually enacted, contained many of the seeds of conflict to come.

Almost all of the conflicts which plagued the legislators divided along labor-management lines, with management favoring the administration approach and labor endorsing the Williams-Steiger version. The important and troublesome issues included provisions relating to the "general duty" requirement imposed upon employers, designation of the "agency'(s)" responsible for enforcement and adjudication, "walkaround" rights of employers and employees during inspection tours, plant closings in cases of "imminent danger," and citation and posting requirements. Fears that employers' rights of due process might be abrogated on the one hand, or that employees might be denied their right to protection or access to relevant information on the other, also hampered agreement in Congress.

Intense hearings were conducted by both the House and Senate Labor Subcommittees. A conference to iron out the differences between the Senate and House bills began in December 1970. Agreement was finally reached between the conferees on the general duty clause, imminent danger clause, and priorities for inspection, investigations, and recordkeeping. The Occupational Safety and Health Act of 1970 was ready for the President's signature and was signed into law on December 29, 1970, taking effect 120 days later. See, generally, Subcommittee on Labor of the Senate Committee on Labor and Public Welfare, Legislative History of the Occupational Safety and Health Act of 1970, 92d Cong., 1st sess.

4. Administrative Organization

In April 1971, the Act took effect, covering approximately fifty-five million employees throughout the country, applying to every employer who had one or more employees and whose business affected interstate commerce. 29 U.S.C. §651 *et seq.* The Act's standards-setting, investigatory, and enforcement powers were combined under the authority of the Secretary of Labor. To discharge this responsibility the Act provided him with an Assistant Secretary of Labor for Occupational Safety and Health who would head the Occupational Safety and Health Administration, known by its acronym OSHA, within the Department of Labor. The Act incorporated the Department of Labor's former safety and health responsibilities under the Walsh-Healey Public Contracts Act, the McNamara-O'Hara Service Contract Act, the Maritime Safety Act, the Construction Safety Act, and certain other statutes pertaining to federally financed or assisted projects.

Under the Act, the Assistant Secretary of Labor was given authority (1) to promulgate safety and health standards, (2) to conduct inspections and investigations, (3) to issue citations and propose penalties, (4) to set abatement times for correcting unsafe or unhealthful conditions, (5) to require employers to keep records of job-related injuries and illnesses, (6) to petition the courts to restrain imminent danger situations, (7) to approve or reject state plans for administering and enforcing state occupational safety and health programs under the Act, (8) to provide information and advice to employers and employees concerning compliance and effective means of preventing occupational injuries and illnesses, and (9) to provide evaluative, consultative, and promotional programs to assist federal agencies in implementing job safety and health programs for federal workers.

Under the Act's general "duties" section, employers were required not only to comply with the occupational safety and health standards promulgated under the Act but to furnish each employee a place of employment which is "free from recognized hazards that are causing or are likely to cause death or serious physical harm. . . ." Employees were also required to comply with standards, rules, and regulations issued pursuant to the Act which are applicable to their own actions and conduct. No sanctions were provided against employees, however, and the courts have emphasized the primacy of the employer's responsibility for assuring compliance because of its "control" over the workplace and over the employment relationship.

A key provision of the Act makes the employer liable for fines the first time his worksite is inspected if violations are disclosed, *i.e.*, the Act employs "first instance" sanctions to foster compliance. The employer is held accountable to comply with all of the Act's requirements *before* the inspector arrives —and advance notice of inspections is usually prohibited. Because OSHA compliance officers are required to issue a citation if they observe a violation, they cannot provide employers with on-site advice or assistance except as part of an inspection. In order to make fine-free, on-site assistance available to employers upon request, OSHA encourages states to maintain safety and health consulting staffs completely independent of their enforcement personnel under agreements by which OSHA reimburses 90 percent of the states' program costs. OSHA compliance officers are permitted to provide advisory opinions *off-site,* if requested.

Several new "entities" were established by the Act to carry out the main functions of the statute: the Occupational Safety and Health Administration, the National Institute for Occupational Safety and Health, and the Occupational Safety and Health Review Commission.

(1) *The Occupational Safety and Health Administration.* The position of Assistant Secretary of Labor for Occupational Safety and Health, the chief decision-maker under the Act, was created to carry out most of the Secretary of Labor's responsibilities under the Act, and to head the Occupational Safety and Health Administration. Organization of OSHA was coordinated on three levels, a national office, ten regional offices, and fifty-four area offices. The National Office in Washington, D.C., contains six directorates (health standards programs, administrative programs, federal compliance and state programs, training, education consultation and federal agency programs, and technical support) and several special offices (toxic substances, physical agents, carcinogen standards, safety standards, electrical engineering and electronic standards, mechanical engineering standards, construction and civil engineering safety standards, maritime safety standards, environmental, inflationary, and economic impact, engineering feasibility and technical assistance, and variance determination). OSHA personnel are also assigned to each of the ten geographical regions. Each region contains several area and district offices and field stations.

One of the most important local OSHA officials is the Area Director. Among the responsibilities of the Area Director is the answering of local information requests, handling employee complaints, scheduling and conducting inspections, issuance of citations and proposed penalties, and conducting of informal conferences.

The Solicitor of Labor represents the Secretary in all OSHA litigation matters, arguing cases before the Occupational Safety and Health Review Commission, appellate courts, district courts, and in other civil litigation. The Solicitor's office also acts informally in the area of legal interpretations and advises the Assistant Secretary on important decisions and policies.

OSHA is responsible for all aspects of the Act's administration and enforcement including the promulgation of standards and regulations, overseeing state plans, and providing programs for training employers and employees. The primary responsibility for direct enforcement of the Act rests with the Compliance Safety and Health Officers, (CSHOs). The main job of these inspectors is to conduct inspections of workplaces in order to determine whether employers are in compliance with the Act. In those inspections that deal with health aspects of the Act, industrial hygienists may accompany or themselves do the inspection.

Section 8(g)(2) of the Act authorizes the Secretary of Labor to prescribe rules and regulations necessary to the administration of the Act. OSHA has issued regulations covering a variety of subjects, including inspections, citations, penalties, recordkeeping, variances, state plans, coverage of employers, access to medical and exposure records, and discrimination complaints. These regulations are contained in 29 CFR 1903—1977.

The Secretary of Labor is required to follow OSHA regulations; failure to do so may result in the vacation of a citation. *Atlantic Marine, Inc. v. Dunlop*, 524 F.2d 476, 3 OSHC 1755 (5th Cir. 1975), *citing Accu-Namics, Inc. v. OSHRC*, 515 F.2d 828, 3 OSHC 1299 (5th Cir. 1975) *cert. den.* 425 U.S. 903 (1976). Regulations, however, to be valid, must be reasonably related to the purposes of the Act and conform to the procedural requirements prior to issuance. *Marshall v. Daniel Construction Co.*, 563 F.2d 707, 6 OSHC 1031 (5th Cir. 1977); *Chamber of Commerce of the United States v. OSHA*, 636 F.2d 464, 8 OSHC 1648 (D.C. Cir. 1980).

The effectiveness of the Act has been influenced not only by the formal rules, regulations, and standards promulgated by OSHA, but also by the issuance of informal policies, procedures, and instructions. OSHA's operating procedures are set forth in a Field Operations Manual (FOM) which contains priorities, procedures, and guidelines for the conduct of inspections, citation preparation, penalty assessment, and other subjects. A companion Industrial Hygiene Manual (IHM) gives similar information for health inspections. The Field Operations Manual is neither a standard nor a substantive rule. It is an agency policy and, accordingly, it has been held that OSHA is not required to follow its own rule-making procedures or those of the Administrative Procedure Act in its promulgation. See *Limbach Co.*, 6 OSHC 1244 (1977).

(2) *The National Institute for Occupational Safety and Health.* The former Bureau of Occupational Safety and Health, within the Department of Health, Education, and Welfare (HEW) was redesignated the National Institute for Occupational Safety and Health (NIOSH), and assigned additional responsibilities for research and experimental programs, industry studies, development of criteria on exposure levels for toxic materials and harmful physical agents, publication at least annually of a list of all known toxic substances (including concentrations at which toxicity is known to occur) and, directly or by grants, assurance of an adequate supply of qualified personnel to carry out the purposes of the Act. The primary responsibilities of NIOSH are to develop and establish recommended safety and health standards, to conduct research experiments and demonstrations related to occupational safety and health, and to conduct education and training programs to provide a needed supply of personnel to carry out the Act's purposes. NIOSH also offers sev-

eral types of technical services to employers and employees (hazard evaluation, technical assistance, accident prevention, industrial hygiene, and medical service). NIOSH regulations are contained in 42 CFR 80, 84–87.

(3) *The Occupational Safety and Health Review Commission.* Adjudicatory authority for enforcement of the Act was assigned to a three-member Occupational Safety and Health Review Commission (the "Commission" or the "Review Commission"), an agency completely independent of the Department of Labor. The Review Commission is an independent agency created by Section 12 of the Act and is a body exclusively exercising adjudicative functions, notwithstanding its placement in the executive branch rather than in the judicial branch. The Act provides that the Commission shall be composed of three members, each of whom is appointed by the President for a six-year term. The Chairman of the Commission is empowered to appoint Review Commission Judges, who hold terms of career tenure, and who hear all cases entered on the Commission docket, rendering the final decision in better than 90 percent of them. Employers are allowed fifteen working days after issuance to contest citations, proposed penalties, and abatement orders issued by the Secretary of Labor; employees or their representatives are allowed fifteen working days to contest the period of time fixed in the citation for the abatement of alleged violations. After an adversary hearing, the Commission issues an order affirming, modifying, or vacating the Secretary's citation or proposed penalty, or directing other appropriate relief. Commission decisions may be further appealed to the courts. Uncontested citations and notices are deemed final orders of the Commission and are not subject to further review. Review Commission procedures are contained in 29 CFR 2200–2400.

(4) *Advisory Committees.* In addition to these agencies, the Act also established a twelve-member National Advisory Committee on Occupational Safety and Health (NACOSH), composed of representatives of management, labor, the occupational safety and health professions, and the public. NACOSH meets at least twice each year to "advise, consult with, and make recommendations to" the Secretary of Labor and the Secretary of Health, Education, and Welfare on matters relating to the administration of the Act.

The Secretary of Labor is also authorized to appoint advisory committees with as many as fifteen members to assist him in his standard-setting functions, and has done so on several occasions (*e.g.*, construction safety, coke oven emissions, cutaneous hazards). A Federal Advisory Council on Occupational Safety and Health (FACOSH), composed of representatives of federal agencies and of labor organizations representing federal employees, has been established by executive order to advise the Secretary of Labor on matters relating to the occupational safety and health programs of federal agencies. The Act also established a temporary commission, now expired, to conduct a two-year study of state worker's compensation laws.

(5) *Bureau of Labor Statistics.* Finally, the responsibility for conducting statistical surveys and collecting injury and illness data, primarily through the

Act's recordkeeping and reporting requirements, is vested in the Bureau of Labor Statistics (BLS).

There are important limitations to the Act's coverage. To the extent that the working conditions of employees are addressed by enforceable regulations issued under the statutory authority of another federal agency, those regulations apply rather than OSHA's. The Act also does not apply to employees of state and local governments. However, to the extent permitted by state law, state occupational safety and health plans must cover state and local government employees in order to receive federal matching grants.

Regarding federal employees, the Act assigned the Secretary of Labor only advisory and evaluative functions. It is the responsibility of the head of each federal agency "to establish and maintain an effective and comprehensive occupational safety and health program which is consistent with OSHA standards." However, since there are no sanctions available to force compliance with OSHA rules, many persons, including labor organizations representing federal employees, have considered this situation to represent a double standard. To correct the situation, three presidents have published executive orders that put occupational safety and health programs of federal agencies under OSHA's jurisdiction.

5. Legislative Involvement Continues

Subsequent to the Act's enactment, the Toxic Substances Control Act[16] (TSCA)—successor to the Federal Insecticide, Fungicide, and Rodenticide Act[17] (FIFRA)—and the Federal Mine Safety and Health Act of 1977[18] (MSHA) have been enacted. These statutes also generally relate to safety and health. TSCA gives the Environmental Protection Agency (EPA) broad authority to regulate the chemical industry, while MSHA, under the authority of the Assistant Secretary of Labor for Mine Safety and Health in the Department of Labor, regulates both the metallic and nonmetallic mining industries (including quarries and open pits). MSHA, a more inspection-intensive statute than OSHA, may overlap jurisdictionally with OSHA in the area of mining-related surface construction projects. TSCA may help OSHA with the task of regulating toxic and harmful substances in the workplace because of restrictions placed thereon prior to entry into the workplace. In addition, the effect of the Resource Conservation and Recovery Act[19] (RCRA) has yet to be determined.

C. STATUTORY PURPOSE

The preamble of the Act, as well as its Section 2(b), states that the Congressional "purpose and policy" is "to assure as far as possible . . . safe and healthful working conditions." Thirteen ways are listed in which the Act seeks to achieve this purpose, including research and development, training, recordkeeping, standards development, and employer and employee responsibilities. While OSHA continually reviews and redefines specific standards, practices, priorities, and approaches, this basic purpose remains constant. In the opinion of Timothy Cleary, Chairman of the Review Commission, the Act's reason for being is the *abatement* of hazards. *American Cyanamid Co.* 8 OSHC 1346 (1980). See, *e.g., Brennan v. OSHRC and Kesler & Sons Constr. Co.,* 513 F.2d 553, 2 OSHC 1668 (10th Cir. 1975); *Dunlop v. Haybuster Mfg.,* 524 F.2d 222, 3 OSHC 1594 (8th Cir. 1975). However, other Commission members and the courts feel that the primary purpose of the Act is the prevention of workplace hazards and resulting injuries. *Farmers Export Co.,* 8 OSHC 1661 (1980). See, *e.g., Bristol Steel & Iron Works, Inc. v. OSHRC,* 601 F.2d 717, 7 OSHC 1462 (4th Cir. 1979); *B & B Insulation, Inc. v. OSHRC,* 583 F.2d 1364, 6 OSHC 2062 (5th Cir. 1978); *Marshall v. Western Electric, Inc.,* 565 F.2d 240, 5 OSHC 2054 (2d Cir. 1977); *Arkansas-Best Freight Systems, Inc. v. OSHRC,* 529 F.2d 649, 3 OSHC 1910 (8th Cir. 1976); *Lee Way Motor Freight, Inc. v. Secretary,* 511 F.2d 864, 2 OSHC 1609 (10th Cir. 1975); *National Realty and Construction Co., Inc. v. OSHRC,* 489 F.2d 1257, 1 OSHC 1422 (D.C. Cir. 1973); *J. M. Martinac Shipbuilding Co. v. Marshall,* 614 F.2d 776, 7 OSHC 2120 (9th Cir. 1980).

OSHA expects to accomplish its goal of assuring safe and healthful working conditions by making noncompliance expensive, both by virtue of the assessment of civil penalties and the imposition of corrective abatement. For it has long been recognized that the civil enforcement provisions of the Act were designed to obtain maximum voluntary compliance with the Act's requirements. *George Hyman Construction Co.,* 5 OSHC 1320 (1977) *aff'd,* 582 F.2d 834, 6 OSHC 1855 (4th Cir. 1978); *Crescent Wharf and Warehouse Company,* 1 OSHC 1219 (1973). Focusing solely on the abatement of the particular hazard cited while ignoring the impact on future voluntary compliance seems contrary to the Act's enforcement scheme and to the continual maintenance of a safe and healthful workplace.

The reason for OSHA's enforcement approach, according to organized labor, is that "industry is interested only in short-term profits and is entirely unfeeling of the problems of the working man and woman. Business responds only when it is hit in the pocketbook." Unfortunately, the past conduct of the corporate world may have led to this characterization, and it continues to haunt and hamper legitimate reform efforts.

Convenience and production are the main reasons usually given by business for not taking immediate action to eliminate a safety or health hazard in the workplace. When the risk of inspection, penalties, and forced compliance under the Act seems slight, the benefits of production and convenience and thus noncompliance may seem to outweigh the cost of immediate abatement of a hazard. But this is not normally the case. In states where the Act is operating, inspections are frequent. In those states which administer their own plans, there are far more inspectors to get the job done, and that means just four things: more inspections, more citations, more penalties, and more abatement. And, for companies in high-hazard or target industries, OSHA's move to concentrate on serious hazards means even more attention from federal and state inspectors.

In addition ther are "hidden" costs associated with the existence of job safety and health hazards, for it goes without saying that an operation where hazards exist is an open invitation for accidents to happen. Any operation that has unnecessary accidents is bound to be less efficient than it could be. Hazards and the accidents they cause are a drain on productivity and efficiency. Employees injured on the job also are able to collect worker's compensation benefits for which the employer pays the premium. As a matter of fact, employees do not even need an accident or unsafe condition or hazard in the workplace to collect—just an injury or illness and proof of compensable loss.

Another hidden expense to be considered is that of potential civil liability through negligence or contract actions. In effect, federal and state job safety and health standards establish a minimum level of safety effectiveness which employers must meet. While employees may not be able to sue their employer directly under OSHA, they may be able to point to these standards and the enforcement action taken as evidence that the employer has not done all that it should where safety is concerned. In some states, employee suits against employers *are* possible, and large awards have been made. In other states, an "exclusive remedy" clause in worker's compensation legislation makes these suits impossible. Complicating matters, however, is the employee option to go after the manufacturer of the equipment, material, or substance involved in the accident. Thereafter, nothing stops the manufacturer from joining the employer in an action for "contributory negligence." Finally, safety and health issues are an increasingly important focal point in the labor relations, collective bargaining, and equal employment opportunity contexts.

It is clear that not only OSHA but other "side effects" of OSHA make noncompliance expensive, providing an incentive for compliance and the accomplishment of the congressional goal of assuring safe and healthful working conditions.

D. BASIC AIMS AND APPROACHES

Under provisions of the Act, the Occupational Safety and Health Administration was created (1) to encourage employers and employees to reduce hazards in the workplace and to implement new or improve existing safety and health programs, (2) to establish "separate but dependent responsibilities and rights" for employers and employees for the achievement of better safety and health conditions, (3) to establish reporting and recordkeeping procedures to monitor job related injuries and illnesses, (4) to develop mandatory job safety and health standards and enforce them effectively, and (5) to encourage the states to assume the fullest responsibility for establishing and administering their own occupational safety and health programs, which must be "at least as effective as" the federal program.[20]

From the beginning, however, OSHA has been surrounded by controversy. The Act was a compromise between conflicting interests, with far-reaching implications in many areas. For example, no one could with certainty anticipate the impact of adopting as mandatory standards under the Act the massive volume of consensus and proprietary standards formulated by various standard-writing organizations as voluntary models for one or more sectors of industry. Nor could one anticipate the implications of language which called upon the Secretary of Labor to promulgate the standard which is "*reasonably necessary*" and assures the "*greatest protection*" of affected employees, and to promulgate the health standard which "*most adequately assures,*" to the extent "*feasible,*" on the basis of the "*best available evidence*" that no employee will suffer "*material impairment*" of health or functional capacity even if such employee has regular exposure to a hazard "*for the period of his working life.*"

Other unknowns concerned the impact of procedures with respect to inspections, discrimination, and imminent danger, the definition of a recognized hazard, the use of the general-duty clause, the likely costs which employers might sustain for payment of penalties and correction of violations, the increasing use of criminal sanctions, and the comparative costs and benefits of the Act to industry and the nation overall.

The need for safety standards was generally accepted, for it was because employers did not voluntarily provide safe work places that OSHA was originally created. But standards relating to the long-term health effects of occupational exposure were thought to lead to a morass. The number of unique substances for which toxic effects information might be available was estimated at approximately 100,000. Moreover, legislative reports repeatedly cited an estimate by the Public Health Service that a new, potentially toxic chemical was introduced into industry every twenty minutes. Requiring industry to take the measures which might be necessary to prevent worker exposure to industrial toxicants was thought to involve enormous costs threatening to close plants and destroy jobs. Just admitting the culpability of certain chemicals and toxic substances for long-term, irreversible, occupational disease might have a traumatic impact on affected workers, and could lead to monumental liability suits and compensation claims. Official cognizance of the relationship between such exposure and occupational disease might result in demands for major new benefit programs similar to the "black lung" program for coal miners, to provide compensation payments and medical services. However, despite these concerns OSHA in the past few years has promulgated comprehensive health standards regulating occupational exposure to vinyl chloride, lead, arsenic, coke oven emissions, cotton dust, and carcinogens.

Imminent-danger procedures also posed a potentially serious problem; for employees and unions could use allegations of imminent danger as a potent bargaining lever. Under the guise of protecting workers, they could either force acceptance of their demands or shut down production. In fact, so threatening was this possibility to administration officials that at a Washington conference of field personnel, an imminent-danger "situation" was narrowly defined as comparable to the situation of "a loaded gun pointed at your head, *after* you see the smoke." The first few years of the Act's existence under the direction of Assistant Secretary George Guenther were spent floundering in an attempt to flush out these problems. Thereafter OSHA achieved a level of credence and was tightened both intellectually and spiritually with the appointment of Dr. Morton G. Corn as Assistant Secretary. OSHA became "results" oriented. However Dr. Corn's tenure during Republican administrations was criticized for not having an organized-labor viewpoint. In the words of Secretary of Labor Ray Marshall after he assumed office: "The tangled history of [OSHA's] first six years illustrates what happens when people are asked to enforce legislation they don't believe in."

Since her appointment in 1977 as Assistant Secretary of Labor for Occupational Safety and Health, Dr. Eula Bingham, who considered organized labor to be her constituency, initiated what has been called a shift to "commonsense priorities" in applying the Act's requirements, including (1) focusing resources on high-risk industries and workplace hazards likely to cause death, serious injury, or irreversible bodily harm, (2) increasing the availability of on-site consultation assistance to employers through state agencies, (3) establishing the New Direction's program for channeling money into training and education; and (4) revoking standards which are not needed, simplifying standards which are unnecessarily complex, and setting performance standards wherever possible. The effect of this shift, however, has been more apparent than real for OSHA still lacks a comprehensive and regulatory strategy conducive to the development of responsive safety and health programs in industry. This is not to say, however, that the shift in priorities has been without any benefit. For example, discretionary or "programmed" inspections are being targeted in high-risk industries such as heavy construction, manufacturing, petrochemicals, grain elevators, and foundries. Priority has also been assigned to standards for health hazards that may cause cancer, leukemia, nerve damage, lung disease, and other irreversible illnesses. On-site consultation services are available upon request to employers in virtually every state through 90 percent federal funding of state consultants, or individual consulting organizations under contract to OSHA.

In addition, OSHA is cooperating closely with other regulatory agencies —principally the Environmental Protection Agency (EPA), the Food and Drug Administration (FDA), the Food Safety and Quality Service (FSQS) of the Department of Agriculture, the Equal Employment Opportunity Commission (EEOC), the Consumer Products Safety Commission (CPSC), the Office of Federal Contract Compliance Programs (OFCCP), and the National Labor Relations Board (NLRB)—to solve problems common thereto in order to achieve the following goals and aims.

1. Social Goals and Aims

Section 2(b) of the Act speaks of the protection of "every working man and woman." This has been interpreted by OSHA to mean that employers must protect *all* workers from *all actual and potential* occupational safety and health hazards, and provide safe and healthful working conditions in *every* workplace. According to the most recent OSHA interpretations this protection would even extend to the unborn potential worker. This is, of course, a cradle-to-grave, strict-liability concept which goes beyond the present and legislates for the hazard-free, risk-free working environment of the future. *See American Cyanamid Co.* C.A. No. 79–5762 (1980).

It would seem, however, that there should be room in this concept for the commonsense approach which recognizes the doctrines of reasonableness, assumption of risk, cost-benefit analysis, contributory negligence, and employee misconduct. For example, virtually every kind of human activity involves some degree of risk. Some of the risks which people incur daily are partly or completely *voluntary* (*e.g.*, smoking). Many other risks that people incur are *involuntary*. People seem to accept a higher level of voluntary risk than involuntary risk. Should not this be factored into any consideration of responsibility under the Act? For whether a benefit is worth an involuntary risk is a determination that obviously cannot be made individually in each situation. It is a societal risk-benefit decision that can and should be made by public officials who have attempted to determine the collective will of the community. The reverse consideration must also be true. If a risk is voluntarily taken, then the individual and not the societal will should prevail. Recently, the Health and Safety Director of the Oil, Chemical, and Atomic Workers' International Union confirmed this when he stated that "industry in the next decade will present workers with information on workplace health hazards and leave it to them to make the choice of whether to take a certain job."

Issues of risk assessment and cost benefit must be addressed for it is essential that there be an identified hazardous condition before OSHA promulgates a regulation restricting exposure to such condition. Moreover, it is not the intent of the Act to require an environment free from any risk. The risk must be significant. It must be recognized that the resources available to protect life in any society are finite. Responsible administration of law and policy must accept the burden of addressing the broader economic and social impacts of each regulatory action.

Finally, should not the development of and compliance with performance standards be emphasized rather than specification standards? (A specification standard "specifies" the type of material, equipment, machinery, or facility that must be used to achieve compliance, *e.g.* ladder rungs must be thirteen inches apart.) Greater emphasis should be placed on a programmatic approach to compliance, focusing on the results to be achieved rather than focusing on the specific means and methods to attain such achievement. The objective must not be to specify how to comply, but to assure that the goal itself is identified and achieved. OSHA should focus on that objective and not limit the means to achieve it. Moreover, since OSHA can inspect no more than 2 percent of the nation's workplaces in any given year, focusing on results will achieve the greatest improvement in employee protection. Is not this the sensible approach when one considers that OSHA's ultimate goal in enforcement of the Act is the attainment of "results" (protection of employees and reduction of workplace injuries, illness, and death) rather than "means" or "methods"?

2. Economic Goals and Aims

Reduction of costs of operation and achievement of more efficient production are sensible business goals, but at times these may be contrary to the philosophy and approach of OSHA with its high-cost, technology-forcing standards and regulations. The Act makes it clear that its goal is to make the workplace as safe as possible. Cost is not mentioned. Is it irrelevant, therefore, that compliance may increase costs of operation and lessen the efficiency of production? Can a price tag be placed on the cost to industry of accidents, injuries, and illnesses? Is it safety and health at *any* cost? It seems that there should be a relationship of the goals and aims of the Act to the concepts of economic feasibility; that is, some sort of a cost-benefit analysis. It also seems, however, that the philosophy behind the new agencies (including OSHA) precludes such a relationship. Their view of their statutory mandate is to cut across industry lines to reduce risks sharply, often to the maximum extent technologically achievable, barring *no* expense. It is this regulatory approach, labeled "zero risk" by its critics, that has come under so much recent fire from economists and businessmen. For the spirit of reform that prevailed in Congress when the Act was passed did not leave room for cost considerations.

Health and safety at any cost carried to its logical conclusion may, however, have socially unacceptable economic implications, "since industries with high accident, injury, and illness risks and rates might not be able to operate." In *"What Price Safety? The Zero-Risk Debate" (Dun's Review,*[21] September 1979), it was stated that "a spirited and intense debate has erupted at the highest levels of government over the degree of safety that society can afford to offer its members." Initiated by economists who warn that the huge cost of complying with government health, safety, and environment standards threatens the economic well-being of the country, this controversy is forcing a fundamental reevaluation of the philosophies behind many of the current domestic programs. Eventually it could lead to a dramatic change in federal policy. At the center of the dispute is a relatively new school of economic thought that holds that physical risk to life from various hazards can be quantified as any other economic unit, and that by using complex mathematical formulas known collectively as *risk-benefit* analyses, an accurate determination can be made as to whether a given safety program is worth the money it costs. Ultimately, its advocates readily concede, such analyses may well lead to an unofficial monetary value being assigned to human life. Predictably, such ideas are generating heavy opposition. Environmental groups, consumer advocates, and labor unions all see the sudden popularity of the new theories as a dangerous trend.

At least one commentator has suggested the advantages of an "injury-tax" system approach to the problem of reducing industrial accidents, injuries, and illnesses, under which the present standards approach should be replaced and employers should be required to pay a fine or monetary penalty for each injury that occurs on the job. See Smith, R.S., *The Feasibility of an Injury Tax Approach to Occupational Safety,* 38 Law and Contemporary Problems 669, Duke University, 1974; *see also* Mendeloff, J., *Regulating Safety: An Economic and Political Analysis of Occupational Safety and Health* (1979).

Conventional wisdom says that safety pays. The report of President Carter's Interagency Task Force on Workplace Safety and Health, entitled "Making Prevention Pay," has found that it often does not, at least in monetary terms. In many industries, the average cost of worker injuries is sufficiently small that management focuses on other cost problems first. The Interagency Task Force report recommended a program of "incentives" to induce employers to comply with federal health and safety standards and also recommended that a more "cost-effective" approach be adopted for reducing workplace injuries and health hazards.

The recommendations are designed to make prevention pay by increasing the effective cost of injuries to high-rate firms and rewarding their low-rate competitors. They would also increase direct financial assistance to high-rate firms in hazardous industries to help them take preventive actions and avoid these higher injury costs.

At the same time, better management of field (regional and area office) enforcement and compliance programs must be required. OSHA should also redirect its enforcement program to target individual firms with known or suspected poor safety and health performance. This is where the serious problems really occur. In doing so, OSHA must consider the concerns of small businesses and provide help and assistance to them, for the management of these businesses are in the most need of support and assistance in correcting safety and health problems. OSHA should also, due to its limited resources, consider the elimination of first-instance penalties unless a serious violation has occurred.

Executive Order 12044 published in 43 Fed. Reg. 12661 (Mar. 24, 1978) requires agencies, such as OSHA, to consider compliance costs, define regulatory need in terms of costs and benefits, and consider less costly alternatives in the development of rules, regulations, and standards, but it does not actually require the agencies to act in accord with the findings in these areas.[22] Such an effort was however, recently undertaken by OSHA with respect to the fire protection standard. 45 Fed. Reg. 60656 (Sept. 12, 1980). A definite commitment to these and related considerations should be required and could possibly be attained through the issuance of an additional Executive Order which would require that the regulator *must* make his decision based on a coordinated review of these interrelated factors, explain to the public how the various factors affect the final decision and consider and apply a risk/cost benefit analysis in both a qualitative and quantitative manner.

3. Technological Goals and Aims

Recently, the social–risk-benefit theory of the Act has been questioned. Although proponents of the theory were aware that continued technological progress was costly, they believed that such costs were reasonable when compared to the benefits. But in the past few years this social understanding has been questioned by some economists. One of them, William D. Rowe, Director of the Institute of Risk Assessment at American University, contends that the rapid growth in technology, the discovery of unforeseen hazards of technology, and growing cynicism about institutions in general have combined, with dangerous results. Not only has the public come to doubt the ability of technology to improve the quality of life, says Rowe, but it also questions the ability of acientists to control technology. The result, he says, has been a wave of "technology anxiety" that borders on anti-risk hysteria.

Should, therefore, good-faith attempts at compliance with the Act in accordance with existing technology be required, or should OSHA be permitted to regulate the quality of life and require the implementation of novel, futuristic, unforeseeable, innovative, and expensive engineering, administrative, or work practice controls in an attempt to further reduce risk and exposure? The answer seems to be yes. According to OSHA, the Act was intended to go beyond the workplace of the present and to consider the working environment of the future, requiring employers to research and develop improvements in existing technology which loom on the horizon. If development and institution of engineering controls will achieve a significant improvement or substantial reduction in employee exposure, must they be implemented even though the resulting exposure levels would still exceed the specified limits in the standards? What if technology achieves only a limited improvement or slight reduction at substantial cost and does not bring an employer into complete compliance with a standard? Must it nevertheless be instituted? Again the answers seem to be yes. OSHA's paternalistic attitude—which considers engineering controls the primary means for achieving compliance and thus preferable to the use of personal protective equipment, since the former are not as subject to the human element as are the latter—may restrict the use of cleaner, more effective means of compliance. Increased use of performance criteria can protect the worker and at the same time allow flexibility and innovation in the methods of protection from identified hazards. Increased innovation in the application of technological and economic resources would reduce compliance costs and permit employers to better use their limited resources while assuring worker protection in the best manner possible.

Along this line, OSHA should shift from a compliance standard emphasis to one which focuses on results and employer motivation, *i.e.*, performance rather than specification targeting. OSHA's enforcement/targeting program must also be redirected to concentrate on high injury rate workplaces that have significant problems from a performance standpoint. Either of two approaches can be taken in this respect: (1) define a safe employer and provide an exemption mechanism for such employers; or (2) target enforcement upon those employers with high incidence of injury and illness rates. Either approach would give recognition to employers with effective safety and health programs, promote voluntary compliance, and at the same time permit innovation in the employment of technological resources.

E. EFFECTIVENESS ASSESSMENT

After almost eight years in existence, there is no documentation that OSHA has had any measurable impact on reducing accidents, injuries, or illnesses in the workplace. The latest available Bureau of Labor Statistics records covering the 1972–77 period demonstrate that private-sector injury and illness incidence rates, considering sampling error, have been on the rise. Occupational Injuries and Illnesses in the United States by Industry—1977, U.S. Dept. of Labor, Bureau of Labor Statistics—Bulletin 2047 (January 1980). Occupational injuries and illnesses occurred at a rate of 9.3 per 100 full-time workers during 1977 (9.2 in 1976, 9.1 in 1975); on the average, one out of every eleven workers in the private economy was injured or made ill while on the job.[23] The United States Chamber of Commerce also reports that since 1971, when OSHA was implemented, the number of serious injuries in America's workplace is up by 25 percent. And the severity of those injuries, measured by total lost workdays, is up by more than 34 percent. Worst of all, OSHA has had almost no positive impact on the number of workplace fatalities.

The National Safety Council (NSC) reports that during the eight-year period prior to 1971 there actually was a steady decline in both the serious-injury and fatality rates. In fact, NSC statistics show steady improvement in industry's prevention of serious injuries and fatalities since World War II. After an initial decline in the fatality rate subsequent to enactment of the Act, fatalities are once again on the rise—the 4,760 fatalities in 1977 is an increase of 20 percent over the 1976 level. The incidence of work days lost per 100 workers due to injury or illness also increased, from 48 days per 100 workers in 1972 to 61.6 days in 1977.[24] There were nearly 5.5 million job-related injuries and illnesses in 1977[25]—an increase of nearly 6 percent from the previous year (there were 5.2 million injuries and illnesses in 1976, a 3 percent increase from 1975). Illnesses continued to account for a relatively small portion of total injury and illnesses cases—about 3 percent.

While it is true that between 1973 and 1977 the case-incidence rate for all industries dropped by more than 15 percent—down from 10.6 to 9.0 injuries per 100 full-time workers, the decrease occurred primarily in less serious cases, those involving no loss of worktime. The rate for lost-workday injuries rose 12 percent. The severity of injuries as measured by the lost-workday incidence rate also increased by 17 percent from 51.2 days to 60.0 days lost per 100 full-time workers.[26] Although, on the whole, for all employers, fatality rates have been decreasing, this is a continuation of a fifty year trend (see Table B in Appendix 1) and seems to have no correlation to the enactment of the Act, which even OSHA admits. But numbers do not specifically explain the effectiveness of OSHA, since many factors affect injury, illness, and fatality rates. See National Safety Council, *Work Accidents,* 1975. As was developed in recent Congressional Oversight Hearings, a decrease in exposure contributed to this improvement; employment declined nearly 2 percent and average hours worked per week went down more than 1 percent. Oversight Hearings on OSHA before the Subcommittee on Health and Safety of the Committee on Education and Labor, House of Regulations, 96th Cong. 2d Sess. 1980.

It seems that enforcement of OSHA's safety and health standards has had little to do with work-related accident, injury, illness, and death rates.[27] OSHA knows little more about what prevents illness, injuries, and death today than it did in 1971. Reference to state worker's-compensation reports filed by employers seem to bear this out. The size and composition of the work force may have more to do with the accident/injury/death rates than anything else. Moreover, and unfortunately, employers now seem to be directing their safety activities to preventing citations, not preventing injuries.

Data prepared by OSHA's Office of Management Data Suppliers indicated that 71% of OSHA inspections in fiscal year 1979 yielded no serious citations and in almost 50% of the inspections no hazards were found—even though a targeting mechanism was directed at high hazard incidents. The Act permits employees and their representatives to submit complaints where they believe a threat of harm or danger exists. While this is and can be an important component of a safety and health program, it has been reported by the General Accounting Office no complaint related violations in 80% of the cases and in only 1% of them were the matters potentially serious. *See How Effective Are OSHA's Complaint Procedures,* GAO, April 1979.[28]

According to McGraw-Hill surveys of industry safety and health expenditures, the cost of industry attempts to achieve compliance with OSHA regulations has been skyrocketing. Furthermore, the figures do not include the formidable litigation expenses incurred by employers contesting citations before the Review Commission, nor do they include the cost of appeals or the cost of retaining counsel to interpret and advise on standards and regulations, all of which add up to sizable business expenses. The United States Chamber of Commerce reported at the Senate Oversight Hearings on the Act that the cost of litigation prevents many employers from contesting citations and recommends reimbursement of litigation costs to successful employer litigants to ensure a more thorough and pragmatic examination of inspection results by OSHA. OSHA also states that citation and standard litigation is a major problem facing the Agency, and labor organizations feel that because of the heavy caseload of contested citations, the Solicitor of Labor is under great pressure to enter into settlements which do not necessarily guarantee protection of workers.

In addition, it is reasonable to assume that to comply with OSHA rules and standards, most of the huge business expenditures are passed on to consumers in the form of higher prices in the marketplace. See *OSHA: Overpriced and under the Gun—a Cost-Benefit Analysis: A Prescription for Reform,* United States Chamber of Commerce, Special Report Congressional Action, November 9, 1979.

In the almost ten years since the Act has been in existence, the labor movement, spearheaded by the AFL–CIO and its Industrial Union Department (IUD) has also become disillusioned with the implementation of the Act and OSHA's commitment to protecting workers' health and safety. Union officials have found themselves thrust into an adversary posture and have spent a great deal of effort in lobbying against a weakening of the federal health and safety program and playing a watchdog role over the Labor Department. The IUD has strongly opposed state takeover of occupational safety and health, on-site consultation, and the low funding and manpower provided by both OSHA and NIOSH. Other unions like United Steelworkers of America, the Oil, Chemical & Atomic Workers, and the United Mineworkers have also opposed several OSHA actions, such as the return of safety and health programs to the state.

The obvious question, then, is "Will increased emphasis on job-site safety and health reduce accidents and illness?" It seems not! Carelessness, negligence, and inadvertence will be present no matter how effectively standards are inforced. The President's Interagency Task Force on Workplace Safety and Health, for one, indicates that compliance alone will not eliminate the problem. According to its recently issued report, compliance with OSHA rules and standards would halt only about 25 percent of accidents. Many industry experts believe that only effective, in-house safety programs will cause a rollback in the accident rate. OSHA's poor performace gives credence to the theory that voluntary safety and health-preservation programs by employers are the only viable solution for protecting workers. Although industry can be aided by government encouragement, incentive, and consultation, the ultimate source of protection in worker safety and health must be the employers *and* the employees themselves.

A compelling reason why employers voluntarily assume safety responsibility themselves is the prohibitive cost of worker's compensation. The lack of OSHA impact supports the theory that most of the on-job injuries are the result of pure accidents. Very few injuries are the result of conditions but are rather the result of acts. You can not enforce skill, you cannot legislate out negligence, you cannot discipline out misconduct, for 85 percent of all occupational accidents are due to unsafe acts by workers and only 15 percent are due to unsafe conditions which the employer can control.

ASSESSMENT FOR THE FUTURE

A recent study by the National Association of Manufacturers (NAM), concluded that OSHA, along with the Environmental Protection Agency (EPA), has been cited countless times as being a major cause of regulation-induced inflation, of lost jobs due to plant closings, and as an obstruction to advanced technology and international competition. Further, this study laments that there have been few significant changes made in the Act in the past ten years other than the yearly laundry list of special-interest exemptions attached to the annual appropriations bills, and that the wholesale repeal of the Act is unlikely, even with the advent of the Reagan Administration. NAM, *Industrial Relations Issues: Management Summary.* The only major exception was the introduction of the Occupational Safety and Health Improvements Bill of 1980 (S.2153), during the 96th Congress, by Senator Richard Schweiker. The bill would have amended Section 4 of the Act to provide exemptions to firms with good safety results, mandated the targeting of OSHA enforcement where most needed, and created incentives for self-enforcement in occupational safety and health. Under this bill a firm would have qualified for an exemption from safety inspections if it had a good safety record, that is, during the preceding year, it incurred no employee deaths resulting from occupational injury and no occupational injuries which required more than one lost workday or no injuries requiring any days off the job.

The bill would also have required the Secretary of Labor to enter into an agreement with state worker's-compensation agencies whereby each agency would submit an annual listing of all employers in that state which had one or more reported occupational injuries during the preceding year. This would include all injuries which resulted in two or more lost workdays. Any workplace not appearing on the list would qualify automatically for an exemption. Workplaces not identified as "safe" through the worker's-compensation data also would qualify for exemption by demonstrating by affidavit that they had no employee deaths caused by occupational injury during the preceding year, and only a low number of lost-workday injuries.

Inspections still would have been permitted at these workplaces, however, (1) to determine the cause of an accident that caused the death or hospitalization of one or more employees, (2) to determine the existence of an imminent danger, (3) to determine whether a violation exists if OSHA was notified of such a violation and was not assured by the employer that the hazard had been corrected, and (4) to determine whether a previously cited violation has been abated properly.

Finally, no employer qualifying for an exemption could have been assessed an OSHA civil penalty for a violation found in the employer's workplace (except for willful failure to abate and for health violations) if the employer maintained an advisory safety committee and a safety-consultation program. OSHA also would have been prohibited from proposing a fine if the employer employed no more than ten workers at any time during the preceding year. Penalties for serious citations would have been limited to $700, and those for nonserious citations to $300.

A "Corporate Crime Bill" (HR 7040) was also introduced which would have made corporations and their managers criminally liable for covering up health and safety problems. A provision to protect corporate "whistle blowers" (employees who turn in their companies) was also included; any employer who fires or fails to hire or promote an employee because he blew the whistle is liable for a penalty or prison term. Like the Schweiker Bill, the Corporate Crime Bill also died with the 96th Congress.

Other recent legislative suggestions for change included the Deckard Bill (HR. 7623) which would have prohibited OSHA from issuing citations for first inspection violations, and the Culver Bill (S. 299) which was actually enacted and signed into the law on September 19, 1980. The Culver Bill requires OSHA to consider the economic impact of its proposed rules, regulations, and standards on small businesses, organizations, and governmental jurisdictions. The bill also requires OSHA to consider alternatives such as the use of performance rather than design standards, exemptions from enforcement and relief from reporting requirements. Other bills which have been introduced include those which would have provided coverage for state and local employees, provided additional assistance to small employers, provided that where violations are corrected within the prescribed abatement period no penalties are assessed, and provided for the repeal of the Act.

Although not involving as significant a change as the legislation described above, OSHA itself, as President Carter noted in his Memorandum chartering the Interagency Task Force, has begun internal reforms to identify the most serious hazards workers face. It recently received public recognition for deleting numerous, unnecessary "Mickey Mouse" standards, even though these standards were rarely if ever enforced and their effect was more apparent than real. Moreover, in the recent past, OSHA's priorities seemed to have turned from the safety area to the health area. As evidence of this turn, standards development has recently focused on cotton dust, acrylonite, lead, benzene, arsenic, coke-oven emissions, and carcinogens. This may be tied to the feeling that society can derive larger gains from an occupational health program geared toward toxic and carcinogenic substances in the workplace. *See* Smith, *The Occupational Safety and Health Act: Its Goals and Its Achievements* (American Enterprise Institute for Public Policy Research, 1976). Organized labor is also increasingly assertive in occupational health issues, including placing high priority on such issues in collective bargaining. *See Business Week,* "The New Activism on Job Health," Sept. 18, 1978. Indeed, most of the complaints now received by OSHA from employees and unions deal with health hazards.

OSHA has not, however, forgotten safety, and has recently given greater attention to occupational safety in its plans for a major revision of job safety standards. Several new or revised safety standards have been announced or issued. A partial explanation for this action may be the recent BLS annual survey which found that, in spite of OSHA's efforts at protecting employees in the workplace: (1) on-the-job fatalities involving companies employing more than eleven workers were 20 percent higher in 1977 than in 1976; (2) the injury rate has been steadily on the rise for the past three years; and (3) lost time due in workplace mishaps is climbing.

These internal reforms are not enough—at least according to three sets of recommendations on OSHA reform that have recently been made public. The first, proposed by the Administrative Conference of the United States, calls upon agencies which have authority to issue mandatory health or safety regulations to draw on knowledge and information available in private organizations that develop voluntary consensus standards. 44 Fed. Reg. 1357, January 5, 1979. The focus of the recommendation is on those groups through which standards are "developed, reviewed, and periodically revised by technical committees of such nongovernmental organizations that follow open and regular procedures, including a process for considering and attempting to resolve negative comments." OSHA participation in or cooperation with those technical committees may result in the development of standards that address safety and health considerations more efficiently and effectively than standards independently formulated by OSHA.

The report of the Senate Governmental Affairs Committee, Vol. VI, "Framework for Regulation," (December 1978) goes well beyond these recommendations. It recommends a number of changes in the agency. It states: "[R]ather than continue on the course of its first 7 years, we would argue, OSHA should be disbanded. Safety and health in the work place would not suffer measurably, significant private and governmental resources would be saved, and an agency perceived primarily as a tool of Government harassment would be eliminated. *Fortunately, these are not the only two courses available, and we would urge that some basic changes be considered (in lieu of abolishing the agency*)." These recommendations, which encourage OSHA to rationalize its interventions and direct itself to areas in which it could be the most productive, include (1) increasing the use of OSHA-promoting incentive mechanisms as an alternative or complement to direct regulation, (2) intervening in the market directly in those areas, and only in those areas, where there is a demonstrated relationship between the means of OSHA's intervention and safety and health, and (3) generating information systematically on the costs and benefits of its regulatory interventions, so as to guarantee that such information receives attention in political and administrative proceedings—that is, legislation by Congress requiring that resource costs be computed.

The final study was prepared by the Interagency Task Force on Workplace Safety and Health. The Task Force was formed by President Carter in August 1977 to consider ways to strengthen the federal role in improving workplace safety and health. Specifically, the Task Force was directed to explore incentives that might supplement direct workplace safety regulations and evaluate government-wide administration of federal workplace safety and health efforts of all federal agencies, including those programs that affect federal employees and the resources devoted to them. In its draft final report dated December 14, 1978, the Task Force issued a long list of recommendations to improve the effectiveness of the Act, to insure greater private-sector participation, and to "reduce serious workplace injuries . . . in a manner which is more cost-effective than the present federal approach," including the following:

(1) Develop and test an enforcement strategy which sends inspectors first to High Injury Rate Establishments (HIRE), and then to establishments most likely to have detectable serious standards violations.
(2) Make more effective use of inspectors by testing a requirement that area directors first respond to employee complaints with a letter to the employer rather than an on-site inspection, except in cases of imminent danger or serious mobile-site hazards.
(3) Begin systematically to identify hazards for which engineering controls can be implemented on normal equipment-replacement cycles instead of by retrofitting.
(4) Give industry the flexibility to adopt different but equally effective safety and health measures.
(5) Streamline OSHA's cumbersome variance procedures.
(6) Develop a penalty system that will assess noncomplying firms a penalty which precisely equals the economic benefits of their noncompliance.

The Task Force also found that perhaps 25 percent of injuries are preventable by enforcement of present OSHA standards, and suggested that 75 percent of injuries can probably not be prevented by present enforcement. It further stated that:

> We also found that concentrating only on compliance, nationwide physical safety standards rather than results would be inefficient due to wide variations of individual workplaces in geographical areas, excessive costs to achieve uniform solutions, and a tendency in some cases to draw resources and attention away from organizational factors which already control significant hazards effectively.

Against this backdrop, the Task Force found that most companies are sincere in their efforts to provide for the safety and health of their workers, yet they are frustrated by punitive and often dictatorial enforcement strategies which emphasize the negative and fail to provide appropriate incentive or encouragement to foster voluntary compliance. In addition, because of the increased costs of regulation, American firms may be "exporting" workplace and environmental hazards by relocating operations to or purchasing processed materials from countries which regulate less stringently. For humanitarian as well as economic reasons, this is not a satisfactory trade-off of costs and benefits.

In spite of all of these suggested avenues of reform, which clearly suggest that some change in OSHA is necessary, Assistant Secretary Bingham vowed that OSHA would not be deterred from carrying out (her view of) the mandate of the Act. She clearly meant to change the way business is done in this county. For four years she succeeded, but with the advent of the Reagan Administration and the anticipated appointment of a new Assistant Secretary, it appears that OSHA's administrative and legislative priorities will be gradually altered and redirected from one of confrontation to cooperation. Given the absence of any demonstrable impact of OSHA on the incidence of injury and illness rates in the workplace as it approaches its tenth anniversary, it appears there is a clear mandate for rethinking, revitalization, and redirection in the utilization of its resources. The first priority of the new Assistant Secretary should be towards administrative modifications in OSHA itself. Areas in which this redirection should be centered, and which can be done *administratively* include the following:

(1) Balance, revitalize, and redirect labor/management input to OSHA through liaison groups as well as advisory committees, such as NACOSH and CACOSH;
(2) Focus on the *prevention* of injury and illness of employees from exposure to workplace hazards rather than on the sanctioning of employers after the occurrence, though the enforcement mechanisms of citations, penalties and abatement. A legal rationale must be developed for maximized on-site consultation. Where enforcement is necessary, better targeting of resources on individual firms on the basis of their accident experience is required, along with exploration and development of injury/illness rate performance standards (as suggested by Section 6 of the Act). OSHA should redirect standards promulgation and enforcement by means of a programmatic approach and strategy;
(3) Acknowledge and hold management responsible for the control of both unsafe conditions and unsafe employee actions;
(4) Initiate consultation and training programs and provide an incentive for the development of employer-based safety and health programs. More external public education is required as well as closer cooperation between governmental agencies and other research groups. This could be done pursuant to Section 21(c)(2) of the Act which requires the Secretary of Labor to consult with and advise employers, employ-

ees, and organizations representing employers and employees as to effective means of preventing occupational injuries and illnesses;

(5) Eliminate first instance sanctions other than for serious or imminent danger violations and redirect the use of the general duty clause to the dictates of its congressional intent;

(6) Strengthen state programs and centralize control over regional and area offices including their enforcement schemes, evaluation of performance, training and development, and personnel to ensure accountability and quality control. Review and revise the FOM, IHOM, and OSHA Instructions;

(7) Avoid conflict with and intrusion into labor relations issues covered by other regulatory agencies, such as the access to medical and exposure records, work refusals and rate retention under the LMRA and MSHA; the regulation of chemical substances under TSCA, RCRA and FIFR; and the genetic testing of employees and the removal/exclusion of susceptible persons from the workplace based upon perceived reproductive hazards under Title VII of the Civil Rights Act of 1964[29] (Title VII). This area must be given meaningful attention, with a view towards both interagency cooperation and in certain instances, deferral; and

(8) Provide that either employers or OSHA should, on a consultation basis, analyze Forms 200 and 100 relative to injuries and illnesses in order to identify needed improvements in hazardous areas and determine if the injuries or illnesses have resulted from unsafe acts or unsafe conditions.

Once OSHA has been administratively tightened, then reform efforts can be directed at the Act itself. Suggested target areas with respect to actual *legislative* changes in the Act, which have frequently been emphasized in congressional oversight and legislative hearings would include the following:

(1) Add to Section 3 of the Act a definition of "feasible" in terms of cost benefit and risk analysis;

(2) Add a new section 4(b) (5) which would circumscribe OSHA interference with employment and labor relations issues under the LMRA, MSHA, and Title VII;

(3) Revise Section 5(b) to provide for complementary employee duties relative to compliance with the Act and possibly revise Section 17 to provide for employee penalties under the Act;

(4) Redefine the abatement terms set forth in Sections 9 and 10 to incorporate the new definition of feasibility set forth in Section 3;

(5) Add a sunset clause through amendment of Section 6 which would set a schedule by which standards would automatically expire (*e.g.*, 10–15 years) unless renewed and affirmatively determined to be both reasonably necessary and appropriate under Section 3(8) of the Act and the criteria set out in Executive Order 12044.

(6) Amend Section 8(f) of the Act, in a manner which rather than speci-

fying exemptions to the conduct inspections, would redirect inspections in terms of proper targeting of hazardous conditions, materials, machines, devices, equipment, and processes. A suggested revision would be to insert a new subsection 1 and change the present subsections 1 and 2 to subsections 2 and 3 and to provide that:

(1) "In order to carry out the purposes of this Section, the Secretary shall conduct both general schedule inspections which shall be comprehensive and special inspections as provided in subsection 2 of this section which shall be limited in scope to the special purpose of the inspection. In making general schedule inspections, the Secretary shall first establish, pursuant to Section 6(b) an administrative plan assuring that prioritization and effective utilization of inspection resources be determined by reliable objective criteria including program criteria and/or incidence criteria. (The criteria would be defined by regulation.)

The combined effect of these suggested administrative and legislative changes should be to begin an era of cooperation rather than confrontation in directing the resources of government, management and labor toward the real issues of fulfilling the purpose of the Act with respect to providing for employee safety and health in the workplace.

G. SOURCE AND REFERENCE MATERIALS

OSHA cannot hope to succeed in its mission of assuring safe and healthful working conditions in the Nation's workplaces without fully informing employers and employees of their rights and duties under the Act. Similarly, employers, employees, and their representatives cannot hope to understand and meet the requirements and demands of the Act without a thorough knowledge of the Act and its aims, programs, policies, practices, and purposes. The source and reference materials listed in the Bibliography are suggested to assist in this regard.

Several Department of Labor, Department of Health and Human Services, and Occupational Safety and Health Administration publications should also be consulted. For example, OSHA publishes many informative pamphlets dealing with various aspects of the Act's operation, such as *Recordkeeping Requirements under the Act, Evaluating Your Firm's Injury and Illness Record, Training Requirements of OSHA Standards, Handbook for Small Businesses, What Every Employer Needs to Know about Recordkeeping,*[30] *Material Safety Data Sheet,* and *Worker's Rights under OSHA.*

OSHA also publishes small manuals in its Safe Work Practices series which deal with such topics as excavating and trenching operations, handling hazardous materials, investigating accidents in the workplace, commercial diving operations, essentials of machine guarding, and ground-fault protection. OSHA also issues a Job Health Hazard series which discusses such topics as carbon monoxide, vinyl chloride, beryllium, toluene diisocyanate, asbestos, inorganic arsenic, and mercury. In its Programs and Policy series OSHA issues such booklets as *All about OSHA; SBA Loans for OSHA Compliance,* and *How OSHA Monitors State Plans.* In its Cancer Alert series it issues booklets concerning health hazards of asbestos, health hazards of inorganic arsenic, and an introduction to occupational cancer. In addition, in its Workplace Rights in Action series it gives answers to common questions concerning job discrimination, safety and health inspections, and health and safety committees. Numerous consumer information leaflets, fact sheets highlighting OSHA programs, are also available. Finally, numerous training materials, including booklets, slides, and tapes, are available at a minimal cost.

OSHA has published a pamphlet entitled *OSHA Publications and Training Materials* which explains how to obtain the information referenced. Normally these publications are available from the Government Printing Office or most OSHA area and regional offices.

Also, important reference sources are the President's Annual Report on Occupational Safety and Health and the records of the Annual Oversight Hearings on the Occupational Safety and Health Act held before the respective House and Senate subcommittees. Of interest also would be OSHA's official monthly magazines, *Job Safety and Health* and *Mine Safety and Health.* Employers, of course, should seek to obtain a copy of the Act and copies of all standards published and promulgated by OSHA. These standards are available by industry from OSHA and are also published in Volume 29 of the Code of Federal Regulations; updates and notices of proposed rule-making are published in the Federal Register. Finally, employers should secure copies of the OSHA Field Operations Manual and Industrial Hygiene Manual. Most of these materials are available from the Occupational Safety and Health Subscription Service which provides standards, interpretations, regulations, and procedures in easy-to-use looseleaf form.

Recent OSHA decisions are compiled and published weekly by the Bureau of National Affairs (BNA) and Commerce Clearing House (CCH).

NIOSH also issues several publications, such as criteria documents and the annual toxicity register. NIOSH has also recently issued a publications catalog. Moreover, the Department of Health and Human Services recently issued an occupational safety and health directory which gives a guide to personnel in NIOSH, OSHA, MSHA, the Review Commission, and state and local agencies dealing with occupational safety and health. The directory lists the names, titles, addresses, and telephone numbers of key personnel in those agencies. The Review Commission issues Rules of Procedure and a guideline to its procedures. Finally, MSHA, like OSHA prepares and issues publications relative to the Mine Act. MSHA decisions are compiled and published by BNA and CCH.

NOTES

1. Most of the early private work was in the area of mining, particularly coal mining.

2. The first federal legislation dealing with occupational safety was enacted in the 1890s, but this pertained only to coal mining operations. Federal involvement in the area of industrial hygiene did not begin until the Act of July 1, 1902, ch. 1378, §§1–10. In 1910, the Federal Bureau of Mines, concerned with health and safety research for underground operations, was established in the Department of Interior. Early state response to job-related accidents and injuries took the form of worker's compensation laws.

3. The early research and study programs apparently provide valuable insights and statistics on the types and frequency of occupational accidents and diseases. The programs were a contributing factor in securing the passage of later health standards and legislation by Congress.

4. During congressional debates on the Act, the House Education and Labor Committee Report noted the disparity in enforcement programs and pointed out that, while uniformity of standards was necessary, the "state response has been minimal." H. Rep. No. 1291, 91st Cong., 2d Sess. (1970).

5. The Davis-Bacon Act, 40 U.S.C. §276(a); the Walsh-Healey Public Contracts Act, 41 U.S.C. §§35–45; and the Longshoremen's and Harbor Workers Compensation Act, 33 U.S.C. §901, *et seq.*, although enacted during this time period, seemed to have had little impact on occupational safety and health. Neither did the Fair Labor Standards Act, 29 U.S.C. §201, *et seq.*, which regulated hours worked through minimum wage and overtime pay guarantees, nor the National Labor Relations Act, 29 U.S.C. §151, *et seq.*, which regulated employer-employee relations.

6. 29 U.S.C. §§141–167.

7. In 1951, Senator Humphrey introduced a bill calling for uniform national health and safety codes and uniform enforcement standards, but legislation was never enacted. S. 2325, 82d Cong., 1st Sess. (1954). Several specialized or limited safety statutes were enacted during the 50s and 60s but they failed to cover a majority of workers.

8. 40 U.S.C. §327, *et seq.*

9. 40 U.S.C. §333, *et seq.*

10. 41 U.S.C. §351, *et seq.*

11. 20 U.S.C. §§784–786, 951, *et seq.*

12. 30 U.S.C. §721, *et seq.*

13. 30 U.S.C. §801, *et seq.*

14. 45 U.S.C. §421, *et seq.*

15. Although injury rates were increasing, fatality rates were decreasing during this period of time. See Table B in Appendix 1 and discussion *infra.*

16. 15 U.S.C. §2601, *et seq.*

17. 7 U.S.C. §136.

18. 30 U.S.C. §801, *et seq.*

19. 42 U.S.C. §6901, *et seq.*

20. When OSHA came into being in 1971, it superseded state job safety and health programs which were operating at that time. The Act included a provision, however, that allowed states to develop their own job safety and health programs, provided their programs were as effective as the federal program. This meant that state standards must be as strict as federal standards; otherwise, OSHA would retain jurisdiction. For most states that developed their own plans, the answer was to adopt federal standards word-for-word as soon as they were promulgated and published. A few states have chosen to adopt some standards which are stricter than their federal

counterparts, but this has been the exception rather than the rule. A few have chosen to write standards to cover areas that the federal program do not cover, and a few have chosen to leave jurisdiction over some areas in the federal program.

21. *Dun's Review* is an influential business publication.

22. Executive Order 12044 also mandates the issuance of clear, simple regulations and requires periodic review of existing regulations and standards to determine if they are still needed to achieve their original regulatory goals. Rulemakers most consider the regulatory burden imposed and reevaluate the applicable technological, economic and other relevant conditions involved in the regulation.

23. Preliminary information indicates that the rate was 9.4 in 1978 and 9.5 in 1979. Occupational Illness and Injuries in 1978: Summary (U.S. Dept. of Labor, Bureau of Labor Statistics, Injury and Illness Incidence Rates, March & November 1980).

24. Preliminary information indicates that the work days lost per 100 workers increased to 63.5 in 1978 and 67.7 in 1979.

25. Preliminary information indicates that there were 5.8 million job-related injuries and illnesses in 1978 and 6.1 million in 1979.

26. Preliminary information indicates that the incidence rate increased to 62.1 for 1978 and 66.2 for 1979.

27. In order to determine whether or not OSHA has been effective it is necessary to predict what injury and illness rates would have been in the absence of the Act, and compare those figures with actual rates. *See* Ginnold, *A View of the Costs and Benefits of the Costs and Benefits of the Job Safety and Health Law,* 24 Monthly Labor Review, (August, 1980). According to R. Nader, M. Green, and M. Waitzman, *Business War on the Law: An Analysis of the Benefits of Federal Health/Safety Enforcement,* 1979, OSHA's enforcement prevented 350 deaths and 40,000–60,000 injuries each year resulting in an annual benefit of 5.1 billion dollars. In addition, some statistics indicate that the Act results in new jobs, a decrease in lost work time, increased productivity, and other benefits. *See* Whiting, *Regulatory Reform and OSHA: Fads and Reality,* 30 Labor L. J. 514 (1979).

28. The Interagency Task Force on Workplace Safety and Health strongly suggests that OSHA implement procedures on a pilot basis similar to those used with respect to the handling of informal complaints, that is, it is OSHA's practice to screen them and use letters to the employer prior to any inspection activity. (OSHA instruction CPL 2.12 dated September 1, 1979). Such an approach is not inconsistent with the provisions of Section 8(f)(1) and may be accomplished administratively. It would also permit a more effective utilization of OSHA's limited resources and direct their focus to serious workplace conditions, while at the same time preventing the utilization of the complaint mechanism for less than legitimate purposes.

29. 42 U.S.C. §2000(e), *et seq.*

30. In *General Motors Corporation, Inland Division,* 8 OSHC 2036, 2038 (1980), the Secretary of Labor attempted to substantiate the employer's recordkeeping obligation by reference to a pamphlet promulgated by the Secretary of Labor entitled "What Every Employer Needs to Know About Recordkeeping," and the administrative law judge at least partially relied on the contents of the pamphlet in finding the employer in violation of 29 CFR 1904.2.

APPENDIX 1 to CHAPTER I

TABLE A
Principal Classes of Accidental Death and Injury, 1903–1975

Cause	Death	Injury
Motor-Vehicle	**46,000**	**1,800,000**
Public non-work	41,900	1,700,000
Work	3,900	100,00
Home	200	10,000
Work	**12,600**	**2,200,000**
Non–motor-vehicle	8,700	2,100,000
Motor-vehicle	3,900	100,000
Home	**25,500**	**4,000,000**
Non–motor-vehicle	23,300	4,000,000
Motor-vehicle	200	10,000
Public	**22,500**	**2,800,000**

Source: National Safety Council Estimates, 1975; Accident Facts, 1976 ed.

TABLE B
Work-related Deaths, 1928–1975, by Number and Rate

Year	Number	Rate	Year	Number	Rate
1928	19,000	15.8	1952	15,000	9.6
1929	20,000	16.4	1953	15,000	9.5
1930	19,000	15.4	1954	14,000	8.7
1931	17,500	14.1	1955	14,200	8.6
1932	15,000	12.0	1956	14,300	8.5
1933	14,500	11.6	1957	14,200	8.3
1934	16,000	12.7	1958	13,300	7.7
1935	16,500	13.0	1959	13,800	7.8
1936	18,500	14.5	1960	13,800	7.7
1937	19,000	14.8	1961	13,500	7.4
1938	16,000	12.3	1962	13,700	7.4
1939	15,500	11.8	1963	14,200	7.5
1940	17,000	12.9	1964	14,200	7.4
1941	18,000	13.5	1965	14,100	7.3
1942	18,500	13.8	1966	14,500	7.4
1943	17,500	13.0	1967	14,200	7.2
1944	16,000	12.0	1968	14,300	7.2
1945	16,500	12.5	1969	14,300	7.1
1946	16,500	11.8	1970	13,800	6.8
1947	17,000	11.9	1971	13,700	6.6
1948	16,000	11.0	1972	14,000	6.7
1949	15,000	10.1	1973	14,300	6.8
1950	15,500	10.2	1974	13,500	6.4
1951	16,000	10.4	1975	12,600	5.9

CHAPTER II
Implementation and Coverage

A. INTRODUCTION

To an employer facing a matter involving an occupational safety and health issue for the first time, the initial inquiry must be "Am I covered?" Thus, before the procedural and substantive aspects of the Act can be explored, the coverage of the Act must be understood.

B. CONSTITUTIONAL BASIS OF COVERAGE

1. The Commerce Clause

The Commerce Clause of the United States Constitution (Article I, Section 8, Clause 3) gives to Congress the power to regulate commerce "among the several States."

The reach of the Commerce Clause extends beyond federal regulation of the channels and instrumentalities of interstate commerce so as to empower Congress to regulate conditions or activities which affect commerce even though the activity or condition may itself not be commerce and may be purely intrastate in character. *Gibbons v. Ogden,* 22 U.S. (9 Wheat.) 1, 195 (1824); *McCulloch v. Maryland,* 17 U.S. (4 Wheat.) 316 (1819); *United States v. Darby,* 312 U.S. 100 (1941); *Wickard v. Filburn,* 317 U.S. 111, 117 (1942); *United States v. Ohio,* 385 U.S. 9 (1966); and *Perez v. United States,* 402 U.S. 146 (1971). It is not necessary to prove that any particular intrastate activity affects commerce if the activity is included in a class of activities which Congress intended to regulate because the class affects commerce. *Heart of Atlanta Motel, Inc. v. United States,* 379 U.S. 241 (1964); *Katzenbach v. McClung,* 379 U.S. 294 (1964); and *Perez v. United States, supra.* Generally speaking, then, the class of activities which Congress may regulate under the commerce power may be as broad and as inclusive as Congress intends, since the commerce power is plenary and has no restrictions placed on it except specific constitutional prohibitions and those restrictions are placed on it by Congress itself. *United States v. Wrightwood Dairy Co.,* 315 U.S. 110 (1942); and *United States v. Darby, supra.* Since there are no specific constitutional prohibitions involved, the issue is reduced to the question "How inclusive did Congress intend coverage to be of the class of activities regulated under the Act?"

2. Extent of Coverage

a. *Statutory*

In view of the "substantial burden" that work-related injuries, illnesses and deaths may place on interstate commerce, Congress found authority to legislate for the Act under the Commerce Clause. In defining an employer in the Act as one "engaged in a business affecting commerce," Congress clearly intended to exercise the full extent of its powers under the Commerce Clause. *Godwin v. OSHRC,* 540 F.2d 1013, 4 OSHC 1603 (9th Cir. 1976); *Brennan v. John J. Gordon Co., Inc.,* 492 F.2d 1027, 1 OSHC 1580 (2d Cir. 1974); *United States v. Dye Construction Co.,* 510 F.2d 78, 2 OSHC 1510 (10th Cir. 1975). Section 2 of the Act provides that:

> The Congress finds that personal injuries and illnesses arising out of work situations impose a substantial burden upon, and are a hindrance to, interstate commerce in terms of lost production, wage loss, medical expenses, and disability compensation payments. . . .
>
> (b) Congress declares it to be its purpose and policy, through the exercise of its powers to regulate commerce among the several States and with foreign nations and to provide for the general welfare, to assure so far as possible every working man and woman in the Nation safe and healthful working conditions and to preserve our human resources.

Congressman William Steiger described the scope of the Act's coverage in the following words during a discussion of the legislation on the floor of the House of Representatives: "The coverage of this bill is as broad, generally speaking, as the authority vested in the Federal Government by the commerce clause of the Constitution" (*Cong. Rec.,* Vol. 116, p.H–11899, Dec. 17, 1970).

> The legislative history, as a whole, clearly shows that every amendment or other proposal which would have resulted in any employee's being left outside the protections afforded by the Act was rejected. The reason for excluding no (private sector)[1] employee, either by exemption or limitation on coverage, lies in the most fundamental of social purposes of the legislation, which is to protect the lives and health of human beings in the context of their employment. [29 CFR 1975.3.]

b. *Regulatory*

Pursuant to Sections 2 and 8(g) (2) of the Act, the Secretary of Labor is authorized to and has promulgated rules and regulations which *inter alia* deal with the scope of the Act's reach under the Commerce Clause: they provide the basis of authority, determine who is an employer, and provide a test for determining what entities are subdivisions of states. *Louisiana Chemical Association v. Bingham,* 496 F. Supp. 1188, 8 OSHC 1950 (W.D. La. 1980).[2] Specifically, the purpose of 29 CFR 1975 is to indicate which persons are covered by the Act as employers, and, as such, are subject to the requirements of the Act. It is not, however, the purpose of these regulations to indicate the legal effect of the Act once coverage is established. To that extent, it is provided in those regulations which are contained in 29 CFR 1975.3 that:

Interpretation of the provisions and terms of the . . . Act must of necessity be consistent with the express intent of Congress to exercise its commerce power to the extent that, "so far as possible, every working man and woman in the Nation" would be protected as provided for in the Act. The words "so far as possible" refer to the practical extent to which governmental regulation and expended resources are capable of achieving safe and healthful working conditions; the words are not ones of limitation on coverage. The controlling definition for the purpose of coverage under the Act is that of "employer" contained in section 3(5). This term is defined as follows:

> (5) The term "employer" means any person engaged in a business affecting commerce who has employees, but does not include the United States or any State or political subdivision of State.

In carrying out the broad coverage mandate of Congress, we interpret the term "business" in the above definition as including any commercial or noncommercial activity affecting commerce and involving the employment of one or more employees; the term "commerce" is defined in the Act itself, in Section 3(3). Since the legislative history and the words of the statute, itself, indicate that Congress intended the full exercise of its commerce power in order to reduce employment-related hazards which, as a whole impose a substantial burden on commerce, it follows that all employments where such hazards exist or could exist (that is, those involving the employment of one or more employees) were intended to be regulated as a class of activities which affects commerce [29 CFR 1975.3(d).]

OSHA's rule-making authority has been held to be as broad as the general remedial purposes of the Act. In *Mourning v. Family Publications Services,* 411 U.S. 356 (1973), the Supreme Court interpreted language in the Truth in Lending Act strikingly similar to that contained in Section 8(g)(2) of the Act:

> Where the empowering provision of a statute states simply that the agency may "make . . . such rules and regulations as may be necessary to carry out the provisions of the Act," we have held that the validity of a regulation promulgated thereunder will be sustained so long as it is "reasonably related to the purposes of the enabling legislation." *Thorpe v. Housing Authority of City of Dupont,* 393 U.S. 2680–281 (1969) (quoting Housing Act of 1937. s81) [411 U.S. at 369].

See also *Whirlpool Corp. v. Marshall,* 435 U.S. 1, 8 OSHC 1001 (1980), where the "refusal-to-work" regulation was upheld by the Court as complementary to the purposes and scheme of the Act. Moreover, the Act allows OSHA to promulgate legislative as well as interpretive rules; as long as the rulemaking complies with the notice and comment procedures specified in the Administrative Procedure Act, 5 U.S.C. §553. *Chamber of Commerce of the United States of America v. OSHA,* 636 F.2d 464, 8 OSHC 1648 (D.C. Cir. 1980);[3] *Cerro Metal Products, Division of Marmon Group, Inc. v. Marshall,* 620 F.2d 964, 8 OSHC 1196 (3d Cir. 1980); *Marshall v. W & W Steel Co.,* 604 F.2d 1322, 7 OSHC 1670 (10th Cir. 1979). Rules and regulations must also be reasonably related to the purposes of the Act. *Marshall v. Daniel Construction Co.,* 563 F.2d 707, 6 OSHC 1031 (5th Cir. 1977).

c. *Procedural*

OSHA program directives are issued to clarify and interpret standards, and prior to October 1978 were classified according to purpose (*e.g.*, the 100 series specified means of compliance, the 200 series contained changes in the Field Operations Manual, the 300 series related to industrial hygiene . . .). A new uniform classification system was established, however, which organized all directives into a single system in which directives are issued either as instructions or notices. An instruction contains information of a single nature; a notice contains interim or temporary information, usually of one year's duration. Directives are also identified by a classification code (*e.g.*, "STD" concerns standards; "CPL" relates to compliance).

OSHA's day-to-day operating procedures are set forth in the Field Operations Manual (FOM) and Industrial Hygiene Manual (IHM). The FOM and IHM provide guidelines and set forth basic policy and instruction to assure uniform enforcement action and contain such matters as inspection procedures, selection of workplaces for inspection, citation issuance, and penalty assessments. Unlike the situation in which the Secretary of Labor's failure to follow the rules and regulations which he has issued may result in the vacation of a citation, the failure to comply with the FOM is not fatal. Compare *Accu–Namics, Inc. v. OSHRC,* 515 F.2d 828, 3 OSHC 1299 (5th Cir. 1975), *cert. denied,* 425 U.S. 903, 4 OSHC 1090 (1976), with *FMC Corp.,* 5 OSHC 1707 (1977).

3. The "Affectation" Doctrine

Jurisdiction (coverage) must be established at the Commission level; primary jurisdiction to determine coverage rests with the Commission. *Marshall v. Able Contractors, Inc.,* 573 F.2d 1055, 6 OSHC 1317 (9th Cir. 1977), *cert. denied,* 439 U.S. 880 (1978).

By its terms the Act covers (1) any employer (2) having employees (3) engaged in a business "*affecting commerce*" but does not include (4) the United States or any state or political subdivision of a state. Section 3(5) of the Act. The "affecting commerce" formula gives the Department of Labor as much jurisdiction under the Act as is constitutionally permissible. *Brennan v. John J. Gordon Co., Inc.,* 492 F.2d 1027, 1 OSHC 1580 (2d Cir. 1974); *United States v. Dye Construction Co.,* 510 F.2d 78, 2 OSHC 1510 (1975). In *Franklin R. Lacy (Aqua View Apartments),* 4 OSHC 115 (1976), *rev'd.;* 628 F.2d 1226, 8 OSHC 2060 (9th Cir. 1980) the court contrasted two of the principal formulations of statutory jurisdiction that have evolved from the congressional history of regulating employment relations:

> If a statute covers businesses "in commerce," a fairly specific showing must be made of a connection between the particular employer regulated and interstate commence. See *e.g.,* Fair Labor Standards Act, 29 U.S.C. §201 *et seq.* (1976). See *Mitchell v. Lublin, McGaughty & Assocs.,* 358 U.S. 207, 211 (1959); *A.B. Kirschbaum Co. v. Walling,* 316 U.S. 517 (1942); *United States v. Darby,* 312 U.S. 100 (1941); *Houser v. Matson,* 447 F.2d 860 (9th Cir. 1971); *Wirtz v. Idaho Sheet Metal Works, Inc.,* 335 F.2d 952 (9th Cir. 1964). This is the interpretation apparently followed by the OSHRC in the instant case, inasmuch as it focused on the allegations and proof concerning whether the employer used materials and tools supplied by interstate manufacturers.

On the other hand, a statute may require only that the particular business "affect" commerce. *See e.g.*, National Labor Relations Act, 29, U.S.C. §§151 *et seq.* OSHA is a statute which employs this formulation. In such cases there is statutory jurisdiction so long as the business is in a class of activity that as a whole affects commerce. *NLRB v. Reliance Fuel Oil Corp.*, 371 U.S. 224 (1963); *Polish Nat. Alliance v. NLRB*, 322 U.S. 643, 647–48 (1944); *NLRB v. Jones & Laughlin Steel Corp.*, 301 U.S. 1, 34–39 (1937). OSHA is an example of this broader jurisidiction, which covers the full reach of Congress' power under the commerce clause.

The application of the "affectation doctrine" by the Secretary of Labor and the Review Commission has differed during the formative stages of the doctrine.

(1) The Secretary of Labor gave broad scope to the jurisdictional language in the Act, based on the fact, that Congress chose "affecting commerce" language rather than limiting coverage to those employers who were themselves "engaged in commerce." 29 CFR 1975, 1–6. *Brennan v. John J. Gordon Co., Inc.*, 492 F.2d 1027, 1 OSHC 1580 (2d Cir. 1974); *Dye Construction Co. v. U.S.*, 510 F.2d 78, 2 OSHC 1510 (10th Cir. 1975)—use of out-of-state supplies.

(2) The Review Commission took a somewhat more limited view of the affectation doctrine, providing employers with a limited jurisdictional defense. In *Ray Morin, Inc.*, 2 OSHC 3285 (1975) a citation was vacated because there was no evidence to show that the employer "purchased the manhole cover (that it was installing) or that the employer regularly used materials that moved in interstate commerce." And, in *Bettendorf Terminal Company and LeClaire Quarries, Inc.*, 1 OSHC 1695 (1974), a company that unloaded and processed sand for sale at a river site downstream from the point where sand was dredged by its parent firm, was not engaged in the continuous process of that excavation and was therefore exempt from the Act's coverage.

The Commission also held that the burden was on the Secretary to show that an employer who contests jurisdiction is subject to the Act, that is, the employer has an effect on commerce. *Leon & Larry Bechard, Brick Contractors*, 1 OSHC 3093 (1972); *Bettendorf Terminal Company and LeClaire Quarries, Inc.*, 1 OSHC 1695 (1974); *Wilshire Terrace*, 1 OSHC 3053 (1973). The Secretary was required to present credible and substantial evidence of an employer's effect on commerce. *Les Mares Enterprises, Inc.*, 3 OSHC 1015 (1975)—land clearing with mere *intent* to grow grapes is not subject to Act; *Anchorage Plastering Company*, 3 OSHC 1284 (1975)—applying stucco to a building; *Cable Car Advertisers, Inc.* 1 OSHC 1446 (1973)—maintenance of place of employment in burned-out ferryboat on a navigable waterway (San Francisco Bay) of United States.

(3) The divergent viewpoints of the Secretary of Labor and the Review Commission have been more or less resolved by two cases, *Les Mares Enterprises*, 3 OSHC 1015 (1975), *rev'd in Godwin v. OSHRC*, 540 F.2d 1013, 4 OSHC 1603 (9th Cir. 1976), and *Anchorage Plastering Co.*, 3 OSHC 1284 (1975), *rev'd in Marshall v. Anchorage Plastering Co.*, 570 F.2d 351, 6 OSHC

1318 (9th Cir. 1978). In *Godwin,* the court held that Congress, in defining employer in Section 3(5) of the Act as one engaged in a business affecting commerce, intended to exercise the full extent of its powers under the Commerce Clause. Therefore the clearing of land for purpose of growing grapes was an integral part of the wine manufacturing process and was covered under the Act; it was significant that the grapes had been neither planted nor harvested. *Anchorage Plastering,* which affirmed that jurisdiction under OSHA is as broad as the scope of the Commerce Clause, held that a company used equipment and materials from out of state, used the telephone and mails, and used a union hiring hall and thus was engaged in business affecting commerce. Based upon these cases, which have been followed by the Commission, it is now extremely easy to prove that an employer is engaged in a business affecting commerce, unless it has terminated operations before the inspection.

Although one of the Commission's Rules of Procedure, 33(a)(2)(i) requires the Secretary to plead and prove jurisdiction, it now seems clear, and the Commission and the courts have held that the Secretary of Labor is not required to prove that an employer is engaged in a business affecting commerce, and therefore subject to the Act's coverage, before proceeding with an inspection. *Marshall v. Able Contractors Inc.,* 573 F.2d 1055, 6 OSHC 1317 (9th Cir. 1978), *cert. denied,* 99 S.Ct. 98 (1978); *Franklin R. Lacy, (Aqua View Apartments)* 4 OSHC 1115 (1976), *rev'd,* 628 F.2d 1226, 8 OSHC 2060 (9th Cir. 1980). The Secretary's mere assertion, however, that an employer is engaged in construction which is "obviously" a business affecting commerce, is insufficient to show it is an employer within the Act's meaning and does not establish jurisdiction. *Steven's Inc.,* 8 OSHC 1269 (1980).

(4) Recent decisions concerning the meaning of "affecting interstate commerce" have continued the expansive view of jurisdictional coverage, holding that a scintilla of evidence that an employer affects commerce is enough to trigger jurisdiction under Section 3(5) of the Act. In *Clarence M. Jones,*— OSHC—(1978) *on review,* No. 77–3676, an apartment owner who hired men to work on his buildings was found to be covered by the Act, since some of the supplies used came from out of state. Similar decisions concerning the use of supplies and equipment from out of state as well as mail and telephone service, which have held that an employer was engaged in a business affecting commerce include: *R. B. Butler, Inc.,* 6 OSHC 1373 (1978); *Finnegan Construction Co.,* 6 OSHC 1496 (1978); *Poughkeepsie Yacht Club, Inc.,* 7 OSHC 1725 (1979), where evidence that a nonprofit corporation, organized for the purpose of maintaining boating facilities for yacht club members dispenses fuel, that members' boats are manufactured outside the state, and that the club marina is situated on a navigable waterway adjoining two states established that the corporation is an employer "affecting commerce." Most recently, the court in *Usery v. Franklin R. Lacy (Aqua View Apartments),* 628 F.2d 1226, 8 OSHC 2060 (9th Cir. 1980) reversed a Commission determination that an employer who employed approximately forty workers for the construction of a fifteen-unit apartment building was not engaged in a business affecting commerce. According to the court, the regulations which implement the Act

state essentially that all employers are covered, with some specific exceptions, such as for those who employ domestic help. See 29 CFR 1975.6. The coverage of the regulations is consistent with the congressional purpose to reach as broadly as constitutionally permissible in regulating employee safety, since nonuniform coverage would give unsafe employers a competitive advantage. Thus it "suffices to say that here the employer hired approximately forty workers for the construction of a fifteen-unit apartment building, and that is a business that affects commerce as a matter of law, for it is within the class of activities that it was Congress' intent to regulate, in extending the Act to employers whose activities in the aggregate affect commerce." See also *Norby & Rocky Norby, Partners,* 4 OSHC 1722 (1974).

C. APPLICATION

The applicability of the Act as set forth in Section 4 thereof is based upon the condition of employment; that is, being an employee[4] employed in the *workplace* of a covered employer.[5] The Act applies to workplaces in all fifty states, the District of Columbia, Puerto Rico, and all United States possessions, including the Virgin Islands, American Samoa, Guam, the Trust Territory of the Pacific Islands, Wake Island, the Outer Continental Shelf Lands as defined in the Outer Continental Shelf Land Act, and the Canal Zone. Section 4 of the Act. *Caribtow Corporation v. OSHRC,* 493 F.2d 1064, 1 OSHC 1592 (1st Cir. 1974), *cert. denied,* 419 U.S. 830, 2 OSHC 1239 (1974); *Pacific International, Inc.,* 7 OSHC 2227 (1979)—work performed in Marshall Islands is covered under the Act.

Congress has not defined the term "workplace" as it is used in Section 4 of the Act, nor has the Secretary of Labor done so in the regulations. However, the definition of the term seems clear and well settled, by judicial and administrative interpretation, encompassing any conceivable location where employees are employed and perform work for a covered employer. Jurisdictionally, however, not every such location would in fact be a "workplace" for purposes of coverage under the Act. For example, it was determined in enacting the Outer Continental Shelf Lands Act (OCSL Act), that Congress intended to exclude OSHA jurisdiction over working conditions/operations conducted on the Outer Continental Shelf. *Pacific International, Inc.* 7 OSHC 2227 (1979). Thus offshore mobile drilling vessels which operate on the high seas would not be workplaces under the coverage of the Act; the United States Coast Guard, the Department of the Interior, and the United States Geological Survey have statutory authority to and do govern the working conditions of seamen and their employment relationships on offshore platforms.[6] *Clary v. Ocean Drilling & Exploration Company,* 609 F.2d 1120, 7 OSHC 2209 (5th Cir. 1980); *Marshall v. Nichols,* 486 F. Supp. 615, 8 OSHC 1129 (E.D. Tex, 1980); *cf., Rosalee Taylor v. Moore-McCormack Lines, Inc.,* 621 F.2d 88, 8 OSHC 1277 (4th Cir. 1980) (Coast Guard regulations apply to seamen working at sea, not to longshoremen working at port.)

In the National Labor Relations Act, 29 U S.C. §151 (NLRA), and Title VII of the Civil Rights Act of 1964, 42 U.S.C. §2000(e) (Title VII), the same jurisdictional formula is used. *NLRB v. Jones & Laughlin Steel Corp.*, 301 U.S. 1 (1937); *Schroepfer v. A. S. Abell Co.*, 48 F. Supp. 88 (D.Md. 1942); *Avigliano v. Sumitomo and Shoji America,*—F.2d—(2d Cir. 1980); *Love v. Pullman Co.*, FEP 423, 430 F.2d 49 (9th Cir. 1970), *rev'd.*, 404 U.S. 522 (1972). But, unlike the NLRA and Title VII, OSHA coverage is not based upon the volume of business or number of employees—each employer is covered. The coverage of the Fair Labor Standards Act, 29 U.S.C. §201 (FLSA), by comparison, is much narrower. *United States v. Darby,* 312 U.S. 100 (1941); *Kirschbaum Co. v. Walling,* 316 U.S. 517 (1942). However, unlike the NLRA, for example, the jurisdictional formula of OSHA would not permit its extraterritorial application to American employees of an American corporation working on foreign soil. See Nothstein and Ayres, *The Multinational Corporation and the Extraterritorial Application of the Labor Management Relations Act,* 10 Cornell Intl. L.J. 1, 35 (1976).[7]

Except for specific exclusions, the Act is applicable to every employer who has *one* or more employees, and, as was previously discussed, is engaged in a business affecting commerce. 29 CFR 1975.4 (a). Thus to be covered under the Act, an employer must have employees at the time of the violation, but not necessarily thereafter. *Price-Potashnick-Codell-Oman,* 6 OSHC 1901 (1978). The entity's activity and employment relationships are the controlling factors in coverage. *Mangus Firearms,* 3 OSHC 1214 (1975)—a "silent partner" who does not manage or control, but works six to seven hours a week is an employee; *Elmer Vath, Painting Contractor,* 2 OSHC 1091 (1974); *R. Colwill Excavating Company,* 5 OSHC 1984 (1977)—a part-owner of a corporation is an employee; *Poughkeepsie Yacht Club, Inc.*, 7 OHSC 1725 (1979)—a nonprofit corporation which is organized for the purpose of maintaining boating facilities for use of yacht club members and which has only one paid employee is a covered employer; *Bloomfield Mechanical Contracting, Inc. v. OSHRC,* 519 F.2d 1257, 3 OSHC 1403 (3d Cir. 1975)—a joint venture can be an employer; *Hydraform Products Corp.*, 7 OSHC 1995 (1979)—a company president is an employee; *Harvey Workover,* 7 OSHC 1687 (1979)—specific criteria are utilized to ascertain whether a person is an independent contractor or an employee.

The interpretive regulations in 29 CFR 1975 provide clarification on coverage with respect to certain employers:

(1) Where a member of a profession, such as an attorney or physician, employs one or more employees, such member comes within the definition of an employer as defined in the Act and is required to comply with its provisions.
(2) Any person engaged in an agricultural activity employing one or more employees comes within the definition of an employer under the Act. However, members of the immediate family of the farm employer are not regarded as employees for the purposes of the definition.
(3) Provided they otherwise come within the definition of the term "employer," Indians and Indian tribes, whether on or off reservations, and non-Indians on reservations, will be treated as employers subject

to the requirements of the Act. But see *Navajo Forest Products Industries,* 8 OSHC 2094 (1980), where an 1868 treaty exempted tribal enterprises. *See also Navajo Tribe v. NLRB,* 288 F.2d 162 (D.C. Cir. 1961), *cert. denied* 366 U.S. 928 (1961); *FPC v. Tuscarora Indian Nation,* 362 U.S. 99 (1960).

(4) The basic purpose of the Act is to improve working environments in the sense that they impair, or could impair the lives and health of employees. Therefore, certain economic tests such as whether the employer's business is operated for the purpose of making a profit or has other economic ends may not properly be used as tests for coverage of an employer's activity. According to the regulations, to permit such economic tests to serve as criteria for excluding certain employers, such as nonprofit and charitable organizations which employ one or more employees, would result in thousands of employees being left outside the protections of the Act. Therefore, any charitable or nonprofit organization which employs one or more employees is covered under the Act. (Some examples of covered charitable or nonprofit organizations would be disaster relief organizations, philanthropic organizations, trade associations, private educational institutions, labor organizations, and private hospitals.)

(5) Churches or religious organizations, such as charitable and nonprofit organizations, are considered employers under the Act where they employ one or more persons in secular activities. As a matter of enforcement policy, the performance of or participation in religious services (as distinguished from secular or proprietary activities whether for charitable or religious-related purposes) will not be regarded as constituting employment under the Act. Any person, while performing religious services or participating in them in any degree is not regarded as an employer or employee under the Act, notwithstanding the fact that such person may be regarded as an employer or employee for other purposes—for example, giving or receiving remuneration in connection with the performance of religious services. Some examples of coverage of religious organizations as employers would be: a private hospital owned or operated by a religious organization; a private school or orphanage owned or operated by a religious organization; commercial establishments of religious organizations engaged in producing or selling products such as alcoholic beverages, bakery goods, religious goods, *etc.;* and administrative, executive, and other office personnel employed by religious organizations. Some examples of noncoverage in the case of religious organizations would be: clergymen while performing or participating in religious services and other participants in religious services, namely, choir masters, organists, other musicians, choir members, ushers, and the like.

Certain employees are specifically not covered even though their employers may be engaged in a business affecting commerce, *e.g.,* self-employed persons, persons engaged in agriculture who are members of the immediate family of the farmer, and domestic household employees. See 29 CFR 1975.4 (b)(2), .6; See also *Mangus Firearms,* 3 OSHC 1214 (1975); *Organized Migrants in Community Action, Inc. v. Brennan,* 520 F.2d 1161, 3 OSHC 1566 (D.C. Cir. 1975). Family members have, however, been considered employees. *Howard M. Clauson Plastering Co.,* 5 OSHC 1760 (1977).

Many cases are concerned with the definition of who is an employee. Control of the employee, that is, the "power to" as distinguished from the "responsibility for," is the most important factor in determining the existence of the employment relationship. The person with the right of control and supervision will almost always be found to be the employer—but this does not preclude a joint control and dual-employer situation. In *Evansville Materials, Inc.,* 6 OSHC 1706 (1978), a lessor of machinery was found to be liable as an employer where the lease included the operator and the lessor controlled the safety of the operator. In *Acchione & Canuso, Inc.,* 7 OSHC 2128 (1980), a general contractor who leased services of an employee-crane operator and a crane to a subcontractor remained the employer because it retained significant control over the employee's activities, paid his wages, had power to fire him and modify employment conditions, and maintained his records, and because the operator considered the general contractor to be his employer. See also *Camden Drilling Co.,* 6 OSHC 1560 (1978); *Lidstrom, Inc.,* 4 OSHC 1041 (1976); *Horn v. C. L. Osborn Contracting Company,* 591 F.2d 318, 7 OSHC 1256 (5th Cir. 1979)—the Act does not impose liability on a general contractor for the safety of its subcontractor's employees; an employer is liable only if his employees are affected; *Cochran v. International Harvester Company,* 408 F. Supp. 598, 4 OSHC 1385 (W.D. Ky. 1975)—the Act does not apply to relations between an owner of a workplace and a worker who was an employer of an independent contractor; *Boston Chimney & Tower Company, Inc.,* 6 OSHC 2190 (1978)—the employer had the right to make job assignments to the employee; *Joseph Bucheit and Sons Company,* 2 OSHC 1001 (1974); *Frohlick Crane Service, Inc. v. OSHRC,* 521 F.2d 628, 3 OSHC 1432 (10th Cir. 1975)—the lessor of a crane was found to be the operator's employer because the lessee relied upon the operator's expertise and gave no specific direction as to the crane's operation; *Sam Hall & Sons, Inc.,* 8 OSHC 2177 (1980)—firm which leased equipment and operators to city that agreed to provide direct supervision at trenching worksite nonetheless shared control over employees with city. In *Griffin & Brand of McAllen, Inc.,* 6 OSHC 1702 (1978) the Commission reversed a judge's holding that the cited corporation was not the employer of migrant workers because their crew leader was an independent contractor. The Commission stated in that case that, while there is no single criterion for determining the existence of an employer-employee relationship, the following factors and principles should be considered in making this determination:

(1) Whom do the workers consider their employer?
(2) Who pays the workers' wages?
(3) Who has the responsibility to control the workers' activities?

(4) Does the alleged employer have the power as distinguished from the authority to control the workers?
(5) Does the alleged employer have the power to fire, hire, or modify the employment condition of the workers?
(6) Do the workers' ability to increase their income depend on efficiency rather than initiative, judgment, and foresight?
(7) How are the workers' wages established?

In addition, the Commission emphasized throughout its decision that the term "employer" under the Act is not limited to employment relationships as defined under common law principles but rather is to be broadly construed in light of the statutory purpose and the economic realities of the relationship at issue. In *S & S Diving Company,* 8 OSHC 2041 (1980) the Commission reaffirmed the criteria for determining employment status found in *Griffin & Brand* and further affirmed that the judge's finding was supported by a preponderance of the evidence.

Even though an employee may be employed by a covered employer, he may nevertheless not be covered by the Act if another aspect of his employment precludes his coverage. For example, certain temporary labor camps in Florida were recently held not to be "workplaces," and therefore employees who worked for employers at such camps were not covered by the Act, 29 C.F.R §1910.142, because neither the employers nor geographical circumstances required the employee to live in the employer-provided housing. *C. R. Burnett & Sons, Inc. and Harlee Farms,*—OSHC—*on review* No. 78–1103 (1979); but see *Walter C. Mehlenbacker,* 6 OSHC 1927 (1978), where the Commission held that the waiver of rental payments for farm laborers who performed work in exchange for housing created an employer-employee relationship rather than a tenant-landlord relationship.

Sale, cessation, or dissolution of business after an inspection does not resolve an employer of its obligation and liability for a violation that occurred when it was a viable business. *Clearwater Veneer Company,* 3 OSHC 1216 (1975); *American Barrel Company,* 5 OSHC 1491 (1977); *Price-Potashnick-Codell-Oman, a Joint Venture,* 6 OSHC 1901 (1978). Moreover, a company in reorganization under the Bankruptcy Act remains subject to the requirements of the Act, including the imposition of penalties. *Penn Central Transportation Company,* 3 OSHC 1856 (1975); *Penn Central Transportation Company,* 4 OSHC 1746 (1976). The Commission has further stated that a federal bankruptcy court lacks jurisdiction to enjoin Commission proceedings to enforce the Act. *Chicago, Rhode Island & Pacific Railroad Company,* 3 OSHC 1694 (1975). In *W. P. Moore (Life Science Products Co.) v. OSHRC,* 591 F.2d 991, 7 OSHC 1031 (4th Cir. 1979), the court held that the company's officers and directors, under a Virginia statute which so provided, incurred personal liability as employers for violations of the Act which occurred in the two and one-half months between the dissolution and reinstatement of the firm's corporate charter. See also *United States v. Pinkston–Hollar, Inc.;*—F. Supp.—, 4 OSHC

1697 (D. Kansas 1976)—a corporate officer charged as an individual with violations of an Act may be an employer.

D. STATUTORY EXCLUSIONS

Specific exclusions from coverage, as set forth in the Act, are limited to the United States and any state or political subdivision of a state (Section 3(5) of the Act). Thus, the working conditions of federal, state, city, and county public sector employees are not covered by the Act.

1. Federal Employees

(1) Although the Act specifically excludes federal agencies and their employees from its jurisdiction, Section 19(a) of the Act requires the head of each federal agency to establish and maintain its own occupational safety and health program, consistent with OSHA's promulgated standards. This requirement has been affirmed and augmented by Executive Order 11612, issued by the President on July 26, 1971, 36 Fed. Reg. 13,891, and reconsidered and reaffirmed in Executive Order 11807 which was issued by the President on September 28, 1974, 39 Fed. Reg. 35,559.

(2) Executive Order 11807 stated that as the nation's largest employer, the federal government should provide a safe and healthful working environment for its employees and is required to maintain detailed safety and health programs. In seeking to ensure this end, the Executive Order required each agency head to designate an agency official who would be responsible for the management and administration of that agency's safety and health program. Executive Order 11807 also provided that the Secretary of Labor, through OSHA, should aid each agency in setting up and administering agency programs. Included among the services which the Secretary of Labor could provide were evaluations of agency working conditions, recommendations of appropriate safety and health standards, inspection (with advance notice) of agencies to detect unsafe or unhealthful working conditions and assist in correcting such conditions. No citations and penalties would issue for federal noncompliance. Federal employees must utilize the administrative safety and health remedies of their agency before they may resort to any other forum for correction of unsafe working conditions. *Rambeau v. Dow*, 553 F.2d 32 (7th Cir. 1977). There is no private or common law right of action implied under Section 19 of the Act. *Federal Employees for Non-Smokers Rights v. United States*, 446 F. Supp 181, 6 OSHC 1407 (D.D.C. 1978), *aff'd*, 598 F.2d 310, 7 OSHC 1634 (D.C. Cir. 1979), *cert. denied*, 444 U.S. 926, 7 OSHC 2238 (1979). Finally, to aid the Secretary of Labor in carrying out the above services, the Federal Advisory Council on Occupational Safety and Health (FACOSH), which was created under an earlier Order, was continued

under Executive Order 11807. See *Articles of Incorporation, Organization of FACOSH* (January 1975).

(3) In addition to Executive Order 11807, one federal agency, the Department of Labor, promulgated implementing and accompanying safety and health provisions for federal employees in 29 CFR 1960. These regulations provide detailed guidelines for carrying out the directives of Executive Order 11807, including guidelines on agency organization, agency occupational safety and health standards, procedures for inspections and abatement, and recordkeeping and reporting of illnesses, injuries, and accidents. The Part 1960 regulations should be consulted for a more thorough view of each agency's safety and health responsibilities. [See *Occupational Safety and Health for the Federal Employee,* Office of Federal Agency Safety Programs].

(4) Many of the above directives received little attention by the federal agencies. Thus the standards in effect may not be "consistent" with the Act's standards as required by Executive Order 11807. Many agencies also failed to set up procedures for processing complaints concerning substandard conditions. See *Safety In the Federal Workplace,* Ninth Report By the Committee on Government Operations (GPO, 1977).

(5) In order to remedy these deficiencies, President Carter issued a new Executive Order 12196, on Feburary 26, 1980 (effective October 1, 1980), which requires federal agencies to comply with the same occupational safety and health standards as are applicable to private sector employees except when the Secretary of Labor approves alternatives. 45 Fed. Reg. 45235, 45 Fed. Reg. 64873. 29 CFR 1960 was revised to reflect the requirements of Executive Order 12196 and to carry out the Secretary of Labor's responsibilities under Sections 19 and 24 of the Act. 45 Fed. Reg. 69796 (October 21, 1980).

For the purposes of the order, the term "agency" means an Executive department, as defined in 5 U.S.C. § 101, or any employing unit or authority of the federal government, other than those of the judicial and legislative branches. Since Section 19 of the Act covers all federal employees, however, the Secretary of Labor shall cooperate and consult with the heads of agencies in the legislative and judicial branches of the government to help them adopt safety and health programs. The directive encourages the formation of joint labor-management occupational safety and health committees to review agency program information, monitor the program's performance, and consult and advise the agency on the operation of the program. In addition, the order requires federal agencies to:

(1) Furnish employees with places and conditions of employment that are free from recognized hazards that are causing or are likely to cause death or serious physical harm.
(2) Operate an occupational safety and health program.

(3) Designate an agency official to be responsible for the management and administration of the agency occupational safety and health program.
(4) Comply with all standards issued under section 6 of the Act, except where the Secretary approves compliance with alternative standards.
(5) Assure prompt abatement of unsafe or unhealthy working conditions. Whenever an agency cannot promptly abate such conditions, it shall develop an abatement plan setting forth a timetable for abatement and a summary of interim steps to protect employees. Employees exposed to the conditions shall be informed of the provisions of the plan.
(6) Assure response to employee reports of hazardous conditions and require inspections within twenty-four hours for imminent dangers, three working days for potential serious conditions, and twenty working days for other conditions. Assure the right to anonymity of those making the reports.
(7) Assure that employee representatives accompany inspections of agency workplaces.
(8) Operate an occupational safety and health-management information system, which shall include the maintenance of such records as the Secretary may require.
(9) Provide safety and health training for supervisory employees, employees responsible for conducting occupational safety and health inspections, all members of occupational safety and health committees where established, and other employees.
(10) Submit to the Secretary an annual report on the agency occupational safety and health program that includes information the Secretary prescribes.
(11) Conduct periodic inspections, and special inspections in response to complaints within a specified time.
(12) Protect employees from discrimination or reprisal for reporting hazards or participating in agency safety and health program activities.
(13) Allow workers to use official duty time to participate in agency programs.

The Order applies to all executive branch agencies except military personnel and "uniquely military" equipment, systems, and operations. A memorandum of understanding between the Office of Management and Budget and the Department of Defense is being drafted which will define the term "uniquely military."

(6) Under the provisions of the Executive Order, the Secretary of Labor will issue rules for agency program operations, continue to evaluate agency programs on a regular basis, conduct unscheduled workplace inspections in agencies that do not establish safety committees, provide technical services to agencies upon request, enter into arrangements with NIOSH for the conduct of inspections and investigations, and prescribe reporting and recordkeeping requirements.

FACOSH is continued under the directive with eight members representing federal agencies. Under Executive Order 11807, FACOSH membership was left to the discretion of the Secretary of Labor.

New duties have been assigned to the General Services Administration (GSA) under the provisions of the Order, including the requirement that the agency initiate a study with the Department of Labor to identify conflicts that may exist in their standards and develop a procedure to resolve any conflicts found.

GSA also will be required to accompany Department of Labor inspectors upon request, assure prompt attention to reports from agencies of unsafe or unhealthy conditions of facilities subject to the agency's authority, and procure and provide safe supplies, devices, and equipment for agencies. The agency also must give priority to prompt hazard abatement when allocating resources.

(7) An example of the understanding between OSHA and a military personnel agency, although predating Order 12196, is the Memorandum of Understanding between the United States Coast Guard and OSHA to establish procedures to increase consultation and coordination between the two "agencies" with respect to matters affecting the occupational safety and health of employees working on the Outer Continental Shelf of the United States. It was agreed in the Memorandum that the Act would apply to working conditions on Outer Continental Shelf Lands, 29 U.S.C §653(a), but would not apply to working conditions with respect to which the Coast Guard or other federal agencies exercise statutory authority to prescribe or enforce safety and health standards. This cooperation was undertaken in order to maximize protection of employees, avoid duplication of effort, and avoid undue burdens in the maritime industry. 45 Fed. Reg. 9142 (February 11, 1980).

(8) A Department of Defense instruction issued April 30, 1980, sets up guidelines for carrying out Executive Order 12196 pertaining to the prevention of occupational illness among federal employees. Instruction 6055.5, "Industrial Hygiene and Occupational Health," establishes uniform procedures to recognize and evaluate workplace health hazards, provides retention periods for medical and environmental surveillance records, and authorizes the publication of a Defense manual which will recommend medical examinations and biological monitoring criteria for selected occupations.

Provisions of the directive apply to both military and civilian personnel employed in the Office of the Secretary of Defense, the Military Departments, the Organization of the Joint Chiefs of Staff, the Unified and Specified Commands, and the Defense Agencies. The instruction was immediately effective.

(9) Efforts have also begun, through introduction of a bill in Congress (HR 826), to put the safety and health of U.S. Postal Service Employees directly under OSHA jurisdiction, in spite of the new Executive Order, due to the fact of its high injury rate in the federal government.

2. State and Local Employees

Employees of states and political subdivisions of states are also excluded from the Act's coverage. Section 3(5) of the Act. A state is defined in Section 3(7) of the Act to include a State of the United States, the District of Columbia, Puerto Rico, the Virgin Islands, American Samoa, Guam, and the Trust Territory of the Pacific Islands.

Any entity created by a state and which constitutes a department or administrative arm of government, or one which is controlled by public officials and responsible to such officials or the general electorate is deemed to be a "state or political subdivision of a state" not within the definition of employer, and, consequently, not subject to the Act as an employer. 29 CFR 1975.5(b). Various factors listed in 29 CFR 1975.5(c) are taken into consideration in determining whether an entity meets the above test (whether it makes a profit, pays taxes, pays the salary of the employees, is administered by a public official). The regulation also gives examples of the types of entities which normally would and would not be regarded as employers under Section 3(5) of the Act. 29 CFR 1975.5.

In *University of Pittsburgh of The Commonwealth System of Higher Education*, 7 OSHC 2211 (1980), the Commission held that the University of Pittsburgh is not a political subdivision and excluded from the Act's coverage because it retains fundamental characteristics of a private institution. Specifically the Commission found that the state controls only one-third of the University's operating budget and appoints only one-third of its board of trustees, and thus the university was not administered by individuals who are controlled by public officials and responsible thereto or to the general electors.

Employees with contractual or other relationships with state or political subsidiaries have usually been found to have insufficient connection to justify exemption. *Gordon Construction Co.*, 4 OSHC 1581 (1976); *Walsh Construction Co.*, 4 OSHC 1116 (1976).

The Act requires that any state which develops and enforces a state plan must include provisions for all employees of public agencies of that state and its political subdivisions to the extent that it is permitted to do so by law. Section 18(c) (6) of the Act. The enforcement agency is directed to provide for and maintain an effective and comprehensive occupational safety and health program for employees of state and local government agencies which is as effective as the standards contained in the approved plan for private employees. Provisions of such programs when adopted, usually provide, however, that public sector employers shall not be subject to any civil or criminal penalty and that public sector employers shall conduct and maintain programs of self-inspection, to be approved and monitored by the enforcement agency. Employees of states which do not have state plans would be denied even this minimal coverage.

3. Small Employers

Recent addendums to annual fiscal appropriateness bills have provided *inter alia* that (1) no penalties may be assessed for first-time, other-than-serious violations unless ten or more violations are detected; (2) farms employing ten or fewer persons, except for migrant labor, are exempt from the Act; and (3) no penalties may be assessed against small employers who have requested consultation services and make good-faith efforts to correct violations.

4. Jurisdiction over "Government" Contractors

Under terms of "Memorandums of Understanding" entered into between OSHA and the various federal governmental executive departments, OSHA inspectors have been given access to government facilities where there are operations conducted by outside independent contractors. For example, OSHA inspectors are now permitted to enter and inspect United States Army ammunition plants. The Army/OSHA Memorandum specifies that access will not be withheld, delayed, or restricted in hazardous operations, but that the plant commander or his designee may require that the OSHA inspectors be briefed. The plant commander may also take other actions if needed to protect life, property, or the safety of plant operations, the Memorandum stated. Advance briefings will be arranged between OSHA and the Army to avoid delays, and inspectors will be required to obtain prior security clearance before inspecting classified areas and processes in the facility. Following inspections, the OSHA compliance officer will apprise the contractor and the plant commander of hazardous conditions and intent to issue a citation. The plant commander will advise OSHA and whether the condition is within Army jurisdiction or if abatement of the condition would result in a greater danger. If jurisdictional disputes arise that cannot be resolved at the plant level, discussions will be held between OSHA-Army officials. In past cases employers under contract with the federal government were not exempt from OSHA unless the other agency prescribed and enforced safety and health standards, *e.g.,* Department of Defense regulations, which were primarily concerned with efficient defense production. *Gearhart-Owen Industries, Inc.,* 2 OSHC 1568 (1975); *Haas & Haynie Corp.,* 4 OSHC 1911 (1976).

E. RELATIONSHIP BETWEEN OSHA AND OTHER FEDERAL AGENCIES

The Act itself is quite straightforward: OSHA has no jurisdiction to regulate working conditions wherever any other federal agency has statutory authority to prescribe safety or health rules or to enforce safety or health rules. If another federal agency has prescribed or enforced such rules, then the Act does not apply. If the agency has not so acted, then OSHA regulations must govern.

1. Exemptory Provision

In the light of the existing safety and health regulations of many other federal agencies, Congress decided to include Section 4(b)(1) of the Act which provides:

> Nothing in this Act shall apply to working conditions of employees with respect to which other Federal agencies . . . *exercise statutory authority* to prescribe or enforce standards or regulations *affecting occupational safety or health.*" [Emphasis added.]

Substantial litigation has arisen regarding what may seem to be a straightforward and easily understandable provision specifically with respect to the underlined language above. The jurisdictional issue has been frequently raised as an "affirmative defense" by employers in OSHA litigation and enforcement. In several cases the employer has succeeded in demonstrating that another federal agency has preempted OSHA regulation. But the employer has the burden to show that the other agency has actually exercised authority over the specific working conditions for which it was cited. *Gold Kist, Inc.*, 7 OSHC 1390 (1979); *Seaboard Coast Line R.R. Co.*, 6 OSHC 1433 (1978); *Delaware & Hudson Railway Company*, 8 OSHC 1252 (1980). Even if an employer prevails with this argument, it does not, however, establish an industry-wide exemption from the Act. *Greyhound Lines, Inc.*, 5 OSHC 1132 (1977). The generally successful OSHA response in this respect has been to argue that the other agency, under its statute, regulates only certain specific working conditions, and that the employer's working conditions are of a different type from those regulated by the other agency. Exemption is thus established only for the employer, the specific working condition, and the hazard involved. Each employer's working conditions must therefore be analyzed separately to see if an exemption is applicable.

The focus, then, in this argument is on the meaning and significance, in the jurisdictional scheme, of the statutory phrase "working conditions." *Southern Railway Co. v. OSHRC*, 539 F.2d 335, 3 OSHC 1940 (4th Cir. 1976), *cert. denied* 429 U.S. 999, 4 OSHC 1936 (1976); but see *Pullman Brick Co.*, 1 OSHC 3045 (1973), where there was a complete exemption from OSHA coverage for an entire industry regulated by another federal agency, rather than for the more narrowly defined regulated working conditions. This concept is discussed in a later portion of this section.

2. Specific Agency Conflicts

A large number of federal agencies have promulgated safety and health standards covering employees in various industries and have thus come into "conflict" with OSHA. The most frequent conflict has occurred between OSHA and the Department of Transportation: *Marshall v. Northwest Orient Airlines*, 574 F.2d 119, 6 OSHC 1481 (2d Cir. 1978); *Penn Central Transportation Co. v. OSHRC*, 535 F.2d 1249, 3 OSHC 2059 (4th Cir. 1976); *Chesapeake & Ohio Railway Co. v. Brennan*, 535 F.2d 1249, 3 OSHC 2071 (4th Cir. 1976); *Southern Railway Co. V. OSHRC*, 539 F.2d 335, 3 OSHC 1940 (4th Cir. 1976) *cert. denied*, 429 U.S. 999, 4 OSHC 1936 (1977); *Baltimore & Ohio Railroad Co. v. OSHRC*, 548 F.2d 1052, 4 OSHC 1917 (D.C. Cir. 1976); *Southern Pacific Transportation Co. v. OSHRC*, 539 F.2d 386, 4 OSHC 1693 (5th Cir. 1976), *cert. denied*, 434 U.S. 874, 5 OSHC 1888 (1977).

Other federal agencies that also have come into conflict with OSHA include:

(1) The Bureau of Alcohol, Tobacco, and Firearms of the Treasury Department. *Mangus Firearms,* 3 OSHC 1214 (1975).
(2) The Department of Defense. *Gearhart-Owen Industries, Inc.,* 2 OSHC 1568 (1975).
(3) The Environmental Protection Agency. *Organized Migrants in Community Action, Inc. v. Brennan,* 520 F.2d 1161, 3 OSHC 1566 (D.C. Cir. 1975).
(4) The Mine Safety and Health Administration. *Aluminum Company of America v. Morton,*—F.2d—, 3 OSHC 1624 (D.D.C. 1975); *Phoenix, Inc.—Legore Quarries Division,* 1 OSHC 1011 (1972); *Pennsuco Cement and Aggregates, Inc.,* 8 OSHC 1378 (1980).
(5) The United States Coast Guard. *Clary v. Ocean Drilling and Exploration Company,* 609 F.2d 1120, 7 OSHC 2209 (5th Cir. 1980)—the Coast Guard has statutory authority over working conditions of seamen, has sufficiently exercised that authority through issuance of regulations, and OSHA has recognized the validity of that jurisdiction; *cf., Prudential Lines, Inc.,* 3 OSHC 1532 (1975); *National Marine Service, Inc. v. Gulf Oil Co.,* 433 F. Supp. 913 (E.D. La. 1977)—a distinction was made between longshoremen and seamen; *Coastal Construction Company,* 7 OSHC 2235 (1979).
(6) The Atomic Energy Commission. *Newport News Shipbuilding & Drydock Co.,* 2 OSHC 3262 (1975).
(7) The Federal Aviation Administration. *Allegheny Airlines, Inc.,* 6 OSHC 1147 (1977) the Review Commission held that OSHA retains authority to enforce its regulations governing specific working conditions not covered by airline safety manuals issued by the Federal Aviation Administration; *Trans World Airlines, Inc.,* 7 OSHC 1047 (1979). The Review Commission held that the Federal Aviation Administration's approval of employer's safety manual established its authority over an employer's air carrier maintenance operation under Section 4(b)(1) of the Act, and barred the Secretary's claim of jurisdiction; *Northwest Airlines, Inc.,* 8 OSHC 1983 (1980)—the Act does not apply to working conditions of airline maintenance personnel while engaged in ground operations involving servicing of aircraft's landing lights because the Federal Aviation Administration has statutory authority over working conditions of these employees by virtue of its power to promulgate rules governing "safety in the air commerce," and has properly exercised this authority by requiring airlines to develop maintenance manuals which include safety instructions for maintenance personnel.
(8) The Department of Agriculture. *Fineberg Packing Co.,* 1 OSHC 1598 (1974); *CIMPL Packing Company,* 2 OSHC 1436 (1974). The Review Commission in these cases upheld the assertion of OSHA jurisdiction, even though the Secretary of Agriculture asserted jurisdiction under the Wholesome Meat Act. The purpose of the Wholesome Meat Act, to protect consumers of meats from adulterated or unwholesome products, was found not to be consonant with the purpose of the Act.

The Review Commission has held that Section 4(b)(1) is not a "jurisdictional" defense but is rather an affirmative type defense, whose burden of proof rests with the employer. The defense must be timely raised, for, unlike a jurisdictional defense, it cannot be raised at any stage of the proceedings. See *Crescent Wharf & Warehouse Co.*, 2 OSHC 1623 (1975). Moreover, an employer who has not exhausted administrative remedies may not be permitted to bring a Section 4(b)(1) jurisdictional challenge in another forum. *Marshall v. Northwest Orient Airlines*, 574 F.2d 119, 6 OSHC 1481 (2d Cir. 1978).

3. Exercise of Authority

In order for the Section 4(b)(1) exemption to apply, the other federal agency with jurisdiction over the employer must not only have statutory authority to adopt regulations affecting job safety and health, but must actually *exercise* that authority. Section 4(b)(1) provides for preemption of OSHA jurisdiction whenever another agency exercises statutory authority to "prescribe *or* enforce" safety and health rules. Predictably, the issue of what constitutes an actual "exercise of authority" by another federal agency over safety and health conditions of employees has produced much litigation.

Recently, the Commission in *Northwest Airlines, Inc.*, 8 OSHC 1982 (1980) explained that Section 4(b)(1) cases fall into four categories relative to the exercise of authority by a "sister" agency:

> First are those cases involving statutes that are concerned solely with the safety and health of particular employees. See *Idaho Travertine Corp.*, 3 OSHC 1535 (1975) (Department of Interior—mine safety). The second category involves situations in which another agency acts to regulate employee safety and health, but the statutory authority serving as the basis of the agency's actions pertains to matters other than safety or health. See *Haas & Haynie Corp.*, 7 OSHC 1911, (1976) (General Services Administration—procurement statute); *Gearhart-Owen Industries, Inc.*, 2 OSHC 1568 (1975) (Department of Defense—procurement statute). Third are situations in which an agency is empowered by statute to regulate an aspect of public safety and health, and its regulations directed toward that end incidentally affect the working conditions of employees in a manner unrelated to the statutory purpose. See *Fineberg Packing Co.*, 1 OSHC 1598 (1974) (Department of Agriculture—Wholesome Meat Act). Fourth are cases in which a statute authorizes an agency to regulate an aspect of public safety or health, and certain employees directly receive the protection the statute is intended to provide. See *Organized Migrants in Community Action, Inc. V. Brennan*, 520 F.2d 1161, 3 OSHC 1566 (D.C. Cir. 1975) (Environmental Protection Agency—Federal Environmental Pesticide Control Act of 1972); *Texas Eastern Transmission Corp.*, 3 OSHC 1601 (1975) (Department of Transportation Office of Pipeline Safety—Pipeline Safety Act); *Mushroom Transportation Co.*, 1 OSHC 1390 (1973) (Department of Transportation—Motor Carrier Safety Regulations); *Southern Pacific Transportation Co.*, 2 OSHC 1313 (1974), *aff'd*, 539 F.2d 386, 4 OSHC 1693 (5th Cir. 1975), *cert. denied*, 434 U.S. 874, 5 OSHC 1888 (1977) (Department of Transportation—Federal Railway Safety Act); *American Airlines, Inc.*, 3 OSHC 1624 (1975) (Federal Aviation Administration—packaging standards for radioactive material); *Magnus Firearms*,

3 OSHC 1214 (1975) (Bureau of Alcohol, Tobacco, and Firearms—regulation of sale and storage of explosives).

It is important to distinguish between the third and fourth categories. In *Fineberg Packing,* the Department of Agriculture exercised its authority to assure the purity of meat produced in a packing facility by regulating sanitary conditions in the plant. These regulations "affected" the conditions under which the packing plant employees worked, but whether or not the regulations added to their health and safety was purely fortuitous, for the Department of Agriculture regulations did not address that objective.

Texas Eastern Transmission is illustrative of the fourth category. There, Congress was concerned with accidents resulting from the transmission and storage of natural gas. Such accidents had affected both members of the public and employees working in the natural gas industry. Thus, in giving the DOT authority to regulate pipeline safety, Congress intended to protect both employees and the general public. Because employees were in the class the statute intended to benefit, the Commission concluded that DOT had the requisite authority to give rise to a section 4(b)(1) exemption.

The purposes behind the Section 4(b)(1) provision are to provide comprehensive coverage to employees, to avoid a duplication of regulatory effort and enforcement among federal agencies, and to prevent a hiatus in the protection of employees. *Taylor v. Moore-McCormack Lines, Inc.,* 621 F.2d 88, 8 OSHC 1277 (4th Cir. 1980); *American Petroleum Institute v. OSHA,* 581 F.2d 493, 6 OSHC 1959 (5th Cir. 1978), *aff'd sub nom., Industrial Union Department v. American Petroleum Institute,* 440 U.S. 906, 8 OSHC 1586 (1980); *Marshall v. Northwest Orient Airlines, Inc.,* 574 F.2d 119, 6 OSHC 1481 (2d Cir. 1978). As stated by the Commission in *Northwest Airlines, Inc., supra:*

> Just as it would be unreasonable to conclude that Congress intended to leave gaps in coverage, it would be similarly unreasonable to conclude that section 4(b)(1) permits the Act to apply to a working condition which is governed by the rules of another agency issued pursuant to statutory authority having the purpose of protecting employees. That is the exact situation that section 4(b)(1) seeks to avoid.

The avoidance of an undue burden on employers of complying with overlapping and often conflicting regulations was not a consideration in enactment of Section 4(b)(1).

Clearly, where an agency *prescribes* a standard or regulation which specifically covers a particular area *and* the agency *enforces* the standard and regulation, OSHA has no jurisdiction and may not issue citations. The Commission has held that preemption is also clearly mandated by Section 4(b)(1) if another agency prescribes *or* enforces such rules or regulations. *Northwest Airlines, Inc., supra.* Moreover, an agency's regulation or standard need not be precisely the same or equally as rigorous or stringent to preclude OSHA from enforcing its standards. Once the Commission has discovered that a rule promulgated by another agency has the force and effect of law, it need not inquire further as to the manner of enforcement. *Northwest Airlines, Inc., supra;*[8] *Mushroom Transportation Co., Inc.,* 1 OSHC 1390 (1973); *American Airlines, Inc.,* 3 OSHC 1624 (1975). Thus Section 4(b)(1) does not permit an inquiry into the stringency of the enforcement powers that have been established for enforcing the rule or permit the Commission to oversee the adequacy of the other agency's enforcement efforts. *Pennsuco Cement and Aggregates, Inc.,* 8 OSHC 1378 (1980)—promulgation of kiln regulations is an actual exercise of authority notwithstanding the fact that MESA (predecessor to MSHA) may have ceased its inspections of kilns, for any oversight of the adequacy of another agency's enforcement activities is beyond the scope of permissible inquiry under Section 4(b)(1); *Organized Migrants In Community Action Inc., supra.* But see *Texas Eastern Transmission Corp.,* 3 OSHC 1601 (1975), in which although another federal agency adopted a regulation there was little likelihood that it would be enforced.

Just as clearly, however, an agency's inchoate authority to regulate or announced notice of intent to regulate occupational safety and health is not enough to provide the exemption, even though the agency actually regulates other aspects of the employer's operations. *Southern Pacific Transportation Co. v. OSHRC,* 539 F.2d 386, 4 OSHC 1693 (5th Cir. 1976), *cert. denied,* 434 U.S. 874 (1977); *Southern Railway Co. v. OSHRC,* 539 F.2d 335, 3 OSHC 1940 (4th Cir. 1976), *cert. denied,* 429 U.S. 999 (1977); *Illinois Central Gulf Railroad Co.,* 6 OSHC 1476 (1978); *Baltimore and Ohio Railroad Company,* 6 OSHC 1072 (1977).

The argument has been made that where an agency has been found to exercise authority over certain working conditions in an industry, Section 4(b)(1) exempts all of that industry's working conditions from OSHA standards.

The Commission has held, however, that Section 4(b)(1) does not provide an "industry-wide exemption" when an agency has exercised partial authority over an industry. *Southern Pacific Transportation Co. v. OSHRC,* 2 OSHC 1313 (1974), *aff'd* 539 F.2d 386, 4 OSHC 1693 (5th Cir. 1976), *cert. denied,* 434 U.S. 874 (1977); *Marshall v. Northwest Orient Airlines, Inc.,* 574 F.2d 119, 6 OSHC 1481 (2d Cir. 1978); *Greyhound Lines, Inc.,* 6 OSHC 1096 (1977); *Belt Railway Co. of Chicago,* 3 OSHC 1612 (1975); *but see Pullman Brick Co., Inc.,* 1 OSHC 3045 (1973). Exemptions only exist for specific working conditions. *Petrolane Offshore Construction Services, Inc.,* 3 OSHC 1156 (1975).

The Review Commission has set up a threefold guideline to determine whether the exemption provided by Section 4(b)(1) of the Act is operative: (1) The sister agency must have statutory authority to regulate the specific working condition[9]; (2) the sister agency must have actually exercised its authority to prescribe and enforce safety and health standards; and (3) the primary policy or purpose of the enabling legislation of the sister agency must be to assure safe and healthful working conditions for the benefit of employees. See *Hermann Forwarding Co.,* 3 OSHC 1253 (1975); *Luhr Brothers, Inc.,* 5 OSHC 1970 (1977).

In *Fineberg Packing Co., Inc.,* 1 OSHC 1598 (1974) the employer, a meat packer who had to comply with sanitation standards imposed under the Wholesome Meat Act by the Department of Agriculture, claimed OSHA was precluded from exercising authority over the working conditions of its employees. The Wholesome Meat Act required the meat packer to constantly wash his floors, while OSHA required that the floors remain dry. The Review Commission found that the real purpose of the Wholesome Meat Act was to protect consumers and any protection afforded to employees was incidental, thus OSHA jurisdiction was not foreclosed. See also *CIMPL Packing Co.*, 2 OSHC 1436 (1974); *Gearhart-Owen Industries, Inc.*, 2 OSHC 1568 (1975).

4. Burden of Proof

Section 4(b)(1) exclusions are considered to be in the nature of a procedural affirmative defense which must be timely raised by the employer. Section 4(b)(1) is no longer considered to be a limitation on the Commission's subject-matter jurisdiction. *Lombard Bros.*, 5 OSHC 1716 (1977); *Idaho Travertime Corp.*, 3 OSHC 1535 (1975).

5. Exhaustion of Remedies

It is well established in administrative law that before a federal court will consider a question of agency jurisdiction, there must have been exhaustion of administrative remedies, so that the agency will have been accorded an opportunity to initially determine whether it has jurisdiction. In *Marshall v. Burlington Northern, Inc.*, 595 F.2d 511, 7 OSHC 1314 (9th Cir. 1979) the court stated that a determination of Section 4(b)(1) preemption, requires an inquiry into complex issues of law and fact, such that courts must defer to agency jurisdiction. See also *Marshall v. Able Contractors, Inc.*, 573 F.2d 1055, 6 OSHC 1317 (9th Cir. 1978), *cert. denied*, 439 U.S. 880 (1978); *Marshall v. Northwest Orient Airlines*, 574 F.2d 119, 6 OSHC 1481 (2d Cir. 1978); *Marshall v. Sauget Industrial Research & Waste Treatment Assn.*, 477 F. Supp. 88, 7 OSHC 1773 (S.D. Ill. 1979)—political subdivision exemption.

This exhaustion requirement also applies to an employer's challenge to an inspection warrant and the manner of inspection conducted pursuant to that warrant. A court will not review such a challenge until the Commission has first ruled on the challenge. *Babcock and Wilcox Company v. Marshall*, 610 F.2d 1128, 7 OSHC 1880 (3d Cir. 1979); *Whittaker Corporation, Berwick Forge & Fabricating Company v. Marshall*, 610 F.2d 1141, 7 OSHC 1888 (3d Cir. 1979). It must be noted, however, that in these cases the employer submitted to the inspection and *then* sought to litigate the matter in court. Direct review of the issuance of a warrant may be obtainable before the inspection by resisting entry, moving to quash, risking contempt, and, if necessary, acting expeditiously to challenge/appeal *before* the warrant is executed. *Babcock and Wilcox, supra.*

Primary jurisdiction to determine questions of OSHA coverage is lodged in the Review Commission, the statutorily created organ for hearing appeals of OSHA violation citations. *Marshall v. Able Contractors, Inc., supra.* In *Burlington Northern, Inc. v. Dunlop*, 595 F.2d 511, 7 OSHC 1314 (9th Cir. 1979) the court stated that "[t]he policy behind the exhaustion doctrine is well served in this case since a determination of [§4(b)(1)] preemption requires an inquiry into complex issues of law and fact. Accordingly, it is proper for a court to defer examination of such difficult questions of agency jurisdiction until a party has fully exhausted its administrative remedies. . . . [T]he agency will be given a chance to apply its expertise." Quoting *Marshall v. Northwest Orient Airlines, Inc.,* 574 F.2d 119, 6 OSHC 1481 (2d Cir. 1978), which held that an employer cannot bring an action before a U.S. Magistrate to quash warrants based upon preemption if it has not exhausted its administrative remedies.

In *The Matter of Establishment Inspection of In Restland Memorial Park,* 540 F.2d 626, 4 OSHC 1485 (3d Cir. 1976), OSHA alleged that the operations of a cemetery require the use of a backhoe, mowers, and trucks, and in view of the alleged safety hazards created thereby, assumed the responsibility of inspection. Restland, however, refused to permit entry, even by warrant, and then moved to quash the warrant contending that operations of the cemetery did not bring it within OSHA's jurisdiction under the interstate commerce clause. The court stated that whether OSHA is authorized to inspect cemeteries depends upon whether Restland is an employer affecting commerce. Restland sought a determination of that issue prior to permitting entry. The court ruled, however, that the Act created in the Commission a forum for adjudications, which is authorized to pass on all factual and statutory defenses against enforcement actions. The decision of the Commission would be subject to review in the Court of Appeals. Moreover, where Congress has designated a specific forum for review of administrative action, that forum is exclusive, and a concerned party must exhaust his administrative remedies prior to seeking relief in the courts.

In matters involving statutory authority for issuance of a warrant and the scope of the Act's coverage (*i.e.,* whether OSHA or the Federal Railroad Administration has the authority to inspect the railroad's premises), one court relying on *Restland,* has held that the Commission must first decide the question, since it was a factual matter requiring agency expertise, even though the employer moved to challenge the warrant *prior* to inspection. The decision would then be subject to review by a court of appeals. *In The Matter of Establishment Inspection of Consolidated Rail Corporation,*—F. Supp.—, 7 OSHC 2015 (W.D. Pa. 1979), *appeal dism'd;*—F.2d—, 8 OSHC 1966 (3d cir. 1980); see also *Marshall v. Sauget Industrial Research and Waste Treatment Association,* 477 F. Supp. 88, 7 OSHC 1773 (S.D. Ill. 1979), where even though a company argued that it was political subdivision of a state rather than an employer, a jurisdictional question of this nature was held to be a factual matter requiring agency expertise such that the proper procedure to challenge jurisdiction was through agency procedures.

Judicial intervention prior to an agency's initial determination of its jurisdiction is appropriate only in extraordinary circumstances where (1) there is clear evidence that exhaustion of administrative remedies will result in irreparable injury; (2) the agency's jurisdiction is plainly lacking; and (3) the agency's special expertise will be of no help on the question of its jurisdiction. *Marshall v. Burlington Northern, Inc.,* 595 F.2d 511, 7 OSHC 1314 (9th Cir. 1979). A challenge to the constitutionality of the Act is often considered such a circumstance. *McGowan v. Marshall,* 604 F.2d 885, 7 OSHC 1842 (5th Cir. 1979).

F. JURISDICTIONAL RELATIONSHIP OF OSHA TO OTHER LAWS

Before the enactment of worker's compensation laws, employees who were injured or made ill could bring negligence actions against their employers. To provide "uniform" compensation for employees, worker's compensation laws were enacted in the states and the District of Columbia, which usually barred negligence actions against their employees. The Act specifically states that it does not apply to worker's compensation laws so as to change or reduce any rights or liabilities of employers or employees thereunder. Section 4(b)(1) of the Act.

But, pursuant to Section 4(b)(2) of the Act, OSHA supercedes the Walsh-Healey Act, the Service Contract Act, the Construction Safety Act, the Longshoremen's and Harbor Worker's Compensation Act, the National Foundation on Arts and Humanities Act, and other federal laws which have "some" relationship to occupational safety and health. The safety and health standards issued or enacted under these federal laws are deemed to be occupational safety and health standards under the Act.

The relationship between OSHA and NIOSH; OSHA and TSCA/RCRA; OSHA and MSHA; OSHA and NLRA; and OSHA and the Discrimination Laws (Title VII, ADEA, and the Vocational Rehabilitation Act) are discussed in detail in other parts of this book. However, in certain of these areas OSHA and other federal agencies have entered into memoranda of understanding with respect to safety and health enforcement, which serve to guide employers and employees in affected industries to determine the jurisdiction of the statutes involved in their operations.

(1) OSHA/EPA/CPSC Interagency Agreement (42 Fed. Reg. 54, 856) October 11, 1977. As the principal regulatory agencies charged with protecting the public and the environment from the effects of toxic and hazardous substances, it was agreed that interagency communication be established to exchange information, and that common/compatible/consistent testing, research, and development criteria will be established as well as compliance and enforcement procedures. OSHA and EPA are presently writing a chemical labeling rule which will make OSHA responsible for chemical identity requirements on labels in the workplace. EPA labels on consumer products will be only hazard warnings, while OSHA labels will list the chemical components of the substance.

(2) OSHA/MSHA Interagency Agreement (44 Fed. Reg. 22, 827) April 17, 1979. The general principle of the interagency agreement is that as to unsafe and unhealthful working conditions on mine sites and in milling operations, the Secretary of Labor will apply the provisions of the Mine Act and the standards promulgated thereunder to eliminate those conditions. However, where the provisions of the Mine Act either do not cover or do not otherwise apply to occupational safety and health hazards on mine or mill sites (*e.g.*, hospitals on mine sites) or where there is statutory coverage under the Mine Act but there exist no MSHA standards applicable to particular working conditions on such sites, then OSHA will be applied to those working

conditions. Also, if an employer has control of the working conditions at the mine site or milling operation and such employer is neither a mine operator nor an independent contractor subject to the Mine Act, the OSH Act may be applied to such an employer. The application of the OSH Act would, in such a case, provide a more effective remedy than citing someone as a mine operator or an independent contractor subject to the Mine Act who does not, in such circumstances, have direct control over the working conditions.

The Agreement provides detailed descriptions of the kinds of operations included in mining and milling and the kinds of ancillary operations over which OSHA has authority.

When any question of jurisdiction between MSHA and OSHA arises, the appropriate MSHA District Manager and OSHA Regional Administrator or OSHA State Designee in those States with approved plans shall attempt to resolve it at the local level in accordance with this Memorandum and existing law and policy. Jurisdictional questions that cannot be decided at the local level shall be promptly transmitted to the respective National Offices which will attempt to resolve the matter. If unresolved, the matter shall be referred to the Secretary of Labor for decision.

In the interest of administrative convenience and the efficient use of resources, the agencies agreed to the following enforcement procedures:

(a) When OSHA receives information concerning unsafe or unhealthful working conditions in an area for which MSHA has authority for employee safety and health, OSHA will forward that information to MSHA.
(b) When MSHA receives information regarding a possible unsafe or unhealthful condition in an area for which MSHA has authority and determines that such a condition exists but that none of the Mine Act's provisions with respect to imminent danger authority or any enforceable standards issued thereunder provide an appropriate remedy, then MSHA will refer the matter to OSHA for appropriate action under the authority of the OSH Act.
(c) When MSHA receives information regarding unsafe or unhealthful working conditions in an area for which OSHA has authority for employee safety and health, MSHA will forward that information to OSHA for appropriate action.
(d) Each agency agrees to notify the other of the disposition of enforcement matters forwarded to it for appropriate action.
(e) OSHA will not conduct general inspections of mine or mill sites except with respect to those areas set forth in this Agreement.

(3) OSHA/NIOSH Interagency Agreement (44 Fed. Reg. 22,834) April 17, 1979. Among other things, the Agreement provides that:

(a) NIOSH and OSHA will jointly establish criteria, document priorities, and develop their development parameters.
(b) NIOSH will provide technical assistance in OSHA standard setting hearings.

(c) NIOSH will respond to health hazard evaluation requests, whether received directly or through OSHA. Upon initiating a health hazard evaluation, it will contact OSHA for relevant information, and if it encounters a situation that may present a serious occupational safety and health hazard, it will notify OSHA immediately.
(d) NIOSH will provide technical assistance to OSHA in compliance investigations, including court actions and administrative proceedings.
(e) NIOSH and OSHA will coordinate training and educational activities and will exchange technical information.
(f) NIOSH will test and certify equipment, instruments, and other devices and OSHA will encourage the use of such certified items.

(4) OSHA/EEOC Interpretive Guidelines (45 Fed. Reg. 7514) February 1, 1980. Interpretive Guidelines on Employment Discrimination and Reproductive Hazards in the Workplace were issued by the EEOC and OFCCP. Prior to issuance, the agencies had extensive consultations and corrdination with OSHA. Although OSHA did not join in the issuance of the guidelines, and, in fact, expressed its concern regarding employment practices which "deny opportunities to any class of workers on the basis of safety and health," its role of consultation and coordination with EEOC and OFCCP will be spelled out in a Letter of Agreement. The guidelines were withdrawn upon review for it was concluded that the most appropriate method of eliminating descrimination in the workplace where there is a potential exposure to reproductive hazards was through investigation and enforcement on a case by case basis.

(5) OSHA/NLRB Memorandum of Understanding (40 Fed. Reg. 26083) June 20, 1975. A procedure for coordinating Section 11(c) litigation under the [OSH] Act and Section 8 litigation under the National Labor Relations Act was issued in order to avoid duplication and reserve the protection of employee rights in the safety and health area. Under the Memorandum, it was recognized that although there may be some safety and health activities which may be protected solely under the OSH Act, many employee safety activities may be protected under both Acts. However, since an employee's right to engage in safety and health activity is specifically protected by the OSH Act and is only generally included in the broader right to engage in concerted activities under the NLRA, it was agreed that enforcement actions to protect such safety and health activities should primarily be taken under the OSH Act rather than the NLRA.

NOTES

1. The Act includes special provisions (Sections 19 and 18 (c)(6)) for the protection of federal and state (public sector) employees to whom the Act's other provisions are made inapplicable under Section 3(5), which excludes from the definition of the term "employer" both the United States and any state or political subdivision of a state.

2. See also Section 8(c)(1)–(3) (authorizing the promulgation of regulations requiring employers to make and preserve various kinds of records); Section 8(e) (concerning regulations governing inspection work sites); and Section 8(g)(2) (a general rule-making provision authorizing such rules as the Secretary of Labor deems necessary to carry out his responsibilities under the Act).

3. OSHA's walkaround pay regulation was at issue in the *Chamber of Commerce* case. Although OSHA characterized the rule as interpretive, the court disagreed holding that "we do not classify a rule as interpretive just because the agency says it is." Interpretive rules "are statements as to what the administrative officer thinks the statute or regulation means"; they only provide "a clarification or explanation of statutory language." Legislative rules, on the other hand, endeavor to *implement* a statute. Thus a rule is interpretive rather than legislative if it is not "issued pursuant to legislatively-delegated power to make rules having the force of law" or if the agency intends the rule to be no more than an expression of its construction of a statute. As such an interpretive rule would be exempt from the notice and comment obligations of the Administrative Procedure Act. The court in *Chamber of Commerce,* however, characterized the walkaround pay regulation as legislative, intended by OSHA to supplement the Act and not simply to construe it. Since it did not comply with the Administrative Procedure Act, it was held to be invalid. See *Whirlpool Corporation, supra,* where the validity of an OSHA regulation protecting an employee's right to refuse to work in an imminent-danger situation was upheld. No issue was raised concerning whether or not the regulation qualified as an interpretive or legislative rule or was promulgated in accordance with the Administrative Procedure Act. The court thus accepted OSHA's designation of the rule as interpretive.

4. The term "employee" means "an employee of an employer who is employed in a business of his employer which . . . affects commerce." Section 3(6) of the Act.

5. The term "employer" means "a *person* engaged in a business affecting commerce who has employees, but does not include the United States or any state or subdivision of a state." Section 3(5) of the Act. The term *person* means "one or more individuals, partnerships, associations, corporations, business trusts, legal representatives, or any organized group of persons." Section 3(4) of the Act. Reference to standards should also be had because the definitions of employer and employee in certain standards may effectively limit OSHA jurisdiction and coverage. *Dravo Corporation v. Marshall,* 613 F.2d 1227, 7 OSHC 2089 (3d Cir. 1980)—the court concluded that "[b]ecause the Secretary has failed in his regulations to state that a structural shop is included with docks and marine railways as a place of maritime employment to which shipbuilding regulations apply, we believe that the structural shop at issue here is not to be held to shipbuilding safety standards."

6. In *Barger and Iwancio v. Mayor & City Council of Baltimore,* 616 F.2d 730, 8 OSHC 1114 (4th Cir. 1980), the court held that the Act, which is incorporated by reference into Maryland law and covers state, municipal, and county employees, does not apply to working conditions of Jones Act (42 U.S.C. §688) seamen, who are employed by the city fire department, since the Coast Guard is empowered to and does regulate the working conditions of seamen. Maryland incorporated into its law the relevant portions of federal OSHA regulations as a whole, including the explicit exclusion of workers under the jurisdiction of federal agencies other than OSHA.

7. The term "commerce" as defined in the Act means trade, traffic, commerce, transportation, or communication among the several States, or between a state and any place outside thereof, or within the District of Columbia, or a possession of the United States (other than the Trust Territory of the Pacific Islands), or between points in the same state but through a point outside thereof. Section 3(3) of the Act.

8. In order to have the force and effect of law, an agency rule must meet two requirements: the agency must have statutory authority to regulate the particular subject, and the rule must be issued in accordance with congressionally established procedures.

9. The breadth of the exemption seems to turn on the meaning of the phrase "working condition". In *Southern Pacific Transportation Co., v. OSHRC,* 539 F.2d 386, 4 OSHC 1693 (5th Cir. 1976) the Secretary of Labor argued that the phrase "working condition" meant "particular discrete hazards" encountered by the employee in the course of his job activities. The employer argued that the phrase meant "the aggregate of circumstances of the employment relationship—that is, the employment itself." The Fourth Circuit refused to adopt the Secretary's view that regulation by another agency of certain hazards only preempts OSHA's authority with respect to those specific hazards. Instead it held that when the other agency regulates certain hazards which occur within an area in which an employee performs his daily tasks, OSHA's authority over *all* hazards in that area is preempted. As stated by the court, ". . . [T]he term 'working conditions' as used in Section 4(b)(1) means *the environmental area in which an employee customarily goes about his daily tasks.* We are further of the opinion that when an agency has exercised its statutory authority to prescribe standards affecting occupational safety or health for such an area, the authority of the Secretary of Labor in that area is foreclosed." (Emphasis added). This would mean, presumably, that if an employee works in an area in which he is exposed to both safety (guarding) and health (lead) hazards and the other agency only enforces safety regulations, then OSHA is precluded from enforcing both its safety and its health regulations. In *Allegheny Airlines, Inc.,* 6 OSHC 1147 (1977), the Commission held that enforcement of FAA rules and safety manuals constituted the exercise of statutory authority, but that OSHA would retain jurisdiction to enforce its regulations concerning specific working conditions in the airline industry not covered by the rules or manuals.

CHAPTER III
State Plans

A. INTRODUCTION

At the time of the passage of the Act, most states had their own occupational safety and health programs. These programs varied widely from state to state, but none had the extensive coverage, standards-setting, enforcement, and other powers given to OSHA in the Act. The Act therefore charged the Secretary of Labor with responsibility to seek the active participation of the states in bringing about safe and healthful working conditions.

B. JURISDICTION

The Occupational Safety and Health Act of 1970 gave to OSHA the mission ". . . to assure so far as possible every working man and woman in the Nation safe and healthful working conditions and to preserve our human resources. . . ." Section 2 of the Act. Temporarily, it preempted all state job safety and health laws, in accordance with the federal supremacy doctrine. As to issue over which no OSHA standard existed (*e.g.*, boilers and elevators) state enforcement was not preempted.

The Act also required OSHA to encourage the states[1] to assume responsibility for employee occupational safety and health by developing and operating their own job safety and health programs, if the programs were "at least as effective as" the federal program. Nothing requires the states to submit a state plan. If they do not they still have jurisdiction over certain safety and health areas (*e.g.*, fire regulations, training and safety information, coverage of state and local government employees). As to other working conditions, states without approved state plans may not assert jurisdiction—OSHA has preempted the field. *Columbus Coated Fabrics, Division of Borden Chemical Co. v. Industrial Commission of Ohio,* 1 OSHC 1361 (S.D. Ohio, 1973), *appeal dism'd.* 498 F.2d 408, 1 OSHC 1746 (6th Cir. 1974); *Five Migrant Farmworkers v. Hoffman,* 136 N.J. Super 242, 345 A.2d 378 (1975). Thus administration of safe and healthful working conditions was not vested in the federal government exclusively. In order to advance this goal of state operation and administration of health and safety, Section 18 of the Act allows the federal government to cease enforcement activity and cede jurisdiction back to the states to run their own job safety and health programs if they meet certain minimum requirements (*e.g.*, the state develops standards comparable to those developed by OSHA and an enforcement plan that meets the requirements of Section 18(c) of the Act).

The effect, then, of a state plan is to transfer jurisdiction from OSHA to the state. But see *Harold Barclay Logging Co., Inc.,* 5 OSHC 1607 (1977), where jurisdiction to enforce the Act on an Indian reservation located within a state was held to lie with the federal government and not with the state of Oregon. With respect to the question of state jurisdiction over private employers operating on federal property, OSHA stated that the state can attempt entry rights under Section 18 of the Act. However, in *Minnesota v. Federal Cartridge Corporation,* 6 OSHC 1287 (D. Minn. 1977), the court said the Minnesota safety and health law did not extend to a Department of Defense plant operated by a private employer on federal land or to conducting inspections of the plant. Since Section 18 of the Act failed to include explicit authorization to states to exercise authority over federal reservations; the state had no jurisdiction over a private employer operating thereon.

State plan enforcement must be at least as effective as OSHA's and may even be more stringent, but no state may exercise wider jurisdiction than OSHA. For example, when OSHA is preempted by another federal agency under Section 4(b)(1), the states are also preempted.

Nothing in OSHA prevents any state agency or court from asserting jurisdiction under state legislation over any occupational safety and health issue with respect to which no standard is in effect under Section 6 of the Act. However, state plans which do not include those safety and health issues covered by the federal program in effect surrender jurisdiction over such issues to OSHA. For example, a state plan may cover all industries except construction. If such is the case, the state surrenders its jurisdiction for safety and health programs in the construction industry to OSHA, and it is then OSHA's obligation to enforce the federal standards for that industry not covered by the state plan. States without approved state plans are preempted by the Act as to all issues covered by OSHA standards. *Display Builders, Inc.,* 1 OSHC 3220 (1973). If state and federal standards differ in a state without an approved plan, the federal standards prevail. *Courier Newsom Express, Inc.,* 1 OSHC 3213 (1973); Sections 18(a) and (b) of the Act. Employers in states *not* having approved plans must therefore be cautious of nonjurisdictional attempts by a state to enforce state standards which have been preempted by OSHA, as occurred in the case of *Columbus Coated Fabrics, Division of Borden Chemical Co. v. The Industrial Commission of Ohio,* 1 OSHC 1361 (S.D. Ohio 1973), *dism'g app.,* 498 F.2d 408, 1 OSHC 1746 (6th Cir. 1974).

C. DEVELOPMENT

Any state desiring to assume responsibility for developing and enforcing its own program of occupational safety and health standards is required to submit a detailed plan to OSHA for approval. Section 18(b) of the Act. The Act provides for the funding of up to 90 percent of the costs of the development of a state plan. If all of OSHA's requirements are met in the state's proposal, the plan will be approved as operational. Section 18(c) of the Act. Once a state plan is approved, OSHA pays up to 50 percent of the program operation cost.

The state plan approval procedures and the regulations applying the provisions of Section 18 of the Act pertaining to the development and enforcement of state standards are contained in 29 CFR 1902 and 1952.

Federal standards merely serve as a starting point for state programs; the Act does not require that state plans be identical to federal OSHA. There is no provision in either the Act or the regulations for a set-off or credit to an employer for payment of any penalty made to a state in connection with a violation identical or similar to one that is subject to an OSHA citation. *Valley Erection Company, Inc.*, 1 OSHC 3074 (1973)—at the time of the assessment of the federal penalty for violation, the New Jersey plan had not complied with the criteria set forth in Section 18(c) or been approved by the Secretary of Labor.

Organized labor consistently opposes development of state plans. It feels that state agencies are not as effective, more lenient, and more amenable to employer pressure than is federal OSHA. Since the Act was passed in large measure because the states did not do a good job in the area of employee safety and health, unions feel that state plans may eventually undermine the Act. See *South Carolina Labor Council v.*—F. Supp.—(D. S.C. 1973), where a union's request for an injunction to bar federal funds for implementing the South Carolina OSHA plan on the basis that it was not "at least as effective as the federal plan" was denied since the necessary conditions to obtain an injunction were not met (*e.g.*, irreparable harm, merit, balancing of relative harm).

Businesses, especially large multistate employers, sometimes do not like state plans because of the difficulty in complying with varying standards in the different states in which they operate. Large companies with plants in more than one state are also having problems with state plans because they feel that state plans present the multistate employer with a confusing variety of standards, enforcement procedures, and interpretations. They must comply with not only federal OSHA, but with many different state plans where their business operates. The multistate employer often is forced to seek relief from enforcement of state standards and procedures which conflict with federal (or other state) standards and procedures, or varying interpretations of federal (or other state) standards and procedures. A multistate employer must strive for uniformity in machinery, equipment, and work procedures if its business is to remain viable. In many cases, in the ordinary conduct of business, equipment is interchanged between various plants of the employer in different states. However, because of conflicting state plan requirements, a multistate employer may often be forced to adopt machinery, equipment, or work procedures solely to the specifications of one particular state, or obtain a multistate variance.

In *AFL-CIO v. Marshall,* 570 F.2d 1030, 6 OSHC 1257 (D.C. Cir. 1978), *remanding,* — F. Supp. —, 6 OSHC 2128 (D.D.C. 1978), the AFL-CIO challenged the validity of "benchmarks" used as criteria by the Secretary in making the determination as to whether to grant or deny federal approval to state plans. States had been required only to assure that they would provide funding and personnel in order to obtain approval of their plans. The court examined the Secretary's criteria and said that although the "at least as effective" criteria could be considered in granting *initial* plan approval under Section 18(c) of the Act, such a standard was not sufficient for realizing the Act's objectives to provide fully effective enforcement. The court required that the Secretary of Labor establish more specific criteria to supplement the "as effective as" criteria in order to assure a fully effective enforcement program for state personnel and funding as a condition for initial and continuing plan approval. These criteria included the specific number of state employees, and the actual amount of state expenditures adequate to effectively administer the plan. OSHA was also required to establish personnel and funding criteria that were part of a written plan to achieve effective enforcement within a given time. According to the court, the "at least as effective" language of Section 18(c)(2) applied only to standards. Sections 18(c)(4) and (5) are phrased in absolute terms; thus, the "at least as effective" test may only be applied on an interim basis. According to the court, there must be a program to realize a *fully* effective enforcement in the forseeable future, and utilization of benchmarks would help assure proper determination in federal approval of state plans. OSHA submitted its staffing and funding levels on April 25, 1980, which required an overall increase in state inspectors from 2393 to 7462 over a five year period.

Many states would be required to appreciably increase staffing totals to meet the levels set in the benchmark plan. As a result, the states alleged impossibility and OSHA petitioned the court for an extension of the time frame to ten years for states to meet the benchmarks. In its submission to the court, OSHA reported that two "expert workgroups" were convened to help OSHA determine how much time is needed to conduct adequate general schedule inspections, and how frequently these inspections should be conducted within various industries.

To obtain "fully effective" benchmarks, OSHA (1) multiplied the panels' recommendations on inspection duration and frequency by the number of establishments in each state, (2) incorporated estimates of time required to complete other-than-scheduled inspections, and (3) determined the number of available man-hours per year in each state.

OSHA noted that the benchmark model projects had less frequent inspections for nonmanufacturing establishments with fewer than ten employees than for large establishments. Coverage of small employers would be supplemented through response to complaints and special emphasis programs directed toward the most hazardous establishments.

The report noted that the Secretary of Labor will propose to amend state plan requirements in 29 CFR 1952 to reflect the benchmark plan. A new subpart will be added to provide that:

(1) Any state submitting a plan for approval must assure OSHA that it will have an adequate number of inspectors consistent with the benchmark plan.
(2) A state must achieve its full benchmark level for final approval under Section 18(e) of the Occupational Safety and Health Act.

(3) A state plan may be withdrawn if, for two consecutive years, the state fails to adhere to the established schedule for meeting its benchmark.

As a further result of the decision, OSHA announced that the provisions of 29 CFR 1902.39(a), which provide that states may petition for an 18(e) determination (final approval of the plan) one year after certification were suspended until the adoption of rules implementing the Court of Appeals decision. Prior to the decision Minnesota and South Carolina had requested final approval of their plans. Thus in the Act's ten year existence, none of the state plans have been given final approval and no such approval is likely to be given in the near future.

D. STATE PLAN APPROVAL

1. Application and Initial 18(b) Determination

Under the Act and regulations, if a state desires to assume "OSHA"-type responsibilities, a plan for development and enforcement of safety and health standards must be submitted to the appropriate OSHA regional office and then to the national office. NIOSH also receives a copy of the plan. Plan approval is essentially a two-step procedure. The state must first submit its plan for an initial determination under Section 18(b) of the Act. A state plan will be approved if it is demonstrated that within three years it will meet all the criteria necessary to become at least as effective as the federal program. The following requirements must be included in the plan: (1) designation of an agency or agencies which will be responsible for administering the plan; (2) provision for development and enforcement of state safety and health standards which are or will be at least as effective as the federal standards; (3) assurance that enabling legislation exists and necessary qualified personnel are available to enforce the standards (in compliance with the benchmarks); (4) assurance that the state will devote adequate funds to the administration and enforcement of its standards; (5) provisions for right of entry and inspection of covered workplaces which are *at least as effective* as provided by OSHA; (6) prohibition against advance notice of inspections; (7) requirement that employers file reports similar to those required by federal OSHA; and (8) coverage of employees of the state and its political subdivisions. [Section 18(c) of the Act; 29 CFR 1902.]

If the Assistant Secretary, after review of the state's submission, determines that the plan satisfies or will satisfy the criteria set forth above, a decision of *initial* approval is issued and the state may begin enforcement of its safety and health standards in accordance with the plan. A state plan may receive initial approval (and eligibility to receive federal funding for plan approval), even though upon submission it does not fully meet the approval criteria, if it contains satisfactory assurance that the state will take the necessary steps to bring the program into conformity with the regulations within three years of commencement of operations. For example, state plans must contain many provisions which correspond or will correspond to federal

OSHA's, *e.g.*, variances, imminent-danger procedure, antidiscrimination. A schedule within which the state must complete these specified developmental steps (*e.g.*, designation or creation of an agency to administer and enforce the plan and enactment of enabling legislation) is issued as part of the initial approval. Hence the reason the plans are known as developmental at this stage.[2]

2. Developmental Stage

The existence and initial approval of a state plan does not immediately divest the Secretary of Labor of jurisdiction to regulate working conditions. Both the state and federal OSHA may exercise concurrent jurisdiction and federal OSHA will retain discretionary enforcement authority for at least a three-year transitional period. During the transitional period, federal authorities will continuously monitor the state's performance to determine whether the state is meeting the commitments contained in its plan and whether the program will ultimately be as effective as the OSHA program. Section 18(e) of the Act. During this period the state will attempt to bring its plan into compliance with federal requirements. This is the developmental stage. Duplicative activity, including enforcement inspections by both federal and state OSHA, may occur during this period, even though the plan is not fully operational. *Robinson Pipe Cleaning Co. v. Department of Labor and Industry,* — F. Supp.—, 2 OSHC 1114 (D. N.J. 1974); *Par Construction Co.,* 4 OSHC 1779 (1976); *Display Builders, Inc.,* 1 OSHC 3220 (1973). Federal enforcement authority will be exercised to the degree necessary to ensure occupational safety and health protection for covered employees as determined by the federal government. The overall purpose of monitoring and evaluation is to provide a substantive basis of review from which a judgment on the performance of a state can be made. The OSHA *Field Operations Manual,* Chapter XVII–"State Jurisdiction and State Plans," sets forth the policy and procedure to be followed in the monitoring and evaluation of approved state plans. See also 29 CFR 1953. All aspects of a state plan are included within the scope of the monitoring and evaluation system, but implicit in the monitoring and evaluation of state performance is the comparison with comparable federal enforcement performance.

Thus during the transition period, state and federal enforcement will overlap and employers will be required to comply with federal as well as state standards.[3] If at any time during the three-year developmental (monitoring) period it becomes apparent that the state's program is not likely to be "at least as effective as" the federal government's, OSHA can revoke the plan. Federal standards would then go back into effect, and federal enforcement activities would resume. *Par Construction Co., Inc.*, 4 OSHC 1779 (1976); *Seaboard Coast Line Railroad Co. and Winston-Salem Southbound Railway Co. v. OSHRC,* 3 OSHC 1767 (1975), *aff'd,* 548 F.2d 1052, 4 OSHC 1917 (D.C. Cir. 1976), *cert. denied,* 434 U.S. 874 (1977). In the latter case, the approval of the North Carolina plan, with express reservation to the federal government of continued enforcement authority within the state, was found to preclude the subsequent action of the state legislature to increase the authority of the state under the plan of approval. The concurrent juristiction could not be eliminated by legislative action. Section 18(e) of the Act gives the Secretary of Labor the authority to determine when enforcement activities will be conducted exclusively by a state. Thus OSHA could conduct an inspection; the fact that the North Carolina plan excluded railroad employees did not prevent federal OSHA from inspecting a railroad.

3. Operational Stage

When a state's program includes certain key elements and has completed all developmental steps specified in the initial approval decision, the plan will receive formal certification of completion of developmental steps, and an "operational status agreement" will be published in the Federal Register.[4] See 29 CFR 1902.34 and 35. Certification provides public notice of completion of an approved plan's developmental stage and announces the beginning of the evaluation and monitoring program to determine on the basis of actual operation, if the state provides a program at least as effective as the federal program. It is the last trial period for state enforcement. Certification does not render judgment as to the adequacy of state performance. The key elements in the final approval of a state plan are:

(1) Enabling legislation substantially in conformity with OSHA regulations,
(2) An operational review appeals system by which employers and employees may contest enforcement action,
(3) Job safety and health standards that are at least as effective as those promulgated by OSHA, and
(4) A sufficient number of competent, qualified, and adequately funded enforcement personnel.

Publication of the certification acknowledging completion of the developmental steps of a state plan will automatically initiate the evaluation of the state plan for purposes of an 18(e) determination. 29 CFR 1902.35.

"Operational" is the term describing the capacity of a state to enforce standards substantially in accordance with OSHA regulations. Only a state that enters into an "operational status agreement" with OSHA is an operational state. Upon the state plan's assumption of operational status, the federal government staff will remain for an additional period of time (one to two years) to continue monitoring the state's performance. 29 CFR 1953 and 1954. The level of federal enforcement authority exercised in a state will be determined by the scope of the operational agreement, *e.g.*, in those areas in which the state is not operative, federal enforcement activity is undiminished.

In *General Motors Corporation, Central Foundry Division*, 8 OSHC 1298 (1980), the effect of a federal-state agreement delineating enforcement authority was reviewed by the Commission. At the time the citation in the case was issued (1974), the federal government exercised exclusive authority for the occupational safety and health of employees in the state of Michigan. Thereafter (1977), Michigan adopted a state plan which was declared operational, pursuant to a federal-state agreement. Subsequently (1978), federal OSHA conducted a reinspection to determine if the violations alleged in the citation had been abated; not being corrected, a "failure to abate" citation was issued. The Commission first held that the employer had standing to raise the validity of the Secretary of Labor's enforcement authority to issue a failure to correct notification to GMC. Thereafter the Commission held that:

> Section 18(e) of the Act gives the Secretary discretion to decide whether and to what extent he should exercise his enforcement authority during the time between initial and final approval of a state plan. The section states that for this interim period, which must extend for at least three years, the Secretary "may, but shall not be required to, exercise his authority under §§8, 9, 13, and 17 with respect to comparable standards promulgated under section 6. . . ." The first question presented was whether in section 1954.3(b), pertaining to the level of federal enforcement once a state plan becomes operational, the Secretary discretionarily relinquished his right to exercise enforcement authority over activities like the failure-to-correct notification issued here. Section 1954.3(b) provides, in part, that "[w]hen a state plan meets all of these guidelines it will be considered operational, and the state will conduct *all enforcement activity* including inspections in response to employee complaints, in all issues where the state is operational," which indicates only that the state will conduct all enforcement activity allocated to it by the federal-state agreement, since the agreement describes the "issues where the state is operational." The language of Section 1954.3(b) should not be read as a grant of exclusive enforcement authority to the state. It merely indicates that a state may conduct enforcement within the entire range of issues in which it is operational. Accordingly, Section 1954.3(b) does not preclude the Secretary from enforcing final orders under the Act. 29 U.S.C. §659(a) (including the issuance of failure-to-correct notifications). The Agreement further indicates that the state will enforce the abatement of conditions that were never the subject of a prior federal citation.

Commissioner Barnako dissented, concluding that the Secretary can initiate only follow-up inspections that involve prior federally-set abatement dates which extend beyond the date the Michigan plan was accorded operational status. If the abatement date did not extend beyond the date the state plan was accorded operational status, the Secretary would have no authority to initiate follow-up inspections but must receive a referral from the state.

At times "spot-check" or "accompanied" monitoring visits are conducted so as to provide on-site observation of the operational impact of the state program on individual establishments and information about the manner in which state personnel follow administrative and enforcement procedures.[5] An "accompanied" visit is the observation by an OSHA monitor of a state safety or health inspection, working with the state inspector for opening to closing conference and drawing violations to his attention. A "spot-check" is simply a reinspection by OSHA of an establishment previously inspected by a state compliance officer or a visit for the purpose of a CASPA investigation (see discussion *infra*). The OSHA "monitor" will hold an opening conference with the employer to discuss the purpose of the visit. A walkaround will be conducted by the OSHA monitor limited to the areas covered by the state compliance officer unless he limited his inspection without good reason. Employers and employees will be given the opportunity to accompany the monitor.

The employer will be informed that no enforcement action will be taken except for imminent danger situations. (FOM, Chapter XVII). However, the employer will also be informed of violations observed during the closing conference and *asked* to correct the hazards. (In the case of imminent danger, the federal inspector can require that the employer make the correction immediately.) A subsequent enforcement inspection will result should any of the violations appear to be serious. However, correction of the hazard by the time of such inspection could remove all cause for citation or penalty. Evidence of correction should be sent to the state by the employer. If the OSHA monitor is denied entrance to the workplace, the procedures for refusal to permit inspection will be followed. (FOM, Chapter XVII).

If the state is carrying out its promises under its plan, OSHA enforcement personnel will be eventually moved out to other areas in the region or to other regions where they are needed, leaving only enough OSHA personnel to monitor the plan. Thus after the state has accomplished the necessary steps for operational status and has operated its program at a fully effective level for at least one additional year (with a two-year maximum period), federal enforcement activity will cease completely in those areas over which the state has jurisdiction. For example, once a state plan is operational, the state recordkeeping and reporting regulations must be followed by employers, rather than the OSHA regulations (if substantially identical to OSHA's), and recordkeeping under such plan regulations would be regarded as compliance with OSHA recordkeeping requirements. 29 CFR 1904.10 and 1952.4. States may, however, require reporting of accidents, illnesses, and injuries regardless of whether a state plan is in effect, since they may assert authority over any issue not covered by a Section 6b5 standard. *P & Z Company, Inc. v. District of Columbia,* 408 A.2d 1249, 8 OSHC 1078 (D.C. 1979); *Harrington v. Department of Labor & Industry,* 163 N.J. Super 595, 395 A.2d 533 (1978)—New Jersey Drilling Water and Toilet Facilities Act not preempted.

4. Final Stage

At the end of the additional monitoring period, OSHA will make the determination required by Section 18(3) of the Act as to whether the state-plan criteria and indices of effectiveness set forth in 18(c) of the Act and in 29 CFR 1902.3–4 are being followed. Final approval of the plan under Section 18(e) of the Act and 29 CFR Part 1902 may not be granted until at least three years after initial approval and until at least one year after completion of developmental steps. The entire approval process, of which the operational stage is the last step, leads to final approval under Section 18(e). During the evaluation and monitoring period (not less than one year after certification), an 18(e) determination will be made as to whether the state program in operation is a fully effective program of enforcement.

When at least one year of 18(e) evaluation is completed and the state is ready for an 18(e) decision (final approval), a Federal Register notice is prepared inviting public comment. 29 CFR 1902.39. This determination to "certify" the state plan is made on the basis of numerous factors set forth in the regulations. 29 CFR 1902.37. Public comment and request for hearing may again be made, after which, within 120 days, OSHA's final decision concerning 18(e) is publicized. 29 CFR 1902.41. Final approval hinges upon a finding that the state program is "as effective as" federal OSHA's. Thereafter, federal standards are no longer applicable and OSHA enforcement authority will be withdrawn within the state except as to those issues which the state has declined to cover. 29 CFR 1902.42–.43. Due to the *AFL-CIO* decision, the eligibility of states for final approval has been suspended, effective January 20, 1978, until the adoption of regulations carrying out the decision of the court.

OSHA approval of a state plan containing different terms than those contained in the OSHA standards (*e.g.*, a state asbestos standard that does not require medical examinations at certain asbestos concentrations) does not constitute a tacit admission that the federal standard includes similar terms. *GAF Corporation v. OSHRC*, 561 F.2d 913, 5 OSHC 1555 (D.C. Cir. 1977); *United Engineers & Constructors, Inc. v. OSHRC*, 546 F.2d 419, 4 OSHC 1960 (3d Cir. 1976).

The U.S. District Courts have jurisdiction over the Secretary of Labor's approval of state plans despite the lack of a specific provision in the Act for judicial review of such approval. *Robinson Pipe Cleaning Co. v. Department of Labor & Industry*,—F. Supp.—, 2 OSHC 1114 (D. N.J. 1974). For example, in *AFL-CIO v. Marshall*, 390 F. Supp. 972, 2 OSHC 1654 *remanding*, 570 F.2d 1030, 6 OSHC 1257 (D.C. Cir. 1978), *on rem.*,—F. Supp.—, 6 OSHC 2128 (D. D.C. 1978), the court refused to sanction disparities in state plan personnel and expenditures and ordered the Secretary of Labor to develop new criteria for personnel and funding for state plans approved under the Act (Sections 18(c)(4)(5)), to establish a schedule for making such criteria effective in currently approved state plans, and to require adherence to such criteria as a condition of future approvals, certifications, and determinations under Section 18(e) of the Act. Moreover, if a state plan is rejected, a state may, pursuant to Section 18(g) of the Act, review the Secretary's decision in the United States Court of Appeals, within thirty days of the decision.

OSHA regulations on state plans and state plan approval are contained in 29 CFR 1902, 1950, 1951, 1952, 1953, 1954, 1955, and 1956.

E. MAINTAINING EFFECTIVENESS

Prior to and after a state plan receives final 18(e) certification and approval, it is required not only to remain at least as effective as the federal program but is also required to adjust as OSHA makes changes in the federal standards. 29 CFR 1953 and 1954. OSHA is required, under Section 18(f) of the Act, to make a continuing evaluation of the operation of approved state plans. This is necessary to ensure that the state plan continues to be "at least as effective as" the federal "on the same issues." Moreover, if OSHA evaluations of a state plan reveal that the state is not meeting the "at least as effective as" requirement of the Act, or that a state has not substantially completed or will not complete the developmental steps of its plan, or that the state has failed to comply substantially with any provision of the plan, OSHA can revoke the 18(e) determination, reintroduce federal enforcement in appropriate areas, and possibly withdraw plan approval.

Interested parties may petition for state plan withdrawal by petitioning the Assistant Secretary of Labor in writing to initiate withdrawal proceedings under Section 18(f) of the Act. The petition must contain a statement of the grounds for initiating a withdrawal proceeding, including facts to support the petition. 29 CFR 1955.5. In Virginia, the Oil, Chemical, and Atomic Workers Union petitioned OSHA to withdraw the Virginia plan approval because it had been denied party status in connection with employee complaints filed against an employer as provided in 29 CFR 1955. The state of Virginia in *Coleman v. Marshall,* C.A. No. 80–0057R (W.D. Va. 1980) filed suit against the Secretary of Labor in response to the threatened withdrawal of federal approval for its state OSHA program. Virginia asked the court to rule that the standard for determining federal approval *or* withdrawal of its state plan be whether it is "at least as effective" as the federal program and not whether it is "fully effective," as decided in *AFL-CIO v. Marshall.* In that case, benchmarks for a state's minimum staffing and funding levels have been proposed by the Labor Department in accordance with requirements for a "fully effective" state program.

After withdrawal of approval of a state plan, such plan shall cease to be in effect and the provisions of the federal Act shall again apply within the state. The regulations set forth in 29 CFR 1955.10–.43 set forth the requirements of the formal proceedings for withdrawal of a plan. Before instituting formal proceedings, a state is given forty-five days to show cause at the hearing why approval should not be withdrawn. An administrative law judge presides over the proceedings; exceptions to his findings may be filed with the Secretary of Labor, whose decision shall be deemed final agency action. A state may appeal to a United States Court of Appeals a decision by OSHA to reject or withdraw approval of its plan. Section 18(g) of the Act. In 1979–1980, OSHA began the formal process of withdrawing approval of the Indiana and Wyoming plans.[6]

A state may at any time voluntarily withdraw its plan by notification in writing to the Assistant Secretary of its reason to do so; the notification will be published in the Federal Register. Withdrawal by a state relinquishes safety and health jurisdiction to OSHA.

Either before or after an 18(e) determination, any interested person may file a complaint with the OSHA Regional Director or his representative against the operation or administration of any aspect of a state plan. 29 CFR 1954.20(a). A complaint about state program administration, called a CASPA, which can be submitted orally or in writing is distinct from workplace or discrimination complaints. The purpose of a CASPA is to give employees and employers the opportunity to bring specific grievances about state performance to OSHA's attention. The complaint should describe the grounds, therefore, and specify the aspects(s) of the administration or operation of the plan which is believed to be inadequate. Some examples include a pattern of delays in processing cases or of inadequate workplace inspections, or the granting of variances without regard to the specifications within the state plans . . . 29 CFR 1954.20(b). If OSHA determines there are reasonable grounds to believe an investigation should be made, it will investigate the complaint as soon as practicable. 29 CFR 1954.20. Where the complaint is found to be valid, OSHA may require appropriate corrective action as a condition of continued plan approval. If the CASPA presents no reasonable grounds for an investigation to be made, the complainant will be notified in writing.

During the period of concurrent federal and state authority as well as after a Section 18(e) determination has been made, the state will be continually evaluated under Section 18(f) as to the manner in which the plan is implemented and modified. 29 CFR 1953. There are four categories of state plan changes which require review: (1) developmental changes required to meet the plan's developmental schedule as approved under section 18(c); (2) federal program changes including, but not limited to, adoption or revision of occupational safety and health standards which require comparable state action; (3) evaluation changes required as a result of federal evaluation of the state operations; and (4) state-initiated changes.

Under 29 CFR 1953.23(a)(1) and 29 CFR 1953.31(a), the OSHA Regional Administrator is required to advise the states of all federal program changes and all changes required as a result of state evaluations, usually written within fifteen days. Within thirty days after such notification, the states must submit a plan supplement indicating the adoption of the change or a schedule for completing the change. The plan supplement must include a timetable for completion of the charge and explain how it will keep the plan at least as effective as the federal plan. Although the regulation does not explicitly state a definite time limit for the state to complete all types of program changes, *standards changes are specifically required to be completed within six months* after the promulgation of the federal standard, and this time frame has generally been applied to other changes. When OSHA changes a standard, the state must make a similar change in its parallel standard in order to ensure that state plans continue to be as effective as the federal plan, unless it can show that there are compelling local reasons which militate against making the change. The regional OSHA office will make the basic decision on whether or not to approve a major change in a state plan. It will then forward its recommendation to the Assistant Secretary for Occupational Safety and Health, who has final approval authority. Upon approval the state may implement its suggested change. When OSHA adopts an emergency standard, states will be given thirty days after its effective date to develop a parallel emergency standard if the state plan covers that issue. OSHA regional offices will review the state's proposal and approve or disapprove it.

When OSHA revokes a standard, the state will not necessarily have to revoke its comparable standard, unles OSHA determines that revocation is essential to maintaining the effectiveness of the state plan.

Whenever there is a proposed change in a state plan, the state must notify OSHA. The proposed change must be consistent with state law and whenever an extended period of time is involved, include a time frame for accomplishing the change. To keep the public informed about changes, the state job safety and health agency is required to make updated copies of its plan available for public inspection. If proposed state standards do not come up to federal requirements, the state will be given thirty days to revise the proposal, or the standards will be rejected. Acceptable proposed state standards are announced in the Federal Register with an opportunity for public comment.

Even in the area of inspection priorities, OSHA may direct states to make changes or follow prescribed procedures. An OSHA program directive (OSHA Instruction STP 2–1.97, August 15, 1979) requires that states with approved plans use the majority of their resources in compliance program activities and that they adopt a high-hazard general inspection schedule list, or develop and submit their own schedule. Moreover, OSHA has retained authority to regulate certain aspects of state recordkeeping and reporting requirements, along with employer petitions for variances and posting. 29 CFR 1952.4, .8–.10. It should also be noted that the OSHA 500 series of program directives, which transmit federal program changes to states, was actively implemented by OSHA with the issuance of numerous directives concerning federal program changes.

F. ON-SITE CONSULTATION

The Act prohibits OSHA from providing preinspection on-site consultation services to employers without triggering normal enforcement activities. OSHA regulations, in fact, require the issuance of citations for all violations which are observed. 29 CFR 1903.11(a). However, almost all states which administer safety and health programs also provide on-site consultation for employers.

Any state (with or without a state plan) may enter into an agreement with OSHA to perform on-site consultation for private sector employers, pursuant to regulations set forth in 29 CFR 1908. A state having an approved state plan under Section 18 of the Act is eligible to participate in such an agreement if its plan does not include provisions for federally funded on-site consultation for private sector employers. Costs will be reimbursed up to 90 percent incurred under an agreement entered into pursuant to the regulations; in those states with plans approved under Section 18 of the Act, reimbursement will be made in accordance with the provisions of Section 23(g) of the Act which authorizes the Secretary of Labor to make grants to the states to assist them in administering and enforcing occupational safety and health programs. In states without plans approved under Section 18 of the Act, no federal reimbursement for on-site consultation will be allowed, although the activity may be conducted independently by a state with 100 percent state funding. 29 CFR 1908.3

On-site consultation visits will be provided only at the request of the employer and shall not result in the enforcement of any right of entry. A consultant is not authorized to make an unscheduled appearance at the workplace of an employer who has not made a request to conduct an on-site consultation or visit. 29 CFR 1980.4(b). While the identification of hazards by a consultant will not mandate the issuance of a citation or penalty, the employer is, nevertheless, required to take action necessary to eliminate a hazard which in the judgment of the consultant represents an imminent danger to employees or which would be classified as a serious violation. The discovery of such a hazard will not initiate any enforcement activity and referral will not take place unless the employer fails to cooperate in the elimination of the identified hazard. 29 CFR 1908.4(a)(3).

Employers may request on-site consultation to assist in the abatement of hazards cited during an OSHA enforcement inspection. However, an on-site consultation visit may not take place after an OSHA inspection until the employer has been notified that no citation will be issued or, if a citation is issued, until those citation items for which consultation is requested have become final orders. 29 CFR 1908.4 (b)(3).

According to the regulations, the consultant retains the right to confer with individual employees during the course of the visit in order to identify and judge the extent of the particular hazards at the workplace. The consultant is required to explain the necessity for this contact to the employer during the opening conference and the employer must agree to this contact before a visit can proceed. 29 CFR 1908.5(c). Employees also may participate in an on-site consultation visit to the extent desired by an employer. Consultants are required in an opening conference to explain to the employer the relationship between on-site consultation and OSHA enforcement activity and explain the employer's obligation to protect employees in the event that certain hazardous conditions are identified.

Activity during an on-site visit will focus primarily on situations described by the employer when the consultation request was made. The consultant will advise the employer of its obligations under the Act and to the extent possible will identify and provide advice on the elimination of hazards, including, in addition to those specified in the employer's request, any other safety or health hazards observed in the workplace during the course of the visit. When a hazard is identified, the consultant shall indicate to the employer his judgment as to whether the hazard is a violation of the Act, and whether it would be classified as "serious" or "other than serious" based on criteria in the OSHA Field Operations Manual. The element of employer knowledge, however, will not be considered. 29 CFR 1908.5(e)(5).

If a hazard is identified or classified as serious, the consultant and the employer shall develop a specific plan to eliminate the hazard, affording a reasonable time for completion of the abatement. If within ten days of the development of the plan the employer disagrees with the period of time established for the elimination of the hazard, the state shall provide an opportunity for discussion.

The employer must take immediate action to eliminate employee exposure to a hazard which in the judgment of the consultant represents an imminent danger to the employees. If he fails to take the action, the consultant must immediately notify affected employees and appropriate OSHA enforcement authorities. 29 CFR 1908.5(f). Moreover, if an employer fails to take action necessary to eliminate a hazard classified as serious, the consultant shall also immediately notify appropriate OSHA enforcement authorities and provide relevant information. OSHA will then make a determination as to whether enforcement activity is warranted.

Upon completion of the consultation visit, a written report will be prepared and sent to the employer. 29 CFR 1908.5(g). The report shall describe the specific hazards and their nature, reference applicable standards, identify the seriousness of the hazards, and suggest means or approaches to their elimination or control.

On-site consultation is conducted independently of enforcement activity, and the identity of employers requesting on-site consultations as well as the file of the consultant's visit is not forwarded or provided to OSHA for use in compliance inspection activities unless the employer fails to take necessary action for protecting employees from hazards found to be imminent danger or serious violations.

An on-site consultation visit already in progress has priority over OSHA compliance inspections and employers should notify the compliance officer of the visit in progress and request delay of the inspection until after the visit is completed. A request for on-site consultation cannot, however, be the basis for the delay of a compliance inspection. Consultants are required to terminate an on-site consultation visit in progress if any of the following kinds of OSHA compliance inspections are about to take place: imminent-danger investigations, fatality/catastrophe investigations, complaint investigations, follow-up inspections, or other critical inspections as determined by OSHA. 29 CFR 1980.6(b)(2).

The advice of the consultant and a consultant's report is not binding on a compliance officer in a subsequent enforcement inspection. The failure of a consultant to point out specific hazards or the commission of other possible errors by the consultant is not controlling upon a compliance officer and will not effect the regular conduct of a compliance inspection, preclude the finding of violations or issuance of citations, or act as a defense to any enforcement action. 29 CFR 1908.6(c)(1), (2) and (3).

In the event of an inspection, the employer is not required to inform the compliance officer of a prior consultation visit, but it may be used to determine the employer's good faith for the purpose of proposed penalties.

G. EXISTENT STATE PLANS

At the present time there are twenty-three state plans in varying degrees of approval and certification:[7] South Carolina, Oregon, Utah, Washington, North Carolina, Iowa, California, Minnesota, Maryland, Tennessee, Kentucky, Alaska, The Virgin Islands, Michigan, Vermont, Nevada, Hawaii, Indiana, Wyoming, Arizona, New Mexico, Virginia, and Puerto Rico. Some of these states enforce primarily federal OSHA standards, while others enforce primarily state OSHA standards. Some states enforce a combination of the two. There is no requirement that approved state plans contain a general duty clause, but many do. See *Green Mountain Power Corp. v. Commissioner of*

Labor & Industry, 136 Vt. 15, 383 A.2d 1046, 6 OSHC 1499 (1978), for an example of litigation procedures under a state-plan general duty clause. Those states which do not have state plans are precluded from enforcing job safety and health standards covering issues which are covered by the federal OSHA programs. Thirteen state plans have been withdrawn: Colorado, Connecticut, Georgia, Illinois, Maine, Mississippi, Montana, New Hampshire, New Jersey, New York, North Dakota, Pennsylvania, and Wisconsin. The remaining states either never submitted plans or submitted plans but failed to pass enabling legislation.

A complete copy of the state plan of a particular state is required to be kept at the office of the state agency administering the plan. 29 CFR 1952.5.

NOTES

1. The term "state" includes the District of Columbia, Puerto Rico, the Virgin Islands, American Samoa, Guam, and the Trust Territory of the Pacific Islands. Section 3(7) of the Act.

2. When a state submits a plan to OSHA for approval, a notice summarizing the plan is published in the Federal Register. Public comment and requests for hearing may be made, but no hearing is required and none is usually held. If OSHA proposes to reject a plan, the state will be provided with an opportunity for a formal hearing. The decision approving or rejecting a plan is published in the Federal Register. 29 CFR 1902 and 1952.

A state plan need not be "fully completed" in order to be approved. Section 18(c)(2) of the Act requires that a plan provide for the *development* and enforcement of safety and health standards which are or *will be* at least as effective as OSHA standards. Based on this authority, OSHA has incorporated the concept of *developmental plans* into state plan regulations. Any developmental plan, that is, a plan not fully meeting the criteria set forth in 29 CFR 1902.3 at the time of approval, must meet the requirements of 29 CFR 1902.2(b). See 29 CFR 1952.3. Satisfactory assurances must be given by the state that it will take the necessary steps to bring the state program into conformity with the criteria within the three-year period immediately following the commencement of the plan's operation. Under 29 CFR 1902.2(b) a developmental period may last no longer than three years. Developmental steps include the specific commitments made by the state to bring the plan into conformity with the approval criteria and the changes required when aspects of the plan which were conditionally approved subject to evaluation have been evaluated and found to be unsatisfactory. *Id.* This concept has withstood court challenge in *Robinson Pipe Cleaning Co. v. Department of Labor & Industry, State of New Jersey,*—F. Supp.—, 2 OSHC 1114 (D. N.J. 1974), where an employee's challenge to a state citation on the basis that the state had not enacted enabling legislation was found to be without merit. A state may incorporate state regulations in effect before OSHA. Since a developmental state plan may have little or no enforcement authority, OSHA enforcement of federal standards may continue for a time following plan approval. As the state adopts standards and increases its enforcement activity, federal involvement will decline.

3. Recordkeeping, reporting, and posting requirements under state plans must, however, meet OSHA requirements. 29 CFR 1959.4, .10.

4. OSHA has developed and used an "operational status agreement" which has the effect of withdrawing active OSHA compliance activity in states demonstrating

effective enforcement during the developmental stage (*e.g.*, enabling legislation, standards, qualified personnel, review/appeals system). OSHA retains jurisdiction over employee complaints and state plan complaints. In *National Steel & Shipbuilding Co.*, 6 OSHC 1131 (1977), execution of such an agreement barred federal enforcement in the areas specified therein.

5. Spot-check monitoring visits of workplaces inspected by state personnel will not be continued pursuant to restrictions set by Congress in the 1980 Appropriations Bill.

6. The procedure for citations and penalties in Wyoming required the use of the criminal court system, with alleged violators entitled to a jury trial. The state also "probably" would need criminal rather than civil search warrants if workplace entry was refused. Criminal warrants are more difficult to obtain than civil warrants.

Other distinctions in the state plan also were attacked by OSHA, including rules prohibiting self-incrimination, which, according to OSHA, "may make it difficult to gain access to records required by OSHA standards." County prosecutors are expected to handle Wyoming cases before criminal judges and juries, while the federal system provides for administrative law judge review of contested cases. Employee participation in criminal proceedings is not guaranteed, as OSHA contends it is under the civil appeals system.

Moreover, in criminal cases violations must be established "beyond a reasonable doubt," while a "preponderance of evidence" is needed for civil cases. Finally, prosecutors "may be reluctant" to take cases to court, since employers convicted of violations would have criminal records. Employers may be hesitant to appeal citations and penalties in the face of possible criminal conviction.

In Indiana the deficiencies alleged included a lack of sufficient industrial hygienics, failure of safety inspectors adequately to identify workplace hazards, and improper calculation of proposed penalties.

7. Eight state plans have completed developmental steps and received certification; fifteen state plans are approved and have completed developmental steps; and thirteen state plans have been withdrawn. No state plan has yet received an 18(e) determination.

CHAPTER IV
Standards

A. INTRODUCTION

1. What Is a "Standard"?

The Act in Sections 2(3) and 6 extends to the Secretary of Labor authority to promulgate *standards* for the improved safety and health of the employees; the Act does not itself contain any standards, nor does it prohibit specific employment practices.

The Act defines the term "occupational safety and health standard" as "a standard which requires conditions, or the adoption or use of one or more practices, means, methods, operations, or processes, reasonably necessary or appropriate to provide safe or healthful employment and places of employment." Section 3(8) of the Act. Put in general terms, a standard is both a rule of conduct for avoiding hazards in the workplace and a legally enforceable obligation governing conditions, practices and operations in order to assure a safe and healthful workplace; it requires employers to take concrete measures to reduce a specific hazard to employees. *Louisiana Chemical Association v. Bingham,* 496 F. Supp. 1188, 8 OSHC 1950 (W.D. La. 1980); *American Industrial Health Council v. Marshall,* 494 F. Supp. 941, 8 OSHC 1789 (S.D. Tex. 1980).[1] In *American Industrial Health Council,* the generic cancer policy, 29 CFR 1990, was challenged in the district court, as nothing more than a set of general regulations which establish procedures and evidentiary rules to govern future enactment of standards. The court rejected this argument finding that:

> Although the cancer policy imposes procedures whereby the Secretary will issue specific standards in the future, it also establishes binding substantive limitations on the Secretary and on industry. See *e.g., Bethlehem Steel Corp v. Occupational Safety and Health Review Commission,* 573 F.2d 157, 6 OSHC 1440 (3d Cir. 1978) ("The purpose of OSHA standards is to improve safety conditions in the working place by telling employers just what they are required to do in order to prevent or minimize danger to employees.") *Id.* at 161.

The court also stated that:

> . . . the generic cancer policy requires that exposure conditions for Category 1 toxic substances shall be limited to the lowest feasible level and that engineering and work practice controls must be adopted as the primary means of compliance.

> *See* [29 CFR] 1990.142; 45 Fed. Reg. at 5286. These requirements are binding; their validity may not be challenged in subsequent proceedings unless the cancer "standard" is amended. *See* 45 Fed. Reg. at 5214. The generic standard further establishes binding criteria for identification, classification, and regulation of potential occupational carcinogens. [29 CFR] 1990.112; 45 Fed. Reg. at 5284. Those criteria are not subject to relitigation. 45 Fed. Reg. at 5214. The court finds [therefore] that the cancer policy is a nationally applicable standard addressed to regulation of toxic substances; further, the determinations underlying the promulgation of the standard are essentially the same as those underlying other standards promulgated by the agency.

The fact that OSHA itself had characterized the policy as a regulation was not persuasive to the court:

> The plaintiffs rely in part on the agency's characterization of the policy as a "regulation". The Court concludes that "regulation" was used in the instant policy much as the term has been used in other contexts involving OSHA "standards", *see, e.g., Nat. Indus. Contractors v. OSHRC,* 583 F.2d 1048, 1050–52, 6 OSHC 1914, 1916–17 (8th Cir. 1978); *National Constructors Ass'n v. Marshall,* 581 F.2d 960, 972, 6 OSHC 1721, 1728 (D.C. Cir. 1978). Moreover, the Secretary concluded that this generic approach to standard-setting is a method which is necessary and reasonable for regulation of occupational carcinogens. 45 Fed. Reg. at 5014; *See* [Sections 3(8), 6(b) (5)]. The Secretary further concluded that the standard at issue is "an occupational safety and health standard within the meaning of the Act." 45 Fed. Reg. at 5015; *See* [Section 3(8)]. In such circumstances, the agency's use of the term "regulation" cannot foreclose a finding that the instant policy is a standard.

Finally, the court considered the effect of the decision in *Whirlpool Corp. v. Marshall,* 435 U.S. 1, 8 OSHC 1001 (1980), wherein the Court reviewed the Secretary's rulemaking power pursuant to [Section 8(g)(2)]. The court held that:

> Plaintiffs assert that the similarity of the Secretary's announcing intended action to be taken against employers in the Whirlpool regulation and the instant generic policy is the decisive factor in determining that the instant policy is a general regulation. The Court finds that assertion unpersuasive. The Whirlpool regulation apparently was promulgated pursuant to the authority of only [Section 8(g)(2)], and the character of the regulation was not at issue in the Whirlpool litigation. *See Usery v. Whirlpool Corp.,* 416 F. Supp. 30, 33, 4 OSHC 1391, 1392 (N.D. Ohio 1976), *rev'd and remanded sub. nom. Marshall v. Whirlpool Corp.,* 593 F.2d 715, 7 OSHC 1075 (6th Cir. 1979), *aff'd.* 63 L.Ed. 2d 154, 8 OSHC 1001 (1980), in which the District Court observed the following: "Pursuant to his authority under the Act to issue regulations, the Secretary promulgated [the challenged regulation]." Id. at 33 [1392]. In contrast, the Secretary promulgated the instant policy pursuant to [8](g)(2).

See also *American Petroleum Institute v. OSHA,*—F.2d—, 8 OSHC 2025 (5th Cir. 1980).

In *Louisiana Chemical Association,* a case involving the rule entitled "Access to Employee Exposure and Medical Records," 29 CFR 1910.20, the court, in holding that the rule was in fact a standard, stated:

> As commonly understood a "standard" provides a means of determining what a thing should be. It is a measure against which a thing may be compared to make

> an immediate determination of whether the thing conforms to established criteria. As used in the definition of an occupational safety and health standard, the word "standard" most rationally denotes a similar meaning with respect to those components of the work environment which affect employee safety. Thus it may be concluded that Congress intended an occupational safety and health standard to establish a measure against which the conditions existing or the practices, means, methods, operations, or processes used in a given work place may be compared for an immediate determination of whether the work place is safe.
>
> * * *
>
> Judicial recognition of the apparent Congressional intent to have standards address hazards may be found in *Industrial Union Dept., A.F.L.-C.I.O. v. American Petroleum Institute,*—U.S.—, [8 OSHC 1586] 48 U.S.L.Week 5022 (1980). . . . In that case a plurality of the court agreed that with regard to toxic substances [Section] 3(8) requires the Secretary [of Labor] to find as a threshold matter that the toxic substance in question poses a significant health risk in the workplace and that a new, lower standard is therefore reasonably necessary or appropriate to provide safe or healthful employment or places of employment. . . .
>
> From this discussion the court has concluded that Congress intended an occupational safety and health standard to possess two qualities: (1) it must address a particular hazard existing in a work environment; and (2) it must establish a measure against which the condition existing or the practices, means, methods, operations or processes used in a work place may be compared for an immediate determination of whether the work place is safe with respect to the hazard addressed by the standard.

By identification of the above two essential qualities of a standard, the court rejected the plaintiff's assertion that a standard must regulate the actual *physical* environment of the workplace. The court felt the word practices in Section 3(8) could reasonably be read to reflect a desire on Congress' part to èxtend to OSHA authority to promulgate standards requiring employer adherence to certain administrative procedures designed to reduce an identified occupational hazard posed by procedures different from those prescribed.

2. Philosophy of Standards

Congress declared that the primary way in which to "assure so far as possible every working man and woman in the Nation safe and healthful working conditions" was to authorize the Secretary of Labor to promulgate mandatory occupational safety and health standards. Section 2 of the Act provides that this authorization resulted from a realization on the part of Congress that industrial safety and health problems are as complex and changing as American industry itself, and cannot be solved by a lengthy list of prohibitions spelled out in a statute. (116 Cong. Rec. Section 17470, Oct. 8, 1970). It was intended that new, detailed standards would be developed and that existing ones would be continually evaluated, refined, and revoked in order to keep abreast of changes in industrial technology. The mechanisms made available to the Secretary of Labor to assure compliance with the standards were civil and criminal penalties and abatement.

3. Categories of Standards

OSHA has promulgated thousands of standards which can be broken down in several different ways, such as by *category*.

OSHA standards fall into four major categories: General Industry, 29 CFR 1910; Maritime (Shipbuilding and Longshoring), 29 CFR 1915–1918; Agriculture, 29 CFR 1928; and Construction, 29 CFR 1926.[2] The Federal Register is the best source of information on standards, since all standards are published therein when promulgated, amended, or deleted. Published standards are also compiled in the Code of Federal Regulations; in addition OSHA publishes the standards in booklet form which are available from the agency at a nominal charge.

Publication in the Federal Register constitutes constructive notice of a regulation or standard, and also creates a rebuttable presumption that the standard or regulation has been validly published and that there has been compliance with all necessary requirements regarding publication. *Leader Evaporator Co., Inc.*, 4 OSHC 1292 (1976). It should be noted, however, that not all standards are published in the Federal Register or the Code of Federal Regulations. Many general industry standards, such as those of the American National Standards Institute (ANSI), National Fire Protection Association (NFPA), American Petroleum Institute (API), American Society of Testing Materials (ASTM), and National Electrical Contractor Association (NECA), among others, are *incorporated by reference* into OSHA standards. See, e.g. 29 CFR 1910.6; 1926.31; 45 Fed. Reg. 44090 (June 30, 1980). Before an agency may incorporate any material by reference into the Code of Federal Regulations (CFR), it must make the material reasonably available to the class of persons affected by it. In *Charles A. Gaetano Construction Corporation*, 6 OSHC 1463 (1978), the Commission stated that publication in the Federal Register creates a *rebuttable presumption* that the "incorporated by reference" standards have been validly published. Materials incorporated by reference are also presumed to be *reasonably available* and to be incorporated with the approval of the Director of the Federal Register in accordance with the limitations on the use of such incorporated by reference materials found at 5 USC Section 522(a)(1).[3] See also *Leader Evaporator Co.*, 4 OSHC 1292 (1976); *Goldkist, Inc.*, 7 OSHC 1855 (1979); *Atlantic & Gulf Stevedores, Inc.*, 3 OSHC 1003 (1975), *aff'd*, 534 F.2d 541, 4 OSHC 1061 (3d Cir. 1976); *Corbin Lavoy d/b/a Empire Boring Co.*, 4 OSHC 1259 (1976); *Ladish Co. Tri-Clover Division*, 8 OSHC 1809 (1980).[4] An employer bears the burden of rebutting at least one of these presumptions in order to prove the invalidity of a standard which has been incorporated by reference. These three presumptions of course are: (1) validity of the promulgation procedures (*e.g.*, not a national consensus standard); (2) proper incorporation; and (3) reasonably available. With respect to reasonable availability, evidence introduced would relate not only to the period of time subsequent to citation, but also would relate to attempts to obtain the standard in order to accomplish compliance before the time of the citation and should include evidence in addition to an employer's own geographically localized efforts to obtain the incorporated material. Unavailability should be demonstrated for employers in general. Each agency that wishes to have material which has been incorporated by reference into the CFR remain effective also must submit annually to the Director of the Federal Register a list of the material and the date of its *last* revision. 1 CFR 51.13. A recent Federal Register notice (45 Fed. Reg. 44090) listed the referenced standards and the corresponding OSHA regulations in which they are to be incorporated. Except for established federal standards in 29 CFR 1910.18, the OSHA regulations themselves do not contain guidelines with respect to the effect of changes in national consensus standards which have been adopted as OSHA standards. The general incorporation by reference regulations set forth in 1 CFR Part 51 would apply. In *Trojan Steel Company,* 3 OSHC 1384 (1975), the oxygen standard 29 CFR 1910.104, was adopted on April 27, 1971, as a national consensus standard within the meaning of Section 3(9) of the Act. The standard was published in the Federal Register on May 29, 1971. The source of the standard was National Fire Protection Assocation (NFPA) No. 566–1965. Between the date of adoption and publication, the annual meeting of the NFPA was held at which NFPA No. 50–1971 was adopted, which superseded the earlier NFPA No. 566–1965. On October 4, 1972, the Secretary of Labor republished the standard. The employer was thereafter cited under 29 CFR 1910.104. At the hearing the employer argued and the administrative law judge held that NFPA No. 566–1965 ceased to be a national consensus standard after May 29, 1971, and that the Secretary of Labor lacked authority to adopt that standard in October, 1972, without first complying with the rule-making procedure prescribed under Section 6(b). Although the Commission was divided in its view as to whether it had the authority to review rule-making actions of the Secretary of Labor and thus did not reach the merits of the issue, Commissioner Cleary posited whether Section 6(b) requires the Secretary to use the procedures set forth therein to keep a rule that has already been adopted because a private organization has changed the private standard upon which the Secretary's rule is based. Commissioner Cleary thought that it did not, for to hold differently:

> would in effect delegate unlawfully rulemaking power to the private organization. Any private organization that has adopted a national consensus standard that was subsequently adopted as an occupational safety and health standard would hold a power to negate the action of the Secretary of Labor.

See also *Kaiser Aluminum & Chemical Corporation,* 8 OSHC 1054 (1979).

Publication of a proposed standard will not relieve an employer of its duty to comply with the existing OSHA standard. Similarly compliance with a private industry standard seems not to be a defense when an OSHA standard is in effect, even though that standard incorporates by reference a prior, revised version of the private industry standard. See *United States Steel Corporation,* 5 OSHC 2073 (1977); *Clover Beef Co.,* 1 OSHC 1254 (1973); *Trojan Steel Company,* 3 OSHC 1384 (1975).

Although the procedures set forth in 5 U.S.C. Section 552(a) and 1 CFR 51 provide that material approved for incorporation by reference has the same legal status as if published in full in the Federal Register, agencies are required to indicate where an employer can obtain each item included in the Federal Register notice of incorporation. The materials approved for incorporation by reference must be available at OSHA offices and would also be available for inspection and copying at the offices of the standards issuing organization and the Office of the Federal Register, 1100 L St., N.W., Washington, D.C. 29 CFR 1910.6; 1926.31.

If the agency wishes to include any amendment to material already approved, the agency must publish a notice to that effect in the Federal Register, and make the amendment available as indicated or as modified in the notice of amendment. Amendments to standards are not properly incorporated until notice is given in the Federal Register, the amendments are filed at the Office of the Federal Register, and made available to the public.

Some standards, such as NFPA provisions, although not incorporated by reference as national consensus standards, may be used to define and interpret cited OSHA standards. *Northern Metal Co.,* 3 OSHC 1645 (1975); *Gold Kist, Inc.,* 7 OSHC 1855 (1979). The ad hoc incorporation of, or reference to private industry standards may be eliminated if the proposed revision of the electrical standards for general industry is any indication. The proposal would end the mandatory reference to the National Electric Code and substitute a smaller text as the OSHA standard. 44 Fed. Reg. 55274 (1979). To avoid present "incorporation-by-reference" problems, employers should endeavor to ascertain which standards apply to their operations (e.g., conduct a standards search), obtain copies of the standards, and familiarize themselves with the requirements imposed thereunder.

In addition to the four major categories of standards, there are specialized *subcategories* of standards contained within the general categories. These subcategories of standards, called "Parts," address particular conditions and hazards associated with differing specialized types of employment. For example, 29 CFR 1926.750, Subpart (R), deals with steel erection; 29 CFR 1910 .401–.440, Subpart (T), contains the standards for commercial diving operations. The diving standards require that anyone employing a diver must insure that the working conditions and work procedures comply with minimum safety and health requirements, since there are special dangers and hazards which are applicable to this employment. For example, the threat of entanglement, trench collapse, pinch-nip, squeeze-points, explosions, and falling objects are greatly amplified when working underwater. There is also difficulty in moving and orienting oneself underwater; there are constraints on communication, and the isolation that is experienced by a diver may create physiological as well as psychological stress. Water currents, low water temperatures, and rough seas also add to divers' problems. Because of hyperbaric conditions, divers must rely completely on life-supporting systems, and on the competence and reliability of persons who operate, monitor, maintain, and repair them. Thus the special nature of commercial diving puts restrictions on activity and work performance. The OSHA standards spell out basic requirements for the protection of the divers and other dive-team members' safety and health. Special qualifications are required, medical examinations are required; special training, briefing programs, communication systems, and recordkeeping requirements are also imposed. See generally "Questions and Answers about the OSHA Standard for Commercial Diving Operations, United States Department of Labor, Occupational Safety and Health Administration," 1979–OSHA 3036. See also *Taylor Diving & Salvage Company, Inc. v. U.S. Department of Labor,* 599 F.2d 622, 7 OSHC 1507 (5th Cir. 1979), where the medical examination provision of the standard, 29 CFR 1910.411, was invalidated as not reasonably necessary or appropriate to provide safe and healthful working conditions in the commercial diving industry, since it imposed upon employers a mandatory job-security provision.

Standards may also be broken down by *characterization* as either horizontal or vertical. Horizontal standards typically apply to several employers within many different industries, while vertical standards apply to very few employers within a particular industry. Finally, standards may be referred to as *specifications* which focus on the particular environment, practice or condition in the work place or *performance* which details the procedures, manner, and acts by which work is to be performed. A performance or programmatic approach to standard compliance is not normally acceptable. Thus, deviation from a specification standard and/or an employer's use of its own safety and health methods is not normally permitted absent the application for a variance. See *Sierra Construction Corp.,* 6 OSHC 1278 (1978).

B. TYPES OF STANDARDS

1. Interim Standards

The Act's objective of assuring so far as possible every working man and woman in the nation safe and healthful working conditions was to be achieved in part by "building upon advances made through employer and employee initiative for providing such conditions." Section 2(b)(4) of the Act. Upon enactment of the Act and within two years of its effective date, the Secretary of Labor was given the authority to adopt/promulgate a large body of pre-existing public and private standards as OSHA standards with only one specific caveat: unless they would "not result in improved safety or health for specifically designated employees." Section (6)(a) of the Act. These interim standards, which were of two types—national consensus standards and established federal standards—were to remain in effect unless and until they were modified or revoked by the Secretary. The power to promulgate interim standards expired on April 28, 1973.

a. *National Consensus Standards*

The term "*national consensus standard*" means any safety and health standard or modification thereof which: (1) has been adopted and promulgated by a nationally recognized standards-producing organization under procedures whereby it can be determined by the Secretary that persons interested and affected by the scope or provisions of the standard have reached a substantial agreement on its adoption; (2) was formulated in a manner which afforded an opportunity for diverse views to be considered; and (3) has been designated as such a standard by the Secretary, after consultation with other appropriate federal agencies. Section 3(9) of the Act; *Noblecraft Industries, Inc. v. Secretary of Labor,* 614 F.2d 199, 7 OSHC 2059 (9th Cir. 1980).

National consensus standards comprise the great bulk of OSHA standards. The "nationally recognized standards-producing organizations" such as ANSI, ASTM and NFPA, which, it was contemplated, would provide the vehicle and major source for standards adoption, did in fact do so and continue to do so. ANSI, for example, is a private organization which utilizes the experience of its members to develop and update safety and health standards. Standards are either drafted by a committee or by an interested group of persons for review by a committee. Upon committee approval, the standards become effective. ANSI works closely with OSHA to develop these standards; in fact, OSHA representatives actually participate on certain ANSI committees.

Since national consensus standards had been [and are] "formulated in a manner which afforded an opportunity for diverse views," and would be more readily accepted by employers, employees, and labor unions, Congress stated that the Secretary could promulgate them without regard to the rule-making requirements of Section 6(b) of the Act or the Administrative Procedure Act. These requirements include notice, hearing, and evaluation of evidence. *Deering Milliken, Inc.,* 6 OSHC 2143 (1978). In accordance with Section 6(a) of the Act, the Secretary of Labor adopted many "interim" national consensus standards, publishing notice to that effect in the Federal Register. (*e.g.,* 36 Fed. Reg. 10466, May 29, 1971). Such promulgation of a national consensus standard is considered to be presumptively valid in the absence of substantial evidence to the contrary. *Kokanee Cedar Sales, Inc.,* 3 OSHC 1530 (1975).

When promulgating a standard pursuant to Section 6(a) of the Act, however, the Secretary of Labor is not permitted to make any *significant* substantive alteration from the source document. *Usery v. Kennecott Copper Corp.,* 577 F.2d 1133, 6 OSHC 1197 (10th Cir. 1977). In *Trojan Steel Company,* 3 OSHC 1384 (1975) a citation for violation of the bulk oxygen storage standard was vacated by an administrative law judge because of a lack of proper promulgation. The "standard" was no longer a national consensus standard when promulgated, thus the Secretary of Labor should have followed the Section 6(b) procedures. The decision was affirmed by an equally divided Commission.

Significant, substantive modifications or alterations of OSHA standards adopted from national consensus standards or the source standards themselves may not be made without resort to the appropriate due process and rule-making procedures. An example of a significant, substantive change would be the alteration of an ANSI standard's requirements from "should" to "shall," making it a mandatory regulation instead of the advisory/directory one denoted. This alteration would affect the validity of the promulgation and the Secretary of Labor would therefore be required to follow the notice-and-comment rule-making procedures prescribed in Section 6(b) of the Act. *Marshall v. Union Oil Company of California,* 616 F.2d 1113, 8 OSHC 1169 (9th Cir. 1980); *Diebold, Inc. v. Marshall,* 585 F.2d 1237, 6 OSHC 2002 (6th Cir. 1978); *Usery v. Kennecott Copper Corp.,* 577 F.2d 1113, 6 OSHC 1197 (10th Cir. 1977); *Marshall v. Pittsburgh-Des Moines Steel Co.,* 584 F.2d 638, 6 OSHC 1929 (3d Cir. 1978). See also *Louisiana-Pacific Corporation,* 4 OSHC 1915 (1976)—the sawguarding standard was validly promulgated as a national consensus standard, under Section 6(a) of the Act; *Boise-Cascade Corp,* 3 OSHC 1804

(1975)—the sawguarding standard was validly promulgated despite the deletion of ANSI headnote; *Clifford B. Hannay & Son,* 6 OSHC 1335 (1978)—employer's compliance with revised NECA standard which differed from earlier standard adopted by OSHA does not preclude finding of violation but is properly characterized as *de minimis.*

In *Bethlehem Steel Corp. v. OSHRC,* 573 F.2d 157, 6 OSHC 1440 (3d Cir. 1978), the court held that a report of an ANSI committee which interpreted an ANSI standard adopted by OSHA is not controlling but is entitled to consideration in the standard's interpretation. And, finally, in *Noblecraft Industries, Inc. v. Secretary of Labor,* 614 F.2d 199, 7 OSHC 2059 (9th Cir. 1980), the court held that the ANSI standard from which the OSHA sawguarding standard, 29 CFR 1910.213(h)(1), was taken was a valid national consensus standard within the meaning of Section 3(9) of the Act, since Congress, aware of ANSI's procedures, specifically approved adoption of ANSI standards as national consensus standards. The court stated, however, that the omission of certain portions of the source standard in OSHA's adoption of 1910.213(h)(1) was material, since those excised portions (a limiting-scope note) excluded certain operations from and limited the scope of the standard. Thus if citations charged a violation of the standard beyond these limitations, they would be invalid and were to be dismissed upon remand, according to the court. But see *Constructora Maza, Inc.,* 6 OSHC 1208 (1977); *Austin Building Co.,* 8 OSHC 2150 (1980); *Boise-Cascade Corp.,* 4 OSHC 1675 (1976).

b. *Established Federal Standards*

An "*established Federal standard*" is defined in Section 3(10) of the Act as any "operative occupational safety and health standard established" by a federal agency (and presently in effect) or contained in a federal law in force on December 29, 1970, the date of enactment. The term "presently in effect," although not defined in the Act, meant those standards in force and effect on the date the Act was enacted and those promulgated prior to the effective date, but subsequent to the date of enactment of the Act. *Lance Roofing Company, Inc.,* 1 OSHC 1501 (1974). Federal laws containing such standards include the Walsh-Healey Public Contracts Act, the Service Contract Act, the Davis-Bacon Act, and the Contract Work Hours and the Safety Standards Act (Construction Safety Act). Examples of such standards include 29 CFR 1926.28(a), which requires the wearing of appropriate personal protective equipment, and 29 CFR 1910.1000 (Table Z) which adopted the threshold limit value (TLV's) for 400 toxic substances developed by the American Conference of Governmental Industrial Hygienists (ACGIH).

The standards adopted by the ACGIH did not fall within the definition of national consensus standards and hence were adopted as established federal standards. *Chlorine Institute, Inc., v. OSHA,* 613 F.2d 120, 8 OSHC 1031 (5th Cir. 1980). In *Rothschild Washington Stevedoring Co.,* 3 OSHC 1549 (1975), the OSHA head protection standard 29 CFR 1918.105(a), was found to be properly promulgated as an established federal standard. Citing *Atlantic & Gulf Stevedores, Inc.,* 3 OSHC 1003 (1975), *aff'd,* 534 F.2d 541, 4 OSHC 1061 (3d Cir. 1976).

Congress itself prescribed that certain established federal standards were *deemed* to be occupational safety and health standards issued under the Act through direct action in Section 4(b)(2) of the Act (presumably regardless of the fact or in addition to the fact that they would also be established federal standards). Congress thus declared that the safety and health regulations originally adopted under five federal statutes (*i.e.*, Walsh-Healey Act, Service Contract Act, Construction Safety Act, Longshoremen's and Harbor Worker's Compensation Act, and National Foundation on Arts and Humanities Act) covering longshoring and shipbuilding industries, federal construction projects, and certain federal contractors and grantees, were to be superseded by and in effect as OSHA standards issued under the Act. It was intended that these standards would apply to industry in general. *Lee-Way Motor Freight, Inc.*, 1 OSHC 1689 (1974), *aff'd*, 511 F.2d 864, 2 OSHC 1609 (10th Cir. 1975); *Coughlan Construction Company, Inc.*, 3 OSHC 1636 (1975). In *Daniel Construction Co.*, 5 OSHC 1005 and 1713 (1977), the effect of "deemed" standards was explained. In that case a challenge to the construction safety standards, 29 CFR Part 1926, was rejected, even though the standards at issue did not become effective on April 28, 1971, the effective date of the Act, but became effective several days later. The Commission stated that these standards became safety and health standards under the Act by operation of Section 4(b)(2), which provides that certain standards in effect on or after the effective date of the Act shall be *deemed* to be safety and health standards under the Act. The "deeming" provision specifically included those standards promulgated under the Construction Safety Act, 40 U.S.C. Section 333. Thus even if the Secretary of Labor was without authority to promulgate the standards at issue because they were not in force on April 28, 1971, the date of the Act, as established federal standards, nevertheless, the Construction Safety Act standards became effective as "deemed" standards by operation of Section 4(b)(2) of the Act.[5]

In *Underhill Construction Corp. v. Secy. of Labor*, 526 F.2d 53, 3 OSHC 1722 (2d Cir. 1975), the court held that a certain regulation, 29 CFR 1926.1050, which established the "effective dates" of construction standards was not in and of itself a standard, could not become an established federal "construction" standard, and was not therefore adopted under the Act by 29 CFR 1910.11 and 12, which defines construction work.[6] See also *Huber Hunt & Nichols, Inc., and Blount Bros. Corp.*, 4 OSHC 1406 (1976). In *Deering Milliken, Inc.*, 6 OSHC 2143 (1978), the Commission held that the failure to incorporate a limiting headnote from a private source standard during its adoption under the Act as an established federal standard did not render the standard invalid, since the headnote was not included in the original standard as adopted under the Walsh-Healey Act. The original headnote was contained in an ACGIH standard for cotton dust. The standard, as promulgated under Walsh-Healey and as adopted by OSHA under Section 6(a) of the Act, referred only to the TLV contained in the ACGIH regulation. Thus the Walsh-Healey standard did not include the headnote. In any event, the Commission said the headnote was only an explanatory note, and, as such, its omission did not constitute a *material* change of the standard, render it ambiguous, or invalidate it.

In *National Industrial Constructors, Inc. v. OSHRC*, 583 F.2d 1048, 6 OSHC

1914 (8th Cir. 1978), it was argued that the Construction Safety Act (CSA) regulations had not been validly promulgated in accordance with the Administrative Procedure Act and consequently were not established federal standards. Although the court rejected this argument by holding that an employer could not challenge the procedure used in adopting standards in enforcement proceedings, citing Section 6(f) of the Act, it did give a detailed account of the procedural requirements attendant upon the promulgation of established federal standards from the Construction Safety Act [*e.g.*, thirty days' publication prior to the standard's effective date was required unless OSHA had good cause for shortening the period. The CSA regulations were only published for ten and seven days; the Secretary of Labor claimed good cause for the shortened procedure.].[7] See also *Marshall v. Union Oil Co. of California*, 616 F.2d 1113, 8 OSHC 1169 (9th Cir. 1980), where a similar challenge to the improper promulgation of the standards was raised and permitted; the court rejected the Eighth Circuit's restrictive reading of Section 6(f) of the Act and thus permitted the employer to challenge the standards' validity in an enforcement proceeding.

The Commission also has the authority to consider issues concerning the validity of a standard, such as whether the Secretary of Labor has acted beyond his statutory authority in adopting a standard. See *Tobacco River Lumber Co.*, 3 OSHC 1059 (1975)—the fact that an employer was not a member of a standards-setting organization and did not participate in the formulation of a standard that was adopted under the Act does not, however, render invalid the application of such standard to the employer; *Huber, Hunt & Nichols, Inc. and Blount Bros. Corp.*, 4 OSHC 1406 (1976). The validity of standards is not a jurisdictional issue that may be raised at any stage of the proceedings. It is basically an affirmative defense and must be raised at the hearing to be considered on review. In *Trojan Steel Co.*, 3 OSHC 1384 (1975), an equally divided Commission vacated a citation in which the Secretary failed to follow rule-making procedures required by Section 6(b) of the Act.

Whenever an established federal standard which has been adopted and incorporated by reference as an OSHA standard is amended, repealed, or modified in whole or in part pursuant to Section 6(b) of the Act and the statute under which it was originally promulgated, the standard shall also be deemed changed for purposes of both that statute and the OSHA regulations. 29 CFR 1910.18, 1911. A change in a standard is defined as an amendment, addition, or repeal in whole or in part. The provisions of 29 CFR 1910.1, .12–.16 provide in this regard that they carry out the directives under Section 6(a) of the Act. See also 1 CFR Part 51; 5 U.S.C. Section 552.

2. Permanent Standards

The Secretary of Labor has broad discretion and authority, pursuant to Section 6(b) of the Act, to initiate actions involving the development, promulgation, modification, or revocation of standards to meet newly recognized hazards or to replace inadequate interim standards. Any standard other than a national consensus or established federal standard which the Secretary of

Labor wishes to adopt can only be promulgated by a different and more burdensome procedure than that provided in Section 6(a) of the Act. Standards adopted under Section 6(a) of the Act are not required to comply with the notice, comment, and publication procedures of the Administrative Procedure Act. Permanent standards, however, established under Section 6(b) of the Act, must go through an informal rule-making process, prescribed by the Administrative Procedure Act. Generally, permanent standards may be promulgated whenever OSHA determines that to do so would serve the objectives of the Act. Section 6(b)(5) of the Act gives more definite guidance relative to the setting of permanent standards for *toxic* substances. A discussion of the procedures for promulgation and adoption of permanent standards occurs in detail in later sections of this chapter.

3. Emergency Temporary Standards

Section 6(c)(1) of the Act provides that an emergency temporary standard (ETS) may be issued, prior to setting a permanent standard, *without notice or hearing,* when the Secretary of Labor determines: (1) that employees are exposed to *grave danger* from exposure to substances or agents determined to be toxic, physically harmful, or from new hazards; and (2) that such emergency standard is necessary to protect employees from such danger. An ETS may be issued because of the existence of certain emergency situations which may and/or would preclude the use of the entire panoply of due process considerations. There must therefore be "a danger of incurable, permanent, or fatal consequences to workers as opposed to easily curable and fleeting effects on their health. . . ." *Florida Peach Growers Assn. v. Dep't of Labor,* 489 F.2d 120, 1 OSHC 1472 (5th Cir. 1974). Thus if this requirement is met, an ETS may be adopted summarily without regard to the APA's rule-making requirement. The courts have held, however, that the extraordinary powers granted to the Secretary of Labor in Section 6(c) of the Act should be exercised delicately and only in those emergency situations which require it. Moreover, the Secretary's determinations in this regard shall be conclusive *only* if supported by substantial evidence in the record considered as a whole. Section 6(f) of the Act. See *Florida Peach Growers Assn. v. United States Department of Labor,* 489 F.2d 120, 1 OSHC 1472 (5th Cir. 1974); *Taylor Diving & Salvage Co. v. U.S. Department of Labor,* 537 F.2d 819, 4 OSHC 1511 (5th Cir. 1976); *Dry Color Mfr's Assn., Inc. v. Dept. of Labor,* 486 F.2d 98, 1 OSHC 1331 (3d Cir. 1973).

Parties affected by the issuance of an ETS may file a request for a stay with the appropriate court of appeals if they feel the standard is improperly promulgated and wish to "avoid" its application during the process of judicial review. In *Vistron Corporation v. OSHA,*—F.2d—, 6 OSHC 1483 (6th Cir. 1978), a motion for a stay of the acrylonitrite ETS pending judicial review was denied because the requirements for obtaining an injunction were not met, but in *Taylor Diving & Salvage Company v. U.S. Dept. of Labor,* 537 F.2d 819, 4 OSHC 1511 (5th Cir. 1976), a stay was granted of the ETS for diving operation, upon a showing by the petitioner that it had good prospects for prevailing on the merits and would suffer irreparable harm if the stay was not granted.

After publication of an ETS, the Secretary must initiate the formal Section 6(b) procedures for promulgation of a permanent standard, for if a permanent standard is not issued within six months, the ETS will expire. An ETS will serve as the proposal for issuance of a permanent standard. One court has held that an ETS may be amended under the same authority, in the same manner, and pursuant to the same criteria that governs its initial issuance. *Florida Peach Growers Assn. v. Brennan,* 489 F.2d 120, 1 OSHC 1472 (5th Cir. 1974). But see *Synthetic Organic Chemical Mfr's Ass'n v. Brennan,* 506 F.2d 385, 2 OSHC 1402 (3d Cir. 1974). The statement of reasons that is required to be issued in conjunction with the promulgation of permanent standards under Section 6(b) of the Act has, however, also been required with respect to the issuance of emergency temporary standards. *Dry Color Mfr's Assoc. v. Dep't. of Labor,* 486 F.2d 98, 1 OSHC 1331 (3d Cir. 1973).[8]

4. Advisory v. Mandatory Standards

After the Act was enacted, the Secretary of Labor adopted a broad range of industry standards (national consensus standards) as occupational safety and health standards. Section 3(9) of the Act defines a national consensus standard as:

> any occupational safety and health standard or modification thereof which has been adopted and promulgated by a nationally recognized standards-producing organization under procedures whereby it can be determined by the Secretary that persons interested and affected by the scope or provisions of the standard have reached substantial agreement on its enactment, which was formulated in a manner which afforded an opportunity for diverse views to be considered, and has been designated as such a standard by the Secretary after consultation with other appropriate agencies.

These standards became upon adoption the primary basis for the new Part 1910 of the Code of Federal Regulations. (Part 1926 was basically a compilation of established federal standards.) Many of them were contained in a safety code originally formulated in 1967 by the American National Standards Institute (ANSI); others were contained in codes originally published by the National Fire Protection Association (NFPA), National Electrical Contractors Association (NECA), and the American Petroleum Institute (API). Pursuant to Section 6(a) of the Act, such standards could be and were promulgated without regard to the procedural requirements provisions of the Administrative Procedure Act. See Senate Report No. 91–1282, reprinted at 1970 U.S. Code Congressional and Administrative News, 5177, 5182. Standards other than national consensus standards (or established federal standards) which the Secretary of Labor wished to adopt could be promulgated only by a different and more burdensome procedure, as provided in Section 6(b) of the act.

Many of the disputes which have developed since the adoption of national consensus standards as OSHA standards relate to whether these standards were mandatory or are merely advisory/directory, serving only as guidelines to industry. The original ANSI standards, for example, distinguished between rules that were mandatory and those that were merely advisory. Section V of the ANSI Code Introduction explained how the two were differentiated:

> Mandatory rules of this Code are characterized by the use of the word "shall." If a rule is of an advisory nature, it is indicated by the use of the word "should" or is stated as a recommendation.

The Secretary of Labor, however, often did not carry over this definition section when he adopted many of the ANSI standards as OSHA regulations. Because he stated an intention to adopt entirely (and only) mandatory regulations, incorporation of such a definition section was considered to be superfluous. In *Marshall v. Anaconda Company, Montana Mining Division,* 596 F.2d 370, 7 OSHC 1382 (9th Cir. 1979), *aff'g,* 6 OSHC 1372 (1978), the court stated that although the Secretary of Labor's interpretation of his own regulation, when affirmed by the Commission, must be accorded substantial weight, citing *Irvington Moore v. OSHRC,* 556 F.2d 431, 5 OSHC 1585 (9th Cir. 1977), it carries much less weight when at odds with the Commission, which it was. Moreover, the court stated, the Secretary's interpretation is less compelling when a regulation has been adopted from a consensus standard. Citing *Bethlehem Steel Corp. v. OSHRC,* 573 F.2d 157, 6 OSHC 1440 (3d Cir. 1978).

The court in *Anaconda,* therefore, concluded that although the Secretary of Labor intended to make certain provisions of the ANSI standard mandatory, the framers of the ANSI code were not so advised, thus the Secretary could not adopt an advisory rule as a national consensus standard and give it mandatory effect. Moreover, the court raised the question of the Secretary's power to substantively amend an ANSI standard without resort to the rulemaking procedure precribed by Section 6(b) of the Act.

The net effect of the *Anaconda* case, then, is that certain provisions in OSHA standards which have been adopted from ANSI are only advisory. For example, the standards contained in 29 CFR 1910.179(b)(2) are only advisory as to cranes installed before August 31, 1971. *Wheeling-Pittsburgh Steel Corp.,* 5 OSHC 1495 (1977), *aff'd sub nom. Marshall v. Pittsburgh-Des Moines Steel Co.,* 584 F.2d 638, 6 OSHC 1929 (3d Cir. 1978). Having recognized that some ANSI standards contain advisory provisions, the Secretary of Labor cannot abrogate to himself the authority to determine that certain of those provisions are mandatory even though considered by ANSI in advisory terms only. See *Usery v. Kennecott Copper Corp.,* 577 F.2d 1113, 6 OSHC 1197 (10th Cir. 1977); *United States Steel Corp.,* 5 OSHC 1289 (1977); *Interlake Inc.,* 8 OSHC 1414 (1980); *General Dynamics Corp.,* 8 OSHC 1360 (1980); *United States Steel Corporation,* 5 OSHC 1289 (1977); *H. K. Ferguson Company, Inc.,* 2 OSHC 1279 (1974); *Westinghouse Electric Corp.,* 6 OSHC 1432 (1978).

The effect of changing the words in an ANSI standard from "should" to "shall," making them mandatory regulations instead of the advisory ones that had been denoted by ANSI, according to the court in *Marshall v. Union Oil Company of California,* 616 F.2d 1113, 8 OSHC 1169 (9th Cir. 1980), *aff'g,* 6 OSHC 1433 (1978), definitely affects the validity of the Section 6(a) promulgation. In *Usery v. Kennecott Copper Corporation,* 577 F.2d 1113, 6 OSHC 1197 (10th Cir. 1977), the court held that 29 CFR 1910.28(a)(3), which requires guardrails on open-sided platforms, was unenforceable as improperly promulgated, rejecting the Secretary of Labor's contention that the change from "should" to "shall" was not significant. The court stated that if the standards were to be adopted during the two-year interim period, which involved the modification of established standards, then a formalized procedure had to be followed, including recommendations from an advisory committee, publication of the proposed rule, and allowance of time for comments and public hearings, if requested. The court held that the Secretary of Labor did not comply with the Act by reason of his failure to adopt the ANSI standard verbatim or by failure to follow the appropriate due process safeguard. The promulgation of the standard with the use of "shall" rather than "should" did not constitute the adoption of the national consensus standard and was therefore unenforceable. The usual procedural due process safeguards accorded to persons who might be adversely affected by governmental regulations were not applied.

Another method by which the Secretary of Labor has attempted to avoid the advisory nature of the many ANSI standards adopted into the Act is simply through interpretation; that is, reading the word *should to mean shall,* even though the language itself is not altered. An example of a standard which contains both terms in its various subparts is 29 CFR 1910.134(b).[9] The Secretary of Labor has frequently cited employers as if the complete standard, and other similarly phrased standards, were couched in mandatory terms. The Commission, however, has rejected such an unfounded application. See *H. K. Ferguson Co., Inc.,* 2 OSHC 1279 (1974); *Matson Terminals, Inc.,* 3 OSHC 2071 (1976); *Edward Heins Lumber Co.,* 4 OSHC 1735 (1976); *McHugh and McHugh,* 5 OSHC 1155 (1977); *A. Prokosch & Sons Sheet Metal, Inc.,* 8 OSHC 2077 (1980).

In *United States Steel Corp., Duquesne Plant,* 8 OSHC 1315 (1980), a citation for a violation of 29 CFR 1910.179(g)(2)(i) was vacated by the Commission because the cranes were in operation prior to August 3, 1971, and the ANSI standard from which the cited standard was derived was advisory only with respect to such cranes. See also *Edgewater Steel Corporation,* 8 OSHC 1314 (1980). In *Consolidated Freightways Corp.,* 5 OSHC 1481 (1977), it was argued that 29 CFR 1910.22(c), which provides that covers and/or guardrails shall be provided to protect personnel from the hazards of open pits . . . had the effect of making the standard hortatory, not mandatory. The Commission, however, rejected this argument.

In one unusual case the use of the word "shall" was found to be an advisory requirement rather than mandatory. 29 CFR 1918.104 provided that an employer shall arrange to make safety shoes available to all employees and shall encourage their use. It was held that this standard required no more than that employers make available and encourage their use; it does not provide that an employer shall require and *insure* their use. *Matson Terminals, Inc.*, 3 OSHC 2071 (1976).

In *Marshall v. Union Oil Company*, 616 F.2d 1113, 8 OSHC 1169 (9th Cir. 1980), the Secretary of Labor argued that the company could not raise the procedural invalidity of a standard as an affirmative defense in an enforcement proceeding, citing *National Industrial Constructors, Inc. v. OSHRC*, 583 F.2d 1048, 6 OSHC 1914 (8th Cir. 1978). In that case the validity of the Secretary of Labor's adoption as an OSHA regulation, 29 CFR Part 1926, a regulation which had been promulgated pursuant to the Construction Safety Act, was challenged. The employer argued that the Construction Safety Act regulations had not been validly promulgated in accordance with the Administrative Procedure Act and thus were not "established federal standards." The Eighth Circuit characterized this attack as a procedural attack on the manner in which 29 CFR 1926 had been promulgated and held that a procedural attack as such, although permissible, was precluded by the enactment of Section 6(f) of the Act, which provides for judicial review of the regulation within sixty days of the regulation's adoption. The Eighth Circuit thus found that although substantive challenges to the validity of regulations may be brought in enforcement proceedings, all procedural challenge must be raised in a pre-enforcement proceeding. The court, however, in *Union Oil* rejected and refused to follow the decision in *National Industrial Constructors*, saying it would not foreclose a challenge of the procedural validity of an OSHA regulation in the absence of express authorization from Congress and that Section 6(f) does not foreclose an employer from challenging the validity of a standard during enforcement proceedings. See also *Atlantic & Gulf Stevedores, Inc. v. OSHRC*, 534 F.2d 541, 4 OSHC 1061 (3d Cir. 1976).

In *Gold Kist, Inc.*, 7 OSHC 1855 (1979), the Commission stated that even if the Secretary of Labor has not adopted an NFPA code section as a standard, he may use it as an aid in interpreting another standard, 29 CFR 1910.36(b)(3), without promulgating that other code as a separate standard. The Commission rejected the argument that the Secretary impliedly and impermissibly tried to adopt by reference to a national consensus standard not adopted in accordance with Section 6(a) of the Act. Reference to the source of a standard, according to the Commission, gives meaning to the standard. Thus it appears that appropriate private industry standards, even though not adopted, may be used to define and interpret the cited OSHA standards.[10]

Disputes have also developed over the use and meaning of the words "and" and "or," but subsequent amendments to the standards involved have generally mooted these questions. *Hoffman Construction Co. v. OSHRC*, 546 F.2d 281, 4 OSHC 1813 (9th Cir. 1976); *Carpenter Rigging & Contracting Corp.*, 2 OSHC 1544 (1974).

C. APPLICABILITY/INTERPRETATION

A substantial body of law has developed concerning whether or not specific employer activities are covered within the scope of certain standards either by the standards' specific applicability or by the interpretations placed upon the standards.

1. Administrative Interpretations

Interpretations of OSHA standards are issued in administrative program directives and field information memoranda, which are collectively called "OSHA Instructions." Like the Field Operations Manual, these "Instructions" are neither substantive rules nor regulations.

In *Empire-Detroit Steel Div., Detroit Steel Corp.,* 4 OSHC 1074, *aff'd,* 579 F.2d 378, 6 OSHC 1693 (6th Cir. 1978), the Commission held that an employer is not entitled to rely on an administrative interpretation of a standard that arguably modifies the requirements of the standard, even though it was incorporated into the abatement provisions of a prior citation. The Secretary of Labor is not estopped from enforcing a standard as written. In *Francisco Tower Service, Inc.,* 3 OSHC 1952 (1976), the Commission stated that an employer may not rely on the interpretation of a standard in a letter from OSHA to another party, since the letter cannot affect an amendment to the standard. For even if it were, it would be ineffective for failure to comply with the standard modification requirements of Section 6(b) of the Act. Neither does it operate as an estoppel, since one cannot raise estoppel against the United States on the basis of a private ruling and there was no evidence that the employer had relied on the letter or even knew of it at the time of inspection. See also, *Holman Erection Co.,* 5 OSHC 2078 (1977); *FMC Corp.,* 5 OSHC 1707 (1977). Similarly, in *United States Steel Corp.,* 2 OSHC 1343 (1974), *aff'd,* 517 F.2d 1400, 3 OSHC 1313 (3d Cir. 1975), the Commission held that the Secretary of Labor could not modify the coal tar pitch volatile standard through an administrative interpretation without resort to the provisions of Section 6(b) of the Act, and that an employer is not entitled to rely on an administrative interpretation, even if published in the Federal Register, that arguably modifies the requirements of the standard, for the Secretary is not estopped from enforcing the standard as written. But see *Belcher-Evans Millwork Co. Inc.,*—OSHC—, 1977–78 OSHD ¶21575 (1977)—if an employer has relied to his detriment on an inspector's interpretation of a standard, due process requires that a citation alleging noncompliance be vacated; *Field Crest Mills, Inc.,* 2 OSHC 1143 (1974).

2. Specific v. General Standards

There are both specific and general standards. There are two types of specific standards: those that have been adopted for particular industries and those that impose specific requirements on certain types of operations. These are differentiated by the appellations vertical and horizontal. The latter standards require something different from and more specific than other standards which may be contained in the same Part of 29 CFR for a certain type of operation or equipment. For example, the guarding requirements for woodworking machines contained in 29 CFR 1910.213 are more specific than the general machine-guarding requirements of 29 CFR 1910.212.

General standards have applicability to general industry and are contained in 29 CFR 1910. Specific standards may be subparts of the general standards or may be contained in totally separate categories (*e.g.*, construction).

As provided in 29 CFR 1910.5(c) (1) and (2), if a specific standard, *e.g.*, a construction or maritime standard or a standard for a special industry contained in 29 CFR 1910 (*e.g.*, Subpart (R)) is specifically applicable to a working condition, then the specific standard is applicable rather than the more general standard which might otherwise be applicable to that working condition. Similarly, a standard which is more specific relative to a working condition, than another standard within the same respective part would be applicable. For example, the requirement regarding machine guards for woodworking saws, 29 CFR 1910.213, is applicable to such saws rather than the more general machine-guarding requirements for power saws. 29 CFR 1910.212.

As stated in 29 CFR 1910.5(c)(1):

> If a particular standard is specifically applicable to a condition, practice, means, method, operation or process it shall prevail over any different general standard which might otherwise be applicable to the same condition, practice, means, method, operation or process.

Moreover, 29 CFR 1910.12(a) states that the standards in 29 CFR 1926 shall apply to every employment and place of employment of every employee engaged in construction work; general industry standards may be applied to employers in the construction industry only to the extent that none of such particular construction standards apply.

The net effect of the discussion above is that general industry standards may only be applied to working conditions when no specific standard relates to the working condition at issue. See *General Electric Company,* 5 OSHC 1448 (1977)—29 CFR 1910.212(a)(3) is not applicable to point of operation guarding relative to a radial cutoff machine since a *more* specific standard, 1910.215(b)(5), requires guards for flying objects, not points of operation on the radial cutoff machine; *Saturn Construction Company, Inc.,* 5 OSHC 1686 (1977); *Matson Terminals Inc.,* 3 OSHC 2071 (1976); *Bristol Steel & Iron Works Inc. v. OSHRC,* 601 F.2d 717, 7 OSHC 1462 (4th Cir. 1979); *Davis/McKee, Inc.,* 2 OSHC 3046 (1974); *Trojan Steel Company,* 3 OSHC 1384 (1975); *Rocksbury Construction Corporation,* 2 OSHC 3253 (1974); and *Rocky Mountain Prestress, Inc.,* 5 OSHC 1888 (1977).

In *Dravo Corp. v. OSHRC,* 613 F.2d 1227, 7 OSHC 2089 (3d Cir. 1980), the court stated that even in areas properly citable under specific maritime standards, an employer may be held to compliance with general industry standards in those particular situations where no specific standard is applicable. The court cited 29 CFR 1910.5(c)(2) which provides:

> (2) On the other hand, any standard shall apply according to its terms to any employment and place of employment in any industry, even though particular

> standards are also prescribed for the industry, as in Subpart B or Subpart R of this part, to the extent that none of such particular standards applies. . . .

The court found that general safety standards complement specific safety standards, of which the set of shipbuilding regulations is one example, by filling the interstices necessarily remaining after promulgation of the specific standards. The Secretary of Labor, it held, cannot be expected to have anticipated every conceivable hazardous situation in promulgating specific standards. It then supported its decision by reference to a Fourth Circuit case which noted:

> Enforcement of general safety standards in situations not covered by specific standards does not render the specific standards meaningless or unnecessary. Specific safety standards insure a minimum level of protection to employees under common conditions widely recognized as potentially hazardous as well as provide employers with a base upon which to establish an adequate safety program. While general safety standards may quite likely cover most conditions addressed by the specific standards, fairness to the employer would require the promulgation of specific safety standards where feasible. . . . Lacking the omniscience to perceive the myriad conditions to which specific standards might be addressed, however, the Secretary, in an effort to ensure the safety of employees as required by the Act, must at times necessarily resort to the general safety standards. *Bristol Steel & Iron Works, Inc. v. OSHRC,* 601 F.2d 717, 7 OSHC 1462 (4th Cir. 1979).

Accordingly, even in areas properly citable under specific standards, the Secretary may at times hold an employer to the general industry standards in those situations where no specific standard is applicable. The aforementioned courts have *only* upheld the application of general safety standards such as those for personal protective equipment standards, where it has been shown that a reasonably prudent employer familiar with the industry would have protected against the hazard by the means specified in the citation. In *Ray Evers Welding Co. v. Secretary of Labor,* 625 F.2d 726, 8 OSHC 1271 (6th Cir. 1980), the court incorporated a reasonableness test as an element of proving a violation of 29 CFR 1926.28(a). It stated that:

> reasonableness is an objective test which must be determined on the basis of evidence in the record. Industry standards and customs are not entirely determinative of reasonableness because there may be instances where a whole industry has been negligent in providing safety equipment for its employees. However, such negligence on the part of a whole industry cannot be lightly presumed . . . it must be proven.

There can be no citation under the general duty clause (Section 5(a)(1) of the Act), if a specific standard applies. *Kaiser Aluminum Co.,* 4 OSHC 1162 (1977); *Mississippi Power & Light Company,* 7 OSHC 2036 (1979); *Armstrong Cork Company,* 8 OSHC 1076 (1980); *Isseles Brothers, Inc.,* 3 OSHC 1964 (1976); *National Realty Construction Company, Inc. v. OSHRC,* 489 F.2d 1257, 1 OSHC 1422, 1423 (D.C. Cir. 1973).

The OSHA Field Operations Manual (FOM) provides that the general duty clause should be used only where there are no specific standards applicable to the particular hazard involved. It also states that 29 CFR 1910.5(f) expressly provides that an employer who is in compliance with a specific standard shall be deemed to be in compliance with the general duty clause insofar as it applies to hazards covered by the specific standard. Any recognized standard created in part by a condition not covered by a standard may, however, be cited under the general duty clause. FOM, Ch. VIII.

3. Standards of Interpretation

The courts dealing with the question of the proper interpretation of standards have basically adopted a bifurcated approach in handling such issues. The courts have generally rejected the contention that regulations should be liberally construed to give broad coverage because of the intent of Congress to provide safe and healthful working conditions for employees. The courts generally feel that an employer is entitled to fair notice in dealing with the government. Like other statutes and regulations which allow monetary penalties against those alleged to have violated them, an occupational safety and health standard must give an employer fair warning of the conduct it prohibits or requires and it must provide a reasonably clear standard of culpability to circumscribe the discretion of the enforcing authority and its agents. *Diamond Roofing Co. v. OSHRC,* 528 F.2d 645, 4 OSHC 1001 (5th Cir. 1976). The court, in *Dravo Corp. v. OSHRC,* 613 F.2d 1227, 7 OSHC 2089 (3d Cir. 1980), citing its decision in *Bethlehem Steel Corp. v. Marshall,* 573 F.2d 157, 6 OSHC 1440 (3d Cir. 1978), stated that the purpose of OSHA standards is to improve safety conditions in the workplace by telling employers just what they are required to do in order to prevent or minimize injury to employees. The responsibility to promulgate clear and unambiguous standards is upon the Secretary. The test is not what he might possibly have intended, but what he said. The court further stated that the Longshoremen's and Harbor Workers' Compensation Act, even though it is given a broad reading in compensation cases in order to provide a remedy for maritime employees, may not be used for interpretative purposes in such a broad sense for OSHA standards for which an employer has been cited. For although courts may be increasingly willing to interpret coverage broadly when an injured person is in need of compensation, the same approach is not followed in cases involving penal sanctions. Citing *Bristol Steel & Iron Works, Inc. v. OSHRC,* 601 F.2d 717, 7 OSHC 1462 (4th Cir. 1979); *Marshall v. Anaconda Co.,* 596 F.2d 377 n. 6, 7 OSHC 1382 (9th Cir. 1979). Several courts, in fact, have urged the Secretary of Labor to rewrite "not easily intelligible" safety and health standards. See *e.g., Westinghouse Electric Corp. v. OSHRC,* 617 F.2d 497, 8 OSHC 1110 (7th Cir. 1980).

The courts thus have usually rejected the Secretary's position that coverage under the Act should be read expansively. For example, with respect to those standards adopted as national consensus standards (ANSI and NFPA) which were not adopted after notice, hearing and evaluation of evidence, but rather were conceived by a nongovernmental agency as a product of its own investigation and research, the Secretary of Labor's views are not as persuasive as they would be if based upon his own review of the problem and a deliberate choice of a satisfactory solution. Relying on the cases of *Skidmore v.*

Swift & Co., 323 U.S. 134, 140 (1944) and *Adamo Wrecking Co. v. United States,* 434 U.S. 275 (1978), the Third Circuit in *Bethlehem Steel Corporation v. OSHRC,* 573 F.2d 157, 6 OSHC 1440 (3d Cir. 1978), in holding that maintenance work on cranes is not covered by a standard which governs their use "under normal operating conditions," stated, in considering the weight to be given an administrative ruling on the standard:

> The weight of such a judgment in a particular case will depend upon thoroughness evident in its consideration, the validity of its reasoning, its consistency with earlier and later pronouncements, and all those factors which give it power to persuade, if lacking power to control.
>
> Unlike *The Budd Co. v. OSHRC,* 513 F.2d 201, 2 OSHC 1698 (3d Cir. 1975), we do not have an authoritative agency interpretation to assist us since the decisions of the Commission are themselves in conflict and inconsistent with the Secretary's position. We rely, therefore, on the plain wording of the standard and conclude that it does exclude maintenance personnel.

The court stated that the Commission should not strain the plain and natural meaning of words in the standard to alleviate an unlikely and uncontemplated hazard. The responsibility to promulgate clear and unambiguous standards is upon the Secretary, for the test is not what he might possibly have intended but what he said. And echoing the statement of many other courts confronted with the question, the court stated that if the language is faulty, the Secretary has the means and the obligation to amend. Citing *Irvington Moore, Division of U.S. Natural Resources, Inc. v. OSHRC,* 556 F.2d 431, 5 OSHC 1585 (9th Cir. 1977); *Diamond Roofing Co. v. OSHRC,* 528 F.2d 645, 650, 4 OSHC 1001, 1005 (5th Cir. 1976).

In the review of Commission decisions which interpret and apply standards, the courts are split on the question of the type and amount of deference to be accorded to the Commission and/or the Secretary of Labor in their interpretations and applications. Although this subject is discussed in another part of the book, the following "blurbs" will present an idea of the thinking of the courts on this matter.

The court in *Diebold Inc. v. Marshall,* 585 F.2d 1327, 6 OSHC 2002 (6th Cir. 1978), stated that the Commission is entitled to great deference in its reasonable interpretations of regulations promulgated under the Act, for historical or factual determinations relating to industrial and technological conditions existing at the time the standard was promulgated are precisely the kinds of determinations the Commission is particularly fitted to make by virtue of its members' education, training or experience. See also *Brennan v. OSHRC (Ron M. Fiegen, Inc.),* 513 F.2d 713, 3 OSHC 1001 (8th Cir. 1975).

Depending upon whether or not the Secretary of Labor and the Commission share congruent views on the Act, its regulations, rules, and standards, deference will vary. For example, several courts state that greater deference should be accorded to the Commission when it agrees with the position of the Secretary of Labor. See *Floyd S. Pike Electrical Contractor, Inc. v. OSHRC,* 576 F.2d 72, 6 OSHC 1781 (5th Cir. 1978).

In *The Budd Company v. OSHRC,* 513 F.2d 201, 2 OSHC 1698 (3d Cir. 1975) and *Clarkson Construction Co. v. OSHRC,* 531 F.2d 451, 3 OSHC 1880

(10th Cir. 1976), the courts also stated that the Secretary of Labor and the Commission should be accorded relatively broad discretion in interpreting the Act, its standards, and its regulations, especially where the Secretary and Commission *are in agreement.*

If the Commission and the Secretary of Labor differ on a legal interpretation, the courts are in disagreement as to the applicable standard of review and deference. Several courts give priority to the views of the Secretary of Labor in this situation. See *GAF Corporation v. OSHRC,* 561 F.2d 913, 5 OSHC 1555 (D.C. Cir. 1977); *Marshall v. Western Electric, Inc.*, 565 F.2d 240, 5 OSHC 2054 (2d Cir. 1977). However, other courts are more willing to accord deference to the Commission's interpretations. *Brennan v. OSHRC (Republic Creosoting Co.*), 501 F.2d 1196, 2 OSHC 1109 (7th Cir. 1974); *Brennan v. OSHRC* (*Gerosa, Inc.*), 491 F.2d 1340, 1 OSHC 1523 (2d Cir. 1974). The Second Circuit and the Sixth Circuit, however, have refused to adopt this distinction and will pay the same deference to the Commission's decisions regardless of whether or not they are in agreement with the Secretary of Labor's position. *General Electric Co. v. OSHRC,* 583 F.2d 61, 6 OSHC 1868 (2d Cir. 1978); *RMI v. Secretary of Labor,* 594 F.2d 566, 7 OSHC 1119 (6th Cir. 1979).

In many other cases, the courts have refused to defer to *either* the Secretary of Labor's interpretation or the Commission's interpretation if those interpretations conflict with the plain meaning of the standard, rule, or regulation in question. See *Westinghouse Electric Corp. v. OSHRC,* 617 F.2d 497, 8 OSHC 1110 (7th Cir. 1980); *Usery v. Marquette Cement Manufacturing Corp.*, 568 F.2d 902, 5 OSHC 1793 (2d Cir. 1977). Applying this standard, for example, the court in *Southwestern Bell Telephone Company v. Secretary of Labor,* 568 F.2d 368, 6 OSHC 1377 (5th Cir. 1978) held that the Commission's interpretation of the telecommunication standard was *not* contrary to its clear and plain meaning.

The Commission itself has also "opined" on the deference issue. In *United States Steel Corp.*, 5 OSHC 1289 (1977), the Commission concluded that the ordinary meaning of "normal operating conditions" set forth in 29 CFR 1910.179(g)(2)(i) prohibiting exposure of employees to live electrical parts does not encompass a situation where maintenance is performed on a crane with the crane out of production. The Commission stated that an interpretation of a standard which permits only unavoidable exposure is consistent with the goal of employee safety. Since the standard referenced no definite legislative history, it had to apply other principles of statutory construction to determine the intent of its drafters. That intent, it held, was primarily conveyed by the ordinary meaning of the language of the standard. It also stated that the Secretary of Labor's own standards distinguish between maintenance and operation of cranes and stated that the Secretary's interpretation is inconsistent with the rule of statutory construction. A meaning should, if possible, be given to every word and phrase of a standard.

The Secretary of Labor had contended that as the promulgator of a standard, its interpretation, if reasonable, must be accepted as controlling, even though that interpretation might not be the most reasonable interpretation of the standard's language. Citing *Brennan v. Southern Contractor Service,* 492

F.2d 498, 1 OSHC 1648 (5th Cir. 1974), the Secretary of Labor viewed the Commission's role as limited to applying Labor's reasonable interpretation of standards to the facts of the controversies it decides.

Rejecting this view, the Commission stated that it itself is the administrative agency with expertise in the area of occupational safety and health, as is the Secretary of Labor. Noting that several courts have held that it is the Commission's interpretations of the Act and the standards promulgated thereunder to which they will accord deference, the Secretary of Labor noted other courts which have accorded deference to its interpretations. The Commission stated, however, in not resolving the conflict, that such question of deference applies only with respect to judicial review of an administrative decision. The criteria for judicial review of a decision by an agency is substantially different from those applied by the agency in reaching its decision. In reaching its decision, the Commission said it must apply the criteria applicable to administrative decision-making as set forth in 29 U.S.C. §659(c), and thus the court decisions cited by the Secretary of Labor had no direct application to its proceedings. The Commission therefore rejected the Secretary of Labor's suggested limitation that its role was merely one to determine whether the Secretary of Labor's interpretation was reasonable. In fact, the Commission listed several cases in which it rejected the Secretary of Labor's interpretation of other standards for not being the most reasonable interpretations. Thus, the Commission said, it would use its expertise to determine for itself the *most reasonable* interpretation of a standard and although it would not ignore an interpretation advanced by the Secretary of Labor, it would use its own independent judgment to determine whether a proposed interpretation is the most reasonable and will apply the usual rules of statutory interpretation in determining the meaning of a standard. Thus in the case at hand, since the language of the standard was ambiguous, it was appropriate to look at the intent of the drafter of the standard to resolve the ambiguity, *e.g.*, ANSI. Since it could not do that, it had to rely on the ordinary meaning of the language of the standard.

In other recent decisions, for example, in *Continental Oil Co.*, 7 OSHC 1432 (1979), the Commission held that the plain meaning of a standard prevails over any other reference or heading that might otherwise limit its meaning. The reliance on a heading to determine the scope and application of a standard is inconsistent with the usual rule of statutory construction. In *Gold Kist, Inc.*, 7 OSHC 1855 (1979), however, the Commission held that NFPA's provisions may be used to define and interpret OSHA standards without promulgating that code as a separate standard.

4. Standards of Applicability

Questions of applicability normally involve the appropriateness of the application of a standard to a particular industry and/or a specific application to an aspect of the operations of that industry. Examples of "standards of applicability" have been broken down by industry.

a. *Construction*

The OSHA regulations at 29 CFR 1926 apply to the construction industry. Construction is defined as "work for construction, alteration, and/or repair, including painting and decorating." 29 CFR 1910.12(b), 1926.13; *Royal Logging Co.,* 7 OSHC 1744, 1747 n.6 (1979). Although the Commission in that case noted that employers not directly engaged in construction have at times been held subject to construction standards if "their operations are an integral part of or intimately involved with the performance of construction work," logging operations could not be so characterized in that case. Since the logging industry is not engaged in construction, the employer could not have violated 20 CFR 1926.28(a) by failing to require that employees wear personal protective equipment on bulldozers equipped with roll-over protective structures, since a logging operation is neither actual construction nor directly related to the performance of construction work. Thus in the circumstances of the case the application of construction standards to a "logger" was incorrect.

In *A. A. Will Sand and Gravel Corp.,* 4 OSHC 1442 (1976), the delivery of material to a construction site was held to constitute construction and the employer was thus subject to 29 CFR 1926 standards. This result was reached since the action of the employees in unloading gravel was considered to be an integral part of construction activities, in that the employee of the cited delivery employer assisted a construction employee in bringing material to the specific work area. See also *Heede International, Inc.,* 2 OSHC 1466 (1975) —dismantling a crane which had been used in building construction was construction work; *Penrod Drilling Co.,* 4 OSHC 1654 (1976)—dismantling of an oil derrick does not constitute construction work within the meaning of 29 CFR 1910.12; *Skidmore, Owings and Merrill,* 5 OSHC 1762 (1978)—construction standards only apply to employers who perform actual construction work or exercise substantial supervision over actual construction or who perform work directly or vitally related to construction work; *Westburne Drilling, Inc.,* 5 OSHC 1457 (1977)—the construction standards were found applicable to the oil and gas mining industry. In *Consumers Power Co.,* 5 OSHC 1423 (1977), employees engaged solely in trimming trees were not engaged in construction and could not be cited for 29 CFR 1926 violations, since tree trimming constituted maintenance work, which was expressly excluded from the coverage of Subpart V of 29 CFR 1926 as evidenced by the remarks on applicability published with the standards in 37 Fed. Reg. 24882 (1972). In *Bechtel Power Corp.,* 4 OSHC 1005 (1976), a construction management firm that supervised and coordinated a construction project involving many contractors made daily inspections of the work in progress, and coordinated the project's safety program, was "engaged in construction work" within meaning of 29 CFR 1910.12, and was therefore subject to the construction standards of 29 CFR 1926. The employer's activity was an integral part of the total construction system at the site, its functions being inextricably intertwined with actual physical labor. The Commission also held that even though OSHA adopted the Construction Safety Act (CSA) standards into the Act, that the limitation of the applicability of the CSA to employers with laborers/ mechanics in their employ did not apply to the scope of construction standards under the Act, since OSHA could properly expand the coverage of such standards to include employees who are not contractors or subcontractors. However, in *John W. McGowan,* 5 OSHC 2028 (1977), *review declined,* 604 F.2d 885, 7 OSHC 1842 (5th Cir. 1979), construction standards were not found to be applicable to an employer who was engaged in pumping crude oil from the earth into tanks and lines and whose only repair activities were normal maintenance of equipment and not remodeling or renovation.

In *Marshall v. Nichols,* 486 F. Supp. 615, 8 OSHC 1129 (E.D. Tex. 1980), OSHA contended that the oil well drilling industry was governed by OSHA construction regulations, 29 CFR Part 1926. However, the court failed to find any OSHA regulations pertaining to the oil well drilling industry and also found that the absence of any participation by the oil well drilling industry in the promulgation of general construction regulations prevented OSHA from applying such regulations to the industry, citing *U.S. v. Frontier Airlines,* 563 F.2d 1008, 1012 (10th Cir. 1977); *H-30, Inc. v. Marshall,* 597 F.2d 234, 7 OSHC 1253 (10th Cir. 1979).

Once coverage under the construction standards has been ascertained, the questions concerning applicability usually center around whether certain equipment is covered under one standard or another, whether guarding requirements apply, whether personal protective equipment requirements are applicable, and whether certain conditions and activities are properly related and/or covered by one standard or another.

(1) *Steel Erection.* The construction safety standard, 29 CFR 1926.28(a), requires than an employer is responsible for requiring the wearing of appropriate protective equipment in all operations where there is an exposure to hazardous conditions or where Part 1926 indicates the need for using such equipment to reduce the hazards to employees. 29 CFR 1926.104 refers to the use of safety belts and life lines while 29 CFR 1926.105 states that safety nets shall be provided when work places are more than twenty-five feet above the ground where the use of ladders, scaffolds, catch platforms, temporary floors, safety lines, or safety belts is *impractical.* The sometimes-apparent conflict between these two standards is particularly apparent when reference is made to the steel erection standards in 29 CFR 1926.750. That standard requires that where steel erection is being done, a tightly planked and substantial floor shall be maintained within two stories or thirty feet; but, when that is not practical and whenever a potential fall distance exceeds two stories or twenty-five feet, nets shall be hung. In certain circumstances, however, the use of safety belts attached to life lines may provide equivalent protection.

The question of when and whether 29 CFR 1926.750, the steel erection standards, apply to and preempt either 29 CFR 1926.28(a) or 105 has been considered in several different cases. The determination of applicability hinges upon whether or not a building is "tiered." If it is, the steel erection standards apply and would preempt 29 CFR 1926.28(a). See, *e.g., Jake Heaton Erecting Company, Inc.* 6 OSHC 1536 (1978)—a single story, loft-like structure with no floors between ground level and roof is not a tiered building; *Havens Steel Company,* 6 OSHC 1564 (1978), *aff'd,* 607 F.2d 493, 7 OSHC 1849 (D.C. Cir. 1979)—the Commission has stated that a tiered building is a multistoried or multileveled building, thus a one-story building warehouse is not a tiered building; *McLean-Behm Steel Erectors, Inc.*, 6 OSHC 2081 (1978), *rev'd,* 608 F.2d 580, 7 OSHC 2002 (5th Cir. 1979)—the steel erection standards, rather than 29 CFR 1926.28(a) are applicable to employees working on an elevated steel erection; *C. F. Braun and Company,* 7 OSHC 1139 (1979)—29 CFR 1926.750(b)(2)(i) or (b)(1)(iii) applies to tiered buildings only and not to open bay structures made of skeletal steel, covered with metal siding, and containing no permanent flooring; however, since the employees installing beams ninety-seven feet above ground were wearing safety belts, the fall protection requirements of 29 CFR 1926.105 were met. In *Wisconsin Associates, Inc.*, 7 OSHC 2083 (1979), it was held that a power plant with floors which do not have the consistency and uniformity of the conventional building is not tiered as interpreted in 29 CFR 1926.750(b)(2)(i). The interpretation of the term "tiered building" is narrow, generally limited to conventional office and apartment buildings where floors are consistent and uniform.

In *Bristol Steel and Iron Works, Inc. v. OSHRC,* 601 F.2d 717, 7 OSHC 1462 (4th Cir. 1979), the employer adopted the position that the Secretary of Labor, in promulgating specific safety standards applicable to steel erection in subpart R Section 1926, precluded the citation of the general construction standards set forth in 1926.28(a) arguing that the entire subject of fall protection was considered in the steel erection standards. In subpart R accordingly the specific standards were exclusive in this area. The court, however, stated that while the Act substantially contemplates specific safety standards promulgated by the Secretary, its purposes are also effectuated by the general safety standards in the general-duty clause which are designed to fill those interspaces. The court therefore concluded that:

> the general safety standard dealing with personal protective equipment found in 29 CFR 1926.28(a) complements the sub-part R specific standards dealing with steel erection by requiring the wearing of appropriate personal protective equipment (where there is a need) for using such equipment to reduce the hazards to the employees. Bristol suggests that its position is supported by 29 CFR 1910.5(c)(1) which provides that a specific standard applicable to a condition shall prevail over any different general standard which might otherwise be applicable thereto. This argument, however, elides the language of 1910.5(c)(2) that any standard shall apply according to its terms even though particular standards are also prescribed for an industry, to the extent that none of such particular standards apply. Were 1910.5(c) rated in the manner Bristol suggests, the Secretary would be prevented from coping with the variety of hazards not covered by these specific standards, and we decline to read it in such a fashion. . . . In determining whether Bristol violated 1926.28(a) the appropriate inquiry is whether under the circumstances a reasonably prudent employer who is familiar with steel erection would have protected against this hazard of falling by the means specified in the citation.

In *Builders Steel Company v. Marshall,* 622 F.2d 367, 8 OSHC 1363 (8th Cir. 1980) the Court held that the steel erection standard 29 CFR 1926.750(b)(2)(ii), which requires installation of flooring thirty feet from the level at which steel erection work is being performed, as opposed to the personal protective equipment standard, 29 CFR 1926.105(a), is applicable to an employer's single-story building where steel erection was being performed at a height of less than thirty feet. The language of the former regulation and related legislative history does not indicate an intention to limit coverage of the regulation to multitiered buildings. Accordingly, the citation for the alleged violation of 29 CFR 1926.105(a) was vacated. See also *Herrick Corporation,* 6 OSHC 1674 (1978)—steel erection standard applies to unguarded floor openings; *Havens Steel Co.,* 6 OSHC 1654, *aff'd.,* 607 F.2d 493, 7 OSHC 1849 (D.C. Cir. 1979); *Ray Evers Welding Co.,* 5 OSHC 1948 (1977), *rev'd on other grounds,* 625 F.2d 726, 8 OSHC 1271 (6th Cir. 1980)—reliance on steel erection standard 29 CFR 1926.750(b) rather than personal protective equipment standard 29 CFR 1926.28(a) in constructing a single-tiered building is inappropriate.

(2) *Trenching/Excavation.* In *Heath & Stich, Inc.,* 8 OSHC 1640 (1980), an employer was cited under 29 CFR 1926.652, the specific trenching requirements, but the employer argued that since the excavation was wider than it was deep, it therefore was not a trench under the standard and that the specific excavation requirements of 29 CFR 1926.651 applied. The Commission rejected the argument because the contention that an excavation with a width greater than its depth cannot be a trench has been rejected in previous cases. The employer then argued that the soil was so stable that even if it were a trench, it need not be protected under the requirements of 1926.652(c). The Commission also rejected that argument, saying it amounted to an impermissible challenge to the wisdom of the standard on which the Commission has no authority to consider that issue. *Austin Bridge Co.,* 7 OSHC 1761 (1979). In *Lloyd C. Lockrem, Inc. v. OSHRC,* 609 F.2d 940, 7 OSHC 1999 (9th Cir. 1979), *remanded (to Commission) and vacated,* 8 OSHC 1316 (1980)—the OSHA excavation standards, 29 CFR 1926.651, were determined to be vague and ambiguous with regard to their applicability to trenches.

In *M. J. Lee Construction Co.,* 7 OSHC 1140 (1979), the Commission interpreted and "equated" the "danger from moving ground" requirement of the excavation standard in 29 CFR 1926.651(c) with the "unstable or soft material" requirement of 29 CFR 1926.652(b), the trenching counterpart to the cited excavation standard, stating that "an excavation dug in unstable soil by definition creates a danger from moving ground."

(3) *Housekeeping.* In *D. Fortunato, Inc.,* 7 OSHC 1643 (1979), the Commission held that garbage, such as paper, coffee cups, and food did not meet the definition of scraps and debris prohibited in a workplace by 29 CFR 1926.25(b) and therefore the violation had to be vacated due to the inapplicability of the standard. See also *H. A. Lott, Inc.,* 6 OSHC 1425 (1978)—form and scrap lumber with protruding nails may constitute debris within the meaning of the housekeeping standard, 29 CFR 1926.25(a), but stacked building materials found in the area could not be classified as debris; *H. P. Foley Electrical Contractor,* 2 OSHC 3214 (1974)—protruding rebars are permanent fixtures and do not constitute debris.

(4) *Personal Protection.* Most often, questions relative to the use of personal protective equipment, such as safety belts, lanyards, life-lines, and nets arise in the context of other specific protections *e.g.*, guardrails, planked floors, etc. See, *e.g.*, *Howard M. Ream, Inc.*, 3 OSHC 1208 (1975); *Warnel Corporation,* 4 OSHC 1034 (1976); *Schiavone Construction Company,* 5 OSHC 1385 (1977); *McKee-Welman Power Gas,* 5 OSHC 1592 (1977); *Linbeck Construction Corp.*, 5 OSHC 1706 (1977); *Builders Steel Co.*, 7 OSHC 1451 (1979); *Universal Roofing and Sheet Metal Company,* 8 OSHC 1453 (1980).

In *Joseph Bucheit & Sons Co.*, 6 OSHC 1640 (1978), the Commission held that the *general* personal protective equipment standard for the construction industry, 29 CFR 1926.28(a), may be applied to require the use of protective devices when a fall distance is less than twenty-five feet, despite the inapplicability of a *specific* personal protective equipment standard, 29 CFR 1926.105(a), to fall distances under twenty-five feet. In *S & H Riggers & Erectors, Inc.*, 7 OSHC 1260 (1979), the Commission held that 29 CFR 1926.28(a) is applicable in preference to 29 CFR 1926.105(a), where a violation for failure to provide safety belts is alleged for employees working on a building forty feet above an unguarded temporary floor.

Other personal protective equipment questions which may arise relate to the use of: (1) eye and face protection: *Camm Industry, Inc.*, 1 OSHC 1564 (1974); *House Wood Product Co.*, 3 OSHC 1993 (1976); *Georgia Quality Masonry, Inc.*, 5 OSHC 2052 (1977); *Walter B. Hawkins d/b/a Walter Hawkins Fruit Co.*, 6 OSHC 1277 (1977); *Stearns-Roger, Inc.*, 7 OSHC 1919 (1979); (2) the use of respiratory protection: *Koppers Co., Inc.*, 2 OSHC 1354 (1974); *Bechtol Power Co.*, 7 OSHC 1361 (1979); *Stabilized Pigment, Inc.*, 8 OSHC 1160 (1980); (3) head protection: *International Terminal Operating Corp. of New England,* 3 OSHC 1831, *aff'd,* 540 F.2d 543, 4 OSHC 1574 (1st Cir. 1976); *Danco Construction Co.*, 3 OSHC 1114 (1975); *General Electric Co.*, 7 OSHC 2183 (1980); (4) foot protection: *Arkansas Best Freight System, Inc. v. OSHRC,* 529 F.2d 649, 3 OSHC 1910 (8th Cir. 1976); *Granson Lumber Co.*, 1 OSHC 1234 (1973); *Sunbeam Corp.*, 4 OSHC 1412 (1976); *American Airlines, Inc.*, 578 F.2d 38, 6 OSHC 1691 (2d Cir. 1978); *United Parcel Service, Inc.*, 577 F.2d 743, 6 OSHC 1588 (6th Cir. 1978); *United Parcel Service of Ohio, Inc.*, 570 F.2d 806, 6 OSHC 1347 (8th Cir. 1978); *Dover Corp., Davenport Machine Tool Company,* 7 OSHC 1237 (1979); (5) hand protection: *Smoke Craft, Inc., v. OSHRC,* 530 F.2d 843, 3 OSHC 2000 (9th Cir. 1976); *Grand Union Co.*, 3 OSHC 1596 (1975); *Great Atlantic and Pacific Tea Company,* 4 OSHC 1699 (1976); *Kroger Co.*, 6 OSHC 1030 (1977); *Owens-Corning Fiberglas Corp.*, 7 OSHC 1291 (1979); and (6) other protective clothing: *Niagara Mohawk Power Corp.*, 7 OSHC 1447 (1979), *Nibb Co.*, 1 OSHC 1777 (1974); *Cerveceria Corona, Inc.*, 6 OSHC 1414 (1978); *Osco Industries Co.*, 8 OSHC 1799 (1980). The word "provide," which is used in many personal protective equipment standards has been controversial and has been defined inconsistently. Compare *Usery v. Kennecott Copper Inc.*, 577 F.2d 1113, 6 OSHC 1197 (10th Cir. 1977)—provide does not mean require the use of—with *Marshall v. Southwestern Industrial Contractors and Riggers Inc.*, 576 F.2d 42, 6 OSHC 1751 (5th Cir. 1978)—provide does mean require the use of.

The requirement that certain personal protective equipment be worn or utilized as set forth in the above cited cases depends in many instances on the finding that a hazard exists, which would require their use. *United Parcel*

Service of Ohio, Inc. v. OSHRC, 570 F.2d 806, 6 OSHC 1347 (8th Cir. 1978); *Duar Iron Co.,* 6 OSHC 1701 (1978); *Walter B. Hawkins d/b/a Walter Hawkins Fruit Co.,* 6 OSHC 1277 (1977); *Kimbob, Inc.,* 6 OSHC 1597 (1978); *Water Works Installation Corp.,* 4 OSHC 1339 (1976); *Hehr International, Inc.,* 7 OSHC 2048 (1979); *Ray Evers Welding Company v. OSHRC,* 625 F.2d 726, 8 OSHC 1271 (6th Cir. 1980); *Pratt and Whitney Aircraft, Division of United Technologies Corporation,* 8 OSHC 1329 (1980). In a recent case, *Turner Communications Corporation v. OSHRC,* 612 F.2d 941, 8 OSHC 1038 (5th Cir. 1980), the court said mere "wearing" of safety belts without tying them does not constitute compliance with 29 CFR 1926.28(a) since the standard requires the *use* as well as wearing of equipment and is not different in that respect from 29 CFR 1910.132(a). See also *Otis Elevator Co. v. Marshall,* 581 F.2d 1056, 6 OSHC 1892 (2d Cir. 1978). Finally, vagueness and/or the failure to provide fair and adequate notice is often alleged as a defense to the requirements set forth in the personal protective equipment standards. See *Cape and Vineyard Division v. OSHRC,* 512 F.2d 1148, 2 OSHC 1623 (1st Cir. 1975); *B&B Insulation, Inc. v. OSHRC,* 583 F.2d 1364, 6 OSHC 2062 (5th Cir. 1978). But the Fourth Circuit in *Bristol Steel and Iron Works, Inc. v. OSHRC,* 601 F.2d 717, 7 OSHC 1462 (4th Cir. 1979) recently stated that

> It would be utterly unreasonable to expect the Secretary to promulgate specific safety standards which would protect employees from every conceivable hazardous condition. . . . The Secretary in an effort to insure the safety of employees as required by the Act must at times resort to general safety standards.

See also *S&H Riggers and Erectors, Inc.,* 7 OSHC 1260 (1979) where the Commission declined to follow the Fifth Circuit's decisions and stated that the test is whether a reasonable person familiar with the actual circumstances surrounding the alleged hazardous condition including any facts unique to a particular industry would recognize a hazard warranting the use of personal protective equipment. According to the Commission, although industry custom and practice are useful points of reference, they are not controlling, for the failure of industry to deal adequately with abatable hazards does not excuse an employer's failure to exercise that degree of care which the law requires. Compliance may require methods of employee protection of a higher standard than industry practice. The Commission's decision in *S&H Riggers and Erectors, Inc.,* was approved by the First Circuit in *General Dynamics Corp. v. OSHRC,* 599 F.2d 453, 7 OSHC 1373 (1st Cir. 1979). But see *Ray Evers Welding Co. v. Secretary of Labor,* 625 F.2d 726, 8 OSHC 1271 (6th Cir. 1980); *McLean-Behm Steel Erectors, Inc. v. OSHRC,* 608 F.2d 580, 7 OSHC 2002 (5th Cir. 1979).

(5) *Cranes, Equipment, and Material.* The decision in *Tri-City Construction Co.,* 7 OSHC 2189 (1980) held that the crane standard was applicable to a caterpillar performing the lifting functions of a crane. Similarly in *Suffolk County Contractors, Inc.,* 8 OSHC 1506 (1980), 29 CFR 1926.550(a)(15)(i), dealing with cranes and derricks was found to be applicable to backhoes used to perform lifting functions usually performed by a crane. In *S. J. Groves & Sons Co.,* 8 OSHC 1767 (1980), the material-handling equipment standard, 29

CFR 1926.602(a)(9)(i), which requires that bidirectional machines be equipped with a horn, was held to be applicable to the employer's scraper, since scrapers are bidirectional machines. Finally, in *Gil Haugan d/b/a Haugan Construction Co.,* 7 OSHC 2004 (1979), the Commission held that 29 CFR 1926.550 applies to any machines used to perform lifting functions usually performed by cranes or derricks; thus distinctions in the specific forms of machines so used are irrelevant as long as the functions are similar.

In *Wisconsin Electric Power Co. v. OSHRC,* 567 F.2d 735, 6 OSHC 1137 (7th Cir. 1977), the court resolved a conflict between the proper application of a standard dealing with mechanical equipment, 29 CFR 1926.952, and one dealing with work on overhead lines, 29 CFR 1926.955. The question was whether a piece of equipment involved in an accident was an electric line truck or lifting equipment; the resolution of the question would determine which standard was properly cited and which requirements thereof would be properly applicable. The court said that determining which of the two standards is more specific depends upon whether the focus is on the kind of equipment or the kind of use to which the equipment is put. See, however, *Horner, Inc. v. Moore,* 8 OSHC 1688 (Ct. App. Ark. 1980), which upheld the applicability of the motor vehicle regulation, 29 CFR 1926.601, to a ready-mix cement truck, which normally operates on highway construction sites not open to public traffic rather than on public streets and highways.

In *Dunlop v. Uriel G. Ashworth,* 538 F.2d 562, 3 OSHC 2065 (4th Cir. 1976), the employer was cited for a violation of 29 CFR 1926.700(a), which incorporated by reference an ANSI standard pertaining to temporary shoring of masonry walls. The OSHA standard, however, only incorporated by reference that portion of the ANSI standard relative to the use of equipment and material in the construction of masonry walls. It did not incorporate the portion on building technique and did not actually direct that shoring be used. It only required that the shoring that was used, if any, meet the applicable ANSI guidelines for equipment and materials. The court, affirming both the ALJ and Review Commission decisions, stated it cannot rewrite the Secretary of Labor's regulation and it would require an unduly strained reading of 29 CFR 1926.700(a) to hold it included the ANSI requirements on building technique. See also *Concrete Construction Co. Inc. v. OSHRC,* 598 F.2d 1031, 7 OSHC 1313 (6th Cir. 1979); *S. J. Groves & Sons Co.,* 7 OSHC 1481 (1979); *Kenneth P. Thompson Co., Inc.,* 8 OSHC 1696 (1980).

(6) *Flat Roofs.* In the decision of *Langer Roofing and Sheet Metal, Inc. v. Secretary of Labor,* 524 F.2d 1337, 5 OSHC 1685 (7th Cir. 1975), the court held that 29 CFR 1926.500(d)(1), which requires perimeter guarding (guardrails) on certain open side floors, by its terms does not apply to flat roofs. See also *Diamond Roofing Company v. OSHRC,* 528 F.2d 645, 4 OSHC 1001 (5th Cir. 1976). However, since 29 CFR 1926 500(d)(1) does not apply to flat roofs, the Commission has stated in several cases that the personal protective requirements of 29 CFR 1926.28(a) and 1926.105(a) are applicable to fall hazards. For example, in *Johns Roofing and Sheet Metal Company,* 6 OSHC 1792 (1978), the Commission stated that 29 CFR 1926.451(u)(3) establishes fall protection requirements for workers on sloped roofs to the exclusion of other standards. Since it is not directed to the same hazard with respect to flat roofs, 29 CFR 1926.28(a) is applicable in the absence of a more specifically applicable standard. See also *Hamilton Roofing Company, Inc.*, 6 OSHC 1771 (1978).

In *Forest Park Roofing Company*, 8 OSHC 1181 (1980), the Commission determined that both 29 CFR 1926.28(a) and 1926.105(a) are applicable to working conditions on flat roofs. In *Everhart Steel Construction Co.*, 3 OSHC 1068 (1975), the term "open-sided floor" in 29 CFR 1926.500(d)(1) was found to encompass roofs and thus the failure to provide perimeter guarding or other protective device/equipment violates the standard. In *Universal Roofing & Sheet Metal Company, Inc.*, 8 OSHC 1453 (1980), the Commission held that a roof cannot be considered a temporary floor within the meaning of 29 CFR 1926.105 (a), that 29 CFR 1926.451(u)(3) does not provide the exclusive fall protection requirements for employees working on roofs, and that 29 CFR 1926.28(a) can appropriately be cited for fall hazards not within the purview of 29 CFR 1926.451(u)(3). Thus compliance with 29 CFR 1926.105(a) is not sufficient; 29 CFR 1926.28(a) is also applicable to the facts even though alleged noncompliance with 29 CFR 1926.105(a) is not a separate violation since the allegations are duplicative under the circumstances of the case. See also *Voegele Company, Inc.*, 7 OSHC 1713, *aff'd*, 625 F.2d 1075, 8 OSHC 1631 (3d Cir. 1980).

b. *Shipbuilding/Longshoring*

In order to ascertain if an employer is in compliance with 29 CFR 1915–1918, it must first be determined if the working conditions and/or workplace are covered by these standards or by the standards of another agency such as the Coast Guard. *Petrolane Offshore Construction Services, Inc.*, 3 OSHC 1156 (1975); *Brown & Root, Inc., Pipeline Div.*, 6 OSHC 1844 (1978); *Pacific International, Inc.*, 7 OSHC 2227 (1979).

In *Dravo Corporation v. OSHRC and Marshall*, 613 F.2d 1227, 7 OSHC 2089 (3d Cir. 1980), the court first noted that an employer is entitled to fair notice in dealing with the government. Like other statutes and regulations which impose penalties, an occupational safety and health standard must give an employer fair warning of the conduct it prohibits or requires. Referring to the definitions set forth in 29 CFR 1916 which deal with shipbuilding and specifically focus upon those terms defining employment, employer, employee, and shipbuilding, the court noted that coverage of the OSHA shipbuilding regulations is limited to maritime work on navigable waters including drydocks, graving docks, and marine walkways. These definitions do not include structural fabrication shops within their scope, because the Secretary of Labor failed to state in his regulations that a structural shop is included within a place of maritime employment to which the shipbuilding regulations apply. Thus the employer is not to be held to shipbuilding safety standards.

In *Seattle Crescent Container Serivce*, 7 OSHC 1895 (1979), an employer was cited for violating the Act for failure to comply with 29 CFR 1918.32(b), which was originally promulgated under the Longshoremen's and Harborworkers' Compensation Act, and were subsequently repromulgated under the Occupational Safety and Health Act as an established federal standard. The Commission stated it was impossible to determine whether the Secretary of Labor intended either to restrict application of the standard to work performed below the deck and between the decks, or, more generally, to protect employees from injuries resulting from working on top of the deck, regardless of their location on the ship. However, implicitly rejecting the argument that standards should be applied and interpreted restrictively, it stated that stronger evidence is needed before it can interpret a standard restrictively to the detriment of the employees' safety.

In *National Steel & Shipbuilding Co. v. OSHRC,* 607 F.2d 311, 7 OSHC 1837 (9th Cir. 1979), the court held that the standard 29 CFR 1916.41(i)(1) is applicable to the situation in which scaffolding is being taken down. The importance of the applicability argument by the employer in that case, which involved a citation for a willful violation, is that if an employer believes in good faith that a standard does not apply, then there may be reason to show that its good-faith belief means it did not knowingly and willfully violate the standard and the Act. The court said that the Commission found that the employer did not act in good faith, since it established it allowed its painter to work on the scaffold for practical reasons, rather than a serious belief that the standard did not apply.

Finally, in *Dravo Corporation v. OSHRC,* 578 F.2d 1373, 6 OSHC 1933 (3d Cir. 1978), the shipbuilding housekeeping standard, 29 CFR 1916.51(c), which requires that slippery conditions on working surfaces and walkways be eliminated as they occur, was held *not* to *apply* if such conditions are caused by bad weather for it would be virtually impossible to so comply. See also *Independent Pier Co.,* 3 OSHC 1674 (1975), which held that the longshoring standard, 29 CFR 1918.105(a), which is applicable to onshore and offshore longshoring activities preempts the general industry standard, 29 CFR 1910.132(a), when working conditions of longshoremen are at issue.

c. *Telecommunications*

In *New England Telephone and Telegraph Company,* 8 OSHC 1478 (1980), the Secretary of Labor and the company disagreed with respect to the application of the telecommunication standard. That is, whether applicability and coverage is to be determined on a geographical or functional basis. The Secretary viewed the standard geographically in that it applies only at telecommunication centers or telecommunications' field installations. The company viewed the standard functionally, that is, it applied where telecommunication workers are performing telecommunication work. The company argued that in view of the applicability of specific standards directed toward telecommunications work, it could not have known that a general industry standard might be applicable to the facts of the case, since the telecommunications standard is the more specific standard and therefore applicable, that is, 29 CFR 1910.268(j)(4)(i) *v.* 29 CFR 1910.180(j)(1)(i). The telecommunication standard requires only a two-foot clearance when working with high-voltage wires, whereas the general industry standard requires a ten-foot distance. Since telecommunications workers are familiar with the hazards and techniques associated with working on or near energized lines it is often necessary for these types of workers to work closer than ten feet to the lines. The Commission found, however, that because telecommunication workers are expected to be trained to work safely at distances closer to energized power lines than most workers, the applicability of the telecommunications standard was intended to hinge primarily on the function and nature of the work operation being performed rather than on the location where the work is being performed. The terms "telecommunication centers" and "telecommunication field installations" reflect the fact that these are types of geographical locations at which telecommunication activities take place, rather than geographically restrict the applicability. Thus the telecommunication standard was found to be applicable. In *Southwestern Bell Telephone Co. v. Secy. of Labor,* 568 F.2d 368, 6 OSHC 1377 (5th Cir. 1978), the court upheld the Review Commission's determination of the applicability of the telecommunication standard, since it was not contrary to the standard's plain meaning.

A related issue, which was raised in the *New England Telephone and Telegraph Company* case, was the question of how an employer is to determine that a specific standard does or does not apply or that a general industry standard would apply instead. This issue, which was also raised in other decisions such as *Kent Nowlin Construction Company v. OSHRC,* 593 F.2d 368, 7 OSHC 1105 (10th Cir. 1979) and *Diebold, Inc. v. Marshall,* 585 F.2d 1327, 6 OSHC 2002 (6th Cir. 1978), in the context of whether a general industry standard would be constitutionally unenforceable under such circumstances is discussed below.

d. *Agriculture*

In *Chapman & Stephens Co.,* 5 OSHC 1395 (1977), the Commission stated that a citrus grower, who is subject to the safety standards for agriculture, 29 CFR Part 1928, is subject to neither the personal protective-equipment requirements of the general industry standard, 29 CFR 1910.132(a), or the construction standard, 29 CFR 1926.28(a). By virtue of 29 CFR 1926.28(a), certain enumerated standards listed in Part 1910 are expressly made applicable to agricultural operations, but 1926.132(a) is not among those listed. Moreover, 29 CFR 1910.21(b) specifically states that, excepting to the extent specified in paragraph (a) thereof, the standards contained in subparts (B) through (S) of 1910 do not apply to agricultural operations. The Commission stated that under proper circumstances, it may be appropriate to apply construction standards to agricultural operations. Having found no standards applicable under construction or general industry and finding no other standard applicable to the facts of the case, the Commission held that the cited hazard was subject to Section 5(a)(1) of the general duty clause, referring to *Whirlpool Corp.,* 5 OSHC 1173 (1977). Farm vehicles are exempted from the requirements of 29 CFR 1910.111(f) relative to tank motor vehicles for transporting ammonia. *Durant Elevator, A Division of Scoular-Bishop Grain Elevator,* 8 OSHC 2187 (1980).

e. *General Industry*

Many of the challenges to general standards allege that their application is improper because a more specific standard applies to the condition, practice, or equipment in issue. This issue occurs most frequently in the application of the guarding standards.

In *Diebold, Inc. v. Marshall,* 585 F.2d 1327, 6 OSHC 2002 (6th Cir. 1978) and *Irvington Moore v. OSHRC,* 556 F.2d 431, 5 OSHC 1585 (9th Cir. 1977), the courts were presented with the question of the apparent conflict between 29 CFR 1910.217(a)(5), which exempts press brakes from the requirements of 29 CFR 1910.212(a)(3)(iii) and 29 CFR 1910.212(a)(3)(ii) itself, which, the Secretary of Labor alleged, required point-of-operation guarding on press brakes. The *Diebold* court stated that the question of whether 29 CFR 1910.212 applies to press brakes is determined by whether its Walsh-Healey predecessor, established federal standard source applied to press brakes, because adoption of established federal and national consensus standards must be accomplished without substantial modification. The Secretary of Labor could not enforceably construe a Section 6(a) standard to impose requirements which the standard's source did not impose. The resolution of that issue depended upon determinations which the court felt the Commission was peculiarly fitted to make by virtue of its members' education, training, and experience and thus, while admitting that 29 CFR 1910.217(a)(5) is inartfully drafted, the court adopted the Commission's interpretation, which held that 29 CFR 1910.212 is applicable to press brakes, despite evidence of contrary industry practice and belief at the time of the standard's adoption, and that the provision in 29 CFR 1910.217 excluding press brakes from coverage of that section did not exempt press brakes from coverage of the general machine-guarding standard at 1910.212. See also *Hobart Corp.,* 5 OSHC 1718 (1977); *General Motors Corp.,* 4 OSHC 1946 (1976)—citations of 29 CFR 1910.212(a)(1) and (3) for inadequate guarding of power presses vacated because specific standard 29 CFR 1910.217 was applicable; *Clark Equipment Co.,* 3 OSHC 1834 (1975)—29 CFR 1910.212 applies to press brakes and requires point-of-operation guarding for them, despite 29 CFR 1910.217(a) (5).

In *Otis Elevator Co. v. OSHRC,* 581 F.2d 1056, 6 OSHC 1892 (2d Cir. 1978) and *Turner Communications Corporation v. OSHRC,* 612 F.2d 941, 8 OSHC 1038 (5th Cir. 1980), the courts interpreted the seeming conflicting requirements of 29 CFR 1910.132(a), which provides that personal protective equipment be *used,* while 29 CFR 1926.28(a) requires that personal protective equipment be *worn.* However, according to the courts, both standards are applicable to and require the *use* of personal protective equipment when necessary to prevent falling; merely wearing and not using personal protective equipment does not comply with either standard.

In *United States Pipe & Foundry Co.,* 6 OSHC 1332 (1978), a coal screw conveyor power switch was properly cited under Section 5(a)(1) of the Act, as the allegedly applicable specific standard, 29 CFR 1910.145(f)(3)(iii) ("Do Not Start" tags), did not apply to the elimination of the hazard after the initial moments during which tags are to be used. In *Greyhound Lines—West v. Marshall,* 575 F.2d 259, 6 OSHC 1636 (9th Cir. 1978), the Court held that open-pit guarding requirements of 29 CFR 1910.22(c) are applicable to vehicle maintenance pits and not only to pits used for handling and storage of materials. Finally, in *Westinghouse Electric Corporation v. OSHRC,* 617 F.2d 497,

8 OSHC 1110 (7th Cir. 1980), the court, after considering that two standards, 29 CFR 1910.94(c) and 1910.107, exist which relate to spray booths and spray rooms (for spray painting), found they were both inartfully drafted, vacated the citation, and remanded the case to the Commission to make a determination of the standards' applicability to the employer's operation; *Highway Motor Corp.*, 8 OSHC 1796 (1980)—employer must prove he falls within exceptions of spray booth standard to avoid applicability; *Ed Jackman Pontiac-Olds, Inc.*, 8 OSHC 1211 (1980)—applicability of spray room vs. spray area standards.

f. *Training Requirements*

Section 2(b)(1) of the Act indicates the need for cooperation between employers and employees in establishing safety programs. In Section 21(c), the Secretary of Labor is required and directed to provide training programs and consultation for employees relative to the prevention of injuries and illnesses. Many OSHA standards also require employers to provide specific safety training for their employees; a failure to provide adequate training may constitute a violation of these standards. See *Brennan v. OSHRC (Gerosa, Inc.)*, 491 F.2d 1440, 1 OSHC 1523 (2d Cir. 1974)—standard requires employer to designate a competent person to inspect and operate; *Ames Crane & Rental Service, Inc. v. Dunlop*, 532 F.2d 123, 4 OSHC 1060 (8th Cir. 1976)—making available instruction rather than giving instruction is not sufficient.

Even in the absence of a specific standard, an employer has an overall duty to train employees adequately for their jobs and take precautionary steps to protect employees from hazards. *Brennan v. Butler Lime & Cement Co.*, 520 F.2d 1001, 3 OSHC 1461 (7th Cir. 1975). The degree of training required will, however, vary depending upon the employee's experience, nature of his job and hazards. See *N. Fields Tree Service, Inc.*, 5 OSHC 1142 (1977). The adequacy of an over-all training and safety program is also important, with respect to the assertion of affirmative defenses such as isolated incident and as an element of good faith.

Many safety and health standards relate to the training of employees, either by requiring a specific training program or requiring that employees be properly instructed. Examples of standards which require employee training programs are 29 CFR 1910.1006(e)(5)(i)–(ii), methyl chloro-methyl ether; 29 CFR 1910.1007(e)(5)(i) 3, 3′-dichlorobenzidine or its salts; 29 CFR 1917.351(d)(1)–(6), gas welding and cutting; and 29 CFR 1926.21(a), which requires the establishment and supervision of programs for the education and training of construction industry employees in the recognition, avoidance, and prevention of unsafe conditions in employment covered by the Act. There are other standards which, although not requiring a formalized training program, do require that employees be properly instructed, such as: 29 CFR 1910.94(d)(9)(i), ventilation; 29 CFR 1910.111(b)(13)(ii), storage and handling of anhydrous ammonia; 29 CFR 1915.36(d)(1)–(4), arc welding and cutting; 29 CFR 1915.82(a)(4), respiratory protection; and 29 CFR 1926.803(a)(2), compressed air.

In *Modern Metal Smiths, Inc.*, 7 OSHC 2132 (1979), an employer was cited for a serious violation of 29 CFR 1910.94(d)(9)(i), which requires that employees working around open-surface tank operations be properly instructed as to job hazards. The citation was based upon a statement that the employer had no training program. The citation was vacated, however, since the standard does not require a training program, but merely requires that employees be properly instructed. In *H. C. Nutting Company v. OSHRC*, 615 F.2d 1360, 8 OSHC 1241 (6th Cir. 1980), an employer was cited for violating 29 CFR 1926.21(b)(2), which requires employers to instruct employees in the recognition and avoidance of unsafe conditions and the regulations applicable to the work environment. It was determined at the administrative level that the employer's safety program fell short of meeting the regulation's requirements since its mere distribution of a safety handbook and the delegation of instruction to supervisors without adequate supervision was insufficient. Addressing the argument that the regulation was unconstitutionally vague, the court stated that the regulation requires an employer to inform employees of safety hazards which are known to a reasonably prudent employer and which are addressed by specific OSHA regulations and that so construed it is not unreasonably vague and does place reasonable duties on an employer. However, the court stated that there is no evidence whatsoever as to what standard of reasonable instruction was required since the Secretary of Labor pointed to no evidence of industry practice or reasonable standards. All that the record revealed was that the OSHA compliance officer was displeased with the employer's safety program, but since there was no substantial evidence showing that the employer violated the regulation against a backdrop of unreasonable industry practice, the citation was vacated. See *John W. McGowan*, 5 OSHC 2028 (1977), *review denied*, 604 F.2d 885, 7 OSHC 1842 (5th Cir. 1979); *Sawnee Electric Membership Corporation*, 5 OSHC 1059 (1977)—Secretary of Labor's failure to show the inadequacy of the employer's safety program, which included monthly safety meetings and on-the-job instructions, required the vacation of a citation for violation of 29 CFR 1926.21(b)(2) for failure to instruct employees in the recognition and avoidance of safety hazards and in safety regulations.

In *National Industrial Constructors, Inc. v. OSHRC*, 583 F.2d 1048, 6 OSHC 1914 (8th Cir. 1978), the court held that 29 CFR 1926.21(b)(2), which requires that employers instruct employees in the recognition and avoidance of unsafe conditions in the applicable work environment sufficiently warns employers of the conduct required, and is not impermissibly vague. The court also noted that the standard requires more than a general safety program, in addition, it requires supervisory personnel to advise employees, especially new employees, of hazards associated with the actual dangerous conduct in which they are presently engaging. See also *Herbert Vollers, Inc.*, 4 OSHC 1798 (1976), permitting a worker who had not been trained to operate a loader tractor violates 29 CFR 1926.20(b)(4), which provides that only qualified personnel shall be permitted to operate machinery and equipment. In this respect see also 29 CFR 1915.10(a)–(b), 1916.10(a)–(b), 1917.10(a)–(b), which require the designation of *competent* persons knowledgeable as to the requirements of the section. Employers are also required to provide adequate supervision to employees. The degree of supervision required will of course depend on several factors, such as the type of employees and the type of work involved. See *Horne Plumbing & Heating Company v. OSHRC,* 528 F.2d 564, 3 OSHC 2060 (5th Cir. 1976); *Brennan v. OSHRC (Hanovia Lamp Division),* 502 F.2d 946, 2 OSHC 1137 (3d Cir. 1974)—although close or constant supervision may be required in some cases, it is unrealistic to expect an experienced technician to be under other than occasional scrutiny. The question of supervision is an important element in the defenses of isolated incidence and employee misconduct, and may also be with respect to general duty clause cases.

Many OSHA standards require that employers provide safe tools and equipment, modify machinery to include safety devices or adopt safety procedures. Employers may also be required to periodically examine, inspect, and measure such equipment and provide personal protective equipment, such as lifelines, safety belts, and hard hats whenever necessary by reason of hazards, processes, the environment, or whenever a standard or reasonable caution so require. See *Ryder Truck Lines, Inc. v. Brennan,* 497 F.2d 230, 2 OSHC 1075 (5th Cir. 1974); *Cape and Vineyard Division v. OSHRC,* 512 F.2d 1148, 2 OSHC 1628 (1st Cir. 1975); *Hoffman Construction Co. v. OSHRC,* 546 F.2d 281, 4 OSHC 1813 (9th Cir. 1976); *Somerset Tire Service, Inc.,* 1 OSHC 1162 (1973). Failure to provide, even though standards only require the use of personal protective equipment, may be a violation of the general duty clause. See *ITO Corp. v. OSHRC,* 540 F.2d 543, 4 OSHC 1574 (1st Cir. 1976); *Atlantic and Gulf Stevedores, Inc. v. OSHRC,* 534 F.2d 541, 4 OSHC 1061 (3d Cir. 1976).

5. Validity and Vagueness

Those courts presented with the question of vagueness, fairness, reasonableness, and due-process challenges to standards have responded negatively to the arguments advanced by OSHA that regulations should be liberally construed to give broad coverage because of the intent of Congress to provide safe and healthful working conditions for employees by saying that an employer is entitled to fair notice in dealing with his government. An occupational safety and health standard must give an employer fair warning of the conduct that it prohibits or requires and must provide a clear standard in order to circumscribe the discretion of the enforcing authority. A regulation cannot be construed to mean what an agency intended, but did not express, especially if a violation or regulation subjects a party to criminal or civil sanctions or penalties. Coverage of an agency regulation, especially one dealing with the penal sanction, should be no broader than what is encompassed in its terms. See generally *Diamond Roofing Co. v. OSHRC,* 528 F.2d 645, 4 OSHC 1001 (5th Cir. 1976); *Bristol Steel & Iron Works, Inc., v. OSHRC,* 601 F.2d 717, 7 OSHC 1462 (4th Cir. 1979); *Marshall v. Anaconda Co.,* 596 F.2d 370, 7 OSHC 1382 (9th Cir. 1979); *Dravo Corp. v. Marshall,* 613 F.2d 1227, 7 OSHC 2089 (3d Cir. 1980). In *Bethlehem Steel Corp. v. OSHRC,* 573 F.2d 157, 6 OSHC 1440 (3d Cir. 1980), the court developed a test for interpretation of the Secretary's OSHA regulations, in that:

> The purpose of OSHA standards is to improve safety conditions in the workplace by telling employers just what they are required to do in order to prevent or minimize danger to employees . . . a responsibility to promulgate clear and unambiguous standards is upon the Secretary. The test is not what he might possibly have intended, but what he said.

However, in *Irvington Moore v. OSHRC,* 556 F.2d 431, 5 OSHC 1585 (9th Cir. 1977), the court held that a reading of regulations which would do violence to the application of a standard that assures the greatest protection for employees can not be reconciled with the general canon of statutory construction that remedial statutes are to be liberally construed in favor of their beneficiaries. Citing *American Smelting & Refining Co. v. OSHRC,* 501 F.2d 504, 2 OSHC 1041 (8th Cir. 1974).

In determining whether a "generalized" standard or regulation gives an employer sufficient warning of the conduct that it prohibits and that its operations are within its scope, courts have held that the question of fairness must be answered in light of the conduct to which the regulation is applied. This has resulted in a variety of interpretations concerning the application of the standard.

a. *The Reasonable Man Test*

In *Diebold, Inc. v. Marshall,* 585 F.2d 1327, 6 OSHC 2002 (6th Cir. 1978), the court held that a combination of the inartful drafting of 29 CFR 1910.217, the power-press guarding standard, in conjunction with the application of the undisputed common understanding and commercial practice relative to press-brake guarding and industry practice denied the employer due process in attempting to apply the general machine-guarding standard, 29 CFR 1920.212, to require point-of-operation guarding for the employer's press brakes. Moreover, in the court's determination, there was a division in the rulings of the Review Commission on the question of such applicability at the time of the citation. The cumulative effect, according to the court, would have been sufficient to lead an employer to believe that press brakes had been specifically exempted from the guarding requirements of 29 CFR 1910.212.

While the court was persuaded that the Commission's interpretation may be correct, that did not lead it to the conclusion that the regulation may be applied, since among the myriad applications of the due-process clause is the fundamental principle that statutes and regulations which purport to govern conduct must give an adequate warning of what they command or forbid. The court stated:

> In our jurisprudence, because we assume that man is free to steer between lawful and unlawful conduct, we insist that laws give the person of ordinary intelligence a reasonable opportunity to know what is prohibited, so that he may act accordingly. . . . Even if a regulation which governs purely economic or commercial activities, if its violation can engender penalties, it must be so framed as to provide a constitutionally adequate warning to those whose activities are governed. . . . Our question therefore is with the question whether the regulation gave Diebold sufficient warning that press brakes were within the scope of its point of operation

guarding requirements. The question is to be answered, of course, "in the light of the conduct to which the regulation is applied." Moreover, the constitutional adequacy or inadequacy of the warning given must be measured by common understanding and commercial practice.

See also *United Parcel Service of Ohio, Inc. v. OSHRC,* 570 F.2d 806, 6 OSHC 1347 (8th Cir. 1978); *Jensen Construction Co. of Oklahoma, Inc., v. OSHRC,* 597 F.2d 246, 7 OSHC 1283 (10th Cir. 1979); *Ray Evers Welding Company v. OSHRC,* 625 F.2d 726, 8 OSHC 1271 (6th Cir. 1980); *Arkansas-Best Freight Systems, Inc. v. OSHRC,* 529 F.2d 649, 3 OSHC 1910 (8th Cir. 1976).

In *Brennan v. OSHRC (Santa Fe Trail Transport Co.),* 505 F.2d 869, 2 OSHC 1274 (10th Cir. 1974), the court also applied the reasonable man standard, vis-à-vis 29 CFR 1910.151(b), asking the question whether the regulation delineates its reach in words of common understanding. In *McLean Trucking Co. v. OSHRC,* 503 F.2d 8, 2 OSHC 1165 (4th Cir. 1974), the court, in upholding 29 CFR 1910.132(a), stated that a standard should be considered in the light of its application rather than on its face; thus the court looked to and applied the reasonable man test.

b. *The Fifth Circuit Test*

In *Ryder Truck Lines, Inc. v. Brennan,* 497 F.2d 230, 2 OSHC 1075 (5th Cir. 1974), the court stated that inherent in the standard is an external and objective test of whether or not a reasonable person would recognize a hazard. So long as the standard affords a reasonable warning of the prescribed conduct in light of the common understanding and practices of the industry, it will pass constitutional muster. In *B & B Insulation, Inc. v. OSHRC,* 583 F.2d 1364, 6 OSHC 2062 (5th Cir. 1978), the court made a finding that 29 CFR 1926.28(a), the personal protective equipment standard, was not unenforceably vague for failure to provide reasonable notice of what was required because the standard must be read "as requiring only those protective measures which knowledge and experience of the employer's industry would clearly deem appropriate under the circumstances." In order to find a violation, it must be shown that the employer's conduct fell below the practice in the industry. This case is interesting because it applied the reasonable man standard in direct reference to persons who would be subject to judgment by that standard. Thus it was necessary to bring in demonstrated industry custom as an outgrowth of the reasonable man standard:

> Where the reasonable man test is used to interpolate specific duties from general OSHA regulations, the character and purposes of the Act suggest a closer identification between the projected behavior of the reasonable man and the customary practice of employers in the industry . . . Where the government seeks to encourage a higher standard of safety performance from the industry than customary industry practice exhibits, the proper recourse is to the standard-making machinery provided in the Act.

In *Power Plant Div., Brown & Root, Inc. v. OSHRC,* 590 F.2d 1363, 7 OSHC 1137 (5th Cir. 1979), the court held that in order to show that a standard is not unconstitutionally vague, the Secretary of Labor must prove that a reasonable employer in the (roofing) industry would have recognized the hazard. It must be shown that a reasonable employer in his industry would have recognized the same conditions as a hazard, requiring the use of safety equipment, the absence of which constitutes the employer's violation. The reasonable employer in the particular industry is generally established by reference to the actual custom and practice in that industry. Compare *Allis-Chalmers Corp. v. OSHRC,* 542 F.2d 27, 4 OSHC 1633 (7th Cir. 1976) 29 CFR 1910.21(a)(1), which requires that scaffolds be provided where work cannot be done safely affords a reasonable warning of prescribed conduct in light of common understanding and practices in the industry; *Jensen Construction Co. of Oklahoma, Inc. v. OSHRC,* 597 F.2d 246, 7 OSHC 1283 (10th Cir. 1979)—when present regulation is viewed in light of conduct to which it is sought to be applied, it is not impermissibly vague; *Cape & Vineyard Div. v. OSHRC,* 512 F.2d 1148, 2 OSHC 1628 (1st Cir. 1975).

c. *The Fifth Circuit Test Rejected*

Several courts have specifically rejected the Fifth's Circuit custom and practice in the industry reasonableness test. In *Voegele Co., Inc., v. OSHRC,* 625 F.2d 1075, 8 OSHC 1631 (3d Cir. 1980), the court held, in a case involving a challenge to 29 CFR 1926.28(a), that the proper test to apply is an objective standard, the reasonably prudent person test—that is, whether a reasonable person familiar with the conditions in the industry would have complied. The court (stating that no other court has adopted the Fifth Circuit's test) rejected the Fifth Circuit's tying of the reasonable person test to measures customarily used in the employer's industry and stated that other courts have evaluated the custom and practice of the industry as one aspect of the reasonable person test, but have refused to limit the reasonable person test to the custom and practice of the industry because such a standard would allow an entire industry to avoid liability by maintaining inadequate safety. See *General Dynamics v. OSHRC,* 559 F.2d 453, 7 OSHC 1373 (1st Cir. 1979); *American Airlines, Inc. v. Secy. of Labor,* 578 F.2d 38, 6 OSHC 1691 (2d Cir. 1978); *Brennan v. Smoke-Craft, Inc.,* 530 F.2d 843, 3 OSHC 2000 (9th Cir. 1976); *Bristol Steel & Iron Works, Inc. v. OSHRC,* 601 F.2d 717, 7 OSHC 1462 (4th Cir. 1979). The Voegele court refused to tie in the reasonable man test to the custom and practice of the employer's industry. The appropriate inquiry to the court was whether, under the circumstances, a reasonably prudent employer, familiar with his industry, would have protected against the hazard. Applying this test, the court upheld the standard, stating that an employer must provide personal protective equipment that a reasonable man would deem sufficient under the circumstances.

In *United Parcel Service of Ohio, Inc., v. OSHRC,* 570 F.2d 806, 6 OSHC 1347 (8th Cir. 1978), the general personal protection standard, at 29 CFR 1910.132(a), was not found to be unenforceably vague as applied to foot protection in the parcel-handling industry since it was reasonable in the circumstances of the particular case. In like respect, the court in *Ray Evers Welding Co. v. OSHRC,* 625 F.2d 726, 8 OSHC 1271 (6th Cir. 1980), stated that the appropriate test relative to a consideration of the "vagueness" of a standard is whether a reasonably prudent employer concerned with the safety of his employees would recognize the existence of a hazardous condition and protect them by the means specified in the citation. See also *Continental Oil Company v. Marshall*, 630 F.2d 446, 8 OSHC 1980 (6th Cir. 1980).

d. *The Plain and Clear Meaning Test*

Other decisions which basically do not fit within either of the above descriptions include the following, which hold that if a standard has a clear and plain meaning, it is not vague and need not be cured by the reasonable person test. *Irvington Moore v. OSHRC*, 556 F.2d 431, 5 OSHC 1585 (9th Cir. 1977)—the regulations are clear and are not unconstitutionally vague; *Austin Commercial v. OSHRC*, 610 F.2d 200, 7 OSHC 2119 (5th Cir. 1979)—the plain and natural meaning of the words of the standard, 29 CFR 1926.251(a)(1), give fair warning of prohibited practice; *Hutton v. Plum Creek Lumber Co.*, 608 F.2d 1283, 7 OSHC 1940 (9th Cir. 1979)—the clear and natural meaning of the standard is not vague, unreasonable, arbitrary, or ambiguous, and it is entitled to a presumption of validity; *Southwestern Bell Telephone Co. v. Secy. of Labor*, 568 F.2d 368, 6 OSHC 1377 (5th Cir. 1978)—the specific application of the telecommunications standard is not contrary to the standard's plain and natural meaning and does not violate due process.

However, if a standard as written is so ambiguous that it is impossible to apply, it is not enforceable. In *Lloyd C. Lockrem, Inc. v. U.S.A.*, 609 F.2d 940, 7 OSHC 1999 (9th Cir. 1979), the court stated:

> The Secretary here contends that his interpretation is reasonable and interprets the standards so as to give effect to the natural and plain meaning of their words. This rule would be easy to apply if it were not for the patent ambiguity in the regulations. Here, the Secretary initially argued to the ALJ that the hole in question was not a trench, since it was wider than 15 feet at the top. However, this argument was abandoned on review in this appeal as the Commission treats the hole as a trench. Even in *D. Federico Co., Inc. v. OSHRC*, 558 F.2d 614, 5 OSHC 1528 (1st Cir. 1977), where the First Circuit held that the excavation standards were applicable to trenches, it also condemned the procedures used by OSHRC and urged it "to straighten up its own trenches lest the employers it prosecutes be able to slip through unscathed." *Federico* at 617.
>
> The Secretary's argument to the contrary notwithstanding, it is far from clear from the natural and plain meaning of its words that the excavation standard, 29 CFR 1926.651(s), applies to trenches. 1926.651 is labeled "Specific Excavation Requirements." While it must be acknowledged that both "excavations" and "trenches" are holes in the ground, there is no indication whatsoever that the "specific excavation requirements" apply to trenches. The fact that the Secretary labeled separately each set of standards and included the word "specific" in each title would indicate that the specific excavation requirements apply to excavations which are not trenches and that only the "specific trenching requirements" apply to the trenches. The maxim *expressio unius est exclusio alterius* applies. Furthermore, the testimony at the hearings clearly demonstrated that other employers in the area were unaware that excavation standards were applicable to trenches. Finally, the evidence supports the view that if the requirements as to excavations would also apply to trenches, it would be impossible to comply with both sets of regulations and still perform the trenching operation.
>
> We note the considerable confusion within the OSHRC and of the ALJ's over

> the question of whether the "specific excavation standards" apply to trenches. We reiterate that an employer should not be held to standards, the application of which cannot be agreed upon by those charged with their enforcement. Regulations cannot be construed to mean what an agency intended but did not adequately express. [*Diamond Roofing Co. v. OSHRC,* 528 F.2d 645, 4 OSHC 1001 (5th Cir. 1976).]

In *Kent Nowlin Construction Company v. OSHRC,* 593 F.2d 368, 7 OSHC 1105 (10th Cir. 1979), a question over whether an employer's earth cavity was a trench or excavation required the vacation of a citation, since the court felt that the employer should not be penalized for a deviation from a standard which is subject to varying interpretations. Administrative regulations cannot be construed to mean what the administration intended but did not adequately express, since an employer is entitled to fair warning of the conduct the regulation prohibits or requires. Similarly, in *Diebold, Inc. v. Marshall,* 585 F.2d 1327, 6 OSHC 2002 (6th Cir. 1978), a question involving the application and interpretation of 29 CFR 1910.217, which excludes press brakes from coverage, and 29 CFR 1910.212, the general machine-guarding standard, was brought into question. The court stated that given the inartful drafting of the standards, neither interpretation concerning the application of the standards could be branded as particularly unreasonable, and although the court was mindful of the great deference owed to the Commission's reasonable interpretations of the Secretary's regulations, it found the regulation so vague in its requirements that its enforcement would violate the due-process clause in the Fifth Amendment. For even a regulation which governs purely economic and commercial activities must be framed so as to provide a constitutionally adequate warning to those whose activities are governed.

It seems clear that employers may raise fairness, due process, vagueness questions, and procedural invalidity as affirmative defenses in enforcement proceedings under Section 11(a) of the Act. However, one court, *National Industrial Constructors, Inc. v. OSHRC,* 583 F.2d 1048, 6 OSHC 1914 (8th Cir. 1978), rejected such a challenge as a procedural attack on the manner in which the standard had been promulgated. The "procedural attack," however, was held to be precluded by Congress by the enactment of Section 6(f) of the Act, which provides for judicial review of a regulation within sixty days of the regulation's adoption. Other courts have rejected this approach, holding that Section 6(f) applies to preenforcement review only and does not foreclose an employer from challenging the validity of a standard during an enforcement proceeding. *Marshall v. Union Oil Co. of California,* 616 F.2d 1113, 8 OSHC 1169 (9th Cir. 1980); *Noblecraft Industries v. Secretary of Labor,* 614 F.2d 199, 7 OSHC 2059 (9th Cir. 1980); *Atlantic & Gulf Stevedores v. OSHRC,* 534 F.2d 541, 4 OSHC 1061 (3d Cir. 1976).

It seems that the Review Commission recently in *Gold Kist, Inc.,* 7 OSHC 1855 (1979); *S & H. Riggers & Erectors, Inc.,* 7 OSHC 1260 (1979), *appeal filed,* (5th Cir. 1979) and *W. J. Lazynski, Inc.,* 7 OSHC 2064 (1979), has adopted the reasonable person familiar with the factual circumstances surrounding the alleged hazardous condition, including factors unique to the particular industry test, relative to fairness, notice, and validity challenge to standards.

The Commission rejects the Fifth Circuit's test and feels that industry custom and practice are useful reference points, but that the standard of care commonly used therein is not conclusive. See also *M. J. Lee Construction Company*, 7 OSHC 1140 (1979)—when the standard is read in light of other applicable provisions, its meaning is sufficiently clear and precise by "definition" to put employers on notice of what the standard requires; impliedly rejecting the understanding of or common knowledge in the industry test.

D. STANDARDS DEVELOPMENT AND ADOPTION

1. Statutory Requirements

The Act authorizes the Secretary of Labor to establish workplace standards aimed at preserving and improving the health and safety of all American employees involved in commerce among the several states. Section 2 of the Act. Most of OSHA's authority and guidance for setting standards comes from Section 6 of the Act. In determining the priority for establishing standards, the Secretary is required to give due regard to the urgency of the need for mandatory safety and health standards for particular industries, occupations, businesses, workplaces . . . and to recommendations by the Secretary of Health and Human Services (*nee,* Education and Welfare). Section 6(g) of the Act.

In relevant part, Section 6(b) of the Act provides as follows:

> The Secretary may adopt, promulgate, modify, or revoke any occupational safety or health standard in the following manner:
>
> (1) Whenever the Secretary upon the basis of information submitted to him in writing by an interested person, a representative of any organization of employers or employees, a nationally recognized standards-producing organization, the Secretary of Health, Education, and Welfare, the National Institute for Occupational Safety and Health, or a State or political subdivision, or on the basis of information developed by the Secretary or otherwise available to him determines that a rule should be promulgated in order to serve the objectives of this chapter, the Secretary may request the recommendations of an advisory committee appointed under section 656 of this title. The Secretary shall provide such an advisory committee with any proposal of his own or of the Secretary of Health, Education, and Welfare, together with all pertinent factual information developed by the Secretary or the Secretary of Health, Education, and Welfare, or otherwise available, including the results of research, demonstrations, and experiments. An advisory committee shall submit to the Secretary its recommendations regarding the rule to be promulgated within ninety days from the date of its appointment or within such longer or shorter period as may be prescribed by the Secretary, but in no event for a period which is longer than two hundred and seventy days.
>
> (2) The Secretary shall publish a proposed rule promulgating, modifying, or revoking an occupational safety or health standard in the Federal Register and shall afford interested persons a period of thirty days after publication to submit written data or comments. Where an advisory committee is appointed and the Secretary determines that a rule should be issued, he shall publish the proposed rule within sixty days after the submission of the advisory committee's recommendations or the expiration of the period prescribed by the Secretary for such submission.
>
> (3) On or before the last day of the period provided for the submission of written data or comments under paragraph (2), any interested person may file with the Secretary written objections to the proposed rule, stating the grounds therefor and requesting a public hearing on such objections. Within thirty days after the last day for filing such objections,

the Secretary shall publish in the Federal Register a notice specifying the occupational safety or health standard to which objections have been filed and a hearing requested, and specifying a time and place for such hearing.

(4) Within sixty days after the expiration of the period provided for the submission of written data or comments under paragraph (2), or within sixty days after the completion of any hearing held under paragraph (3), the Secretary shall issue a rule promulgating, modifying, or revoking an occupational safety or health standard or make a determination that a rule should not be issued. Such a rule may contain a provision delaying its effective date for such period (not in excess of ninety days) as the Secretary determines may be necessary to insure that affected employers and employees will be informed of the existence of the standard and of its terms and that employers affected are given an opportunity to familiarize themselves and their employees with the existence of the requirements of the standard.

Once the standard is promulgated or the Secretary makes any rule, order, or decision, he must include a statement of the reasons for such action in the Federal Register. Section 6(e) of the Act. OSHA's judicial review provision, Section 6(f) of the Act, mandates that "[t]he determination of the Secretary shall be conclusive if supported by substantial evidence on the record considered as a whole."

The Secretary of Labor may promulgate standards in either of two ways: as an emergency temporary standard under Section 6(c) of the Act or as a permanent standard under Section 6(b) of the Act. (The two-year authority to promulgate national consensus standards or established federal standards under Section 6(a) of the Act expired on April 29, 1973). In *Florida Peach Growers Ass'n., Inc. v. Secretary of Labor,* 489 F.2d 120, 1 OSHC 1472 (5th Cir. 1974), the court said that the key to the issuance of an emergency temporary standard is the necessity to protect employees from a grave danger, but that *after* issuance of such a standard, the Secretary of Labor *must* set in motion procedures for promulgation of a permanent standard which must issue within six months of the publication of the ETS. An ETS may be promulgated without the necessity of observing the notice and comment provisions of the Administrative Procedure Act. By contrast, a *permanent standard* which is issued in order to serve the general objectives of the Act requires promulgation procedure similar to informal rule-making under the Administrative Procedure Act, 5 U.S.C. Section 553. Whenever the Secretary of Labor *modifies or revokes* either a permanent or ETS standard, he must follow certain rule-making procedures set forth below and include a statement of reasons for such actions. See *AFL-CIO v. Brennan,* 530 F.2d 109, 3 OSHC 1820 (3d Cir. 1975)—vis-à-vis a comparison of the review process for permanent standards promulgated pursuant to Section 6(b) of the Act with national consensus standards promulgated under Section 6(a); *Florida Peach Growers Ass'n. Inc. v. Dept. of Labor,* 489 F.2d 120, 1 OSHC 1472 (5th Cir. 1974)—amendment of emergency temporary standards promulgated under Section 6(c). Correction of ministerial mistakes—if the mistakes were inadvertent and the power to correct was not used as a guise to change a previous position because the wisdom of that decision appears doubtful in light of changing policies—may be made without necessity for resort to the public notice and comment procedures. *Chlorine Institute, Inc. v. OSHA,* 613 F.2d 120, 8 OSHC 1031 (5th Cir. 1980); 29 CFR 1911.5—minor changes in standards permitted. The promulgation procedures of permanent standards are set forth in summary form below.

a. *Publication*

Publication of a proposed standard in the Federal Register is required. If an advisory committee is appointed, the proposed standard must be published within *sixty* days of submission of the Committee recommendations. Section 6(b)(2) of the Act. If an advisory committee is not appointed, the Act nevertheless requires publication of the proposed rule or standard in the Federal Register. *Florida Peach Growers, supra.*

b. *Hearing*

Publication is followed by a *thirty*-day comment period. Objections or written comments may be submitted within the thirty-day period following the proposed standard's publication. Interested persons must be afforded an opportunity to submit written data or comments or file obligations and request a public hearing on the proposal. *Taylor Diving & Salvage Company Inc. v. Dept. of Labor,* 599 F.2d 622, 7 OSHC 1507 (5th Cir. 1979); *Bethlehem Steel Corp. v. Dunlop,* 540 F.2d 157, 4 OSHC 1239 (3d Cir. 1976)—industrial slings standard vacated and remanded for lack of adequate rule-making notice. Frequently the comment period will be extended to afford interested parties sufficient time to submit their views on the standard.

If a hearing is requested, the Secretary of Labor must publish a notice in the Federal Register specifying the standard or portion thereof objected to and set a time and place for a hearing within *thirty* days of the close of the comment period. Section 6(b)(2) and (3) of the Act. The hearing is a "notice and comment" type, as provided in 5 U.S.C. Section 553 of Administrative Procedure Act and is held in accordance with administrative regulations published in 29 CFR 1911. *Associated Industries of New York State, Inc. v. Department of Labor,* 487 F.2d 342, 1 OSHC 1340 (2d Cir. 1973). The Administrative Procedure Act in 5 U.S.C. Section 553 merely requires that an administrative agency give interested persons an opportunity to participate in rule-making through the submission of written data, views, or arguments, with or without opportunity for oral presentation; it is an informal rather than a formal rule-making procedure. *California Citizens Band Association v. U.S.,* 375 F.2d 43 (9th Cir. 1967), *cert. denied,* 389 U.S. 844 (1968). In *Molina v. Marshall,*—F. Supp.—, 7 OSHC 1405 (D.D.C. 1979), an action was filed challenging the Secretary of Labor's notice of revocation deleting the Employment and Training Administration's housing standards for temporary agricultural workers, as adopted in 29 CFR 1910.142, from the OSHA standards. The court stated that although Section 553(b) and (c) of the Administrative Procedure Act provides that a general notice of proposed rule-making be published in the Federal Register, including, among other things, either the terms or substance of the proposed rule, a description of the subjects and issues involved, and after such notice has been given, the agency must give interested persons an opportunity to participate in the rule-making through the submission of written data, views, or arguments. That section does not require that interested parties be provided precise notice of each aspect of the regulations eventually adopted or deleted. The court stated that notice is sufficient if it affords interested parties a *reasonable opportunity* to participate in the rule-making process.

The hearing is informal and legislative in type and is presided over by an administrative law judge. 29 CFR 1911. The administrative law judge makes no decision on the issues and does not even make recommendations or conclusions, but has the power to conduct a full and fair hearing. Participants in the hearing are permitted to present written recommendations, which become part of the record that is certified to the Secretary.

c. *Promulgation*

The Secretary "must" either promulgate a permanent standard or decide not to issue one within *sixty* days of the hearing, if one was requested, or within sixty days after the period for filing objections has expired. Section 6(b)(4) of the Act. This timetable normally is not adhered to by the Secretary of Labor, since the D.C. Circuit Court has held that the statutory deadline of Section 6(b) gives the Secretary discretion to alter priorities and reallocate resources when in good faith he determines that other priorities demand timetable adjustments; it is not, therefore, mandatory. *National Congress of Hispanic American Citizens v. Brennan,* 425 F. Supp. 900, 3 OSHC 1564 (1975), *remanded,* 554 F.2d 1196, 5 OSHC 1255 (D.C. Cir. 1977), *on remand,*—F. Supp.—, 6 OSHC 2127 (D.D.C. 1978), *rev'd,* 626 F.2d 882, 7 OSHC 2029 (D.C. Cir. 1979). This same procedure also governs the modification and revocation of standards.

d. *Effective Date*

The Secretary of Labor must include the date upon which the new standard becomes effective and may also publish a provision delaying the effective date of the standard up to *ninety* days so that employers and employees will be informed of the standard's existence and understand its terms and requirements. Specific requirements may be phased in at later dates. A delay in an effective date beyond the ninety-day period, or a delay for reasons other than orientation and education, may possibly be considered to be an unauthorized change in the standard.

e. *Statement of Reasons*

Whenever a standard is promulgated, a statement of reasons for such action must be published in the Federal Register articulating the rationale for promulgating the standard in question. Section 6(e) of the Act. The Third Circuit held in *Dry Color Mfr's Ass'n. Inc. v. Dept. of Labor,* 486 F.2d 98, 1 OSHC 1331 (3d Cir. 1973) that Section 6(e) of the Act precludes OSHA from articulating nothing more than a "conclusory statement of reasons"—particularly when there is a "voluminous factual record." This requirement also applies to an ETS. In *Associated Industries of New York State Inc. v. Department of Labor,* 487 F.2d 342, 1 OSHC 1340 (2d Cir. 1973), the court struck down an OSHA standard where the agency did not explain its dismissal of industry objections which "challenged the priority of the requirement on various grounds." Observing that OSHA "must take reasonable steps to enable [the courts] to carry out the task Congress has imposed on them," the court stated:

> When Congress has required us to review standards established by the Department—as we see it, under the substantial evidence test—we must insist on something more than *ipse dixit*. . . . In a case where a proposed standard under OSHA has been opposed on grounds as substantial as those presented here, the Department has the burden of offering *some* reasoned explanation. *Id* at 354 (Emphasis in original.)

The asbestos standard, for example, was remanded because the Secretary of Labor failed to explain his reasons for setting one effective date for all industries and requiring only a three-year record retention for monitoring results. *Industrial Union Dept., AFL-CIO v. Hodgson,* 499 F.2d 467, 1 OSHC 1631 (D.C. Cir. 1974). The medical examination sections of the original fourteen carcinogens standard were also remanded because the Secretary of Labor failed to explain the absence of specific tests that were recommended by the advisory committee. *Synthetic Organic Manufacturers Chemical Mfr's Ass'n v. Brennan,* 503 F.2d 1155, 2 OSHC 1159 (3d Cir. 1974), *cert. denied,* 420 U.S. 973, 2 OSHC 1654 (1975).

f. *Regulatory Analysis*

The Secretary is required to prepare a regulatory analysis of significant regulations under Executive Order 12044. 43 Fed. Reg. 12661 (1978). Regulations requiring an analysis are, at minimum, those that will result in: (1) an annual effect on the economy of $100 million or more; or (2) a major increase in costs or prices for individual industries, governments, or regions. The analysis includes a heavy emphasis on evaluation of alternatives and requires justification for choosing one over the others.

g. *Substantive Requirements*

In addition to the above-described procedural requirements for the promulgation of standards, the Act imposes certain additional substantive requirements on the promulgation of standards, depending upon the type of standards involved. For example, Section 6(a) of the Act provides that for national consensus standards and established federal standards, they may be promulgated *unless* it is determined that the promulgation of such a standard would not result in improved safety or health for specifically designated employees.

The term "standard," as defined in Section 3(8), means a practice, means, method, operation, or process reasonably necessary or appropriate to provide safe or healthful employment and places of employment. In addition, those standards dealing with toxic materials or harmful physical agents, must

> . . . adequately assure, to the extent feasible, on the basis of the best available evidence, that no employee will suffer material impairment of health or functional capacity even if such employee has regular exposure to the hazard dealt with by such standard for the period of his working life. [Section 6(b)(5) of the Act.]

These substantive requirements, as indicated in the phrases to the *extent feasible, reasonably necessary or appropriate, best available evidence,* and *material impairment* will be discussed later in this Chapter in the context of challenges to standards.

2. Regulations

Section 6(b) of the Act provides that the Secretary may, by rule, promulgate, modify, or revoke any occupational safety and health standard, and thereafter specifies the manner in which such may be accomplished. In 29 CFR Part 1911, OSHA has set forth the rules of procedure for promulgating, modifying, or revoking standards under Sections 6(b)(1), (2), (3), and (4) of the Act so as to insure uniformity of the standards to be enforced and in order to avoid a multiplicity of rule-making proceedings.

Any interested person may file with the Assistant Secretary of Labor, Occupational Safety, and Health a written petition for the promulgation, modification, or revocation of a standard. The petition should be accompanied by or include the proposed rule or standard desired, a statement of reasons, and the intended effect. 29 CFR 1911.3. In any particular proceeding, the Assistant Secretary of Labor, upon reasonable notice, may prescribe additional or alternative procedural requirements in order to expedite the conduct of the proceedings or for any other good cause. 29 CFR 1911.4. The procedural regulations for the promulgation of standards are divided into three categories: construction, other, and emergency.

a. *Construction Standards*

In the promulgation, modification, or revocation of standards applicable to construction work, as defined in 29 CFR 1910.12(b), the Assistant Secretary of Labor shall first consult with the Advisory Committee on Construction Safety and Health. Thereafter, the Committee shall submit to the Assistant Secretary its recommendations regarding the rule to be promulgated within the period prescribed by the Assistant Secretary of Labor which in no event shall be longer than two hundred seventy days from the date of initial consultation. 29 CFR 1911.10.

Within sixty days after the submission of the Committee's recommendation, or after expiration of the period prescribed, whichever is earlier, the Assistant Secretary, if he determines that a rule should be issued, shall publish in the Federal Register a notice of proposed rule-making, including: (1) the terms of the proposed rule; (2) a reference to Section 6(b) of the Act and to Section 107 of the Construction Work, Hours, and Safety Standards Act; (3) an invitation to interested persons to submit comments within thirty days after publication of the notice; (4) the time and place for an informal hearing to be commenced, not earlier than ten days following the end of the period for written comments; (5) a requirement for filing an intention to appear at the hearing together with a statement of position; (6) a designation of presiding officer; and (7) any other appropriate provisions. Any interested person who files an intention to appear in accordance with 29 CFR 1911.10(b) shall have a right to participate at the informal hearing. In lieu of the procedure prescribed, the Assistant Secretary may, in certain circumstances, follow the procedure prescribed in 29 CFR 1911.11 for informal hearings.

b. *Other Standards*

The Assistant Secretary may promulgate, modify, or revoke standards applicable to employments other than those in construction work by requesting the recommendations of an Advisory Committee appointed under Section 7 of the Act. Thereafter, but no later than two hundred seventy days, the Committee shall submit to the Assistant Secretary its recommendation with regard to the rule to be promulgated within the period prescribed by the Assistant Secretary. Thereafter, the Assistant Secretary shall publish in the Federal Register a notice of proposed rule-making.

Where an Advisory Committee has been consulted and the Assistant Secretary determines that a rule should be issued, the notice shall be published within sixty days after the submission of the Committee's recommendations or the expiration of the period prescribed, whichever date is earlier. The notice shall include: (1) the terms of the proposed rule; (2) a reference to Section 6(b) of the Act and to the appropriate section of any particular statute applicable to the employments affected; (3) an invitation to any interested persons to submit comments within thirty days after publication of the notice; (4) the time and place of an informal hearing on the proposed rule to be held not earlier than ten days from the last day of the period for written comments; and (5) any other appropriate provisions with regard to the proceeding. 29 CFR 1911.11.

A provision is also set in the rules for interested persons to file objections to the proposed rule and request an informal hearing on the objections. If objections are filed, they must include the name and address of the objector, they must be submitted on or before the thirtieth day after the date of publication of the notice of proposed rule-making, and must specify with particularity the provision of the proposed rule to which objection is taken. Objections must state the grounds therefore, must be separately stated and numbered, and must be accompanied by a summary of the evidence proposed to be adduced at hearing. 29 CFR 1911.11(c). Within thirty days after the last date for filing objections, if objections are filed in compliance with the above requirements, the Assistant Secretary *shall,* and in any other case *may,* publish in the Federal Register a notice of informal hearing containing the statement of time, place, nature of the hearing, reference to the authority under which the hearing is held, specification of the provisions of the proposed rule which have been objected to, specification of the issues on which the hearing is to be held, designation of the presiding officer, and any other appropriate provisions. Any objection or request for hearing on the proposed rule and any interested person who files an intention to appear shall be entitled to participate in the hearing.

c. *Emergency Standards*

Whenever an emergency standard is published, pursuant to Section 6(c) of the Act, the Assistant Secretary must commence a proceeding under Section 6(b) of the Act within six months and the standard as published must serve as the proposed rule. An emergency standard promulgated pursuant to Section 6(c) of the Act shall be considered issued at the time when it is officially filed in the Office of the Federal Register. 29 CFR 1911.12. The Assistant Secretary may consult an Advisory Committee and, if so, he shall afford interested persons an opportunity to inspect and copy any recommendations of the Advisory Committee within a reasonable time before the commencement of any informal hearing which may be held and before the termination of the period for submission of written comments. Section 6(c) requires that a permanent standard must be promulgated following the rule-making proceedings and within six months after the publication of the emergency standard.

d. *Hearings*

The regulations provide and the legislative history of Section 6 of the Act indicates that Congress intended informal rather than formal rule-making procedures to apply to standards proceedings. See Conference Report, H. Report, 91–1765, 91st Cong. Second Session 34 (1970). The informality of the proceedings is also suggested by the fact that Section 6(b) of the Act permits the making of a decision on the basis of written comments alone, the use of advisory committees, and the inherent legislative nature of the tasks involved. For these reasons, the proceedings conducted pursuant to Sections 29 CFR 1911.10 or .11 are informal.

Section 6(b)(3) provides an opportunity for a hearing on objections to proposed rule-making and Section 6(f) provides in connection with the judicial review of standards that the determinations of the Secretary of Labor shall be conclusive if supported by substantial evidence in the record as a whole. Although these sections have not been read as requiring a rule-making proceeding within the meaning of the last sentence of U.S.C. Section 553(c) requiring the application of the formal requirements of U.S.C. Section 556 and 557, they do suggest a congressional expectation that the rule-making would be on the basis of a record to which a substantial evidence test may be applied. The oral hearing is legislative in type, but fairness may and often does require an opportunity for cross-examination on crucial issues. Thus the presiding officer, the administrative law judge, is empowered to permit cross-examination under such circumstances. The hearing is normally reported verbatim, and the transcript made available to any interested person on such terms as the presiding officer may provide.

The presiding officer at a hearing shall have powers necessary or appropriate to conduct it fairly and fully, including the power to regulate the course of the proceedings, to dispose of procedural requests and objections, to confine the presentation to the issues specified in the notice, to permit cross-examination of any witness, to take official notice of material facts not appearing in the record, and to keep the record open for a reasonable period of time to receive written recommendations and supporting reasons, data, and views.

Upon completion of the hearing, all submissions and exhibits and post-hearing comments and recommendations are certified. Within sixty days after the expiration of the period provided for the submission of written data, views and arguments on a proposed rule on which no hearing is held, or within sixty days after the certification of the record of a hearing, the Assistant Secretary shall publish in the Federal Register either an appropriate rule promulgating, modifying, or revoking a standard or a determination that such a rule should not be issued upon consideration of all relevant matters presented in written submission at any hearings held.

The determination that a rule should not issue may be accompanied by an invitation for the submission of additional data, view, or arguments, in which event an appropriate rule or other determination shall be made within sixty days following the end of the additional time period. Any rule or standard adopted pursuant to 29 CFR 1911.18(a) shall incorporate a concise general statement of its basis and purpose. This statement is not required to include specific and detailed findings and conclusions of the kind customarily associated with formal proceedings, but it should show the significant issues which have been faced and the rationale for their resolution.

Where an advisory committee has been consulted in the formulation of a proposed rule, the Assistant Secretary may seek its advice as to the disposal of the proceeding.

A rule promulgating, modifying, or revoking a standard, or a determination that such a rule should not be promulgated shall be considered issued at the time when the rule or determination is officially filed in the office of the Federal Register.

e. *Procedural Requirements*

The Act extends to the Secretary of Labor the authority to promulgate rules and regulations. *Louisiana Chemical Association v. Bingham,* 496 F. Supp. 1188, 8 OSHC 1950 (W.D. La. 1980); *American Industrial Health Council v. Marshall,* 494 F. Supp. 941, 8 OSHC 1789 (S.D. Tex. 1980). For example, Section 8(c)(1–3) authorizes the promulgation of regulations requiring employers to make and preserve various kinds of records. Section 8(e) authorizes regulations governing inspection of worksites. Section 8(g)(2) is a general rule-making provision authorizing promulgation of rules and regulations which are deemed necessary to carry out the responsibilities under the Act.

The Act, however, does not contain any indication of the proper forum in which judicial review may be sought of administrative action promulgating rules and regulations. Thus judicial review of such action is governed by relevant provisions of the Administrative Procedure Act, 5 U.S.C. Section 553 *et seq.* Under the Act, therefore, a court would have to determine whether some independent jurisdictional basis existed in which would enable it to entertain such an action. In *Louisiana Chemical Association, supra,* and *American Industrial Health Council, supra,* challenges were filed concerning OSHA's generic cancer "policy" and the "rule" regarding access to employee exposure and medical records. In those actions, the plaintiffs raised in the first instance the question whether the Secretary of Labor's actions were rules and regulations or constituted a standard. While both courts stated that an agency's own interpretation and label is indicative but not dispositive, the courts in both instances found that these promulgations were standards rather than rules and therefore they did not have jurisdiction to consider the validity thereof. Jurisdiction thus lies in the court of appeals.

In *Chamber of Commerce of the United States of America v. Occupational Safety and Health Administration,* 636 F.2d 464, 8 OSHC 1648 (D.C. Cir. 1980) the legitimacy of OSHA's walkaround pay regulation was challenged. The court held that even though the regulation was not a standard, OSHA violated the requirements of the Administrative Procedure Act by issuing the regulation without publishing a general notice of proposed rule-making in the Federal Register and without allowing interested parties to submit their comments before the regulation's adoption, pursuant to 5 U.S.C. Section 553. OSHA had argued in the case that it did not afford notice or opportunity for comment since the regulation was an *interpretative* rule exempt from the notice and comment obligation of the Administrative Procedure Act. The court, however, rejected this argument, stating that a rule is interpretative rather than legislative *only* if it is not issued pursuant to legislatively delegated power to make rules having the force of law or if the agency intends the rule to be no more than an expression of its construction of a statute or rule. Interpretative rules are statements as to what an administrative officer thinks the statute or regulation means and only provides clarification of the statutory language. The court stated that the walkaround pay regulation does not merely explain the statute, it endeavors to implement the statute and is therefore a legislative rule, and must comply with the Administrative Procedure Act requirements.

3. Advisory Committees

Whenever the Secretary, upon the basis of information he develops or information submitted to him in writing by an interested person, a representative of any organization of employers or employees, a nationally recognized standards-producing organization, the Secretary of Health, Education, and Welfare, or the National Institute for Occupational Safety and Health, determines that a rule should be promulgated in order to serve the objectives of the Act, the Secretary may request the recommendations of an advisory committee appointed under Section 7 of the Act. The Secretary shall provide such an advisory committee with any proposals of his own or of the Secretary of Health, Education, and Welfare, together with all pertinent factual information developed by the Secretary of Labor or the Secretary of Health, Education, and Welfare, or otherwise available, including the results of research, demonstrations, and experiments. An advisory committee shall submit to the Secretary its recommendations regarding the rule to be promulgated within ninety days from the date of its appointment or within such longer or shorter period as may be prescribed by the Secretary, but in no event for a period which is longer than two hundred and seventy days. The caveat is, of course, that such actions must be in accordance with the statute and requirements of the Administrative Procedure Act.

Section 7(a) of the Act establishes a National Advisory Committee on Occupational Safety and Health consisting of twelve members appointed by the Secretary of Labor, four of whom are designated by the Secretary of Health, Education, and Welfare, and composed of representatives of management, labor, occupational safety and occupational health professions, and the public. One of the members serves as chairman. The committee, known as NACOSH, is required under Section 7(a) of the Act to consult and advise and make recommendations to the Secretary of Labor on matters related to the administration of the Act.

Under Section 7(b) of the Act, an advisory committee may be appointed by the Secretary of Labor to assist in standard setting functions under Section 6 of the Act. Each such committee shall consist of not more than fifteen members and shall include as a member a designee of the Secretary of Health, Education, and Welfare, and shall include persons able to present the viewpoint of employers, workers, and representatives of safety and health agencies of the state. Such advisory committee may also include other persons who are qualified by knowledge and experience to make useful contributions to the work of the committee. The Secretary is also authorized under Section 7(c) of the Act to use the services, facilities, and personnel of state agencies and to employ experts or consultants for organizations.

a. *Advisory Committees on Standards*

29 CFR 1912 prescribes the policies and procedures governing the composition and functions of advisory committees which are appointed and authorized under Section 7(b) of the Act to assist the Assistant Secretary of Labor in carrying out standards-setting duties under Section 6 of the Act. There are two types of committees: one is a *continuing committee,* which has been or may be established from time to time to assist in the development of standards in areas where there is frequent rule-making and the use of temporary (*ad hoc*) committees is impractical; the other is an *ad hoc committee,* which is established to render advice in particular rule-making proceedings. 29 CFR 1912.2.

Under the authority of Section 7(b) of the Act and the provisions of the Federal Advisory Committee Act, the Secretary of Labor may appoint specialized *standing* (continuing) advisory committees to assist him with his safety and health responsibilities. Each committee may not exceed fifteen members and shall include at least one member appointed by the Secretary of Health, Education, and Welfare. Membership shall include an equal number of management and employee representatives, as well as one or more state health and safety agency members and such other persons as the Secretary of Labor may designate. There have been several *standing* advisory committees, such as construction safety, agriculture, and longshoring.

Standards (*ad hoc*) advisory committees are also established expressly under the authority of Section 7(b) of the Act. They also may be appointed by the Secretary of Labor to assist him in his standard-setting functions under Section 6 of the Act. Their makeup and composition is limited to fifteen members, with equal representation of both management and labor, as well as public members. These committees are temporary, designed to deal with a single problem and serve until they make recommendations on the standard they were assigned to consider. There have been several such *standards* advisory committees including asbestos, noise, fourteen carcinogens, labeling, coke oven emissions, and cutaneous hazards.

29 CFR 1912.3 applies to the Advisory Committee on Construction Safety and Health established under Section 107 of the Contract Work Hours and Safety Standards Act, commonly known as the Construction Safety Act. Section 107 requires the Secretary of Labor to seek the advice of the Advisory Committee in formulating construction standards, which would be published in 29 CFR 1926. Due to the fact that the Construction Safety Act would apply to much of the work covered under the Act whenever occupational safety or health standards for construction activities are proposed, the Assistant Secretary is often required to consult the advisory committee.

29 CFR 1912.3–.43 sets forth the general rules and regulations for advisory committees; it is provided *inter alia* in the regulations for the issuance of reports of the composition of the representation on Section 7(b) committees, the terms of committee members, both *ad hoc* and *continuing* (29 CFR 1912.11 and 1912.10), the operation and termination of advisory committees, including the notice of meetings, attendance, quorum requirements and minutes, Freedom of Information Act requests, and the availability and costs of transcripts.

Of specific importance is the provision in 29 CFR 1912.36 relative to the advice of advisory committees. That section provides that approval by a majority of all members of an advisory committee is encouraged for rendering advice or making recommendations. A failure to secure a majority is not a reason for refusing to give advice to the Assistant Secretary, but the Assistant Secretary must be informed of concurring and/or dissenting views. The advisory committee is required to submit to the Assistant Secretary its recommendations within ninety days from the date of the commencement of its assigned task or within such longer or shorter period otherwise prescribed.

b. *National Advisory Committee on Occupational Safety and Health*

Section 7(a) of the Act established the National Advisory Committee on Occupational Safety and Health (NACOSH), whose basic function is to advise, consult with, and make recommendations to the Secretary of Labor and the Secretary of Health, Education, and Welfare on matters relating to the general administration of the Act. Advisory committees appointed under Section 7(b) of the Act, by contrast, have a more limited role; they are exclusively concerned with assisting the Assistant Secretary in his standards-setting functions under Section 6 of the Act. 29 CFR 1912.5. There is one exception: the Advisory Committee on Construction Safety and Health discussed above provides assistance in both standards-setting and policy matters arising under the Construction Safety and Health Act. The regulations provide in 29 CFR 1912.5(c) that to the extent the Advisory Committee on Construction Safety and Health renders advice to the Assistant Secretary on general policy matters, its activities should be coordinated with those of NACOSH.

29 CFR 1912(a)(1)–(14) sets forth the procedures to be used by NACOSH in fulfilling its responsibilities. The Committee is a continuing advisory body of twelve members: two represent management, two represent labor, two represent the occupational health professions, two represent the

occupational safety professions, and four members represent the public. One of the public members is designated by the Secretary of Labor to serve as Chairman. The Secretary of Health, Education, and Welfare designates the two members of the occupational health professions and two of the members who represent the public. The regulations set forth in 29 CFR 1912(a) provide for the terms of membership of the committee, meetings, quorum, advice and recommendations, forms, etc. Any advice or recommendations shall be made or given with the approval of a majority of all the Committee members present.

The Chairman, however, shall include in any report of such advice or recommendations the concurring and/or dissenting views, as well as any abstentions and absences. Any member may also submit his own advice and recommendations in the form of individual views with respect to any matter considered by the committee.

Meetings of advisory committees shall be in the presence of an official or employee of the federal government pursuant to 29 CFR 1912(a)(10) and detailed minutes of such meeting shall be prepared.

The Federal Advisory Council on Occupational Safety and Health (FACOSH) is utilized in the public sector as an avenue to submit relevant occupational safety and health material for consideration in standards setting.

c. *Utilization and Legal Significance*

If OSHA determines that promulgation of a specific standard is needed, any of several advisory committees may be called upon to develop specific recommendations. Section 7(b) of the Act sets forth the procedures by which the Secretary may appoint an advisory committee when promulgation of a standard is contemplated. A procedural irregularity occurred in one case, however, when the Secretary appointed an advisory committee *after* the issuance of an emergency temporary standard for 4, 4′ methylene bis (2-chloroaniline) (MOCA). Reviewing the matter, the Third Circuit held that the requirements of Section 7(b) of the Act should prevail over the dictates of Section 6(c). *Synthetic Organic Chemical Manufacturers Association v. Brennan (SOCMA II),* 506 F.2d 385, 2 OSHC 1402 (3d Cir. 1974), *cert. denied,* 423 U.S. 830, 3 OSHC 1559 (1975). Thus in *SOCMA II* the Secretary issued an emergency temporary standard and then appointed an advisory committee to make recommendations on the upcoming permanent standard. The Secretary thereafter published a proposed permanent rule before the advisory committee's recommendations were submitted. Had the Court found that utilization of the advisory committee was of no legal significance in this situation, the Secretary's actions would have been acceptable. Instead, the Court invalidated the subsequently promulgated standard for the following reason:

> [T]he procedures followed here did not comply with [Section 6]. The proposed rule was published before the advisory committee had submitted its report. Consequently, the parties were not given adequate time to submit comments or to prepare for the hearing after the committee's work was completed.

Underlying the procedural differences between the two sections of the Act involved in the court's decision was a conflict in their respective rationales. Section 6(b) of the Act permanently exists to provide procedural safeguards in the predominantly informal rule-making setting of the Act. On the other hand, Section 6(c) of the Act temporarily allows for a rapid response to emergency situations and protects employees by assuring the most prompt possible issuance of permanent standards consistent with reasoned consideration. The Third Circuit simply concluded that procedural safeguards were more important:

> The courts should not permit temporary emergency standards to be used as a technique for avoiding the procedural safeguards of public comment and hearings required by [Section 6(b)]. Especially where the effects of a substance are in sharp dispute, the promulgation of standards under [Section 6(b)] is preferable since the procedure for permanent standards is specifically designed to bring out the relevant facts. [quoting *Dry Color Manufacturers' Association v. Dept. of Labor,* 486 F.2d 98, 1 OSHC 1331 (3d Cir. 1973).]

There are at least three purposes for advisory committee consultation: (1) to enable Department of Labor officials to take advantage of the expertise of committee members in formulating a wide variety of standards; (2) to allow the persons who will be affected by the standards to participate in their promulgation; and (3) to enable those same persons to understand and comply with the standards once promulgated. A fourth reason, identified by the Third Circuit in *SOCMA II,* is that the prepromulgation recommendations of an advisory committee will have a beneficial organizing and educative effect during the public comment and hearing procedure that follows.

The D.C. Circuit, like the Third Circuit, also considers advisory committee functions to be legally significant. In *National Constructors Association v. Marshall,* 581 F.2d 960, 6 OSHC 1721 (D.C. Cir. 1978), in connection with the issuance of ground-fault protection standards, 29 CFR 1910.309(c) and 1926.400(h), the court noted that OSHA may not bypass a standards advisory committee in the event it uses one in the standards formulation process:

> Under the Department's [OSHA's] own promulgation regulation, as explained by its statutory models, CSA and OSHA, advisory committee consultation should, but in this case did not, consist of something more than a single and brief rest stop on the route between a tentative proposal of one construction health and safety standard, and the final promulgation of another, superficially related, but substantively quite different, standard. Because the deviation from the requisite procedures is so great, we cannot be sure that the Assistant Secretary would have promulgated the modified ground fault protection standard in its current form had he used the proper process. Accordingly, the modified ground fault protection standard must be remanded to the Department so that it may engage in the appropriate consultation with the Advisory Committee.
>
> On the other hand, because we do not find that the standard as promulgated is illegal, nor, upon our initial review of the record, do we see any glaring deficiencies in the evidence supporting that standard, we feel that a minimum ninety-day remand of the record—during which the regulations shall remain in effect—will suffice to allow the committee to be convened and to issue any recommendation it chooses on the assured grounding/GFCI alternative standard.

The court most likely reached this result because under the Construction Safety Act, standards promulgation requires consultation with the advisory committee by the Secretary of Labor; thus the Secretary will normally attempt to satisfy both CSA and OSHA requirements by submitting any proposed standard to the Advisory Committee on Construction Safety and Health.

The Fifth Circuit has apparently adopted a different viewpoint, at least with respect to the issuance of an ETS. In *Florida Peach Growers Ass'n., Inc. v. Brennan,* 489 F.2d 120, 1 OSHC 1472 (5th Cir. 1974), the Court concluded that the rapid response to emergencies allowed by Section 6(c) of the Act outweighed the procedural safeguards of Section 6(b). In rejecting the argument that the Secretary can only *modify* an emergency temporary standard by adhering to a Section 6(b) proceeding, the Court noted:

> It is inconceivable that Congress, having granted the Secretary the authority to react quickly in fastbreaking emergency situations, intended to limit his ability to react to developments subsequent to his initial response. Although the Secretary decided here that his initial standards were too broad, in other circumstances he might first issue standards too narrow to protect the employees against the grave danger envisioned by the Act. In such a case, adherence to [Section 6(b)] procedures would not be in the best interests of employees, whom the Act is designed to protect. Such lengthy procedures could all too easily consume all of the temporary standard's six month life. [*Id* at 127.]

See also *National Roofing Contractors Ass'n v. Brennan,* 495 F.2d 1294, 1 OSHC 1667 (7th Cir. 1974), *cert. denied,* 419 U.S. 1105, 2 OSHC 1447 (1975) which held that the Secretary of Labor's failure to appoint representatives of the roofing industry to the advisory committee on the "catch platform" standard, 29 CFR 1926.451(u)(3), did not violate the requirements of Section 7(b) of the Act, which requires that an advisory committee be representative of employers as well as employees and public representatives, absent a showing of "specific prejudice" suffered by roofing contractors.

4. Legal Significance of NIOSH

Recommendations for standards may also come from the National Institute for Occupational Safety and Health (NIOSH), which has been established by the Act as an agency of the Department of Health, Education, and Welfare (Health and Human Services). Section 22 of the Act. The Act provides in this respect:

> Whenever the Secretary, on the basis of information submitted to him in writing by an interested person, a representative of any organization of employers or employees, a nationally recognized standards-producing organization, the Secretary of Health, Education and Welfare, the National Institute for Occupational Safety and Health . . . determines that a rule should be promulgated in order to serve the objectives of this Act . . . the Secretary shall provide such an advisory committee with any proposals of its own or the Secretary of Health, Education and Welfare together with all pertinent factual information developed by the Secretary or the Secretary of Health, Education and Welfare or otherwise available including the results of research, demonstrations and experiments. [Section 6(b)(1).]

NIOSH has no authority to promulgate or enforce standards but it is, among its other functions, directed to "develop criteria dealing with toxic materials and harmful physical agents and substances which will prescribe exposure levels that are safe for various places of employment." Section 20(a) (3) of the Act. The Secretary of Labor is directed to promulgate toxicity standards reflecting the same considerations, with the added requirement that he must consider the element of *feasibility*. Some employers have argued from this similarity that the Secretary of Labor is bound by the NIOSH toxicity criteria in promulgating standards, to the extent that the standards are otherwise feasible. The District of Columbia Circuit in *Industrial Union Dept. AFL-CIO v. Hodgson,* 499 F.2d 467, 1 OSHC 1631 (D.C. Cir. 1974), has rejected this argument and held that the NIOSH criteria are merely advisory:

> With respect to the former, the statute directs NIOSH to develop criteria documents that describe safe levels of exposure, and the Secretary is to promulgate standards that insure that employees are protected. The language employed by Congress in these two mandates is essentially identical except that the Secretary must consider elements of feasibility. From this similarity petitioners argue that the determinations of NIOSH are meant to be conclusive on the question of what exposure levels adequately protect health, and that the Secretary may deviate from the NIOSH document only to the extent dictated by feasibility.
>
> The Act merely says that the Director of NIOSH shall immediately forward recommended standards to the Secretary without specifying how the Secretary is to use them, but the procedure prescribed for the formulation of standards militates against petitioners' position.
>
> It is the Secretary, rather than NIOSH, who conducts hearings and receives the comments of interested persons. The Secretary may also appoint a special advisory committee to assist him in his standard-setting functions, and receive recommendations from it, as it did here.
>
> The Act, or so it seems to us, must be taken as contemplating that the Secretary may consider all of this information as well as that received from NIOSH.

The Commission also has rejected a similar argument in a decision basically affirmed by the District of Columbia Court of Appeals. *GAF Corp.,* 3 OSHC 1686 (1975), *rev. denied,* 561 F.2d 913, 5 OSHC 1555 (D.C. Cir. 1977):

> We have previously held that the Secretary is not bound by NIOSH recommendations. *Industrial Union Department v. Hodgson,* 499 F.2d 467, 1 OSHC 1631 (1974). And in this case the Secretary's departure from the NIOSH recommendation is reasonable. NIOSH noted that concentrations as low as 1.2 fibers per cubic centimeter have been known to cause serious diseases. Furthermore, NIOSH noted, the state of knowledge concerning asbestos-related diseases is such that no exposure standard other than zero would assure freedom from such diseases. NIOSH criteria at III–23, III–9. Hence the Secretary acted reasonably and on substantial evidence in requiring medical examinations for those exposed to any concentration of airborne asbestos rather than limiting examinations to those exposed to more than 1 fiber per cubic centimeter.

NIOSH follows a detailed procedure in formulating recommendations or criteria for standards. It reviews all existing literature on the subject, conducts its own research, and may even conduct inspections of workplaces and question employers and employees. Sections 8 and 20 (b) of the Act. The role and function of NIOSH in the occupational safety and health environment is discussed in chapter XVI.

5. Hearings

The rules and regulations set forth in 29 CFR 1911.15 state and the legislative history of Section 6 of the Act indicates that Congress intended an informal rather than formal rule-making procedure to apply to the standard-setting process. Conference Report, H. Rep. No. 91–1765, 91st Cong. 2d Sess. 34 (1970). The regulations further state that the informality of the proceedings is also suggested by the fact that Section 6(b) of the Act permits the making of a decision on the basis of written comments alone unless an objection to a rule is made and a hearing is requested. Thus proceedings for standard-setting held pursuant to 29 CFR 1911.10 and 1911.11 can be and have been informal.

Section 6(b)(3) of the Act provides an opportunity for a hearing on and objections to proposed rule-making. Section 6(f) provides in connection with the judicial review of standards that the determinations of the Secretary shall be conclusive if supported by *substantial evidence in the record as a whole.* The rules provide that although these sections are not read as providing a rule-making proceeding which meets formality requirements, they do suggest an expectation that rule-making would be conducted on the basis of a record to which a substantial evidence test may be applied.

The hearing is legislative in type but fairness may require an opportunity for cross-examination on crucial issues. 29 CFR 1911.15(a)(3). Thus the rules provide in 29 CFR 1911.15(b) that although the hearing provided shall be informal and legislative in type, the requirements specify that there should be a hearing examiner, the hearing examiner shall provide an opportunity for cross-examination, the hearing should be reported verbatim, and the transcript should be made available to interested parties. The hearing officer shall have powers necessary to conduct a full and fair hearing. 29 CFR 1911.16. Upon completion of the oral presentations, a transcript, written submissions, exhibits, and comments certified by the hearing officer shall be submitted.

Within sixty days after the expiration of the period provided for submission of written comments or data on a proposed rule on which *no hearing* is held or within sixty days after the certification of the record of a hearing, the Assistant Secretary of Labor shall publish in the Federal Register an appropriate rule modifying, promulgating, or revoking the standard or a determination that such a rule should not issue. 29 CFR 1911.18. A determination that a rule should not issue may be accompanied by an invitation for the submission of additional data in which event an appropriate rule or other determination may be made within sixty days following the end of that period allowed for comments.

Any rule or standard adopted shall incorporate a general statement of its basis and purpose. A rule promulgating, modifying, or revoking a standard or a determination that a rule should not be promulgated shall be considered issued at the time when the rule or determination is officially filed in the Office of the Federal Register. The time of the official filing in the Office of the Federal Register is established for the purpose of determining timeliness or lateness of petitions for judicial review. 29 CFR 1911.18.

6. Outside Petitions to Adopt Standards

In *National Congress of Hispanic American Citizens (El Congreso) v. Brennan,* 425 F. Supp. 900, 3 OSHC 1564 (1975), *remanded,* 554 F.2d 1196, 5 OSHC 1255 (D.C. Cir. 1977), *on remand,* —F. Supp.—, 6 OSHC 2157 (D. D.C. 1978), *reversed and remanded,* 626 F.2d 882, 7 OSHC 2029 (D.C. Cir. 1979), the members of El Congreso filed an action to compel the Secretary of Labor to promulgate safety and health standards for the agricultural industry, specifically identifying standards governing field sanitation, farm safety equipment, roll-over tractor protection, personal protective equipment, nuisance dust and noise. (El Congreso had previously petitioned OSHA for the promulgation of these standards). Subsequently, the agency promulgated a roll-over tractor protection standard 29 CFR 1928.51, and a farm machinery guarding standard, 29 CFR 1928.57. El Congreso argued that, by not issuing the other standards within a designated time after he had begun action on them, the Secretary of Labor had abused his discretion and had unlawfully withheld and unreasonably delayed agency action in violation of Section 6(b)(1)–(4) of the Act. The district court, holding that mandatory time frames are triggered once the Secretary begins action on a standard, ordered the Secretary to publish a final farm machinery guarding standard, and to proceed, according to the time frames of Section 6(b), with publishing final protective equipment and field sanitation standards. Since the district court believed that any departure from the statutory timetable violated the Act, it did not examine the Secretary's criteria for setting the priorities that led to delay in issuing the particular standards in dispute, did not reach the question of whether that delay constituted an abuse of discretion, and made no finding regarding the relative need for a field sanitation standard or any other standard.

On appeal, the Court of Appeals reversed, stating in its 1977 opinion that the mandatory language of the Act did not negate the "implicit acknowledgement that traditional agency discretion to alter priorities and defer action due to legitimate statutory considerations was preserved." It pointed out that the Secretary may "rationally order priorities and reallocate his resources at any rule-making stage" so long as "his discretion is honestly and fairly exercised." It ordered the district court to require the Secretary to file a report on the situation with regard to each proposed standard, including timetables for their development. Finally, the Court stated that if the district court was satisfied with the sincerity of the Secretary's effort, it should hold the case for further reports; but that if it was not so satisfied, it should act accordingly.

On December 26, 1978, in a memorandum opinion and final order, the district court concluded its consideration upon remand with respect to the standard for field sanitation. Finding that "the Secretary had not established any criteria which would enable the Court to determine that the agency had acted in a rational manner," the district court concluded that the Secretary's refusal to complete a field sanitation standard was inconsistent with the requirements of the Act and the mandate of the court, and that, therefore, the Secretary "must be directed to complete development of a field sanitation standard . . . as soon as possible." The Secretary appealed and sought and obtained a stay of the district court order. The court of appeals again considered the matter and held that:

> Finally, since the decision of our court in *Hispanic I*, it has been clear that the timetables set forth in [Section 6](b) are not etched in stone, that they do not so circumscribe the discretion of the Secretary that his failure to act within their limits is, not by itself, an abuse by him of his discretionary powers. So long as his action is rational in the context of the statute, and is taken in good faith, the Secretary has authority to delay development of a standard at any stage as priorities demand.[30]
>
> [30] We noted in our opinion in *Hispanic I* that [Section 6](b) does envision a situation where, once the Secretary has sought the recommendation of an advisory committee regarding a proposed standard, consideration of the standard moves through several statutory steps, each with a maximum time frame. The specific maxim within which action must be taken are: advisory committee consideration, 270 days; publication of the recommended standard by the Secretary, 60 days; comments by interested parties, 30 days; specification of the time and place for a hearing, 30 days; promulgation of the standard or, alternatively, a determination that it shall not be issued, 60 days. . . . In *Hispanic I*, we held that this timetable was not mandatory.[11]

The court thus found that OSHA had established criteria used in setting its priorities, which were found to be rational; it was not the court's function to substitute its judgment for OSHA's where OSHA has reasonably exercised its discretion; OSHA's action was not arbitrary, capricious, or an abuse of discretion.

Thus according to the court the Secretary may delay development of a standard beyond the statutory timetables when, in good faith, he determines that other priorities demand an adjustment. Where delay is necessary, however, it is not enough for the Secretary, having initiated the standards-setting process, merely to state that the standard will not be developed within the period covered in his next planning period. Those seeking the protection of a standard are entitled to some assurance that its development has not been inadvertently sidetracked, and they are also entitled to a good faith representation from the Secretary regarding his reasonable expectation as to when the standard will be forthcoming.

The court found, however, in *Hispanic II* that the Secretary of Labor acted within the bounds of his discretionary powers in setting the priorities for his department. There was no finding that he had acted in bad faith.

In *Textile Workers Union of America, AFL-CIO v. Marshall*,—F. Supp.—, 5 OSHC 1591 (D.D.C. 1977), an action to compel the issuance of a standard, although rendered partially moot by the Secretary of Labor's issuance of such standard, was retained under the court's jurisdiction. Moreover, questions concerning the delay in initiating the rule-making procedures and discovery requests in respect thereto were also held pending the outcome. Finally, in

Molina v. Marshall,—F. Supp.—, 7 OSHC 1405 (D.D.C. 1979), OSHA published a notice in the Federal Register revoking certain housing standards for temporary agricultural workers, adopted other standards, and changed enforcement policy. Twenty-four migrant farm workers challenged the notice as invalid in that it failed to comply with the notice and comment provisions of the Administrative Procedure Act. The plaintiffs contended that proper notice was not given because the defendants confused the housing standards section of 20 CFR 620 with the enforcement of those standards, also contained in that section. They argued that it was the specific standards themselves and not the enforcement of these standards which were to be changed and set forth in a single rule. Therefore, they asserted that the deletion of the entire Part 620 violated the notice requirements of the Administrative Procedure Act.[12] The court held:

> Section 553(b) of the APA does not require that interested parties be provided *precise* notice of each aspect of the regulations eventually adopted or deleted. Instead, notice is sufficient if it affords interested parties a reasonable opportunity to participate in the rulemaking process.
>
> * * *
>
> Although lacking in perfect clarity, the earlier proceedings constituted minimal statutory notice that the Department of Labor was attempting to achieve unanimity regarding the regulations, encompassing both standards and enforcement.

In like respect is the recent decision in *Public Citizens Health Research Group v. Marshall,* 485 F. Supp. 845, 8 OSHC 1009 (D.D.C. 1980), which was an action to compel the Secretary of Labor to promulgate regulations which would compel employers to inform employees of the level and identity of toxic substances in the workplace. The action was based on Section 8(c)(3), which requires employers to maintain accurate records of employee exposures to potentially toxic materials and Section 6(b)(7), which requires standards promulgated thereunder to use labels or warnings to apprise employees of hazards to which they are exposed. The plaintiffs had petitioned for development of such a standard and were informed it was being developed, but more than two years after the petition none had been issued. Considering the plaintiffs' claims, the court stated:

> First, section 8(c)(3) of the Act states that the Secretary "shall issue regulations requiring employers to maintain accurate records of employee exposures to potentially toxic materials or harmful physical agents *which are required to be monitored or measured under* [*Section 6 of the Act*]." [Emphasis added.] By its plain language, this provision only requires the Secretary to issue recordkeeping regulations for specific substances which are the subject of a rule promulgated pursuant to Section 6. Absent such a rule, the toxic material would simply not be one which is "required to be monitored" under Section 6 of the Act. Because the mandatory portion of Section 8(c)(3) is limited to specific materials which are the subject of regulation under Section 6, plaintiffs' contention that the statute compels the Secretary to promulgate generic regulations is without a basis in law.
>
> Likewise, Section 6(b)(7) merely states, "Any standard promulgated *under this subsection* shall prescribe the use of labels or other appropriate forms of warn-

ing. . . ." (Emphasis added.) The words "this subsection" refer to subsection (b) of Section 6; this particular provision is the source of the Secretary's rulemaking authority for specific toxic materials and other hazardous agents. The Court reads the phrase "this subsection" as a limitation on the mandatory scope of Section 6(b)(7). In other words, labeling is required when a specific rule on a dangerous substance is issued, but generic labeling is not compelled.

* * *

Plaintiffs' complaint, as its heart, rests upon an interpretation of Sections 6(b)(7) and (8(c)(3). The Court, however, is unable to accept plaintiffs' reading of these statutes and it finds that the compulsory portion of these provisions is limited in scope to substances which are the subject of a specific regulation under Section 6(b). Simply put, the Secretary has no duty to perform the acts which plaintiffs have made the subject of litigation. Because the Secretary is not compelled to issue the generic regulations requested in plaintiffs' rulemaking petition, the complaint here must be dismissed for failure to state a claim.

7. New Developments

a. *Diving Operations*

The emergency temporary standard for diving operations was stayed prior to its effective date by the court in *Taylor Diving and Salvage Co., Inc. v. United States Department of Labor,* 537 F.2d 819, 4 OSHC 1511 (5th Cir. 1976). Subsequently, a permanent standard was promulgated by OSHA and was codified in 29 CFR 1910.401–441. Again, the Fifth Circuit in *Taylor Diving and Salvage Company, Inc. v. United States Department of Labor,* 599 F.2d 622, 7 OSHC 1507 (5th Cir. 1979), vacated certain portions of the standard dealing with employer obligations to provide comprehensive medical examinations to divers in order to assure they are medically fit to perform their assigned tasks in a safe and healthful manner. The court concluded that the medical examination procedure was not reasonably necessary or appropriate to provide safe or healthful working conditions in the commercial diving industry. In effect, the medical examination provision imposed upon employers a mandatory job security provision, and the court felt that OSHA was not authorized to regulate job security as it attempted to do and interfere with the proper maintenance of employee/employer relations.

b. *Coke Oven Emissions*

A new health standard governing employee exposure to coke oven emissions was promulgated in 29 CFR 1910.1029, which prescribes particular controls and procedures to reduce coke oven employees' exposure in specified regulated areas to toxic emissions in concentrations no greater than 0.15 mg. of the benzene-soluble fraction of the total particulate matter (BSFTPM) per cubic meter of air (0.15 mg/m^3) present during the production of coke averaged over an eight-hour time period. Soon after issuance a challenge was filed. The court to which the challenge was directed in *American Iron & Steel Institute v. OSHA,* 577 F.2d 825, 6 OSHC 1451 (3d Cir. 1978) upheld: (1) the Secretary of Labor's factual determination that coke oven emissions are carcinogenic and that no safe level of exposure exists and that such determination is supported by substantial evidence; (2) that the exposure limits for coke oven emissions meet the technological feasibility requirement of the Act; (3) that the standard is not economically infeasible; and (4) that the decision to combine the performance standard for specific controls by requiring implementation of specific work practices and engineering controls for reduction of exposure to emissions together with the performance standard that includes an exposure limit was supported by the record in a reasoned analysis. However, the court stated that although the Secretary can impose a standard which requires an employer to implement technology looming on the horizon and is not limited to issuing a standard solely based on fully developed technology, the Act does not permit an affirmative duty to be placed upon an employer to research and develop new technology, holding the research and development provision standard invalid and unenforceable.

c. *Benzene*

OSHA promulgated a new health standard limiting occupational exposure to benzene, 29 CFR 1910.1028, which required employers to assure that no employee is exposed to an airborne concentration of benzene in excess of 1 part benzene per 1 million parts of air (1 ppm) averaged over an eight-hour day. Also, it required employers to assure that no employees are exposed to dermal contact with liquid benzene and required employers to assure that caution labels are affixed to all containers of products containing benzene and that the labels remain affixed when the product leaves the employers' workplace.

In *American Petroleum Institute v. OSHA*, 581 F.2d 493, 6 OSHC 1959 (5th Cir. 1978), the court, relying on the obligation set forth in Section 3(8) of the Act to enact only standards that are reasonably necessary or appropriate to provide safe or healthful workplaces and determining that Section 6(b)(5) of the Act requires that standards dealing with toxic materials be feasible, held that OSHA must determine whether the benefits expected from the standard bear a reasonable relationship to the costs imposed thereby and finding that OSHA failed to provide an estimate supported by substantial evidence of the expected benefits derived from lowering the permissible exposure limit for benzene from 10 ppm to 1 ppm, thus making it impossible to assess the reasonableness of the relationship between costs and benefits required, the court set the exposure limit aside. The court also set aside the provision prohibiting dermal contact with liquid benzene but upheld the prohibition against employers' removing warning labels from containers of benzene and benzene products when those containers leave the workplace. In *Industrial Union Department, AFL-CIO v. American Petroleum Institute*, 440 U.S. 906, 8 OSHC 1586 (1980), the Court affirmed the Fifth Circuit's holding that Section 3(8) of the Act requires that the Secretary of Labor find as a threshold matter that the "regulated" toxic substance poses a significant health risk in the workplace and that a new lower standard is therefore "reasonably necessary or appropriate" to provide safe or healthful employment in places of employment. The court also stated that unless and until such a finding is made, it was not necessary to address the further question whether there must be a reasonable correlation between costs and benefits or whether OSHA and promulgation of standards may go as far as technologically and economically possible to eliminate the risk.

d. *Cotton Dust*

A new permanent health standard, 29 CFR 1910.1043, limiting occupational exposure to cotton dust, sets forth permissible exposure limits for each manufacturing operation and industry that exposes workers to cotton dust. OSHA set: (1) a 200 microgram per cubic meter (200 $\mu g/m^3$) as the permissible exposure limit for lint-free, respirable cotton dust in yarn manufacturing; (2) a 750 $\mu g/m^3$ for slashing and weaving operations in the cotton industry; and (3) a 500 $\mu g/m^3$ for all other processes in the cotton industry and for all nontextile industries that expose its workers to cotton dust. In *AFL-CIO v. Marshall,* 617 F.2d 636, 7 OSHC 1775 (D.C. Cir. 1979), the court upheld OSHA's decision in this stringent regulation, which would reduce even the occurrence of reversible acute stage of disease. The court also held that that standard was both technologically and economically feasible and rejected the argument that OSHA must conduct a formal cost-benefit analysis before promulgating the standard. Moreover, the court upheld medical removal transfer and wage guarantee provisions of the standard as within OSHA's authority to adopt methods reasonably necessary to insure safe working conditions, impliedly adopting a job-security provision. The Supreme Court recently granted *certiorari* in the case and it is expected the forthcoming decision will specifically address the cost-benefit analysis question.

e. *Lead*

OSHA recently issued a new standard designed to protect employees from exposure to airborne lead in the workplace. The standard restricts employee exposure to metallic lead, inorganic lead compounds, and organic lead soaps, and applies to almost all workplaces covered under the general industry standards. 29 CFR 1910.1025. The general scheme of the standard resembles earlier OSHA lead standards, but sets a permissible exposure limit (PEL) of 50 $\mu g/m^3$ and an action level of 30 $\mu g/m^3$. To determine whether exposure to any workplace exceeds the PEL or the action level, the employer must use environmental monitoring to measure airborne lead at least every six months or whenever changes in operations may alter lead exposures, and must warn employees whenever airborne lead exposure exceeds the PEL. Affected employers must meet the 50 $\mu g/m^3$ PEL immediately through some combination of engineering controls, work practices, or administrative controls and supplemental respirators. Under one of the most controversial portions of the standard, whenever biological monitoring reveals a worker has an abnormally high blood lead level or whenever medical surveillance reveals that a worker may suffer actual physical impairment from lead exposure, the employer must remove the employee from the workplace. Responding to challenges that were filed, the court in *United Steelworkers of America, AFL-CIO-CLC v. Marshall,*—F.2d—, 8 OSHC 1810 (D.C. Cir. 1980), upheld most aspects of the standard, remanding to OSHA for reconsideration only the question of the feasibility of the standard for a number of the affected industries, rejecting many procedural arguments advanced by industries in challenge to the standard. The court also upheld the standard's medical removal provision,[13] and upheld a provision of the standard which grants both OSHA and NIOSH access to employee medical records and the disclosure of limited medical information to labor unions. The court found that OSHA in the lead standard met Section 3(8) of the Act's threshold test by proving significant risk.

f. *Arsenic*

The permanent standard for inorganic arsenic, 29 CFR 1910.1018, reduces permissible worker exposure to 10 micrograms per cubic meter of air. Organic arsenic compounds and arsine are excluded from the standard and the exposure limits for those substances remain at the levels of 500 micrograms and 200 micrograms respectively. In addition to the eight-hour time-weighted exposure limit of 10 micrograms, the standard mandates an action level of 5 micrograms and also requires the use of respirators in plants with exposure levels higher than 500 micrograms and monitoring in all workplaces where inorganic arsenic is present. Challenges to the arsenic standard were filed in the Ninth Circuit in *Asarco, Inc. v. Marshall,* C.A. No. 78–1959, (9th Cir. 1979), but the court announced that it would not rule upon the validity of the standard until the issuance of the decision in the benzene case. Thus the court stayed certain provisions of the standard scheduled to go into effect before the expected date of the Supreme Court's decision, including those involving engineering and administrative controls, temperature controls, positive air pressure, and filtered air supply for lunchrooms.

g. *Generic Cancer Policy*

The final OSHA comprehensive policy for regulating cancer-causing chemicals and other substances in the workplace entitled "Identification, Classification and Regulation of Potential Occupational Carcinogens" established the criteria and procedures under which substances will be regulated by OSHA. According to the "policy," potential occupational carcinogens will be identified and classified on the basis of human epidemiological studies and/or carcinogenesis bioassays on animals. Carcinogens are to be classified into two categories depending upon the nature and extent of available scientific evidence linking the substance to the causing of cancer. Category 1 substances are viewed by OSHA as presenting a grave danger to employee health. Other toxic substances which are linked to cancer scientifically but which present a lesser risk fall into Category 2. OSHA also lists a candidate's list of potential occupational carcinogens. Upon promulgation, the standard was challenged in the Fifth Circuit, the D.C. Circuit, and the District Court for the Southern District of Texas, which declined jurisdiction because the policy was a standard. *American Industrial Health Council v. Marshall,* 494 F. Supp. 941, 8 OSHC 1789 (S.D. Tex. 1980).

h. *Access to Medical Records*

On May 21, 1980, OSHA issued a new regulation governing access by employees, unions, and OSHA to employee medical records, employee exposure records, and analyses of employee exposure and medical records applicable to virtually every workplace in the nation except for agricultural employers. 29 CFR 1910.20 Access to medical records is required whether or not exposure to a toxic or harmful subject exceeds permissible limits imposed by OSHA and whether or not the substance is in fact even subject to OSHA regulations. The regulations require that employers preserve employee medical records for at least the duration of an employee's employment plus thirty years, and exposure records be preserved for thirty years, and analyses of exposure records for at least thirty years. The "regulation" was immediately challenged in the Fifth and D.C. Circuits by both AFL-CIO and United States Chamber of Commerce, as well as in the western District of Louisiana, which declined jurisdiction on the basis that the rule is actually a standard. *Louisiana Chemical Association v. Bingham,* 496 F. Supp. 1188, 8 OSHC 1950 (W.D. La. 1980).

i. *Fire Protection*

The OSHA Fire Protection standard, 29 CFR 1910.35–.38, .107–.109, .155–.165, was issued on September 12, 1980, and applies to all places of employment except for the maritime industry, construction, and agriculture. The standard features performance-oriented provisions with appendices furnishing various guidelines for employer compliance in achieving worker protection. Highlights of the new standard include requirements for the design and installation of portable and fixed fire suppression equipment, fire detection systems, and local fire and emergency alarm systems, recognition of the differences between incipient-stage firefighting hazards and advanced-interior structural firefighting hazards, where there is a greater need for personal protective equipment and training. Specifications for fire alarm testing requirements enhance the flexibility for combating small fires by reducing reliance on portable fire extinguishers and allowing instead a hose system and allowing the use of alternative employee alarm systems, such as blinking lights and touch systems in addition to audible systems.

j. *Other Standards*

Other health standards recently issued included dibromochloropropane (DBCP) and acrylonitrile, which were not challenged in judicial review. On November 28, 1978, almost 1,000 general and specific industry standards thought to cover remote or *de minimis* safety standards were revoked by OSHA.

Construction industry standards and those general industry standards that have been identified as applicable to construction were recently published together as part of OSHA's effort to develop a single, comprehensive set of regulations for construction sites. A proposed amendment of the electrical safety standards of Subpart S of Part 1910 is intended to substantially simplify and update, and also places relevant requirements of the National Electrical Code (NEC) directly into the text of the regulations, making it unnecessary for employers to refer to the NEC to determine their obligations and unnecessary to continue to incorporate the NEC by reference. OSHA has also indicated that it will reopen the record with respect to the noise standard (hearing conversation and engineering/administrative controls), and has also initiated rule-making for a machinery and electrical safety-lock-out tagging standard, a marine terminal facilities standard, conveyors standard, floor and wall openings standard, grain handling facilities standard, tunnelling standard, low pitched roofs standard, confined spaces standard, ionizing radiation standard, and a standard on labeling and requiring of identification of hazardous or toxic chemicals in the workplace. Also, the beryllium standard was reported to be in the final stages.

E. REQUIREMENTS FOR ENVIRONMENTAL IMPACT AND ECONOMIC ANALYSIS STATEMENTS

1. Environmental Impact Statements

Section 102 of the National Environmental Policy Act of 1969 (NEPA) 42 U.S.C. 4221, *et seq.*, (1969), requires that environmental impact statements (EIS) be prepared by federal agencies when significant federal action, such as the promulgation of a safety or health standard would "significantly affect" the quality of the human environment. See also Executive Order 11514 (1970). More particularly Section 102(2)(B) of NEPA requires federal agencies to develop methods and procedures in consultation with the Council on Environment Quality, which will insure that presently unquantified environmental amenities and values will be given appropriate consideration in decision-making. This requirement has been held applicable to permanent standards issued by OSHA under Section 6(b), but a limited exception has been held to exist for EIS. For to require the issuance of an environmental impact statement before the promulgation of an emergency temporary standard would impair the purpose of Section 6(c) of the Act, to provide speedy protection from grave dangers to the health of employees. *Dry Color Manufacturers Assn. v. Department of Labor,* 486 F.2d. 98, 1 OSHC 1331 (3d Cir. 1973).

The *initial determination* of whether a statement is required is made by OSHA's Director of the Office of Standards. 29 CFR 1999.3. The Director is the official responsible for the preparation and circulation of the impact statement (draft and final) and performance of all other duties required by NEPA relating to the administration of the Act. If there is a potential that a proposed action may significantly affect the human environment or that the environmental impact of the action is likely to be highly controversial, an impact statement must be prepared. A statement shall also be prepared if it is reasonable to anticipate a cumulatively significant impact on the environment from a number of actions which individually have a limited environmental effect.

Where the Director determines that an impact statement is required, either initially or after a previous determination that one was not required, a short statement indicating the reasons for such a decision and soliciting comments that may be helpful in preparing a draft impact statement shall be *published in the Federal Register* as soon as practicable after the determination is made. The statement shall note that: (1) a copy of the draft impact statement, once it is prepared, will be available to any member of the public who requests an opportunity to comment on it; (2) any person or agency submitting comments on the draft impact statement to the Director must at the same time forward five copies of the comments to the Council on Environmental Quality; and (3) a forty-five day period will be allowed for the submission of comments on the draft impact statement. If the Director determines that an impact statement is not required for a proposed action, he shall set forth in writing the decisions and the reasons for it. The Director shall also maintain a list of proposed actions for which impact statements are under preparation, which list shall be revised at regular intervals, and which will be periodically published in the Federal Register.

After an initial determination of the need for an impact statement, such statement must be prepared and made available to the public within ninety days before final action is taken on the standard. 29 CFR 1999.4. The *draft impact statement* must include the following information: probable environmental impact of the standard, alternatives to the standard, the relationship between local, short-term uses of man's environment and the maintenance of long-term productivity, irreversible or irretrievable commitments of resources which would become involved if the standard were to be implemented, and the relationship of proposed action to land-use plans, policies, and controls.

At least forty-five days after a draft statement has been made available, a *final impact statement* must be published in the Federal Register. 29 CFR 1999.5. The final environmental impact statement must contain an analysis of all "significant comments." All "substantive comments" received on the draft impact statement (or summaries thereof) must be attached to the final statement.

The regulations set forth in 29 CFR 1999 need not be followed in an emergency situation in which the promulgation of a standard having a significant environmental impact is required. 29 CFR 1999.6. The Director must, however, consult the Council on Environmental Quality as soon as possible regarding alternative procedures to be followed.

2. Economic Regulatory Analysis

Executive Order 11821 (1974), revised by Executive Order 11949 (1977), provided that any regulations or standards promulgated by an executive branch agency must be preceded by an evaluation of the standard's inflationary impact. This requirement also specified that the final standard must be accompanied by a statement certifying that the inflationary impact of the standard has been evaluated. On March 23, 1978, Executive Order 12044 was issued with the aim of improving existing and future government regulations; it superseded Executive Order 11821.[14] The purpose of the "new" Executive Order was to insure that regulations shall be as simple and clear as possible, achieving legislative goals effectively and efficiently. Regulations were not to impose unnecessary burdens on the economy, on individuals, on public or private organizations, or on state and local governments. In order to give the public adequate notice of impending government regulations, agencies were directed to publish at least semiannually a "regulatory agenda" of significant regulations under development or review. At a minimum each published agenda was to describe the regulations being considered by the agency, the need for and the legal basis for the action being taken, and the status of regulations previously listed on the agenda.

Agencies are required under the Executive Order to give the public an early and meaningful opportunity to participate in the development of agency regulations. Agencies are directed to consider a variety of ways to provide this opportunity, including publishing advanced notice of proposed rule-making, holding open conferences or public hearings, sending notices of proposed regulations to publications likely to be read by those affected, and notifying interested parties directly. Agencies are also required to give the public at least sixty days to comment on proposed significant regulations.

With respect to "significant" regulations, each agency is required to approve such regulations before they are published for public comment in the Federal Register. At a minimum this means that the agency must determine that (1) the proposed regulation is needed, (2) the direct and indirect effects of the regulation have been adequately considered, (3) alternative approaches have been considered and the least burdensome of the acceptable alternatives have been chosen, (4) public comments have been considered and an adequate response has been prepared, (5) the regulation is written in plain English and is understandable to those who must comply with it, (6) an estimate has been made of the new reporting burdens or recordkeeping requirements necessary for compliance with the regulation, (7) the name, address, and telephone number of a knowledgeable agency official is included, and (8) a plan for evaluating the regulation after its issuance has been developed.

Agencies were also directed to establish criteria for identifying which regulations were in fact significant and were required to consider, among other things: (1) the type and number of individuals, businesses, organizations, and governmental units affected; (2) the compliance and reporting requirements likely to be involved; (3) the direct and indirect effects of the regulation, including the effect on competition; and (4) the relationship of the regulation to those of other programs and agencies. Some of the regulations identified as significant were, according to the Executive Order, considered to possibly have major economic consequences to the general economy, for individual industries, geographical regions, or levels of government. For these regulations agencies were required to prepare a regulatory analysis. Such analysis must involve a careful examination of alternative approaches early in the decision-making process. The following requirements govern the preparation of regulatory analyses:

(1) Agency heads must establish criteria for determining which regulations require regulatory analysis, which shall insure that the regulatory analysis is performed for all regulations which will result in (1) an annual effect on the economy of $100,000,000 or more or (2) a major increase in costs or prices for individual industries, levels of government, or geographic regions.
(2) Agency heads shall establish procedures for developing the regulatory analysis and obtaining public comment. Each regulatory analysis shall contain a succinct statement of the problem, a description of the major alternative ways of dealing with the problem considered by the agency, and an analysis of the economic consequences of each of these alternatives in a detailed explanation of the reasons for choosing one alternative over the other. Agencies shall also include in their public notice of proposed rules an explanation of the regulatory approach that has been selected or is favored and a short description of the other alternatives considered. A statement of how the public may obtain a copy of the draft regulatory analysis shall also be included. Finally, agencies shall prepare a final regulatory analysis to be made available when the final regulations are published.

Regulatory analysis was not required under Executive Order 12044 in rule-making proceedings pending at the time the order was issued if an economic impact statement had already been prepared in accordance with Executive Order 11821.

The final aspect of the Executive Order directs agencies to review their existing process for developing regulations and revise it as needed to comply with and implement the Executive Order. The Order, however, does not apply to regulations issued in accordance with the formal rule-making provisions of the Administrative Procedure Act 5 U.S.C. Sections 556 and 557 or regulations issued in response to an emergency, but in such a case the agency is required to publish in the Federal Register a statement of the reasons why it is impractical or contrary to the public interest of the agency to follow the procedures of the Executive Order.

F. JUDICIAL REVIEW OF THE VALIDITY OF STANDARDS

Direct judicial review of occupational safety and health standards promulgated by OSHA is specifically authorized by Section 6(f) of the Act. The courts have also held that challenges to the validity of standards may also be brought within the context of enforcement proceedings under Section 11(a) of the Act, even though that section addresses judicial review of factual findings and is silent on procedural and legal issues. *Atlantic & Gulf Stevedores, Inc. v. OSHRC,* 534 F.2d 541, 4 OSHC 1061 (3d Cir. 1976); *Marshall v. Union Oil Co. of California,* 616 F.2d 1113, 8 OSHC 1169 (9th Cir. 1980); *Noblecraft Industries, Inc. v. OSHRC,* 614 F.2d 199, 7 OSHC 2059 (9th Cir. 1979); *Contra National Industrial Constructors, Inc.,* 583 F.2d 1048, 6 OSHC 1941 (8th Cir. 1978).

Thus the scope of review in a Section 11(a) case is coextensive with that of a Section 6(f) case; the validity of OSHA standards and regulations may be challenged in enforcement proceedings as well as in direct petitions for review. However, in a Section 6(f) proceeding OSHA has the affirmative burden to demonstrate the reasonableness of a standard, while in an enforcement proceeding invalidity is an affirmative defense to a citation and the employer must bear the burden of proof on the issue. *Atlantic & Gulf Stevedores, supra.* The applicable scope of review of a standard's validity may also differ depending upon whether it is the Secretary of Labor's issuance of a standard or the Commission's interpretation and application of a standard which is in issue; that is, an enigmatic scope of review versus a narrow scope of review. *RMI Company v. Secretary of Labor,* 594 F.2d 566, 7 OSHC 1119 (6th Cir. 1979); *Brennan v. OSHRC (Canrad Precision Industries),* 502 F.2d 946, 2 OSHC 1137 (3d Cir. 1974); *American Petroleum Institute v. OSHA,* 581 F.2d 493, 6 OSHC 1959 (5th Cir. 1978).

1. "Standing" to Contest

The validity of any OSHA standard promulgated under Section 6 of the Act is subject to pre-enforcement review. Any "person" who is *adversely affected or aggrieved* by an OSHA standard may file a petition challenging its validity in the appropriate United States Court of Appeals at any time prior to the *sixtieth* day after its issuance, promulgation, or modification. Section 6(f) of the Act. In order to have standing in a Section 6(f) challenge, the "person" must normally have participated in the administrative standard-setting proceedings. See *National Industrial Constructors v. OSHRC,* 583 F.2d 1048, 6 OSHC 1914 (8th Cir. 1978); *Aqua Slide 'N Dive Corp. v. Consumer Product Safety Comm'n,* 569 F.2d 831 (5th Cir. 1978). However, persons also may challenge the validity of standards during enforcement proceedings. *Atlantic & Gulf Stevedores, Inc. v. Dunlop,* 534 F.2d 541, 4 OSHC 1061 (3d Cir. 1976); *Marshall v. Union Oil Co. of California,* 616 F.2d 1113, 8 OSHC 1169 (9th Cir. 1980); *Noblecraft Industries, Inc. v. OSHRC,* 614 F.2d 199, 7 OSHC 2059 (9th Cir. 1979). Even when persons are satisfied with a standard, a protective challenge urging judicial approval can be filed in a friendly hospitable circuit. The question of whether such a petitioner would have a standing under Section 11(a) of the Act has not yet been addressed.

Unless otherwise ordered, the filing of a petition for judicial review does not operate as a stay of enforcement action under the challenged standard. See *Florida Peach Growers Ass'n., Inc. v. Brennan,* 489 F.2d 120, 1 OSHC 1472 (5th Cir. 1974). An application for a stay must first have been filed with OSHA. Then, in order to obtain a temporary stay, a petitioner must show that: (1) he is likely to prevail on the merits; (2) he will be irreparably injured without a stay; and (3) a stay would not cause substantial harm to the public interest. *Taylor Diving & Salvage Co., Inc. v. Department of Labor,* 537 F.2d 819, 4 OSHC 1511 (5th Cir. 1976); *United Steelworkers of America, AFL-CIO v. Marshall,* C.A. No.79–1048 (D.C. Cir. 1979).

For an interesting discussion and court review of a whole panoply of alleged procedural infirmities (bias of decisionmaker, use of outside consultants, separation of functions) attendant upon the promulgation of a standard, see *United Steelworkers of America, AFL-CIO v. Marshall*,—F.2d—, 8 OSHC 1810 (D.C. Cir. 1980).

2. Appropriate Court for Review

The United States Courts of Appeal have exclusive statutory jurisdiction over OSHA's rule-making proceedings. *Synthetic Organic Chemical Manufacturers Assn. (SOCMA) v. Brennan*,—F. Supp.—, 1 OSHC 1403 (D.D.C.1973). In *American Industrial Health Council v. Marshall*, 494 F. Supp. 941, 8 OSHC 1778 (S.D. Tex. 1980), the court was presented with the question of the proper characterization of OSHA's generic cancer policy. In finding that the policy was actually a standard the court stated:

> Jurisdiction for review of Section 6(b) standards is vested in the courts of appeals by Section 6(f) of the Act. Accordingly, the issue before the Court is whether the cancer policy is a Section [6(b)] standard, as defendants contend it is, or a regulation promulgated pursuant to Section [8](g)(2), as plaintiffs contend. Inasmuch as Section [8] does not vest jurisdiction to review rules and regulations in a particular court, if the generic cancer policy is a regulation, it properly is reviewable in this Court.[2] See *In re School Board of Broward County, Florida*, 475 F.2d 1117, 1119 (5th Cir. 1973).
>
> [2] Defendants point out that the Act specifically provides for review in district courts of only two actions: (1) when the Secretary charges an employer with discriminating against an employee for exercising his rights under the Act. [Section 11] (c)(2); and (2) when the Secretary seeks an injunction against conditions or practices in any place of employment which present an imminent danger of death or serious physical harm, [Section 13](a),(b). The Court finds, however, that in the absence of a specific statutory grant of jurisdictional power regarding regulations, they also would be reviewable in the district courts, pursuant to the Administrative Procedure Act. *In re School Board of Broward County, Florida*, 475 F.2d 1117, 1119 (5th Cir. 1973).

Finding that the generic cancer policy imposes enforceable legal obligations on employers, the court concluded that it was a standard within the meaning of Section 3(8) of the Act, and accordingly jurisdiction lies in the court of appeals.

The case of *United Steelworkers of America v. Marshall*, 592 F.2d 693, 7 OSHC 1001 (3d Cir. 1979), presented the problem of the appropriate court to review an OSHA action when petitions for review have been filed "simultaneously" in multiple courts. Both industry and union representatives were aware that the proposed lead standard would be filed in the Office of the Federal Register on November 13, 1978, and both sides anticipated filing a petition for review pursuant to Section 6(f) of the Act. Each side, therefore, had a representative at the Office of the Federal Register when it opened for business on November 13, 1978. Aware of the provisions of 28 U.S.C. Section 2112(a), each side had prepared in advance a petition for review for filing in the court of appeals of its choice. (OSHA regulations are considered promulgated and issued when they are filed with the Federal Register. 42 Fed. Reg. 65,166 [1977]). Counsel for Lead Industries Association chose the Fifth Circuit, while counsel for the Steelworkers chose the Third Circuit. When at 8:45 a.m. the OSHA standard was stamped "filed" at the Office of the Federal Register, counsel for the Steelworkers, having arranged for an open telephone line to a public telephone near the entrance to the Third Circuit Clerk's Office, instructed its representative in Philadelphia to file the Steelworkers' petition for review. The Third Circuit Clerk's office marked the petition filed at 8:45 a.m., Eastern Standard Time. Counsel for the Lead Industrial Association, having arranged for an open line to a telephone near the entrance of the Fifth Circuit Clerk's office in New Orleans, instructed its representative there to file their petition for review. It was marked filed at 7:45 a.m., Central Standard Time. Thus the court records of the respective clerk's offices reflected a dead heat in the race to the courthouse.

The governing statute provides that "[i]f proceedings have been instituted in two or more court of appeals with respect to the same order the . . . commission . . . concerned shall file the record in that one of such courts in which a proceeding with respect to such order was first instituted." 28 U.S.C. Section 2112(a). On November 29, 1978, the Solicitor of Labor informed the chief judges of both courts by letter that in the Labor Department's view there was no court of first filing. The letter called to the courts' attention the additional provision in 28 U.S.C. Section 2112(a) that "[f]or the convenience of the parties in the interest of justice such court may thereafter transfer all the proceedings with respect to such order to any other court of appeals," and suggested consultation between the respective courts as to which of the two should make a determination regarding venue for the review of the lead standard.

The petitions for review were referred to panels in each court. Following the procedure adopted by the District of Columbia Circuit in *American Public Gas·Ass'n v. FPC,* 555 F.2d 852 (D.C. Cir. 1976), the Third Circuit panel conferred with the Fifth Circuit panel, which on December 14, 1978, filed a *per curiam* opinion on the LIA petition for review as follows:

> Having conferred with the judges of the Third Circuit, and by agreement with them, we defer proceedings pending a decision by the Third Circuit Court of Appeals designating the forum to consider and decide this matter.

On December 20, 1978, the Third Circuit heard argument on the question of proper forum, during which counsel for LIA advanced several arguments in support of its contention that the case should be heard in the Fifth Circuit. The court stated after considering the requests advanced by both parties concerning the time of filing that:

> Unlike race tracks, courts are not equipped with photoelectric timers, and we decline the invitation to speculate which nose would show as first in a photo finish.

The court then adopted a rule that in the absence of extraordinary circumstances, the official notations of time of filing are conclusive. When those notations show a simultaneous filing the agency should proceed, as did the Labor Department here, to notify both courts, who by agreement will determine which one will determine venue "[f]or the convenience of the parties in the interest of justice." Then, upon consideration of the convenience factor the court decided that:

> ... three possible institutional considerations occur to us. One is the relative expertise of a given court of appeals in the area of law under review. We think that Congress, by providing for review in the circuit wherein the petitioner resides or has his principal place of business, [Section 6(f)], has implicitly determined that OSHA litigation should not be concentrated in a single "expert" court. Thus we think it would be improper to speculate that any circuit court of appeals is more expert in OSHA matters than another. *Accord, Public Service Comm'n v. FPC, supra,* 472 F.2d at 1272–73. A second institutional consideration is the relative state of the dockets of those courts to which the case might be relegated. But since no court which is likely to be convenient to counsel has an uncrowded docket, that consideration is essentially neutral. A third institutional consideration is the desirability of concentrating litigation over closely related issues in the same forum so as to avoid duplication of judicial effort. In her November 29, 1978 letter, the Solicitor of Labor informed us that on November 21, 1978 LIA filed in the District of Columbia Circuit a petition to review ambient air quality standards for lead issued pursuant to the Clean Air Act by the Environmental Protection Agency (EPA). (43 Fed. Reg. 46246 October 5, 1978). Since it seems likely that the District of Columbia litigation will involve issues with respect to the health need for limiting exposure to lead similar to those raised in the LIA and Steelworkers petitions pending here and in the Fifth Circuit there appears to be a strong institutional interest in having these petitions considered in the District of Columbia forum. That institutional interest would not in our view suffice to justify a serious imposition of inconvenience upon those counsel most closely involved in the litigation. But in this instance, the District of Columbia is obviously a convenient forum.

The case was thus transferred to the District of Columbia Circuit, which decided the case in decision entitled *United Steelworkers of America, AFL-CIO-CLC v. Marshall,* —F.2d—, 8 OSHC 1810 (D.C. Cir. 1980). A similar forum-shopping or race to the courthouse situation occurred in the challenges to OSHA's "Access to Medical and Exposure Records" and generic carcinogen standards. In the former the Fifth Circuit deferred to the D.C. Circuit despite the fact that it would be hearing an appeal from a district court's decision in the same matter, *American Petroleum Institute v. OSHA,*—F.2d—, 8 OSHC 2025 (5th Cir. 1980); in the latter it determined that it was the proper forum rather than the D.C., Third or Seventh Circuit.

The race to courthouse problem that exists under Section 6(f) would not normally arise under Section 11(a) because only one person is aggrieved in such an adjudication.

3. Statutory Authority

The Secretary of Labor's authority to promulgate OSHA standards derives from Section 6 of the Act. This authority has been delegated to the Assistant Secretary of Labor for Occupational Safety and Health, the chief executive officer of OSHA. The Act defines in Section 3(8) an occupational safety and health standard as:

> a standard which requires conditions, or the adoption or use of one or more practices, means, methods, operations, or processes, reasonably necessary or appropriate to provide safe or healthful employment and places of employment.

In promulgating standards dealing with toxic materials or harmful physical agents, Section 6(b)(5) further provides:

> The Secretary, in promulgating standards dealing with toxic materials or harmful physical agents, under this subsection, shall set the standard which most adequately assures, to the extent feasible, on the basis of the best available evidence, that no employee will suffer material impairment of health or functional capacity even if such employee has regular exposure to the hazard dealt with by such standard for the period of his working life.

By virtue of these *two* substantive, statutory requirements and/or restrictions, depending upon one's characterization, the Secretary of Labor can [only] effectuate the will of Congress as expressed therein. At the same time, the Secretary must be ever mindful of complying with the Act's procedural requirements such as notice, publication, and statement of reasons.

The power of an administrative officer or board to administer and enforce a federal statute and to prescribe rules and regulations to that end is not the power to make law, for no such power can be delegated by Congress —but the power to adopt regulations to carry into effect the will of Congress is expressed by the statute. A rule or regulation which does not do this, but operates to create a rule out of harmony with the statute, is a nullity. *Lynch v. Tilden Produce Co.,* 265 U.S. 315 (1931); *Miller v. United States,* 294 U.S. 129 (1936). It is within these guidelines that OSHA must promulgate occupational safety and health standards.

Judicial review of occupational safety and health standards must therefore include "a review of whether the Secretary exercised his decisionmaking power within the limits imposed by Congress." *American Petroleum Institute v. OSHA,* 581 F.2d 493, 6 OSHC 1959 (5th Cir. 1978), *aff'd,* 440 U.S. 906, 8 OSHC 1586 (1980).

4. The Statement of Reasons

Section 6(e) of the Act states that "whenever the Secretary promulgates any standards, . . . he shall include a statement of the reason for such action, which shall be published in the Federal Register." *Dry Color Mfr's Ass'n Inc. v. Dept. of Labor,* 486 F.2d 98, 1 OSHC 1331 (3d Cir. 1973); *Assoc. Industries of New York State Inc. v. Dept. of Labor,* 487 F.2d 342, 1 OSHC 1340 (2d Cir. 1973); *Industrial Union Dept. AFL-CIO v. Hodgson,* 499 F.2d 467, 1 OSHC 1631 (D.C. Cir. 1974). The statement-of-reasons requirement of Section 6(e) complements the requirement of Sections 6(b) and (c) that certain factual findings be made by the Secretary of Labor prior to the promulgation of a permanent or an emergency temporary standard. Congress, for example, contemplated that emergency temporary standards would be developed and issued without the benefit of ordinary standard-setting procedures involving public comment and open hearings in the interest of permitting rapid action to meet emergencies. But, by the language of Section 6(e), it clearly refused to eliminate the general requirement of articulation of the reasons for such action as an essential safeguard to emergency temporary standard-setting. The failure of the Secretary to meet (or adequately meet) this requirement has resulted in the invalidation of several standards.

The statement of reasons requirement is designed to serve several general functions: it provides an internal check on arbitrary agency action by insuring that prior to taking action, an agency can clearly articulate the reasons for its decision; it makes possible informed public criticism of a decision by making known its underlying rationale; and it facilitates judicial review of agency action by providing an important part of the record of the decision. Basically, then, OSHA in the promulgation of standards must state reasons demonstrating that it has given consideration to the factors relevant under the Act; findings of fact and policy determinations may be included but are not absolutely required in every statement of reasons.

The statement-of-reasons requirement of Section 6(e) complements the requirement of Section 6(c)(1) that certain factual findings be made by the Secretary of Labor prior to promulgation of an emergency temporary standard. In *Dry Color Mfr's Ass'n v. Department of Labor,* 486 F.2d 98, 1 OSHC 1331 (3d Cir. 1973), the Secretary of Labor promulgated an ETS for exposure to fourteen carcinogenic chemicals. The only statement of reasons contained in the standard consisted of a conclusion that the chemicals were carcinogens and that conditions necessary for issuance of an ETS were met. The court found this to be inadequate:

> In the context of a voluminous factual record, such a conclusory statement of reasons places too great a burden on interested persons to determine and challenge the basis for the standard, and makes possible in any subsequent judicial review the use of *post hoc* rationalizations that do not necessarily reflect the reasoning of the agency at the time the standard was issued.
>
> Congress contemplated that emergency temporary standards would be developed and issued without the benefit of ordinary standard-setting procedures involving public comment and open hearings in the interest of permitting rapid action to meet emergencies. But, by the language of subsection 6(e), it clearly refused to eliminate the general requirement of articulation of the reasons for such action as an essential safeguard to emergency temporary standard-setting.

According to the court a statement of reasons for an ETS in order to satisfy Section 6(e) must indicate

> . . . which data in the record is being principally relied on and why that data suffices to show that the substances covered by the standard are harmful and pose a grave danger of exposure to employees. This could have been accomplished here by a brief statement in the notice that certain scientific data (citing the record documents) showed that DCB and EI produced cancer in rodents and supported the conclusion that they were therefore carcinogenic in man. Second, the statement of reasons failed to offer any explanation as to why this particular standard is necessary to protect the employees from such exposure. We do not mean to say that every procedure must be justified as to every substance, type of use or production technique. But we do read subsection 6(e) as requiring at least a general explanation as to why the procedures prescribed were chosen in light of the rec-

ommendations of scientific experts and other governmental bodies, the types of industrial practices with these chemicals, and the alternative kinds of regulations considered by OSHA.[15]

The District of Columbia Circuit in *Industrial Union Department, supra,* has given an excellent summary of what the "statement of reasons" (relative to the issue of permanent standards) should contain:

> What we are entitled to at all events is a careful identification by the Secretary, when his proposed standards are challenged, of the reasons why he chooses to follow one course rather than another. Where that choice purports to be based on the existence of certain determinable facts, the Secretary must, in form as well as substance, find those facts from evidence in the record. By the same token, when the Secretary is obliged to make policy judgments where no factual certainties exist or where facts alone do not provide the answer, he should so state and go on to identify the considerations he found persuasive.

The statement of reasons requirement also applies to the promulgation of standards from national consensus standards, for Section 6(b)(8) of the Act provides:

> Whenever a rule promulgated by the Secretary differs substantially from an existing national consensus standard, the Secretary shall, at the same time, publish in the Federal Register a statement of the reasons why the rule as adopted will better effectuate the purposes of this chapter than the national consensus standard.

The requirement of this section is analogous to, yet greater than that in Section 6(e) which is generally applicable to standard-setting; it must be more detailed, indicating why a modification is necessary to increase employee safety and health. In *AFL–CIO v. Brennan,* 530 F.2d 109, 3 OSHC 1820 (3d Cir. 1975), the court stated:

> Although the statement of reasons quoted in note 26 *supra* in most respect satisfies the requirements of *Synthetic Organic I,* in this instance, because the Secretary's standard substantially differs from a national consensus standard, we have indicated that Sec. 6(b)(8) imposes a specific additional requirement. The Secretary must disclose the reasons why his rule will: better effectuate the purposes of the Act. We have found that the six reasons listed by him are supported by substantial evidence. We do not, however, find that they adequately disclose why his rule will better effectuate OSHA's purposes. . . . Nowhere does the Secretary discuss the reason why a partial departure from the national consensus standard would be inappropriate. Nowhere does he discuss the possibility of utilizing the variance procedures of Sec. 6(b)(6), for relief in specific instances of infeasibility. There may be adequate reasons for rejecting either approach, but we do not know them. Unless they are disclosed we cannot properly perform our reviewing function.

The court remanded the matter to OSHA for preparation of a more complete statement of reasons.

5. Substantial Evidence

OSHA is a self-contained statute in the sense that it does not depend upon the Administrative Procedure Act for specification of the procedures to be followed in the promulgation of standards. The Act provides that the process of promulgating a standard is to be initiated by the publication of the proposed rule or regulation. Thereafter a notice and comment period follows, and finally publication and promulgation of the final rule or regulation. The determination of the Secretary of Labor in the promulgation of a standard shall be conclusive if supported by substantial evidence in the record as a whole. Thus, OSHA has the burden of establishing the need for a proposed standard's issuance. Persons affected by the issuance of a standard may "challenge" it by the process of judicial review. Section 6(f) of the Act provides in this regard that:

> Any person who may be adversely affected by a standard issued under this section may at any time prior to the sixtieth day after the standard is promulgated by filing a petition challenging the validity of such standard with the United States Court of Appeals for the Circuit wherein such person resides or has his principal place of business, for a judicial review of such standard. . . . The filing of such petition shall not unless otherwise ordered by the court, operate as a stay of the standard. The determinations of the Secretary shall be conclusive if supported by substantial evidence in the record considered as a whole.

The OSHA rule-making procedure is characteristically informal, as contemplated by Section 4 of the Administrative Procedure Act and it was so understood by Congress. However, as indicated in the regulations, 29 CFR 1911.15, the provisions, even though described as legislative in type, provide for an oral hearing and provide for certain elements such as cross-examination and preparation of a transcript normally associated with formal rule-making. Moreover, the substantial evidence test, which is applicable to the Secretary's determinations, according to several courts, has customarily been considered and utilized in adjudicatory proceedings or formal rule-making. *Camp v. Pitts,* 411 U.S. 138 (1973). The Act is one of a number of recent Congressional statutes, however, that designates the stringent, substantial evidence test for judicial review of rule-making. See also *Consumer Products Safety Act,* 15 U.S.C. Section 2060(c); *Toxic Substances Control Act,* 15 U.S.C. Section 2618(c)(1)(B)(i). The substantial evidence test provides for a more rigorous scrutiny than the usual arbitrary and capricious test normally applicable to informal rule-making.

Section 6(f) of the Act provides that "[t]he determinations of the Secretary shall be conclusive if supported by *substantial evidence* in the record considered as a whole." The requirement of substantial evidence on the record considered as a whole means that reviewing courts must take into account not just the evidence that supports the agency's decision but also the countervailing evidence. *AFL-CIO v. Marshall,* 617 F.2d 636, 7 OSHC 1781 fn. 44 (D.C. Cir. 1979). The substantial evidence test may at times seem to both coincide with and come into conflict with the "best available evidence" requirement of Section 6(f) of the Act. For example, in *American Petroleum Institute v. OSHA,* 581 F.2d 493, 6 OSHC 1959 (5th Cir. 1978), the principal argument advanced in attacking the benzene standard's promulgation was that both substantial evidence and the best available evidence did not substantiate that the reduction of the PEL for benzene from 10 ppm to 1 ppm was reasonably necessary or appropriate . . . or that there was a possibility of absorption of benzene through the skin, which might cause cancer. The court agreed with the argument advanced in this respect and set aside the challenged portion of the standard, stating:

> In light of unrefuted testimony on the ready availability of conclusive evidence on the subject, OSHA's choice to rely on old and inconclusive evidence that there is a possibility of absorption of benzene through the skin which might cause cancer is in clear disregard of the congressional directive as to the kinds of evidence OSHA is required to consider. Therefore, the provision of the standard prohibiting dermal contact with liquid benzene cannot stand on the present record.

In *United Steelworkers of America v. Marshall,*—F.2d—, 8 OSHC 1810 (D.C. Cir. 1980), however, the court rejected the argument that the best available evidence imposes any substantive burden on OSHA (see discussion, *infra*). Of necessity, then, since the courts have intertwined the two statutory restrictions on the promulgation of standards, a discussion of the best available evidence requirement will be undertaken in conjunction with the application of the substantial evidence standard. The legislative history of the Act indicated that standards were intended to be promulgated pursuant to informal rule-making procedures, and then reviewed by the less demanding "arbitrary and capricious test." See Subcomm. on Labor of the Senate Comm. on Labor and Public Welfare, 92nd. Cong., 1st Sess., *Legislative History of the Occupational Safety and Health Act,* at 1189 (Comm. Print 1971); 29 CFR 1911.15. In view of this conflict between the apparent intent of Congress and the clear language of the Act, the courts reviewing OSHA's issuance of standards have been forced to fashion their own tests for determining questions of validity. For there would be significant problems in attempting to apply wholesale the *traditional* substantive evidence test in reviewing OSHA standards resulting from informal rule-making. *American Petroleum Institute, supra; Associated Industries of New York State, Inc. v. United States Department of Labor,* 487 F.2d 342, 1 OSHC 1340 (2d Cir. 1973); *Florida Peach Growers Association, Inc. v. United States Department of Labor,* 489 F.2d 120, 1 OSHC 1472 (5th Cir. 1974); *Industrial Union Department, AFL-CIO v. Hodgson,* 499 F.2d 467, 1 OSHC 1631 (D.C. Cir. 1974); *Synthetic Organic Chemical Manufacturers Association v. Brennan,* 503 F.2d 1155, 2 OSHC 1159 (3d Cir. 1974), *cert. denied,* 420 U.S. 973, 2 OSHC 1654 (1975). In attempting to resolve this conflict, however, the courts have not reached a consistent, uniform approach. It is appropriate, then, to review the significant decisions of the courts to determine just what standard, if any, is utilized in this regard.

A blanket arbitrary and capricious test for review has been rejected, of course, because of the language in Section 6(f) of the Act. See *Associated Industries of New York State, Inc. v. Department of Labor,* 487 F.2d 342, 1 OSHC 1340 (2d Cir. 1973); *Industrial Union Department, AFL-CIO v. Hodgson,* 499 F.2d 467, 1 OSHC 1631, (D.C. Cir. 1978).

In *Florida Peachgrowers Association, Inc. v. Department of Labor,* 489 F.2d 120, 1 OSHC 1472 (5th Cir. 1974), the emergency temporary standard for pesticides was challenged. The government suggested that the arbitrary and capricious test, and not the substantial evidence test (which it argued was applicable only to Section 6(b) standard proceedings), was the appropriate standard of review for an exercise of quasi-legislative discretion in informal rule-making. The court stated, however, that although the substantial evidence test is anomalous in the context of informal rule-making, and would not apply where rule-making was governed solely by the APA, here rule-making was *not* governed solely by the APA. Moreover, Section 6(f) (the substantive evidence test) refers to a standard issued under Section [6] which would also include emergency temporary standards. The court stated, however:

> Thus, it seems clear that even with the required substantial evidence test, our review basically must determine whether the Secretary carried out his essentially legislative task in a manner reasonable under the state of the record before him.
>
> Alternatively the government argues that if the substantial evidence test is to be applied, its application should be less rigorous than would be the case were a fully developed adversary record before us. Force of circumstances supports this contention to a degree. The type of administrative proceeding and the form of record it produces will influence the method of review. As Judge Friendly noted in *Associated Industries:* it may well be that the controversy is semantic in some degree, at least in the context of informal rulemaking . . . Commentators have suggested that in the class of cases in which the ground for challenging the agency action is the inadequacy of its evidentiary basis, it is difficult to imagine a decision having no substantial evidence to support it which is not arbitrary, or a decision struck down as arbitrary which is in fact supported by substantial evidence, and that the true significance of the substantial evidence rule is in limiting the agency to the confines of the public record . . . While . . . there may be cases where an adjudicative determination not supported by substantial evidence . . . [and] would not be regarded as arbitrary or capricious, in the review of rules of general applicability made after notice and comment rulemaking, the two criteria do tend to converge.

The problem then centers not on how to apply the "substantial evidence" test to factual findings subject to evidentiary development, but rather how to review legislative-like policy judgments. With respect to the former, the substantial evidence standard provided in the statute would clearly be applicable. See, *e.g., Industrial Union Department, AFL-CIO v. Hodgson,* 499 F.2d 467, 1 OSHC 1631 (D.C. Cir. 1974); *American Iron & Steel Institute, et al. v. OSHA,* 577 F.2d 825, 6 OSHC 1451 (3d Cir. 1978); *Texas Independent Ginners Ass'n v. Marshall,* 630 F.2d 398, 8 OSHC 2205 (5th Cir. 1980). Policy choices, though not so susceptible to verification or refutation on the record, must be scrutinized nevertheless. Although the courts may differ in their articulation of the standard of review of these policy judgments, they have at least required the Secretary of Labor's action to be consistent with the language and purpose of the Act. Thus in pragmatic terms, the Courts of Appeals have said that the essential requirement is that the Secretary of Labor must demonstrate that he has engaged in "reasoned" decision-making.

In *Industrial Union Dept., AFL-CIO v. Hodgson,* 499 F.2d 467, 1 OSHC 1631 (D.C. Cir. 1974), the court characterized the standard-making procedure of the Act as characteristic of informal rule-making contemplated by Section 4 of the Administrative Procedure Act, 5 U.S.C. Section 553. Even though it is legislative in type, the procedure contains some elements which are normally associated with the adjudicatory or formal rule-making model. The court indicated that apparently this is necessary because of the necessity of having a record to which the statutorily mandated substantial evidence test could be meaningfully applied by a reviewing court. The court noted, however, that the substantial evidence test has customarily been directed to adjudicatory proceedings or formal rule-making:

> The hybrid nature of OSHA in this respect can be explained historically, if not logically, as a legislative compromise.

The question for the court, then, was whether, by virtue of the anomalous combination of formal and informal rule-making, the determinations in question are of the kind to which substantial evidence review can appropriately be applied. The government, in its argument, suggested that a proper accommodation could be effected by construing the Act to require a substantial evidence review of factual determinations, while weighing the inferences of policy drawn from those facts in terms of their freedom from arbitrariness or irrationality. The court, however, stated that it did not believe that this approach would affect the rigorousness of review to the extent the government seems to suppose or that the industry petitioners purport to fear. The court did agree, however, that the record was of a mixed nature since in some degree it approaches the form of one customarily conceived of as appropriate for substantial evidence review and in other respects it does not.

The court recognized that in a statute like the Act where the decision-making vested in the Secretary is also legislative in character, there will be areas where explicit findings are not possible and a decision will essentially be a prediction based upon pure legislative judgment. Thus the record contains elements of both legislative policy determination and adjudicative resolution of facts. The court, therefore, determined:

> Regardless of manner in which the task of judicial review is articulated, policy choices are not susceptible to same type of verification or refutation by reference to the record as are some factual questions. Consequently, the courts' approach must necessarily be different no matter how the standards of review are labeled. That does not mean that such decisions escape exacting scrutiny, for, as this court has stated in a similar context, this exercise need be no less searching and strict in its weighing of whether the agency has performed in accordance with the congressional purposes, but, because it is addressed to different materials, it inevitably varies from the adjudicatory model. The paramount objective is to see whether the agency, given an essentially legislative task to perform, has carried it out in a manner calculated to negate the dangers of arbitrariness and irrationality in the formulation of rules for general application in the future. . . .

What the court in effect adopted as the standard of review for OSHA rule-making was a "non-arbitrary and irrational reasonableness" test. The court thus construed the substantial evidence review in the following terms:

> What we are entitled to at all events is a careful identification by the Secretary when his proposed standards are challenged, of the reasons why he chooses to follow one course rather than another. Where that choice purports to be based on the existence of certain determinable facts, the Secretary must, in form as well as

> substance, find those facts from evidence in the record (substantial evidence test). By the same token, when the Secretary is obliged to make policy judgments where no factual certainties exist, or facts alone do not provide the answer, he should so state and go on to identify the considerations he found persuasive.

Thus the court rejected the claim that the substantial evidence test should apply only to factual determinations, for such a view would leave the policy judgments behind a challenged standard reviewable only under the arbitrary and capricious test. The court concluded that under the Act, policy inferences, along with factual determinations, must not escape exacting scrutiny, even though the inferences cannot be strictly verified.

In *Associated Industries of New York State v. Department of Labor,* 487 F.2d 342, 1 OSHC 1340 (2d Cir. 1973), the court was asked to apply a standard of review less severe than the substantial evidence test to the "minority sanitation" standard. The Secretary of Labor in that case again pointed to what it considered an anomaly in subjecting the Act's informal notice and comment rule-making, which produces a regulation that is essentially legislative in nature, to the substantial evidence test, which, according to the Administrative Procedure Act, would only apply to determinations resulting from formal adjudication. It was also argued that the final sentence of Section 6(f) requires only that the Secretary's "determinations" be supported by substantial evidence and that the regulation as a whole need not be so supported; the clear implication being that Section 6(f) imposes a substantial evidence test only upon findings of fact ("determinations") which are announced in the promulgation of a rule. The court rejected this argument, stating that there is no question but that the determinations to which the substantial evidence test of Section 6(f) applies, whatever that may mean in the context, include the overall policy decision to promulgate the rule. However, the court ruled, in applying the substantial evidence test to both factual findings as well as policy decisions, that:

> While we have felt constrained to determine and sustain the applicability of the substantial evidence test, it may well be that the controversy is semantic in some degree, at least in the context of informal rulemaking, and that it lacks the dispositional importance that respondents imply. [Referring to a case in which it applied the arbitrary and capricious standard] it is hard to see in what respect we would have treated the question differently if we had been applying a substantial evidence test . . . While we still have a feeling that there may be cases where an adjudicative determination not supported by substantial evidence within the test of *Universal Camera Corp. v. NLRB,* 340 U.S. 474 (1951), would not be regarded as arbitrary and capricious . . . in the review of rules of general applicability made after notice and comment rulemaking, [however] the two criteria do tend to converge.

The true significance of the substantial evidence rule, according to the court as applied to the rule-making procedures of the Act, is to limit the agency to the confines of the record.

In *AFL-CIO v. Marshall,* 617 F.2d 636, 7 OSHC 1775 (D.C. Cir. 1978), the court reaffirmed its decision in *Industrial Union Department* and attempted to harmonize it with the *Associated Industries of New York* decision, stating that while it recognized that an anomaly exists between applications of the two scopes of review vis-à-vis standards' promulgation, its task is to insure that the promulgated regulations result from a process of *reasoned* decision-making. It stated that the task of a reviewing court in this respect is to insure that the agency: (1) acted within the scope of its authority; (2) has followed the procedures required by statute and by its own regulations; (3) has explicated the bases for its decision; and (4) adduced substantial evidence in the record to support its determinations:

> The reviewing court's task under the Act is to provide a careful check on the agency's determinations without substituting its judgment for that of the agency. Congress apparently created an uneasy partnership between the agency and the reviewing court to check extravagant exercises of the agency's authority to regulate risk.
>
> Our role in this partnership is to insure that the regulations resulted from a process of reasoned decision making consistent with the agency's mandate from congress.

The court stated, however, that the meaning of substantial evidence in this context is problematic, for factual proof about particular health risks may not be substantial in the traditional sense, since OSHA must rely on predictions about possible future events and extrapolations from limited data and may have to fill gaps in knowledge with policy considerations. Congress recognized this problem by authorizing OSHA to promulgate rules on the basis of the best *available* evidence—OSHA's mandate thus empowers it to act even when the evidence is incomplete when the best *available* evidence indicates a serious threat to the health of the workers. The court stated, however, when entrusted with the rigorous "substantial evidence" review, it must examine not only OSHA's factual support but also the "judgment calls" and reasoning that contributed to its final decision.[16] Otherwise the agency's claim of ignorance would clothe it with unreviewable discretion. Thus the court stated it must examine both the factual evidence and the policy considerations, set forth in the record:

> This requires explication of the assumptions underlying predictions or extrapolations, and of the basis for its resolution of conflicts and ambiguities. In enforcing these requirements, the court does not reach out to resolve controversies over technical data. Instead, it seeks to insure public accountability. Explicit explanation for the basis of the agency's decision not only facilitates proper judicial review but also provides the opportunity for effective peer review, legislative oversight and public education. This requirement is in the best interest of everyone, including the decision makers themselves. If the decision making process is open and candid, it will inspire more confidence in those who are affected. Further, by opening the process to public scrutiny and criticism, we reduce the risk that important information will be overlooked or ignored. Instructed by these ends, this court on review will combine supervision with restraint.

The Third Circuit Court of Appeals also declined to apply an across-the-board substantial evidence test to the ethyleneimine standard in *Synthetic Organic Chemical Mfr's Assn. v. Brennan (SOCMA I)*, 503 F.2d 1155, 2 OSHC

1159 (3d Cir. 1974), *cert. denied,* 420 U.S. 973, 2 OSHC 1654 (1975). The court, noting that there is compounded confusion concerning the standard of review, said that it was clear that the application of a substantial evidence test to informal rule-making was a deliberate legislative compromise between House and Senate versions in the bill which became law, referring to the legislative history discussed in Judge Friendly's opinion in *Associated Industries of New York.* Referring to prior decisions where it was suggested that the difference between a substantial evidence test and the more traditional abuse of discretion test in the context of informal rule-making may be largely semantic and that even in applying the substantial evidence test, a court's review is basically to determine whether the Secretary carried out his legislative task in a manner reasonable under the stated record before him. The court noted, then, in its review of the ethyleneimine (EI) standard, that there were determinations of both factual matters and nonfactual policy decisions involved, which raised the problem of the court's role in reviewing the nonfactual policy decisions:

> In *Industrial Union Department, AFL-CIO v. Hodgson,* 499 F.2d 467 (D.C. Cir. 1974) Judge McGowan distinguished those rulemaking issues which were susceptible of resolution on a factual record as to which the court would review by substantial evidence and those issues which by their nature were not capable of such resolution but represented policy judgments. As to the latter, Judge McGowan suggested that the Secretary's policy judgment would be affirmed if he indicated that factual certainty was not possible and identified rationally the controlling considerations. . . . This case is a good illustration of the difficulty of attempting to measure the legislative policy decision against a factual yardstick. OSHA's position with respect to EI is bottomed on an extrapolation from data gathered from two animal studies . . . but the extrapolation of that determination from animals to humans is not really a factual matter. . . . It seems to us that what the Secretary has done in extrapolating from animal studies to humans is to make a legal rather than a factual determination. . . . This is in the nature of a recommendation for prudent legislative action. This raises the problem of our role under Section 6(f) in reviewing the Secretary's nonfactual policy decisions.

The court then set out a five-step process for judicial review of OSHA standards under Section 6(f) of the Act, which include the following:

(1) Determining whether the Secretary's notice of proposed rule-making adequately informed interested persons of the action taken;
(2) Determining whether the Secretary's promulgation adequately sets forth reasons for his action;
(3) Determining whether the statement of reasons reflects consideration of factors relevant under the statute;
(4) Determining whether presently available alternatives were at least considered; and
(5) Determining if the Secretary's determination is based in whole or in part on factual matters subject to evidentiary development, whether substantial evidence in the record as a whole supports the determination.

Since the question challenged in the ethyleneimine (EI) standard was the validity of the Secretary of Labor's extrapolation from animal data, which was "a legal rather than a factual determination," "in the nature of a recommendation for prudent legislative action," the Third Circuit, upon review, in accordance with the process outlined above refused to invalidate the standard:

> Lastly, we hold that there does exist substantial evidence in the record as a whole to support the Secretary's finding that EI is carcinogenic in rats and mice and in the absence of evidence of carcinogenicity in humans, the Secretary properly waived the only available alternatives. We therefore reject the petitioner's challenge to the EI standards except insofar as they relate to laboratories as outlined above.

The court did, however, vacate and remand the portion of the standard which applied special provisions to research laboratories because OSHA had failed to give sufficient public notice concerning the special provisions. See also *Synthetic Organic Chemical Mfr's. Ass'n v. Brennan (SOCMA II),* 506 F.2d 385, 2 OSHC 1402 (3d Cir. 1974), *cert. denied,* 423 U.S. 830, 3 OSHC 1559 (1975).

The five-part *SOCMA I* test received further refinement in *AFL–CIO v. Brennan,* 530 F.2d 109, 3 OSHC 1820 (3d Cir. 1975), in which the newly promulgated mechanical power presses standard was reviewed. While the standard in *SOCMA I* had been consistent with the national consensus standard upon which it was promulgated, the mechanical power presses standard under review in *AFL–CIO v. Brennan,* by its deletion of the "no hands in dies" requirement, differed substantially from the national consensus standard. The petitioner argued in that case that the scope of review in such a situation should be far less circumscribed than under Section 6(f), which provides that determinations shall be supported by substantial evidence. Section 6(b)(8) of the Act provides however that:

> Whenever a rule promulgated by the Secretary differs substantially from an existing national consensus standard, the Secretary shall, at the same time, publish in the Federal Register a statement of the reasons explaining why the rule as adopted will better effectuate the purposes of the [Act] than the national consensus standard.

The petitioner further stated that when the Secretary departs from a national consensus standard his rule-making is deprived of the presumption of validity afforded by Section 6(f). The court thus had to initially determine whether to review the action by the *SOCMA I* standards or by a stricter standard. The court held that Section 6(b)(8) did not intend to impose a higher burden of proof or to require a more stringent standard for judicial review than would otherwise apply to a standard, since the language of Section 6(b)(8) speaks in terms of a statement of reasons but not in terms of factual support for those reasons. The language of Section 6(b)(8) is more directly related to the generally applicable statement of reasons requirement contained in Section 6(e) than to the evidentiary standard requirement of Section 6(f). The purpose of Section 6(b)(8) is to require the Secretary to make particular reference to his reasons. Before he may adopt any standard, the Secretary must, under Section 6(e), state reasons demonstrating that he has given the consideration to facts relevant under the statute, referring to *SOCMA I,* 503 F.2d 1155, 2 OSHC 1159 (3d Cir. 1974) and *Dry Color Manufacturers Association, Inc. v. Department of Labor,* 486 F.2d 98, 1 OSHC 1331 (3d Cir. 1973), but if this standard differs substantially from an existing national consensus standard, he must go further and show what reasons there were for the departure. If the reasons for the departure are based in whole or in part on factual matters susceptible of evidentiary development, they must as in any other case be supported by substantial evidence in the record as a whole. Thus Section 6(b)(8) is no more than a particularization of the general statement of reasons requirement of Section 6(e), and that whether or not a permanent standard differs substantially from an existing national consensus standard, the Secretary's promulgation of such standard is reviewable in a manner outlined in *SOCMA I.*

While the mechanical power presses standard promulgated by the Secretary satisfied the five *SOCMA I* tests (and was supported by substantial evidence in the record as a whole), the Third Circuit remanded the case to OSHA, directing that a more complete statement of reasons be prepared to explain why the new standard *would better* effectuate the purposes of the Act than the national consensus standard, as requested by Section 6(b)(8).

In one of the most controversial standards cases yet decided, the Second Circuit, utilizing the "nonarbitrary-and-irrational reasonableness" test formulated by the D.C. Circuit in *Industrial Union Dept. AFL–CIO v. Hodgson,* 499 F.2d 467, 1 OSHC 1631 (D.C. Cir. 1974), upheld the permanent standard regulating exposure to vinyl chloride in *Society of the Plastics Industry, Inc. v. OSHA,* 509 F. 2d 1301, 2 OSHC 1496 (2d Cir.), *cert. denied,* 421 U.S. 992, 2 OSHC 1195 (1975). The court stated that the traditional substantial evidence test is almost impossible of application where the Secretary of Labor's decision-making is essentially legislative in character; policy choices are not susceptible to the same type of verification as factual questions. The court's approach to such determination must necessarily be different no matter how standards of review are labeled. Thus the court must determine whether the agency given an essentially legislative task to perform has carried it out in a manner calculated to negate the dangers of arbitrariness and irrationality in the formulation of rules for general application in the future. Since the Secretary's extrapolation of animal data for carcinogenicity was a policy determination rather than a factual conclusion, the Second Circuit only examined the reasonableness of the Secretary's actions. Taking judicial notice of the deaths of thirteen workers within a three-year period from overexposure to vinyl chloride, the Court found the standard to be overwhelmingly reasonable.

In *American Iron & Steel Institute v. OSHA,* 577 F.2d 825, 6 OSHC 1451 (3d Cir. 1978), the court in deciding against the Institute on a challenge to the coke oven emissions standard quoted from *SOCMA I,* applied its five-step process and stated that:

> As we examine the regulation at issue in this case, it is imperative to distinguish between determinations bottomed on factual matters and non-factual, legislative-

like policy decisions. It is only the former that we subject to the "substantial evidence" test. The evidence in support of a fact-finding is "substantial" when "from it [the evidence] an inference of the fact may be drawn reasonably." B. Schwartz, *Administrative Law* 595 (1977). In such a case, the reviewing court must uphold the finding "even though [it] would justifiably have made a different choice had the matter been before it de novo." *Palmer v. Celebrezze,* 334 F.2d 306 (3d Cir. 1964), quoting *Universal Camera Corp. v. Labor Board,* 340 U.S. 474, 488 (1950).

However, with respect to such decisions as establishing a 0.15 $\mu g/m^3$ exposure level, based on evidence that coke oven emissions are carcinogenic at any level of exposure, the court found that such decision was a policy judgment made on the basis of the *best available evidence* as to what the industry could achieve in an effort to best protect its employees. The court stated that this was not a factual determination for which it needed substantial evidence in the record to support. Thus the court said that even though it might have drawn different inferences from the information before the Secretary, this conclusion was *reasonably drawn* from the record and must be upheld; reading the record to find support for the conclusion is the proper reviewing function. The court concluded that because judicial review of legislativelike decisions runs the risk of becoming arbitrary supervision and revision of the Secretary's efforts to effectuate the legislative purposes, it would remand only those provisions of the standard which were not reasoned decisions, but rather left "nagging questions as to the reason and rationale for the Secretary's particular choices." The court held therefore that:

(1) The Secretary of Labor did not act *capriciously* (in his legislative-type policy decision) in adopting the standard for coke oven emissions without considering a single steel company's evidence that environmental cofactors shielded its employees from carcinogenic effects of such emissions, since studies relied on by the Secretary had much larger statistical base than company's study;
(2) The Secretary of Labor's factual determination, made during a standard-setting process, that coke oven emissions are carcinogenic and that no safe level of exposure exists, is supported by *substantial evidence* and, therefore, is upheld; and
(3) The Secretary of Labor's *policy decision* to require implementation of specific work practices and engineering controls for reduction of exposure to coke oven emissions, together with a performance standard that includes exposure limit, is *supported by the record and by reasoned analysis,* and, therefore, is upheld.

In the recent decision in the lead case, *United Steelworkers of America v. Marshall,* —F.2d—, 8 OSHC 1810 (D.C. Cir. 1980), the court as it had in *AFL-CIO v. Marshall,* 617 F.2d 636, 7 OSHC 1775 (D.C. Cir. 1979), dealt again with the criteria for judicial review appropriate to "hybrid" rule-making, incorporating its analysis in the cotton dust case as the established and proper interpretation of the scope of review for OSHA cases.[17] The court noted as it did in the cotton dust case that:

> We do not pretend to have the competence or the jurisdiction to resolve technical controversies in the record or where the rule requires setting a numerical standard to second guess an agency decision that falls within a zone of *reasonableness*. . . . Rather, our task is to *insure public accountability* by requiring the agency to identify relevant factual evidence, to explain the logic and policies underlying any legislative choice, to state candidly any assumptions on which it relies, and to present its reasons for rejecting significant contrary evidence and argument. Generalization cannot usefully take us further. [Emphasis added.]

Applying this test in the specific context of Section 6(b)(5), the court was troubled by the vague, invocational rhetoric contained therein: *e.g.*, "most adequately assures," "best available evidence," "material impairment"—especially in the context of feasibility. Although not specifically addressing what any of these terms actually meant (except that it stated "most adequately" means "best"), the court found that OSHA had more than substantial evidence on these issue; that the evidence in support of its correlations, conclusions, and determinations, though controverted, was substantial. In such a situation it determined that it must defer to the reasonable and conscientious interpretations of the agency. For example, the court noted that the "best available evidence rule," as construed in the cotton dust case [*AFL–CIO v. Marshall*], and by the Supreme Court in the benzene case [*Industrial Union Dep't, AFL–CIO v. American Petroleum Institute*], requires OSHA to act immediately to protect workers as best it can, without waiting for scientific certainty. The court declined to read the phrase as setting an unprecedented evidentiary burden on the agency to show not only that its evidence was substantial, but also was the best it could possibly have presented.

Finally, in an important decision, the Fifth Circuit in *American Petroleum Institute v. OSHA*, 581 F.2d 493, 6 OSHC 1959 (5th Cir. 1978), *aff'd,*—U.S. —, 8 OSHC 1586 (1980), set aside all provisions of the OSHA standard for employee exposure to benzene [one part per million parts of air (1 ppm)] including the labeling and the dermal contact provisions. In that case, the court noted that several courts have pointed out problems involved in attempting to apply the traditional substantial evidence test in review of OSHA standards resulting from informal rule-making, noting that the problems center not only on how to apply that test to factual findings subject to evidentiary development, but also on how to review legislative-like policy judgments. The court noted that with respect to factual findings, the substantial evidence standard provided in the statute clearly is applicable. Policy choices, though not so susceptible to verification or refutation by the record must be scrutinized, nevertheless; although the courts have differed in their articulation of the standard to review these policy judgments, they have required the Secretary's action to be consistent with the statutory language and purpose of the Act. The court then stated that its review of such determination must also be whether the Secretary carried out his essentially legislative task in a manner *reasonable under the state of the record* before him and within the limits imposed by Congress; including a review of whether the decision-making power was exercised within the limits imposed by Congress. See also *Texas Independent Ginner's Ass'n v. Marshall*, 630 F.2d 398, 8 OSHC 2205 (5th Cir. 1980) which discussed the substantial evidence test in the context of the reasonable necessity requirement of Section 3(8) and the material impairment requirement of Section 6(b)(5). The court found that OSHA relied on assumptions rather than findings of significant risk of harm as required by the Act. The court referred to the language of the Act for its standard of review and found that *both* factual and policy determinations of the Secretary of Labor must be based on substantial evidence. Furthermore, policy determinations must be consistent with the statutory language and purpose, as well as be reasonable.

[The Fifth Circuit again had an opportunity to consider the requirements of Section 3(8) in *Texas Independent Ginner's Ass'n v. Marshall,* 630 F.2d 398, 8 OSHC 2205 (5th Cir. 1980), and reaffirmed its prior holding in *American Petroleum Institute,* that it requires a cost-benefit analysis:

> The agency's determination that the cost bears a reasonable relationship to the benefit, although a policy determination to a large degree, also is subject to the substantial evidence standard. For such a policy determination to be supported by substantial evidence, sufficient facts must be available to render a policy judgment rational; the factual premises underlying that judgment must be supported by substantial evidence; and the judgment must be rationally related to those factual premises so that the regulatory manner is reasonable. As with other determinations, OSHA must make specific findings of these elements of reasonable necessity. *See Industrial Union Department, AFL-CIO v. American Petroleum Institute,* —U.S. at — no. 45, 100 S.Ct. at 2863 [8 OSHC at 1597] no. 45 and must include its evidence in the record.

Applying this standard, the court concluded that OSHA assumed and conclusorily asserted that significant worker protection would result from promulgation of the standard; an assumption without supporting evidence is not sufficient to meet the reasonable necessity requirement of the Act.]

Holding that by defining an occupational safety and health standard in Section 3(8) as one requiring conditions reasonably necessary or appropriate to provide safe and healthful places of employment and that Congress recognized that safety and health resources are not unlimited, the court found that OSHA had not made a valid determination that reducing the permissible exposure level of benzene from 10 ppm to 1 ppm was *reasonably necessary* to protect workers from the risk of leukemia nor had it used "*the best available*" evidence to make the invalid determination it did reach. In addition, OSHA failed to conduct a cost-benefit analysis, thus the benzene standard was not supported by substantive evidence:

> Moreover, OSHA was required to determine whether the benefits expected from the standard bear a reasonable relationship to the costs imposed thereby. Substantial evidence does not support OSHA's conclusions that benefits are likely to be appreciable. Without an estimate of benefits supported by substantial evidence, OSHA is unable to justify a finding that the benefits to be realized from the standard bear a reasonable relationship to its one-half billion dollar pricetag. . . . OSHA's failure to provide an estimate of expected benefits for reducing the permissible exposure limit, supported by substantial evidence, makes it impossible to assess the reasonableness of the relationship between expected costs and benefits. This failure means that the required support is lacking to show reasonable necessity for the standard promulgated. Consequently, the reduction of the permissible

exposure limit from 10 ppm to 1 ppm and all other parts of the standard geared to the 1 ppm level must be set aside.

In affirming the decision of the Fifth Circuit in *American Petroleum Institute,* the Supreme Court plurality stated that, given the conclusion that the Act empowers the Secretary of Labor to promulgate standards only where a significant risk of harm exists, the issue then becomes how to define and allocate the burden of proving the significance of risk where knowledge is imperfect and the precise quantification of risk is therefore impossible:

> The Agency's position is that there is substantial evidence in the record to support its conclusion that there is no absolutely safe level for a carcinogen and that, therefore, burden is properly on industry to prove, apparently beyond a shadow of a doubt, that there is a safe level for benzene exposure. We disagree.
>
> As we read the statute, the burden was on the Agency to show, on the basis of *substantial evidence,* that it is at least more likely than not that long-term exposure to 10 ppm of benzene presents a significant risk of material health impairment. . . . In this case OSHA did not even attempt to carry the burden of proof. The closest it came to making a finding that benzene presented a significant risk of harm in the workplace was its statement that the benefits to be derived from lowering the permissible exposure levels from 10 to 1 ppm were likely to be appreciable. The Court of Appeals held that this finding was not supported by substantial evidence. Of greater importance, even if it were supported by substantial evidence, such finding would not be sufficient to satisfy the agency's obligation under the Act. [Congress intended the agency to bear the normal burden of establishing the need for proposed standard]. . . . Contrary to the Government's contentions, imposing a burden on the Agency of demonstrating a significant risk of harm will not strip it of its ability to regulate carcinogens, nor will it require the Agency to wait for death to occur before taking any action. First, the requirement that a significant risk be identified is not a mathematical straightjacket. . . . Although the Agency has no duty to calculate the exact probability of harm, it does have an obligation to find that a significant risk is present before it can characterize a place of employment as "unsafe." Second, OSHA is not required to support its finding that a significant risk exists with anything approaching scientific certainty. Although the agency's finding must be supported by substantial evidence, Section 6(b)(5) specifically allows the Secretary to regulate on the basis of the best available evidence. So long as they are supported by a body of reputable, scientific thoughts, the agency is free to use conservative assumptions in interpreting the date with respect to carcinogens, risking error on the side of overprotection rather than underprotection. . . . As several courts of appeals have held, this provision requires a reviewing court to give OSHA some leeway where its findings must be made on the frontiers of scientific knowledge.

In effect then the court held that OSHA had the initial burden to show that the present exposure level was not adequate to protect the employees.

The Court, however, stated that while the agency must support its finding that a certain level of risk exists by substantial evidence, it recognized that its determination that a particular level of risk is "significant" will be based largely on policy considerations, that is, Section 6(b)(5) permits the Secretary of Labor to regulate on the basis of the best available evidence. It need not be calculated with mathematical precision. The Court stated, however, "at this point it has no need to reach the issue of what level of scrutiny a reviewing court should apply to the latter type of determination."[18] Nor did the Court reach the cost-benefit issue.

In Justice Powell's concurrence, he further stated that a determination that benefits justify costs is largely a policy judgment. Thus when a court reviews such judgments under the "substantial evidence" standard mandated by the Act, the court must determine whether the agency has carefully identified the reasons why it chooses to follow one course rather than another as the most reasonable method of affectuating the purposes of the applicable law:

> But the record before us contains neither adequate documentation of this conclusion [that the substantial costs of the benzene regulation are justified] or any evidence that OSHA waived the relevent consideration. The agency simply announced its finding of cost-justification without explaining the method by which it determined that the benefits justified the costs and economic effects. No rational system of regulation can permit its administrators to make policy judgments without explaining how their decisions effectuate the purposes of the governing law, and nothing in statute authorizes such laxity in this case. . . .
>
> [Moreover if there] is no substantial evidence supporting OSHA's determination . . . then OSHA has failed to show that its regulation rests on the best available evidence. . . . If OSHA's finding of significant risks is not supported by substantial evidence, then OSHA has failed to show that regulation is reasonably necessary to provide for safe and healthful workplaces.

Thus the Court at least in concurrence seems to adopt the standard of review set forth by the Court *Industrial Union Department.*

The dissent seemed to feel that the Court's decision amounted to an almost *de novo* review of both questions of fact and legislative/regulatory policy, scrutinized under a substantial evidence test, making no distinction between the two types of decision-making. The dissent stated that while the Secretary's "determinations," whether factual or policy, must be upheld and supported by substantial evidence on the record considered as a whole, such review must be ultimately deferential. In addition:

> The plurality is insensitive to three factors which in my view make judicial review of occupational safety and health standards under the substantial evidence test particularly difficult. First, the issues reach a high level of technical complexity. . . . . Second, the factual issues with which the Secretary must deal are frequently not subject to any definitive resolution. . . . Third, when the question involves a determination of an acceptable level of risk, the ultimate decision must necessarily be based on considerations of policy as well as empirically verifiable facts. Factual determinations can at most define the risk in some statistical way; judgment whether that risk is tolerable cannot be based solely on a resolution of the facts.
>
> The decision to take action in conditions of uncertainty [bears] little resemblance to the sort of empirically verifiable factual conclusions to which the substantial evidence test is normally applied. Such decisions were not intended to be unreviewable; they too must be scrutinized to assure that the Secretary has acted reasonably and within the boundaries set by Congress, but a reviewing court must be mindful of the limited nature of its role. It must recognize that the ultimate decision cannot be based solely on determination of the facts, and that those factual conclusions that have been reached are ones which the courts are ill equipped to resolve on their own.

Under this standard of review, the decision to reduce the permissible exposure level to 1 ppm was well within the Secretary's authority.

Thus the various courts of appeal and the Supreme Court, faced with the conflict between Section 6(f) and the apparent intent of Congress to relax formal rule-making requirements in the area of OSHA standards have essentially adopted a hybrid scope of review which boils down to this: The Secretary's legislative policy determinations are "at least" subject to the test of "reasoned decision-making" or "reasonableness." The Secretary's reasons for promulgating a standard must be supported by findings of fact, and will be reviewed pursuant to a substantial evidence test. The courts will look to the record to determine if there is substantial evidence. The limited and hybrid scope of review highlights the importance to parties of making the most compelling case for their position at a hearing before the Secretary. Under this hybrid scope of review, standards reveal their desired flexibility, and aggrieved parties are provided with some basis from which to attack misconceived findings of fact by the Secretary.

See however, *National Congress of Hispanic American Citizens (El Congress) v. Marshall,* 626 F.2d 882, 7 OSHC 2029 (D.C. Cir. 1979), in which the court applied the arbitrary, capricious, and abuse-of-discretion test to the Secretary of Labor's criteria in the setting of priorities for the development of standards.

An employer may challenge a standard in an *enforcement* action under Section 11 of the Act even though it had neither sought review originally when the standard was promulgated nor sought a variance. *Arkansas Best Freight System, Inc. v. OSHRC,* 529 F.2d 649, 3 OSHC 1910 (8th Cir. 1976); *Atlantic & Gulf Stevedores Inc. v. OSHRC,* 534 F.2d 541, 4 OSHC 1061 (3d Cir. 1976).

However, the proper standard of review related to challenges to standards (*i.e.,* the Secretary of Labor's interpretation and application) during enforcement proceedings is the *arbitrary and capricious* test, not the more stringent substantial evidence test that applies to preenforcement review of standards, since matters primarily of legislative policy are involved. *Atlantic & Gulf Stevedores, Inc. v. OSHRC,* 534 F.2d 541, 4 OSHC 1061 (3d Cir. 1976); *GAF Corporation v. OSHRC,* 561 F.2d 913, 5 OSHC 1555 (D.C. Cir. 1977).

6. Reasonable Necessity

The Act in Section 2(b)(3) of the Act, authorizes the Secretary of Labor to promulgate occupational safety and health standards, which are defined in Section 3(8) to mean a standard which requires conditions or the adoption or use of one or more practices, means, methods, operations, or processes *reasonably necessary or appropriate* to provide safe or healthful employment and places of employment. The requirement that standards be reasonably necessary or appropriate is also referenced obliquely in the Section of the Act dealing with the promulgation of standards concerning toxic materials, in which the Secretary is required to set the standard which most adequately assures to the extent feasible, on the basis of the *best available evidence* that no employee will suffer material impairment of health or functional capacity. The Secretary is also directed to take into account research, demonstrations, experiments, and such other information as may be *appropriate*. The question which has recently arisen, and now has seemingly been resolved, is whether Section 3(8) has any independent legal significance in the standard-setting process.

In *Taylor Diving & Storage Co., Inc. v. U.S. Dept. of Labor*, 599 F.2d 622, 7 OSHC 1507 (5th Cir. 1979), the court was asked to determine whether the medical examination procedure contained in 29 CFR 1910.411 of the commercial diving operations standard was reasonably necessary or appropriate to provide safe or healthful working conditions in the commercial diving industry. In light of the Act's language and legislative history, the Secretary of Labor's admitted purpose for promulgating the standard, and the practical effect of the standard on the employer's hiring practices, the court concluded that the Secretary was not authorized to enact 29 CFR 1910.411.

Looking first to the language of the Act and its legislative history, the court noted that Congress created the Act for the sole purpose of protecting the health and safety of workers and improving physical working conditions on employment premises. *Brennan v. OSHRC (Pearl Steel Erection Company)*, 488 F.2d 337, 1 OSHC 1429 (5th Cir. 1973). The court then noted that the Secretary is authorized only to promulgate regulations which are reasonably necessary or appropriate to achieve that goal.

In promulgating 29 CFR 1910.411, however, OSHA's choice of medical examination procedures was controlled by its "cognizan[ce] of the employees' countervailing rights to be protected in their choice of occupation." In its effort "not to create, through a health and safety standard, a situation which restricts entry into a profession or allows employees to be dismissed for a cause which is less than substantial," OSHA sought to achieve a "balance between the need for a mandatory medical examination and the employee's right to a thorough medical assessment." In effect, then, 29 CFR 1910.411 attempted to impose upon employers a mandatory job security provision. The court stated, however, that OSHA is simply not authorized to regulate job security as it has done here; it is not reasonably necessary or appropriate to the purposes of the Act. "When adopting OSHA, Congress deliberately sought to achieve job safety while maintaining proper employee employer relations." *Marshall v. Daniel Construction Co.*, 563 F.2d 707, 6 OSHC 1031 (5th Cir. 1977), *cert. denied*, 439 U.S. 826, 99 S. Ct. 216, 6 OSHC 1957 (1978). The court also stated that a regulation enacted in the singleminded pursuit of an absolutely risk-free workplace regardless of cost or limitation of job opportunities might well be struck down.

A permanent standard for occupational exposure to benzene was promulgated on February 3, 1978, published on February 10, 1978, and was immediately challenged by the American Petroleum Institute and other industry groups. *American Petroleum Institute v. OSHA*, 581 F.2d 493, 6 OSHC 1959 (5th Cir. 1978), *aff'd*, 440 U.S. 906, 8 OSHC 1586 (1980). (A stay pending judicial review was granted as is authorized by Section 6(f) of the Act.) The challenge alleged in part that the proposed reduction of the permissible exposure limit from 10 ppm to 1 ppm was not *reasonably necessary* or *appropriate* to provide a safe workplace. The petitioners in that case argued that by defining standards in Section 3(8) of the Act as being reasonably necessary to provide a safe or healthful place of employment, that Congress recognized that safety or health resources are not limited, and required OSHA to determine in the decision-making process the extent to which standards would benefit workers and decide whether the projected benefits justify the cost of compliance with the standard. The petitioners supported their contention by reference to the decision in *Aqua Side 'N Dive Corp. v. Consumer Product Safety Commission,* 569 F.2d 831 (5th Cir. 1978). OSHA denied in its response that the "reasonably necessary" language imposed any substantive obligation on it in promulgating standards, attempting to distinguish the *Aqua Slide 'N Dive* case on the basis that the "reasonably necessary" language in the Consumer Product Safety Act (which was involved in that case) appeared in the sections dealing with the agency's process of setting standards, whereas the "reasonably necessary" language in the Act appeared in the definitions defining the type of standard which may be promulgated.

The Fifth Circuit declined to view the precisely similar requirements of the two Acts differently or to read the words of Section 3(8) out of the Act and held that OSHA is obligated to enact only standards that are reasonably necessary or appropriate to provide safe or healthful workplaces. If a standard does not fit the definition of Section 3(8) it is not one OSHA is authorized to enact.

Moreover, in evaluating the reasonable necessity of a standard (and once that determination is made), the court held that the agency must show that a hazard exists, that its regulation will reduce the risk of a hazard, and it must assess the expected benefits in light of the burdens to be imposed by the standard. In conclusion, the court stated that although the agency does not have to conduct an elaborate cost-benefit analysis, it does have to determine whether the benefits expected from the standard bear a reasonable relationship to the cost imposed by the standard. The Act does not give OSHA the unbridled discretion to adopt standards designed to create absolutely risk-free workplaces, regardless of cost, for that would not comport with the requirements imposed that standards be reasonably necessary or appropriate. The court stated:

> To the contrary, that section requires standards to be feasible, and it contains a number of pragmatic limitations in the form of specific kinds of information OSHA must consider in enacting standards dealing with toxic materials. Those include "the best available evidence," "research, demonstrations, experiments, and such other information as may be appropriate," "the latest available scientific data in the field," and "experience gained under this and other health and safety laws." Moreover, in standards dealing with toxic materials, just as with all other occupational safety and health standards, the conditions and other requirements imposed by the standard must be "reasonably necessary or appropriate to provide safe or healthful employment and places of employment." . . . Without an estimate of benefits supported by substantial evidence, OSHA is unable to justify a finding that the benefits to be realized from the standard bear a reasonable relationship to its one-half billion dollar price tag.

It must be made clear at this point that this cost-benefit analysis is also inextricably intertwined with a determination of economic feasibility as well as a determination of reasonable necessity. Moreover, according to the court, such a finding must be supported in the record. Thus the court concluded that OSHA's failure to provide an estimate of expected benefits for reducing the permissible exposure limits for benzene, supported by substantial evidence, made it impossible to reassess the reasonableness of the relationship between expected costs and benefits. The court stated, therefore, that this failure means that the required support is lacking to show a reasonable necessity for the standard, and set aside the reduction of the permissible exposure limit. The court noted that it would not attempt to reconcile its decisions with the decisions of other courts upholding standards, since they did not address what Congress meant by requiring conditions imposed to be reasonably necessary. See *Industrial Union Dept, supra; Society of the Plastics Industry, supra; American Iron & Steel Institute, supra.* Shortly after the Fifth Court rendered its decision in the *American Petroleum Institute* case, the Supreme Court granted *certiorari;* its decision is discussed below.

Shortly after the promulgation of the benzene standard, a new permanent health standard, limiting occupational exposure to cotton dust was promulgated by OSHA on June 19, 1978. Again, the standard was immediately challenged by several industry groups and the United States Court of Appeals for the District of Columbia Circuit considered these challenges. The court in *AFL–CIO v. Marshall,* 617 F.2d 636, 7 OSHC 1775 (D.C. Cir. 1979), stated that although the actual test for economic feasibility has yet to be fully developed by the courts, costs cannot be "prohibitively expensive." Yet when confronted with the argument that a cost-benefit or cost-effectiveness analysis is mandated by the "reasonably necessary or appropriate" language in the Act's definition of health and safety standards, the court held that a formalized cost-benefit and cost-effectiveness analysis is not required by the Act's provisions. Holding that such an analysis is not mandated in explicit terms of the statute, the court rejected the basis of the Fifth Circuit's decision in *American Petroleum Institute* as unpersuasive and misguided.[19]

The decisions of *American Petroleum Institute* and *AFL-CIO v. Marshall* set up a clear conflict between the Fifth Circuit and the D.C. Circuit over the meaning, application, and authority of the "reasonable necessity" language of Section 3(8) of the Act.

In *Industrial Union Department, AFL-CIO v. American Petroleum Institute,* 440 U.S. 906, 8 OSHC 1586 (1980), the United States Supreme Court "resolved" a portion of the apparent conflict between the Fifth and D.C. Circuit concerning the "reasonable necessity" requirement of the Act, agreeing with the Fifth Circuit's holding that Section 3(8) of the Act requires the Secretary of Labor to find as a *threshold* matter that the proposed regulation of a substance through promulgation of a standard poses a significant health risk to a workplace, that the current exposure limit leaves workers subject to that risk, *and* that a new lower standard (exposure limit) is therefore reasonably necessary or appropriate in that the risk can be eliminated or lessened by lowering the limit:

> For we think it is clear that Section 3(8) does apply to all permanent standards promulgated under the Act and that it requires the Secretary, before issuing any standard, to determine that it is reasonably necessary and appropriate to remedy a significant risk of material health impairment. Only after the Secretary has made the threshold determination that such a risk exists with respect to a toxic substance, would it be necessary to decide whether Section 6(b)(5) requires him to select the most protective standard he can consistent with economic and technological feasibility, or whether, as respondent argued, the benefits of the regulation must be commensurate with the costs of the implementation. Because the Secretary did not make the required threshold finding in this case, we have no occasion to determine whether costs must be weighed against benefits in an appropriate case.[20]

The Secretary of Labor had also argued that apart from a minimal requirement of rationality, Section 3(8) imposed no limits on OHSA's power to promulgate standards and that with respect to toxic substances and harmful physical agents, under Section 6(b)(5), that OSHA is required to impose standards that either guarantee that workplaces are free from any risks of material health impairment or that comes as close as possible to doing so without ruining entire industries. The court, however, stated that it is clear that the Act was not designed to require employers to provide absolutely risk-free workplaces whenever it is technologically feasible to do so, so long as the cost is not great enough to destroy an entire industry. Rather, the language and structure of the Act as well as the legislative history indicated that it intended to require the elimination as far as feasible of significant risks of harm:

> By empowering the Secretary to promulgate standards that are reasonably necessary or appropriate to provide safe or healthful places of employment, the Act implies that before promulgating any standard, the Secretary must make a finding that the workplaces in question are not safe. Safe is not the equivalent of risk free. There are many activities that we engage in every day—such as driving a car, or even breathing city air—that entails some risk of accident or material health impairment; nevertheless, few people would consider these activities unsafe. Similarly, a workplace can hardly be considered unsafe unless it threatens the workers with significant risk of harm.
>
> Therefore, before he can promulgate any permanent health or safety standard the Secretary is required to make a threshold finding that a place of employment is unsafe—in the sense that significant risks are present and can be eliminated or lessened by a change in practices. This requirement applies to permanent standards promulgated by Section 6(b)(5) as well as to other types of permanent standards. For there is no reason why Section 3(8)'s definition of a standard should not be deemed incorporated by reference into Section 6(b)(5).

The court stated that its interpretation of Sections 3(8) and 6(b)(5) is supported by reference to other provisions of the Act, in particular, Section 6(g) and Section 6(b)(8). For example, Section 6(b)(8) requires that when a Secretary substantially alters an existing consensus standard he must explain how a new rule better effectuates the purposes of the Act; Section 6(g) provides in part that in determining the priority for establishing a standard, the Secretary should give due regard to the urgency of the need.

The court also stated that the legislative history supports the conclusion that Congress was concerned not with absolute safety but with the elimination of significant harm. And Congress amended Section 6(b)(5) to make it perfectly clear that it does not require the Secretary to promulgate standards that would assure a risk-free workplace.[21] The only issue which the court did not address was the cost-benefit question (in fact Justice Powell alone in his concurring opinion took a partial step toward approving the assertion that OSHA must attempt a quantitative comparison of costs and benefits to prove a standard is reasonably necessary under Section 3[8]). This issue, however, will be addressed in the upcoming review by the court of the cotton dust decision.

United Steelworkers of America v. Marshall,—F.2d—, 8 OSHC 1810 (D.C. Cir. 1980) was the first decision subsequent to *Industrial Union Department* to consider the question of reasonable necessity under Section 3(8) in light of the Supreme Court's decision. Although the court stated that the plurality decision requires that before OSHA creates a new exposure limit for a toxic agent, it must first find as a threshold matter that the toxic substance in question poses a significant health risk in the workplace and that a new lower standard is therefore reasonably necessary or appropriate to provide safe and/or healthful employment and a place of employment, the lack of a majority opinion in the complex overlapping among the five separate opinions leaves the Supreme Court's view of OSHA's statutory responsibilities in some doubt. While the Secretary of Labor must examine a current exposure level of a toxic substance in order to determine whether it leaves workers subject to a significant risk and whether that risk can be eliminated or lessened by a lowering of the level, it is the agency itself that determines the existence of a significant risk. Thus a court's review is limited to reviewing the agency's finding for substantial evidence support. In making the difficult judgment as to what level of harm is unacceptable the agency may rely on its own sound considerations for policy as well as hard factual data:

> In creating the new lead standard OSHA has clearly met the Section 3(8) threshold test of proving significant harm described by the American Petroleum Institute plurality. . . . As characterized by [that decision] the sole basis for the new 1 ppm level for benzene was the likelihood of a significant number of leukemia deaths at any higher level. In the plurality's view OSHA had failed to produce any reliable evidence on the number of deaths from leukemia at 10 ppm or below. By contrast the lead standard does not rest on the possibility of predicting the specific number of deaths from a single disease. Rather in formulating the lead standard, OSHA found evidence of a wide variety of significant harms that proved scientifically observable at specific blood lead levels.

The court noted, however, that, having carried its burden under Section 3(8), OSHA still must meet the demands of Section 6(b)(5), which requires that the Secretary must set the standard which most adequately assures to the extent feasible on the basis of best available evidence that no employee will suffer an impairment of health or functional capacity. . . . Although the court found that the language was neither precise nor artful and contains a rather troublesome phrase—"most adequately assure"—it assumed that Congress used that phrase to mean "best." However, the court noted that there are also two other limitations which Section 6(b)(5) places on OSHA's ability to protect workers. The first of these is that OSHA has a mandate to eliminate only a material impairment of health and safety; the word "material" is not defined in the Act. In upholding OSHA's determination to prevent subclinical effects of lead exposure, the court held that OSHA need not wait until symptoms of a disease appear but can act to *reduce the risk* of serious material impairment. With respect to the second limitation, that OSHA must carry out its decision on the basis of the best available evidence, the court recognized that in an area of scientific uncertainty, the agency has broad discretion to form the best possible solution. Thus, like the material reduction aspect, the court deferred to the reasonable and conscientious interpretations of the agency:

> In reviewing a numerical standard we must ask whether the agency's members are within a *zone of reasonableness,* not whether its members are precisely right. . . . Where the agency presents scientifically respectable evidence which the petitioner can continually dispute with rival and we will assume equally respectable evidence, the court must not second-guess the particular way the agency chooses to weigh the conflicting evidence or resolve the dispute. As with OSHA's choice of a blood lead level goal [*e.g.,* material impairment], so with the choice of a PEL [*e.g.,* best available evidence] we affirm the agency's chosen number as lying within a zone of reasonableness.

7. Feasibility

The Act mentions feasibility only once, as a factor relevant in the formulation of permanent standards for toxic materials and harmful physical agents. Section 6(b)(5) of the Act provides in this respect that:

> the Secretary, in promulgating standards dealing with toxic materials or harmful physical agents under this subsection, shall set the standards which most adequately assure, to the *extent feasible,* on the basis of the best available evidence, that no employee will suffer material impairment of health or functional capacity even if such employee has regular exposure to the hazard dealt with by the standard for the period of his working life.

Neither the statute nor the legislative history defines feasibility or explains how the limitations on safety and health are to be determined—that has been left to administrative and judicial development. Feasibility has been clearly construed to include both technological and economic considerations. *Industrial Union Dept. AFL-CIO v. Hodgson,* 499 F.2d 467, 1 OSHC 1631 (D.C. Cir. 1974); *American Iron & Steel Institute v. OSHA,* 577 F.2d 825, 6 OSHC 1451 (3d Cir. 1978); *AFL-CIO v. Marshall,* 617 F.2d 636, 7 OSHC 1775 (D.C. Cir. 1979); *RMI Company v. Secretary of Labor,* 594 F.2d 566, 7 OSHC 1119 (6th Cir. 1979); *Turner Co. v. Secretary of Labor,* 561 F.2d 82, 5 OSHC 1790 (7th Cir. 1977); *United Steel Workers of America v. Marshall,*—F.2d—, 8 OSHC 1810 (D.C. Cir. 1980). The overall feasibility problem to be determined by courts and administrative agencies is whether the safety and health protection that a standard provides warrants the technological difficulty and economic cost of compliance and under what circumstances an industry or individual employer's technological and economic hardship is entitled to consideration.

It appears that the Secretary of Labor's decisions with respect to technological and economic feasibility are policy judgments made on the basis of the best available evidence as to what industry could achieve in an effort to best protect its employees. The courts have not normally characterized this decision as a factual determination which needs substantial evidence in the record for support. Rather, the Secretary of Labor's task in this respect combines elements of legislative policy determinations and factual findings. Thus even though a court might have drawn different inferences from the information, the conclusion of the Secretary is normally upheld if it is reasonably drawn from the record and/or can be characterized as reasoned decision-making. See *Industrial Union Department AFL–CIO v. Hodgson,* 499 F.2d 467, 1 OSHC 1631 (D.C. Cir. 1974); *Society of Plastics Industry, Inc. v. OSHA,* 509 F.2d 1301, 2 OSHC 1496 (2d Cir. 1974), *cert. denied,* 421 U.S. 992, 3 OSHC 1195 (1975); *American Iron and Steel Institute v. OSHA,* 577 F.2d 825, 6 OSHC 1451 (3d Cir. 1978).

a. *Economic Feasibility*

While some observers have felt that the Act was intended to protect workers regardless of the economic impact on employers, most commentators have agreed that a standard which is prohibitively expensive is not "feasible." As the District of Columbia Circuit stated in *Industrial Union Department, AFL-CIO v. Hodgson* 499 F.2d 467, 1 OSHC 1631 (D.C. Cir. 1974), the Secretary of Labor must properly consider economic factors in determining questions of feasibility. The court cited the remarks of Senator Javits who added the "feasibility" phrase to the Act:

> As a result of this amendment, the Secretary, in setting standards, is expressly required to consider the feasibility of proposed standards. This is an improvement over the Daniels Bill, which might be interpreted to require absolute health and safety in all cases, regardless of feasibility, and the Administration Bill, which contains no criteria for standards at all. [Senate Report No. 9–1282, 91st Congress, Second Session, page 58; Legislative History at 197.]

The court stated that the thrust of Senator Javits's remarks would seem to be that practical considerations can temper protection requirements. Thus the court stated:

> Congress does not appear to have intended to protect employees by putting their employers out of business either by requiring protective devices unavailable under existing technology or by making financial viability generally impossible. . . . Standards may be economically feasible even though, from the standpoint of the employer, they are financially burdensome and affect profit margins adversely. Nor does the concept of economic feasibility necessarily guarantee the continued existance of any individual employer. It would appear to be consistent with the purposes of the Act to envisage the economic demise of an employer who has lagged behind the rest of the industry in protecting the safety and health of employees and is consequently financially unable to comply with the new standard as quickly as other employers. . . . As the effect becomes more widely spread within an industry, the problem of economic feasibility becomes more pressing.

Thus according to the court the practical question relative to economic feasibility is whether the standard threatens the competitive stability of an industry, or whether any intra- or inter-industry discrimination in the standard might ruin such stability or lead to an undue concentration.

Similarly, in *RMI Company v. Secretary of Labor,* 594 F.2d 566, 7 OSHC 1119 (6th Cir. 1979), the Court echoed the decision in *Industrial Union Department* and stated that even though the statute or regulations neither specifically describes economic considerations which are relative to the implementation of a standard and speak only in terms of feasibility, a consensus has developed among the courts that the term "feasibility" should encompass both technological and economic aspects. The court further stated that in many cases the Secretary of Labor should weigh the estimated costs of compliance against the benefits reasonably expected therefrom in promulgating or enforcing the regulations. The decisions in *American Petroleum Institute v. OSHA,* 581 F.2d 493, 6 OSHC 1959 (5th Cir. 1978), *aff'd,* 440 U.S. 906, 8 OSHC 1586 (1980); *American Iron & Steel Institute v. OSHA,* 577 F.2d 825, 6 OSHC 1451 (3d Cir. 1978); *UPS of Ohio, Inc. v. OSHRC,* 570 F.2d 806, 6 OSHC 1347 (8th Cir. 1978); *AFL-CIO v. Brennan,* 530 F.2d 109, 3 OSHC 1820 (3d Cir. 1975); *Arkansas-Best Freight Systems, Inc. v. OSHRC,* 529 F.2d 649, 3 OSHC 1910 (8th Cir. 1976); and *ITO Corp. v. OSHRC,* 540 F.2d 543, 4 OSHC 1574 (1st Cir. 1976), have all basically reached this same result (with more or less emphasis on cost-benefit) based both upon the Congressional intent derived from the statutory scheme of the Act and the language and the legislative history of the Act.

A more difficult decision which courts must face in determining the question of feasibility is how one defines and applies the term "economic feasibility." As stated previously, the court in *Industrial Union Department* has set forth the basic role economic considerations are to play in standards review:

> There can be no question that OSHA represents a decision [by Congress] to require safeguards for the health of employees even if such measures substantially increase production costs. . . . But practical considerations can temper protective requirements. Congress does not appear to have intended to protect employees by putting their employers out of business—either by requiring protection devices unavailable under existing technology or by making financial viability generally impossible. . . .
>
> Standards may be economically feasible even though from the standpoint of employers they are financially burdensome and affect profit margins adversely. Nowhere does the concept of economic feasibility necessarily guarantee the continued existence of individual employers. It would appear to be consistent with the purposes of the Act to envisage the economic demise of an employer who has lagged behind the rest of the industry in protecting the health and safety of the employees and is consequently financially unable to comply with the new standards as quickly as other employers.

The thrust of the court's decision is simply that standards may be economically feasible even though they are financially burdensome and affect profit margins adversely. For as the Fifth Circuit in *Taylor Diving and Salvage Co. Inc., v. Dept. of Labor,* 537 F.2d 819, 4 OSHC 1511 (5th Cir. 1976) aptly said:

> Indeed, a regulation enacted in the single-minded pursuit of absolutely risk-free workplaces regardless of cost or limitation might well be struck down; it would certainly be ill-advised. On the other hand, "[s]tandards may be economically feasible even though, from the standpoint of employers, they are financially burdensome and affect profit margins adversely."

The considerations involved in making such "profitability/feasibility" determinations were discussed, of course, in the *Industrial Union Dept.* decision:

> This qualification [of economic feasibility] is not intended to provide a route by which recalcitrant employers or industries may avoid the reforms contemplated by the Act. . . . It would appear to envisage the economic demise of an employer who has lagged behind the rest of the industry in protecting the health and safety of employers and is consequently financially unable to comply with new standards as quickly as other employers. As the effect becomes more widespread within an industry, the problem of economic feasibility becomes more pressing. For example, if the standard requires changes that only a few leading firms could quickly achieve, delay might be necessary to avoid increasing the concentration of that industry.

In *American Federation of Labor v. Brennan,* 530 F.2d 109, 3 OSHC 1820 (3d Cir. 1975), and *American Iron & Steel Institute v. OSHA,* 577 F.2d 825, 6 OSHC 1451 (3d Cir. 1978), the court reiterated the *Industrial Union Department* statement that Congress did not intend to eliminate all hazards to industrial employees at the price of crippling an industry or rendering it extinct, however, it upheld the coke oven standard in the latter case by saying that although it would have an adverse impact on the industry, it was not persuaded that its implementation would precipitate anything approaching the "massive dislocation"—"imperil-the-existence-of" standard which would characterize an economically infeasible standard. In ruling as it did, the court attached significance to the United Steelworkers Union's strong support for the standard, reasoning that the Union would not support a standard the effect of which would be to put its members out of work.

In *AFL-CIO v. Marshall,* 617 F.2d 636, 7 OSHC 1775 (D.C. Cir. 1979), the court, while admitting that the actual test for determining economic feasibility has not yet been developed by the courts, agreed that the cost cannot be prohibitively expensive, for Congress did not intend OSHA to make financial viability impossible for a regulated industry. Even if a few older and smaller firms must bear a disproportionate financial burden and are forced to shut down, however, the standard is not necessarily economically infeasible; reiterating its statement in *Industrial Union Dept.:*

> It would appear to be consistent with the purposes of the Act to envisage the economic demise of an employer who has lagged behind the rest of the industry in protecting the health and safety of employees and is consequently financially unable to comply with the new standards as quickly as other employers.

In arriving at this result and affirming the cotton dust decision, the court *specifically rejected* the claim that under the economic feasibility requirement OSHA must demonstrate that the benefits of the standard are in proportion with the cost which it imposes. A cost-benefit analysis is not required and OSHA is constrained only by the limits of feasibility and declined to read the "reasonably necessary and appropriate" language contained in the definition of safety and health standards Section 3(8) of the Act, as requiring a cost-benefit analysis:

> The OSH Act constrains [OSHA's] regulation of dangerous substances only by the limits of feasibility. . . . We also find that no additional constraint is imposed by the Act's definition of a health or safety standard as reasonably necessary or appropriate to provide safe or healthful employment. . . . Other statutory schemes explicitly require such particular kinds of analyses. In the Clean Air Act for example Congress required the Environmental Protection Agency to perform a cost-benefit analysis before prohibiting the manufacture or sale of a fuel or fuel additive which endangers public health or welfare. In the OSH Act in contrast, Congress itself struck the balance between cost and benefits in the mandate to the agency. Section 6(b)(5) unequivocally mandates OSHA to set the standard which most adequately assures, to the extent feasible, on the best available evidence that no employee will suffer material impairment of health or functional capacity. . . . In contrast to the Act for which Congress contemplated a cost-benefit requirement, the legislative history of the OSH Act contains no reference to this kind of economic analysis.

The court, in specifically rejecting the argument that Section 3(8) requires a cost-benefit analysis, referred to the fact that other statutes such as the Federal Hazards and Substances Act, 15 U.S.C. Section 1261, Consumer Products Safety Act, 15 U.S.C. Section 2058(c)(2)(A) and the Toxic Substances Control Act, 15 U.S.C. Section 1605(a) all require a showing of unreasonable risk prior to regulation. According to the court, the Act does not have such a requirement. The court therefore rejected the decisions of the courts in *Turner Co. v. the Secretary of Labor,* 561 F.2d 82, 5 OSHC 1790 (7th Cir. 1977) and *American Petroleum Institute v. OSHA,* 581 F.2d 493, 6 OSHC 1959 (5th Cir. 1978), which required a cost-benefit analysis.[22] See also *Florida Peach Growers Ass'n., Inc. v. Dept. of Labor,* 489 F.2d 120, 1 OSHC 1472 (5th Cir. 1974)—the court recognized that the promulgation of any OSHA standard will depend upon a balance between the protection afforded by the requirement and the effect upon economic and market conditions.

In *American Petroleum Institute v. OSHA,* 581 F.2d 493, 6 OSHC 1959 (5th Cir. 1978), the court specifically upheld the concept of a cost-benefit analysis and developed a three-part test for determining economic feasibility (the first two of which are reviewable under the substantive evidence rule, the third, being a policy judgment, must be consistent with the purpose of the Act): (1) is the estimate of expected benefits supportable by substantial evidence; (2) is the estimate of expected costs supportable; and (3) do the benefits bear a reasonable relationship to the costs. According to the court, OSHA failed to prove even the first test with respect to the benzene standard, that benefits from the standard would be appreciable. Since inadequate evidence of expected benefits existed, the cost-benefit analysis was not reached:

> Although [Section 61](b)(5) requires the goal of attaining the highest degree of health and safety protection for the employee, it does not give OSHA the unbridled discretion to adopt standards designed to create absolutely risk free workplaces regardless of cost. To the contrary, that section requires standards to be feasible, and it contains a number of pragmatic limitations in the form of specific kinds of information OSHA must consider in enacting standards dealing with

> toxic materials. Those include the "best available evidence," "research, demonstrations, experiments and such other information as may be appropriate," "the latest available scientific data. . . . Moreover, in standards dealing with toxic materials just as with other occupational safety and health standards conditions and other requirements imposed by the standard must be reasonably necessary or appropriate. . . .
>
> Before it regulates, the agency must show that a hazard exists and that its regulation will reduce the risk from the hazard, for "no [occupational safety and health] standard would be expected to impose added costs or inconvenience . . . unless there is reasonable assurance that the frequency or severity of injuries or illnesses will be reduced. . . . Most importantly for today's case, *Aqua Slide* also requires the agency to assess the expected benefits in light of the burdens to be imposed by the standard. Although the agency does not have to conduct an elaborate cost-benefit analysis, it does have to determine whether the benefits expected from the standard bear a reasonable relationship to the costs imposed by the standard.
>
> The only way to tell whether the relationship between the benefits and costs of the benzene standard is reasonable is to estimate the extent of the expected benefits and costs. OSHA did this with respect to costs. . . . However, OSHA disclaimed an obligation to balance these costs against expected benefits. Without an estimate of benefits supported by substantial evidence, OSHA is unable to justify a finding that the benefits to be realized from the standard bear a reasonable relationship to its one-half billion dollar price tag.
>
> OSHA's assumption that the standard is likely to result in benefits is not unsupported.

The requirement, however, that an agency determine the level of benefits and assess expected benefits in light of burdens imposed by the standard, that is, requiring it to determine whether the benefits expected from the standard bear a reasonable relationship to the cost imposed by the standard, were construed by the court to fall under the Section 3(8) language of the Act rather than the economic feasibility of Section 6(b)(5) of the act.[23] The court noted that OSHA in promulgating the benzene standard determined that the cost would be feasible since it would not threaten the financial welfare of affected firms or the economy. However, it disclaimed any obligation to balance these costs against the expected benefits. This, in the court's determination, was not sufficient, for "feasibility" means that once the costs were determined to be feasible in the *Industrial Union Dep't* context, they must then be balanced against benefits in the context of Section 3(8) of the Act:

> OSHA's failure to provide an estimate of expected benefits for reducing the permissible exposure limit, supported by substantial evidence, makes it impossible to assess the reasonableness of the relationship between expected costs and benefits. This failure means that the required support is lacking to show reasonable necessity for the standard promulgated. Consequently, the reduction of the permissible exposure limit from 10 ppm to 1 ppm and all other parts of the standard geared to the 1 ppm level must be set aside.

Although the United States Supreme Court affirmed the decision in *American Petroleum Institute,* it did not do so on the cost-benefit analysis question, and specifically reserved ruling on that matter until a later decision.

The Third Circuit has recognized that the economic infeasibility of OSHA regulations may be asserted as an affirmative defense in enforcement actions as well in *Atlantic Gulf Stevedores, Inc. v. OSHRC,* 534 F.2d 541, 4 OSHC 1061 (3d Cir. 1976). The employer there contended that the longshoring hard-hat standard was economically infeasible, and hence invalid, because attempts at enforcement would provoke a wildcat strike by employees. However, the court found that the petitioners had failed to establish the infeasibility of the challenged regulation, since they did not show that they had taken steps to discipline or discharge employees who defied the standard. Furthermore, employers may have other legal remedies available to them according to the court:

> We must face squarely the issue whether the Secretary can announce, and insist on employer compliance with a standard which employees are likely to resist to the point of concerted work stoppages. To frame the issue in slightly different terms, can the Secretary insist that an employer in the collective bargaining process bargain to retain the right to discipline employees for violation of safety standards which are patently reasonable and are economically feasible except for employee resistance?
>
> We hold that the Secretary has such power. [T]he entire thrust of the Act is to place primary responsibility for safety in the work place upon the employer. That, certainly, is a decision within the legislative competence of Congress. In some cases, undoubtedly, such a policy will result in work stoppages. But as we observed in *AFL-CIO v. Brennan, supra,* the task of weighing the economic feasibility of a regulation is conferred upon the Secretary. He has concluded that stevedores must take all available legal steps to secure compliance by the longshoremen with the hardhat standard.

Recently other courts have also adopted the economic infeasibility defense. In *RMI Company v. Secretary of Labor,* 594 F.2d 566, 7 OSHC 1119 (6th Cir. 1979), the court, in affirming the Review Commission's finding of technological feasibility of noise controls, remanded the case to permit taking of additional evidence on economic feasibility and the costs of engineering controls versus benefits to the employees' health and safety, as they related to an alleged violation of the noise standard. 29 CFR 1910.95(b)(1). See also *Turner Company v. Secretary of Labor,* 561 F.2d 82, 5 OSHC 1790 (7th Cir. 1977); *ITO Corp. v. OSHRC,* 540 F.2d 543, 4 OSHC 1574 (1st Cir. 1976); *Arkansas-Best Freight Systems, Inc., v. OSHRC,* 529 F.2d 648, 3 OSHC 1910 (8th Cir. 1976); *Southern Colorado Pre-Stress Co. v. OSHRC,* 586 F.2d 1342, 6 OSHC 2032 (10th Cir. 1978). The economic feasibility test adopted by several circuits in the enforcement context has not been limited to determining whether an employer could afford the control. Rather the costs of the proposed controls are to be balanced against the proposed benefits flowing therefrom in order that the resources are allegated in priority to the degree of harm established. The Review Commission has adopted the following standard in comport therewith:

> [s]tandard requires engineering and administrative controls where are economically and technically feasible. Controls may be economically feasible even though they are expensive in increased production . . . but they will not be required without regard to the costs which must be incurred and the benefits they will achieve. All the relevant costs and benefit factors must be weighed.

Continental Can Company, 4 OHSC 1541 (1976); *Sampson Paper Bag, Inc.,* 8 OSHC 1515 (1980).

b. *Technological Feasibility*

In addition to economic feasibility, the issue of tecnnological feasibility has become important recently, and promises to remain so in the area of occupational safety and health. A major aspect of the issue of technological feasibility, of course, is whether employers can be required to implement novel and/or expensive technological changes to deal with newly discovered occupational hazards. In *Industrial Unions Departments AFL-CIO v. Hodgsons, supra,* the court held that it remains the duty of the Secretary of Labor in the enforcement of the Act to protect workers, even in situations where existing methodology or research are deficient. The leading court decision in this area, however, is *Society of Plastics Industry, Inc. v. OSHA,* 509 F.2d 1301, 2 OSHC 1496 (2d Cir. 1975), where the court stated, in reviewing the technological feasibility issues under a "reasoned decision-making" standard that:

> We cannot agree with petitioners that the standard is so clearly impossible of attainment. It appears that they simply need more faith in their own technological potentialities, since the record reveals that, despite similar predictions of impossibility regarding the emergency 50 ppm standard, vast improvements were made in a matter of weeks, and a variety of useful engineering and work practice controls have yet to be instituted.
>
> In the area of safety, we wish to emphasize, the Secretary is not restricted by the status quo. He may raise standards which require improvements in existing technologies or which require the development of new technology, and he is not limited to issuing standards based solely on devices already fully developed.

Under this view OSHA can force industry to develop and diffuse new technology; so long as it presents substantive evidence that companies acting vigorously and in good faith can develop the technology, OSHA can require industry to meet exposure levels never previously attained.

In *AFL–CIO v. Brennan,* 530 F.2d 109, 3 OSHC 1820 (3d Cir. 1975), the court reiterated the oft-stated view that Congress meant the Act to be technology forcing and added:

> Although we hold that the Secretary may, consistent with the statute, consider the technological feasibility of a proposed occupational health and safety standard promulgated pursuant to Sec. 6(a), we agree with the Second Circuit in *Society of Plastics Industry, Inc. v. OSHA, supra,* that, at least to a limited extent, *OSHA* is to be viewed as a technology-forcing piece of legislation. Thus, the Secretary would not be justified in dismissing an alternative to a proposed health and safety standard as infeasible when the necessary technology looms on today's horizon.

The Third Circuit has in fact characterized the Act as a technology-forcing piece of legislation in three cases. *Atlantic & Gulf Steelworkers, Inc. v. OSHA,* 534 F.2d 541, 4 OSHC 1061 (3d Cir. 1976); *AFL–CIO v. Brennan,* 530 F.2d 109, 3 OSHC 1820 (3d Cir. 1975) and *American Iron & Steel Institute v. OSHA,* 577 F.2d 825, 6 OSHC 1451 (3d Cir. 1978). Although recognizing that the Secretary of Labor is constrained by the requirements of feasibility, the court stated that a standard is not infeasible if the necessary technology looms on the horizon. This means, at the very least, that OSHA can impose a standard which only the most technologically advanced plants in the industry have been able to achieve—even if only in some of their operations, some of the time. However, in *American Iron and Steel Institute* the court placed a limit on the Act as a technology-forcing legislation when it stated that while OSHA may raise standards which requires employers to implement technology looming on the horizon and is not limited to issuing a standard based solely upon technology fully developed today, it may not place an affirmative duty on employers to research and develop new technology.

In *AFL-CIO v. Marshall,* 617 F.2d 636, 7 OSHC 1775 (D.C. Cir. 1979), the court stated that judging technological feasibility of a particular agency is beyond the expertise of the the judiciary, especially where the assessment involves predictions of technological changes. Thus its task on review is finding whether it has sufficient support for the feasibility determination in record. The Act only requires the agency to develop standards based upon the best *available* evidence and not necessarily the best evidence:

> In the area of safety, we wish to emphasize, the Secretary is not restricted by the status quo. He may raise standards which require development of new technology, and he is not limited to issuing standards based solely on devices already fully developed.

c. *Concepts of Proof*

In *United Steelworkers of America v. Marshall,*—F.2d—, 8 OSHC 1810 (D.C. Cir. 1980), which involved a review of the standard for occupational exposure to airborne lead in the workplace, the court felt that although previous standards-review decisions had provided guidance on the meaning of feasibility, there still was a question as to how OSHA was to prove feasibility within the rather generous constraints of the substantial evidence test, given a court's necessary deference to the agency when it is making an essentially legislative policy. Thus the court stated it must decide what to require of OSHA in the matter of types and quantity of data, precision of cost estimates, and certainty of prediction.

As for technological feasibility, the court stated it knows it could not require of OSHA anything like certainty since technology-forcing assumes the agency will make highly speculative projections about future technology. A standard is obviously not infeasible solely because OSHA has no hard evidence to show that the standard has been met. More to the point, a court cannot require OSHA to prove with any certainty that industry will be able to develop the necessary technology or even to identify the single technological means by which it expects industry to meet a given permissible exposure level. OSHA can force employers to invest all reasonable faith in their own capacity for their own technological innovation and can thereby shift to industry some of the burden of choosing the best strategy for compliance. OSHA's duty is to show that modern technology has at least conceived some industrial strategies or devices which are likely to be capable of meeting the permissible exposure level and which the industries are generally capable of adopting.

The court stated that its view finds support in the statutory requirement that OSHA act according to the best available evidence. OSHA cannot let workers suffer while it awaits the "Godot" of scientific certainty. It can and, therefore, must make reasonable predictions on the basis of credible sources of information, whether data is from existing plants or expert testimony. The court stated, therefore, that its role is to insure that the agency has developed substantial evidence in meeting that task, however:

> The ironic truth is that "technology-forcing" makes the agency's standard of proof somewhat *circular*. Since the agency must hazard some predictions about experimental technology, it may not be able to determine success of new means of compliance until the industry implements them. . . . Conversely, OSHA or the courts may discover that a standard is infeasible only after industry has exerted all good faith efforts to comply. . . . Both the agency in issuing, and the court, in upholding a standard under this principle obviously run the risk that an apparently feasible standard will prove technologically impossible in the future. But, while we address below the complexities of this issue, we note the general agreement in the courts that such flexible devices as variance proceedings can correct erroneous predictions about feasibility when the error manifests itself, and thus can reduce this risk.

The court stated that proving economic feasibility presents different problems. Although OSHA need not specifically weigh industry costs against worker benefits, nevertheless, the agency must provide a reasonable assessment of the likely range of costs of its standard and the likely effects of these costs on the industry. The practical question is what form that assessment must take. In the cotton dust case, the court stated that the agency need not engage in massive data collection with its own staff, but could rely on data and estimates produced by consultants or even by industry itself. OSHA can, in effect, produce its own estimate by modifying the consultant's or industry's estimates where it finds specific faults in their calculations:

> In relying on such revised estimates, however, OSHA must do more than generally allege that the independent estimate is too high. Rather it must carefully identify the sources and assess the magnitude of any overestimation. . . . OSHA can work from the outside party's data and estimates so long as it is thorough and precise in explaining its revisions. In short, the agency can rely on the best available evidence and the court demands only substantial evidence. We do not require the agency to establish the standard's economic feasibility in a particular way.

In addressing the circularity problem, the court stated that the problem is essentially that it cannot know if a standard is feasible until it knows exactly what it expects of employers, (*e.g.*, the ambiguous means of compliance in the lead standard). There should be a clear line between construing the precise terms of the demands OSHA places on industry and assessing the feasibility of those demands. The court recognized, however, that some cases have blurred this line and have justified their deference to OSHA's feasibility findings, at least in part, by discovering flexibility in OSHA's regulatory scheme that renders the standards less stringent than they may first appear. The court found, however, that these apparently very reasonable decisions do not address a serious analytical problem in the standards.

For example, it noted that two cases, in upholding OSHA's standards as feasible, have pointed to the availability of variance proceedings as means by which companies can gain relief from standards which prove too demanding. *American Iron & Steel Institute, supra; Society of Plastic Industries, supra.* Another court has noted that even if industry cannot find the technology to meet the permissible exposure level, that OSHA simply requires companies to install new engineering and work-practice controls to the extent feasible and then allows them to use respirators to meet the permissible exposure limit (PEL) when feasible engineering and work-practice controls prove insufficient to meet the PEL by themselves—again circular reasoning. If employers need only use engineering or work-practice controls which prove feasible, what meaning is there to the feasibility requirement and how could a standard ever be infeasible under this approach?

The references to variances also gloss over important details in the Act. For example, permanent variances from a standard are only available to employers whose workplaces are as safe and healthful as those which would prevail if they fully complied with the standard. Variances of this type are useless to an employer who claims he can find no practical way of meeting the health and safety demands of an OSHA standard, assuming, of course, that respirators are not found to be as safe and healthful as engineering and work-practice controls. Permanent variances would be generally irrelevant to the whole scheme of compliance with the standards.

The other type of variance—the temporary variance—is used if an employer cannot comply with a standard because of the unavailability of professional or technical personnel or of materials and equipment . . . or because necessary construction or alteration of facilities cannot be completed by the effective date. Although this type of variance appears to grant some relief to an employer who is technologically incapable of meeting a standard, such a variance is very temporary. Relief is generally no more than six months to one year, although sometimes renewals can effectively extend the variance to three years. However, OSHA's predictions about feasibility may turn out to be so faulty as to make even three years' relief insufficient.

In an attempt to resolve the circularity problem, the court first stated that the feasibility test for a standard that only expects feasible improvements from employers may appear circular, but reasonable construction of such a standard avoids circularity. The cases, according to the court, have apparently treated the standards as creating a general presumption of feasibility for an industry, that is, a company or an industry could not simply refuse to pursue engineering or work practice controls by asserting their infeasibility. Rather it would have to attempt to install controls to the limits of contemporary, technical knowledge and of its own financial resources. Traditional review of feasibility in this context would have meaning because the court would have to find substantial evidence to justify this presumption—evidence that the technical knowledge to meet the PEL without relying on respirators was available or would likely be available when the deadline arrives, and that enough firms could afford to so meet the PEL that the makeup and structure of the industry would survive. Thus the court would do a preliminary test on feasibility in any preenforcement challenge to rule-making. The court stated, however, that it reserves the power to test feasibility in reviewing the denial of the temporary variance or where an employer found such a variance insufficient in judicial review of an enforcement proceeding, alleging the affirmative defense that the standard had proved generally infeasible, even if the court had earlier found otherwise on preenforcement review. In any of these proceedings, however, the employer would bear the burden of proof.[24]

The court then attempted to set forth a systematic analysis of the issue of feasibility and means of compliance and relief from impractical standards. When affected parties petition for pre-enforcement review of an OSHA standard, the standard of course must pass a preliminary test of feasibility:

> First, within the limits of the best available evidence, and subject to the court's search for substantial evidence, OSHA must prove a reasonable possibility that the typical firm will be able to develop and install engineering and work-practice controls that can meet the PEL in most of its operations. OSHA can do so by pointing to technology that is either already in use or has been conceived and is reasonably capable of experimental refinement and distribution within the standards' deadlines. The effect of such proof is to establish a presumption that industry can meet the PEL without relying on respirators, presumptions which firms will have to overcome to obtain relief in any second inquiry into the feasibility in any of the proceedings we discuss below.
>
> Insufficient proof of technological feasibility for a few isolated operations within an industry, or even OSHA's concession that respirators will be necessary in a few such operations, will not undermine this general presumption in favor of feasibility. . . .
>
> Second, as for economic feasibility, OSHA must construct a reasonable estimate of compliance costs and demonstrate a reasonable likelihood that these costs will not threaten the existence or competitive structure of an industry, even if it does portend disaster for some marginal firms.
>
> To protect industry from the risk of wasteful guesses by the best means of compliance, we can also expect OSHA to show that new technology, even if it might fall short of meeting the PEL, will nevertheless significantly reduce the lead exposure.

There always remains, however, the chance that an OSHA standard, after appearing generally feasible on review of the rule-making, will ultimately prove generally infeasible. Thus there is a scheme of devices by which employers can gain relief from such problems in the future, and by all of which devices employers can revert to respirators as a means of compliance when engineering and work-practice controls prove insufficient. First, for employers who need as much as three years to meet the deadline in a standard, the temporary variance proceeding is available. Second, an employer who is cited for failing to meet the standard in a particular operation and who believes the standard has proved technologically infeasible for that operation can claim this specific infeasibility as a defense in an enforcement proceeding. See *e.g., Marshall v. West Point Pepperell, Inc.*, 588 F.2d 979, 7 OSHC 1035 (5th Cir. 1979). Thus an OSHA standard would remain subject to a second test of feasibility with respect to *special* difficulties in certain operations. Moreover, OSHA may also face a second test of feasibility on a claim that a standard is *generally* infeasible for an entire industry:

> First, we see nothing to prevent an employer from raising a defense in an enforcement proceeding that the standard has proved infeasible for all similar companies not just his own. The alleged infeasibility may be technological if the firm is typical or is large enough to prove that it has made all possible efforts to develop engineering or work-practice controls to meet the PEL. Or it might be economic where the defending firm can use its own experience to prove that no employer could afford to meet the standard. . . . Second, where employers believe that a standard originally determined to be feasible had proved infeasible they could petition the agency to initiate a new rulemaking to revise the standard. . . . [W]here time does demonstrate the infeasibility of a standard once proved as feasible, so that the very predicate for the original rule and its statutory basis have disappeared, a court may well be able to deny the agency any discretion to refuse a new rulemaking. There is no hard rule of *res judicata* or estoppel barring employers from obtaining relief from a standard which time proves impossible to meet.

Having established the general principles for analyzing feasibility, the court considered the particular problems OSHA faced in determining the lead standards feasibility and upheld the standard with respect to some industries, but remanded it to OSHA for other industries.

Some standards, such as asbestos (29 CFR 1910.93(a)) and noise (29 CFR 1910.95) require, as an additional element of proof of a violation (by OSHA), that (engineering and administrative) controls be technologically feasible; that is, it must be shown that feasible engineering or administrative controls exist to abate the condition. Where such controls do not exist to lower employee exposure to permissible levels, or to effect a *significant* reduction in employee exposure, then the standards require employers to implement personal protection programs for all affected workers (*e.g.*, ear plugs, respirators). *Marshall v. West Point Pepperell, Inc.*, 588 F.2d 979, 7 OSHC 1034 (5th Cir. 1979); *RMI Company v. Secretary of Labor,* 594 F.2d 566, 7 OSHC 1119 (6th Cir. 1979); *Southern Colorado Pre-Stress Co. v. OSHRC,* 626 F.2d 1342, 6 OSHC 2032 (10th Cir. 1978); *Diversified Industries Division Independent Stone Company v. OSHRC,* 618 F.2d 30, 8 OSHC 1107 (8th Cir. 1980).

See generally Berger and Risken, *Economic and Technological Feasibility in Regulating Toxic Substances Under the Occupational Safety and Health Act,* 7 Ecol. L.Q. 285 (1978), quoted in *Industrial Union Dept., AFL-CIO v. American Petroleum Institute,* 440 U.S. 906, 8 OSHC 1586 (1980).

8. Stay of Standards

The issuance of a stay pending judicial review is authorized by Section 6(f) of the Act. In order to obtain a stay pending appeal, it must be demonstrated that (1) there is a substantial likelihood that the applicants will prevail on the merits, (2) they will suffer irreparable harm if the stay is denied, (3) the issuance of the stay will not substantially harm other parties to the proceeding, and (4) the issuance of the stay will not interfere with the public interest. See *Vistron Corporation v. OSHA,* —F.2d—, 6 OSHC 1483 (6th Cir. 1978); *Taylor Diving & Salvage Company, Inc. v. U.S. Dept. of Labor,* 537 F.2d 819, 4 OSHC 1511 (5th Cir. 1976); *Chlorine Institute, Inc. v. OSHA,* 613 F.2d 120, 8 OSHC 1031 (5th Cir. 1980).

G. ADMINISTRATIVE REVIEW OF STANDARDS

The Commission has held that affected persons may challenge the validity (but not the wisdom) of standards during enforcement proceedings before the Commission. See *Charles A. Gaetano Construction Corp.*, 6 OSHC 1463 (1978); *Daniel Construction Co.*, 5 OSHC 1713 (1977); *Modern Automotive Service, Inc.*, 1 OSHC 1544 (1973); *Kennecott Copper Corp.*, 4 OSHC 1400 (1976); *cf.*, *Van Raalte Co. Inc.*, 4 OSHC 1151 (1976) and *Fabricraft, Inc.*, 7 OSHC 1540 (1970). The Commission has also decided vagueness and lack of fair notice challenges to standards. *Gold Kist, Inc.*, 7 OSHC 1855 (1979); *American Airlines, Inc.*, 7 OSHC 1980 (1979). The Commission also has the authority to determine the feasibility of standards, in the enforcement context. *Sampson Paper Bag Co.*, 8 OSHC 1515 (1980); *Wheeling Pittsburgh Steel Corp.*, 7 OSHC 1581 (1979); *Continental Can Corp.*, 4 OSHC 1541 (1976).

NOTES

1. See also *Northwest Airlines, Inc.*, 8 OSHC 1989 (1980), where the Commission stated: "As used in the Act, a 'standard' is a substantive rule containing a requirement 'reasonably necessary or appropriate to provide safe or healthful employment.' [Section 3](8). A 'regulation' is a rule governing matters such as posting of notices, recordkeeping, and conduct of inspections. *E.g.*, [Section 8](c)(1)–(3), [Section 8] (g)(2). We shall use the word 'rule' to refer to both 'standards' and 'regulations.' "

2. In February, 1979, a single set of regulations was developed by OSHA with the advice of the Advisory Committee on Construction Safety & Health and published in the Federal Register for the use of the construction industry. The new set of regulations contains the entire text of 29 CFR Part 1926—Construction, and certain general industry standards contained in 29 CFR Part 1910 which have been identified as applicable to construction work.

3. The Administrative Procedure Act provides at 5 U.S.C. Section 522(a)(1) that "[e]xcept to the extent that a person has actual and timely notice of the terms thereof, a person may not in any manner be required to resort to, or be adversely affected by, a matter required to be published in the Federal Register and not so published. For the purposes of this paragraph, matter reasonably available to the class of persons affected thereby is deemed published in the Federal Register when incorporated by reference therein with the approval of the Director of the Federal Register.

4. It should also be noted that the Secretary of Labor is not required to physically introduce an ANSI standard into evidence at a hearing if there has been a valid incorporation by reference; it may be officially noticed. 44 U.S.C. Section 1507; *Atlantic & Gulf Stevedores, Inc.*, 3 OSHC 1003 (1975).

5. Section 4(b)(2) of the Act provides that "[t]he safety and health standards promulgated under the Act of June 30, 1936, commonly known as the Walsh-Healy Act (41 U.S.C. 35 *et seq.*), the Service Contract Act of 1965 (41 U.S.C. 351 *et seq.*), Public Law 91–54, Act of August 9, 1969 (40 U.S.C. 333), Public Law 85–742, Act of August 23, 1958 (33 U.S.C. 941), and the National Foundation on Arts and Humanities Act (20 U.S.C. 951 *et seq.*) are superseded on the effective date of corresponding standards, promulgated under this Act, which are determined by the Secretary to be more effective. Standards issued under the laws listed in this paragraph and in effect on or after the effective date of this Act *shall be deemed* to be occupational safety and

health standards issued under this Act, as well as under such other Acts (emphasis added)."

6. 29 CFR 1910.11 provides: "(a) The provisions of this Subpart B (Adoption and Extension of Established Federal Standards) adopt and extend the applicability of established Federal standards in effect on April 28, 1971, with respect to every employer, employee, and employment covered by the Act.

"(b) It bears emphasis that only *standards (i.e.,* substantive rules) relating to safety and health are adopted by any incorporations by reference of standards prescribed elsewhere in this chapter of this title. Other materials contained in the referenced parts are not adopted. Illustrations of the types of materials which are not adopted are there. The incorporations by reference of Parts 1915, 1916, 1917, 1918 in Sections 1910.13, 1910.14, 1910.15, and 1910.16 are not intended to include the discussion in those parts of the coverage of the Longshoremen's and Harbor Workers' Compensation Act or the penalty provisions of the Act. Similarly, the incorporation by reference of Part 1926 in Section 1910.12 is not intended to include references to interpretative rules having relevance to the application of the Construction Safety Act, but having no relevance to the application of the Occupational Safety and Health Act."

29 CFR 1910.12 (Construction Work) provides: "(a) Standards. The standards prescribed in Part 1926 of this chapter are adopted as occupational safety and health standards under Section 6 of the Act and shall apply, according to the provisions thereof, to every employment and place of employment of every employee engaged in construction work. Each employer shall protect the employment and places of employment of each of his employees engaged in construction work by complying with the appropriate standards prescribed in this paragraph.

"(b) Definition. For purposes of this section, 'construction work' means work for construction, alteration, and/or repair, including painting and decorating.

"(c) Construction Safety Act distinguished. This section adopts as occupational safety and health standards under Section 6 of the Act the standards which are prescribed in Part 1926 of this chapter. Thus the standards (substantive rules) published in Subpart C and the following subparts of Part 1926 of this chapter are applied. This section does not incorporate Subparts A and B of Part 1926 of this chapter. Subparts A and B have pertinence only to the application of Section 107 of the Contract Work Hours and Safety Standards Act (the Construction Safety Act). For example, the interpretation of the term 'subcontractor' in paragraph (c) of §1926.13 of this chapter is significant in discerning the coverage of the Construction Safety Act and duties thereunder. However, the term 'subcontractor' has no significance in the application of the Act, which was enacted under the Commerce Clause and which establishes duties for 'employers' which are not dependent for their application upon any contractual relationship with the Federal Government or upon any form of Federal financial assistance."

Similar provisions are included in 29 CFR 1910.13, .14, .15, and .16 for ship repairing, shipbuilding, ship breaking, and longshoring. Special provisions are also contained for certain air contaminants. 29 CFR 1910.19.

7. The Review Commission's decision in *Daniel Construction Co.,* 5 OSHC 1005 (1977), which held that a shorter period of publication was justified because of the Secretary's statement of good cause, was criticized in *National Industrial Constructors, Inc.,* on the basis that the cases defining the scope of good cause have required some showing of emergency or immediacy that does not appear to have been alleged by the Secretary in that action.

8. The court in *Dry Color Manufacturers Ass'n v. Department of Labor,* 486 F.2d 98, 1 OSHC 1331 (3d Cir. 1973), discussed the emergency temporary standard there under

review in the framework of the substantial evidence test but did not rule on the question, because it vacated the standard for lack of an adequate statement of reasons.

9. 29 CFR 1910.134(b) provides, *inter alia:* "(b) Requirements for a minimal acceptable program. (1) Written standard operating procedures governing the selection and use of respirators *shall* be established.

"(2) Respirators *shall* be selected on the basis of hazards to which the worker is exposed.

"(3) The user *shall* be instructed and trained in the proper use of respirators and their limitations.

"(4) Where practicable, the respirators *should* be assigned to individual workers for their exclusive use.

"(5) Respirators *shall* be regularly cleaned and disinfected. Those issued for the exclusive use of one worker *should* be cleaned after each day's use, or more often if necessary. Those used by more than one worker *shall* be thoroughly cleaned and disinfected after each use.

"(6) Respirators *shall* be stored in a convenient, clean, and sanitary location.

"(7) Respirators used routinely *shall* be inspected during cleaning. Worn or deteriorated parts *shall* be replaced. Respirators for emergency use such as self-contained devices *shall* be thoroughly inspected at least once a month and after each use.

"(8) Appropriate surveillance of work area conditions and degree of employee exposure or stress *shall* be maintained.

"(9) There *shall* be regular inspection and evaluation to determine the continued effectiveness of the program.

"(10) Persons *should* not be assigned to tasks requiring use of respirators *unless* it has been determined that they are physically able to perform the work and use the equipment. The local physician *shall* determine what health and physical conditions are pertinent. The respirator user's medical status *should* be reviewed periodically (for instance, annually).

10. 29 CFR 1910.39, for example, provides that the entire subpart is promulgated from NFPA 101–1970, Life Safety Code.

11. The Court in *Hispanic I* had also held that "[s]ince the Congress left such open-ended discretion in the Secretary at many key points in the Act, including Section 6(g), we find implicit acknowledgment that traditional agency discretion to alter priorities and defer action due to legitimate statutory considerations was preserved. If the Secretary may rationally order priorities and reallocate his resources at the point when Section 6(b)(1) through (4) becomes applicable, he should be able to do so at any rule-making stage, so long as his discretion is honestly and fairly exercised. To hold otherwise would encourage refusals to initiate rulemaking and create incentives to deceive which anyone, including El Congreso, decries. As we see it, there is no sense in proceeding completely through the rule-making process in accordance with El Congreso's claim that it is mandatory only to end up with the Secretary issuing a notice that the standard is not adopted."

12. Section 553(b) of the APA provides that a general notice of proposed rulemaking shall be published in the Federal Register and shall include, among other things, either the terms or substance of the proposed rule or a description of the subjects and issues involved. After such notice has been given, the agency must give interested persons an opportunity to participate in the rule-making through the submission of written data, views, or arguments. 5 U.S.C. Section 553(b)(3) and (c).

13. One of the most vigorously contested issues in the lead standard was the provision for medical removal protection (MRP) which supplements the medical surveillance provisions of the standard. All employers under the standard are required

to measure the lead-air content of all workplaces to determine whether employees suffer exposure to lead above the action level. If any workers are exposed to air lead above the action level for more than thirty days a year, the employer must offer a program of biological monitoring, including exhaustive blood measurements, other laboratory tests, and detailed medical examinations. If these tests and examinations reveal that a worker has a high blood-lead level or some ailment attributable to lead exposure, the employer must remove the worker from the high-exposure workplace. The standard sets out the criteria for and terms of this removal in great detail. OSHA phased in the removal standard gradually, with three preliminary stages requiring removal at progressively lower blood-lead and air-lead levels. For each stage of the removal standard, OSHA has also designated the reduced blood-lead level which the employees must achieve before they can return to the jobs from which they were removed. Finally, the employer must remove an employee when a final medical examination is made that the lead exposure at the employee's workplace threatens his health; the employer cannot return the employee to the workplace until a final medical determination is made that the exposure in the workplace no longer places the employee's health at risk.

The employer enjoys the discretion to place a removed worker in a low-exposure job in the same plant, or at a job at a nonlead facility, or even to simply retain him at his high-exposure job for a smaller number of hours per week to reduce his time-weighted average exposure level below the action level. If these actions fail, however, the employer must lay off the worker. But, most important, whatever form removal takes, the employer must maintain the worker's earnings and seniority rights during the removal for a period of up to eighteen months and must restore him to all the rights of his original job status when he becomes medically able to return. The court upheld the medical removal and wage guarantee provisions as reasonable, economically feasible, and reasonably necessary and appropriate. See also *AFL-CIO v. Marshall,* 617 F.2d 636, 7 OSHC 1775 (D.C. Cir. 1979), which upheld similar, but not as expansive, provisions in the cotton dust standard.

14. Under Executive Order 11821, guidelines on the preparation of inflationary impact statements from the Office of Management and Budget required OSHA to analyze the *inflationary* effect of the proposed standard upon the market, consumers, and businesses, compare the benefits derived from the proposed action with estimated costs and inflationary impacts through quantitative analysis where possible, and review alternatives to the proposed standard with consideration for the possible costs, benefits, risks, and inflationary impact compared with those for the proposed standard.

15. In reviewing standards under Section 6(f) of the Act, the Courts have concentrated on assuring that OSHA's "statement of reasons" gives a rational explanation of the conclusions it has reached, the significant issues confronted and an articulated rationale for their solution, the areas of uncertainty and dispute, the alternative resolutions of those areas, and the policies used to choose between alternatives. See *Synthetic Organic Chemical Mfr's. Ass'n v. Brennan,* 503 F.2d 1155, (3d Cir. 1974), *cert. denied,* 420 U.S. 973; *rehearing denied,* 423 U.S. 886 (1975); *Industrial Union Dept. AFL-CIO v. Hodgson,* 499 F.2d 467 (D.C. Cir. 1974); *Associated Industries of NY., Inc. v. Dept. of Labor,* 487 F.2d 342 (2d Cir. 1973). See generally *Florida Peach Growers Ass'n v. Dept. of Labor,* 489 F.2d 120 (5th Cir. 1974); *Citizens to Preserve Overton Park, Inc. v. Volpe,* 401 U.S. 402 (1971).

16. The court referred to the fact that Congress recognized the problem by authorizing OSHA to promulgate rules on the basis of the best available evidence. The best available evidence in an area of changing technology and incomplete scientific data may leave gaps in knowledge that require policy judgments in construing a health

and safety standard. In recent statutes, such as the Toxic Substance Control Act, Congress has defined "best available evidence" to mean any matter in the rule-making record.

17. The court stated that its role is to ensure that the regulations resulted from a process of reasoned decision making consistent with the agency's mandate from Congress. A court entrusted with the rigorous substantial evidence review must examine not only OSHA's factual support but also the judgment calls and reasoning that contribute to its final decision. Otherwise, an agency's claim of ignorance would clothe it with unreviewable discretion.

18. The court did state, however, that in making the difficult judgment as to what level of harm is unacceptable, the agency may rely on its own sound "considerations of policy" as well as hard factual data. Moreover, the agency need not present a precise quantification of risk where even on the best available scientific evidence such precision is impossible, and, so long as OSHA relies on reputable scientific authority, it may employ conservative assumptions in assessing the data before it, erring on the side of worker safety. Justice Stevens, in concurrence, stated, however, that the Act was not intended to require employers to provide absolutely risk free workplaces. It was only intended to require the elimination, as far as feasible, of significant risks.

19. The industry petitioners also argued that the effort to reduce cotton dust exposure to the lowest feasible level was not "reasonably necessary or appropriate," since certain symptoms of bysinosis, *e.g.*, at the *acute* stage, do not themselves constitute a material impairment of health. The court declined to respond to this issue in the significant risk context but rather stated that the agency's mandate requires it to protect workers' health even before the resolution of all medical and scientific uncertainties about the particular health risk. Moreover, although it did not define material impairment, the court also said there was *adequate/considerable* support for the choice of the agency's dust control strategy to protect workers from bysinosis, which causes material impairment in the *chronic* stage.

Significantly, however, the court used Section 3(8)'s "reasonably necessary and appropriate language" to justify the medical transfer and wage guarantee provisions of the standard.

20. The court stated that it could not accept the argument that Section 3(8) was totally meaningless. The Act authorizes the Secretary of Labor to promulgate three different kinds of standards: national consensus standards, permanent standards, and temporary emergency standards. The only substantive criteria given for two of these, national consensus standards and permanent standards, for safety hazards not covered by Section 6(b)(5) is set forth in Section 3. The court stated that while it is true that Section 3 is entitled definitions, that fact does not drain each definition of substantive content, for otherwise there would be no purpose in defining the critical terms of the statute. Moreover, if the definitions were ignored, there would be no statutory criteria at all to guide the Secretary of Labor in promulgating either national consensus standards or permanent standards other than those dealing with toxic materials and harmful physical agents. Although the Court may be correct in its assessment of the significance to be given to the term "necessary," it failed to consider the significance of the disjunctive "or," and it failed to address the meaning of the term "appropriate," which in common parlance would require a somewhat less restrictive threshold determination of risk than "necessary."

21. The dissenting opinion stated that "when the question involves determination of the acceptable level of risk, the ultimate decision must necessarily be based on considerations of policy as well as empirically verifiable facts. Factual determinations can at most define risk in some statistical way; the judgment whether that risk is

tolerable cannot be based solely on a resolution of the facts." The plurality agreed. Thus while OSHA must support its finding that a certain level of risk exists by substantial evidence it recognized that its determination that a particular level of risk is "significant" will be based largely on policy considerations. The court stated, however: "At this point we have no need to reach the issue of what level of scrutiny a reviewing court should apply to the latter type of determination."

22. The court noted that the industry petitioner's reliance on the *American Petroleum* case was unpersuasive, since the court in that decision based its holding on an entirely different statutory scheme in *Aqua Slide 'N Dive Corp. v. Consumer Product Safety Commission.* The court noted, however, that *Aqua Slide 'N Dive* involved the Consumer Product Safety Act, which employs the unreasonable risk language that courts have construed to require a balancing test. The court noted that no such language appeared in the OSH Act under which the provisions of the cotton dust standard were issued. *See also United Steelworkers of America v. Marshall,*—F.2d—, 8 OSHC 1810 (D.C. Cir. 1980).

23. The Fifth Circuit had addressed similar concerns when it interpreted the reasonable necessity criterion of the Consumer Product Safety Act as requiring the Commission to take a hard look, not only at the nature and severity of the risk, but also at the potential the standard has for reducing the severity and frequency of injury and the effect the standard would have on the utility, cost, or availability of the product.

Before [an agency] regulates, [the court held] it must show that a hazard exists *and that* its regulation will reduce the risk from the hazard, for "no [occupational safety and health] standard would be expected to impose added costs or inconvenience . . . unless there is reasonable assurance that the frequency or severity of injuries or illnesses will be reduced." More importantly, that case, *Aqua Slide 'N Dive Corp. v. Consumer Product Safety Comm'n,* 569 F.2d 831 (5th Cir. 1978), also required the agency to assess the expected benefits in light of the burdens to be imposed by the standard. Although the agency does not have to conduct an elaborate cost-benefit analysis, it does have to determine whether the benefits expected from the standard bear a reasonable relationship to the costs imposed by the standard.

24. Thus, since the presumption of feasibility remains rebuttable, in pre-enforcement review the court would not expect OSHA to prove the standard *certainly* feasible for *all* firms at *all* times in *all* jobs, but it would have to justify the presumptions and the attendant shift in burden, with reasonable, technological, and economical evidence and analysis.

CHAPTER V
Variances

A. INTRODUCTION

The variance is a device frequently utilized in areas of pervasive agency regulation such as zoning, environmental, and health law. Variances allow persons affected by the regulation to obtain exceptions from its application.

In the occupational safety and health area, there are occasions when the specification standards cannot be met by employers. In other cases, the protection already afforded by an employer to his employees is equal to or better than what could be afforded by strict compliance with the OSHA standards. Attention will be focused on the employer's practices, therefore, in the agency's decision to grant or deny a variance.

Thus if an employer believes that the immediate enforcement of an occupational safety or health standard will cause irreparable harm to his business or workplace, if he lacks the means to comply with the standard, or if he can prove that his facilities or methods of operation provide employee protection that is at least as effective as that required by OSHA, then an application for a variance may be requested from the Assistant Secretary of Labor for Occupational Safety and Health. Variance application must be submitted in accordance with the "Rules of Practice for Variance, Limitations, Variations, Tolerances, and Exemptions" contained in 29 CFR 1905. No particular form is prescribed for applications for variances, but they must be clearly legible and in writing. A model variance application letter has been published by OSHA that sets forth the necessary information to be contained therein and is attached as Appendix 1. A typewritten original and six copies must be filed.

B. TYPES OF VARIANCES

The Act references four types of variances and four separate procedures for obtaining these variances.

1. Temporary Variance

A temporary variance may be requested when an employer needs additional time to comply with a standard. Section 6(b)(6)(A) of the Act.

Recognized reasons justifying the inability to presently comply with a standard include: (1) the unavailability of professional or technical personnel; (2) the unavailability of necessary materials or equipment; and (3) the impossibility of completing necessary construction or alterations within the time allowed. The employer must clearly establish that he is taking all available steps to safeguard his employees against the hazards covered by the standard, and that he has put in force an effective program for compliance with the standard as soon as practicable. Economic hardship is not a consideration in the qualification for a temporary variance.

A temporary variance may only remain in effect for the period of time needed by the employer to achieve compliance with the standard, which cannot exceed one year. The variance order may be renewed twice, each time for one hundred eighty (180) days. The application for renewal must be filed at least ninety (90) days prior to the expiration date of the previous order.

An interim order may be sought by an employer to provide relief pending a decision on the application for a temporary variance. 29 CFR 1905.10(c).

2. Permanent Variance

A permanent variance may be requested when an employer feels that he can prove, by a preponderance of the evidence, that "the conditions, practices, means, methods, operations, or processes used would provide employment and a place of employment which, although different, are as safe and healthful as those required by the standard from which a variance is sought." Section 6(d) of the Act. See also 29 CFR 1905.11(b)(4). A permanent variance allows an employer to accomplish the goals of a standard through means other than those which OSHA has required, *e.g.*, performance versus specification standard. Generally, an employer would apply for such a variance so that he may use less expensive, more effective, or more convenient means of compliance, and accomplish the same result.

Permanent variances can only be granted after a hearing, with notice to employees. The variance order must indicate the methods and operations the employer will use to afford equivalent protection. In lieu of a variance, an employer may be granted relief in the form of a clarification of a standard, which is recognized as and given the same effect as a variance.

The order granting a permanent variance can be modified or revoked at any time after six months from the date of its issuance, upon application by the employer, employees, or OSHA. 29 CFR 1905.13.

An interim order may be sought by an employer, to provide relief while the application for a permanent variance is pending. 29 CFR 1905.11(c).

3. National Defense Variance

A national defense variance can be granted if OSHA determines that it is necessary to avoid serious impairment of the national defense. Section 16 of the Act; 29 CFR 1905.12(b)(4). Such a variance may not remain in effect for more than six months duration without notification to affected employees and an opportunity for a hearing. No action by a state under a plan shall be inconsistent with action by OSHA under Section 16 of the Act. 29 CFR 1952.8. The procedure for Section 16 variances is set forth in 29 CFR 1905.12.

4. Research or Experimental Variance

A research or experimental variance may be useful if the employer is involved in health and safety experiments and may be granted if either the Secretary of Labor or the Secretary of Health, Education, and Welfare certifies that (1) such a variance is necessary to permit an employer to participate in an approved experiment and (2) the experiment is designed to demonstrate or validate new and improved techniques to safeguard the health or safety of employees. Section 6(b)(6)(C) of the Act.

C. THE VARIANCE APPLICATION PROCESS

1. The Application

Although no particular form is prescribed, an application for any type of variance should be carefully drafted—legibly and in writing—in order to avoid delay and possible rejection by OSHA.

The information which should be included in *any* type of variance application consists of: (1) the name and address of the employer; (2) the addresses of the places of employment for which a variance is requested; (3) identification of the particular regulation or standard from which a variance is sought; (4) a statement describing the method used or proposed to be used to protect employees from the hazard covered by the standard; (5) a statement that the employer has informed affected employees of the request for a variance; and (6) a description of *how* employees have been informed of the application (*e.g.*, posting of the application) and of their right to petition for a hearing. Where the variance is applicable to employment in more than one state (including at least one state with an approved state plan) and involves a standard identical to a state standard, the employer must certify that he has not filed for a variance on the same material facts and same employment under the state plan, must include a side-by-side comparison of the standards involved and must state whether any citations have been issued or are pending for violations of the state standard.

An *application for a temporary variance* must include the following additional information: (1) a statement, supported by qualified persons with first-hand knowledge, setting forth the reasons the employer is unable to comply with the standard; (2) a detailed statement of when the employer expects to be able to comply with the standard, and the steps which have been or will be taken to come into compliance; and (3) a statement of the facts upon which the employer relies to establish that he is unable to comply with the standard, that all available steps are being taken to protect employees against the hazards covered by the standard, and that an effective program has been adopted to enable the employer to come into compliance as quickly as practicable. 29 CFR 1905.10(b).

The *application for a permanent variance* must also include a statement describing the conditions, practices, means, methods, operations, or processes used or proposed to be used and detailing how the proposed alternative would provide as safe a place of employment as if the employer were in compliance with the standard from which a variance is sought. 29 CFR 1905.11(b).

Other kinds of information, which, while not required, are helpful and should be included in the application are: (1) an explanation of why the alternative methods the employer proposes to utilize will be as safe as practices required by the standard from which a variance is sought; (2) statements of safety and health experts, accident reports, and employee health histories relating to the specific occupational safety and health issue; (3) confidential information or trade secrets, which, if supportive of the application, should be included but must be clearly marked and a special request made that this material not be copied or communicated to other parties or agencies; and (4) certain kinds of economic information, which, even though technically not considered in determining the propriety of the variance, may be helpful in putting the application into proper perspective (*e.g.*, a description of the particular plant, facility, or process, the cost of abatement and safety equipment, the number of employees involved, the payroll, the number of customers involved, and additional costs which will be incurred as a result of compliance).

If an application for a variance does not technically conform to the above-stated requirements, the Assistant Secretary may deny and return the application without prejudice to the filing of another application. Notice of the denial, accompanied by a brief statement of grounds, will be given to the applicant. If the application is sufficient, a notice of its filing will be published in the Federal Register. This notice includes the terms of the application, a reference to the kind of variance requested, and an invitation to interested persons to comment. It also provides information to affected employers, employees, and appropriate state authorities of their right to request a hearing. OSHA will send the employer a copy of the published notice.

2. Notice and Hearing

Employees must be notified of any requests for variances. This notification may be accomplished by serving a copy of the application on the authorized representative of the employees, if any, by posting a copy of the application on the employee bulletin board, or by placing an article in the employer's newsletter. If the employees or their representative can show that they were not given proper notice, the application may be rejected by OSHA.

OSHA has a timetable for action on adequate variance applications providing for final decisions to be made normally within one hundred twenty (120) days. This time period provides for thirty days from receipt of the application by OSHA to publication of the notice in the Federal Register. Thirty (30) days are then given for public comments. If no comments are received, OSHA schedules a final order to be sent for publication forty-five (45) days later. If comments are received and no hearing is requested, the decision is made in sixty (60) days.

If the variance is approved, the order granting it is sent to the employer and also published in the Federal Register. If the application is denied for any reason, the employer is given prompt notice of that denial with an explanation of the grounds for the denial. The employer must then submit a new variance application containing new information, or, if he wants to contest the denial, he may request a hearing. It must be emphasized that a variance application must set forth an alternative method of compliance or it will be denied as procedurally inadequate within fifteen days,[1] and if there is any potential hazard involved in the operation or workplace as indicated by the variance application, the employer will be referred to the OSHA area office for guidance. All notices of a denial of a variance must be posted for employees to read. If a variance request is denied for a workplace where no citation was involved previously, OSHA will conduct a compliance inspection within thirty days of the denial, but the inspection will cover only the areas included within the variance denial.

An affected party may request and participate in a hearing on a variance application within the time specified in the notice of application. 29 CFR 1905.15. A notice of hearing will then be sent to all affected parties. The hearing which is held before an administrative law judge is a formal adversary proceeding, with sworn testimony, exhibits and cross-examination conducted under the provisions of the Administrative Procedure Act.[2] A record is made of the hearing and all parties are given an opportunity to submit any additional information. The administrative law judge then issues a decision which becomes final within twenty days after it is sent to the parties involved, unless exceptions are taken by way of appeal to the Assistant Secretary of Labor for Occupational Safety and Health. The hearing examiner's decision is not operative pending a decision on the appeal to the Assistant Secretary. 29 CFR 1905.20–.30.

The entire record of the proceeding is sent to the Assistant Secretary of Labor for review whenever a party files exceptions to the administrative law judge's finding of fact, conclusions of law, or terms of the decision. 29 CFR 1905.28–.30. The Assistant Secretary may affirm, modify, or set aside the decision of the hearing examiner and include a statement of reasons therefor. There is no specific timetable for review of the judge's decision after exceptions are filed, but they are usually given prompt attention. Only a decision by the Assistant Secretary is deemed final agency action for purposes of judicial review.

Judicial review of the Assistant Secretary of Labor's decision on the variance application is not specifically provided in either the Act or the regulations. There are, however, two potential avenues of judicial review: (1) after all administrative remedies have been exhausted, an aggrieved party may bring an appeal pursuant to the Administrative Procedure Act, 5 U.S.C. §704, (*i.e.*, the Secretary must have completed his review of the administrative law judge's decision[3]); and (2) pursuant to the doctrine enunciated in *Scanwell Laboratories, Inc. v. Shaffer*, 424 F.2d 859 (D.C. Cir. 1970), in which it was held that, where a *prima facie* showing of illegal agency action is made, a plea that such action is reserved to agency discretion will not be allowed to deny judicial review. In other words, the Assistant Secretary of Labor must follow the pertinent regulations. See also *Service v. Dulles*, 354 U.S. 363 (1957).

The employer's workplace may well be inspected before the variance is granted. OSHA representatives investigating variances are required, pursuant to OSHA Instruction STD–6, to notify employers of workplace hazards observed during the investigation *and* report the hazards to their Area Director for follow-up compliance action. Unless the hazard is abated immediately, the employer will be required to provide "interim protective measures" until it can be corrected. The Area Director will be informed of the method or plan of abatement, and is empowered to take "whatever compliance action is deemed necessary" to determine that the hazard is or will be corrected.

Variance investigations will continue to take place only with the employer's previously obtained consent, and will be limited to areas concerned with the variance request. Variance representatives will not have authority to issue citations and must advise the employer and the authorized employee representative of the investigation in advance, according to the OSHA Instruction.

A compliance inspection may also be conducted within thirty days of a denial of a variance request if a citation was involved previously. The employer, however, would be informed before the inspection and a survey would involve only the operation or process involved in the variance application. Inspections are also required for research/experimental variances and temporary variances. OSHA requires variance inspections before making decisions on variance requests, specifically, those involving hazardous materials, toxic and carcinogenic substances, explosives, and electrical equipment. Also, if an employee requests a hearing on the merits of a variance application, a variance inspection will be made of the situation or operation within fifteen days. OSHA also makes inspections on variance applications where first-hand examination is necessary to obtain further information.

3. The Interim Order

An interim order may be requested by an employer either at the same time or after the application for a variance has been made; the mere filing of the application does not excuse compliance with the Act. In the former situation, the interim order may be attached to the application for either a permanent or a temporary variance. 29 CFR 1905.10(c) and .11(c). Since an employer is still exposed to the possibility of inspections, further citations, and penalties for violations of the standard from which a variance is sought during the time period involved in the variance application and approval process, the interim order can become quite important, since it excuses compliance until final action is taken on the variance. An interim order will be effective until a decision is rendered on the variance application itself. 29 CFR 1905.10(c) and 11(c). In fact, the Review Commission has stayed abatement requirements, pending the Secretary of Labor's action on a permanent variance application. *Deemer Steel Casting Co.*, 2 OSHC 1577 (1975). It has also been held that an employer could not be required to abate during the pendency of its variance application. *Ensign Electric Division, Harvey Hubbell, Inc.*, 1973–74 OSHD ¶ 18,261(1974); but see *Midvale-Heppenstall Co.*, 1 OSHC 1155 (1973), which held that OSHA's denial of an application for a variance precluded the extension of the abatement period.

The application for an interim order should contain: (1) a statement of the facts which support the granting of an order, (2) a statement of the protective steps taken by the employer to guard employees from the hazards covered by the standard, and (3) assurances that employee safety will not be threatened if the interim order is granted.

Interim orders can be granted expeditiously and *ex parte* by the Assistant Secretary even while the employer is contesting a citation based on the standard from which the variance is sought. See *Bethlehem Steel Corp.*, 1 OSHC 3087 (1972), where the Review Commission withdrew a citation because an interim order had been granted during the contest period; *cf.*, *Star Textile and Research, Inc.*, 2 OSHC 1697 (1975), where a citation was vacated because a permanent variance from the applicable standard was granted while the citation was being contested.

A hearing is not required for an interim order, although affected parties must be given notice of the application and may request a hearing. The Secretary of Labor may convene a conference, however, at which the applicants and affected parties will seek to resolve any differences.

If OSHA denies the request for an interim order, the employer will be promptly notified of the reason for that denial. If the interim order is granted, the employer and other parties concerned are served with a copy of the order, and the terms of the order will be published in the Federal Register. The employer must inform his or her employees of the order by giving a copy to the authorized employee representative and by posting a copy of the order at the location where notices are customarily posted.

4. Federal and State Variances from Identical Standards

If an employer's plant or operation is located in a state with an approved state plan, the variance application should be made to the state authority. If the employer's facility is in a state with no OSHA approved plan, the variance application should be submitted to OSHA. An employer with operations in more than one state, however, may be subject to standards under the Act as well as standards under approved state plans. If the state(s) standards are identical to a federal standard addressed to the same hazard, the employer may elect to apply to OSHA for a variance, pursuant to the provisions of 29 CFR 1905. According to OSHA regulations, state and federal variances are mutually recognizable whenever the respective standards are identical and are addressed to the same hazard. 29 CFR 1952.9.

Employers, therefore, have the option of applying for a variance from identical standards in either a federal or state forum and also may apply for, and OSHA may issue, a multistate variance.[4] 29 CFR 1905.13 and 1952.9. The variance application, wherever filed, must certify, however, that it has not been previously filed, denied, modified, or revoked in another forum, for any action by one authority has a *res judicata* effect on the other.[5] Filing a variance application with either a federal or state agency precludes any further consideration of the matter by the other agency (29 CFR 1952.9(c). Where a federal variance has been granted with multistate applicability, such variance shall be deemed an authoritative interpretation of the employer's compliance obligations with respect to the state standard.

5. Modification, Revocation, and Renewal

An affected employer or employee may apply for a modification, renewal, or revocation of a permanent variance order issued under Section 6(b)(6)(A), 6(d) or 16 of the Act at any time during the first six months of issuance. The application must contain certain information, such as a description of the action and reasons therefore and a statement that employees have been informed. Hearings, if requested, follow the same rules as those for an original variance application. If approved, the decision is published in the Federal Register. Temporary or research experimental variances may be renewed twice, but the request for extension renewal must be filed ninety days before the expiration date of the order, and interim renewals may not remain in effect for longer than one hundred eighty days. Defense variances may not be extended beyond six months without giving affected employees notice. 29 CFR 1905.13.

D. EFFECT OF VARIANCE APPROVAL

Every final action granting a variance must be published in the Federal Register, and must specify the alternative to the standard involved which the particular variance permits. 29 CFR 1905.6. A variance is not retroactive; it has only prospective effect from the date of issuance. Furthermore, approval of a variance does not automatically obviate a citation previously issued for violation of the standard in question. Thus while variance approval may render the issues in contest moot, a pending request for a variance does not affect an employer's liability for violation of a standard which existed at the time of the inspection. *Star Textile & Research, Inc.*, 2 OSHC 1697 (1975); *Stoughton Body, Inc.*, 3 OSHC 1359 (1975). *Safeway Stores, Inc.*, 2 OSHC 1439 (1974). OSHA also may, in its discretion, decline to entertain a variance application while citation proceedings (state or federal) are pending against the employer with respect to the standard at issue. 29 CFR §1905.5. In such event, an employer should contest the citation or challenge the appropriateness of the abatement. If this is not possible (*e.g.*, the contest period has run or a final order was entered), a petition for modification of abatement should be filed.

An employer who has been cited for violation of a standard may not be able to seek relief from that citation by applying for a variance, since the regulations permit OSHA to decline a variance application in such circumstances. *George A. Hormel & Co.*, 2 OSHC 1190 (1974), *den. pet. for reconsid.*, 2 OSHC 1282 (1974); *Western Steel Mfg. Co.*, 4 OSHC 1640 (1976); 29 CFR 1905.5. However, a variance proceeding which may change future methods of compliance may warrant an extension of the abatement period and may, upon approval, cause dismissal of the issues in contest. *Deemer Steel Casting Co.*, 2 OSHC 1577 (1975); *Star Textile & Research, Inc.*, 2 OSHC 1697 (1975). Alternately, an employer should also consider filing a notice of contest challenging the appropriateness of the abatement period or if the citation has resulted in the entry of a final order, filing a petition for modification of abatement (PMA) under Section 10(c) of the Act.

In order to establish and prove the affirmative defenses of infeasibility, impossibility of performance, greater hazard, or jurisdiction of another agency, an employer must first establish that an application for a variance was made (and rejected) or would have been inappropriate under the circumstances. *Atlantic & Gulf Stevadores, Inc. v. OSHRC*, 534 F.2d 541, 4 OSHC 1061 (3d Cir. 1976); *Noblecraft Industries, Inc. v. Secretary of Labor*, 614 F.2d 199, 7 OSHC 2059 (9th Cir. 1980); *Diebold, Inc. v. Marshall*, 585 F.2d 1327, 6 OSHC 2002 (6th Cir. 1978); *General Electric Co. v. Secretary of Labor and OSHRC*, 576 F.2d 558, 6 OSHC 1541 (3d Cir. 1978); *Meilman Food Industries*, 5 OSHC 2060 (1977); *"AH" Metal Fabricators, Inc.*, 5 OSHC 1403 (1977); *Bellmonte Finishing Co., Inc.*, 6 OSHC 1416 (1978). It is the Commission's policy to reject these defenses where an employer fails to prove the unavailability of a variance. *General Electric Co. v. Secretary of Labor and OSAHRC*, 576 F.2d 558, 6 OSHC 1541 (3d Cir. 1978); *Irwin Steel Erectors, Inc. v. OSAHRC*, 574 F.2d 222, 6 OSHC 1601 (5th Cir. 1978); *Hood Sailmakers, Inc.*, 6 OSHC 1206 (1977). See also *Sierra Construction Corp.*, 6 OSHC 1278 (1978), ruling that an employer's use of his own signaling system does not eliminate the need for compliance with posting requirements or absent an application for a variance from the standard. But see *Fred Wilson Drilling Company, Inc. v. Marshall*, 624 F.2d 38, 8 OSHC 1921 (5th Cir. 1980).

NOTES

1. A variance application which is considered procedurally inadequate (*e.g.*, does not state an alternative method of compliance or does not identify steps to safeguard employees) will be denied within fifteen days.

2. The regulations provide that a party may, at least twenty days prior to the date of the hearing, move for a summary decision in its favor. The hearing examiner may

then in his discretion set the matter for argument. If no genuine issue of material fact is found to have been raised, the hearing examiner may issue a decision. The other party may then file exceptions which will be reviewed by the Assistant Secretary. 29 CFR 1905.40–.41.

3. A hearing examiner's decision is not operative, pending a decision on the appeal to the Assistant Secretary of Labor. Only a decision by the Assistant Secretary is deemed to be final agency action for purposes of judicial review. A decision by a hearing examiner which becomes final for purposes of appeal to the Assistant Secretary of Labor is not deemed final agency action for purposes of 5 U.S.C. §704.

4. Information in addition to that set forth above, the Variance Application Process, should be included when a multistate variance is filed: (1) side by side comparison of the standards; (2) certification that employer has not applied for the variance with the state agency; and (3) statement regarding any pending citations for violation of the same standard. 29 CFR 10(b)(11) and 11(b)(8).

5. "The filing with, as well as granting, denial, modification, or revocation of a variance request or interim order by either authority (Federal or State) shall preclude any further substantive consideration of such application on the same material facts for the same employment or place of employment by the other authority." 29 CFR 1952.9(c).

APPENDIX 1 to CHAPTER V
Model of an Application for Variance from Standards Adopted by the Occupational Safety and Health Administration

The following model of an application for a variance from a standard or standards issued by the Occupational Safety and Health Administration was prepared by OSHA. It includes all required information for a variance application listed in the regulations that appear on the preceding pages.

Assistant Secretary of Labor
for Occupational Safety and Health
United States Department of Labor
Washington, D.C. 20210

Dear Sir/Madam:

Pursuant to Section 6(d) of the Occupational Safety and Health Act, 29 U.S.C. 655, ________________(Employer/Applicant) respectfully requests a permanent variance from the requirements of 29 CFR 1910, concerning the ________ on the ________________of a gantry crane.

(1) Employer/Applicant.

(2) Place of employment.

(3) Employer/Applicant has a gantry crane which operates beside a slip for unloading shallow draft barges. It currently is used only for unloading salt barges, usually about twice a year. Because of the location of the operation and the experience we have had with high-wind conditions, Employer recognizes the importance for monitoring wind conditions. As a result, our plant has two wind-indicating devices with visual and automatic recording devices. We have a well-established plan for monitoring federal Weather Bureau reports and emergency plans of action for pending high winds. The key wind-indicating device is located in the Shift Superintendent's office. A Shift Superintendent is on duty around the clock, seven days a week, and is responsible for the plant. He keeps a close watch on the wind conditions and also monitors weather forecasts at least

three times during each eight-hour period. A written record is kept of the weather conditions. If there are any changes in wind conditions or anticipated changes due to approaching bad weather, the wind indicators and weather forecast are monitored continuously. Because of these well-established policies and the limited use of the gantry crane, we feel that our plan would afford a safer operation. In addition, we feel that the installation of a wind-indicating device on the gantry crane would not be feasible because of conditions which tend to cause malfunction. Our past experience has verified this consistently. The gantry crane is very close to the molten sulfur pits and salt water. The close proximity to these agents creates a corrosive condition which would make the wind indicator inoperative soon after installation. We feel our monitoring and early warning system is more effective and safe than is the use of an infrequently used and difficult-to-maintain wind velocity instrument on the crane.

(4) Due to the corrosive conditions present around the gantry crane, ______ feels that our established procedures would give more reliable information than an indicator on the crane. Therefore, we feel that this would insure the safety of the crane and operator better by following our procedures.

Respectfully submitted,

____

Manager

(5) This is to certify that a copy of the variance from 29 CFR 1910.179(b)(4) requested by ________________has been posted in an appropriate place for employees to examine. Also, an explanation of the employees' right to send comments or petition for a hearing are explained on an attached sheet. A copy of the variance application was also given to the authorized employee representative.

Posted on ____________, 19_____

____

Manager

(6) NOTICE TO ALL EMPLOYEES

The attached form is a copy of the variance application submitted by ______ for a variance from 29 CFR 1910.179(b)(4). This is to inform you that you may request a hearing with the Assistant Secretary of Labor for Occupational Safety and Health, United States Department of Labor, according to 29 CFR 1905.15, Request for Hearing, if you object to the variance. This request must be filed in writing.

____

Manager

(7) ________________ also respectfully requests that an interim order be granted to allow operation of the gantry crane without the indicator until action is

taken on the variance application. We feel that the safety of the employees will be guaranteed because we do have wind indicators being monitored at our plant and the gantry operator would be promptly informed before wind velocities rose or approached a dangerous level.

____

Manager

CHAPTER VI
Rights and Duties

A. INTRODUCTION

In 1970, Congress enacted the Act "to assure so far as possible every working man and woman in the nation safe and healthful working conditions and to preserve our human resources." To accomplish this purpose, Congress established dual employer duties under OSHA. First, Congress conferred upon the Secretary of Labor authority to establish specific standards for safe and healthful conditions of employment. Each covered employer has the duty of complying with all safety and health standards promulgated by OSHA which are applicable to the employer's place of employment. Second, in all cases not covered by a specific standard, an employer has a duty to furnish to each of his employees employment and a place of employment which are free from recognized hazards that are causing or likely to cause death or serious physical harm to his employees. Employees, too, have an obligation to comply with the Act; however, this obligation is neither enforced nor enforceable. Employer rights guaranteed under the Act are less comprehensive than are the duties; on the other hand, employee rights are more substantive than the correlative duties.

While the approach utilized in the other chapters of the book is to provide a detailed and complete explanation of the topic and contents thereof, in this chapter the discussion will of necessity be in outline form with little or no legal support since this subject is detailed throughout the other chapters. Reference will, of course, be made to those chapters of the book where the matter is treated substantively.

B. WHO ARE EMPLOYERS AND EMPLOYEES?

Section 4(a) of the Act provides that the obligations and responsibilities set forth in the Act are applicable with respect to employment performed in a "workplace." The term "workplace," although it is not defined in the Act, except obliquely by reference to Section 8(a)(1), has been given broad judicial interpretation. *REA Express, Inc. v. Brennan*, 495 F.2d 822, 1 OSHC 1651 (2d Cir. 1974); *Clarkson Construction Company v. OSHRC*, 531 F.2d 451, 3 OSHC 1880 (10th Cir. 1976). Basically, the term is defined as any area subject to an employer's control.

The determination of whether a particular party is an employer or an employee is largely a factual question. *W. J. Barney Corp.*, 4 OSHC 1119 (1976). The definitions set forth in Sections 3(4), (5) and (6) of the Act are legal niceties, but offer no substantive guidance. According to Commission and court precedent, the principal factor in determining whether one is an employer is who has the ability to abate the violate condition or otherwise protect or control the employees or their working condition. In this respect the Review Commission will look generally to see (1) who the employees consider to be their employer, (2) who pays their wages, and (3) who is responsible for controlling the employees' activities. *Gordon Construction Co.*, 4 OSHC 1581 (1976); *Lidstrom, Inc.*, 4 OSHC 1041 (1976); *Giamoi Bros., Inc.*, 5 OSHC 1831 (1977); *Manpower Temporary Services, Inc.*, 5 OSHC 1803 (1977). The officers and directors of a Virginia company which manufactured the pesticide Kepone were found to be personally liable as "employers" for violations of the Act cited at the company's plant during the ten-week period between dissolution and reinstatement of the firm's corporate charter. *W. P. Moore v. OSHRC and Marshall,* 591 F.2d 991, 7 OSHC 1031 (4th Cir. 1979).

The determination of whom is considered to be an employee is similarly left to Commission and court interpretation. Common law, contractual, and statutory concepts have been rejected; but it appears that one factor in the determination of employee status is the performance of work for an employer. It has been determined that supervisors, foremen, stockholders, owners, and officers can be employees for purposes of the Act. *FEC, Inc.*, 1 OSHC 3043 (1972); *Magnus Firearms,* 3 OSHC 1214 (1975).

C. DUTIES OF EMPLOYERS

1. Compliance with the Act

The basic source of OSHA obligations imposed upon employers is Section 5(a) of the Act which provides that each employer: (1) shall furnish to each of his employees employment and a place of employment which are free from recognized hazards that are causing or are likely to cause death or serious physical harm to his employees, and (2) shall comply with occupational safety and health standards promulgated under this Act. OSHA assures compliance by means of conducting inspections, from which citations, monetary penalties, and abatement requirements may result. The Act's purpose is to eliminate and protect employees from hazardous conditions by means of this compliance obligation.

a. *General Duty Clause*

Section 5(a)(1) of the Act, known as the general duty clause, requires that each employer covered by the Act furnish to each of his employees employment and a place of employment which are free from recognized hazards that are causing or likely to cause death or serious physical harm to his employees. *National Realty and Construction Co. v. OSHRC,* 489 F.2d 1257, 1 OSHC 1422 (D.C. Cir. 1973); *American Smelting & Refining Co. v. OSHRC,* 501 F.2d 504, 2 OSHC 1041 (8th Cir. 1974); *Georgia Electric Company v. Marshall,*

595 F.2d 309, 7 OSHC 1343 (5th Cir. 1979). Being remedial and preventive in nature the Act's general duty clause has been liberally construed in favor of employees "whom it was designed to protect." *Southern Railway Co. v. OSHRC,* 539 F.2d 335, 3 OSHC 1940 (4th Cir. 1976).

The clause obligates employers to provide a safe workplace free from recognized hazards and to avoid and/or eliminate hazardous conditions not covered by a specific standard. Where a specific standard covers a particular hazard or condition, the general duty clause may not be used as basis for citation. *American Smelting & Refining Company v. OSHRC,* 501 F.2d 504, 2 OSHC 1047 (8th Cir. 1974); *Dan J. Brutger,* 4 OSHC 1745 (1976).

Citations based on the general duty clause are limited to serious violations, including willful or repeated violations. Compare in this respect Section 17(k) of the Act which defines a serious violation with Section 5(a)(1).

Abatement of a general duty clause hazard must be feasible (that is, the method of abatement must be such as to materially reduce the injury that would result from the condition cited as violative), and the burden of proving feasibility rests with the Secretary of Labor. If this burden is met, then the employer must show that the use of the methods established by the Secretary would cause consequences so adverse as to render their use infeasible.

Finally, as in all other alleged violations of the Act, exposure to a hazard must be shown, but "access" to the zone of danger by employees is deemed sufficient for citation. Substantive aspects of the general duty clause are discussed in Chapter VII.

b. *Specific Standards*

Section 5(a)(2) requires that employers comply with the thousands of specific standards promulgated under the Act and incorporated by reference into the Act. The aspects of a violation of a specific standard include the following: (1) the terms of the standard must have been violated by a condition existing at the worksite; (2) there must be employee exposure or access to the zone of danger; and (3) there must be actual or constructive employer knowledge of the hazard or violation. *Brennan v. Alsea Lumber Co.,* 511 F.2d 1139, 2 OSHC 1646 (9th Cir. 1975); *REA Express, Inc. v. Brennan,* 495 F.2d 822, 1 OSHC 1651 (2d Cir. 1974). Employers should indeed be familiar with all OSHA standards applicable to their operations. They should also become aware of specific posting, labeling, and other particular requirements (*e.g.,* provide medical examinations, monitor the workplace, give employee training) which may be required by those standards. One approach recommended in this respect is to conduct a standards search or utilize the OSHA or private on-site consultation service. Employers may not deviate from strict compliance with standards unless a variance has been obtained.

2. Supervision and Training

Normally, the act or failure to act of a supervisor is considered to be the act of the employer. Knowledge on the part of a supervisor is normally imputed to the employer and where a supervisor's order requires employees to carry out duties under hazardous conditions, the employer will normally be held strictly liable.

Supervisors are required to do more than simply orally warn employees of the existence of a hazard; they may also have an obligation to see that a hazardous condition is eliminated. The degree of supervision required, however, may vary from situation to situation.

An employer, through his supervisors, cannot fail to properly instruct, train, and supervise its employees and/or cannot hide behind its lack of knowledge concerning their dangerous working practices. *Danco Construction Company v. OSAHRC,* 586 F.2d 1243, 6 OSHC 2039 (8th Cir., 1978). This is especially true if an employer cannot provide its employees with continuous, on-site supervision.

Employers are also required to ensure that employees have proper tools and personal protective equipment and that the equipment is used, not merely provided. Machinery may be required to be guarded or modified to ensure safety. Tools, machinery, and equipment must be properly maintained and in safe, good working condition. Although OSHA under Section 21(c) of the Act is directed to provide training programs and consultation services for employers, employees, and labor organizations about preventing workplace injuries and illnesses, private training and instruction is of primary importance. Chapters IV, VIII, XII, and XVII discuss various aspects of an employer's obligations vis-à-vis its employees.

3. Recordkeeping, Reporting, and Posting

Numerous requirements are imposed upon employers to assure an adequate flow and dissemination of occupational safety and health information to both employees and OSHA, *e.g.,* posting of certain informative materials in the workplace such as the workplace poster, summaries of petitions for variance, citations, and other notices.

Employers must record and report certain occupational injuries, illnesses, and deaths, and must notify employees of their rights under certain standards rules and regulations. Some standards also require that employers keep records of employee exposure to potentially toxic materials or harmful physical agents. Other standards require that employees be given access to their medical records, exposure records, and analyses of employee medical and exposure records. To the same extent, certain standards require that records must be maintained and preserved for certain periods of time.

Reporting duties are also contained in the Act. For example, employers are required to report to OSHA within forty-eight hours the occurrence of an accident which is fatal to one or more employees or which results in the hospitalization of five or more employees. Finally, some employees may be required to participate in the BLS annual survey. Employer recordkeeping, recording, and posting obligations are discussed in Chapter VII.

4. Nondiscrimination

Section 11(c) of the Act prohibits disciplinary actions against employees for engaging in certain protected activities, such as instituting a complaint or testifying under the Act. Chapter XV discusses the discrimination protection afforded by the Act.

5. Abatement and Penalties

If hazardous locations or conditions are discovered either in enforcement or other proceedings, employers must abate, guard, or remove them. This duty is heighted when health hazards such as exposure to lead, benzene, coke oven emissions, and asbestos are involved. Periodic monitoring and testing may be required to determine and, if necessary, eliminate the presence of toxic or hazardous substances. In addition to abatement, penalties may be assessed for a violation of the Act. Chapter X contains a discussion on this aspect of an employer's obligation.

D. DUTIES OF EMPLOYEES

Section 5(b) of the Act requires that employees shall comply with occupational safety and health standards and all rules, regulations, and orders issued pursuant to the Act which are applicable to their own actions and conduct. Despite the language of the Act, employees are not subject to any legal sanctions for failure to observe applicable standards or rules or for employee misconduct. *Atlantic & Gulf Stevedores, Inc. v. OSHRC,* 534 F.2d 541, 4 OSHC 1061 (3d Cir. 1976). Rather, the matter of employee compliance is one left to the collective bargaining process. The practical significance to employers of the Section 5(b) duty of employees is small, although the Commission, courts, and labor organizations have acknowledged an employer's right to discipline employees for violation of their duty.[1]

Notwithstanding the language of the Act, the OSHA *Field Operations Manual* states that where an employee's misconduct results in a violation of the Act, the employer will only be answerable when "demonstrably feasible measures" existed for reducing its incidence. Accordingly, an employer is not strictly liable for the acts of employees. *National Realty & Constr. Co. v. OSHRC,* 489 F.2d 1757, 1 OSHC 1422 (D.C. Cir. 1973). Generally then, to avoid liability under the Act for the acts of its employees, an employer has to show that its employee's act (breach of duty) was unpredictable and unforeseeable. Chapter XVII contains information on an employer's relationship with its employees in the labor relations context; Chapter XII reviews this relationship in the context of affirmative defenses to OSHA citations.

E. RIGHTS OF EMPLOYERS

1. Inspections

Section 8(a) of the Act, which provides for the inspection of workplaces by OSHA compliance officers or inspectors, states:

> In order to carry out the purposes of this Act, the Secretary, upon presenting appropriate credentials to the owner, operator, or agent in charge, is authorized:
> i. to enter without delay and at reasonable times any factory, plant, establishment, construction site, or other area, workplace or environment where work is performed by an employee of an employer; and
> ii. to inspect and investigate during regular working hours and at other reasonable times, and within reasonable limits and in a reasonable manner, any such place of employment and all pertinent conditions, structures, machines, apparatus, devices, equipment, and materials therein, and to question privately any such employer, owner, operator, agent or employee.

Although employers are required to submit to OSHA inspections, they have the initial right to require the inspector to present his credentials and to otherwise inspect at reasonable times, in a reasonable manner, and within reasonable limits. Employers have no right to receive advance notice of an inspection.

Warrantless inspections to enforce the provisions of the Act have been held to be unconstitutional within the meaning of the Fourth Amendment; OSHA may thus be required by the employer to obtain a search warrant prior to conducting an inspection of a company's premises. *Marshall v. Barlows, Inc.,* 436 U.S. 307, 6 OSHC 1571 (1978).

a. *Opening Conference*

The OSHA inspector must conduct an opening conference with the employer or employer representative. 29 CFR 1903.7(a). At the conference, the purpose and procedures for conducting the inspection will be described and discussed. The employer will be given copies of the Act, applicable standards, and the employee complaint if one has been lodged. Failure to conduct an opening conference will not be grounds for vacating a subsequently issued citation unless prejudice can be shown.

b. *Walkaround*

Employer representatives (different representatives may accompany any inspector during different phases of inspection) are permitted to accompany the inspector during workplace inspections. Section 8(e) of the Act. The "walkaround" right is mandatory and the failure to accord this right may require a vacation of citation if prejudice evolves.

c. *Trade Secrets*

The employer has the right to designate those areas of his establishment which contain and which may reveal trade secrets during an inspection. Employers may request that an employee representative entering a trade secret area be an employee that works in that area or one authorized by the employer. All information (photographs, samples, and other material) will be labeled "Confidential Trade Secrets" by the inspector. Trade secret information will not be disclosed except as prescribed by Section 15 of the Act. Trade secrets include confidential formulas, patterns, devices, or information which give the employer a competitive advantage over those who do not possess the information.

d. *Closing Conference*

After the inspection tour the inspector will hold a closing conference with the employer representative. 29 CFR 1903.7(f). The employer will be informed of all conditions and practices which may constitute violations as well as the applicable standards involved. The inspector will not indicate any proposed penalties; only the Area Director has that authority. Usually, however, the employer will be informed as to the enforcement action which may be taken. Although a hazard may not violate a particular standard or the general duty clause, it may still be brought to the employer's attention at the closing conference.

Employee representatives may, but generally do not attend the same closing conferences as the employer representative; separate closing conferences may be held. The failure to hold a closing conference will not be grounds for vacating a citation unless prejudice can be shown.

The procedural and substantive aspects of the inspection are discussed in Chapter IX.

2. Citations

After the inspector makes a report to the OSHA Area Director, he will determine which citations, if any, will be issued, and which penalties and abatement, if any, will be proposed. Citations inform the employer and employees of the regulations, rules, and standards which have been violated and of the time set for their abatement. The employer will receive citations and notices of proposed penalties by certified mail promptly, usually within four to six weeks after the inspection, but not more than six months thereafter. Citations must be written with particularity. The employer must post a copy of each citation at or near the place the violation occurred for three days or until the violation is abated, whichever is longer. The issuance of citations is discussed in Chapters X and XI.

3. Informal Conference

Subsequent to the issuance of a citation, the employer may request an informal conference with the Area Director to discuss the alleged violations. *Note:* This right does not stay the fifteen-working-day notice of contest period, or the abatement requirements of the citation.

The informal conference may be beneficial to the employer as a means of settlement. The employer may be represented by counsel at the informal conference. If the fifteen-day notice of contest period has expired and an employer has filed a notice of contest, the Regional Solicitor's Office must be consulted before a conference will be held and a representative from that office will attend any authorized conference.

The informal conference is discussed in Chapter XI.

4. Contest of Citations and Hearing

An employer who believes a citation and/or accompanying penalty to be improper may, within fifteen working days after receipt of the citation, file a notice of contest with the Area Director. (Employers who wish to challenge only the reasonableness of the abatement requirements of a citation may wish to file a Petition for Modification of Abatement). Sections 10(a) and (c) of the Act. Shortly after receipt of the notice of contest, the Solicitor of Labor will file a complaint and the employer will respond by filing an answer in which it can, indeed, must raise affirmative defenses. Subsequently the matter will be set down for hearing.

An employer who contests a citation has a right to a fair and impartial hearing before an administrative law judge. Prior to the hearing the employer may exchange information and obtain discovery from the Solicitor of Labor, file subpoenas and FOIA requests in preparation for the proper presentation of his case, and may participate in prehearing or settlement conferences. If the case is not settled, at the hearing the employer will be afforded the right to present witnesses, exhibits, and other evidence in support of his position. Upon the conclusion of the hearing, the judge will issue a decision.

Procedural aspects of an employer's contest are discussed in Chapters XI, XII, and XIII.

5. Review by the Commission and Courts

An employer may request review of an adverse administrative law judge's decision by the Review Commission. Whether review will be granted is within the discretion of the Commission, and, consequently, is not strictly a matter of right. Unlike a request for review to the Commission, the right to appeal a final order is of the Review Commission to the appropriate United States Court of Appeals is a prerogative of the employer. Section 11(a) of the Act. Chapter XIII discusses the employer's rights in this regard.

6. Standards Promulgation

Prior to standard issuance and promulgation by OSHA, employers may submit data, information, and comments for its consideration. Once a proposed standard is published an employer may (1) submit data to support or to discredit the proposal, (2) confer with OSHA officials regarding the standard to explain industry's viewpoint, (3) submit comments which examine and analyze the issues involved, and (4) request and participate in a public hearing on the standard. Due process, notice, and fairness are required before an employer can be held to the requirements of a standard.

Any person who is adversely affected by a standard may file a petition challenging its validity at any time prior to the sixtieth day after issuance. The petition must be filed with the United States Court of Appeals wherein the person resides or has his principal place of business. Section 6(f) of the Act. Standards may also be challenged in enforcement proceedings.

Chapters IV and XIII discuss an employer's rights with respect to standards promulgation.

7. Seek Variance

An employer may seek a variance from a particular standard where he needs additional time to comply therewith or where he can show his existing practices or operations are as safe or healthful as those required by the standard. Variances are discussed in Chapter V.

8. Discipline Employees

Disciplinary measures may be taken by an employer to enforce compliance with the Act or in response to an employee's refusal to comply with safety rules and regulations. 29 CFR 1977.6. Discipline for these reasons will not be considered discrimination within the meaning of Section 11(c) of the Act. According to a recent Supreme Court decision, employers may not discipline employees if they walk off the job because of potentially unsafe working conditions, provided that the employees' refusal is reasonable and based upon objective facts and circumstances. *Whirlpool Corporation v. Marshall,* 435 U.S. 1, 8 OSHC 1001 (1980).

9. Consultation Services

In order to educate employers and to help them achieve voluntary compliance with the Act, on-site consultative services are available in certain instances. Consultation services are discussed in Chapter III.

10. NIOSH

Employers may request the assistance of NIOSH to conduct health-hazard evaluations or to provide technical services. Chapter XVI contains a detailed discussion of NIOSH's functions.

F. RIGHTS OF EMPLOYEES

1. Request Inspection and File Complaints

Any employee or representative of employees who believe that a violation of a safety or health standard exists that threatens physical harm or that an imminent danger exists may complain to OSHA (and retain anonymity) and request an inspection by giving notice to the Secretary of Labor or his authorized representative of such violation or danger. Section 8(f)(1) of the Act. If an inspector fails to issue a citation, an employee has the right to obtain reivew if he has provided a written statement of alleged violations. Section 8(f)(2). The Act does not, however, allow an employee to appeal to the Commission if OSHA refuses to inspect or issue a citation in response to an employee's complaint. Chapter IX discusses employees' rights to this respect.

2. Protection against Discrimination

Section 11(c) of the Act provides that employees may not be discharged or otherwise disciplined for utilizing the Act's procedures, instituting a proceeding under the Act, testifying under the Act, or filing a complaint under the Act. Any employee who feels he has been a victim of discrimination may file a complaint with the Secretary of Labor. Chapter XV discusses the substantive aspects of OSHA discrimination.

3. Participation in Inspection

The Act gives employees, as well as employers, the right to participate in "walkaround" inspections. Section 8(e) of the Act. Before an OSHA inspector proceeds through a workplace to check for safety and health hazards, the employer will appoint his representative to accompany the inspector. An authorized representative of the employees is also given the opportunity to accompany the OSHA inspector. If employees are represented by a collective bargaining representative, the union ordinarily would designate the employee representative. Similarly, if there is a plant safety and health committee, the employee members of that committee would ordinarily designate the representative. Where neither employee group exists, the employee representative may be selected by the employees themselves, but not by the OSHA inspector or the employer. Employee accompaniment is utilized to identify where and how various toxic substances are used or hazardous conditions maintained, and otherwise generally help OSHA conduct an inspection.

Where no authorized employee representative is chosen, the OSHA inspector is required to consult privately with a reasonable number of employees concerning safety and health matters in the workplace. Employees have the right to respond to questions the OSHA inspector may ask, and may bring to his attention any conditions that they believe violate the safety and health regulations of the Act. They also have the right to observe monitoring and sampling conducted by the inspector. Section 8(c)(3) of the Act. Employees are also encouraged to request participation in opening, closing, and informal conferences.

Chapter IX sets forth information relative to employee roles during inspections.

4. Compensation for Walkaround Time

OSHA issued an interpretative regulation of Section 11(c), the antidiscrimination provision of the Act, which required payment to employees for normal work time spent accompanying OSHA representatives conducting workplace inspections, including time spent participating in opening and closing conferences. 29 CFR 1977.21.

The courts have held, however, that time spent by employee representatives during a walkaround is *not* "compensable hours worked" under the Fair Labor Standards Act, since it is for the employees' benefit, it is voluntary, and the employer neither requires, supervises, or controls the employee during this time. *Leone v. Mobil Oil Corp.*, 523 F.2d 1153, 3 OSHC 1715 (D.C. Cir. 1975); *Chamber of Commerce of the United States v. OSHA*, 636 F.2d 464, 8 OSHC 1648 (D.C. Cir. 1980). Chapters IX and XV discuss walkaround pay "rights."

5. Imminent Danger

An employee has a right to objectively determine, in fact, that a workplace practice or condition presents an imminent-danger situation and to refuse to work upon making such a determination. Employees may thus request OSHA to immediately inspect an employer's premises when they believe an imminent danger exists at the workplace and provide the inspector with information during the inspection. Employees also have a ("limited") right to refuse to work in conditions of "perceived" imminent danger and according to some courts, to be paid for the time lost if an imminent-danger situation was in fact determined to exist. When OSHA fails to seek immediate abatement of an imminent danger, employees or their representative may seek a writ of mandamus in federal district court to compel the Secretary of Labor to act. Section 13 (d) of the Act.

6. Information, Notice, and Party Status

Employees, whether represented or unrepresented, have a right to receive (or have posted at the place where the violation occurred) certain information during enforcement proceedings, including copies of the citation issued to the employer, copies of the notice of contest filed by the employer, and copies of the pleadings and other documents filed prior to, during, and after the hearing. Sections 8(c)(1) and 9(b) of the Act. Under Section 10(c), affected employees may elect to participate in hearings as parties or interveners. Election of party status entitles the employees or their authorized representative to meaningful participation in certain aspects of the contest process, including settlements. Employees must, however, notify the Commission of their desire to participate.

Interested employees or their representatives may also petition the Commission to intervene in particular cases. The right to intervene will be granted when sufficient interest is shown and when participation will be helpful to a determination of the issues without causing undue delay in the proceedings. Chapters XI and XIII discuss employee rights in this regard.

7. Contest Abatement Period

Employees may contest the "reasonableness" of any abatement period set forth in a citation. Section 10(c) of the Act. Employees may also contest an employer's petition for modification of abatement. Recent Commission decisions which seem to expand the scope of this right are discussed in Chapters XI and XIII.

8. Access to Information

Certain standards, such as the Access to Employee Medical and Exposure Record Standard, and the general standards on recordkeeping and reporting as well as specific standards on toxic and hazardous substances afford specific rights relative to employee access to the information contained therein. Chapter VII discusses these employee access rights.

9. Participation in Standards Development and Variance Procedure

Pursuant to Sections 10(c), 6(b) and (f) employees, like employers, may participate in all proceedings relevant to standards promulgation or variance approval, and challenge such standards in the courts, as outlined in Chapters IV and V.

10. Request Health Hazard Evaluation

Under Section 20(a)(6) of the Act employees may request NIOSH to conduct a health-hazard evaluation of their place of employment. Chapter XVI sets forth the employee rights in this regard.

NOTE

1. A pamphlet entitled *Workers' Rights Under OSHA* published by the Department of Labor states in this respect: "There are no legal sanctions for employees who violate

OSHA standards or in other way obstruct the purpose of the Act. Congress refrained from interfering in the traditional employer-employee relationship. However, the Act specifically requires you to comply with all safety and health rules and regulations that apply to your own actions and conduct on the job. You are expected to comply with all internal or administrative safety and health regulations issued by your employer, including the wearing or using of protective gear and equipment. Failure to do so may result in your employer's reprimanding, suspending, or dismissing you."

CHAPTER VII
Recordkeeping, Reporting, and Posting

A. INTRODUCTION

Prior to the enactment of the Act, accident, illness, and injury reporting varied significantly among the states, and what was available was usually kept in a worker's compensation context. Congress included specific recordkeeping and reporting provisions in the Act in order to provide OSHA and NIOSH with a source of information to aid in enforcement. Posting provisions were also provided to aid workers in their understanding of the Act and its protections.

B. STATUTORY AUTHORITY

The Act contains specific authorization for OSHA to promulgate regulations on recordkeeping, reporting, and posting. See *e.g., Louisiana Chemical Association v. Bingham,* 496 F. Supp. 1188, 8 OSHC 1950 (W.D. La. 1980).[1] OSHA believes that maintenance of records of illness and injuries are necessary to carry out the purpose of the Act and to assist compliance officers in their enforcement responsibilities (inspections and investigations). Records also provide the basis for statistical programs for injury and illness incidence rates. Additionally they are used to aid employers and employees in evaluating the success of their occupational safety and health activities, and in developing information and identifying "hazards" which cause or may cause occupational injury and illness in the workplace.

Section 8(c) of Act provides in pertinent part:

> (c)(1) Each employer shall make, keep and preserve, and make available to the Secretary [of Labor] or the Secretary of Health, Education and Welfare, such records regarding his activities relating to this Act as the Secretary, in cooperation with the Secretary of Health, Education, and Welfare, may prescribe by regulation as necessary or appropriate for the enforcement of this Act or for developing information regarding the causes and prevention of occupational accidents and illnesses. In order to carry out the provisions of this paragraph such regulations may include provisions requiring employers to conduct periodic inspections. The

Secretary shall also issue regulations requiring that employers, through posting of notices or other appropriate means, keep their employees informed of their protections and obligations under this Act, including the provisions of applicable standards.

(c)(2) The Secretary, in cooperation with the Secretary of Health, Education and Welfare, shall prescribe regulations requiring employers to maintain accurate records of, and to make periodic reports on, work-related deaths, injuries and illnesses other than minor injuries requiring only first aid treatment and which do not involve medical treatment, loss of consciousness, restriction of work or motion, or transfer to another job.

(c)(3) The Secretary, in cooperation with the Secretary of Health, Education and Welfare shall issue regulations requiring employers to maintain accurate records of employee exposures to potentially toxic materials or harmful physical agents which are required to be monitored or measured under section 6. Such regulations shall provide employees or their representatives with an opportunity to observe such monitoring or measuring, and to have access to the records thereof. Such regulations shall also make appropriate provisions for each employee or former employee to have access to such records as will indicate his own exposure to toxic materials or harmful physical agents. Each employer shall promptly notify any employee who has been or is being exposed to toxic materials or harmful physical agents in concentrations or at levels which exceed those prescribed by an applicable occupational safety and health standard promulgated under section 6, and shall inform any employee who is being thus exposed of the corrective action being taken.

Section 8(g) of the Act also provides:

(g)(1) The Secretary of [of Labor] and [the] Secretary of Health, Education and Welfare are authorized to compile, analyze, and publish, either in summary or detailed form, all reports or information obtained under this section.

(g)(2) The Secretary [of Labor] and the Secretary of Health, Education and Welfare shall each prescribe such rules and regulations as he may deem necessary to carry out his responsibilities under this Act, including rules and regulations dealing with the inspection of an employer's establishment.

Section 20(a)(5) of the Act is also a very important source of statutory authority to promulgate recordkeeping requirements. It provides authority for NIOSH to make inspections:

(a)(5) The Secretary of Health, Education and Welfare, in order to comply with his responsibilities under paragraph (2) [Section 20(a)(2)], and in order to develop needed information regarding potentially toxic substances or harmful physical agents, may prescribe regulations requiring employers to measure, record, and make reports on the exposure of employees to substances or physical agents which the Secretary of Health, Education and Welfare reasonably believes may endanger the health or safety of employees. The Secretary of Health, Education and Welfare also is authorized to establish such programs of medical examinations and tests as may be necessary for determining the incidence of occupational illnesses and the susceptibility of employees to such illnesses. Nothing in this or any other provision of this Act shall be deemed to authorize or require medical examination, immunization, or treatment for those who object thereto on religious grounds, except where such is necessary for the protection of the health or safety of others. Upon the request of any employer who is required to measure and record exposure of employees to substances or physical agents as provided under this subsection, the Secretary of Health, Education and Welfare shall furnish full financial or other assistance to such employer for the purpose of defraying any additional expense incurred by him in carrying out the measuring and recording as provided in this subsection.

Finally, Section 24(a) and (e) provide that the Secretary shall develop and maintain an effective program of collection, compilation, and analysis of occupational safety and health statistics on injuries and illnesses. Employers are required to file such reports with the Secretary as are prescribed by regulation.

Enactment of the foregoing statutory requirements has resulted in the promulgation of regulations which require maintenance of both general and special records, posting of these records and other documents, and extensive litigation concerning an employer's obligation under and compliance with recordkeeping requirements. With this framework in mind it is helpful to review the applicability of the recordkeeping requirements.

C. REGULATIONS

Pursuant to the statutory authority granted in the Act, OSHA in cooperation with the Department of Health, Education, and Welfare (NIOSH), has promulgated recordkeeping, reporting, and posting regulations. *Louisiana Chemical Association, supra.* These regulations are published in the Code of Federal Regulations at 29 CFR 1903 and 1904. Employers, unless exempted, are required to comply with these regulations.

The Bureau of Labor Statistics is responsible along with OSHA for compiling, developing and maintaining the recordkeeping, reporting, and posting program.

D. COVERAGE

All employers covered by the Act are required to maintain in each establishment records of recordable occupational injuries and illnesses, unless specifically exempted. Even federal, state, and local government agencies, which are not covered by the Act, may be required to keep records. Sections 18(c)(7) and 19(a)(3) and (5) of the Act. The detailed "Occupational Safety and Health Administration Regulations for Recording and Reporting Occupational Injuries and Illnesses" are codified in 29 CFR 1904.

1. Who Is a "Covered" Employer?

The definition of a "covered" employer is contained in Section 3(5) of the Act, that is, "a person engaged in a business affecting commerce who has employees but does not include the United States[2] or any state or political subdivision of a state." This definition of "employer," however, is actually too limited in describing who must comply with reporting and recordkeeping requirements. An element of *control* is also involved. See *Dayton Tire and Rubber Co. v. OSHRC,* 539 F.2d 242, 4 OSHC 1497 (D.C. Cir. 1976), where an employer was held responsible for keeping records of employees provided by Manpower, a temporary personnel referral service, since the employer had control over the temporary employees; *Elmer R. Vath, Painting Contractor,* 2 OSHC 1091 (1974), where the employer of a painter who hired a helper (to be paid by and under the control of the painter) was required to report the helper's death since the painter was not an independent contractor; *Petrolane Offshore Construction Services, Inc.,* 3 OSHC 1156 (1975); *Joseph Bucheit & Sons Co.,* 2 OSHC 1001 (1974) an injury to a loaned employee is not recordable as to the regular employer.

2. Exemptions From Coverage

a. *Applicability*

Coverage is limited to those employers specified in Section 4 of the Act. 29 CFR 1904.1. Federal agencies and state agencies acting under Section 274 of the Atomic Energy Act of 1954 as amended are not covered under the general recordkeeping requirements. But see Sections 18 and 19 of the Act; 29 CFR 1960. 3–.14.

b. *Jurisdictional*

Coverage is restricted where other federal agencies regulate the employer's reporting and record keeping. Section 4(b)(1) of the Act. This provision is controlling even if the other federal agency does not have recordkeeping or recording requirements which are as stringent as OSHA's (*e.g.,* Department of Transportation's accident reporting requirements). See *Secretary of Labor v. Southern Pacific Transportation Company,* 2 OSHC 1313, *affd,* 539 F.2d 386, 4 OSHC 1693 (5th Cir. 1976), *cert. denied,* 434 U.S. 874 (1977). The other federal agency must, however, specifically regulate the employer's recordkeeping and reporting. General requirements will not be sufficient to divest OSHA of authority since recordkeeping requirements are substantive rules. See *Bettendorf Terminal Company and LeClaire Quarries, Inc.,* 1 OSHC 1695 (1974); but *cf., Belt Railway Co. of Chicago,* 3 OSHC 1612 (1975); *Seaboard Coast Line RR Co.,* 3 OSHC 1758 (1975). Notwithstanding another agency as regulator of an employer's reporting and recordkeeping, OSHA can require the posting of the OSHA poster. *Chicago, Rock Island and Pacific Railroad Co.,* 3 OSHC 1694 (1975). See also *P &Z Co. v. District of Columbia,* 408 A.2d 1249, 8 OSHC 1078 (D.C. 1979)—the injury reporting requirements of the District of Columbia Industrial Safety Act is not preempted by the Act.

3. Specific Recordkeeping Exemptions

A small employer which employed no more than ten (10) full or part-time *employees* at any time during the calendar year immediately preceding the current calendar year is exempt from and need not comply with any of the recordkeeping requirements *except* that it must under 29 CFR 1904.8 report fatalities or multiple hospitalization accidents, must maintain a log and summary of occupational injuries and illnesses under 29 CFR 1904.2(1), and must maintain records if it is selected by the Bureau of Labor Statistics (BLS) to participate in the annual statistical survey of occupational injuries and illnesses. 29 CFR 1904.15. Also, state safety and health laws may require small employers to keep injury and illness records.

If an employer has more than one establishment with a combined number of more than ten (10) employees, records must be kept for all individual establishments.

Other employers exempted would be farmers employing family members (records must be kept for nonfamily employees), certain personnel supplied by temporary employment services, employers of domestics (in the employer's private residence), and employers in religious activities with respect to the conduct of religious services or rites. Records of injuries and illnesses occurring to employees while performing secular activities must be kept. Recordkeeping is also required for employees of private hospitals, schools, orphanages, and commercial establishments owned or operated by religious organizations.

Employers covered under other acts pursuant to the Section 4(b)(1) exemption may claim an OSHA recordkeeping exception. Employers covered by MSHA are also exempt from OSHA recordkeeping.

Even though OSHA's regulatory approach encompasses three separate functions: standard specification, standard enforcement, and recordkeeping/reporting, the Act's preemption provision, Section 18, applies only to standards promulgated under the Act; the Act's reporting requirement is not covered by the preemption since it is not considered a standard. States may continue their reporting requirements regardless of whether a state plan is in effect since they may assert authority over any issue not covered by a §655 standard. Thus injury-reporting requests of the District of Columbia Industrial Safety Act are not preempted. *P & Z Company, Inc. v. District of Columbia,* 408 A.2d 1249, 8 OSHC 1078 (D.C. 1979).

E. GENERAL RECORDKEEPING AND REPORTING REQUIREMENTS

Only two of the forms discussed below must be maintained by employers *and* shown to OSHA compliance officers during inspections, the Log and the Supplementary Record (OSHA Form Nos. 200 and 100), which are available at BLS regional offices. OSHA Form No. 200 replaces two separate forms that were previously required (Forms 100 and 102). In states with state plans, additional recordkeeping requirements may be applicable. 20 CFR 1904.1 and 1952.4. Compliance with the recordkeeping requirements of an approved state plan constitutes compliance with OSHA recordkeeping requirements. 29 CFR 1952.4. Records must be established (and maintained) on a calendar year basis. 29 CFR 1904.3. Federal and state agencies are also required to maintain records. Sections 19(a) and (d), 18(8) of the Act.

1. The Log and Summary of Occupational Injuries and Illnesses

Each employer must prepare and maintain in each establishment a log and summary of all recordable occupational injuries and illnesses (the "Log"), OSHA Form No. 200,[3] for that establishment and enter each recordable injury and illness on the Log as early as practicable, but no later than six (6) working days after receiving information that an injury or illness has occurred. 29 CFR 1904.2. An employer who is on a centralized recordkeeping system or who uses data processing equipment may maintain the Log of Occupational Injuries and Illnesses at a place other than at each establishment. If the Log is maintained elsewhere, a copy updated to within forty-five calendar days must be present at all times at each establishment.

All recordable injuries and working illnesses must be recorded within six working days from the time the employer learns of them. The employer is given discretion, however, to reasonably determine whether or not an injury is recordable. *LaBiche's, Inc.*, 2 OSHC 3110 (1974). If the employer determines the injury or illness is work-related, it is recordable. An injury or illness is work-related if it is directly related to the workplace or to the occupational environment of an employee. If no illnesses or injuries occur, the employer has no obligation to maintain a Log. *Taysom Construction Company,* 2 OSHC 1606 (1975); *P.M.F. Inc.*, 1 OSHC 3258 (1973); *Intermountain Block and Pipe Corporation,* 1 OSHC 3145 (1972). In *General Motors Corporation, Inland Division,* 8 OSHC 2037 (1980), an employee contracted bronchitis due particularly to her smoking. The bronchitis was also possibly aggravated by exposure to toluene diisocyanate (TDI) in the workplace. Two other employees also had respiratory problems possibly related to TDI exposure. The employees were transferred to other departments. None of the incidents were reported on the Log and Summary of Occupational Injuries and Illnesses. OSHA Form No. 100 (now No. 200). The employer argued it was only required to record those occupational illnesses caused by the occupational environment,[4] at best the TDI exposure merely aggravated an existing condition. According to the Secretary of Labor, a broad interpretation of the employer's obligation to record occupational illnesses was substantiated by looking at the legislative history of the Act and the interpretations of the regulations promulgated by the Secretary of Labor in a pamphlet, "What Every Employer Needs to Know About Recordkeeping."[5]

The Commission stated that the legislative history of Section 8(c)(1) of the Act shows a clear intent to interpret the recordkeeping request broadly.

Moreover, a broad definition of occupational injuries and illnesses is set forth in 29 CFR 1904.12(c),[6] thus:

> The illnesses in this case were sufficiently serious to be recordable. Each illness resulted in lost workdays and a transfer to another job. This places the incidents squarely under the definition in 29 CFR 1904.12(c)(3). The only question that remains is whether the illnesses were sufficiently connected to the employees' occupation so that they had to be recorded.
>
> Respondent has argued that it was only responsible to record illnesses which it found were directly caused by the occupational environment. We reject this narrow interpretation of the recording obligation.
>
> As our discussion of the legislative history of the Act indicates, the primary purpose of the recording obligation is to develop information for future scientific use. Moreover, Congress showed a clear preference for overreporting injuries and illnesses rather than underreporting them. Given this background, employers must record illnesses in which the occupational environment either was a contributing factor to the illness or aggravated a preexisting condition. Furthermore, employers must record illnesses, when, as in this case, there is medical evidence from personal physicians linking the illness and the occupational environment and no other medical evidence contradicts this demonstrated relationship.

The Commission also held that the recordkeeping violation could not be characterized as *de minimis* since its effect on employee safety and health was not trifling.

OSHA does not normally provide employers with recordkeeping forms. If OSHA does not provide the employer with Form 200, the employer must maintain a private log equivalent—which is as readable and comprehensible to a person not familiar with it. *V. O. Hegsted d/b/a Challenger Supply*, 1 OSHC 1484 (1973). For the failure of OSHA to furnish forms is no excuse for an employer's failure to maintain records. *Puterbaugh Enterprises, Inc.*, 2 OSHC 1030 (1974). A lack of time to complete the forms is similarly not a defense. *Automotive Products Corp.*, 1 OSHC 1772 (1974).

OSHA recordkeeping and reporting requirements may differ from those under state worker's compensation laws. A state worker's compensation form, therefore, may or not be a sufficient alternative to OSHA Form No. 200, depending on the contents. The New York form was held sufficient in *Jenny Industries, Inc.*, 2 OSHC 3308 (1975); the Virginia form was held insufficient in *William C. Bradley, Inc. of Virginia*, 1 OSHC 3041 (1973); the New Hampshire form was held insufficient in *New Hampshire Provision Company, Inc.*, 1 OSHC 3071 (1974).

An employer who wishes to maintain records in a manner different from that required must petition and secure prior approval by OSHA for a "Recordkeeping Exception." 29 CFR 1904.13. Affected employees or their representatives are afforded the opportunity to comment on the petition.

Whether or not deviation in format will result in a citation depends upon the effect on comprehensibility. *Cf., Novo Corporation, Export Packing Division*, 1 OSHC 3076 (1973); *Automatic Products Corporation*, 1 OSHC 1772 (1974). Although an employer's obligation to keep records is unconditional, where the violation is merely technical or *de minimis* and bears no direct relationship to employee safety or health, the Review Commission has normally not assessed penalties. *Mantua Mfg. Co.,* 1 OSHC 3070 (1973); *P. M. F., Inc.,* 1 OSHC 3258 (1973).

2. The Supplementary Record

In addition to completing OSHA Form No. 200, an employer must prepare a Supplementary Record, OSHA Form No. 101,[7] of any *recordable injuries or illnesses* at the same time, that is, within six working days after receiving information that a recordable case of illness or injury has occurred. 29 CFR 1904.4. The Supplementary Record shall be completed according to the detail prescribed in the instructions to OSHA Form 101. If a Log is not required, neither is a Supplementary Record. See *Taysom Construction Co., supra.*

While the information in the Supplementary Record must comport with the contents of OSHA Form No. 101, deviations from the official format are much less serious than with the Log. Worker's compensation, insurance, and other reports may be acceptable alternative records if they contain the information required by Form 101. 29 CFR 1904.4. Private forms which have been held insufficient as Form 200 (*né* Form 100) equivalents have sufficed as Form 101 alternatives. *Automotive Products Corporation,* 1 OSHC 1772 (1974) (insurance reports).

The employer must maintain private forms if OSHA does not supply him with official supplementary records. See *Puterbaugh Enterprises, Inc., supra.*

3. The Annual Summary

The Annual Summary, based on the information contained in the Log, must be completed by February 1, beginning with calendar year 1979. 29 CFR 1904.5. The last page of OSHA Form No. 200 is used in presenting the Annual Summary. The Annual Summary of the 1977 calendar year's occupational injuries was required to be posted on OSHA Form No. 102. The Annual Summary consists of a copy of the year's totals[8] from Form No. 200 plus the following information therefrom: year covered, company name, establishment name and address, signature, title, and date. The fact that the Annual Summary is prepared in longhand rather than typewritten precludes a citation for failure to compile under 29 CFR 1904.5(a). *Simmons, Inc.,* 6 OSHC 1157 (1977). If no recordable injuries or illnesses occurred in the year, zeroes must be entered on the totals lines. *LaBiche's, Inc.,* 2 OSHC 3110 (1974). The Annual Summary shall contain a certification that it is true and complete. *Stowe Canoe Company,* 4 OSHC 1013 (1976).

The Annual Summary *must be posted* annually, from February 1 until March 1, in a conspicuous place or the place where employee notices are usually posted. Failure to post a copy may result in the issuance of citations and assessment of penalties pursuant to Sections 9 and 17 of the Act. *Simmons, Inc.,* 6 OSHC 1157 (1977).

4. Material Safety Data Sheet

Those employers in the ship repairing, 29 CFR 1915.57(c), shipbuilding, 29 CFR 1916.57(c), and ship breaking, 29 CFR 1917.57(c), industries which use *hazardous materials*[9] must ascertain the hazards and obtain certain items of information prior to use of such materials and record such information on a Material Safety Data Sheet, OSHA Form No. 20, or an essentially similar form. The Data Sheet is required to be completed only on materials which, when used without special precautions and controls, will constitute a health hazard to employees. The Data Sheet is not required to be maintained except by employers in maritime operations, but can be useful as a fact sheet on the properties and potential hazards of a chemical. The pertinent information required includes: the chemical name of the material; physical data; fire and explosion hazard data; reactivity data; spill, leak, and clean-up procedures; special protection and handling information. Where information is not available from published sources or by normal test procedures, the fact is indicated by recordation of the word "unknown"; if information is inapplicable to the material, the initials "n.a." are inserted.

Some employers use the Data Sheet as the vehicle for communicating information about potentially hazardous substances to employees. Employers request suppliers to fill out the form in order to assess the risks associated with the chemical. Manufacturers of chemicals may supply these forms in response to such a request. Although possibly providing protection under OSHA, this action could lead to potential liability under tort and contract laws. See discussion Chapter XIV, *infra.*

In addition, under the recent OSHA regulation on access to medical and exposure records, material safety data sheets are included within those records to be made available to employees. 29 CFR 1910.20.

5. Retention of Records

All of the required OSHA forms (including Form No. 200 and its predecessors No. 100 and No. 102) must generally be *retained* in each establishment for a period of five years following the end of the calendar year to which they relate. 29 CFR 1904.6. Where the ownership of an establishment changes hands, the new employer must preserve prior records for the remainder of the five-year period. The new employer has responsibility for maintaining records and filing reports *only* for the period during which he owns the establishment. 29 CFR 1904.11. He need not update the records of the predecessor employer.

6. Maintenance of Records

Failure to maintain the required records may result in the issuance of a citation and assessment of penalty as provided for in Sections 9, 10 and 17 of the Act. 29 CFR 1904.9. *Howard P. Foley Co.*, 5 OSHC 1501 (1977); *Kensington*

West Corp., 3 OSHC 1519 (1975). In *McDowall Co.*, 1 OSHC 3021 (1972) the Review Commission vacated a penalty where the failure to maintain was a result of a misinterpretation of the regulations; *General Motors Corp., Delco Air Conditioning Division,* 5 OSHC 1044 (1977)—in order to prove violation of the regulation requiring an employer to maintain the Supplementary Record, OSHA must show there were injuries/illness that existed and were not recorded. Typical penalties for recordkeeping and reporting violations of the Act are as follows: (1) failure to maintain the Log or Supplemental Record —$100; (2) failure to compile and maintain the Summary of Occupational Illnesses and Injuries—$100; and (3) failure to timely report an occurrence of a reportable employment accident—$400. (FOM-Chapter XII). *Chrysler Corporation,* 7 OSHC 1578 (1979). Citations and penalties of up to $10,000 or six months in prison may, however, be issued for: (1) knowing falsification of records submitted to OSHA; (2) failure to maintain the OSHA forms; and (3) failure to file fatality or multiple hospitalization reports.

Falsification of OSHA records is a criminal offense that can result in imprisonment for up to six months, a fine up to $10,000, or both. Section 17(g) of the Act. Records maintained by an employer and reports submitted pursuant to and in accordance with the requirements of an approved state plan under Section 18 of the Act shall be regarded as compliance with 29 CFR 1904.

7. Location of Records

Ordinarily, records must be maintained at each establishment (workplace). The reverse side of form OSHA Form No. 200 defines the term "establishment." It is also defined in 29 CFR 1904.12(g). See *McDowall Co.*, 1 OSHC 3021 (1972), "when an employer has more than one establishment, a different set of records must be maintained at each one." Employers with activities that are physically dispersed, such as construction and transportation, must keep the records of injuries and illnesses of employees engaged in such activities at the place where the employees report each day. *Parnon Construction, Inc.*, 5 OSHC 1232 (1977); *Adler & Neilson Co., Inc.*, 5 OSHC 1130 (1977); *Otis Elevator Co.*, 3 OSHC 1736 (1975).

If employees do not regularly report to the same place (*e.g.*, traveling salesmen), records may be maintained at a central place for each group of employees regularly supervised by the same person. These are basically employees who do not work in fixed establishments. If records are so maintained centrally, two conditions must be met: (1) address and telephone number of the place where the records are kept must be available at the worksite; and (2) there must be personnel available at the central place during normal business hours to provide information from the records. See, *e.g.*, *E. J. Albrecht Co., Inc.*, 5 OSHC 1126 (1977).

a. *Definition of "Establishment" under the Act*

"Establishment" is defined as a single physical location where business is conducted or where services or industrial operations are performed (for example: a factory, mill, store, hotel, restaurant, farm, movie theater, ranch, or bank). Where distinctly separate activities are performed at a single physical location (such as contract construction activities operated from the same physical location as a lumberyard), each activity shall be treated as a separate establishment and joint records may not be maintained. 29 CFR 1904.12(g)(1).

For firms engaged in activities which may be physically dispersed (agriculture, construction, transportation, communications) records may be maintained at the place to which employees report each day. 29 CFR 1904.12(g)(2).

In the *construction industry,* for example, the actual performance of work is accomplished at dispersed locations. OSHA records may, in compliance with 29 CFR 1904.12, be maintained at an established central place where the employee reports to work each day, *i.e.,* the construction site rather than the main administrative office. *Truland Corporation,* 6 OSHC 1896 (1978); *McDowall Co.,* 1 OSHC 3021 (1972); *Parnon Construction, Inc.,* 5 OSHC 1232 (1977); *Adler & Neilson Co., Inc.,* 5 OSHC 1130 (1977); *Otis Elevator Company,* 3 OSHC 1736 (1975). Until a "recordable" illness or injury occurs on the job site, though, records need not be maintained there.

Records for personnel who do not primarily report or work at a single establishment, and who are not generally supervised in their daily work, such as traveling salesmen and engineers shall be maintained at the location from which they are paid or the base from which personnel operate to carry out their activities. 29 CFR 1904.12(g)(3).

b. *Exception to the Location Rules*

A narrowly-drawn exception to the general rule requiring record maintenance at each establishment applies to OSHA Form No. 200 (Log). The exception will be granted under the following circumstances:

(1) There is available at the place where the Log is maintained sufficient information to complete the Log to a date within six working days after receiving information that a recordable case has occurred. . . .
(2) At each of the employer's establishments, there is available a copy of the Log which reflects separately the injury and illness experience of that establishment complete and current to date within forty-five calendar days. 29 CFR 1904.2(b).

If these conditions are met, an employer may maintain the Log at a place other than the establishment or by means of data-processing equipment or both. This is particularly a good idea for large companies with centralized locations or a computerized data system.

Employees who work in physically dispersed operations, a situation sometimes seen in the construction industry, for example, may not report to any fixed establishment on a regular basis. The employer may satisfy the establishment requirement under such circumstances by:

(1) Maintaining the records for each operation or group of operations which is subject to common supervision (field superintendent/supervisor) in an established central location;
(2) Having available at each worksite the address and telephone number of the central location;
(3) Providing personnel during normal business hours at the central location to dispense information from the records. 29 CFR 1904.14.

8. Access to Records and Information

a. *In General*

Section 8(c)(1) of the Act provides that employers shall make available such records as OSHA, in cooperation with NIOSH, shall prescribe by regulation and as are necessary, for either enforcement or for the development of information regarding the causes and prevention of occupational accidents and illnesses. The Act also provides that OSHA, in cooperation with NIOSH, shall prescribe regulations which require employers to maintain records of and make periodic reports on work-related deaths, injuries, and illnesses.

As a general rule, none of the basic records (Forms No. 200 and 101) required to be maintained by the OSHA regulations (29 CFR 1904.7) must be submitted to either OSHA or to NIOSH, but each must be available at the employer's establishments for inspection and copying by various agency officials including:

(1) OSHA compliance officers conducting inspections under Section 20(b) of the Act;
(2) Bureau of Labor Statistics representatives preparing statistical compilations under Sections 18 and 24 of the Act;
(3) NIOSH representatives carrying out research responsibilities under Section 20(b) of the Act; and
(4) State representatives making inspections; and

Present employees, former employees, and their designated representatives (*e.g.*, labor organizations) are also entitled to *access* to the Log, the Supplementary Record, and the Annual Summary of Occupational Injuries and Illnesses, for inspection and copying in a reasonable manner and at reasonable times. 29 CFR 1904.7(b)(1). The regulations specify, however, that nothing precludes employees or their representatives from collectively bargaining to obtain access to information relating to occupational injuries and illnesses in addition to that provided for in 29 CFR 1904.7

b. *OSHA Instruction*

OSHA Instruction STD 1–24.1 (February 2, 1979) provides that if an employer refuses access to records to a compliance officer, an administrative subpoena may be issued *and* the employer may be cited with a violation of the Act. OSHA Instruction STD 1–24–1 affects both the FOM and IHM concerning reporting and recordkeeping procedures during inspections. It has also been held that OSHA has authority to request and receive employer records of employee exposure which are necessary to the performance of its legislative as well as enforcement functions under the Act. In *Marshall v. American Olean Tile Co.*, 489 F. Supp. 32, 8 OSHC 1136 (E.D. Pa. 1980), the court compelled an employer to produce designated records requested by OSHA, pursuant to an administrative subpoena issued pursuant to Section 8(b) of the Act.

c. *Regulations*

In Section 8(c)(3) of the Act, OSHA, in cooperation with NIOSH, is required to issue regulations requiring employers to maintain records of employee exposures to potentially toxic materials or harmful physical agents which are required to be monitored or measured under Section 6 of the Act. Under these regulations, affected employees shall be provided with information regarding the toxic effects, conditions of exposure, and precautions for safe use of all hazardous materials in the establishment, by means of labeling or other forms of warning where such information is prescribed by a standard. If employees are exposed to harmful materials in excess of levels set forth in occupational safety and health standards promulgated under Section 6 of the Act, the affected employees must be so notified by the employer, and the employer must also inform the employees thus exposed what corrective action is being taken. Upon request, employees must be given access to records of their *own* history of exposure to toxic materials or harmful physical agents which are required to be monitored or measured and recorded. Finally, if a standard requires monitoring or measuring hazardous materials or harmful physical agents, employees must be given the opportunity to observe such monitoring or measuring.[10]

The regulations which were issued by OSHA implementing the provisions of Section 8(c)(3) of the Act are contained in 29 CFR 1910.20. Employee medical and exposure records were thus made available to specified persons in a specific circumscribed manner. However, OSHA issued a new rule, effective August 21, 1980, which greatly increases the access to medical and exposure records set forth in 29 CFR 1910.20 and which substitutes therefore virtually unrestricted access for employees, OSHA and collective bargaining agents to the records in question. 45 Fed. Reg. 35212 (May 23, 1980).

Pursuant to the "new" regulations, employers who make, maintain, contract for, or have access to exposure records or medical records and analyses (compilation of data including research) of employees exposed to toxic substances or harmful physical agents,[11] whether or not the records are related to specific occupational safety and health standards,[12] are required to preserve and retain them, for at least the duration of the affected employee's employment with the employer plus thirty years in the case of medical records and thirty years for exposure records and analyses, unless a specific occupational safety and health standard provides a different period of time.[13]

The standard provides that once employee exposure and medical records are created for *any* reason they must be preserved; it does not however require that an employer create such records. The records do not have to be within the employer's physical control as long as the employer has access to them either in-house or on a contractual basis *(e.g.,* in a physician's or insurance company's office). The standard provides in this respect that "the concept of employee access encompasses situations in which any of the employer's officers, employees, agents or contractors have physical control or access to records even though they are not generally available to all officers, employees, agents and contractors."

The regulation, as issued, applies to all employers in the maritime, construction, and general industries. A proposed rule was also recently published for agricultural employers which would provide access to records similar to that in general industry. 29 CFR 1928.20; 45 Fed. Reg. 35298 (May 23, 1980).

(1) *Purpose.* The purpose of the regulation is to make data contained in exposure and medical records available to both employees and OSHA in order to promote the recognition of workplace hazards and the subsequent reduction of occupational disease. Moreover, these records, according to the standard's preamble, are relevant to OSHA's statutory functions in the following ways:

(1) Certain health standards contain medical surveillance and recordkeeping requirements; access to employee medical records is viewed as necessary to verify employer compliance.

(2) OSHA enforcement of discrimination complaints under Section 11(c) of the Act would be assisted in those situations where an employer has altered a worker's job status for "purported" medical reasons and the worker then alleges that this action resulted from a retaliatory or discriminatory intent; access to relevant employee medical records may be important to ascertain the legitimacy of employer's actions.

(3) Access to records may be relevant to the type of enforcement action OSHA initiates against an employer or to prove the appropriateness of an enforcement action. For example, serious violations under Section 17(k) of the Act require an element of actual or constructive employer knowledge; the content of employer medical records could document a pattern of disease sufficient to give an employer actual knowledge of the hazard involved.

(4) The question of whether an employer willfully violated an OSHA standard may also be influenced by the content of employee medical records, willful violations may involve evidence of an employer's knowledge of both the requirements of an OSHA regulation and the factual circumstances underlying the hazardous working condition, which may be satisfied by a pattern of disease related to occupational exposure to a toxic substance which is documented by employee medical records.

(5) Medical records could influence whether or not a general duty clause violation occurred under Section 5(a)(1) of the Act. In addition to the knowledge element, being traced either to industry or common sense recognition of the hazard, a recognized hazard element of the general duty clause can be proved by actual employer knowledge of the hazard. *Brennan v. OSHRC (VyLactos Laboratories, Inc.)*, 494 F.2d 460, 1 OSHC 1623 (8th Cir. 1974). Employee medical records could document a pattern of disease, constituting actual knowledge of a hazard or a hazardous working condition.

(6) In some instances medical practice itself constitutes a violation of an OSHA standard. Employee medical records could help document such instances.

(7) Employee medical records are relevant to the rulemaking process, where OSHA health standards are established.

(8) Employee medical records could be relevant to imminent-danger situations and could demonstrate that a particular employee, in light of his current health status, faces an imminent danger of disease or death from present working conditions. Content of employee medical records could precipitate initiation of these procedures.

(9) Employee medical records could be relevant in the situations where concern arises over the etiology of a newly discovered disease pattern.

(10) Employee medical records can be relevant to compliance investigations on the efficacy of controls (such as respirators) on worker exposure to toxic chemicals.

(11) Access to medical records may be necessary in situations where an employer conducts medical surveillance but fails to evaluate the medical data.

(2) *General Access.* The records which must be maintained under the regulation shall be made available within a reasonable time (no later than fifteen days after the request is made), place and manner without cost to current and former employees, employees being transferred or assigned to work areas where there will be exposure to toxic substances or harmful physical agents, designated representatives of employees (in certain circumstances) and OSHA. NIOSH is excluded from the scope of the standard, since it has independent legal authority to seek access to employee medical and exposure records. "Access" is defined in the regulation as the "right and opportunity to examine and copy." The right of access varies dependent upon whether access is being sought to medical or exposure records. Nothing in the regulation is intended to preclude employees and collective bargaining agents from bargaining to obtain access to information in addition to that provided therein. An unanswered question related to a request for access to exposure or medical records which were prepared in anticipation of litigation and would normally be covered by the attorney work product doctrine. It seems not to be the intent of the standard to include these records, but a literal reading of the standard conceivably may require an opposite conclusion.

(3) *Exposure Records.* Each employer shall, upon request, assure the access of OSHA, each employee, *and* his designated representative to *employee exposure records*[14] relevant to the employee. (The term exposure is not defined in the regulation; providing records upon request therefore, may be considered an admission of exposure). For the purpose of the regulation, exposure records relevant to the employee consist of:

(1) Records of the employee's past or present exposure to toxic substances or harmful physical agents;
(2) Exposure records of other employees with past or present job duties or working conditions related to or similar to those of the employee;
(3) Records containing exposure information concerning the employee's workplace or working conditions; and
(4) Exposure records pertaining to workplaces or working conditions to which the employee is being assigned or transferred.

Such access is required whether or not exposure to a substance exceeds OSHA's PEL's, or, in fact, whether the substance is even regulated by OSHA. Although the Act provides that an employee shall have access to exposure records that indicate his own exposure to substances, the new regulations grant such access to designated employee representatives. It also gives employees access to exposure records of similarly situated employees in the workplace.

(4) *Medical Records.* Each employer shall, upon request, assure the access of each employee to *employee medical records*[15] of which the employee is the physician, nurse, or other health care personnel or subject, except where a physician representing the employer believes that direct employee access to the information contained in the records regarding a specific diagnosis of a terminal illness or a psychiatric condition could be detrimental to the employee's health. In such a situation the employer may inform the employee that access will only be provided to a designated representative of the employee having specific written consent, and by the employee's request for direct access to this information only. Each employer shall also, upon request, assure the access of each designated representative to the employee medical records of any employee who has given the designated representative specific *written* consent. (Within hours of the standard's announcement, the AFL-CIO asked the U.S. Court of Appeals for the District of Columbia Circuit to review it, challenging the presumption that a confidential relationship exists between company doctors and employees). OSHA has the right of access to employee medical records whether or not the employee has given authorization, subject to compliance with the procedural safeguards discussed below.

(5) *Analyses.* Each employer shall, upon request, assure the access of each employee and his designated representative to each *analysis*[16] using exposure or medical records concerning the employee's working conditions or workplace. Whenever access is requested to any analysis which reports the contents of employee medical records by either direct identifier (name, address, social security number, payroll number, etc.) or by information which could reasonably be used under the circumstances indirectly to identify specific employees (exact age, height, weight, race, sex, date of initial employment, job title, etc.), the employer shall assure that personal identifiers are removed before access is provided. If the employer can demonstrate that removal of personal identifiers from an analysis is not feasible, access to the personally identifiable portions of the analysis need not be provided.

(6) *OSHA Access.* Each employer shall, upon request, assure the immediate access of OSHA representatives to employee exposure and medical records and to analyses using exposure or medical records. Rules of agency practice and procedure governing OSHA access to employee medical records are contained in 29 CFR 1913.10.[17]

Whenever OSHA seeks access to personally identifiable employee medical information it is required—unless specific written consent of the employee is obtained pursuant to 29 CFR 1910.20(e)(2)(ii)—to present to the employer a written access order pursuant to 29 CFR 1913.10(d). Upon receipt the employer shall prominently post a copy of the written access order and its accompanying cover letter for at least fifteen working days. If deemed appropriate, a written access order may constitute, or be accompanied by, an administrative subpoena. Copies of the access order shall be presented to the employees' designated representative prior to entry. Provision is also made for employer, employee, and designated representative objections to access orders.

As part of every workplace health inspection, OSHA inspections are required to routinely investigate whether employers are complying with the medical and exposure records access standard under 29 CFR 1910.20, according to a yet "unpublished" OSHA Instruction. If the standard is applicable, the inspector should determine if medical and exposure records exist. Should the employer indicate that they do not exist, the inspector should verify this by interviewing "several employees or employee representatives" as to whether monitoring or medical examinations were conducted, it added.

An employer's failure to preserve medical records in violation of 1910.20(d) should normally be classified as serious according to the Instruction.

(7) *Employee Information.* The employer is required to inform employees exposed to toxic substances or harmful physical agents upon first entering into employment and at least annually thereafter of the existence, location, and availability of the records covered by the regulation, the person responsible for maintaining and providing access to them, and the right of access to the records.[18] Copies of the standard, its appendices, and informational material must also be provided and distributed. 29 CFR 1910.20(g).

(8) *Transfer of Records.* Whenever an employer ceases to do business, it must transfer all records subject to the standard to its successor who is required to receive and maintain them. If no successor is available, employees shall be timely notified and given access to the records at least three months prior to business cessation. Employers must transfer the records to NIOSH if a specific standard requires they be kept or otherwise notify NIOSH of impending disposal. 29 CFR 1910.20(h).

(9) *Trade Secrets.* Nothing in the standard precludes an employer from deleting trade secret data from records as long as the employee or designated representative is notified of the deletion. Alternative information sufficient to permit employees to identify where and when exposure occurred shall be provided if necessary, including chemical names/identities, exposure levels, and employee health status data. 29 CFR 1910.20(f).

Challenges to the "rule" were filed at both the district and circuit court (Fifth and District of Columbia) levels. In *American Petroleum Institute v. OSHA*,—F.2d—, 8 OSHC 2025 (5th Cir. 1980), the court ceeded jurisdiction to the D.C. Circuit. In *Louisiana Chemical Association, supra,* the court dismissed the district court action on the basis that the rule was actually a standard and thus the district courts lacked subject matter jurisdiction to entertain the issue. Section 6(f) of the Act provides that preenforcement judicial review of a standard may be entertained only in a United States Court of Appeal.

In the lead standard which was recently upheld in *United Steelworkers of America, AFL–CIO–CLC v. Marshall*,—F.2d—, 8 OSHC 1810 (D.C. Cir. 1980), the court approved a provision which requires employers to maintain a variety of health records on employees, including exposure monitoring, medical surveillance, and medical removals. 29 CFR 1910.1025(n). Access to these records is available to OSHA, NIOSH, employees, and their authorized representatives (union). In upholding this provision, which, unlike the "standard" on access to medical and exposure records contained in 29 CFR 1910.20, gives unions *direct* access to certain employee medical records (*i.e.*, medical removal records), the court stated:

> The recordkeeping requirements of the standard, however, contain a special section entitled "Medical Removals." This section requires employers to maintain records on all removed employees and states that these records "shall include:"
>
> (A) The name and social security number of the employee;
> (B) The date on each occasion that the employee was removed from current exposure to lead as well as the corresponding date on which the employee was returned to his or her former job status;
> (C) A brief explanation of how each removal was or is being accomplished; and
> (D) A statement with respect to each removal indicating whether or not the reason for the removal was an elevated blood lead level.
>
> Section 1910.1025(n)(3)(ii). OSHA argues that the "medical removal records" mentioned in Section (n)(4)(ii) refer only to this limited information.
>
> The bare language of the standard unaided by the agency's construction, cannot resolve the ambiguity. Section (n)(4)(ii) does not specifically refer to Section (n)(3)(ii). That latter section, moreover, says that medical removal records "shall include" the limited facts listed there, not that they shall contain only those facts.
>
> We agree with LIA that if the standard allows the unions, without the employee's permission, to examine the intimate results of physician examinations—information to which the employee and the government do have rightful access—it may violate the statute and the Constitution.
>
> * * *

> Under OSHA's construction, however, disclosure of "medical removal records" to labor unions probably avoids statutory and constitutional problems . . .

Thus the court held that disclosure of limited, nonconfidential medical (removal) information is lawful.

d. *Disclosure of Information*

The Secretaries of Labor and Health, Education, and Welfare are authorized to compile, analyze, and publish all reports or other information pertaining to OSHA inspections, investigations, or recordkeeping. Section 8(g)(1) of the Act. Employers, however, have the right to require that certain information, such as trade secrets, be kept confidential and not subject to disclosure. Certain other information, such as personally identifiable employee medical information, is also subject to confidentiality strictures. See discussion *supra.*

e. *Inspection of Records*

Prior to the walkaround inspection and most likely during the opening conference, the OSHA compliance officer is instructed to inspect all records required by 29 CFR 1904 to ascertain compliance with the recordkeeping requirements. These records include the Log (OSHA Form No. 200), the Supplementary Record (OSHA Form No. 101), and other records which the employer may be required to keep under the Act such as toxic substance records. [FOM, Chapter V, Section (D)(5)]. The compliance officer is also directed to determine if posting requirements specified in 29 CFR 1903.2(a) and 1904.5(d) are met. Medical records required under 20 CFR 1910.93 will also be examined. The OSHA *Field Operations Manual* also directs that the compliance officer

> . . . should ask [at the opening conference] to see the minutes of safety committee meetings, if the employer has no objections, to better understand accident patterns and identify problem areas. The CSHO must keep in mind [however] that the only records the employer must reveal are those required by the Act, OSHA regulations and the standards. [FOM, Chapter 5, Section (D)(2)(b)]

An as yet unpublished OSHA Instruction provides that inspectors should routinely investigate whether employers are complying with the access rules under 29 CFR 1910.20. In most instances serious citations will be issued for violation of the standard. Moreover, according to the FOM, Ch. VIII, willful violations may also be appropriate if the employer committed an intentional violation of the standard.

f. *Obtaining Recordkeeping Forms*

OSHA does not automatically mail recordkeeping forms to employers each year. Employers must request the forms from OSHA or use acceptable alternative records.

g. *BLS Statistical Surveys*

Under Section 24 of the Act, employers may be selected to participate in surveys of occupational injuries and illnesses conducted by the Commissioner of Labor Statistics (BLS). 29 CFR 1904.20. Upon receipt of a survey form, the employer must complete and return it in accordance with the instructions. 29 CFR 1904.21. Even those employers qualifying for the *small employer* recordkeeping exception may be selected to participate in state statistical programs under approved state plans. Nothing in any state plan affects the duties of employers to complete and submit statistical report forms. Failure to submit the form may result in a citation and penalty. Form No. 200S (an extension of the Log and Summary) became effective in 1979 and replaced OSHA Form No. 103, previously used for this purpose.

h. *Variations from Recordkeeping Requirements*

An employer who wishes to maintain records in a manner different from that required may submit a petition to the Regional Director of the Bureau of Labor Statistics wherein the establishment involved is located. 29 CFR 1904.13(a).

Petitions may be granted if the alternate recordkeeping procedure "will not hamper or interfere with the purposes of the Act and will provide equivalent information." The variation may be subject to specific conditions and "for cause" revocation. 29 CFR 1904.13(f).

Petitions must include the following:

(1) The name and address of the applicant;
(2) The address of places of employment involved;
(3) Reasons for seeking an exception;
(4) A description of the different recordkeeping procedures;
(5) A statement that the applicant has informed affected employees of the petition and of the employees' right to submit written data, views, or arguments concerning the petition (the applicant should post a summary of the petition in each establishment at the place where employee notices are customarily posted);
(6) A list of the states in which the applicant's establishments are located and the number of establishments in each such state. 29 CFR 1904.13(c).

Petitions when granted and the reasons for such grant are published in the Federal Register. 29 CFR 1904.13(g).

F. SPECIFIC RECORDKEEPING REQUIREMENTS

The OSHA *Industrial Hygiene Manual* requires that upon inspection, the industrial hygienist shall determine the extent of the employer's recordkeeping program. This not limited to OSHA required records but is to be extended to that information pertinent to the inspection. IHM, Chapter 1, Section (E)(3)(F). The industrial hygienist is directed to determine if records pertaining to employee exposure and medical records are being preserved in accordance with 29 CFR 1910.20, 43 Fed. Reg. 31330 (1978). Information concerning the monitoring and medical programs shall also be reviewed. Where a specific standard has provisions for employee access to the records, the industrial hygienist shall determine whether the results of environmental measurements and medical examinations are accessible to the affected employees. Information concerning the employer's medical program shall also be requested. In newly issued standards, rules on maintenance and access to exposure and medical records have been promulgated and proposed. Provisions of certain standards have recently been revised to comply with this general requirement. Employers are required to record and maintain specific information concerning the following standards, substances, incidents, and exposures.

1. Toxic and Hazardous Substances (29 CFR 1910.20)

Where employers are subject to standards regulating exposure to toxic substances, special recordkeeping provisions are encountered. Since the Act does not describe exactly what must be recorded or in what form recorded information should be compiled, employers must check specific standards very carefully. Section 8(c)(3) of the Act mandates the issuance of regulations which require employers to maintain accurate records of employee exposure to potentially toxic materials or harmful physical agents which are required to be monitored or measured under Section 6 of the Act. These regulations are generally set forth in 29 CFR 1910.20, discussed *supra.*

2. Asbestos (29 CFR 1910.1001)

Under the old standard, employers were required to maintain records of any personal or environmental monitoring of employee exposure to asbestos. 29 CFR 1910.93(a)(i)(1). The District of Columbia Circuit Court of Appeals, however, held that the three-year record retention period required under this regulation was arbitrary and capricious. *Industrial Union Department, AFL–CIO v. Hodgson,* 499 F.2d 467, 1 OSHC 1631 (D.C. Cir. 1974). The Department of Labor revised the standard effective July 1, 1976. Under the revised standard, records of personal or environmental monitoring must be maintained for a period of at least twenty (20) years, and must be made accessible to OSHA and NIOSH.

The standard was revised again due to the promulgation of the OSHA final rule on access to exposure and medical records. Thus employee exposure records required by the standard shall be provided upon request to employees' designated representatives and OSHA, in accordance with 29 CFR 1910.20(a)–(e) and (g)–(l). Records of medical examinations shall be provided, upon request, to employees' designated representatives and OSHA again in accordance with 29 CFR 1910.20. These records shall also be provided upon request to NIOSH. Any physician who conducts a medical examination required by the standard shall furnish to the employer of the examined employee all the information specifically required by this paragraph and any other medical information related to occupational exposure to asbestos fibers.

3. Cotton Dust (29 CFR 1910.1043)

The cotton dust standard requires employer preparation and maintenance of records of exposure measurements. The required records include name and job classification of employees measured, details of the sampling and analytic techniques, weight of the dust collected, results, type of respiratory protection worn, and exposure levels. The standard also requires records of medical surveillance. These include names of employees, the physician's written opinion, employee medical complaints, a copy of the results of the examination, and the information furnished by the employer to the examining physician.

Employee exposure measurement records and employee medical records required by this paragraph shall be provided upon request to employees, designated representatives, and OSHA, in accordance with 29 CFR 1910.20(a)–(e) and (g)–(i). The employer shall also comply with any additional requirements involving the transfer of records set forth in 29 CFR 1910.20(h).

4. Ionizing Radiation (29 CFR 1910.96)

Employers must maintain records of the radiation exposure of all employees for whom personal monitoring is required. Every monitored employee must be advised of his individual exposure level on at least an annual basis, and the monitoring records must be maintained for an indefinite period. 29 CFR 1910.96(n). Upon the request of a former employee, a report of the employees to radiation as shown in the records maintained by the employer must be supplied within thirty (30) days.

5. Carcinogens (29 CFR 1990.151)

Prior to 1978, employers subject to carcinogen standards were required to maintain monitoring, measuring, and personal records, as well as records of medical examinations. 29 CFR 1910.93. This standard was removed and its provisions incorporated in other standards.

The new cancer policy provides in the "model standard" section that employers will be required to establish and maintain certain monitoring, exposure, and medical surveillance records for specified periods of time, depending upon the identification and classification of the occupational carcinogen involved and the outcome of the rule-making with respect thereto. Employee exposure measurement records and medical records shall be provided upon request to employees, designated representatives, and OSHA in accordance with 29 CFR 1910.20(a)–(e) and (g)–(i). Provisions on the transfer of records when an employer ceases to do business are also contained in the standard. The employer also must comply with additional transfer requirements set forth in 29 CFR 1910.20(h).

6. Benzene (29 CFR 1910.1028)

Employers must establish and maintain records of all measurements of environmental and biological (exposure) monitoring and medical surveillance for at least forty (40) years or the duration of an employee's employment plus twenty (20) years, whichever is longer. Access to the records is required for employees, former employees, NIOSH, and OSHA. Provisions for transfer of records are also provided. Employee exposure measurement records and employee medical records required under the standard shall be provided upon request to employees, designated representatives and OSHA in accordance with 29 CFR 1910.20(a)–(e) and (g)–(i). Employers shall also comply with additional requirements involving the transfer of records set forth in 29 CFR 1910.20(h).

7. Coke Oven Emissions (29 CFR 1910.1029)

The employer must establish and maintain records of all measurements taken to monitor employee exposure to coke oven emissions and for all employees subject to medical surveillance. The records must be maintained for at least forty (40) years or for the duration of employment plus twenty (20) years, whichever is longer. Records of employee medical examinations must also be maintained.

Employee exposure measurement records and employee medical records required by the standard shall be provided upon request to employees, designated representatives and OSHA in accordance with 29 CFR 1910.20(a)–(e) and (g)–(i). Employers shall also comply with any additional requirements involving the transfer of records set forth in 29 CFR 1910.20(h).

8. Lead (29 CFR 1910.1025)

The final lead standard requires employers to complete and maintain records of all monitoring and employee exposure measurements. The records required include name and job classification of employees measured, details of the sampling and analytic techniques, results, and type of respiratory protection worn. The standard also requires maintainance of records of medical surveillance (biological monitoring and medical exam results). These include names of employees, the physician's written opinion, and a copy of the results of the examination. These records must be kept for forty (40) years or for the duration of employment plus twenty (20) years, whichever is longer.

The standard also contains a recordkeeping requirement concerning removals of employees from exposure to lead pursuant to the medical removal protection program of the standard.

Environmental monitoring, medical removal, and medical records required by the standard shall be provided upon request to employees, designated representatives, and OSHA in accordance with 29 CFR 1910.20(a)–(e) and (g)–(i). Medical records shall be provided in the same manner as environmental monitoring records. Employer shall also comply with any additional requirements involving the transfer of records set forth in 29 CFR 1910.20(h).

9. Inorganic Arsenic (29 CFR 1910.1018)

Employers must maintain accurate records of exposure monitoring and of all employees subject to medical surveillance for at least forty (40) years or for the duration of the employees employment plus twenty (20) years, whichever is longer. Records required by the standard shall be provided upon request to employees, designated representatives, and the Assistant Secretary in accordance with 29 CFR 1910.20(a)–(e) and (g)–(i). Employers shall also comply with any additional requirements involving the transfer of records set forth in 29 CFR 1910.20(h).

10. Vinyl Chloride (29 CFR 1910.1017)

Records of required monitoring and measuring and personnel rosters shall be maintained for not less than thirty (30) years; medical records shall be maintained for the duration of employment of each employee plus twenty (20) years or thirty (30) years, whichever is longer. Reports must be made to OSHA with respect to regulated areas and employee exposure in excess of the permissable exposure limit set forth in the standard.

Employers must report emergencies and pertinent facts to OSHA within twenty-four (24) hours.

Records of required monitoring and measuring and medical records shall be provided upon request to employees, designated representatives, and OSHA in accordance with 29 CFR 1910.20(a)–(e) and (g)–(i). These records shall be provided upon request to NIOSH. Authorized personnel rosters shall also be provided upon request to OSHA and NIOSH. In the event that the employer ceases to do business and there is no successor to receive and retain his records for the prescribed period, these records shall be transmitted by registered mail to NIOSH, and each employee individually notified in writing of this transfer. The employer shall also comply with any additional requirements set forth in 29 CFR 1910.20(h).

11. Acrylonite (29 CFR 1910.1045)

Records of materials exempted under the standard must be maintained, as well as exposure monitoring records and medical surveillance records which must be kept for at least forty (40) years or the duration of employment plus twenty (20) years, whichever is longer. Records required by this section shall be provided upon request to employees, designated representatives, and OSHA in accordance with 29 CFR 1910.20(a)–(e) and (g)–(i). Records required by the standard shall be provided in the same manner as exposure monitoring records. The employer shall also comply with any additional requirements involving the transfer of records set forth in 29 CFR 1910.20(h).

Reporting of information in regulated areas, emergencies, emergency plans, and compliance program is also provided.

12. Commercial Diving Operations (29 CFR 1910.440)

Reports on diving-related illnesses or injuries, decompression sickness, repair, and modifications to equipment are to be maintained. The records and documents required by the standard must be provided upon request to employees, designated representatives, and OSHA in accordance with 29 CFR 1910.20(a)–(e) and (g)–(i). Safe practice manuals, depth-time profiles, recordance of dives, decompression procedure assessment evaluations, and records of hospitalizations shall be provided in the same manner as employee exposure records or analyses using exposure or medical records. Equipment inspections and testing records, which pertain to employees, shall also be provided upon request to employees and their designated representatives.

After the expiration of the retention period of any record required to be kept for five years, the employer shall forward such records to NIOSH. The employer shall also comply with any additional requirements with respect to the transfer of records set forth in 29 CFR 1910.20(h).

13. 1, 2-dibromo-3-chloropropane [DBCP] (29 CFR 1910.1044)

The employer is required to maintain records of exposure monitoring and employee medical surveillance for at least forty (40) years or the duration of employment plus twenty (20) years, whichever is longer. Access to the records shall be provided upon request to employees, designated representatives, and OSHA in accordance with 29 CFR 1910.20(a)–(e) and (g)–(i). Provisions on the transfer of records if the employer ceases to do business as set forth in 29 CFR 1910.20(h) is also provided.

14. Emergency Temporary Standard, pursuant to Section 6(c) (29 CFR 1990.152)

The employer shall assure that all records required to be maintained by the standard be made available upon request to OSHA and NIOSH for examination and copying. Employee exposure measurement records and employee medical records required shall be provided upon request to employees, designated representatives, and OSHA in accordance with 29 CFR 1990(a)–(e) and (g)–(i)

15. Other Toxic Substances

Various recordkeeping and reporting requirements are also set forth in the following "toxic and hazardous substances" standards: (1) alpha-Naphthylamine—29 CFR 1910.1004; (2) methyl chloromethyl ether—29 CFR 1910.1006; (3) 3, 3–dichlorobenzidine (and its salts)—20 CFR 1910.1007; (4) bis-chloromethyl ether—29 CFR 1910.1008; (5) beta-Naphthylamine—29 CFR 1910.1009; (6) benzidine—29 CFR 1910.1010; (7) 4–Aminodiphenyl—29 CFR 1910.1011; (8) Ethyleneimine—29 CFR 1910.1012; (9) beta-Propiolactone—29 CFR 1910.1013; (10) 2–Acetylaminofluorene—29 CFR 1910.1014; (11) 4–Dimethylaminoazobenzene—29 CFR 1910.1015; and (12) N–Nitrosodimethylamine—29 CFR 1910.1016; and (13) 4–nitrobipheny–1—29 CFR 1910.1003.

The records required by these standards to be maintained shall be provided upon request to employees, designated representatives, and OSHA in accordance with 29 CFR 1910.20(a)–(e) and (g)–(i). These records shall also be provided upon request to NIOSH.

16. Other Requirements

In addition to specific recordkeeping requirements imposed upon employers with respect to toxic substances, employers are also required to record and maintain information under other standards, *inter alia:* (1) cranes and derricks (reports, load rating, inspections, conditions of ropes)—29 CFR 1910.179, 180, 181; (2) fire extinguishers—29 CFR 1910.157; (3) mechanical power presses—29 CFR 1910.217; (4) respiratory protection—29 CFR 1910.134; (5) forging machines—29 CFR 1910.218; (6) compressed air—29 CFR 1926.803; (7) hazardous materials—29 CFR 1910.106–108; (8) ground fault protection—29 CFR 1926.954; (9) ship repairing—29 CFR 1915.57, shipbuilding—29 CFR 1916.57 and ship breaking—29 CFR 1917.57; and (10) explosives—29 CFR 1926.900.

Failure to maintain the required records may result in a citation. *Overhead Door Corporation,* 7 OSHC 1280 (1979); *Everglades Sugar Refinery, Inc.,* 7 OSHC 1410 (1979).

SPECIFIC REPORTING REQUIREMENTS

1. Fatality and Multiple Hospitalization Reports

All employers subject to the Act must report within forty-eight (48) hours[19] thereof any employment accident which results in the *death* of *one* (1) or more employees or which causes the *hospitalization*[20] of *five* (5) or more employees. See Section 8(c)(2) of the Act; 29 CFR 1904.8. There is no small employer "reporting exception."

The report can be *oral or written* and the regulations explicitly authorize the telephone and telegraph as appropriate means for transmission. An Area Director's receipt of actual notice through a television news broadcast has even been held sufficient; as is notification by newspaper. *Englehaupt Construction Corp.,* 1 OSHC 3049 (1972); *Fort Worth Enterprises, Inc.,* 2 OSHC 1103 (1974); *L. R. Brown, Jr., Painting Contractor,* 3 OSHC 1318 (1975). The required information includes the circumstances of the accident, the number of fatalities, and the extent of the injuries. In states with state plans, the report is made to the state agency with enforcement responsibilities.

Additional reports concerning the accident, in writing or otherwise, may be required. Failure to file a report may result in the issuance of a citation and assessment of penalties. See *Brennan v. Southern Contractors Service,* 492 F.2d 498, 500 n. 3, 1 OSHC 1648 (5th Cir. 1974), where the Fifth Circuit affirmed a fine of $200 for an employer's failure to report a fatal injury. As with any citation, however, not every failure to report will result in the assessment of penalties. The citation was affirmed but a $200 proposed penalty was vacated in *The Anderson Company,* 2 OSHC 3024 (1974). The Secretary bears the burden of proving that a reportable occupational accident has occurred. It is equally clear that where the evidence does not prove there was a "work-related" fatality, the alleged violation will be vacated. *Brilliant Electric Signs, Inc.,* 1 OSHC 3222 (1973)—death apparently due to a heart attack.

Section 9(c) of the Act does not bar the issuance of a citation more than six (6) months after the occurrence of violation when OSHA's failure to discover the violation within six (6) months was due to the employer's failure to report the fatality as required by 20 CFR 1904.8. *Yelvington Welding Service,* 6 OSHC 2013 (1978).

2. Reports of Power Press Injuries

Employers must report *all* "recordable" injuries to employees operating mechanical power presses (point-of-operation accidents), within thirty days of the accident, either to OSHA's Directorate of the Office of Standards Development or to the agency administering an approved state plan. The report must include (29 CFR 1910.217(g)):

(1) The name of the employer and the address of the establishment where the accident occurred;
(2) The employee's name, the injury sustained, and the task being performed;
(3) The type of clutch used on the press;
(4) The type of safeguards being used;
(5) The cause of feeding;
(6) The type of feeding;
(7) The means used to actuate the press stroke;
(8) The number of operators provided with controls and safeguards.

H. POSTING REQUIREMENTS

1. General Rules on Posting

In addition to certain records discussed *supra*, any notices (*e.g.*, citations,[21] notices of contest,[22] notices of hearing, petitions for modification of abatement date (PMA), variance applications, access orders for employee medical records, lead exposure signs, cotton dust warning signs, coke oven cancer hazard signs) furnished by OSHA that inform employees of their rights and obligations under the Act, must be posted at each establishment by the employer, in a conspicuous place or places where employee notices are customarily posted. See, *e.g.*, Sections 9(b) and 10(c) of the Act; Commission Rule 7 and 34; see also 29 CFR 1903.16, .14(c) and .2; and 2200.34(c)(1); *Ed Miller & Sons, Inc.*, 2 OSHC 1132 (1974). Failure to post such notices and/or certify that posting was accomplished may prejudice an employer's rights under the Act. *Car and Truck Doctor, Inc.*, 8 OSHC 1767 (1980). The requirement that an employer display an OSHA poster does not violate the First Amendment right of freedom of speech. *Lake Butler Apparel Co. v. Secretary of Labor*, 519 F.2d 84, 3 OSHC 1522 (5th Cir. 1975).

"Establishment" is defined as a single physical location where business is conducted or where services or industrial operations are performed (*e.g.*, factory, store, hotel, restaurant, movie theater, farm, bank, warehouse, office). Where distinctly separate activities are performed at a single physical location, each activity is treated as a separate physical establishment, with separate notices posted. Where employers are engaged in activities which are physically dispersed, such as agriculture, construction, or transportation, the notice shall be posted at the location to which employees report each day or if they do not usually work at or report to a single establishment, such notice shall be posted at the location from which they operate to carry out their activities.

For example, 29 CFR 2200.34(c)(1) (Commission Rule 34), sets forth the specific requirements for notifying employees that a PMA has been filed by an employer. A copy of the petition must "be posted in a conspicuous place where all affected employees will have notice thereof or near each location where the violation occurred." The rule also requires that the petition remain posted for a period of ten days. Absent this notice, affected employees may be deprived of their right to oppose an employer's request for an extension of the abatement date or to otherwise participate in proceedings which result from an employer's PMA. *Auto Bolt & Nut Company*, 7 OSHC 1203 (1979); *Brockway Glass Company*, 6 OSHC 2089 (1978).

Completion of abatement (or vacation of citation) does not require a posting, rather, at that time the posted notice of contest may be removed. 29 CFR 1903.16(b). The failure of an employer to post a notice informing employees of their right to participate in Commission proceedings may require dismissal of the notice of contest. *Caribtow Corp.*, 1 OSHC 1503 (1973), *aff'd*, 493 F.2d 1064, 1 OSHC 1592 (1st Cir. 1974), *cert. denied*, 419 U.S. 830, 2 OSHC 1239 (1974).

OSHA posting requirements are different from the recordkeeping provisions: whereas the recordkeeping provision may not apply to certain employers (railroads), the posting provision may. *Chicago, Rock Island & Pacific Railroad Co.*, 3 OSHC 1694 (1975).

2. The OSHA Poster and Annual Summary

Each employer must display in each establishment a poster (the OSHA poster) which informs employees of the protections and obligations under the Act. See Section 8(c)(1) of the Act; 29 CFR 1903.2(a)(1). Failure to post the OSHA poster constitutes a violation of the Act and may result in issuance of a citation. *Intermountain Block & Pipe Corp.*, 1 OSHC 3145 (1972); *Automotive Products Corp.*, 1 OSHC 1772 (1974). The posting requirement has been held not to violate the First Amendment right of freedom of speech. *Lake Butler Apparel Co. v. Secretary of Labor*, 519 F.2d 84, 3 OSHC 1522 (5th Cir. 1975). Where a state has an *approved* poster informing employees of their protections and obligations, such poster, when posted, shall constitute compliance with the posting requirements of the Act. 29 CFR 1903.2(a)(2).

The Annual Summary, that is, the last page of OSHA Form No. 200, must be posted yearly at the time specified in the regulations. The failure to post the Annual Summary may result in a violation of 29 CFR 1904.5(d)(1). *Simmons, Inc.*, 6 OSHC 1157 (1977).

3. Place of Posting; Definition of "Establishment"

The regulations require that the OSHA poster be displayed at the employer's establishment at a location to which the employees report every day. 29 CFR 1903.2(b). As applied to the posting requirement, the definition of "establishment" is very broad. 29 CFR 1903.2(b). For example, a temporary construction worksite has been called an establishment such that the posting of notices was required under the Act. *Intermountain Block & Pipe Corporation*, 1 OSHC 3145 (1972); but see *San Juan Construction Company, Inc.*, 3 OSHC 1445 (1975). Chapter VII of the FOM provides for posting of notices on construction sites at the trailer or office if employees report there on a daily basis; in addition, a copy must be posted at any other location where employees are required to report daily. If there is no obvious place to post, the employer must furnish a suitable place in a conspicuous location adjacent to the worksite and protect the notice from the weather.

In *Western Waterproofing, Inc.*, 7 OSHC 1499 (1979), the employer was a roofing and waterproofing contractor with its principal place of business in Dallas. The worksite in question was located in Albuquerque, New Mexico. The employees assigned to the Albuquerque location were regularly stationed in Dallas but did not report to the Dallas headquarters during the term of the six-week Albuquerque assignment. There was no OSHA poster at the Alburquerque site at the time of the inspection. A poster was, however, properly posted at the Dallas headquarters warehouse where the employees regularly gathered, and a poster furnished by the OSHA compliance officer was posted at the Albuquerque site immediately subsequent to his inspection.

The regulation required the employer to post the notice in "each establishment." The regulation defined an "establishment" as "a single physical location where business is conducted or where services or industrial operations are performed." 29 CFR 1903.2(b). This definition, standing alone, would bring the Albuquerque jobsite under the regulation. However, 1903.2(b) later qualifies the definition as follows:

> Where employers are engaged in activities which are physically dispersed, such as agriculture, construction, transportation, communications, and electric, gas and sanitary services, the notice or notices required by this section shall be posted at the location to which employees report each day. Where employees do not usually work at, or report to, a single establishment, such as longshoremen, travelling salesmen, technicians, engineers, etc., such notice or notices shall be posted at the location from which the employees operate to carry out their activities.

Thus the definition of "establishment" distinguishes between situations where employers maintain multiple worksites but with each employee reporting to a particular location each day, and those where each employee frequently changes the physical location at which he works. In the former situations, the notice must be posted at each location; in the latter, it need only be posted "at the location from which the employees operate to carry out their activities."

Although the regulation lists "construction" as a type of activity for which the notice must be posted at the location to which employees report each day, some construction work falls under the category described in the latter part of the definition. See, *e.g., A. A. Will Sand & Gravel Corp.* 4 OSHC 1442 (1976) (driver who delivered and helped unload gravel at construction site held engaged in construction work); *Heede International, Inc.*, 2 OSHC 1466 (1975) (employee dismantling crane that was used on construction site held engaged in construction work). The listing of "construction" in the former part of the definition is, therefore, for purposes of illustration; it does not provide a mandatory requirement that each employer engaged in construction work post a notice at each site at which its employees may work. Thus simply because the employees in *Western Water Proofing* were engaged in construction work did not prove that the Albuquerque location was an "establishment" within the meaning of the cited regulation.

In this case, both parts of the definition partially described the nature of the employees operation. Although the employees reported to Albuquerque each day during the six-week period, they were normally assigned to various job sites from the Dallas headquarters, which can, therefore, be considered the "location from which (they) operate to carry out their activities." In other words, the regulation did not draw a sharp distinction between worksites at which the notice must be posted, and those that are so temporary that a notice is not required. Whereas in this case, either part of the definition is arguably applicable, it was appropriate to look to whether the purpose of the regulation was fulfilled in deciding whether an employer violated it. Here the employees had a full opportunity to read the OSHA poster, which was properly displayed at the Dallas headquarters where they regularly reported, thus fulfilling the purpose of the regulation. The Review Commission, therefore, concluded that the company did not violate 29 CFR 1903.2(a)(1).

Places and the time periods of posting for other notices are set forth as follows: (1) citation—post immediately upon receipt at or near place where violation occurred, to remain posted for 3 days or until violation is abated or vacated whichever is later; (2) notice of contest—posted immediately upon receipt at location of citation until disposition of matter; (3) PMA—10 days if not objected to (otherwise until settled) at location where citation is posted; (4) Annual Summary—in a conspicious place from February 1 to March 1 of each year; (5) labels and signs—continuously as required by standard at safe distance from area and on the materials involved; and (6) petition for recordkeeping exceptions—continuously where OSHA poster is posted.

4. Multiple Worksites

When the employer is engaged in activities which require that operations be physically dispersed, the notice must be posted at the location where the employees *report* to work each day. *Par Construction, Inc.*, 4 OSHC 1799 (1976).

Whether or not the employees actually see the notice is irrelevant for purposes of compliance with this regulation. As long as the notices are conspicuously posted at the place where the employees report, the employer should not be cited. *Danco Construction Co.*, 3 OSHC 1114 (1975).

5. Multiple Reporting Sites

If the employees do not report to a single establishment, then the notices must be posted at the *base* from which the employees operate to carry out their activities. The failure to post the poster at a job site is not a violation where posters were maintained at the main headquarters and field offices, since there is nothing permanent at the job site on which the poster could be posted. *Kent Nowlin Construction Co., Inc.*, 5 OSHC 1051 (1977).

The permanence of the structure is often determinative of whether something is a "base" at which notices must be posted. See *BI-CO Pavers, Inc.*, 2 OSHC 3142 (1974), where the lack of permanent shelters at a worksite meant that a "base" was not established.

6. How to Post Notices or Posters

Posting must be in a "conspicuous" place at the establishment, or in places where employee notices are customarily posted or maintained. The employer must "take steps to insure that such notices are not altered, defaced, or covered by other material" (29 CFR 903.2(a)(1)). Destruction of poster and wall by a tornado did not absolve employer from the duty to post. *Custom Fabricators, Inc.,* 3 OSHC 1459 (1975). Merely showing the poster to employees does not fulfill the employer's duty to post in a conspicuous place. *Kinney Steel,* 3 OSHC 1453 (1975).

7. Failure to Post

The OSHA Field Operations Manual directs compliance officers to determine, upon inspection, if posting requirements have been met in accordance with 29 CFR 1903.2(a) and 1904.5(d). Failure to comply with the posting regulations may result in issuance of citations and proposed penalties. *Intermountain Block & Pipe Corporation, supra; McDowell Co., surpa.* But the temporary removal of the poster during yearly painting and clean-up accompanied by subsequent replacement will be substantial compliance with 29 CFR 1903.2(a). *Glendale Mills, Inc.,* 6 OSHC 1361 (1978).

An employer cannot be cited, however, unless OSHA has supplied him with the posters or other notice, *i.e., Woerfel Corp.,* 1 OSHC 3299 (1974); *Oak Lane Diner,* 1 OSHC 1248 (1973); *Ira Holliday Logging Co., Inc.,* 1 OSHC 1200 (1973).

Failure to post a notice of contest and provide certification of posting may result in dismissal of the notice of contest. *Car and Truck Doctor, Inc.,* 8 OSHC 1767 (1980).

Certain set penalties (gravity based) will normally be proposed for violations of the posting regulations: (1) failure to post notice of the Act furnished by OSHA—$100; (2) failure to post citations—$500 for *each* citation; and (3) failure to post a copy of the Summary of Occupational Injuries and Illnesses —$200. Violations of posting and recordkeeping requirements which involve the same document are grouped for penalty purposes. FOM, Chapter XII. However, in *Cornett & Weaver,* 7 OSHC 1665 (1979), the failure to post the OSHA poster drew only a $1.00 penalty. In *Stowe Canoe Co.,* 4 OSHC 1012 (1976), no penalty was assessed for failure to post the annual summary. The Secretary bears the burden of establishing that employees do not report to the place where notices are posted. *Davis-McKee, Inc.,* 2 OSHC 3046 (1974).

8. Time of Posting

From February 1 to March 1, each establishment having eleven or more employees must post in areas where notices to employees are customarily posted a copy of the year's total count of job-related injuries and illnesses that occurred in their establishments during the previous calendar year. The last page or right-hand portion of OSHA Form No. 200 with only the yearly totals, establishment identification, and certification may be used for this purpose. The form must be posted in the place or places where notices to employees are customarily displayed.

Establishments having no injuries or illnesses during the year must enter zeros on the totals line, sign, and post the form.

The OSHA poster must be posted at all times.

9. Specific Posting Requirements

a. *Imminent Danger*

If when notified of an imminent-danger situation by an OSHA inspector, the employer fails to eliminate the danger or remove employees from the area and give assurances that the danger will be eliminated, the inspector, with the approval of the OSHA Area Director, shall post a "notice of alleged imminent danger." Section 13 of the Act; FOM Chapter IX(c)(4).

b. *Toxic and Hazardous Substance*

Standards regulating exposure to toxic substances or harmful physical agents (*e.g.*, asbestos, cotton dust, benzene, coke oven emissions, 4-Nitrobiphenyl, lead, acrylonite, inorganic arsenic) require the posting of caution signs, labels, and warnings.

c. *Signs, Tags, Labels, and Warnings*

These are required to be posted with respect to the following standards, *inter alia:* (1) accident prevention (exits, slow moving vehicles, danger, caution) 29 CFR 1910.145; (2) respirators 29 CFR 1910.134; (3) powered platforms 29 CFR 1910.66; (4) crane operating instructions 29 CFR 1926.550(a)(2); (5) fire and explosion prevention (no smoking) 29 CFR 1926.151–.152; (6) conveyors 29 CFR 1926.555; and (7) explosives 29 CFR 1926.903. Program directives such as No. 100–24 concerning the "Caution Labeling of Radial Saws," 29 CFR 1910.213(h)(5), may impose additional obligations with respect to posting. Failure to post signs, labels, or tags will subject an employer to a violation of the standard or even to the general-duty clause. *Cajun Company,* 6 OSHC 1804 (1978); *United States Pipe and Foundry Company,* 6 OSHC 1332 (1978); *Hillsdale Lumber & Mfg., Inc.,* 5 OSHC 1281 (1977).

In *Sierra Construction Corp.,* 6 OSHC 1278 (1978) the Commission held that an employer's use of its own signaling system did not relieve it from compliance with 29 CFR 1926.550(a)(4) posting requirements, absent application for a variance. Moreover, the incorrect citation of a subsection of a standard did not prejudice the employer since the substance of the alleged violation, absence of a poster, was basically the same. The failure to post was also not a *de minimus* violation since the hazard addressed in the standard was not trifling.

I. GUIDELINES FOR DETERMINING RECORDABILITY

Employers have the responsibility for determining whether an injury is recordable.

OSHA recordkeeping and reporting requirements differ from those established under the various state worker's compensation laws. Thus employ ers must not substitute worker's compensation criteria in determining whether or not a case should be recorded for OSHA. Worker's compensation rules may require employers to record more or less cases than the OSHA rules. How then does an employer determine whether an injury or illness is occupational or work-related and thus recordable?

Basic recordkeeping concepts and guidelines are included with instructions on the back of OSHA Form No. 200. They are also set forth in 29 CFR 1904.12(c). This section summarizes the major recordkeeping concepts and provides additional information to aid in keeping records accurately.

An injury or illness is considered to be occupational if it is directly related to the workplace and occurs in the work environment, which is defined as any area on or off the employer's premises, *e.g.*, worksite, company cafeteria, or company parking lot or results from a work accident or exposure to environmental factors, materials, equipment used by employees to perform their jobs, or associated with employment. An employee sporting event paid for by the employer may, for example, be considered the work environment. The work environment surrounds the worker wherever he or she goes—in official travel, in dispersed operations, or along regular routes (*e.g.*, sales representative, pipeline worker, vending machine repairer, telephone line worker). Once an injury or illness is determined to be occupational, then the employer must determine if it is recordable.

A recordable occupational injury or illness[23] for purposes of OSHA Form No. 200 includes the following:

(1) Fatalities, regardless of the time between injury/illness and death, must be recorded (and reported to OSHA within 48 hours).
(2) All diagnosed work-related illnesses reported to the employer must be recorded.
(3) All work-related injuries (nonfatal cases) if they require medical treatment rather than first aid or involve a loss of consciousness, restriction of work or motion, employment termination, or transfer to another job must be recorded.
(4) Lost workday cases, other than fatalities, that result in lost workdays must be recorded. A "lost workday" is defined as: the number of days (consecutive or not) after, but not including, the day of injury or illness during which the employee would have worked but could not do so; that is, could not perform all or any part of his normal assignment during all or any part of the workday or shift, *because of* the occupational injury or illness. 29 CFR. §1904.12(f).

Recordable and nonrecordable injuries are distinguished by the treatment provided; *i.e.*, if the injury required medical treatment, it is recordable; if only first aid was required, it is not recordable. See, Chart II, *infra*. However, medical treatment is only one of several criteria for determining recordability. Regardless of treatment, if the injury involved loss of consciousness, restriction of work or motion, transfer to another job, or termination of employment, the injury is recordable.

Medical treatment is treatment administered by a physician or by registered professional personnel under a physician's standing orders. The fact that first aid is administered by a physician or by registered professional personnel does not make such first aid a medical treatment case. The kind of treatment, rather than the person who administers it, determines whether the remedy is first aid or medical treatment. *La Biches, Inc.*, 2 OSHC 3110 (1974). The following are considered to involve medical treatment and must be recorded for a work-related injury.

- Antiseptics applied on second or subsequent visit to a doctor or nurse.
- Burns of a second or third degree.
- Butterfly sutures.
- Compresses, hot or cold, on second or subsequent visit to a doctor or nurse.
- Cutting away dead skin (surgical debridement).
- Diathermy treatment.
- Foreign bodies, removal if embedded in eye.
- Foreign bodies, if removal from wound requires a physician because of depth of embedment, size or shape of object(s) or location of wound.
- Infection, treatment for.
- Prescription medications used.
- Soaking, hot or cold, on second or subsequent visit.
- Sutures (stitches).
- Whirlpool treatment.
- X-ray which is positive.

First-aid treatment is one-time treatment of minor cuts, scratches, burns, and the like. It also includes a follow-up visit for purposes of observation. First aid may be administered by a physician, registered professional personnel, or others. The following are considered to involve only first-aid treatment and need not be recorded if the work-related injury does not involve loss of consciousness, restriction of work or motion, or transfer to another job.

- Antiseptics, application of, on first visit to a doctor or nurse.
- Bandaging on any visit to a doctor or nurse.
- Burns of first degree.
- Compresses, hot or cold, on first visit to a doctor or nurse only.
- Foreign bodies, not embedded, irrigation of eye for removal.
- Foreign bodies, removal from wound by tweezers or other simple techniques.
- Nonprescription medications, use of.
- Observation of injury on second or subsequent visit.
- Ointments applied to abrasions to prevent drying or cracking.

Other procedures which are not considered medical treatment.

- Tetanus shots, initial or boosters along.
- Hospitalization for observation (no treatment other than first aid).
- X-ray which is negative.

Injuries are to be recorded as soon as they are determined to be work-related (requiring medical treatment rather than first aid), *Chrysler Corp.*, 7 OSHC 1578 (1979); illnesses are to be recorded as of the date of diagnosis or the first day of absence due to such illness. Few alleged violations concerning failure to record "recordable injuries or illnesses" have been issued, but see *H. Wilson Corp.*, 1 OSHC 3031 (1972); *Intermountain Boock & Pipe Corp.*, 1 OSHC 3145 (1972); *La Biches, Inc.*, 2 OSHC 3110 (1974).

NOTES

1. The Act is silent, however, as to the proper forum wherein judicial review of such action may be sought. *Id.* As used in the Act, a "standard" is a substantive rule containing a requirement "reasonably necessary or appropriate to provide safe or healthful employment." Section 3(8) of the Act. A "regulation" is a rule governing matters such as posting of notices, recordkeeping, and conduct of inspections. *E.g.*, Section 8(c) of the Act; *Northwest Airlines, Inc.*, 8 OSHC 1982 (1980).

2. Federal agencies, while not subject to the Act, are subject to requirements under 29 U.S.C. §668. States and their political subdivisions may be subject to requirements under approved state plans.

3. The information required includes: (1) date of injury or onset of illness; (2) employee's name; (3) department and occupation/workplace; (4) description of illness or injury; (5) fatality, if any; (6) lost workdays; (7) job transfer or termination; and (8) changes in illness or injury.

4. The instructions to OSHA Form No. 100 require that all occupational illnesses be recorded and define occupational illness: "Occupational illness of an employee is any abnormal condition or disorder, other than one resulting from an occupational injury, *caused* by exposure to environmental factors associated with employment. It includes acute and chronic illness or diseases which may be caused by inhalation, absorption, ingestion, or direct contact."

5. The following question and answer from the pamphlet were pertinent to the case:

Q. What are the reporting requirements of pre-existing physical deficiencies so far as the OSH Act is concerned?

A. None. However, each case which involves aggravation of pre-existing physical deficiency must be examined to determine whether or not the employee's work was a contributing factor. If a work accident or exposure in the work environment contributed to the aggravation, the case is work related. It must be recorded if it meets the other requirements of recordability.

6. 29 CFR 1904.12(c) provides: "(c) 'Recordable occupational injuries or illnesses' are any occupational injuries or illnesses which result in: (1) Fatalities, regardless of the time between the injury and death, or the length of the illness; or (2) Lost workday cases, other than fatalities, that result in lost workdays; or, (3) Nonfatal cases without lost workdays which result in transfer to another job or termination of employment, or require medical treatment (other than first aid) or involve: loss of consciousness or restriction of work or motion. This category also includes any diagnosed occupational illnesses which are reported to the employer but are not classified as fatalities or lost work day cases."

7. The information required includes: (1) employer's name, address, and location; (2) injured or ill employee's name, address, age, occupation, sex, department, and social security number; (3) place of accident/exposure (on/off premises of employer), how accident occurred, what employee was doing when it occurred; (4) description of injury and illness/body part affected; (5) object or substance which injured employee; (6) date of injury or diagnosis of illness; (7) did employee die; (8) name and address of physician and hospital; and (9) name and position of preparer of report and date of report.

8. The following information must be recorded: (1) total number of cases; (2) number of deaths; (3) lost workday cases; (4) nonfatal cases without lost workdays; and (5) terminations or permanent transfers.

9. A hazardous material has one or more of the following characteristics: (1) has a flashpoint below 140° F, closed cup, or subject to spontaneous heating; (2) has a threshold limit value below 500 ppm for gases and vapors, below 500 mg/m^3 for fumes, and below 25 mp/ft^3 for dusts; (3) has a single dose oral LD below 500 mg/kg; (4) is subject to polymerization with the release of large amounts of energy; (5) is a strong oxidizing or reducing agent; (6) causes first-degree burns to skin in short time exposure or is systemically toxic by skin contact; or (7) in the course of normal operations, may produce dusts, gases, fumes, vapors, mists, or smokes which have one or more of the above characteristics.

10. On written request to NIOSH, an employee has the right to obtain a determination of whether or not a substance found or used in the workplace is harmful. 42 CFR 85.

11. "Toxic or harmful physical agent" means a chemical substance, physical agent, or physical stress which is: (1) regulated by federal law or rule due to a hazard to health; (2) is listed on NIOSH's Registry of Toxic Chemical Substances; (3) has a material-safety date sheet available; or (4) has in human, animal, or other biological testing shown positive evidence of being an acute or chronic health hazard.

12. The Act is concerned only with exposures to toxic materials or harmful physical agents which are required to be monitored or measured under Section 6, that is, a substance regulated by an OSHA standard. The new regulations, however, contain no such limitation; in fact, they specifically indicate that the substance need not be one that is regulated by OSHA.

13. Health insurance claims maintained separately from an employer's medical program need not be maintained for any specified period. Material-safety data sheets concerning the identity of a substance or agent need not be retained for any specified period as long as some record of the agent or substance identifies where/when it was used and is retained for at least thirty years. Background data to environmental monitoring and measuring need be retained for only one year as long as the results, methods used, and their summary are retained for at least thirty years.

14. Employee exposure records include the following information: environmental monitoring or measuring, including personal, area, wipe, grab, or other sampling plus methodology, calculations, and other background data relevant to the interpretation of results; biological monitoring results which directly assess the absorption of a substance or agent by body systems; material-safety data sheet or any other record which reveals the identity of a toxic substance or harmful physical agent.

15. "Employee medical records" means a record concerning the health status of an employee made or maintained by a technician, including: medical and employment questionnaires and histories; results of medical examinations; laboratory tests; and medical opinions on employee medical complaints (but excluding normally discarded physical specimens, records on health insurance claims if maintained separately from

medical records, and records concerning voluntary employee assistance programs), whether or not the records are related to an OSHA standard.

16. "Analysis" is defined as any compilation of data, research, statistical, or other study based at least in part on information collected from individual employee exposure or medical records or information collected from health insurance claims records, provided that either the analysis has been reported to the employer or no further work is currently being done by the person responsible for preparing the analysis.

17. Due to the substantial personal privacy interests involved, OSHA authority to gain access to personally identifiable employee medical information (*i.e.*, employee medical information accompanied by either direct identifiers (name, address, social security number, payroll number, and so forth), or by information which could reasonably be used in the particular circumstances indirectly to identify specific employees (*e.g.*, exact age, height, weight, race, sex, date of initial employment, job title, *etc.*,) will be exercised only after it has made a careful determination of its need for this information, and only with appropriate safeguards to protect individual privacy. Once this information is obtained, OSHA examination and use of it will be limited to only that information needed to accomplish the purpose for access. Personally identifiable employee medical information will be retained by OSHA only for so long as needed to accomplish the purpose of access, will be kept secure while being used, and will not be disclosed to other agencies or members of the public except in narrowly defined circumstances. 29 CFR 1913.10 establishes procedures to implement these policies in all requests by OSHA to examine or copy personally identifiable medical information, whether or not pursuant to the access provisions of 29 CFR 1910.20(e). This regulation does not apply to OSHA access to, or the use of, aggregate employee medical information or medical records on individual employees which is not in a personally identifiable form, nor does it apply to records required by 29 CFR 1904, to death certificates, or to employee exposure records, including biological monitoring records treated under 29 CFR 1910.20(c)(5) or by specific occupational safety and health standards as exposure records. Moreover, the section does not apply where OSHA compliance personnel conduct an examination of employee medical records solely to verify employer compliance with the medical surveillance recordkeeping requirements of an occupational safety and health standard or with 29 CFR 1910.20.

18. The summary of the regulation states that employee access to exposure and medical information is important to the effectiveness of certain employee rights established by the Act, to wit: (1) access is "crucial" to the effectiveness of the Section 8(f)(1) right of employees and their representatives to complain to OSHA about safety and health problems and to obtain a prompt inspection of the workplace; (2) access "gives meaning" to the Section 8(c) right of employees and their representatives to accompany OSHA inspectors during walkaround tours in order to identify where and how various toxic substances are used and which operations generate the greatest exposure and otherwise to help OSHA conduct a thorough inspection; (3) access will enable workers to exercise their twin rights under Section 10(c) to contest the reasonableness of abatement periods and to participate as parties in Review Commission proceedings; (4) access to records will enhance the Section 20(a)(6) right of workers to request a NIOSH health-hazard evaluation; and (5) the knowledge learned by workers due to access to medical and exposure records will heighten the impact of employee training and education programs funded under section 21(c) of the Act.

19. OSHA has proposed to amend 29 CFR 1904.8 to require that employers report a fatality or multiple hospitalization accident within eight hours. OSHA also proposed that employers report all deaths resulting within six months of an accident

within eight hours of the time the employer becomes aware of the death. 44 Fed. Reg. 59,560 (October 16, 1979). Most employers responding to the proposal opposed it as being burdensome and inflexible.

20. OSHA Instruction CPL 2.43 clarifies the meaning of hospitalization as to be sent to a hospital, to go to a hospital, or to be admitted to a hospital or equivalent medical facility, regardless of the treatment provided or the length of the stay.

21. A citation must remain posted until the violation is abated or for three working days, whichever is later. When a citation is amended both the original and amended version must be posted. OSHA Notice CPL 2–1.16. The filing of a notice of contest does not affect an employer's posting responsibility, unless and until an order is issued by the Commission vacating the citation. 29 CFR 1903.16.

22. Pursuant to Commission Rule 7, when the employer receives notice of "docketing" of its notice of contest or petition for modification of the abatement period, the employer must post, where the citation is required to be posted, a copy of the notice of contest and a notice informing such affected employees of their right to party status and of the availability of all pleadings for inspection and copying at reasonable times. A notice in the following form shall be deemed to comply with this paragraph:

[Name of employer]

Your employer has been cited by the Secretary of Labor for violation of the Occupational Safety and Health Act of 1970. The citation has been contested and will be the subject of a hearing before the Occupational Safety and Health Review Commission. Affected employees are entitled to participate in this hearing as parties under terms and conditions established by the Occupational Safety and Health Review Commission in its Rules of Procedure. Notice of intent to participate should be sent to:

Occupational Safety and Health Review
Commission
1825 K Street, N.W.
Washington, DC 20006

All papers relevant to this matter may be inspected at

[A place reasonably convenient to employees,
preferably at or near workplace].

Where appropriate, the second sentence of the above notice will be deleted, and the following sentence will be substituted:

> The reasonableness of the period prescribed by the Secretary of Labor for abatement of the violation has been contested and will be the subject of a hearing before the Occupational Safety and Health Review Commission.

A copy of the notice of hearing to be held before the Judge must also be posted.

23. An occupational injury is any injury, such as a cut, fracture, sprain, or amputation, which results from a work accident or from an exposure involving a single incident in the work environment. An occupational illness is any abnormal condition or disorder, other than one resulting from an occupational injury, caused by exposure to environmental factors associated with employment. The work environment is defined as the physical location, equipment, materials processed or used, and kinds of operations performed in the course of an employee's work, whether on or off the employer's premises.

CHAPTER VIII
The General Duty Clause

A. INTRODUCTION

It was intended that the occupational safety and health standards promulgated pursuant to the authority set forth in Section 6 of the Act would be the principal method of enforcement and compliance under the Act. Section 5(a)(2) of the Act, dubbed the "specific duty clause," provides that employers shall comply with all such standards promulgated under the Act. In addition, in those circumstances in which standards had not been adopted or were not applicable to a given situation, it was intended that Section 5(a)(1), known as the "general duty clause," would provide protection. Section 5(a)(1) provides:

> Each employer shall furnish to each of his employees employment and a place of employment which are free from recognized hazards that are causing or are likely to cause death or serious physical harm to his employees.

Unlike specific standards which are normally phrased in absolute terms, the general duty clause is actually a catchall. Violations of the general duty clause will, however, subject an employer to more or less the same penalties as violations of specific standards. Application of the general duty clause has been held not to be an unconstitutional delegation of legislative power and not to be void for vagueness. *REA Express Inc. v. Brennan,* 495 F.2d 822, 1 OSHC 1651 (2d Cir. 1974); *Bethlehem Steel Corp. v. OSHRC,* 607 F.2d 1069, 1 OSHC 1833 (3d Cir. 1979).

B. NATIONAL REALTY AND THE GENERAL DUTY CLAUSE ELEMENTS

The first major court decision to consider the construction of the general duty clause was *National Realty and Construction Company v. OSHRC,* 489 F.2d 1257, 1 OSHC 1422 (D.C. Cir. 1973). That seminal decision has set forth the definitive basis for the interpretation and application of the general duty clause to this date. The citation in the case charged that the employer had "permitted" the existence of a condition which constituted a recognized hazard that was likely to cause death or serious physical harm to its employees. The condition arose when one of the employees stood as a passenger on the running board of a piece of construction equipment which was in motion.

The word "permit" in the circumstances of the case was read to suggest a "failure to prevent" the incident rather than a knowing authorization of the conduct. Since no specific standard proscribed this conduct the employer was cited for a violation of the general duty clause.

The court stated that under the general duty clause the Secretary of Labor must prove the following elements: (1) that the employer failed to render its workplace *free* of a hazard which was (2) recognized *and* was (3) causing or likely to cause death or serious physical harm. Addressing the issue of the existence of the hazard, the first element, the court stated that the hazard involved was the dangerous activity of riding heavy equipment. According to the court the record clearly contained substantial evidence to support the Commission's finding that the hazard was recognized. The court then discussed the term "recognized":

> An activity may be a "recognized hazard" even if the defendant employer is ignorant of the activity's existence or its potential for harm. The term received a concise definition in a floor speech by Representative Daniels when he proposed an amendment which became the present version of the general duty clause:
>
>> A recognized hazard is a condition that is known to be hazardous, and is known not necessarily by each and every individual employer but is known taking into account the standard of knowledge in the industry. In other words, whether or not a hazard is recognized is a matter for objective determination; it does not depend on whether the particular employer is aware of it.
>
> 116 Cong. Rec. (Part 28) 38377 (1970). The standard would be the common knowledge of safety experts who are familiar with the circumstances of the industry or activity in question. The evidence below showed that both National Realty and the Army Corps of Engineers took equipment riding seriously enough to prohibit it as a matter of policy. Absent contrary indications this is at least substantial evidence that equipment riding is a "recognized hazard."

The court also found that the recognized hazard was likely to cause death or serious physical harm, the third element of the general duty clause. In this respect the court stated that:

> Presumably any given instance of equipment riding carries a less than 50 percent probability of serious mishap, but no such mathematical test would be proper in construing this element of the general duty clause. See Morey, *The General Duty Clause of the Occupational Safety and Health Act of 1970,* 86 Harvard Law Review 988, 997 to 998 (1973). If evidence is presented that a practice *could* eventuate in a serious physical harm upon other than a freakish or utterly implausible concurrence of circumstances, the Commission's expert determination of likelihood should be afforded considerable deference by the courts. For equipment riding, the potential for injury is indicated on the record by Smith's death and, of course, by common sense [Emphasis added.]

The remaining question for the court was whether the employer rendered its workplace "free" of the hazard, the first element of proof under the general duty clause case. In this respect the meaning of the term "free" was presented as an instance of first impression. The court stated that a workplace cannot be just reasonably free of a hazard or merely as free as the average workplace in the industry. Although both the Senate and House

Committee reports on the Act stated that the general duty clause incorporated "common law principles," the court noted that the standard of care imposed on employers by the general duty clause was not characterized in terms of reasonableness. Rather, the court found that the reports spoke of a "general and common duty to bring *no* adverse effects to the life and health of their employees throughout the course of their employment." See House Report No. 91–1291, Committee on Education and Labor, 91st Cong., 2d Sess., 14–15 (July 9, 1970); Senate Report No. 91–1282, 91st Cong., 2d Sess., 9 (Oct. 6, 1970). Overtones of reasonableness were to be found only in the Act's definition of a serious violation. (See discussion *infra.*) The court felt that employers must take more than merely *reasonable* precautions, for the safety of employees follows from the control which employers exert over their conduct and working conditions. However, the court stated that Congress did not intend that the general duty clause should impose strict liability. The duty was to be an achievable one.[1] In this respect the court referred to the Act's purpose "to assure so far as possible every working man and woman in the Nation safe and healthful working conditions." In support for this position the court again relied on House Report No. 91–1291, which stated:

> An employer's duty under Section 5(a)(1) is not an absolute one. It is the Committee's intent that an employer exercise care to furnish a safe and healthful place to work and to provide safe tools and equipment. This is not a vague duty, but is protection for the worker from preventable dangers.

The court construed from this language that the very word *duty* implied an obligation capable of *achievement. Restatement (Second) of Torts,* Section 4 (1965). The court also construed the language of the Congress to be consonant with its intent:

> [O]nly where the recognized hazard in question *can be* totally eliminated in a workplace. A hazard consisting of conduct by employees, such as equipment riding, cannot however, be totally eliminated. A demented, suicidal or willfully reckless employee may on occasion circumvent the best conceived and most vigorously enforced safety regime. . . . This seeming dilemma is, however soluble within the literal structure of the general duty clause. . . . The employer's duty is, however, disqualified by the simple requirement that it be achievable and not be a mere vehicle for strict liability.

Thus the court concluded that Congress intended only to require the elimination of preventable hazards; Congress did not intend unpreventable hazards to be considered recognized under the clause. The court then stated as an example that although a *generic* form of hazardous conduct, such as equipment riding, may be recognized, unpreventable instances of it may not be, and thus the possibility of their occurrence at a workplace was not inconsistent with the workplace being free of recognized hazards.

The court then attempted to draw forth on the criteria of preventability, stating that it "draws content from the informed judgement of safety experts." Hazardous conduct is not preventable if it is so idiosyncratic and implausible in motive or means that conscientious experts *familiar with the industry* would not take it into account in prescribing a safety program. Nor is misconduct preventable if its elimination would require methods of hiring, training, monitoring, or sanctioning workers which are either so untested or so expensive that safety experts would substantially concur in thinking the methods infeasible. However, the court cautioned that:

> This is not to say that a safety precaution must find general usage in the industry before its absence gives rise to a general duty violation. The question is whether a precaution is recognized by safety experts as feasible, not whether the precaution's use has become customary. Similarly a precaution does not become infeasible merely because it is expensive. But if adoption of the precaution would clearly threaten the economic viability of the employer, the Secretary should propose the precaution by way of promulgated regulations, subject to advance industry comment, rather than through adventurous enforcement of the general duty clause. Finally, in an abundance of caution we emphasize that an instance of hazardous employee conduct may be considered preventable even if no employer could have detected the conduct, or its hazardous character, at the moment of its occurrence. Conceivably, such conduct might have been precluded through feasible precautions concerning the hiring, training, and sanctioning of employees.

Finally, the court stated (and in doing so engrafted a *fourth* element onto the general duty clause) that in order to establish a violation of the general duty clause, hazardous conduct need not actually have occurred, for feasibly curable inadequacies may sometimes be demonstrated before employees have acted dangerously. At the same time, however, the court noted that the actual occurrence of hazardous conduct is not, by itself, sufficient evidence of a violation even when the conduct had led to injury. The record must therefore additionally indicate that demonstrably *feasible* measures would have *materially reduced* the likelihood that such misconduct would have occurred. The court stated that to assure that violations issue only upon careful preparation, the Secretary of Labor must specify the particular steps a cited employer should have taken to avoid citation and demonstrate the feasibility and likely utility of those measures. Since the Secretary of Labor failed to demonstrate the feasibility and likely utility of the particular measures, the employer in the case before the court should have taken, it did not shoulder its burden of proof, and the record lacked substantial evidence of a violation.

A similar and equally extensive discussion of the application and interpretation of the general duty clause in the context of state plan enforcement is *Green Mountain Power Corporation v. Commissioner of Labor & Industry,* 383 A.2d 1046, 6 OSHC 1499 (S. Ct. Vt. 1978).

C. FAIRNESS CHALLENGES

Often citations issued under the general duty clause are challenged on the basis of a due process rationale, the argument being that fair notice has not been afforded an employer as to what conduct was prohibited under the Act. Although recognizing that the citation in *National Realty* was deficient, the court determined that the citation and pleadings were not so misleading as to foreclose the Secretary of Labor from litigating the citation or as to prejudice the employer in defending it. The court also noted that ambiguities in the citation could be cured at the hearing itself. The court stated that the operative factor is whether fair notice was afforded and that this follows from the familiar rule that administrative pleadings are very liberally construed and very easily amended. Thus the court stated:

> The rule has particular pertinence here, for citations under the 1970 Act, are drafted by non-legal personnel, acting with necessary dispatch. Enforcement of the Act would be crippled if the Secretary were inflexibly held to a narrow construction of citations issued by his inspectors. Allowing subsequent amendment of a citation's charges will not disturb the central function of the citation, which is to alert a cited employer that it must contest the Secretary's allegation or pay the proposed fine.

However, in *National Realty* it was noted that where the Secretary of Labor, in the absence of a specific regulation, elects to proceed under a general safety standard, as a matter of fundamental fairness, it is

> [o]nly by requiring the Secretary at his hearing to formulate and defend his own theory of what a cited defendant should have done that the Commission and the courts assure evenhanded enforcement of the general duty clause. . . . To assure that citations issue only upon careful deliberation, the Secretary must be constrained to specify the particular steps a cited employer should have taken to avoid citation, and to demonstrate the feasibility and likely utility of those measures.

See *Baroid Division of NL Industries, Inc.*, 7 OSHC 1466 (1979).

The courts, since *National Realty,* have held that since the general duty clause is so broad the evidence to support a charge of its violation should be specific in detail. *Brennan v. OSHRC (Canrad Precision Industries),* 502 F.2d 946, 2 OSHC 1137 (3d Cir. 1974). However, as the court in *Babcock and Wilcox Company v. OSHRC,* 622 F.2d 1160, 8 OSHC 1317 (3d Cir. 1980), stated:

> We have no doubt that the employer was put on notice as to the nature of the Secretary's complaint. Citations must give fair notice to the employer so that it understands the charge being made and has an adequate opportunity to prepare and present a defense. Citations, however, are prepared by inspectors who are not legally trained and who should act with dispatch. For these reasons citations should not be as tightly construed as other pleadings—a grand jury indictment, for example.

This same standard has been applied in cases dealing with the enforcement of specific standards which are relatively vague in their compliance requirements. For example, 29 CFR 1926.28(a) and 29 CFR 1910.132(a) require the wearing of *appropriate/necessary* personal protective equipment where there is a need for using such equipment to reduce the hazards to the employees. Thus in cases such as *B & B Insulation, Inc. v. OSHRC,* 583 F.2d 1364, 6 OSHC 2062 (5th Cir. 1978); *Cape and Vineyard Division of New Bedford Gas and Edison Light Company v. OSHRC,* 512 F.2d 1148, 2 OSHC 1628 (1st Cir. 1975); *American Airlines, Inc. v. Secretary of Labor,* 578 F.2d 38, 6 OSHC 1691 (2d Cir. 1976); *McLean Trucking Co. v. OSHRC,* 503 F.2d 8, 2 OSHC 1165 (4th Cir. 1974); *Ray Evers Welding Company v. OSHRC,* 625 F.2d 726, 8 OSHC 1271 (6th Cir. 1980); and *Voegele Company v. OSHRC,* 625 F.2d 1075, 8 OSHC 1631 (3d Cir. 1980), the courts "generally" have held:

> Since the language of this type of standard is similar in breadth to the general duty clause language, analogies are also appropriate to that clause. Employers have challenged these types of regulatory provisions (1926.28(a) and 1910.132) as unconstitutionally vague because the regulations fail to provide adequate notice to the employer of what conduct is prohibited. In order to uphold the regulations in the face of such a constitutional attack, the first test of the regulation ([1]) has been held to imply an objective standard—the reasonably prudent person test. . . .
>
> Whether a reasonable person familiar with the conditions in the industry would have instituted more elaborate precautions? . . .
>
> We are presented with the question of whether a reasonable person familiar with the factual circumstances surrounding the allegedly hazardous condition, including any facts unique to this particular industry, would recognize a hazard warranting the use of personal protective equipment. There are actually three factors to evaluate under this standard: recognition of a hazard, feasibility of alternatives, and whether the alternatives would create a greater hazard. The burden of proof rests with the Secretary to prove all elements of a violation of a general duty safety standard.

The Commission, however, disagrees with the foregoing analysis to the extent that it incorporates the "feasibility and likely utility" element of a Section 5(a)(1) charge into the Secretary of Labor's burden of proving noncompliance with a specific standard. *S & H Riggers and Erectors, Inc.*, 7 OSHC 1260 (1979); *Austin Building Co.*, 8 OSHC 2150 (1980).

D. INAPPLICABILITY OF SPECIFIC STANDARDS

The OSHA Field Operations Manual (FOM) provides that, during the course of an inspection, the inspector's primary concern should be the determination of whether the employer is in compliance with the several safety and health standards promulgated under the Act. FOM, Ch. VIII. However, his attention is also required to be directed to whether the employer is complying with the general duty clause. The FOM provides, however, that the general duty provisions should be cited in inspections *only* where there are no specific standards applicable to the particular hazard involved. In this respect 29 CFR 1910.5(f) expressly provides that an employer who is in compliance with a specific standard shall be deemed to be in compliance with the general duty clause insofar as it applies to the same condition, practice, means, method, operation, or process covered by the specific standard. Any "recognized" hazard created in part by a condition not covered by a standard may, however, be cited under the general duty clause.

The courts and the Commission have also adopted this rationale, stating that the use of the general duty clause is improper and should not be available when a specific standard has been promulgated which covers the hazards involved. *American Smelting and Refining Company v. OSHRC*, 501 F.2d 504, 2 OSHC 1041 (8th Cir. 1974); *R. L. Sanders Roofing Company*, 7 OSHC 1566 (1979); *Central City Roofing Company*, 4 OSHC 1286 (1976); *U.S. Pipe & Foundry Co.*, 6 OSHC 1332 (1978); *Pratt & Whitney Air Craft, Division of United Technologies Corp.*, 8 OSHC 1332 (1980); *Issek Brothers*, 3 OSHC 1964 (1976); *Mississippi Power & Light Co.*, 7 OSHC 2036 (1979); *Austin Building Company*,

8 OSHC 2150 (1980); *A. Prokosch & Sons Sheet Metal, Inc.,* 8 OSHC 2077 (1980).

Often, however, the Secretary of Labor may cite an employer in the alternative, alleging both a violation of a specific standard and the general duty clause. *Henkels & McCoy, Inc.,* 4 OSHC 1502 (1976). Even if this is not permitted, liberal amendment of the citation and complaint to conform to the evidence discovered and presented is generally permitted. *Dunlop v. Ashworth,* 538 F.2d 562, 3 OSHC 2065 (4th Cir. 1976); *Usery v. Marquette Cement Mfg. Co.,* 568 F.2d 902, 5 OSHC 1793 (2d Cir. 1977).

E. PLACE OF EMPLOYMENT

Employers are required under the Act to furnish to their employees, employment and *a place of employment* which are free from recognized hazards. In *REA Express, Inc. v. Brennan,* 495 F.2d 822, 1 OSHC 1651 (2d Cir. 1974), one employee was electrocuted and another employee was injured in the employer's circuit room when they were standing on a damp concrete floor four to five feet from cut live wires. The employer argued that since the circuit room was generally off-limits to it employees, it could not be considered a place of employment.

The court noted, however, that since the Act required the employer to furnish *each* employee a place of employment free from recognized hazards, then, even though the circuit room was not open to all employees but only to certain authorized employees, and even though it was kept locked and generally unavailable, it was a place of employment. The court noted that the high-voltage circuit room obviously presented a hazard which was recognized. To exclude the room from coverage of the Act would exonerate the employer from providing occupational protection and safety in the most hazardous area of its plant. Similarly, in *Sugar Cane Growers Cooperative of Florida,* 4 OSHC 1320 (1976), the employer was cited for violation of the general duty clause with respect to the transportation of employees in trucks without seating facilities, since the Commission found that the truck itself constituted a place of employment. Thus it appears that any workplace where *one* of the employer's employees is present may be considered a place of employment. *Clarkson Construction Co. v. OSHRC,* 531 F.2d 451, 3 OSHC 1880 (10th Cir. 1976)—a shoulder of a public highway is a workplace.

F. RECOGNIZED HAZARD

A recognized hazard is a condition that is known to be hazardous, not necessarily by each and every individual employer, but is known taking into account the standard of knowledge in the industry. In other words, whether or not a hazard is recognized is a matter for objective determination, and such determination does not necessarily depend on whether the particular employer is aware of it. See 116 Cong. Rec., Part 28377 (1970). Either an employer's actual or constructive knowledge may be utilized to determine "recognition."[2] As the court in *Brennan v. OSHRC (Vy Lactos Laboratories, Inc.)*, 494 F.2d 460, 1 OSHC 1623 (8th Cir. 1974), stated: the constructive knowledge standard would be determined by reference to the common knowledge of safety experts who are familiar with the circumstances of the industry or activity in question. See *National Realty, supra; Brennan v. OSHRC (Republic Creosoting Company)*, 501 F.2d 1196, 2 OSHC 1109 (7th Cir. 1974). This aspect of the concept of "recognized" hazard will be discussed in detail below.

The concept of "recognized" includes by its own terms the element of preventability, since Congress intended to require only the elimination of preventable hazards; that is, the employer's obligation under the Act should be an achievable one. To that end the court in *National Realty* stated that the Act does not impose strict liability upon employers. Although the concepts of assumption of risk and contributory negligence are not applicable in dealing with citations under the general duty clause, there may be instances where accidents will result from negligence or bad judgment and not from an employer's neglect of its general duty under the Act. In such instances, the hazard would not be foreseeable or preventable, and compliance with the general duty clause would not be achievable. Thus it would be unfair to impose a strict liability concept upon employers. In this respect the court in *Canrad Precision Industries, supra,* stated:

> Certainly these [hazards] are recognized by any industry utilizing high voltage electricity. The Congress apparently did not intend to impose strict liability in this sense. The purpose clause of the Act declares it to be the Congressional purpose "to assure *so far as possible* every working man and woman in the Nation safe and healthful working conditions." . . . The italicized language suggests that the duty imposed by Section 5(a)(1) must be a duty which is capable of achievement. Moreover the definition of serious violations in Section 17(k) of the Act is qualified by the requirement that the employer with the exercise of reasonable diligence know of the presence of violation. Thus, we do not construe the general duty clause of Section 5(a)(1) as imposing strict liability for a Section 17(k) penalty for the results of idiosyncratic, demented or perhaps suicidal self-exposure of employees to recognized hazards.

See also *Brennan v. OSHRC (Hanovia Lamp Division)*, 502 F.2d 946, 2 OSHC 1137 (3d Cir. 1974); *Marshall v. L. E. Myers Company*, 589 F.2d 270, 6 OSHC 2159 (7th Cir. 1978); *Atlantic and Gulf Stevedores, Inc. v. OSHRC*, 534 F.2d 541, 4 OSHC 1061 (3d Cir. 1976); *Getty Oil Company v. OSHRC*, 530 F.2d 1143, 4 OSHC 1121 (5th Cir. 1976); *Richmond Block, Inc.*, 1 OSHC 1505 (1974). Thus in order to establish a recognized hazard, the Secretary of Labor must prove that the occurrence of an incident is reasonably foreseeable. *Brown & Root, Inc.*, 8 OSHC 2140 (1980); *Pratt & Whitney Aircraft*, 8 OSHC 1329 (1980).

Moreover, when we focus upon whether a hazard is recognized, it must be remembered that although a generic form of the hazard may be recognized, specific unpreventable instances of it may not. See *National Realty, supra; Canrad Precision Industries, supra.* But, in *Champlin Petroleum Company v. OSHRC*, 593 F.2d 637, 7 OSHC 1241 (5th Cir. 1979), the court stated that a general duty clause citation may be supported by evidence which shows the preventability of the generic hazard, if not the particular instance. Similarly, the court in *Vy Lactos* observed:

> The basic weakness of the . . . rationale is that it addresses itself to the foreseeability of the incident as it actually occurred rather than the foreseeability of the general hazard. . . .

See also *Boeing Company, Wichita Division,* 5 OSHC 2015 (1976); *Beaird-Poulan, a Division of Emerson Electric Co.,* 7 OSHC 1229 (1979)—it is the general hazard, not the cause of a particular accident, that is relevant in determining the existence of a recognized hazard; *Brown & Root, Inc.,* 8 OSHC 2140 (1980)—it is the hazard, not a specific incident which resulted in injury, which is relevant in determining the existence of the recognized hazard.

In *American Smelting and Refining Company, supra,* it was argued that the general duty clause was not intended to cover recognized hazards that could be detected only by testing devices. The employer asserted that "recognized" only meant recognized directly by human senses without the assistance of any technical instruments. In support of this argument, the employer relied upon a statement made by Representative Steiger just prior to the House roll call enacting the Act, which said:

> The conference bill takes the approach of this House to the general duty requirement that an employer maintain a safe and healthful working environment. The conference-reported bill recognizes the need for such a provision where there is no existing specific standard applicable to a given situation.
>
> However, this requirement is made realistic by its application only to situations where there are recognized hazards which are likely to cause or are causing serious injury or death. Such hazards are the type that can readily be detected on the basis of basic human senses. Hazards *which require technical or testing devices to detect them are not intended to be within the scope of the general duty requirement.* It is expected that the general duty requirement will be relied upon infrequently and that primary reliance will be placed on specific standards which will be promulgated under the act. 116 Congressional Record 11899, December 17, 1970 [Emphasis added.]

The court stated that although Steiger's statement was clear concerning his interpretation of recognized hazard, it did not follow that his interpretation was correct. The court, therefore, relying upon language and an amendment introduced by Senators Dominick and Javits, which substituted the words "recognized hazards" for "readily apparent hazards" (3 U.S. Code Congressional & Administrative News at 522 (1970); 2 U.S. Code Congressional & Administrative News at 5177, 5222 (1970)), rejected the employer's argument. Specifically, the court noted that Senator Javits's remarks stated in this respect:

> As a result of this amendment the general duty of employers was clarified to require maintenance of a work place free from recognized hazards. This is a significant improvement over the Administration bill, which requires employers to maintain the work place free from readily apparent hazards. That approach would not cover non-obvious hazards discovered in the source of inspection.

The court also stated that no previous court decision had specifically interpreted the meaning of recognized hazard. (In *Brennan v. OSHRC,* [*Vy*

Lactos Laboratories, Inc.], 494 F.2d 460 1 OSHC 1623 (8th Cir. 1974), for example, it was held that recognized hazards include an employer's actual knowledge and the generally recognized knowledge of the industry as well, but it did not interpret the term "recognized" any further.) The *American Smelting* court thus rejected the employer's argument concerning the definition of recognized hazard, since the court felt it denoted a broader meaning than readily apparent:

> We further think that the purpose and intent of the Act is to protect the health of the workers and that a narrow construction of the general duty clause would endanger this purpose in many cases. To expose workers to health dangers that may not be emergency situations and to limit the general duty clause to dangers only detectable by the human senses seems to us to be a folly. Our technological age depends upon instrumentation to monitor many conditions of industrial operations and the environment. Where hazards are recognized but not detectable by the senses, common sense and prudence demand that instrumentation be utilized. The health of the workers should not be subjected to such a narrow construction.

See also *Edgewood Construction Co.*, 2 OSHC 1485 (1975).

The next major issue for resolution in the context of the definition of a recognized hazard was whether a failure to take certain precautions could in and of itself be a recognized hazard. In *Beaird-Poulan, a Division of Emerson Electric Company*, 7 OSHC 1229 (1979), the Commission said "no," stating that a recognized hazard cannot be defined in terms of the absence of appropriate abatement measures; that question would properly be considered under the element of rendering a workplace free of the hazard. Rather, it stated that a recognized hazard is a condition or practice in the workplace that is known by the industry in general or by the employer in particular to be hazardous, citing *Empire Detroit Steel Division, Detroit Steel Corporation v. OSHRC*, 579 F.2d 378, 6 OSHC 1693 (6th Cir. 1978). But this resolution appears to conflict with the decisions of the courts and the Commission itself. See *Republic Creosoting Company, supra,*—the recognized hazard was the failure to properly stack windrowed piles of logs; *Champlin Petroleum Company v. OSHRC*, 593 F.2d 637, 7 OSHC 1253 (5th Cir. 1979)—the recognized hazard was the failure to have handles on valves to prevent the escape of hot oil. *Magma Copper Company v. Marshall*, 608 F.2d 373, 7 OSHC 1993 (9th Cir. 1979)—a lack of non-interchangeable nose fittings was the recognized hazard; *Royal Logging Company*, 7 OSHC 1744 (1979)—the nonuse of seatbelts to prevent rollovers on earth-moving equipment was the recognized hazard.[3]

In addition to showing that a hazard was recognized, and in order to establish that an employer was in violation of the general duty clause, the Secretary of Labor must first prove that one or more employees were exposed to or had *access* to the recognized hazard. *Baroid Division of N. L. Industries, Inc.*, 7 OSHC 1466 (1979); *Continental Oil Co.*, 6 OSHC 1816 (1977), *aff'd*, 630 F.2d 446, 8 OSHC 1980 (6th Cir. 1980). For implicit in the proof of the three elements of the general duty clause is the necessity of establishing employee exposure at the site of the hazardous condition. *Grossman Steel and Aluminum Corporation*, 6 OSHC 2020 (1978); *Arizona Public Service Co.*, 1 OSHC 1369

(1973); *Fry's Tank Service, Inc. and Cities Service Oil Co.*, 4 OSHC 1515 (1976), *aff'd*, 577 F.2d 126, 6 OSHC 1631 (10th Cir. 1978). However, an injury resulting from exposure to the hazard is not necessarily required. Although the occurrence of death or serious injury may be relevant to prove a violation of the general duty clause, the Act is violated when a recognized hazard is present regardless of whether or not an injury occurs. *Allis-Chalmers Corp. v. OSHRC*, 542 F.2d 27, 4 OSHC 1633 (7th Cir. 1976); *REA Express, Inc. v. Brennan*, 495 F.2d 822, 1 OSHC 1651 (2d Cir. 1974); *Bethlehem Steel Corp. v. OSHRC*, 607 F.2d 871, 7 OSHC 1802 (3d Cir. 1979); *Norman R. Bratcher Co.*, 1 OSHC 1152 (1973); *Vy Lactos, supra; Dale M. Madden Construction, Inc.*, 1 OSHC 1030 (1972).

Finally, as discussed above, in order to prove that a hazard is recognized it must be shown that the employer had or should have had notice thereof, either by actual or constructive knowledge, due to the custom and practice in the industry. Moreover, this knowledge must be established by substantial evidence. *Georgia Electric Co. v. Marshall*, 595 F.2d 309, 7 OSHC 1343 (5th Cir. 1979); *Magma Copper Company v. Marshall*, 608 F.2d 373, 7 OSHC 1893 (9th Cir. 1979); *Marshall v. Cities Service Oil Company*, 577 F.2d 126, 6 OSHC 1631 (10th Cir. 1978); *Penrod Drilling Co.*, 4 OSHC 1654 (1976).

1. Actual Knowledge

In *Empire Detroit Steel Division, Detroit Corp. v. OSHRC*, 579 F.2d 378, 6 OSHC 1693 (6th Cir. 1978), the court determined that the hazard was recognized, since:

> Petitioner does not dispute that water entrapped by molten slag is a hazard. Because Petitioner admits knowledge of the dangerous potential of this condition, the hazard constitutes a recognized hazard.

In *Continental Oil Co. vs. OSHRC*, 630 F.2d 446, 8 OSHC 1980 (6th Cir. 1980), the Sixth Circuit, affirming the Commission, determined that Continental recognized the hazard posed by overflows of "this inherently dangerous petroleum product" because Continental was aware of a similar overflow explosion at another tank farm which like Continental used an unmanned delivery system. In *Ringland-Johnson, Inc. v. Dunlop*, 551 F.2d 1117, 5 OSHC 1137 (8th Cir. 1977), an employer was found to have violated the general duty clause, since it knowingly allowed its employees to work on a scaffold which had gridwork with gaps large enough for an employee to fall through. In *Royal Logging Co.*, 7 OSHC 1744 (1979), the Commission found that actual knowledge was established, since:

> Respondent's witnesses including its own employees and those engaged in logging operations in other companies, knew of rollovers. Additionally the evidence showed that Respondent provided much of its equipment with seat belts and encouraged their use which is evidence the Respondent recognized the existence of a hazard.

In *Mobil Oil Company*, 3 OSHC 2014 (1976), the Secretary of Labor attempted to establish the existence of actual knowledge on the basis of the employer's maintenance of an operating manual which contained procedures for lighting "off-gas" fired process heaters; these procedures themselves constituted the recognized hazard. The employer, however, successfully rebutted this argument. The Secretary of Labor, however, attempted to change his legal theory of violation at the briefing stage, from one of inadequate procedures to one of inadequate implementation. The court rejected this "amendment" on the basis of fairness, in that it was unfair to decide a case on a legal theory or set of facts not presented at the hearing. See also *Pratt & Whitney Aircraft Division of United Technologies Corp.*, 8 OSHC 1331 (1980)—the employer had a past history of chemical spills; *American Smelting, supra*—evidence of such actualities as death or injury is admissable to indicate actual knowledge; *Brown & Root, Inc.*, 8 OSHC 2140 (1980)—the employer instructed its employees to secure themselves and their tools from falling; *Young Sales Corp.*, 5 OSHC 1564 (1977); *American Gypsum Co.*, 6 OSHC 1900 (1978); *Owens-Corning Fiberglass Corp.*, 7 OSHC 1291 (1979); *Wilson Freight*, 5 OSHC 1692 (1977); *Copperweld Steel Co.*, 2 OSHC 1602 (1978); *Williams Enterprises of Georgia, Inc.*, 7 OSHC 1572 (1979); *The Coca-Cola Bottling Co.*, 7 OSHC 1609 (1979); *Advance Specialty, Inc.*, 3 OSHC 2072 (1976).

Finally, in *Magma Copper Co. v. Marshall*, 608 F.2d 373, 7 OSHC 1893 (9th Cir. 1979), the court reviewed a Commission determination that the employer had actual knowledge since its supervisors were aware of the cited problem. While the court stressed that proof of an employer's actual knowledge of a hazard is sufficient to prove that a hazard was recognized,[4] *citing Usery v. Marquette Cement Mfg. Co.*, 568 F.2d 902, 5 OSHC 1793 (2d Cir. 1977); *Titanium Metal Corp. v. Usery*, 579 F.2d 536, 6 OSHC 1973 (9th Cir. 1978); and *Brennan v. OSHRC (Vy Lactos)*, 595 F.2d 460, 1 OSHC 1623 (8th Cir. 1974), the court held that the Commission failed to prove this knowledge by substantial evidence. The employer in *Magma* argued that although it took certain safety precautions which were different from those "required" by OSHA, this action should not be equated with knowledge that its safety precautions were insufficient. It asserted that it was the Secretary of Labor's burden to demonstrate that the industry would have deemed its safety system inadequate. The court agreed, stating in cases involving the application of "vague" specific standards, the Secretary must carry such a burden, using a standard of a "reasonably prudent man familiar with the practices in the industry." Similarly, in general duty clause cases, there must be an adequate warning to an employer before he is charged with actual knowledge and a violation of the Act based on that actual knowledge. The court held that

> . . . where evidence of the employer's actual knowledge is relied on as proof that a hazard is recognized, under the general duty provision of the Act, the Secretary has the burden of demonstrating by substantial evidence that the employer's safety precautions were unacceptable in this industry or a relevant industry.

The court thus adopted the *General Dynamics Corp. v. OSHRC*, 599 F.2d 453, 7 OSHC 1373 (1st Cir. 1979), standard of a reasonably conscientious safety expert familiar with the pertinent industry and found that there was no evidence of that sort shown. See *Upshur Rural Electric Cooperative Corporation*, 7 OSHC 1573 (1979).

This standard of a "reasonably conscientious safety expert" has also been used in conjunction with the interpretation/application of "vague and ambiguous" specific standards to obviate concerns over a lack of specificity and fair notice to an employer. For example, in *Cape and Vineyard Division v. OSHRC,* 512 F.2d 1148, 2 OSHC 1528 (1st Cir. 1975), and *Brennan v. Smoke Craft, Inc.,* 530 F.2d 843, 3 OSHC 2000 (9th Cir. 1975), the OSHA standard relative to the use of personal protective equipment was involved. The courts held in those cases that, in order for the Secretary of Labor to prove a violation, it first required a showing that such equipment was appropriate or necessary under the circumstances. The court stated that a regulation without internally ascertainable standards does not provide a constitutionally adequate and fair warning to an employer of the conduct proscribed, unless read to penalize only that conduct unacceptable in light of the common understanding and experience of those working in the industry. Thus, in determining whether an employer violates a standard, the appropriate inquiry is whether under the circumstances a reasonably prudent employer familiar with the industry would have protected against the hazard by the means specified in the citation. See *Bristol Steel and Iron Works, Inc. v. OSHRC,* 601 F.2d 717, 7 OSHC 1462 (4th Cir. 1979); *Diebold, Inc. v. Marshall,* 585 F.2d 1327, 6 OSHC 2002 (6th Cir. 1978). The Fifth Circuit has limited this test somewhat, however, stating that the standard must be read as requiring only those protective devices which the knowledge and experience of the *employer's* industry would clearly deem appropriate under the circumstances. See *B. & B Insulation v. OSHRC,* 583 F.2d 1364, 6 OSHC 2062 (5th Cir. 1978); *Ryder Trucklines, Inc. v. Brennan,* 497 F.2d 230, 2 OSHC 1075 (5th Cir. 1974). Other courts, however, have refused to limit the reasonable person test to the custom and practice of the industry, because "[s]uch a standard would allow an entire industry to avoid liability by maintaining inadequate safety." *General Dynamics v. OSHRC,* 599 F.2d 453, 7 OSHC 1373 (1st Cir. 1979); *Voegele Co. v. OSHRC,* 625 F.2d 1075, 8 OSHC 1631 (3d Cir. 1980).

Whatever the test, it is clear that in order to satisfty due process requirements the courts have applied some form of the reasonable man test to both the general duty clause and the various general safety standards found in the regulations.

2. Constructive Knowledge

A recognized hazard may be shown not only by an employer's actual knowledge but also by reference to the general knowledge of the industry as well, if there is an accepted standard in the industry. Recognition may thus take into account the custom and practice in a relevant industry, for if an employer is ignorant of the hazard, the industry must generally guard against it.

Whether or not a hazard is recognized is a matter for objective determination, even where constructive knowledge is relied upon. Recognition, if it is to be determined by constructive knowledge, must be made by reference to an appropriate industry, and the finding must be supported by substantial evidence. In *H-30, Inc. v. Marshall,* 597 F.2d 534, 7 OSHC 1253 (10th Cir. 1979), the Secretary of Labor attempted to apply standards from other industries to the devices used by the employer, on the theory that they were similar mechanically or in purpose. However, the Court rejected this, stating that Congress included in the general duty clause the concept that *the industry* recognized certain hazards and had met them with generally accepted standards:

> This of course, contemplates some sort of consensus in this general acceptance and recognition which cannot necessarily be extrapolated from one industry to another, especially so where there is nothing to show that the second group has not generally recognized the hazard and has adopted no standards. Some general similarity and purpose of the devices is not enough to carry over recognition from one industry to another.... There is no indication in the record before us that the oil well drilling industry had any part or was consulted in the development of the construction industry standards.

See *Usery v. Kennecott Copper Corp.,* 577 F.2d 1113, 6 OSHC 1197 (10th Cir. 1978); *Republic Creosoting Co., supra; Sun Shipbuilding & Drydock Co.,* 3 OSHC 1413 (1975). However, in *R. L. Sanders Roofing Co.,* 7 OSHC 1566 (1979), the Commission held that the recognition of a hazard by the construction industry in general would constitute recognition by the various member trades within that industry, since employers whose employees engage in construction owe a duty to those employees to protect them from those hazards recognized by the construction industry in general.

The next question, of course, is how constructive knowledge of a recognized hazard is established by reference to the employer's industry. Obviously, evidence of industry custom and practice may be shown by actual witnesses' testimony and evidence. For example, the court stated in *Getty Oil Co. v. OSHRC,* 530 F.2d 1143, 4 OSHC 1121 (5th Cir. 1976):

> The uncontroverted evidence in the record shows that failure to pressure test a pressure vessel before activation is a universally-recognized hazard in the oil industry and that omitting such tests was likely to cause serious injury and created an extremely high probability of rupture ensuing harm in the instant case. Thus it is clear that the hazard at issue was both recognized and likely to cause serious harm, as well as preventable.

Similarly, in *Magma Copper Company v. Marshall,* 608 F.2d 373, 7 OSHC 1893 (9th Cir. 1979), it was noted that

> The Secretary sought to prove by expert testimony that explosions due to the lack of non-interchangeable hose fittings in oxygen systems was a recognized hazard.
>
> But as *Magma* points out the Secretary's expert witness had no experience in the smelting or refining industry but testified primarily about the use of portable oxygen units in hospitals and in welding work.... It was stipulated by the parties that the evidence offered by the Secretary did not apply directly to a strap or plait system; it was offered only by analogy.... But Magma's conduct, (particularly in view of the nature of the hazard with which it is charged) must be judged against the standards of a relevant industry. There was no expert evidence introduced by the Secretary as to the practice of the smelting and refining industry, the industry of which Magma is a part.... There is not substantial evidence in this record to support the Secretary's claim that it carried its burden by expert testimony.[5]

In *H-30, Inc. supra,* the court found that the Secretary of Labor had failed to demonstrate that in this industry it considered a practice to be a recognized hazard:

> The agency did not present a witness to show that the hazard was recognized. The drilling industry is large and active but no evidence was offered.

See *Schriber Sheet Metal and Roofers, Inc. v. OSHRC,* 597 F.2d 78, 7 OSHC 1246 (6th Cir. 1979); *Baroid Division of N. L. Industries, Inc.,* 7 OSHC 1466 (1979)—the description between separators and busters summarizes the uniform unequivocal testimony of the other witnesses who have extensive experience in oil drilling; *Continental Oil Co. v. OSHRC,* 630 F.2d 446, 8 OSHC 1980 (6th Cir. 1980)—the evidence established that other farms, not necessarily all tank farms, utilized either manned delivery systems, high-level alarms, or both; *The Boeing Co., Wichita Division,* 5 OSHC 2014 (1977); *Niagara Mohawk Power Corporation,* 5 OSHC 1919 (1977).

In the absence of actual testimony, recognition of a hazard by means of constructive knowledge of industry custom or practice may be shown, for example, by reference to OSHA standards, ANSI standards, NFPA standards, NFC standards, and manufacturer's warnings. In *Titanium Metals Corporation of America v. Usery,* 579 F.2d 536, 6 OSHC 1873 (9th Cir. 1978), for example, the fact that a fire hazard posed by titanium generally and by the accumulation of dust and fines in particular was recognized by both the industry and petitioner was shown by reference to the National Fire Code, NFPA No. 481–1972. Moreover, recognition was shown by the adoption of a safety program at the employer's plant, which allegedly attested to the hazard posed by titanium as being one that was recognized.[6] In *Atlantic Sugar Association,* 4 OSHC 1355 (1976), it was shown that an employer violated Section 5(a)(1) by failing to require employees to remain seated while being moved in trucks:

> It is found that the evidence presently of record is sufficient to prove a violation of Section 5(a)(1) of the Act. It is enough here to find that by means of correspondence, industry meetings and highly publicized accidents, there is sufficient evidence that the sugar cane industry and respondent had knowledge that transporting field workers in a standing position in trucks without adequate seating or substantial rear barriers was hazardous to the workers.

Similarly, in *Fry's Tank Service, Inc., and Cities Service Oil Company,* 4 OSHC 1515 (1976), *aff'd,* 577 F.2d 126, 6 OSHC 1631 (10th Cir. 1978), it was stated:

> Both Fry's and Cities were aware of the hazard that hydrogen sulfide presented in oil fields producing sour crude. Additionally at least Cities was aware of other hazards that could be encountered by entering the tank, including hydrocarbon vapors and oxygen deficiency. Because it know of these hazards, Cities instructed its employees that they were not to enter tanks on the Stultanberg lease. Cities also made tank entry by its employees unnecessary by hiring independent contractors to do all work requiring entry into tanks. In other locations where its employees were required to enter confined spaces, it was Cities' policy to take precautions such as testing the atmosphere for entry, using breathing equipment where necessary, using safety lines, and having at least one worker outside the vessel in constant communication with the workers inside. There is no indication in the

record that the prohibition against entering the tanks on the Stultanberg lease was ever breached except on this occasion. Fry also had a policy against employees entering tanks. Mr. Swalley, Fry's president, testified that he instructed his employees not to enter tanks and had so instructed Thach.

Remember, substantive evidence is required to show that a hazard is recognized by the employer or its industry. *Penrod Drilling Co.*, 4 OSHC 1655 (1976).

G. RENDER A WORKPLACE "FREE"

Section 5(a)(1) requires the employer to free its workplace of recognized hazards. The Act, however, does not specify the manner by which recognized hazards are to be removed from the worksite. Accordingly, the court in *National Realty, supra,* stated that Congress intended to require the elimination only of preventable hazards. Rendering a workplace free of recognized hazards, therefore, does not mean the elimination of *all* hazards, for Congress did not intend unpreventable hazards to be considered recognized. Construing this requirement of the Act with respect to the employer's obligation to render a workplace free, the court in *Empire Detroit Steel Division, Detroit Steel Corporation v. OSHRC,* 579 F.2d 378, 6 OSHC 1693 (6th Cir. 1978), stated that a violation of the general duty clause occurs whenever an employer fails to take reasonable precautionary steps to protect its employees from *reasonably foreseeable* recognized hazards. The court in *L. E. Meyers Company, supra,* similarly stated:

> The Secretary has been unable to demonstrate that Meyers failed to render the worksite "free" of the hazard. Actual occurrence of hazardous conduct is not by itself sufficient evidence of a violation, even when the conduct has led to injury. The record must additionally indicate demonstrably *feasible* measures would have materially reduced the likelihood that such misconduct would have occurred.

Similarly, the court in *Titanium Metals, supra,* stated:

> While it is beyond dispute that an accident need not occur for a violation of Section 5(a)(1) properly to be found, . . . conversely it has been held that while the occurrence of injury may be relevant to prove a violation it is not conclusive.

Finally, in *Bethlehem Steel Corporation, supra,* the court stated:

> Much of Bethlehem's argument concerning the way in which the accident may have occurred is misdirected, for although the occurrence of a death or serious injury may be relevant to proving a violation of a general duty clause, the statute is violated when a recognized hazard is maintained, whether or not an injury occurs.

It would appear, therefore, that an employer may not be found to have violated the general duty clause and to have failed to render its workplace free of a hazard, notwithstanding the occurrence or nonoccurrence of an accident or injury, *if* the hazard itself was *unforeseeable.* See *John R. Jurgensen Co.,* 7 OSHC 1251 (1979). The court in *Babcock and Wilcox, supra,* also stated in this regard that foreseeability is an element of a Section 5(a)(1) violation, stating specifically:

> Moreover, either the employer must be aware of the hazard's existence, or if he is ignorant of it industry must generally guard against it. In other words, the hazard must be reasonably foreseeable.

In *Fry's Tank Service, and Cities Service Oil, supra* the Commission stated that although it is possible to determine in retrospect how an accident could have been prevented, in order to show a violation it must be shown that the exposure of employees to the "recognized hazard" was reasonably foreseeable prior to the time such exposure occurred. Otherwise, an accident is not preventable by the employer, and the purposes of the Act are not served by finding the employer in violation.[7]

Similarly, the Commission in *Pratt and Whitney Aircraft Division of United Technologies Corp.*, 8 OSHC 1329 (1980), stated that:

> The evidence demonstrates the reasonable foreseeability of a serious incident . . . (the neutral term incident is used to avoid the connotation of chance frequently associated with the term accident in common usage). . . . The occurrence of an incident need not be likely. It is enough if an incident is reasonably foreseeable.

See also *Marquette Cement Manufacturing, supra; R. L. Sanders Roofing Co.*, 7 OSHC 1566 (1979);[8] *Republic Creosoting Co., supra.*

In *General Dynamics Corporation, Quincy Shipbuilding Division v. OSHRC,* 599 F.2d 453, 7 OSHC 1373 (1st Cir. 1979), the court indicated, however, that a violation of the general duty clause may exist even though an accident may occur because of unforeseeable events. The Commission in *R. L. Sanders Roofing Company, supra,* also stated that a fall hazard may be recognized within the meaning of Section 5(a)(1) and the workplace not rendered "free" even if the unusual chain of events in the case was not foreseeable. In *Vy Lactos Laboratories,* the court also indicated:

> The second basic weakness of the hearing examiner's rationale is that it addresses itself to the foreseeability of the incident as it actually occurred rather than the foreseeability of the general hazard. A violation occurs whenever an employer fails to take reasonable precautionary steps to protect his employees from reasonably foreseeable recognized hazards.

Finally, in *Magma Copper Company, supra,* the court stated that it was not necessary to address the question of the employer's duty to eliminate only those hazards that are foreseeable and preventable, since it found no substantial evidence that the employer's safety precautions were unacceptable in the industry or relevant industry. See also *California Stevedore and Ballast Company v. OSHRC,* 517 F.2d 986, 3 OSHC 1174 (9th Cir. 1975).

An interesting sideline of the foreseeability discussion centers on the question of causation. In *Bethlehem Steel Corporation, supra,* the court stated that its task on review was *not* to look for a proximate cause relationship between the accident which preceded the inspection and the specific violation charged, but to determine whether there was a recognized hazard to the safety of employees. See also *The Boeing Co., Wichita Division,* 5 OSHC 2014 (1977)—whether a violation occurred (and a recognized hazard existed) does not depend on the causes of the particular accident; *Beaird-Poulan, supra*—it is the general hazard, not the cause of a particular hazard, that is relevant in determining the existence of a recognized hazard. However, the court in *Champlin Petroleum Company, supra,* specifically stated that in the particular accident involved therein there may have been no direct causal connection between the lack of a handle (which was the alleged hazard set forth in citation) and the occurrence of a flash fire; *Nugent Service, Inc.*, 4 OSHC 1075 (1976).

In order to render its workplace free of recognized hazard, an employer is required to take certain precautionary steps to protect its employees from reasonably foreseeable, that is, recognized hazards. *Empire Detroit Steel Division, Detroit Steel Corp., supra.* As stated in *General Dynamics Corporation, supra:*

> It is undisputed that an unsupported webframe constitutes a recognized hazard in the industry. The question then becomes *what precautionary steps a conscientious safety expert familiar with the pertinent industry* would take to avoid the occurrence of the hazard (*i.e.* render its workplace free of the hazard).

In *Magma Copper Company v. Marshall,* 608 F.2d 373, 7 OSHC 1893 (9th Cir. 1979), the court held that it was the Secretary of Labor's burden to demonstrate by substantial evidence that the industry would have deemed the employer's safety system and precautions inadequate or unacceptable. In *Bristol Iron & Steel Workers, Inc. v. Marshall,* 601 F.2d 717, 7 OSHC 1462 (4th Cir. 1979), the court held that the record must demonstrate (in order to show a violation of the general duty clause) that a *reasonably prudent employer* under the circumstances and familiar with the industry would have protected against the hazard *by the means specified* in the citation, *rather than by the means used* by the cited employer.[9]

The issue that has generated the most controversy recently in Section 5(a)(1) cases is by what standard does an employer determine whether and what precautions would and should have been taken against a particular hazard. The court in *Titanium Metals Corporation, supra,* obliquely answered by stating:

> We have held that where an employer's safety record is the only evidence offered concerning whether a reasonably prudent man would have found that further precautions against a particular hazard were necessary where certain precautions already existed, the Commission's order vacating a citation for violation of Section 5(a)(1) is not supported by substantial evidence.

The Commission's position relative to what must be shown has recently and clearly been demonstrated. In *Continental Oil Company,* 6 OSHC 1814 (1978), *aff'd,* 630 F.2d 446, 8 OSHC 1893 (6th Cir. 1980), the Commission stated that:

> Though the level of hazard *recognition* required by the general duty clause is measured against industry employer knowledge . . . an abatement order under Section 5(a)(1) may require that work practices and safety precautions be upgraded by a feasible level which is above that considered customary or reasonable by an industry.

As the basis for this statement the Commission relied upon *National Realty, supra,* which stated that the hazard must be one which is preventable and defined preventability as follows:

> This is not to say that a safety precatuion must find general usage in an industry before its absence gives rise to a general duty violation. The question is whether a precaution is recognized by safety experts as feasible, not whether the precaution's use has become customary.

The Commission reaffirmed its position in *Sanders Roofing Company, supra,* stating that an employer's abatement duty under Section 5(a)(1) to render its workplace free of hazard is not controlled by the practices of its industry but is dependent *solely* upon a showing that there were feasible steps it could have taken to abate the hazardous condition. See also *Ford, Bacon and Davis Construction Corporation,* 6 OSHC 1910 (1978); *Williams Enterprises, Inc.,* 7 OSHC 1250 (1979); *Beaird-Poulan, supra*—compliance with the Act may require methods of protection of a higher standard than industry practice; that is, the question is whether a precaution is recognized by safety experts as feasible, not whether the precaution's use has become customary.

The courts are in general agreement with the Commission concerning the requirements placed upon employers to take precautions to eliminate or to free the workplace of a recognized hazard. In *General Dynamics Corporation, Quincy Shipbuilding Division, supra,* the court stated that the administrative law judge committed error in focusing on the accident as it occurred rather than whether the employer had taken proper precautionary steps, including adequate safety training, to prevent the occurrence of the hazardous condition. A violation of the general duty clause may exist if the employer failed to take precautionary steps to protect its employees from the occurrence of a general hazard. The court, relying on the case *Cape & Vineyard Div. of New Bedford Gaŝ v. OSHRC,* 512 F.2d 1148, 2 OSHC 1628 (1st Cir. 1978), which discussed the standard of care relative to the requirement that employers provide protective equipment, then set forth the following test relative to the utilization of precautions:

> Consequently *Cape & Vineyard* must be read as holding that knowledge of the existence of a hazardous situation must be determined in light of the common experience of an industry, *but* that the extent of precautions to be taken against a known hazard is that which a conscientious safety expert would take. . . . Furthermore we cannot agree with Qincy's position that the measure of the adequacy of its safety program should be that of the industry. Such a standard would allow an entire industry to avoid liability by maintaining inadequate safety training. The purpose of the Act is to require all employers to take all feasible steps to avoid industrial accidents. While the definition of a "recognized hazard" should be made in reference to industry knowledge, by virtue of the definition of the word "recognized," we cannot accept a standard for the precautions which should be taken against such a hazard which is any less than the maximum feasible. . . . In the present case it is undisputed that an unsupported web frame constitutes a recognized hazard in the industry. The question then becomes what precautionary steps a conscientious safety expert would take to avoid the occurrence of the hazard.

See *Magma Copper Company, supra; Continental Oil Company supra; Voegele Co., Inc. v. OSHRC,* 630 F.2d 446, 8 OSHC (3d Cir. 1980); *Ray Evers Welding Co. v. OSHRC,* 625 F.2d 726, 8 OSHC 1273 (6th Cir. 1980); *General Dynamics Corp. v. OSHRC,* 599 F.2d 453, 7 OSHC 1371 (1st Cir. 1979).

Any precautions which are required to be taken, however, must be feasible. A precaution, according to the court in *National Realty, supra,* does not become infeasible merely because it is expensive. But if adoption of the precaution would threaten the economic viability of the employer, that would render it infeasible. If, however, hazards are created by a proposed abatement method or suggested precautions, this evidence is relevant in determining feasibility, but is properly a part of the employer's case in rebuttal. As the Review Commission stated in *Royal Logging Company, supra:*

> Thus complainant need only prove that an abatement method exist that would provide protection against the cited hazard. The burden then shifts to the employer to produce evidence showing or tending to show that the use of the method or methods established by Complainant will cause consequences so adverse as to render their use infeasible. Furthermore because feasibility is an issue, the greater hazards defense cannot apply in a case such as this arising under Section 5(a)(1). If the employer demonstrates that the means of abatement specified by Complainant create other hazards such as they are considered infeasible, the citation must be vacated for failure of proof by Complainaint; there is no need for the employer further to demonstrate the unavailability of other means of abatement or the appropriateness of a variance application.

Abatement efforts subsequent to the issuance of a citation do not negate or excuse an employer's failure to comply with the Act. *Whirlpool Corporation,* 8 OSHC 2243 (1980).

H. "CAUSING OR LIKELY TO CAUSE"

Section 17(k) of the Act, which sets forth the definition of a serious violation, provides that:

> [a] serious violation shall be deemed to exist in a place of employment if there is a substantial probability that death or serious physical harm could result from a condition which exists or from one or more practices, means, methods, operations or processes which have been adopted or are in use, in such place of employment unless the employer did not and could not with the exercise of reasonable diligence know of the presence of the violation.

The phrase "causing or likely to cause" in Section 5(a)(1), on the other hand, is not specifically defined, and the legislative history of the Act provides no interpretative evidence.

The FOM states, however, that because of the virtually equivalent language used in Section 17(k) and Section 5(a)(1) of the Act in describing serious and general duty clause violations, respectively, an other-than-serious citation should not be issued for violations based on the general duty clause. Citations based on the general duty clause should be and are limited to alleged serious violations.

In order to prove a serious violation of the Act, it need not be likely that an accident will occur; it is only necessary to prove that if such an accident occurred, the probable result will be serious injury. As discussed below, a violation of the general duty clause may be proven in a similar manner. Neither the general duty clause nor Section 17(k) requires that an accident occur or that any actual death or physical injury occur as a precondition for a violation to exist. See *Vy Lactos Laboratories, supra; American Smelting and Refining Corporation, supra; Babcock and Wilcox Co., supra.* However, even though a *prima facie* case under the general duty clause does not require the showing of death or serious physical harm, evidence of such an actuality may be relevant and admissible to indicate actual knowledge by the employer of the existence of a recognized hazard. However, the court in *L. E. Myers, Company, supra,* stated that actual occurrence of hazardous conduct is not by itself sufficient evidence of violation, even when the conduct has led to an injury.[10]

Similarly, the court in *Titanium Metals Corporation, supra,* has stated that an accident need not occur for a violation of the general duty clause properly to be found; conversely, it has been held that occurrence of injury may be relevant to proving a violation but is not conclusive. However, in *Empire Detroit Steel Division, supra,* the court stated that the third element of *National Realty* was established when it was shown that water entrapped by molten slag had killed one employee and seriously injured other employees. In *Georgia Electric Company v. Marshall,* 595 F.2d 309, 7 OSHC 1343 (5th Cir. 1979), the court held that the Commission properly found that the hazard was one that was likely to cause death or serious physical harm in that:

> The hazard associated with the reversed control was such as to be likely to cause death or serious physical harm. A death actually occurrred from the reversed control. This is convincing evidence of the seriousness of the hazard. An accident was always possible. A loading control was used to raise the poles. The poles were of steel and weighed between 700 and 800 pounds. The possibility always existed that the control could have been moved in the wrong direction and caused a pole to fall on employees. In such an event the resulting accident would have caused death or serious physical harm. . . . The seriousness of the violation of the general duty clause has also been satisfactorily proven. The evidence supporting the finding that the hazard was likely to cause death or serious bodily harm supports the similar requirement of [Section 17(k) of the Act] that there be a "substantial probability" that death or serious physical harm could result from the questioned condition.

In this respect, it is important to note that it is not necessary to show that an accident has occurred to prove a violation of the general duty clause; it is enough to show that had one occurred it was likely that death or serious physical harm to an employee would have resulted. See *Babcock and Wilcox Company, supra; Pratt and Whitney Aircraft Division, supra.* It is only necessary to prove that if such an accident occurs the probable result will be serious injury. The issue, then, is not whether an accident is likely, but, assuming one has occurred, whether it is likely that serious physical harm would have resulted. *Andy Anderson,* 6 OSHC 1595 (1978); *Usery v. Marquette Cement Mfg. Co.,* 568 F.2d 902, 5 OSHC 1793 (2d Cir. 1977); *Usery v. Hermitage Concrete Pipe Co.,*

584 F.2d 127, 6 OSHC 1886 (6th Cir. 1978); *Shaw Construction, Inc. v. OSHRC,* 534 F.2d 1183, 4 OSHC 1427 (5th Cir. 1976); *L. E. Myers Company, supra; Storey Electric Co. v. OSHRC,* 553 F.2d 357, 5 OSHC 1285 (4th Cir. 1975). The Commission in *Pratt & Whitney Aircraft, supra,* stated that the occurrence of an "incident" need not be likely; it is enough if an incident is reasonably foreseeable.[11] The Commission further stated that

> Moreover, the courts have admonished that the general duty clause is applicable even where the probability of an incident is less than 50 percent. In applying the likely to cause element of the general duty clause it is improper to apply mathematical tests relating to the probability of a serious mishap occurring, given the Act's prophylactic purpose to prevent employee injuries. Citing *Allis Chalmers Corp. v. OSHRC,* 542 F.2d 27, 4 OSHC 1633 (7th Cir. 1976).

Also relevant in this respect is the decision of *Illinois Power Company v. OSHRC,* 632 F.2d 25, 8 OSHC 1512 (7th Cir. 1980), in which the court stated that:

> The company's argument stems in part from a misapprehension of the kind of evidence necessary to show that a hazard is likely to cause death or serious physical harm or to show that a violation is serious. Looking primarily to a balance of need for safety in the work place against the interests of the industry to operate without undue interference, the statutory standard does not imply any mathematical test of probability. *Titanium Metals Corp. v. Usery,* 579 F.2d 536, 6 OSHC 1873 (9th Cir. 1978). Rather the Act is concerned with preventability and, therefore, with possible hazards and potential danger. *Usery v. Marquette Cement Manufacturing Co.,* 568 F.2d 902, 5 OSHC 1793 (2nd Cir. 1977). As a result, although the fact of an accident may not be sufficient to prove the likelihood of an injury, it is at least *prima facie* evidence of a likelihood and the rest may be supplied by common sense or an understanding of physical law. It is settled moreover that if evidence is presented that a practice could eventuate a serious physical harm upon other than a freakish or utterly implausible concurrence of circumstances, the Commission's expert determination of likelihood should be accorded considerable deference by the court. Finally, the language in Section 17(k) requiring substantial probability that death or serious physical harm could result in order to find a serious violation refers not to the probability that an accident will occur but to the probability that an accident having occurred death or serious physical injury could result. . . . The very fact of the injury meanwhile is powerful evidence of a likelihood of injury and while not dispositive is sufficient under the prevailing standards to give rise to a prima facie case that a violation has occurred.

Thus the interpretation that Section 5(a)(1), which requires that the workplace be kept free of recognized hazards likely to cause death or serious physical harm, can be shown regardless of whether an accident is likely to occur (or conversely is not necessarily shown just because an accident has occurred) is supported by the decision of those courts and the Commission which have considered it. *Marquette Cement Manufacturing Co. supra; Republic Creosoting Co., supra; Canrad Precision Industries, supra; R. L. Sanders Roofing Company, supra; Royal Logging Company, supra; Kelsey-Hayes Company,* 5 OSHC 1550 (1977).

I. MATERIALLY REDUCE

Once it is established that there is a recognized hazard which is likely to cause serious physical harm and that the employer has failed to render its workplace free of those hazards by taking reasonable precautions, it must be further and additionally shown that *demonstrably feasible measures would have materially reduced* the likelihood that such misconduct or injury would have occurred. In this respect, as in all other respects, the substantial evidence standard is applicable. *Ryder Truck Lines, Inc., v. Brennan,* 497 F.2d 230, 2 OSHC 1075 (5th Cir. 1974); *Brennan v. OSHRC (Alsea Lumber Co.),* 511 F.2d 1139, 2 OSHC 1646 (9th Cir. 1975). In *Champlin Petroleum Company, supra,* the court stated that the general duty obligation is not designed to impose absolute liability or *respondeat superior* liability for employee negligence. Rather, it requires the employer to eliminate only feasibly preventable hazards. Moreover, in this respect the court stated that:

> It is the Secretary's burden to show that demonstrably feasible measures would materially reduce the likelihood that such injury as that which resulted from the cited hazard would have occurred. . . . The Secretary must specify the particular steps the employer should have taken to avoid citation and he must demonstrate the feasibility and likely utility of those measures.

In that case, the court vacated the citation, since the Secretary of Labor did not show that the use of handles on pipeline bleeder valves or the effective communication of the employer's rule against opening the valves without handles would have *materially reduced* the hazard that caused the injuries, even though these precautions may have been utilized and may have been necessary according to the Secretary of Labor to render the workplace "free" of hazard.

In *Canrad Precision Industries, supra,* the court noted that the case before it differed from *National Realty* in two respects:

> In *National Realty* the Commission had upheld the Secretary's citation whereas here it was vacated. In *National Realty* there was no evidence bearing upon the existence of feasible measures for reducing the employee misconduct in question. Here there is testimony by the Secretary's expert that the employer could have had two technicians in attendance during the test, and that Gallagher's supervisor could have examined the progress of the test more frequently. . . . There is direct testimony by the expert.

The court in *Titanium Metals Corporation, supra,* also stated that the employer's obligation to render its workplace free from recognized hazards must be reconciled with a reasonable limitation placed on such duty; that is, it must eliminate only those hazards which are foreseeable and preventable. As the court stated:

> In the context of the Secretary's burden of establishing a violation of Section 5(a)(1), this means that the record must indicate that demonstrably feasible measures would have materially reduced the likelihood that such injury would have occurred. Put another way, the Secretary must be constrained to specify the particular steps a cited employer should have taken to avoid citation, and to demon-

> strate the feasibility and likely utility of those measures. . . . Unlike the case in *National Realty*, in this instance the Secretary offered ample evidence concerning what steps were open to petitioner which could and should have been taken to *eliminate or substantially reduce* the hazards of fires emanating from or fueled by accumulations of titanium dust and fines.

See also *L. E. Myers Company, supra; Atlantic and Gulf Stevedores, Inc., supra; Hanovia Lamp, supra;*[12] *Empire Detroit Steel Corp., supra; Getty Oil Company, supra; Babcock and Wilcox Company, supra; R. L. Sanders Roofing Company*, 7 OSHC 1566 (1979); *Emerson Electric Co., Edwin L. Wiegand Div.*, 7 OSHC 1441 (1979); *Pratt & Whitney Aircraft, supra; Beaird-Poulan, supra.*

The general duty clause must therefore be applied in a manner consistent with the realities of the employer's operation. *Bethlehem Steel Corp., supra.*

In *Royal Logging Company*, 7 OSHC 1744 (1979), the Commission also stated that in addition to showing the existence of a recognized hazard it is incumbent upon the Secretary of Labor to demonstrate that there are feasible methods which would substantially reduce or eliminate the incidence of hazard. The Commission dealt with the question of what happens when hazards may be created by a proposed method of abatement and concluded that this evidence is relevant in determining the feasibility but would be part of the employer's case in rebuttal. That is, the Secretary of Labor need only prove that an abatement method exists which would provide protection against the cited hazard. The burden would then shift to the employer to produce evidence showing or tending to show that the use of the method or methods established by the Secretary of Labor will cause consequences so adverse as to render their use infeasible. The Commission continued:

> Furthermore, because feasibility is an issue, the greater hazards defense cannot apply in a case such as this arising under Section 5(a)(1). If the employer demonstrates that the means of abatement specified by Complainant create other hazards such that they are considered infeasible, the citation must be vacated for failure of proof by Complainant; there is no need for the employer further to demonstrate the unavailability of other means of abatement or the inappropriateness of a variance application. . . . In the instant case the record establishes that the wearing of seat belts can subject employees to injury from jill pokes, sweepers and other materials. In the opinion of Respondent's witnesses, death can occur from being struck by a jill poke or sweeper and, therefore, the hazard of jill pokes and sweepers entering the cab is equivalent in severity to the hazard of roll-over. . . . The creation of an additional hazard commensurate with the hazard the abatement method is intended to prevent is sufficient to establish the infeasibility of the abatement method. We find, therefore, that safety belts are not feasible where employees are exposed to the entry of foreign material into the cabs of their equipment.

See also *S & H Riggers and Erectors, Inc.*, 7 OSHC 1260 (1979).

J. EXISTENCE OF A SERIOUS VIOLATION

As previously stated, because of the virtually equivalent language used in Section 17(k) and Section 5(a)(1) of the Act, describing serious and general duty clause violations, respectively, other-than-serious citations may not be issued for violations based on the general duty clause. Citations based on the general duty clause are limited to alleged serious violations (and can include repeated and willful violations). In order to prove a serious violation, Section 17(k) of the Act provides that two elements must be shown: (1) a substantial probability that the death or serious physical harm could result; (2) unless the employer did not and could not with the exercise of reasonable diligence know of the presence of the violation. (For example, in *Baroid Division of NL Industries, Inc.*, 7 OSHC 1469 (1979), the Commission stated that an employer with notice that a hazard *may* exist must make reasonable efforts to ascertain if, in fact, the hazard *does* exist. Concluding in that particular case that the employer could have known, through the exercise of reasonable diligence, of the existence of the hazard and the conditions surrounding it, the citation was affirmed.) See *Getty Oil Co., supra*—employer failed to exercise reasonable diligence to determine if tank was pressure tested.

The requirement of showing a substantial probability that death or serious physical harm could result is basically identical to the showing that a hazard is causing or is likely to cause serious physical harm under Section 5(a)(1). Thus proof under Section 5(a)(1) would indeed be proof under this aspect of Section 17(k). However, Section 17(k) also requires that an employer have actual or constructive knowledge, as does the general duty clause. However, the actual or constructive knowledge under the general duty clause goes toward whether there is a recognized hazard. The actual or constructive knowledge under Section 17(k) goes toward whether or not the employer knew of the presence of the violation. Thus under the general duty clause, knowledge must be shown of the *hazard;* under Section 17(k), knowledge must be shown of the *violation.* Knowledge of the existence of the hazard does not always equal knowledge of the existence of the violation. The courts have been clear that knowledge of the violation does not mean knowledge of the requirements of the law but rather that the condition itself is a violation of the Act. In *National Realty, supra,* the court stated:

> With respect to the general duty clause itself, the Commission can ameliorate the fair notice problem by attending carefully to the statutory definition of a "serious violation":
>
> > [A] serious violation shall be deemed to exist in a place of employment if there is a substantial probability that death or serious physical harm could result from a condition which exists, or from one or more practices, means, methods, operations, or processes which have been adopted or are in use, in such place of employment unless the employer did not, and could not with the exercise of reasonable diligence, know of the presence of the violation.
>
> . . . Where the hazard involved is a form of hazardous conduct by employees, an employer's safety program is in "serious" violation of the general duty clause only if (1) the misconduct involves a substantial risk of harm and is substantially probable under the employer's regime of safety precautions, or (2) the employer, with the exercise of reasonable diligence, could have known that its safety program failed the standards of the clause by failing to preclude the occurrence or preventable misconduct. If either condition applies, it is hardly unfair for the Commission to assume that the defendant-employer had at least constructive notice that the

> law required more than was being done. Only if a violation is serious is a penalty necessarily imposed.

In *Republic Creosoting Company, supra,* the court entwined the finding of a serious violation with an aspect of foreseeability, stating that the issue before it was whether an employer using reasonable diligence would have foreseen the danger in question:

> The Commission rejected the Secretary's argument on the theory that the Act does not require that a new employee always be trained in proper procedures for a task simply because he is required to be present at the place of the operation in question in which he is not a participant. We agree with the Commission's interpretation of the Act. The Act clearly requires that, for a serious violation citation to be sustained, the danger must be one of which the employer knew or with reasonable diligence could have known. Whether training is necessary will depend on other factors such as the experience of the employees in a particular field of work, extent of the employees' participation in the operation in question, and the complexity and danger involved in the operation. Where an employee is directly participating in a job, the employer may well, as the Commission noted, have a duty under the Act to instruct him on the safe procedure for handling the job. On the other hand, the Commission accurately recognized that training may be unnecessary for an employee who is wholly associated with the operation in question and who would not be foreseeably exposed to danger.

The Secretary of Labor's burden of proof under Section 5(a)(1) is greater than his burden of proving a violation of a specific standard under Section 5(a)(2), *(e.g.,* feasibility). *S & H Riggers & Erectors, Inc.,* 7 OSHC 1260 (1979); *Austin Building Co.,* 8 OSHC 2150 (1980); *Forest Park Roofing Co.,* 8 OSHC 1181 (1980). This burden may be sustained by introduction of substantial evidence of a probative nature.

K. SPECIFIC APPLICATIONS

Although the Commission has stated that the failure to take precautions cannot in and of itself constitute a recognized hazard, it is clear that the failure to provide certain training, instructions, or equipment may indeed be considered a recognized hazard. Moreover, the adequacy of safety practices, training, and equipment is important in determining whether a workplace is free of a recognized hazard. See *Marshall v. Cities Service Oil Company,* 577 F.2d 126, 6 OSHC 1631 (10th Cir. 1980)—failure to provide rescue equipment for purposes of rescuing employees of independent contractors; *Marshall v. L. E. Myers Company,* 589 F.2d 270, 6 OSHC 2159 (7th Cir. 1978)—failure to use preform grips on tension cables; *General Dynamics Corporation, Quincy Shipbuilding Division v. OSHRC,* 599 F.2d 453, 7 OSHC 1373 (1st Cir. 1979)—the employer failed to take proper precautionary steps, including adequate safety training to prevent the occurrence of hazardous conditions; *Brennan v. OSHRC (Canrad Precision Industries),* 502 F.2d 946, 2 OSHC 1137 (3d Cir. 1974)—the employer failed to provide adequate supervision of its employees; *H-30, Inc., supra*—the employer permitted employees to ride long champs; *Republic Creosoting Co., supra*—windrowed railroad ties were "improperly" stacked; *Getty Oil Company, supra,*—failure to pressure test a fluid booster tank.

L. DEFENSES

While most of the affirmative defenses listed in Chapter XII are theoretically available in cases involving violations of the general duty clause, the fact that an employer had, or through its supervisor (imputation) had, no knowledge of the hazard or the violation, or that employees committed the violation contrary to employer instructions *(e.g.,* isolated occurrences and/or employee misconduct), are particularly applicable defenses to general duty clause citations. If these defenses are proven, they would insulate an employer from liability under the Act. However, if the defenses are raised, it puts the employers' training and supervision generally into issue. See *United States Sugar Corporation,* 5 OSHC 1655 (1977); *Edwin F. Cillessen, General Contractor,* 5 OSHC 1570 (1976); *William R. Davis and Son, Inc.,* 6 OSHC 2000 (1978); *Jensen Construction Co.,* 7 OSHC 1477 (1979). However, in *Barker Brothers, Inc.,* 6 OSHC 1282 (1978), the Commission stated that the employee misconduct or isolated incident defense was not available therein, since the instructions given to the employees were inadequate. As the Commission stated:

> An employer must do more than give generalized instructions to its employees to work safely; it must specifically instruct its employees when and how to take necessary safety precautions. . . . It's not sufficient for the employer to leave the decision to use protective devices to the discretion of the employees, as Barker Brothers did.

See also *Southwest Central Rural Electric Cooperative Corporation,* 7 OSHC 2086 (1979). For an employer is not an insurer and need not take steps to prevent hazards which are not generally foreseeable, including the idiosyncratic behavior of an employee; the employer's duty under the general duty clause must be achievable. See *General Dynamics Corporation, supra; Magma Copper Company, supra; Titanium Metals Corporation of America, supra; Brennan v. Butler Lime & Cement Co.,* 520 F.2d 1011, 3 OSHC 1461 (7th Cir. 1975). Consequently, an employer can defend a charge that it violated the general duty clause of the Act on the basis that it took all necessary precautions to prevent the occurrence of the employee misconduct. *Horne Plumbing & Heating Co. v. OSHRC,* 528 F.2d 564, 3 OSHC 2060 (5th Cir. 1976).

See also *Royal Logging Company,* 7 OSHC 1744 (1979), in which the Commission stated that the greater hazard defense does not apply in cases arising under the general duty clause, but that evidence concerning creation of hazards by the proposed abatement method will be relevant in determining feasibility. But see *Westinghouse Electric Corp.,* 4 OSHC 1952 (1976).

NOTES

1. This is not to say that an employer's statutory responsibility for a hazard vanishes, or is even diminished, because the hazard was directly caused by an employee. The Act provides "that employers and employees have separate but dependent re-

sponsibilities and rights with respect to achieving safe and healthful working conditions." An employer has a duty to prevent and suppress hazardous conduct by employees, and this *duty is not qualified* by such common law doctrines as assumption of risk, contributory negligence, or comparative negligence.

2. In the OSHA Field Operations Manual (FOM) a hazard is defined as recognized if it is a condition that is of common knowledge or general recognition in the particular industry in which it occurs and is detectable by means of the senses (sight, smell, touch, and hearing) or if it is of such wide general recognition as a hazard in the industry that, even if it is not detectable by means of the senses, there are generally known and accepted tests for its existence which should make its presence known to the employer. For example, the excessive concentration of a toxic substance in the air would be a recognized hazard if it could be detected through the use of measuring devices. FOM, Ch. VII.

3. In *R. L. Sanders Roofing Co.*, 7 OSHC 1566 (1979), the Commission stated: "It is well established that falling is a hazard recognized throughout the construction industry. The specific question to be answered here is whether the hazard of falling 13 feet from the edge of a flat roof is recognized."

4. In *Getty Oil Company, supra,* the court stated that Getty could not be found "guilty" of a serious violation unless it did not and could not with the exercise of reasonable diligence know the presence of the violation (referencing Section 17(k) of the Act). Since it was undisputed that the company did not have actual knowledge of the fact that the vessel had not been pressure tested, the sole issue was whether Getty could have discovered that fact by reasonable diligence, in order to support a finding of a serious violation. Since the court in *Getty* stated a serious violation must be supported by substantial evidence, the citation must be vacated.

5. An employer's conduct must, under the constructive knowledge test, by judged against the standards of a *relevant* industry. *Brennan v. Smoke Craft, Inc.*, 530 F.2d 843, 3 OSHC 2000 (9th Cir. 1975).

6. In *Bethlehem Steel Corporation v. OSHRC*, 607 F.2d 871, 7 OSHC 1802 (3d Cir. 1979), recognition was established by reference to a standard promulgated by the American National Standards Institute (ANSI) referencing the speed of operation of cranes; *Betten Processing Corp.*, 2 OSHC 1724 (1975)—in evidence was a handbook of industrial loss prevention, an insurance-industry publication for general usage, and recommendations with respect to cranes; *Beaird-Poulan, A Division of Emerson Electric Company,* 7 OSHC 1225 (1979)—a National Safety Council data sheet on magnesium was used to demonstrate recognition; *Pratt & Whitney Aircraft Div., supra*—a treatise entitled "Sax, Dangerous Properties of Industrial Materials;" *R. L. Sanders Roofing Company,* 7 OSHC 1566 (1979)—OSHA standards for the construction industry which addressed falls of thirteen feet and a publication prepared by the National Roofing Contractors Association entitled "Protection of Roofers from Falling Risks;" *American Smelting & Refining, supra*—a chapter of a learned treatise entitled "Industrial Lead Poisoning in Industrial Hygiene and Toxicology," relative to the hazards of lead exposure; *Upshur Rural Electric Cooperative Corp.*, 7 OSHC (1979)—manual on safe work practices published by Texas A & M University; *Young Sales Corporation,* 7 OSHC 1298 (1979)—notices on a manufacturers brochure were warnings informing users of dangers. But see *Toms River Chemical Corporation,* 2 OSHC 3063 (1974)—a NIOSH criteria document could not be used to establish a general duty clause violation; *Triangle References Inc.*, 1 OSHC 3050 (1973)—the National Safety Council's Accident Prevention Manual is not evidence of a recognized hazard.

7. It should be noted that an employer's general duty only extends to his *own* employees; the multiemployer principle applicable to violations of specific standards does not apply to Section 5(a)(1) violations. The citations in *Fry's Tank Service* alleged that both "employers" violated Section 5(a)(1) because of the exposure of their employees to a confined space incapable of supporting human life. There was no question that the employees were so exposed and that both employers recognized the hazards involved. Thus the element of hazard recognition necessary to find a Section 5(a)(1) violation was present. Both employers, however, claimed that, by prohibiting their employees from entering the tanks, they had fulfilled the duty placed on them by the Act with respect to the hazard, and that they should therefore not be found liable. The Commission stated: "With respect to Fry's, we agree with the Judge's finding that a violation occurred. An employer's duties under the Act run to experienced as well as inexperienced employees. . . . All employees must receive supervision concerning the hazards inherent in their employment. *Brennan v. Butler Lime and Cement Co.*, 520 F.2d 1011, 3 OSHC 1461 (7th Cir. 1975).

"Cities also had a policy against its employees entering tanks on the Stoltenberg Lease. Even though its employees did enter the tank on this occasion, Complainant and OCAW do not argue that its basic policy was inadequate. Instead, they argue that despite this policy, Cities should have foreseen the possibility that its employees might enter the tank in a rescue attempt, and should have provided for this eventuality. Thus the gravamen of the charge against Cities is that it should have had rescue equipment, such as breathing equipment and lifelines, present at the site so that a rescue could have been conducted without endangering its employees. Judge Riehl found that, since Cities knew that somebody was to enter the tank, the possibility that its employees might also enter the tank in order to effect a rescue was also foreseeable, and that rescue equipment should therefore have been provided. In reaching his conclusion, the Judge relied on the Commission decision in *Aro, Inc.*, 1 OSHC 1453, (1973).

"In *Aro,* an employee entered an enclosed vessel in which the atmosphere had not been tested for oxygen sufficiency. The employee was overcome due to a lack of oxygen, and other employees tried to rescue him even though no rescue equipment was available. The Commission held that the failure to test the atmosphere before the first employee entered was a violation of Section 5(a)(1). The Commission also held that, under the circumstances, the possibility of a rescue attempt was reasonably foreseeable, and that the failure to furnish rescue equipment was a separate Section 5(a)(1) violation. The only distinction between this case and *Aro* is that, in this case, the employer charged with failing to furnish rescue equipment is not the same one that committed the violation which led to the necessity for the rescue attempt.

"A necessary element of a Section 5(a)(1) violation is that the employer must have been able to foresee the hazard to *its* employees. *Brennan v. OSHRC (Republic Creosoting Co.),* 501 F.2d 1196, 2 OSHC 1109 (7th Cir. 1974). The Judge found that, because Cities knew that somebody was going to enter the tank, the possibility that Cities' employees might enter the tank in a rescue attempt was reasonably foreseeable. But the need for a rescue arose, not simply because Thach entered the tank, but because he entered the tank under conditions which constituted a Section 5(a)(1) violation on Fry's part. Thus, in order to determine whether the entry of Cities' employees was reasonably foreseeable, we must inquire whether Cities could reasonably have foreseen that Fry's would perform the work in such an unsafe manner as to constitute a violation of the Act and give rise to the need for a rescue attempt.

"Normally, when an employer hires an independent contractor to perform certain work, it relies on the contractor's expertise to perform the work correctly. Generally, such an arrangement will be entered into because the contractor has greater expertise than the hiring employer in performing the work involved. We agree with Cities that it would constitute an unreasonable burden for an employer to be required to anticipate every violation an independent contractor might commit which results in a rescue attempt by its own employees. In general, the hiring employer is only responsible when it has reason to foresee that the independent contractor might perform the work in an unsafe manner.

"The record does not establish that Cities should reasonably have foreseen the violation committed by Fry's."

8. In this case, there were two separate and distinct hazards: falling and being burned by hot tar. The particular chain of events came about from the almost simultaneous occurrence of two "accidents," with the injury from each aggravated by the other. While the fall itself may have been precipitated by a burn and the fall injuries compounded by subsequent burns, the Commission stated that it is enough that the fall hazard involved was foreseeable, even if the unusual chain of events was not foreseeable.

9. Accordingly, an employer may use *any* method that renders its work site free of the hazard and is not limited to those methods suggested by the Secretary. It follows that an employer may defend against a Section 5(a)(1) citation by asserting that it was using a method of abatement other than the one suggested by the Secretary. However, the burden is then on the employer to establish that its abatement method is as effective as the one suggested by the Secretary. See *Beaird & Poulan, A Division of Emerson Electric Co., supra; Brown & Root, Inc., supra.* As stated by the Commission in *Brown & Root:* "Where, as here, Respondent asserts that it had an established work rule designed to prevent the violation, Respondent must show that the work rule was in fact adequate to prevent the violation and was adequately communicated to its employees. Respondent must also show that it had taken steps to discover violations of this rule and had effectively enforced the rule in the event of infractions." *Cf. Jensen Construction Co.,* 7 OSHC 1147 (1979)—elements needed to prove the affirmative defense of unpreventable employee misconduct under Section 5(a)(2) of the Act.

10. The court stated, in *Brennan v. OSHRC (Republic Creosoting Co.),* 501 F.2d 1196, 2 OSHC 1109 (7th Cir. 1974), that the lack of an accident could have been fortuitous, but a high rate of accidents with resulting injuries from the same cause is often given some weight as being symptomatic of a dangerous situation.

11. The Commission used the neutral term "incident" to avoid the connotation of chance frequently associated with the term "accident." The Commission previously addressed the meaning of the term "causing or . . . likely to cause" serious physical harm in numerous cases prior to its decisions in this case and *R. L. Sanders Roofing Co., supra.* To the extent that those decisions are inconsistent with the holdings in *Sanders* and this case, they were overruled. See, *e.g., Norman R. Bratcher Co.,* 1 OSHC 1152 (1973); *Betten Processing Corp.,* 2 OSHC 1724, (1975); *Nugent Service, Inc.,* 4 OSHC 1075, (1976); *Arizona Public Service Co.,* 1 OSHC 1369, (1973).

12. The court in *Atlantic & Gulf Stevedores,* referencing its *Hanovia Lamp* decision, stated that an employer could be held answerable for a violation resulting from employee misconduct only when demonstrably feasible measures exist for materially reducing its incidence, noting that while *Hanovia Lamp* involved a citation for violations of the general duty clause, the case before it involved citations for violation of a specific safety standard. The court declined to bifurcate the statute in such a manner as to hold employers to a higher standard of care under specific regulations than under the general duty clause. The court held that the *Hanovia Lamp* standard governing employer responsibility applies to specific standards to the same extent as to the general duty clause.

CHAPTER IX
Inspections, Compliance, and Enforcement

A. INTRODUCTION

In order to determine if employers are in compliance with the standards, regulations, rules, and orders (or the general duty clause) promulgated under the Act, and to carry out the purposes of the Act, agents of the Secretary of Labor (or his counterpart in state plans), called inspectors or compliance officers, are authorized to conduct inspections and investigations of workplaces or other places of employment.[1] The National Institute for Occupational Safety and Health (NIOSH), and the Mine Safety and Health Administration (MSHA), may also inspect an employer's workplace for research or compliance purposes, but their authority is discussed in later parts of this book. This part of the book will review the procedural aspects of an OSHA inspection from the arrival of an inspector at an employer's workplace through the issuance of a citation.

B. STATUTORY AND REGULATORY AUTHORITY

Section 8(a) of the Act provides, with respect to the conduct of inspections:

> In order to carry out the purposes of this Act, the Secretary, upon presenting appropriate credentials to the owner, operator, or agent in charge is authorized—
> (1) to enter without delay and at reasonable times any factory, plant, establishment, construction site, or other area, workplace, or environment where work is performed by an employee of an employer; and
> (2) to inspect and investigate during regular working hours and at other reasonable times, and within reasonable limits and in a reasonable manner, any such place of employment and all pertinent conditions, structures, machines, apparatus, devices, equipment, and materials therein, and to question privately any such employer, owner, operator, agent, or employee.

29 CFR 1903 contains the OSHA regulations relative to the conduct of inspections. Specifically, 29 CFR 1903.1 provides in pertinent part:

> The Act authorizes the Department of Labor to conduct inspections, and to issue citations and proposed penalties for alleged violations. The Act, under Section

> 20(b), also authorizes the Secretary of Health, Education, and Welfare to conduct inspections to question employers and employees in connection with research and other related activities. . . . The purpose of this part 1903 is to prescribe rules and to set forth general policies for enforcement of the inspection, citation, and proposed penalty provisions of the Act.

29 CFR 1903.3 sets forth the authority for conducting inspections, basically echoing Section 8(a) of the Act, but also providing that an inspector may review records required by the Act and its regulations and other records which are directly related to the purpose of the inspection.

Every establishment covered by the Act is subject to inspection by OSHA Compliance Safety and Health Officers (Inspectors)[2] during regular working hours and at other reasonable times. Only in "special circumstances" will an inspector attempt to inspect a workplace at times other than during the employer's regular working hours. FOM, Ch. V.D.1.a. In *Kinney System, Inc.*, 7 OSHC 1428 (1979), it was argued that an inspection conducted after 5:30 P.M. was not made during reasonable hours. The evidence established, however, that since the usual practice of the business was to operate in several consecutive day and night shifts, the term "regular working hours" should be construed to mean the working period comprising all of the work shifts. These situations will occur rarely, but may happen, for example, where an inspection during working hours would increase the danger or reduce the effectiveness of the inspection. 29 CFR 1903.6(a)(2).

Employers should not expect any advance notification of an OSHA inspection. Indeed, Section 17(f) of the Act contains a specific prohibition against prior notification:

> Any person who gives advance notice of any inspection to be conducted under this Act, without authority from the Secretary or his designees, shall, upon conviction, be punished by a fine of not more than $1,000 or by imprisonment for not more than six months, or by both.

See also 29 CFR 1903.6(c). The major purpose behind this statutory provision against advance notice is to "prevent employers from creating a misleading impression of conditions in an establishment." FOM, Ch. V.C.1.a.

Nevertheless, there are some exceptions to this general prohibition. Even under these circumstances, however, an employer will receive advance notice, by telephone, not more than twenty-four hours in advance of the inspection. 29 CFR 1903.6(b); FOM, Ch. V.C.2.6. The situations in which OSHA will contact an employer in advance of an inspection are as follows:

(1) In cases of apparent imminent danger, to enable the employer to abate the danger as quickly as possible;
(2) In circumstances where the inspection can most effectively be conducted after regular business hours or where special preparations are necessary for an inspection;
(3) Where necessary to assure the presence of representatives of the employer and employees or the appropriate personnel needed to aid in the inspection; and
(4) In other circumstances where the Area Director determines that the

giving of advance notice would enhance the probability of an effective and thorough inspection.

29 CFR 1903.6(a). See FOM, Ch. V.C.1.c. When the employer receives advance notice of an inspection, it is his responsibility to notify any authorized employee representative of the impending inspection, although the inspector will notify the representative of the employees if requested to do so by the employer and if the employer furnishes the necessary information. 29 CFR 1903.6(b); FOM, Ch. V.C.2.d. See also Section 8(e) of the Act; 29 CFR 1903.8.

As long as an employer continues to operate his business, OSHA may conduct inspections. The Secretary of Labor is not required to prove that an employer is engaged in a business affecting commerce subject to coverage of the Act before proceeding with an inspection. *Marshall v. Able Contractors, Inc.*, 573 F.2d 1055, 6 OSHC 1317 (9th Cir. 1978), *cert. denied*, 439 U.S. 880, 6 OSHC 1957 (1978). Thus, even where there has been an adjudication of bankruptcy, if the employer is still functioning as an operating entity, OSHA may inspect and enforce the Act. *Chicago Rock Island & Pacific Railroad Company*, 3 OSHC 1694 (1975); *In Re Penn Central Transportation Co.*, 347 F. Supp. 1356, 2 OSHC 1379 (E.D. Pa. 1972), *enf'd*, 535 F.2d 1249, 3 OSHC 2059 (4th Cir. 1976); *Penn Central Transportation Company*, 3 OSHC 1856 (1976). The financial status of an employer would, however, be a relevant consideration in the imposition of penalties. *Colonial Craft Reproductions, Inc.*, 1 OSHC 1063 (1973). See FOM, Ch. V.D.1.f. Primary jurisdiction to determine any questions of employer coverage under the Act is lodged in the Review Commission. *In The Matter of Restland Memorial Park*, 540 F.2d 626, 4 OSHC 1485 (3d Cir. 1976); *Lance Roofing Co. v. Hodgson*, 343 F. Supp. 685, 1 OSHC 1012 (M.D. Ga. 1972).

The same holds true in situations involving strikes, work stoppages, picketing, or other labor disputes at a workplace. In these circumstances, however, the inspector has an obligation to "take care to avoid involvement in the dispute while at the same time carrying out his responsibilities in determining whether the employer is complying with the Act." FOM, Ch. V.D.1.g. The inspector is instructed to contact the OSHA Area Director if he is unsure whether to proceed with the inspection. In *Amoco Oil Company v. Marshall*, 496 F. Supp. 1234, 8 OSHC 2030 (S.D. Tex. 1980), the Secretary of Labor sought to secure a warrant that ordered entry onto the employer's premises of not only an OSHA inspector but also of an employee-representative who represented striking Oil, Chemical and Atomic Worker employees. These "employees" were separated from the company and not present on company premises by reason of their strike status, both at the time of the accident that gave rise to the inspection and at the time of the inspection itself. The question presented to the court was whether, for purposes of determining the coverage of the Act, striking employees had the same right to be an accompanying representative as they would ordinarily have before their strike status. The court concluded that, because of the employees' absence both during the time of the accident and during the inspection, they really had little interest in the inspection itself and denied enforcement of the warrant.

Because OSHA regulations require the issuance of a citation for all citations observed by an inspector (even if abatement is immediate), OSHA cannot and does not provide pre-inspection consultation at the employer's workplace. However, states may and do provide consultation services, and these services are discussed in Chapter III. The normal consultation visit will not result in inspection activity by OSHA, since the consultant may not issue citations. However, OSHA must be informed of apparent violations if an imminent danger or unabated serious violation is discovered.

C. INSPECTION PROCEDURES

OSHA's operating procedures for the conduct of inspections are set forth in the Field Operations Manual (FOM), which contains guidelines and directives on such matters as inspection priorities, citation preparation, the characterization of violations, and assessment of penalties. The FOM procedures are basically applicable to the general safety inspection; the Industrial Hygiene Manual (IHM) contains similar procedures for the conduct of health inspections. Copies of both the FOM and the IHM may be obtained through OSHA's subscription service. OSHA also publishes additional memorandums and directives for guidance of its compliance personnel, called "OSHA Instructions."

While the Secretary of Labor is required to follow its regulations in the enforcement of the Act, *Atlantic Marine, Inc. v. OSHRC,* 524 F.2d 476, 3 OSHC 1755 (5th Cir. 1975), noncompliance with the FOM or IHM is not fatal, since these publications are agency policy and thus are directory only. *Combustion Engineering, Inc.*, 5 OSHC 1943 (1977); *FMC Corp.,* 5 OSHC 1707 (1977); *Limbach Co.*, 6 OSHC 1244 (1977).

An inspector is requested to conduct inspections according to the procedures set forth in the FOM or IHM in such a manner so as not to expose himself unnecessarily to hazards, recording all pertinent facts with respect to apparent violations and hazards. Although OSHA inspectors initially were poorly trained, lacked formal safety and health education, and viewed their mission as a holy one, concerned with citing employers with violations rather than elimination of work place hazards, much progress has been made in increasing both the quantity and quality of inspectors. *See* Whiting, *OSHA's Enforcement Policy,* 31 Lab.L.J. 259 (1980). Inspectors must now have professional safety and health credentials, training programs have been approved and their number has been increased (996 safety officers and 579 industrial hygienists as of May 1980). Basic information which is needed for enforcement purposes, and thus the information the inspector will attempt to obtain during the inspection, is set forth in the FOM as follows:

(1) Apparent violations observed.
(2) Serious, other-than-serious, or *de minimis.*
(3) Location in the establishment.
(4) Units (number, names, identification, etc.).
(5) Diagram where required.

(6) Interview with employees (name, title, assignment, or job) as applicable.
(7) Hazards corrected at time of inspection.
(8) Instrument readings.
(9) Time violation observed.
(10) The number and kinds of employees, *e.g.*, carpenters, machine operators, affected by a hazard which may be the subject of a citation. The inspector should make every effort to identify employees exposed to a hazard through admissions of the representative of their employer because at subsequent enforcement proceedings the inspector may testify as to statements made to him by such representative. On the walkaround the inspector should ask the employer's representative the name and employer of any employee exposed to a workplace hazard and then ask the employee to confirm his name and employer.
(11) Duration of condition.
(12) Employer/employee comments.
(13) All measurements where required.

Health inspections will follow the same basic procedures as safety inspections but will differ from safety inspections in that they are more complex and time-consuming and generally will involve the input of an industrial hygienist, who will accompany the inspector. The IHM sets forth general industrial hygiene policy for the conduct of health inspections. IHM, Ch. 1. Health inspections are normally scheduled in accordance with the priorities and procedures specified generally in the FOM. The industrial hygienist is instructed, as in normal workplace inspections, to conduct an opening conference, at which he or she will request process flow charts and plant layouts relative to the inspection. An examination will then be made of workplace records pertinent to the inspection. Thereafter, the health industrial hygienist will make a walkaround to identify potential health hazards in the workplace, recording relevant information concerning potential exposure to chemical substances or physical hazards (*e.g.*, he will observe employees activities to see if they are engaged in stationary or transient activities, survey engineering controls, collect screening samples, and record information relative to employee exposure to health hazards). Material safety data sheets will be reviewed and collected where available and appropriate in order to identify any hazardous materials and toxic substances which are present in the workplace. Information concerning the employer's medical program shall be requested, as well as his education, training and recordkeeping program, and his emergency procedures. The industrial hygienist will also determine whether a program for engineering and administrative controls exists and will determine whether an effective personal protective-equipment program exists. Finally, the industrial hygienist must determine whether sampling is required by utilizing the information collected during the walkaround.

If sampling is necessary, a sampling strategy will be developed by considering potential chemical and physical hazards, the number of samples to be taken, and the operations and locations to be sampled. Representative jobs must be selected for sampling and personal sampling devices prepared accordingly. Employees with the *highest* expected exposures at specific operations should be monitored. It is not necessary, however, according to the IHM, to monitor every employee that may be overexposed. All sampling equipment shall be checked and calibrated according to the procedures prescribed in the IHM. The IHM states that although it is not essential that the industrial hygienist continuously observe each employee being monitored, an account must be made for each monitored employee's movements and duties and each area of the establishment which may significantly affect the total exposure. The industrial hygienist should remain at the workplace while samples are being collected. Employers with the capability of doing so should consider running "side-by-side" sampling, which could later be used as a basis for comparison, in the event a citation is issued. It should be remembered, however, that if the inspector's sampling data does not turn out or is lost, OSHA may be able to subpoena the employer's data.

If an employee refuses to wear a sampling device and another employee who is similarly exposed cannot be sampled, the industrial hygienist is instructed to collect the sample by holding the collection device in the breathing zone of the employee or by any other means which provides a representative sample of the employee's exposure. If the employer has instructed the employees not to wear sampling equipment, the industrial hygienist shall inform the Area Director. In *Hutton v. Plum Creek Lumber Co.,* 608 F.2d 1283, 7 OSHC 1940 (9th Cir. 1979), the employer had a policy forbidding employees to wear measuring devices in aid of an OSHA inspection. The court stated that the federal courts are without power in the absence of specific legislative authorization to order an employer to require its employees to wear such devices. But see *Establishment Inspection of Keokuk Steel Castings, Division of Cast Metals Corporations,* 493 F. Supp. 842, 8 OSHC 1730 (S.D. Iowa 1980). Area samples may be taken to identify exposure sources relative to employee exposure.

After completion of the sampling, the general FOM procedure for closing conferences shall be followed, and an explanation of available inspection results shall be given to the employer, along with general guidelines on correction procedures and interim methods of control.

The IHM gives a detailed description and classification of violations and sets forth the procedures for the issuance of citations. The IHM, like the FOM, also covers such matters as the format in which citations are to be structured, discusses considerations such as mixture formulas, that is, whether substances have additive or synergistic effects, and considers the employer's work practices and personal hygiene requirements *(e.g.,* ingestion hazards, skin effects, carcinogenic substances, use of wipe samplings, biological monitoring). Prior to the completion of the inspection, the industrial hygienist is instructed to carefully investigate the source or cause of the observed hazards to determine whether some type of engineering, administrative, or work practice control or combination thereof may be applicable which would reduce employee exposure. Administrative controls are defined as any procedure which significantly limits daily exposure by control or manipulation of the work schedule. The use of personal protective equipment is not considered a means of administrative control. A work practice control is defined as the action of the employees which would result in the reduction of exposure through such methods as the effective use of engineering controls, sanitation and hygiene practices, or other changes in the way the employee performs the job. Engineering controls consist of substitution, isolation, ventilation, and equipment modification. Substitution may involve process change, equipment replacement, or material substitution. Isolation results in the reduction of a hazard by providing a barrier around the material, equipment, process, or employee. This barrier may consist of a physical separation or isolation by distance. Ventilation controls mean what they state, ventilation. Equipment modification results in increased performance or change in character, such as the application of sound-absorbing material.

Feasibility is defined in the IHM as the existence of general, technical knowledge of materials or methods which are available or adaptable to specific circumstances and which can be applied with a reasonable probability that employee exposure to occupational safety or health hazards will be reduced.

Technological feasibility is to be determined by reference to several sources listed in the IHM. Economic feasibility (cost of correction) is generally not considered to be a factor in the issuance of a citation.

The IHM states that whenever feasible engineering, administrative, or work practice controls can be instituted and yet are not sufficient to reduce exposure to or below the permissible exposure limits, they shall be used nontheless to reduce exposure to the lowest practical level. A determination that engineering controls are not feasible should not be made until consultation has been made with the OSHA national office. Finally, the IHM, like the FOM, specifies procedures for following and setting a reasonable time for abatement of the health hazard violations and for specifying abatement plans to correct the problem and for implementation and approval of petitions for modification.

D. INSPECTION PRIORITIES

An estimated five million workplaces employing sixty-five million employees are subject to inspection by OSHA.[3] In order for OSHA to inspect those workplaces which present the greatest threats to employee well-being, OSHA has established a system of inspection priorities, which is contained in the FOM.

1. Imminent Danger

Top priority has been assigned to imminent-danger situations. An imminent danger is defined as any condition where there is a reasonable certainty that a danger exists that can be expected to cause death or serious physical harm immediately or before the danger can be eliminated through normal enforcement procedures. Section 13(a) of the Act; FOM, Ch. IX.A.1. The Field Operations Manual defines "serious physical harm" as

> that type of harm which could cause permanent or prolonged impairment of the body or is the type of harm, which while not impairing the body on a prolonged basis, could cause such temporary disablement which would warrant in-patient treatment at a hospital.

FOM, Ch. IX.A.2.a. With regard to health hazards, an imminent-danger situation exists where there is a reasonable expectation that toxic substances or hazardous materials are present and that an employee's exposure to them will cause irreversible harm to a degree which would shorten life or cause a reduction in physical or mental efficiency, despite the fact that these harms may not be manifested immediately. FOM, Ch. IX.C.2.

When OSHA receives a complaint of an imminent danger and there appears to be a reasonable basis for the complaint, or where the situation is otherwise brought to the attention of OSHA, an inspection will be conducted immediately, usually within twenty-four hours. FOM, Ch. IX.B.1.c.(1). As noted above, the regulations also provide that in some situations OSHA will provide notice to the employer in advance of any inspection, in order to enable the employer to eliminate the danger as quickly as possible. 29 CFR 1903.6; FOM, Ch. IX.C.2.b.

It is the duty of the inspector at the site of an imminent-danger situation to encourage the employer to do whatever is possible to eliminate the danger. No Notice of Alleged Imminent Danger shall be prepared and no imminent-danger proceedings instituted, if voluntary elimination of the danger is immediately accomplished by permanent correction of the condition, or if the employer gives satisfactory assurance that he will not permit employees to work in the area of danger until the danger is permanently eliminated. Otherwise, the inspector shall post an Imminent Danger Notice and recommend court action. 29 CFR 1903.13; FOM, Ch. VIII.

The Notice of Alleged Imminent Danger does not constitute a citation of alleged violations or a notice of proposed penalty. It is only a notice that an imminent danger is believed to exist and that the Secretary of Labor will be seeking a court order to restrain the employer from permitting employees to work in the area of the danger until it is eliminated.

After an imminent danger has been found, (an) appropriate citation(s) shall be completed in accordance with the procedures contained in FOM, Chapters VIII and X. Penalties will be proposed in accordance with the procedures applicable to the violations involved, even though the employer immediately eliminates the imminence of the danger and initiates abatement steps. 29 CFR 1903.13.

Where a court has issued an injunction in an imminent-danger situation, a follow-up inspection shall take place immediately after the court order has been issued, to determine if the employer is complying with the terms of the order. Where no court proceeding has been initiated, because the imminence of the danger has been voluntarily eliminated by removal of affected employees or by the issuance and use of personal protective equipment, but permanent correction of the condition has not been achieved at the time of the inspection, a prompt citation shall be issued, and a follow-up inspection shall be conducted on the date set for abatement.

Imminent-danger situations have become intertwined with discrimination situations under Section 11(c) of the Act and are discussed in Chapter II of the book as well. See *e.g., Whirlpool Corp. v. Marshall,* 435 U.S. 1, 8 OSHC 1001 (1980).

2. Accidents and Fatalities

OSHA gives second priority to the investigation of catastrophes, fatalities, and accidents resulting in the hospitalization of five or more employees. Accidents of this nature must be reported to OSHA within forty-eight hours after their occurrence. 29 CFR 1904.8. Where accidents do not result in death or injury but do receive significant publicity, occur frequently, occur in an industry which is the subject of a special emphasis program, or involve extensive property damage and could have resulted in an enormous loss of life or many injuries, OSHA will investigate on a priority basis. FOM, Ch. IX.B.3.b; Ch. XVI.A.2. The fact that an accident has occurred, however, does not require that the Secretary of Labor conduct a general inspection of the entire workplace. *Truax & Hovey Drywall Corporation,* 6 OSHC 1654 (1978).

3. Complaints

Third priority is given to the investigation of valid complaints of alleged unsafe or unhealthful working conditions filed by employees or their authorized representatives. Complaints may be formal or informal; 29 CFR 1903.11(a) provides that employees may request an inspection of a workplace by giving notice of an alleged violation to the Area Director or Inspector. (According to OSHA, Congress recognized the harassment potential of employee complaints but enacted the provisions nevertheless. 45 Fed. Reg. 65916, October 3, 1980.) Any such notice shall: (1) be reduced to writing; (2) allege that a violation of the Act exists in the workplace; (3) set forth with reasonable particularity the grounds upon which it is based *(e.g.,* specify a condition or practice that is considered hazardous); and (4) be signed by the employee or representative of the employees. Although "employee" normally is defined as a present employee, it may also be a former employer *(e.g.,* where he has been discharged in violation of Section 11(c) of the Act or is still exposed to the workplace hazards by virtue of his employment on a multi-employer worksite). A copy of the complaint shall be provided to the employer or his agent by the inspector at the time of the opening conference, except, upon the request of the person giving the notice, the name or names of the complaining employees may be deleted therefrom.

If, upon receipt of the notice, the Area Director determines that the complaint meets the four formality requirements set forth in 29 CFR 1903.11(a) and that there are reasonable grounds to believe that the alleged violation exists, he will cause an inspection to be made as soon as practicable to determine if such a violation exists in accordance with the inspection priorities of FOM, Ch. IV. A complaint which does not meet the formality requirements will be investigated pursuant to the provisions set forth in FOM, Ch. VI, discussed below.

Prior to or during an inspection of the workplace, any employee or representative of employees in such a workplace may notify the inspector of any violation of the Act which he has reason to believe exists in the workplace. 29 CFR 1903.11(c). Any such notice shall comply with the requirements of paragraph (a) of the section. The regulations specify that inspections shall not be limited to matters referred to in the complaint but rather may encompass the entire establishment (unless there has been a very recent inspection such that only a partial inspection is necessary). FOM, Ch. VI; 29 CFR.

Because experience has taught OSHA that the investigation of substantially all the complaints which were filed with it would result in considerable backlog and in an inability to adequately concentrate the agency's resources in high-hazard areas (and because some complaints are filed as a result of harassment), OSHA has become more selective in its screening of complaints. The criteria by which OSHA will evaluate complaints is contained in an appendix to OSHA Instruction CPL 2.12A, issued on September 1, 1979. Thus OSHA will normally conduct inspections in response to formal complaints, except where (1) an evaluation indicates no reasonable grounds to believe that a violation exists, (2) the complaint involves a condition which has no direct or immediate relationship to safety and health, (3) a recent inspection or other evidence indicates that the hazard is not present or has been abated, or (4) the complaint is not within OSHA's jurisdiction.

OSHA is required under Section 8(f)(1) of the Act to investigate "as soon as practicable" complaints where it determines there are "reasonable grounds to believe that a violation or danger exists." See also 29 CFR 1903.11(b). Naturally, complaints alleging imminent-danger situations are afforded the highest priority and shall be investigated immediately regardless of whether the formality requirements are met. FOM, Ch. VI.B. With the exception of imminent-danger situations, which will be investigated immediately, complaints will usually be investigated between three and twenty working days after a complaint is filed, depending upon the potential seriousness of the hazard involved.

In 29 CFR 1903.12 it is provided that, if the Area Director determines that an inspection is not warranted because there are no reasonable grounds to believe that a violation exists with respect to a complaint filed under 29 CFR 1903.11, he shall notify the complaining party in writing of such determination. The person who filed the complaint may then obtain an informal review of such determination by submitting a statement of position with the Assistant Regional Director, providing the employer with a copy of such statement by certified mail. The employer may then submit an opposing written statement of position with the Assistant Regional Director. Thereafter, an informal conference may be held in which both the complaining party and employer may present their views; the Assistant Regional Director shall then make a determination relative to the appropriateness of conducting an investigation.

If the Area Director determines that an inspection is not warranted because the "formality" requirements of 29 CFR 1903.11(a) have not been met, he shall notify the complainant in writing of such determination. Such a determination shall be without prejudice to the filing of a new complaint which meets the requirements of 1903.11(a). In situations involving "informal" complaints, *i.e.*, those where the formality requirements of the regulations are not met or where the complaint is *not made by an employee* or his authorized representative, OSHA may advise the employer by letter that a complaint has been received. This letter will ask the employer to respond by explaining why there is no violation or what steps have been taken to correct any possible violations. If the employer's response is satisfactory, no investigation will be conducted, except for occasional follow-up inspections, which will occur in no more than one out of every ten situations. OSHA Instruction CPL 2.12A. If the response is not satisfactory, an investigation may be conducted.

Complaints that do not meet the formality requirements of 29 CFR 1903.11 *(e.g.*, oral complaints in person or by telephone) will not be given the priority of a formal complaint. Moreover, the complainant will not be given the additional rights accorded a formal complaint. OSHA will give such persons an opportunity to formalize their complaints as it will do for persons who file written complaints that do not meet all of the formality requirements. If a complainant decides not to formalize his complaint, it will then not be given the priority of a formalized complaint. Complaints from persons who are not employees or their representatives are treated as informal complaints. OSHA Instruction CPL 2.12(A), issued September 1, 1979, sets forth OSHA's procedure for screening "informal complaints." It provides that OSHA will use letters to advise employers of the filing of the complaint and of the action OSHA requires to remedy the alleged violation. Concurrently, the complainant will be notified and asked to notify the OSHA Area Director if no action has been taken by the employer within the prescribed period of time. If the employer corrects the hazard or responds that no correction is necessary, the case will be closed. The authority of OSHA to conduct inspections based upon informal employee complaints has been approved. *Burkart Randall, Division of Textron, Inc. v. Marshall*, 625 F.2d 1313, 8 OSHC 1467 (7th Cir. 1980); *Dravo Corp. v. Marshall*, 578 F.2d 1373, 6 OSHC 1824 (3d Cir. 1978).

Section 11(c) of the Act provides that no person shall discriminate against an employee who has filed a complaint or instituted proceedings under the Act. See 29 CFR 1903.11(d).

4. Special Inspection Programs

Fourth in priority are Special Program Inspections, which focus on industries with high rates of accidents or illnesses. These special-emphasis programs periodically are aimed at specific high-hazard target industries, occupations, or health substances. Industries are selected for special emphasis inspections on the basis of the severity of their death, injury, and illness incidence rates or employee exposure to toxic or hazardous substances. The Target Industry Program (TIP), for example, focused on five "hazardous" industries (marine cargo handling, roof and sheet metal work, meat products, transportation and lumber). After the TIP was retired, OSHA established the National Emphasis Program (NEP), which was characterized as a "general administrative plan" for the enforcement of the Act and which focused on foundries. *The Fountain Foundry Corporation,*—F. Supp.—, 6 OSHC 1885 (S.D. Ind. 1978). (Other special programs included the target health hazard program and the trenching and excavation program). OSHA disbanded the NEP and now focuses on industries with serious health and safety problems in its Hazard Analysis and Countermeasures Evaluation System. On November 1, 1980, OSHA issued a new system for conducting program inspections, which will replace its prior scheduling system. OSHA Instruction CPL 2.25A. As with the previous system, 95 percent of the programmed inspections will be conducted in high hazard industries. The selection of a particular employer in a category of employment is based on an "annual field operator's program plan projection" made at the area office. Establishments within a category are selected from an establishment list for that category and placed on an inspection cycle. Whether an establishment is considered high hazard is based on its standard industrial classification (SIC).

5. Routine Inspections

At the bottom of OSHA's priority list, but the most numerous and frequent, are "routine" inspections. Routine inspections are somewhat random in nature and may occur at any type of establishment at any time, although emphasis is given to high-hazard industries and industries with high accident rates. FOM, Ch. IV.B.3.d. OSHA Instruction CPL 2.25 attempts to allocate 95 percent of programmed, routine inspection resources to high-hazard industries, determining the viability of such inspections by the employer's standard industrial classification (SIC) and size. Factors considered in scheduling such routine program inspections are the establishment's location, the nature of the establishment, the time required for the inspection, the availability of sampling and other investigative equipment, the weather, the determination whether the establishment has recently been inspected, and special circumstances such as a labor strike.

6. Follow-up Inspections

In order to assure that abatement has been effected, the Area Director will schedule follow-up inspections as necessary. Follow-up inspections, which will normally be considered within the basic priority activity of the initial inspection, shall normally be conducted after an employer has been cited for a serious, willful, or repeated violation or when an employer has been subject to a court order in an imminent-danger situation, unless the inspector is certain the hazard has been corrected.

Follow-up inspections may be conducted during the fifteen-day notice of contest period for purposes of determining whether abatement has been achieved, provided the employer has not actually filled such a notice; however, it will not be conducted until after the earliest date for correction set forth in the citation. See *D. Fortunato, Inc.*, 2 OSHC 1446 (1974)—reinspection one day after employer's receipt of citation was invalid. A failure to abate citation will not be issued until after the contest period for the underlying citation has expired and it has been determined that the employer has not filed a notice of contest. FOM, Ch. VIII.

Where the employer has been cited for a number of violations with varying correction dates, the Area Director will normally arrange a schedule of follow-up inspections so as to minimize the number of follow-up inspections that will be necessary.

Follow-up inspections are normally conducted in cases of serious, willful, and repeated violations; the Area Director will normally use his discretion in scheduling follow-up inspections in the case of other-than-serious violations.

The primary purpose of the follow-up inspection is, of course, to determine if the cited hazard has been corrected. In addition, the inspector is instructed to determine if the citations have been posted in accordance with 29 CFR 1903.16 and if notices of contest or applications for variance have been posted, as applicable. Such determination shall be by observation. However, if posting time has elapsed, employee interviews may be used.

If violations noted in the citation have been corrected in a satisfactory manner, the inspector, after consulting with the Area Director, will mark the file accordingly. If the employer has not corrected the alleged violations, a notice of failure to correct and additional proposed penalties may be considered appropriate.

7. Inspection Frequency

An employer can expect objections to the reasonableness and frequency of inspections to meet with little success. The Third Circuit has, for example, rejected an employer's contention that OSHA abused its discretion in conducting twenty-three inspections, which it claimed amounted to nothing more than harassment. Because the inspections were initiated based on employee complaints, the court held that OSHA had little choice but to investigate. *Dravo Corp. v. Marshall,* 613 F.2d 1227, 7 OSHC 2089 (3d Cir. 1980). See also *Adams Steel Erection, Inc.*, 7 OSHC 1594 (1979), where the Review Commission took into account the hazardous nature of the industry in determining there was no harassment. Indeed, the fact that there has been no employee complaint prior to an inspection does not in itself constitute evidence of OSHA's bad faith or of harassment. *Robbins Electric Co., Inc.,* 5 OSHC 1688 (1977). With the above considerations in mind, we turn to the conduct of the inspection itself. The annual Department Labor and Department of Health, Education and Welfare Appropriations Act contains certain limitations on and exemptions from OSHA enforcement, such as: the assessment of penalties for first instance other than serious violations unless 10 or more violations are cited; farmers with 10 or fewer employees except those with migrant labor camps were exempted; restriction of enforcement activity in recreational, hunting, fishing, or shooting areas; inspection of employers with 10 or fewer employees in industries with 3 digit SIC injury and illness rates of less than 7 for 100 employees. These limitations do not apply in response to complaints for willful violations, in response to accidents or imminent dangers, in response to health hazards or to the investigation of discrimination complaints.

E. ARRIVAL AT WORKSITE AND PRESENTATION OF CREDENTIALS

An employer's first contact with OSHA typically occurs when an inspector actually appears at the workplace to begin an inspection. Workplace is defined as the place where work is performed by an employee of the employer. Section 8(a)(1) of the Act. A cement truck has been considered a workplace as long as it was used in furtherance of an employer's business. *West Allis Lime & Cement Co.,* 2 OSHC 1453 (1974); *Getter Trucking Co.,* 1 OSHC 1743 (1974). The first thing the inspector is required to do upon arriving at the workplace is to present "tactfully" his or her official Department of Labor credentials to the owner, operator, or agent in charge of the workplace. Section 8)a) of the Act; 29 CFR 1903.7(a); FOM, Ch. V.D.1.b. OSHA recommends that employers always insist upon seeing the inspector's credentials, which bear the inspector's photograph and serial number and which can be verified by contacting the nearest OSHA office. Normally, the inspector's credentials will be legitimate. In rare cases, however, the credentials or the inspector himself may be suspect. If an employer has a good reason to doubt the inspector's credentials *(e.g.,* arrival after regular working hours), such doubt may constitute valid grounds to refuse entry. The refusal of entry is not normally considered inconsistent with good faith. *Troxels, Inc.,* 2 OSHC 1377 (1979). However, it is not advisable to refuse such entry before an attempt is made to contact the OSHA area office.

If the inspector observes violations during his approach to the workplace and prior to the presentation of his credentials, these observations can legitimately form the basis of any subsequently issued citation. *Accu-Namics, Inc.,* 1 OSHC 1751 (1974), *aff'd,* 515 F.2d 828, 3 OSHC 1299 (5th Cir. 1975), *cert. denied,* 425 U.S. 930, 4 OSHC 1090 (1976). In its decision, the court concentrated on the fact of the requested *exclusion* of the evidence obtained in the inspection, stating that:

> Because *Accu-Namics* has shown no harm resulting from the Secretary's alleged violations of the Act, we decline to adopt any sweeping exclusionary rule that would exclude the facts regarding the violation. . . . There has been no violation of the employer's Fourth Amendment right here; the jobsite was on a public street. There is no showing that the inspector looked where he had no right to look, nor

> that he filched information to which he was not entitled, had he shown his credentials.

See also *Hartwell Excavating Co. v. Dunlop,* 537 F.2d 1071, 4 OSHC 1331 (9th Cir. 1976); *Hoffman Construction Co. v. Dunlop,* 546 F.2d 281, 4 OSHC 1813 (9th Cir. 1976); *S & H Riggers & Erectors, Inc.,* 8 OSHC 1173 (1980); *Kokanee Cedar Sales, Inc.,* 3 OSHC 1530 (1975). It has thus been held that the failure of an inspector to present his credentials to the employer before commencing an inspection, and even the subsequent failure to afford the employer an opportunity to accompany the inspector on the actual inspection tour, do not warrant the suppression of any evidence obtained by the inspector, *absent* the showing of prejudice to the employer. *Marshall v. Western Waterproofing Co.,* 560 F.2d 947, 5 OSHC 1732 (8th Cir. 1977); *Pullman Power Products, Inc.,* 8 OSHC 1930 (1980); *Environmental Utilities Corp.,* 5 OSHC 1195 (1977). The requirement that an inspector present his credentials pursuant to Section 8(a) of the Act has been relied upon to consider the merits of employer challenges to the validity of such inspection without the necessity of ruling on Fourth Amendment constitutional grounds. See *Marshall v. C.F. & I. Steel Corp.,* 576 F.2d 809, 6 OSHC 1543 (10th Cir. 1978). But see *Western Waterproofing Co., supra,* which held that where there has been neither a Fourth Amendment violation *nor* prejudice to an employer as a result of a failure of an inspector to present his credentials to an employer, suppression of evidence is not justified under Section 8(a) of the Act.

Most of the cases dealing with the issue of an inspector's observation of violations prior to the presentation of his credentials have dealt with the situation in which a workplace was open to public view. In such situations, it has not been held error or prejudicial to an employer's rights for the inspector to take pictures of violative conditions, even using a telephoto lens or other "sense-enhancing" device. Moreover, such action has not been considered to be an unlawful search and seizure in violation of the Fourth Amendment or a violation of Section 8(a) of the Act. In *Minotte Contracting and Erection Corporation,* 6 OSHC 1369 (1978), photographs taken by an inspector from a public roadway prior to his presentation of credentials were admitted into evidence, since the conduct of the inspection did not violate the Fourth Amendment in that the employer had no reasonable expectation of privacy for those operations which were open to the public view. See also *Laclede Gas Company,* 7 OSHC 1874 (1979)—photographs which showed violations at the employer's workplace were admissible as a result of the plain view doctrine where the compliance officer took the photographs with a telephoto lens while on a highway near the site, despite his failure to present his credentials prior to the taking of the photographs. *Titanium Metals Corporation of America,* 7 OSHC 2172 (1980).

It must also be noted that contrary to instructions contained in the FOM, it has been held an inspector is not required actually to observe a violation of the Act in order to establish a violation. *Combustion Engineering, Inc.,* 5 OSHC 1943 (1977); *Otis Elevator Co.,* 6 OSHC 2048 (1978); *Dixie Mills Co.,* 2 OSHC 3121 (1974); *Kenneth P. Thompson Co.,* 8 OSHC 1696 (1980)—determination of employee access to a hazard was made without observation of actual exposure.

However, in two other cases identification of a contaminant by a sense of smell was not found sufficient cause for the issuance of a citation. *Equity Supply Co.*, 1 OSHC 1082 (1973); *Apex Construction Inc.*, 2 OSHC 3025 (1974).

If, after his arrival at the workplace, the inspector is unable to locate the owner or operator in order to present his credentials, the inspector will endeavor to identify the agent in charge. Depending upon the nature of the workplace, this agent may be a supervisor, gang boss, or even the senior member of the work crew who has apparent or actual authority. *Hartwell Excavating Co. v. Dunlop*, 537 F.2d 1031, 4 OSHC 1331 (9th Cir. 1976). In *Andy Anderson, d/b/a Andy Anderson Irrigation and Construction*, 6 OSHC 1595 (1978), it was held that a foreman or manager or other person working in a managerial capacity with no person superior in authority at a worksite is a representative or an agent of the employer within the meaning of Section 8(e) of the Act. If the identity of the agent in charge cannot be ascertained, the inspector is then required to attempt to contact the employer by telephone in order to request the presence of the owner, operator, or agent in charge. In no case, however, will such initial confusion unduly delay the inspection. The inspector is operating under the express instruction that the "inspection should not be delayed unreasonably in order to await the arrival of the employer's representative." FOM, Ch. V.D.1.b(2). In *Northern Metal Company*, 3 OSHC 1645 (1975), the inspector was admitted to the employer's workplace by a guard and thereafter was unable to locate any other company officials. The inspector conducted the inspection and was accompanied by a foreman during which certain violations of the Act were observed and noted. Following the inspection, a closing conference was held with a company official. The employer argued that the credentials requirement of Section 8(a) of the Act was not met, but the Commission rejected this argument, stating that substantial compliance with the terms of the Act was sufficient, noting that the inspector had presented credentials to the person on the shift, establishing his authority. The Commission further noted that the inspector was well-known at the employer's place of business. See also *Tobacco River Lumber Co.*, 3 OSHC 1059 (1975).

WARRANT REQUIREMENTS

If the inspector's credentials are in order, the employer is faced with the decision of whether or not to allow an inspection to be conducted. The inspector at this point in time will either have no warrant or will have a warrant which has been obtained *ex parte* (without prior notification to the employer). When an inspector encounters a refusal to permit entry upon presenting proper credentials or an employer initially allows him to enter but thereafter refuses to permit an inspection, the inspector is instructed that he should tactfully advise the employer that the Act (Section 8(a)) provides for an inspection and he should endeavor to ascertain the reason for such refusal. If the employer persists in his refusal, the inspector is required to advise him as follows: "Due to your refusal to comply with Section 8(a) of the Occupational Safety and Health Act of 1970 I will be compelled to seek appropriate legal process to order your compliance." FOM, Ch. V. The inspector is then instructed to leave the premises and immediately report to the Area Director, who will immediately consult with the Regional Administrator and the Regional Solicitor, who will thereafter take appropriate action.

When permission to enter is not clearly given, the inspector is instructed to make every effort to clarify the employer's intent; if the inspector has any doubt as to whether the employer intends to permit an inspection, he should not proceed. If the employer is hesitant or absents himself for a substantial period of time, the inspector shall use professional judgment to determine whether to proceed with the inspection or delay until the Area Director is consulted. The inspector should not engage in argument or discussion concerning the refusal but may inquire as to the reason. The inspector may answer reasonable questions presented by the employer relative to the scope of inspection and its purpose but should avoid giving any impression of insistence or intimidation concerning the right to inspect. If the employer inquires about his right to obtain a warrant, the inspector should truthfully answer.

Where entry has been allowed but the employer interferes with or limits an important aspect of the investigation, the inspector is instructed to proceed as outlined above. Examples given by the FOM are the refusal to permit the walkaround, the examination of records which are essential to the inspection or the restriction of the inspection to a particular part of the premises. In *Usery v. Godfrey Brake & Supply Service, Inc.*, 545 F.2d 52, 4 OSHC 1843 (8th Cir. 1976), the employer, after verifying the inspector's credentials, refused to permit an inspection until a lengthy questionnaire was completed. The court stated that once verification has been obtained, the "use of a questionnaire as a condition of inspection was an unreasonable and arbitrary request not permitted by the Act and was used for the purpose of and patently designed to delay the inspection of the premises."

Although no search warrant or other process is expressly required by the Act, the United States Supreme Court, in *Marshall v. Barlow's Inc.*, 436 U.S. 307, 6 OSHC 1571 (1978), held that the Fourth Amendment's prohibition against warrantless searches and seizures applies to inspections conducted pursuant to the Act. It is thus clear that an employer has a right to require an inspector to obtain a search warrant prior to permitting an inspection of the workplace.

An employer may not, however, voluntarily permit the inspector to inspect the workplace and then later contest the constitutionality of the inspection conducted without a warrant. Voluntary consent to a search obviates any need for OSHA to obtain a warrant, and the consent prevents the employer from subsequently asserting that the search somehow violated his expectation of privacy. An inspection which is conducted with the consent of an employer does not violate the Fourth Amendment, even in the absence of a warrant. *Stephenson Enterprises, Inc. v. Marshall*, 578 F.2d 1021, 6 OSHC 1860 (5th Cir. 1978); *Milliken & Company v. OSHRC*, 605 F.2d 1201, 7 OSHC 1700 (4th Cir. 1979)—an inspector's mere announcement that he is at a workplace for the purpose of conducting an inspection, without more, is not coercive and does not change the voluntary nature of an employer's consent to be searched; thus the employer's acquiescence in the inspection constitutes consent. However, an employer who permits an inspection under protest is not considered to have consented to the inspection or waived his objections, which may be applicable under the Fourth Amendment either to the validity of the inspection or to the exclusion of evidence which has been obtained as a result of the inspection. See *Blocksam & Co. v. Marshall,* 582 F.2d 1122, 6 OSHC 1865 (7th Cir. 1978); *Weyerhaeuser & Company v. Marshall, supra.*

Prior to the Supreme Court's decision in *Barlow's, Inc.,* the lower courts had split on the question of whether nonconsensual, warrantless inspections were constitutionally suspect. In *Brennan v. Buckeye Indus., Inc.,* 374 F. Supp. 1550, 1 OSHC 1703 (S.D. Ga. 1974), the court held that Section 8(a) of the Act provided the statutory authorization for conducting warrantless searches, since the statutory restrictions mandating that inspections only be conducted "during regular working hours and at other reasonable times, and within reasonable limits and in a reasonable manner" was a sufficient limitation on "unreasonable" inspections, thereby eliminating the necessity for a warrant. The court held that the Supreme Court's exception from search warrant requirements for "pervasively regulated business," set forth in *United States v. Biswell,* 406 U.S. 311–316 (1972), and for "closely regulated" industries "long subject to close supervision and inspection," set forth in *Colonnade Catering Corp. v. United States,* 397 U.S. 72, 74, 77 (1970), *(e.g.,* liquor and firearms), should be extended to industries regulated by OSHA.

Other courts, however, rejected the *Buckeye* view of the Fourth Amendment's strictures. In *Brennan v. Gibson's Products, Inc.,* 407 F. Supp. 154, 3 OSHC 1944 (E.D. Tex. 1976), *vacated and remanded* 584 F.2d 668, 6 OSHC 2092 (5th Cir. 1978), a three-judge district court parted company with the court in *Buckeye Industries,* holding that warrantless inspections were indeed unconstitutional. The court in *Gibson's Products* expressly refused to extend the *Colonnade-Biswell* exception. The court went on to hold that Section 8(a) of the Act could be construed to authorize the issuance of warrants by United States magistrates or other judicial officers under the probable cause standards appropriate to administrative searches.

Finally, a three-judge district court concurred in the reasoning of *Gibson's* and held, in *Barlow's Inc. v. Usery,* 424 F. Supp. 437, 4 OSHC 1887 (D. Idaho 1976), *aff'd sub nom., Marshall v. Barlow's Inc.,* 436 U.S. 307, 6 OSHC 1571 (1978), that the Court's holding in *Buckeye Industries* which extended the *Colonnade-Biswell* exceptions to all industries regulated by OSHA was erroneous. However, the court in *Barlow's Inc.* refused to construe Section 8(a) of the Act as authorizing the issuance of search warrants. It held that the inspection provisions of OSHA were unconstitutional and enjoined enforcement of the Act. While the case was pending before the United States Supreme Court, Justice Rehquist stayed the injunction as to all workplaces except that of Barlow's itself. Upon review, the Supreme Court affirmed the judgment of the three-judge district court. The Court noted that the Fourth Amendment warrant requirements apply to commercial buildings as well as to private homes and that the Fourth Amendment's protections extend not only to criminal investigations, but also to civil searches.

The Court rejected the Secretary of Labor's contention that OSHA inspections should come within the *Colonnade-Biswell* exceptions, noting that those cases "represent responses to relatively unique circumstances" where there is such a history of government oversight and regulation that no businessman could have a reasonable expectation of privacy. Also rejected was the Secretary's argument that warrantless inspections were essential to enforcement of the Act. The Court noted that "the great majority of businessmen can be expected in normal course to consent to inspections without warrant," 436 U.S. at 316, and that there was no evidence of any "widespread pattern of refusal." *Id.* The Court went on to say that in those cases where a businessman does insist upon a warrant, regulations have already been promulgated which would enable OSHA to obtain compulsory process. See 29 CFR 1903.4. To the Secretary's argument that the advantages of surprise searches would be lost, the Court held that is was not "apparent why the advantages of surprise would be lost if, *after* being refused entry, procedures were available for the Secretary to seek an *ex parte* warrant to reappear at the premises without further notice to the establishment being inspected." 436 U.S. 319–320. In a footnote the Court noted that the Secretary of Labor had the statutory authority to promulgate a regulation expressly providing for *ex parte* warrants.

The Court further held that OSHA's entitlement to inspect will not depend upon its demonstration of probable cause to believe that conditions in violation of the Act exist on the premises to be inspected in order to obtain a warrant or other process, with or without prior notice:

> Probable cause in the criminal law sense is not required. For purposes of an administrative search such as this, probable cause justifying the issuance of a warrant may be based not only on specific evidence of an existing violation but also on a showing that "reasonable legislative or administrative standards for conducting an . . . inspection are satisfied with respect to a particular [establishment]." [*Camara v. Municipal Court, supra,* at 538.] A warrant showing that a specific business has been chosen for an OSHA search on the basis of a general administrative plan for the enforcement of the Act derived from neutral sources such as, for example, dispersion of employees in various types of industries across a given area, and the desired frequency of searches in any of the lesser divisions of the area, would protect an employer's Fourth Amendment rights. We doubt that the consumption of enforcement energies in the obtaining of such warrants will exceed manageable proportions.

According to this passage from the opinion, in order to establish probable cause for the issuance of a warrant, OSHA must establish that there is a reasonable, objective basis for selection of the particular workplace. The Court in so holding rejected the argument that any incremental protections afforded the employer's privacy by the necessity for a warrant were so marginal that they failed to justify the administrative burdens that would be imposed upon OSHA. In the Court's view, a warrant was necessary to provide assurances that a neutral officer had found that the inspection in question was reasonable under the Constitution, authorized by statute, and issued pursuant to an administrative plan containing specific neutral criteria. Significantly, the Court also noted that an important function of a warrant would be to "advise the owner of the scope and objects of the search, beyond which limits the inspector is not expected to proceed."

The Court thus concluded that the Act was unconstitutional insofar as it purported to authorize inspections without a warrant or its equivalent. The Court did note, however, that its holding "should not be understood to forbid the Secretary from exercising the inspection authority conferred by Section 8 of the Act pursuant to regulations and judicial process that satisfied the Fourth Amendment. . . . Of course, if the process obtained here, or obtained in other cases under revised regulations, would satisfy the Fourth Amendment, there would be no occasion for enjoining the inspections authorized by Section 8(a)." 436 U.S. at 325 n.23.[4]

The decision of the Supreme Court in *Barlow's* left in its wake the important question of determining when "probable cause" exists to issue an administrative search warrant. It also raised an additional consideration. Before an employer decides to request an inspection warrant from an inspector, he should certainly consider and take into account the relative ease by which the inspector will be able to obtain a warrant and the impression that such a demand would make on the attitudes of OSHA in the conduct of the inspection. For example, where an employer has had a fatal accident or catastrophe at its workplace, OSHA will normally and easily be able to obtain a warrant. A refusal by the employer to allow an inspector to inspect without a warrant will undoubtedly be viewed as nothing more than a delaying tactic. An employer's request that an inspector delay inspection until later in the day or until the following day and subsequent attempts on the part of the employer to negotiate the terms and the scope of inspection may also be considered a denial or refusal of entry by an employer and would thus permit an inspector to seek a warrant to search an employer's workplace. See *The Marmon Group, Inc., Cerro Metal Products Division,*—F. Supp.—, 7 OSHC 1327 (M.D. Pa. 1979). Indeed, if the inspector, who is refused entry without a warrant, subsequently obtains a warrant, returns to the workplace, and ascertains from employees that the employer used the extra time needed to obtain the warrant to "clean up," the inspector still may cite the employer and the citation may, according to OSHA, be considered "willful" in some cases. Moreover, assuming that there are *valid* OSHA regulations authorizing the issuance of an *ex parte* warrant before entry is denied, OSHA's records of employers who have demanded warrants in the past will probably result in an employer only being able to delay an inspection once. For, in subsequent inspections, OSHA would probably secure an *ex parte* warrant in advance.

1. General "Probable Cause" Standards

The *Barlow's* decision determined that probable cause may be established in two ways: by evidence of an existing violation or by application of reasonable administrative or legislative standards. Whether or not probable cause sufficient to justify the issuance of an inspection warrant is present in any given case depends, of course, on the specific facts of that case as presented to a court in a warrant application. Jurisdiction to consider OSHA's warrant applications lies with the magistrates of the United States District Courts.[5] *Marshall v. Pool Offshore Co.*, 467 F. Supp. 978, 7 OSHC 1179 (W.D. La. 1979); *Marshall v. Reinhold Construction Inc.*, 441 F. Supp. 685, 6 OSHC 2195 (M.D. Fla. 1977).

The courts, however, have outlined some general standards. In order to establish probable cause for an OSHA inspection warrant, the courts have required the Secretary to demonstrate a reasonable, objective basis for the selection of a particular workplace for inspection. Thus an administrative plan derived from neutral sources as required by *Marshall v. Barlow's*, if properly demonstrated in the warrant application, will justify the issuance of a warrant. Employee complaints and prior violations may also be used to establish the existence of a reasonable, objective basis for selection of a particular workplace for inspection. A sufficient lapse of time between inspections or the occurrence of an accident or injury has also been found sufficient as probable cause. The district court or magistrate's role in reviewing the warrant application is to ensure that the factual reasons given for the decision by OSHA to inspect a particular workplace are not clearly erroneous and that the general policy considerations used by OSHA are not arbitrary or an abuse of administrative discretion.

Based on the Supreme Court's specific holding in *Marshall v. Barlow's* that "probable cause in a criminal law sense is not required," the courts have held that probable cause is to be judged according to a relaxed or flexible standard of administrative probable cause, whether the warrant is sought for an inspection to be conducted pursuant to a legislative or administrative plan or is based on an employee complaint. The administrative probable-cause standard requires that an inspection be reasonable; the public interest in the inspection must outweigh the invasion of privacy which the inspection entails. See *Camara v. Municipal Court*, 387 U.S. 523, 536–39 (1967); *Marshall v. North American Car Co.*, 476 F. Supp. 698, 7 OSHC 1551 (M.D. Pa. 1979); *aff'd* 626 F.2d 320, 8 OSHC 1722 (3d Cir. 1980). In determining whether this standard is satisfied in any case, a court may consider *only* the evidence presented to the court or magistrate at the time of the application for the inspection warrant. See *Burkart Randall Division of Textron, Inc. v. Marshall*, 625 F.2d 1313, 8 OSHC 1467 (7th Cir. 1980); *Marshall v. W & W Steel Co.*, 604 F.2d 1322, 1326, 7 OSHC 1670, 1672 (10th Cir. 1979); *Plum Creek Lumber Co. v. Hutton*, 608 F.2d 1283, 1287, 7 OSHC 1940, 1942 (9th Cir. 1979); *Gilbert & Bennett Mfg. Co.*, 589 F.2d 1335, 6 OSHC 2151 (7th Cir. 1979), *cert. denied*, 444 U.S. 884, 7 OSHC 2238 (1979).

It would appear that three basic classes of information may generally be used to establish probable cause:

(1) General information about the employer's industry;
(2) General information about the employer and its workplace; and
(3) Specific information about the employer's working conditions.

The first two classes of information are relevant to the conduct of routine inspections, and the third is relevant to special inspections. With respect to the first two classifications, most of the courts who have considered the question have ruled, for example, that probable cause for the issuance of a warrant could be based upon the NEP, the "worst first" list, or some other special-emphasis program which is compiled in a geographical area by multiplying the injury rate, based upon BLS figures, for the employer's standard industrial classification (SIC). However, even though probable cause for an inspection may be based upon such information, the Secretary of Labor is still required to indicate how the "target" employer compares with other companies who are included in such a program. See *Reynolds Metals Co. v. Secretary of Labor,* 442 F. Supp. 195, 5 OSHC 1964 (W.D. Va. 1977); *In Re Mine Equipment Co.,*—F. Supp.—, 7 OSHC 1185 (E.D. Mo. 1979), *vacated on other grounds,* 608 F.2d 719, 7 OSHC 1907 (8th Cir. 1979); *Marshall v. Weyerhaeuser Company,* 456 F. Supp. 474, 6 OSHC 1920 (D. N.J. 1978).

With respect to information about the specific employer and its workplace, an important consideration in determining whether the employer has been reasonably selected for inspection (in addition to the receipt of a complaint) is the length of time since the employer had last been inspected. See *Reynolds Metals Co. v. Secretary of Labor,* 442 F. Supp. 195, 5 OSHC 1965 (W.D. Va. 1977). Also important is the employer's history of prior violations. See *Marshall v. Northwest Orient Airlines, Inc.,* 574 F.2d 119, 6 OSHC 1481 (2d Cir. 1978); *Pelton Casteel Inc. v. Marshall,* 588 F.2d 1182, 6 OSHC 2137 (7th Cir. 1978). But see *In Re Urick Property,* 472 F. Supp. 1193, 7 OSHC 1497 (W.D. Pa. 1979), although the employer had not been inspected in a year and a half and had a history of past violations citations, and employee complaints, a warrant was not issued, since there was no indication why the employer was chosen from among other foundries in the foundry inspection program.

2. Specific Evidence of Existing Violations

Where a warrant application has been based solely upon evidence of an existing violation(s), for example, an employee complaint, the courts have found that it is not absolutely necessary that the complaint itself be attached to the warrant application, as long as the "evidence" is brought to the court's attention. *North American Car Co., supra.* In this respect however, the courts have demanded more than a mere conclusory statement in the application that employee complaints have been received by OSHA; this is deemed to be insufficient. *Weyerhaeuser Co. v. Marshall,* 592 F.2d 373, 378, 7 OSHC 1090, 1093 (7th Cir. 1979); *Central Mine Equipment Co., supra; Burkart Randall Division of Textron, supra.* That is, the courts have required that OSHA at least set out the substance of the employee complaints so that the judicial officer issuing the warrant may exercise independent judgment as to whether or not an inspection is justified, rather than acting merely as a rubber stamp validating a decision already made. *Weyerhaeuser Co. v. Marshall, supra.*

There should also be an indication that employee complaint inspections are part of an overall administrative program. *Northwest Airlines, Inc., supra.* In *Burkhart Randall, supra,* the court found that a sworn warrant application submitted by the inspector alleging specific unsafe and unhealthful working conditions was adequate to establish the requisite degree of probable cause without requiring that the complaining employees also swear to the truth of their statements, since OSHA had determined from the information it received that a reasonable basis existed for believing violations existed. The court noted, however, that the application was not a model of specificity and clarity:

> Burkart, in essence, objects to the failure of the warrant application to demonstrate the underlying factual basis for the complainants' allegations. Under *Gilbert & Bennett,* however, the Magistrate need not be presented with enough evidence to allow him to verify to reevaluate the observations of the employee-informants. The application in this case presented to the Magistrate the substance of the complaints received by OSHA,[8] in sufficient detail for the Magistrate to evaluate OSHA's decision to conduct an inspection based on these complaints. It thus provided "specific evidence of an existing violation":
>
> > The affidavit established particularized probable cause because it provided the magistrate with the underlying factual data giving rise to the compliance officer's belief that a violation existed.
>
> *In the Matter of the Establishment Inspection of Marsan Co.,*—F. Supp.—, 7 OSHC 1557 (N.D. Ind. 1979).
>
> ---
>
> [8] Plaintiff suggests that the warrant application be found fatally defective because it does not state that the particular compliance officer applying for the warrant personally received the employee complaints or had personal knowledge of the circumstances surrounding their receipt. Plaintiff cites *Marshall v. Horn Seed Co.,* 7 OSHA (BNA) 1182 (W.D. Okla. 1979), in support of this proposition. In *Horn Seed,* however, the application also failed to indicate that the complaints received by OSHA came from employees of the subject employer. *Id.* at 1185. In addition, the compliance officer there specifically testified at a show cause hearing that she had not received the complaints and did not know who had. *Id.* at 1183. Therefore, *Horn Seed* does not hold that an application's failure to state that the affiant personally received the employee complaint referred to in the application, without more, renders the application fatally defective. To the extent that *Horn Seed* can be read to support such a position it is inconsistent with the probable cause standard described in *Gilbert & Bennett, supra,* 589 F.2d at 1339–40 (6 OSHC at 2153–54).

Although a warrant application based upon an employee complaint must be more than mere boilerplate, the judicial officer need not be presented with such specific evidence to allow a complete verification or reevaluation of the allegation of the complaining employee. In *Gilbert & Bennett Mfg. Co., supra,* for example, the application presented to the magistrate set forth the substance of the complaints received by OSHA (plus an affidavit from an inspector) in sufficient detail for the magistrate to evaluate OSHA's decision to conduct an inspection based on these complaints. The warrant application thus provided "specific evidence of an existing violation," since the attached affidavit of the inspector established particularized probable cause, because it provided the magistrate with the underlying factual data giving rise to the compliance officer's belief that a violation existed. See also *Weyerhaeuser Co. v. Marshall, supra.* A signed employee complaint with supporting affidavits, presented with a warrant application, would, however, be the best evidence of probable cause, *Marshall v. W & W Steel Co.,* 604 F.2d 1322, 7 OSHC 1670 (10th Cir. 1979), but cause may be found even based upon an anonymous employee complaint or where the complaint's contents are summarized in the warrant application. *Gilbert & Bennett Mfg. Co., supra; In Re Marsan Company, supra; North American Car Company, supra.*

Section 8(a) of the Act seems probable cause for the issuance of a search warrant may be established upon information received from any reasonably reliable and credible source. It does not actually require that information be supplied by someone currently employed by the employer. *Central Mine Equipment Co.*,—F. Supp.—, 7 OSHC 1185 (E.D. Mo. 1979), *vacated on other grounds*, 608 F.2d 719, 7 OSHC 1907 (8th Cir. 1979). In *Millcon Corporation*, 7 OSHC 1926 (1979), the court held that Section 8(f)(1) of the Act, which requires that the inspection be based upon a complaint of an employee or representative of an employee, includes complaints which came from a former employee. In *Robberson Steel Company*, 6 OSHC 1430 (1978), the Commission held that a complaint by a person who was no longer an employee at the time of the inspection, although contrary to the terms of Section 8(f)(1), was nevertheless valid if otherwise in accord with Section 8(a) of the Act. See also *In the Matter of Establishment Inspection of Marsan Co.*,—F. Supp.—, 7 OSHC 1557 (N.D. Ind. 1979)—the complaint was filed by a union on behalf of its employee members. Similarly, an inspector's personal observation of possible violations has established probable cause. *Marshall v. Miller Tube Corp. of America*,—F. Supp.—, 6 OSHC 2042 (E.D. N.Y. 1978).

The warrant application must, of course, state more than the fact that an employee (or other) complaint has been received giving OSHA reason to believe that an inspection was necessary. It must outline, with some degree of detail and specificity, the nature and contents of the employee complaint. *In Weyerhaeuser Co. v. Marshall*, 452 F. Supp. 1375, 6 OSHC 1811, (E.D. Wis. 1978), *aff'd*, 592 F.2d 373, 7 OSHC 1090 (7th Cir. 1979), the court stated:

> We believe that the Secretary's claim that the warrant application established probable cause by supplying specific evidence of an existing violation at the Manitowik plant must likewise fail as the magistrate was given no clue as to what the nature of the alleged violation might be. The affidavit is unrelieved boiler plate, stating merely that the Secretary had received a written complaint from an employee and that the Secretary had determined that there were reasonable grounds to believe that a violation of the Act existed. . . . Furthermore, in view of the ease with which OSHA can insert in the application the few additional facts which will satisfy the "specific evidence" manner of establishing probable cause (for example the application in *Gilbert and Bennett Manufacturing Company*, 589 F.2d 1335, 6 OSHC 2151 (7th Cir. 1979) where OSHA presented facts elucidating the nature of the employee's complaint), we see no reason to uphold the boiler plate application submitted in this case.

Similarly, the court in *Burkart Randall, supra* at 1472, stated:

> Plaintiff also argues that the warrant application is fatally deficient in terms of detail and specificity. In support of this contention, plaintiff cites cases holding inadequate to establish administrative probable cause applications which state only that a complaint has been received giving OSHA reason to believe an inspection is necessary. See, *e.g.*, *Weyerhaeuser Co. v. Marshall*, 452 F. Supp. 1375, 1379–80, 6 OSHC 1811, 1814 (E.D. Wis. 1978), *aff'd*, 592 F.2d 373, 378, 7 OSHC 1090, 1093 (7th Cir. 1979); *In the Matter of: Establishment Inspection of: Northwest Airlines, Inc.*, 437 F. Supp. 533, 536, 5 OSHC 1848, 1849 (E.D. Wis. 1977), *aff'd*, 587 F.2d 12, 13, 6 OSHC 2070, 2071 (7th Cir. 1978). These cases, however, do not advance the inquiry in this case. The warrant application at issue here went beyond the conclu-

> sory statement that complaints had been received: it described the contents of the complaints. As this court stated in *Weyerhaeuser, supra,* only a "few additional facts" need be added to a conclusory application to "satisfy the 'specific evidence' manner of establishing probable cause." 592 F.2d at 378 (7 OSHC at 1093) [footnote omitted]. The application before us contains more than a "few additional facts" and is not properly compared to the boilerplate applications presented in *Weyerhaeuser* and *Northwest Airlines.* On the contrary, it is comparable in terms of specificity and detail to the warrant applications upheld in *Gilbert & Bennett, supra,* 589 F.2d at 1339, 6 OSHC at 2153 and *Marsan Co., supra,* 7 OSHC at 1559.

See also *Northwest Airlines, Inc.*, 437 F. Supp. 533, 536, 5 OSHC 1848, 1849 (E.D. Wis. 1977), *aff'd,* 587 F.2d 12, 13, 6 OSHC 2070, 2071 (7th Cir. 1978); *Marshall v. Horn Seed Co.,*—F. Supp.—, 7 OSHC 1182 (W.D. Okla. 1979); *West Point Pepperell v. Marshall,* 496 F. Supp. 1178, 8 OSHC 1954 (N.D. Ga. 1980)—evidence which fails to establish the validity of employee complaints concerning a respirator program and which shows that an employer had instituted a comprehensive respirator program and did not discriminate against employees who were not able to wear respirators requires rejection of a warrant, since there is no probable cause to base an inspection with respect to specific violations.

The fact that the warrant application is based upon employee complaints which are not formal, *i.e.,* not in writing and signed by the complainant, as provided for in Section 8(f)(1) of the Act, will not necessarily invalidate a warrant. Section 8(f)(1) merely identifies a particular set of circumstances under which OSHA must inspect in response to an employee complaint. Although it is not clear whether or not OSHA must inspect in response to all complaints, formal or informal, there seems to be no question that OSHA may inspect based on an informal complaint. *Dravo Corp. v. Marshall,* 5 OSHC 2057 (W.D. Pa. 1977), *aff'd,* 578 F.2d 1373, 6 OSHC 1824 (3d Cir. 1978); *Aluminum Coal Anodizing Corp.,* 1 OSHC 1508 (1974); *Burkart Randall Division of Textron, Inc. v. Marshall,* 625 F.2d 1313, 8 OSHC 1467 (7th Cir. 1980).

With regard to those warrant applications which are not based on a general administrative or legislative plan, *Barlow's* requires that there be "specific evidence of an *existing* violation." 436 U.S. at 320, 6 OSHC at 1575. Thus the length of time between the filing of the complaint (or other basis for inspection) and the application for a warrant may become very important. Employers have successfully challenged the information contained in warrant applications as being stale. In *Marshall v. Pool Offshore Co.,* 467 F. Supp. 978, 7 OSHC 1179 (W.D. La. 1979), an inspector's affidavit which failed to state any specific evidence of an *existing* violation or to set forth any facts from which a court could conclude that an inspection is required by reference to reasonable legislative or administrative standards did not establish probable cause to support the issuance of an inspection warrant. Due to the nature of most allegations of health and safety standard violations, it is unlikely, however, that the conditions giving rise to employee complaints will cease to exist merely because of a passage of time. Usually, affirmative action is necessary to correct those conditions. Thus, the courts seem willing to issue a warrant for an OSHA inspection despite the fact that the application is based on information several months old. *Burkart Randall Division of Textron, Inc. v.*

Marshall, 625 F.2d 1313, 8 OSHC 1467 (7th Cir. 1980)—6 months; *Central Mine Equipment Co.,* 7 OSHC 1185 (E.D. Mo.), *vacated on other grounds,* 7 OSHC 1907 (8th Cir. 1979)—8 months. But see *Marshall v. Sun Petroleum Products Co.,* 622 F.2d 1176, 8 OSHC 1422 (3d Cir. 1980)—an employer will not be required to defend a stale citation. It must be noted in this respect that the lapse of a considerable amount of time between inspections may establish probable cause in and of itself. *Weyerhaeuser Co. v. Marshall,* 456 F. Supp. 474, 6 OSHC 1920 (D. N.J. 1978).

Several courts, including two courts of appeal, have taken the position that inspection warrants issued in response to specific employee complaints must be limited in scope to the subject matter of the complaints, if at all possible, since no probable cause has been shown for a search into other areas of the workplace. *Marshall v. North American Car Co.,* 476 F. Supp. 698, 7 OSHC 1551 (M.D. Pa. 1979) *aff'd,* 626 F.2d 320, 8 OSHC 1722 (3d Cir. 1980); *Marshall v. Trinity Indus., Inc.,*—F. Supp.—, 7 OSHC 1851 (W.D. Okla. 1979); *Whittaker Corporation v. OSHA,*—F. Supp.—, 6 OSHC 1492 (M.D. Pa. 1978), *dismissed* 610 F.2d 1141, 7 OSHC 1638 (3d Cir. 1979); *Central Mine Equipment Co.,*—F. Supp.—, 7 OSHC 1185 (E.D. Mo.), *vacated on other grounds,* 608 F.2d 719, 7 OSHC 1907 (8th Cir. 1979); *Urick Foundry,* 472 F. Supp. 1193, 7 OSHC 1497 (W.D. Pa. 1979). The Eighth Circuit, in *Central Mine Equipment Co.,* concluded that the scope of an OSHA inspection should be a factor considered by the judicial officer in the balancing process to be used in determining whether probable cause exists in a particular case. The court felt that the scope of such an inspection should be limited whenever possible. However, the Seventh Circuit (and one decision by the Third Circuit) has parted company with the Eighth and the Third Circuits on this issue and has taken the position that, in the absence of extraordinary circumstances, an application for a warrant to conduct a general wall to wall inspection should be allowed, even though it is based on specific employee complaints directed to a specific area. *Burkart Randall supra.; Dravo Corp. v. Marshall,*—F. Supp.—, 5 OSHC 2057 (W.D. Pa. 1977), *aff'd,* 578 F.2d 1373, 6 OSHC 1824 (3d Cir. 1978); *Gilbert & Bennett Mfg. Co.,*—F. Supp.—, 5 OSHC 1375, (N.D. Ill. 1977), *aff'd,* 589 F.2d 1335, 6 OSHC 2151 (7th Cir.), *cert. denied,* 444 U.S. 884, 100 S.Ct. 174 (1979). See also Rothstein, *OSHA Inspections After Marshall v. Barlow's, Inc.,* 1979 Duke L.J., 63, 96.

In order to meet the requirements of specific evidence of an existing OSHA violation, the warrant application need not be based on employee complaints alone. Indeed, one court has recently held that administrative probable cause may properly be founded solely upon newspaper reports of a serious accident occurring at an employer's workplace. *Federal Clearing Die Casting Co.,*—F. Supp.—, 8 OSHC 1635, 1636 (N.D. Ill. 1980). An OSHA warrant request will also normally be granted in situations involving a followup inspection. Thus, where an employer had been found in violation of the Act at one facility and had moved some of the operations which had led to the OSHA citation to a new location, probable cause was found for an OSHA inspection of the new plant to determine whether or not the conditions which violated the Act had been corrected. *Pelton Casteel, Inc. v. Marshall,*—F.2d—, 6 OSHC 2137 (7th Cir. 1978).

3. Reasonable Legislative or Administrative Standards

Pursuant to *Barlow's,* courts also seem willing to find probable cause to justify the issuance of an administrative warrant based upon a showing that a specific employer was chosen for an investigative search on the basis of "a general administrative plan for the enforcement of the Act derived from neutral sources. . . ." *Weyerhaeuser Co. v. Marshall,* 592 F.2d 373, 378, 7 OSHC 1090, 1093 (7th Cir. 1979); *Marshall v. Weyerhaeuser Co.,* 456 F. Supp. 474, 6 OSHC 1902 (D. N.J. 1978)—approval of OSHA's worst-first inspection program, as applied to the particular establishment in a neutral manner. In *Gilbert & Bennet Manufacturing Co., supra,* the court held that probable cause for a NEP inspection was shown through the use of industry-wide accident statistics.

> Paragraph 2's bare assertion that the desired inspection is "part of an inspection and investigation program . . . authorized by section 8(a) of the Act" does not indicate that the program was based on neutral criteria. Nor did the addition of the phrase "to assure compliance with the Act in the foundry business" eliminate this insufficiency. Without a more definite statement of the neutral criteria being utilized by the Secretary in selecting Chromalloy for inspection, it was impossible for the magistrate to form a conclusion of reasonableness. This finding comports with the criticism leveled by the Supreme Court in *Barlow's* at the warrant application in that case, which contained similar language as here. The Court stated: "the program was not described . . . or any facts presented that would indicate why an inspection of Barlow's establishment was within the program." 436 U.S. at 323 n.20. See also *Matter of Northwest Airlines, Inc.,* No. 77–2268 (6 OSHC 2070) (7th Cir., Nov. 8, 1978); *Dunlop v. Hertzler Enterprises, Inc.,* 418 F. Supp. 627, 629 n. 3 (4 OSHC 1569) (D. N.M. 1976).
>
> Paragraph 9 supplied additional pertinent information, namely that the inspection was part of a "National-Local plan designed to achieve significant reduction in the high incidence of occupational injuries and illnesses found in the metal-working and foundry industry." No other information was before the magistrate. And no other can be considered by this court in passing on the validity of the warrant. *Aguilar v. Texas,* 378 U.S. 108, 109 n.1 (1964). Thus, the instant warrant was plainly supported by probable cause in the *Camara/Barlow's* sense since Chromalloy was selected for inspection not as the result of the "unbridled discretion" of a field agent, but rather, pursuant to "a National-Local plan" designed by agency officials for the purpose of reducing the high incidence of occupational injuries and illnesses found in the metal-working and foundry industry.

In the context of inspection justification vis-à-vis a "general administrative plan," the judicial officer is required to determine that there is a reasonable legislative or administrative inspection program and that the inspection of a particular employer fits within that program. *Northwest Airlines, Inc.,* 587 F.2d 12, 6 OSHC 2070 (7th Cir. 1978). As with an inspection based upon employee complaints, noted above, a warrant application which contains only conclusory allegations that the inspection is reasonable or is authorized by law is clearly insufficient and fails to describe the program in which the inspection fits. *Weyerhaeuser Co. v. Marshall,* 592 F.2d 373, 7 OSHC 1090 (7th Cir. 1979); *In Re Urick Foundry,* 472 F. Supp. 1193, 7 OSHC 1497 (W.D. Pa. 1979)—with respect to warrants issued pursuant to an administrative plan, the warrant application must describe how the neutral/specific criteria resulted in the selection of the particular employer—the fact that the employer has a history of past citations and inspections and had not been recently inspected is not sufficient unless other employers under the plan do not have similar records.

The courts have also held that national emphasis programs taken in conjunction with statistics that show high illness or injury rates would normally constitute the neutral administrative plan for enforcement of the Act. However, OSHA must explain why a particular employer is chosen rather than other similarly situated employers. As stated by the court in *Northwest Airlines, Inc.:*

> The question of whether section 8(f)(1) constitutes "a general administrative plan . . . derived from neutral sources," *Barlow's, supra* at 1825, should not, however, divert us from the true problem in this case, which is the total lack of any description of the "inspection and investigation program" referred to in paragraph 3 of the affidavit. In *Barlow's* the Supreme Court stated in reference to a warrant application virtually identical to paragraph 3 of the application in the instant case that "the program was not described . . . or any facts presented that would indicate why an inspection of Barlow's establishment was within the program." [98 S.Ct. at 1826 n.20.] This passage makes it clear that the magistrate has two functions to perform under *Barlow's* second method of establishing probable cause: (1) He must determine that there is a reasonable legislative or administrative inspection program and (2) he must determine that the desired inspection fits within that program.
>
> The affidavit here provides insufficient information for the magistrate to perform the first of these functions, so *a fortiori* he is unable to perform the second. Paragraph 3's bare assertion that there is an "inspection and investigation program" does not permit the magistrate to ascertain that there is in fact a program or what the nature of such a program might be, let alone whether it is reasonable. It is only in the Secretrary's briefs that we are told the "program" of paragraph 3 refers to the alleged section 8(f)(1) "program" of inspecting pursuant to employee complaints. But the identification of the program must take place in the warrant application itself in order to enable the magistrate to perform his first function under *Barlow's.* Accordingly, we hold that the warrant application here was insufficient to support the issuance of a warrant as it failed to describe or identify the program which allegedly supported the desired inspection.

In *Central Mine Equipment Company,*—F. Supp.—, 7 OSHC 1185 (E.D. Mo.), *vacated,* 608 F.2d 719, 7 OSHC 1907 (8th Cir. 1979), the failure of the warrant application to describe adequately how specific criteria of OSHA's plan for inspection-site selection resulted in the employer's designation for inspection deemed it insufficient to establish probable cause. See also *Marshall v. Multi-Cast Corp.,*—F. Supp.—, 6 OSHC 1486 (W.D. Ohio 1978). Thus evidence must be presented with the warrant application which shows that the administrative standards are being applied to the respective employer in a neutral manner.

This is not to say that OSHA must supply such detailed information to the judicial officer as to convert "a simple warrant request into a full blown hearing." *Marshall v. Chromalloy American Corp.,* 589 F.2d 1335, 1342, 6 OSHC 2151, 2156 (7th Cir. 1979). The affidavit in support of the warrant application need only set forth the necessary neutral criteria used by OSHA in selecting a particular employer for inspection. In the *Chromalloy* case the warrant application contained a brief description of a national-local plan designed to reduce the high incidence of occupational injuries and illnesses in the metal-working and foundry industry. The court found that the warrant was supported by probable cause and not issued based on "unbridled discretion of a Field Agent." See also *Fountain Foundry Corp. v. Marshall,*—F. Supp.—, 6 OSHC 1885 (S.D. Ind. 1978); *Pfister & Vogel Tanning Co., Division of Beatrice Foods, Establishment Inspection of,* 493 F. Supp. 351, 8 OSHC 1693 (E.D. Wis. 1980)-(NIOSH).

Because of the nature of an investigation based on a general administrative plan, the exact location and nature of the conditions to be investigated at an individual workplace cannot be known before entering the particular establishment. Thus, courts in these situations will allow the warrant to be fairly broad in scope, possibly encompassing the entire workplace. *Marshall v. Chromalloy,* 589 F.2d at 1343, 6 OSHC at 2157; *Gilbert & Bennett, supra.*

4. Harassment and Misrepresentation

In several cases, employers have alleged that an inspection has been conducted or requested due to harassment. However, such arguments have almost uniformly been rejected by the courts (and the Commission). For example, in *The Marmon Group, Cerro Metal Products Division,*—F. Supp.—, 7 OSHC 1327 (N.D. Pa. 1979), the employer argued that the broad inspection was unjustified because it was part of a pattern of harassment and in a sense was an abuse of process. The court disagreed and stated that repeated exchanges between inspector and employer do not establish bad motive or harassment. In *Dravo Corporation v. OSHRC,* 613 F.2d 1227, 7 OSHC 2089 (3d Cir. 1980), the employer also argued that inspections which lead to citations were the product of harassment, complaining that there were twenty-three inspections ordered of its facilities while competitors were inspected one, two, or three times and the Secretary's continued inspections constituted an abuse of its discretion. The court, however, stated:

> Admittedly the respondent's premises have been inspected more than would normally be expected. However, for the most part, the inspections were initiated at the instigation of complaints by respondent's employees. Under the provisions of Section 8(f) of the Act it would appear that the Complainant had little choice but to investigate the complaints, albeit he did have discretion with respect to the scope of the inspections. Given the broad scope of prosecutorial discretion it is concluded that the evidence does not warrant vacation of the citations.

In *Quality Stamping Products Co.,* 7 OSHC 1285 (1979), the Commission held that the motive of a complainant is *not* relevant to a determination of whether there is probable cause to inspect. Moreover, OSHA's duty is "to determine whether there is cause to inspect an employer's workplace rather than attempt to discern a complainant's motive." *Aluminum Coil Anodizing Corp.,* 1 OSHC 1508 (1974); *Hydroform Products Corp.,* 7 OSHC 1995 (1979).

Section 8(a)(2) of the Act requires, however, that inspections be conducted within reasonable limits and in a reasonable manner; if not, and if the employer is harassed or prejudiced in the preparation and presentation of his case, the evidence obtained from the inspection may be suppressed or the citation vacated. See *Robbins Electric Company, Inc.*, 5 OSHC 1658 (1977); *Adams Steel Erection, Inc.*, 7 OSHC 1549 (1979); *Electrocast Steel Foundry, Inc.*, 6 OSHC 1562 (1978)—case remanded to produce evidence that the inspection was conducted in a positive and reasonable manner. A warrant application must therefore indicate that the inspection will be conducted at a reasonable time and in a reasonable manner. See *Empire Steel Manufacturing Co. v. Marshall,* 437 F. Supp. 873, 5 OSHC 1819 (D. Mont. 1977), where the period of time specified in the warrant had expired and the inspection had not been authorized for a time other than normal business hours. See also *In Re Central Mine Equipment Co.,*—F. Supp.—, 7 OSHC 1185 (E.D. Mo. 1979), *vacated on other grounds,* 608 F.2d 719, 7 OSHC 1907 (8th Cir. 1979).

Finally, in *Marshall v. Milwaukee Boiler Manufacturing, Inc.*, 626 F.2d 1339, 8 OSHC 1923 (7th Cir. 1980), evidence in the Secretary of Labor's application for a warrant to inspect the employer's workplace contained a misrepresentation or misleading statements on the injury rate in the employer's industry in relationship to others in the metropolitan area where it was located, as well as on the programmatic authority for selection of the employer's worksite for inspection. Such misrepresentation did not, however, bar the affirmance of the validity of the warrant, because the record failed to establish that the misrepresentations were intentionally or recklessly made or that OSHA's selection of the employer's worksite on the basis of no prior inspection, history, and industry injury rates on a statewide basis was unreasonable. See also *Pelton Casteel, Inc., v. Marshall,*—F.2d—, 6 OSHC 2137 (7th Cir. 1978).

5. Ex Parte Warrants

A major problem area, which now appears to be on the verge of resolution, regards the issuance of *ex parte* warrants. In *Cerro Metal Products v. Marshall,* 467 F. Supp. 869, 7 OSHC 1125 (E.D. Pa. 1978), *aff'd,* 620 F.2d 964, 8 OSHC 1196 (3d Cir. 1980), a district judge ruled, in November 1978, that OSHA did not have the regulatory authority to seek warrants on an *ex parte* basis. The court based its opinion on language contained in the *Barlow's* decision.[6] See also *Marshall v. Huffhines Steel Co.,* 478 F. Supp. 986, 7 OSHC 1850, 7 OSHC 1910 (N.D. Tex 1979); *contra S. D. Warren, Division of Scott Paper,* 481 F. Supp. 491, 7 OSHC 2010 (D. Me. 1979); *Amoco Oil Co. v. Marshall,* 496 F. Supp. 1234, 8 OSHC 2030 (S.D. Tex. 1980). In response, OSHA attempted to clarify its regulations, 29 CFR 1903.4, in an interpretation issued on December 22, 1978, to make clear its authority to seek *ex parte* warrants. 43 Fed. Reg. 59839. Although the Tenth Circuit ruled that the regulation was valid and that compliance with the notice and comment procedures of the Administrative Procedure Act were not necessary, *Marshall v. W & W Steel Co.,* 604 F.2d 1322, 7 OSHC 1670 (10th Cir. 1979), the Third Circuit, in affirming the original *Cerro Metal Products* case, flatly held that the

December 1978 regulation was *interpretive* in nature and was therefore invalid because of the agency's failure to use the notice and comment rule-making procedures. In *United States Chamber of Commerce v. OSHA,* 636 F.2d 464, 8 OSHC 1648 (D.C. Cir. 1980), the court distinguished between legislative and interpretive rules and regulations, stating that while Section 8(g)(2) of the Act empowers the Secretary of Labor to prescribe rules and regulations that are necessary to carry out his responsibility and that such provision allows the Secretary to promulgate legislative as well as interpretive rules, only interpretive rules are exempt from the notice and comment obligations of 4 U.S.C. §553(b) of the Administrative Procedure Act. In that case, the court invalidated the walkaround pay requirement, because the regulation was not interpretive in nature but rather legislative in nature.[7] The court found that OSHA attempted through the regulation to supplement the Act and not simply to construe it; therefore, the regulation must be treated as a legislative rule and must meet the notice and comment provisions of the Administrative Procedure Act.

Instead of attempting to resolve the conflict between the circuits relative to *ex parte* warrants in an appeal to the Supreme Court, OSHA decided simply to follow the notice and comment procedures required by the Administrative Procedure Act and issued a proposed amendment to 29 CFR 1903.4. 45 Fed. Reg. 33652 (May 20, 1980). The comment period ended on July 21, 1980, and the final version of the regulations was announced on October 3, 1980. The regulations provide that:

> (a) Upon a refusal to permit a compliance safety and health officer, in the exercise of his official duties, to enter without delay and at reasonable times any place of employment or any place therein, to inspect, to review records, or to question any employer, owner, operator, agent or employee, in accordance with [29 CFR] 1903.3 or to permit a representative of employees to accompany the compliance safety and health officer during the physical inspection of any workplace in accordance with [29 CFR] 1903.8, a safety and health officer shall terminate the inspection to other areas, conditions, structures, machines, apparatus, devices, equipment, materials, records or interviews concerning which no objection is raised. The compliance safety and health officer shall endeavor to ascertain the reason for such refusal, and shall immediately report the refusal and the reason therefor to the Area Director. The Area Director shall consult with the Regional Solicitor who shall take appropriate action including compulsory process, if necessary.

The regulation also provides, in subpart (b) thereof, that compulsory process may be sought and obtained in advance of an attempted inspection or investigation if in the judgment of the Area Director or Regional Solicitor circumstances exist which make such preinspection process desirable or necessary. The regulation lists several examples of circumstances in which it may be desirable to seek a warrant in advance and includes (but is not limited to) situations in which:

> (1) The employer's past practice either implicitly or explicitly puts the Secretary on notice that a warrantless inspection will not be allowed;

(2) An inspection is scheduled far from the local office, and procuring a warrant prior to leaving to conduct the inspection would avoid, in case of a refusal of entry, the expenditure of significant time and resources to return to the office, obtain a warrant, and return to the worksite;

(3) An inspection includes the use of special equipment, or when the presence of an expert or experts is needed in order to conduct the inspection properly, and procuring a warrant prior to an attempt to inspect would alleviate the difficulties or costs encountered in coordinating the availability of such equipment or expert.

The term "compulsory process" is defined in the regulation as meaning the institution of any appropriate action, including an *ex parte* application for an inspection warrant or its equivalent. An *ex parte* inspection warrant is the preferred form of compulsory process in all circumstances where compulsory process is relied upon to seek entry to a workplace under the regulation.[8]

Prior to promulgation of the rule, OSHA published the proposed rule and invited the public to submit written data, views, and arguments with respect to the issues involved therein. Many of the comments requested that public hearings be held on the proposed rule, but OSHA determined that public hearings would not be held, since the regulation was, in its understanding, not an occupational safety and health standard, as defined in Section 3(8) of the Act,[9] and therefore Section 6(b) of the Act did not apply. Additionally, OSHA held that hearings were not required by the Administrative Procedure Act when, as here, an agency was engaged in informal rulemaking. See 45 Fed. Reg. 65916 (Oct. 3, 1980). OSHA also determined that no commentator was unable adequately to present its views through written submissions. Moreover, no regulatory analysis was required, OSHA said, since the proposed regulation did not meet criteria for a major action laid down in Executive Order 12044. See, however, *Louisiana Chemical Association v. Bingham,* 496 F. Supp. 1188, 8 OSHC 1950 (W.D. La. 1980), which held that the "regulation" set forth in 29 CFR 1910.20, entitled "Access to Employee Exposure and Medical Records," is an occupational safety and health standard defined by Section 3(8) of the Act; and *American Industrial Health Council v. Marshall,* 494 F. Supp. 941, 8 OSHC 1789 (S.D. Tex. 1980),[10] which held that OSHA's generic cancer policy was also a standard. Pursuant to the holdings of these two cases, it is possible and it may be argued that the *ex parte* warrant regulation would indeed be a standard and subject to the hearing provisions of the Administrative Procedure Act.

It appears, however, that, notwithstanding such an argument, the final regulation should be sufficient to withstand challenges to OSHA's right to obtain *ex parte* warrants for inspection, at least according to the dicta in *Barlow's.* Indeed, the court in *Stoddard Lumber Co. v. Marshall,* 627 F.2d 984, 8 OSHC 2055 (9th Cir. 1980), held that the Secretary of Labor has the authority to seek *ex parte* warrants under the regulation as originally issued *and* as amended. Moreover, the court held that, regardless of whether the regulation was legislative (which would subject it to the notice and consent procedure) or interpretive in nature, the conclusive test was whether the inspection selection process had a substantive impact, that is, whether sufficient confusion and controversy were generated to require expertise to resolve the issue. Without such impact, the notice and comment procedure need not be followed.

6. Exhaustion of Remedies

Once an OSHA inspection warrant has been issued by a judicial officer, it may still be possible for an employer to challenge the sufficiency and validity of the warrant, if it appears that the warrant was not issued according to the standards outlined above. Courts have afforded judicial review to employers who resist entry, expeditiously moving to quash the warrant, and if necessary thereafter expeditiously filing an appeal. *Babcock & Wilcox, supra.* OSHA has taken the position, however, in several district court cases, that any challenges to the sufficiency and validity of an *issued* warrant should first be raised through the administrative process before the Review Commission. OSHA's position is based on the exhaustion of remedies doctrine. The circuit courts are split on this issue. The First, Third, Eighth, and Ninth circuits have required exhaustion,[11] holding that an employer has an adequate remedy in the ongoing enforcement proceedings before the Commission (plus the opportunity is provided to develop the factual record), since the issue of a warrant's validity can be raised in the appeals court after a final Commission order. *Quality Products, Inc.*, 592 F.2d 611, 7 OSHC 1093 (1st Cir. 1979); *Babcock & Wilcox Co.*, 610 F.2d 1128, 7 OSHC 1880 (3d Cir. 1979); *In Re Central Mine Equipment Co.*, 608 F.2d 719, 7 OSHC 1907 (8th Cir. 1979); *Marshall v. Burlington Northern, Inc.*, 595 F.2d 511, 7 OSHC 1314 (9th Cir. 1979); *Whittaker Corp., Berwick Forge & Fabricating Co., Div. v. Marshall,* 610 F.2d 1141, 7 OSHC 1888 (3d Cir. 1979). See also *Sauget Industrial Research and Waste Treatment Association,* 477 F. Supp. 88, 7 OSHC 1773 (S.D. Ill. 1979) —an employer must exhaust its administrative remedies (*e.g.*, including the raising of a defense that it is not an employer) in administrative proceedings before it can come to the federal district court; *Hayes-Albion Corporation v. Marshall,*—F. Supp.—, 8 OSHC 1581 (N.D. Ohio 1980); but see *Marshall v. Huffhines Steel Company,* 478 F. Supp. 986, 7 OSHC 1850, 7 OSHC 1910 (N.D. Tex. 1980)—the Secretary's argument that an employer must submit to an inspection pursuant to an *ex parte* search warrant and then challenge the Secretary's entitlement to that warrant before the Commission was rejected, because the Secretary failed to show that the Review Commission provides an adequate forum in which an employer can contest a search *before* it takes place.

The Seventh Circuit has held that exhaustion of administrative remedies is not necessary. *Weyerhaeuser Co. v. Marshall,* 592 F.2d 373, 7 OSHC 1090 (7th Cir. 1979). See also *Baldwin Metals Co. v. Marshall,*—F. Supp.—, 7 OSHC 1403 (N.D. Tex. 1979), *appeal filed,* (5th Cir.); *Marshall v. Nichols,* 486 F. Supp. 615, 8 OSHC (E.D. Tex., 1980)—no exhaustion required where OSHA acted beyond its statutory authority in issuing a subpoena. There is strong support for the exhaustion position on this point since it is now clear that the Review Commission has authority to and will consider challenges to inspection warrants. *Chromalloy American Corp.*, 7 OSHC 1547 (1979).

7. Exclusionary Rule

The Circuit Courts of Appeal are also split on the question of whether or not the exclusionary rule applies to OSHA enforcement proceedings. Under the exclusionary rule, any evidence obtained pursuant to an invalid search under the Fourth Amendment would be considered illegally seized and could not be used to form the basis of an OSHA violation. Compare *Todd Shipyards Corp. v. Secretary of Labor,* 586 F.2d 683, 6 OSHC 2122 (9th Cir. 1978) (exclusionary rule should not be applied) with *Savina Home Industries, Inc. v. Secretary of Labor,* 594 F.2d 1358, 7 OSHC 1154 (10th Cir. 1979) (exclusionary rule should be applied). It should be noted that the exclusionary rule does not apply retroactively to bar evidence obtained in pre-*Barlow's* inspections. *Meadows Industries, Inc.,* 7 OSHC 1709 (1979); *Poughkeepsie Yacht Club, Inc.,* 7 OSHC 1725 (1979); *Savina Home Industries, Inc. v. Secretary of Labor,* 594 F.2d 1358, 7 OSHC 1154 (10th Cir. 1979).

8. Consent

Naturally, an employer may not challenge a warrantless inspection of an employer's workplace if it is conducted with the employer's consent. Consent to a warrantless inspection may either be expressly given or implied from a failure to object to an inspection. *Stephenson Enterprises, Inc. v. Marshall,* 578 F.2d 1021, 6 OSHC 1860 (5th Cir. 1978); *Granite State Minerals, Inc.,* 6 OSHC 2019 (1978); *C. F. & I. Steel Corp. v. Marshall,* 576 F.2d 809, 6 OSHC 1543 (10th Cir. 1978). An interesting question which has not yet been addressed is presented by multiemployer worksite situations. Where the general contractor consents to the inspection, yet one of the subcontractors refuses entry, or, conversely, if a warrant was presented and the general contractor permitted entry and a subcontractor refused entry onto its portion of the site, the question arises as to what is the proper enforcement procedure. The decision in *Marshall v. Western Waterproofing Company, Inc. v. OSHRC,* 560 F.2d 947, 5 OSHC 1732 (8th Cir. 1977), addressed this issue—somewhat. In that case, inspectors investigating an accident were unable to contact the employer's agent in charge of the workplace but did talk to a person in control of the premises, the building manager, and were given permission to inspect the premises. The employer filed a notice of contest and moved to suppress the evidence gained from the inspection on the grounds that the inspectors violated the Fourth Amendment and that there was a failure to comply with Sections 8(a) and 8(e) of the Act. The court denied the employer's claims, stating:

> Generally a person who exercises control over a premises may consent to a search and evidence gathered in that search may be used against persons who did not consent. *United States v. Matlock,* 415 U.S. 164 (1974). . . . In *Matlock* the Supreme Court held that a search was valid where permission to search was obtained from a third party who possessed common authority over or other sufficient relationship to the premises or effects sought to be inspected. The court went on to define common authority as being where there is mutual use of the property by persons

> having joint access or control for most purposes, so that it is reasonable to recognize that any of the co-inhabitants has the right to permit the inspection in his own right and that the others have assumed the risk that one of their number might permit the common area to be searched. . . . Accordingly, Western lacked standing to raise Fourth Amendment claims since valid consent to entry upon the areas had been given by others who were in control.

It is possible, however, to distinguish the *Western Waterproofing* situation from the question posed above, since the court stated that the complaining employer could have had no reasonable expectation of privacy, since the matter investigated was exposed to public view and consequently there could be no Fourth Amendment violation. The court also found that since there was no Fourth Amendment violation, and that even if there was a violation of Section 8(a), the employer's ability to defend its case on the merits was not prejudiced.

Note: Where an employer allows an OSHA inspection "under protest" when faced with an inspector bearing a warrant issued by a magistrate, the employer is not considered to have consented to the inspection and may challenge the validity of the warrant in later proceedings. *Weyerhaeuser Co. v. Marshall,*—F. Supp.—, 6 OSHC 1811 (E.D. Wis. 1978), *aff'd,* 592 F.2d 373, 7 OSHC 1090 (7th Cir. 1979). In order to assure the availability of this defense in later proceedings, the employer should retain evidence of his protest (*e.g.,* signed receipt from inspector that the inspection was conducted under protest).

An employer who, when presented with a warrant, refuses to permit entry, may be risking a citation for civil contempt in pursuing such a course of conduct.[12] The purpose of a civil contempt order is to deter frivolous warrant challenges, coerce compliance with the underlying order, and/or compensate the complainant for loss sustained by disobedience. An employer's failure to advance rational grounds to support its disobedience of a warrant will normally result in a finding of civil contempt. *Federal Clearing Die Casting Co.,*—F. Supp.—, 8 OSHC 1635 (N.D. Ill. 1980). Civil contempt may be defended on the ground that the underlying order was erroneously issued. *Blocksom and Co., supra.* The better approach to risking contempt may be to either file a "pre-emptive" injunction or motion to quash or to allow the inspection to proceed under protest, challenging the constitutionality of the action before the Review Commission by filing motions to suppress the evidence obtained. *Quality Products, Inc.,* 592 F.2d 611, 7 OSHC 1093 (1st Cir. 1979); *Consolidated Rail Corporation,*—F.2d—, 8 OSHC 1966 (3d. Cir. 1980); *Miller Tube Corp., supra.* Appeal to the courts is accorded as a matter of right from adverse Commission decisions. In *The Matter of Worksite Inspection of S. D. Warren, Division of Scott Paper,* 481 F. Supp. 491, 7 OSHC 2010 (D. Me. 1979), before OSHA had made its application to obtain a search warrant, the employer had informed the court that it desired to participate in the warrant proceedings. The court stated that an employer does not have a right to appear and contest the issuance of an OSHA inspection warrant. Since doing so would frustrate the congressional policy as expressed in the Act and would seriously impede the efficiency of the inspection, such participation was not considered necessary to protect the employer's rights.

Finally, in *Marshall v. Shellcast,* 592 F.2d 1369, 7 OSHC 1239 (5th Cir.

1979), the court dismissed the Secretary of Labor's request for an injunction ordering an employer to submit to an inspection, since Section 8(e) of the Act does not give the Secretary of Labor authority to seek such injunctive relief.

G. THE OPENING CONFERENCE

Once the inspector has been admitted to the employer's workplace, either by the employer's consent or pursuant to an inspection warrant, a brief opening conference will be held, at which time the inspector should explain the purpose, basis, and scope of the inspection. The purpose, of course, is to ascertain if the employer is in compliance with the Act; the scope of the inspection will include employee interviews, physical inspection, records review, and a consideration of discrimination complaints. (As noted above, any inspection warrant should clearly state the scope of the inspection to be conducted.) The opening conference is mandated by the regulations. 29 CFR 1903.7(a). The FOM warns inspectors that they should be alert to attempts by employers to correct violations during the opening conference. FOM, Ch. V. For, as the Secretary of Labor argued in *Barlow's, Inc.*,

> The risk is that during the interval between an inspector's initial request to search a plant and his procuring a warrant following the owner's refusal of permission, violations of this later type could be corrected and thus escape the inspector's notice.

The opening conference may be held jointly with employer and employee representatives, or, if an objection is stated, separate opening conferences may be held. An employee representative can be a collective bargaining representative or a member of any other group which represents employees or an individual himself. OSHA Instruction, CPL 2.23A.

The employer will be given copies of the applicable safety and health standards and the Act, as well as a copy of any employee complaint that may be involved. The failure to give a copy of the Act or standards at the opening or closing conference is not considered to be a fatal defect to the conduct of an inspection. *Edward Hines Lumber Co.*, 4 OSHC 1735 (1976). If a copy of the complaint is not volunteered, the employer should request it. If, however, the employee has so requested, his or her name will not be revealed.

The employer will be asked to select an employer representative to accompany the CSHO during the inspection tour. Moreover, an authorized employee representative will also be given an opportunity to accompany the inspector.[13] If the employees at the workplace are represented by a recognized collective bargaining representative, the union ordinarily will designate the employee representative who will accompany the inspector. Similarly, if there is a plant safety committee, the employee members of that committee will designate the employee representative, in the absence of a recognized bargaining representative. 29 CFR 1903.8. The Act does not require that there be an employee representative for each inspection. However, where there is no authorized employee representative, the inspector will normally consult with a reasonable number of employees concerning safety and health matters in the workplace. 29 CFR 1903.8(b).

The normal opening conference will rarely exceed one hour in length. FOM, Ch. V.D.2. Where the inspection is being conducted pursuant to an allegation of the presence of imminent danger or following a serious accident, the opening conference will be abbreviated in order more quickly to conduct the inspection. FOM, Ch. XVI.D.2.e. During the opening conference the inspector will ask to inspect all safety and health records that the employer is required to keep pursuant to 29 CFR 1904; FOM, Ch. V.D.5.a. The employer must show the inspector these required OSHA forms, which include: (1) Log and Summary of Occupational Injuries and Illnesses (OSHA Form 200); (2) Supplementary Record (OSHA Form 101); (3) Occupational Injuries and Illness Survey (OSHA Form 103) (if the employer was selected by the Bureau of Labor Statistics to participate); and (4) other records required to be kept, such as medical examinations of employees, toxic substance reports, and air contaminant readings. FOM, Ch. V.D.5.a.; 29 CFR 1903.3(a); 29 CFR 1910.1000. The inspector is required routinely to investigate whether any medical or exposure records are kept by the employer and to determine whether employers are complying with the medical- and exposure-records access regulation under 29 CFR 1910.20. If the employer indicates that such records do not exist, the inspector should verify this by interviewing several employees or employee representatives as to whether medical examinations or exposure monitoring have been conducted. An employer's failure to preserve medical records in violation of the standard is, according to an OSHA Instruction, classified as serious if the records pertain to serious health effects or potential serious health effects resulting from exposure to toxic substances or harmful physical agents. The OSHA Instruction also gives examples of other situations which should be cited as serious violations. Where the inspection uncovers evidence that the employer knowingly disposed of relevant employee records, the employer will be cited for a willful violation of the Act. FOM, Ch. VIII; OSHA Instruction (to be published).

The inspector will also request to see the minutes of safety committee meetings, checklists, and safety program plans. The employer is not obligated to show these to the inspector. FOM, Ch. V.D.5.a(6). The inspector will also determine if there is compliance with posting requirements under 29 CFR 1903.2(a) and 1904.5(d) (*e.g.*, OSHA poster, citations, Annual Summary).

At the opening conference (or before if necessary), the employer should explain to the inspector all of the safety and health rules of the workplace to be inspected. The inspector is required to obey these rules and practices and must wear any protective clothing and equipment required. 29 CFR 1903.7(c); FOM, Ch. V.B.4. Moreover, if the employer's workplace has areas where there is a potential for exposure to communicable diseases, the Field Operations Manual provides that "[t]he CSHO may be refused entry . . . without current, valid immunization proof." FOM, Ch. V.B.5.a. Normally, however, OSHA will have anticipated this problem and arranged for an inspector with the proper immunization to conduct the inspection.[14] Although there are no reported cases, it appears that an inspector is required to provide his own personal protective equipment; the employer has no obligation to do so. See 29 CFR 1903.7(c); FOM Ch. V.B.(2), (4); IHOM Ch. XII. Theoretically, an employer could deny entry to an inspector who lacked the proper personal protective equipment upon arrival at the workplace.

At the opening conference, the employer should inform the inspector of any areas of the workplace which may contain trade secrets. The inspector is obligated under Section 15 of the Act to treat any information obtained during the inspection, which contains or which might reveal a trade secret, as confidential. The regulations provide that:

> At the commencement of an inspection, the employer may identify areas in the establishment which contain or which might reveal a trade secret. If the compliance safety and health officer has no clear reason to question such identification, information obtained in such areas, including all negatives and prints of photographs, and environmental samples, shall be labeled "confidential—trade secret" and shall not be disclosed except in accordance with the provisions of Section 15 of the Act.

29 CFR 1903.9(c). This information may only be disclosed to other officers or employees of OSHA or when relevant in any proceeding under the Act. If the inspector or any other OSHA employee violates these regulations, he is subject to a penalty of $1,000 or one year in prison or both and shall be removed from employment. 29 CFR 1903.9(b). The regulations further provide that if the employer so requests, the authorized representative of employees for the purposes of the inspection shall be an employee of the area which contains the trade secrets or an employee authorized by the employer to enter that area. If there is no such employee, the inspector is required to speak with a number of employees who work in that area concerning matters of safety and health. 29 CFR 1903.9(d).

It was formerly the view of the Review Commission that OSHA could only use federal employees to inspect areas of an employer's workplace which contained trade secrets if OSHA could not show that it was impossible to locate a federal expert capable of performing the required inspection. The issue arose with regard to the use of a specialized expert who was utilized by OSHA to determine whether engineering controls to reduce noice levels were feasible. *Reynolds Metals Co.*, (I) 3 OSHC 1749 (1975); *Reynold's Metals Co.*, (II) 6 OSHC 1667 (1978). However, the Review Commission has reversed itself and, in a split decision, overruled *Reynolds* (I & II), holding that an employer may not restrict entry into areas containing trade secrets to federal experts only. *Owens-Illinois, Inc.*, 6 OSHC 2162 (1978). The Commission has held that an employer's trade secrets can be adequately protected by a protective order and a provision in OSHA's contract with the outside expert that would allow the inspected employer an enforceable third-party beneficiary interest. See also *Kaiser Aluminum & Chemical Corp.*, 7 OSHC 1486 (1979).

Finally, if the employer's workplace contains areas which contain information which has been declared classified by a federal agency for national security reasons, the inspector must have the appropriate security clearance. 29 CFR 1903.3(b); FOM, Ch. IV.C.1.

H. INSPECTION TOUR (WALKAROUND)

At the conclusion of the opening conference, the inspector will normally proceed through the workplace, along with employer and employee representatives, inspecting work areas for compliance with OSHA standards and the General Duty Clause. The inspector will also determine at this time whether the employer has complied with the posting requirements of 29 CFR 1903.2(a) and 29 CFR 1904.5(d). These regulations require the display of an OSHA poster, which informs employees of their rights and obligations under the Act, the Annual Summary (OSHA 102), and any previous citations which have not become a final order or been abated. FOM, Ch. V.D.5.b.

The right to accompany the inspector on the inspection tour (more commonly known as the "walkaround") is guaranteed under Section 8(e) of the Act for both the employer and a representative of the employees:

> Subject to regulations issued by the Secretary, a representative of the employer and a representative authorized by his employees shall be given an opportunity to accompany the Secretary or his authorized representative during the physical inspection of any workplace under subsection (a) for the purpose of aiding such inspection. Where there is no authorized employee representative, the Secretary or his authorized representative shall consult with a reasonable number of employees concerning matters of health and safety in the workplace.

See also 29 CFR 1903.8; FOM, Ch. V.D.3.

Normally, the selection of the employer representative is routinely handled and poses no problem of consequence. See *ITO Corporation of Baltimore*, 6 OSHC 2058 (1978)—employer acquiesced in walkaround procedure. *Note:* Walkaround rights may be waived by an employer if, for example, he should be offered but declines to accompany an inspector during a tour of the worksite. *Jones Oregon Stevedoring Co.*, 3 OSHC 2067 (1976). However, on a multiemployer worksite significant problems may arise.

1. Construction Sites

The standards published as 29 CFR Part 1926 apply to every employment and place of employment of every employee engaged in construction work, including noncontract construction work. The term "construction work" means work for construction, alteration, and/or repair, including painting and decorating. 29 CFR 1926.13; *Royal Logging Co.*, 7 OSHC 1744 (1979); *A. A. Will Sand and Gravel Corp.*, 4 OSHC 1442 (1976).

When he arrives at the construction site, the inspector is instructed to contact the "prime" or general contractor's representative in charge of the job; usually, this will be the superintendent or project manager. The inspector shall advise this individual that the purpose of his visit is to make an inspection to determine compliance with the requirements of the Act. Normally, there will be several subcontractors at the site. In such cases, the individual in charge shall be asked to identify them and to provide the name of the individual in charge of each subcontractor's operations at the site. This person shall also be requested to immediately notify such individuals of the inspection and to ask them to assemble in the general contractor's office or some other suitable place in order to discuss the inspection with the inspector. The inspection shall not be postponed or substantially delayed because of the unavailability of one (or more) subcontractor's representatives.

The inspector is required to advise all of the construction employers that a closing conference will be held with each of them following the complete inspection and to request that each of them arrange to have a representative available.

At the opening conference, or at some other suitable time during the inspection, the inspector should ascertain who is responsible for providing such special services as common sanitation, eating facilities, and first aid which are available to all employees on the worksite. Even though arrangements have been made for one subcontractor or the general contractor to provide common services, each employer is responsible for his own employees in this regard. Any or all of the employers can be cited for lack of such services.

If the inspection is being conducted as a result of a complaint or complaints received, the inspector is required to furnish a copy of the complaint(s) as follows:

(1) A copy of every complaint, including complaints against subcontractors, should be provided to the general contractor.
(2) A copy of every complaint against the general contractor should, if possible, be provided to every subcontractor whose employees may be exposed to the alleged hazard.
(3) A copy of every complaint against a subcontractor should be provided to that subcontractor and, if possible, to others whose employees may be exposed to the alleged hazard.

Each employer is entitled to select an authorized representative to accompany the inspector during the inspection. Similarly, the employees of each employer have the right to select an authorized representative for this purpose. If the job is unionized, then the labor organization representing the employees would select the authorized employee representative. In situations where there is no union, the unorganized employees shall have the opportunity to select a representative who shall accompany the inspector during the inspection. If no representative is chosen, the inspector would normally interview a reasonable number of employees to determine whether hazards exist. A reasonable number of employees should include at least some employees of each employer and each craft on the job.

The main difficulty in implementing the walkaround provisions on construction sites derives from the fact that in the usual situation there will be numerous employers on the job; the result might be that if all employers and groups of employees selected a different representative to accompany the inspector on the inspection, the group participating in the inspection would be so large that work on the worksite might be disrupted, and the effectiveness of the inspection would be diminished. If possible, the inspector will allow representatives of every employer on the worksite to accompany him on the walkaround. Because of the physical problems inherent in such an undertaking, this right will usually be limited to those portions of the worksite under the control of the particular employer. FOM, Ch. VII.D.3(a). On certain construction sites, however, there may be numerous employers on the worksite, causing significant logistical difficulties. In these situations, the inspector will attempt to encourage both the employers and the employees to select a limited number of representatives for the purpose of accompanying him on the walkaround. In extreme cases, the inspector may determine that the inspection is not being conducted in an effective manner or that work is being significantly disrupted and that the walkaround will be discontinued. FOM, Ch. VII.4.

Situations may arise in which a subcontractor on a multiemployer worksite is not given an opportunity to accompany the inspector on the walkaround because, for example, on the day of the inspection no representative of that employer is present. Usually in these circumstances, the general contractor or the owner of the worksite will have had a representative attend the opening conference and accompany the inspector on the walkaround. While the Review Commission had been of the opinion that the Act's walkaround provision contained in Section 8(e) is merely directory, *Kokanee Cedar Sales, Inc.*, 3 OSHC 1530 (1979), the Seventh Circuit has held that the provision is mandatory and that if the OSHA inspector does not give *all* employers an opportunity to accompany him, then a citation based on the inspection *may* be invalid. *Chicago Bridge & Iron Co. v. Review Commission,* 537 F.2d 371, 4 OSHC 1181 (7th Cir. 1976). In that case, the worksite covered over twenty-six acres and there were over fifty subcontractors involved in the project. The inspector only permitted a representative inspection party to accompany him. The cited employer, however, was not included in the walkaround party. Even though it held that the walkaround provision was mandatory, the Seventh Circuit refused to void the citation, because it held that there had been "substantial compliance" with the employer's walkaround rights and that there had been no prejudice to the cited employer as a result of the inspection. In *Marshall v. Western Waterproofing Co., Inc.*, 560 F.2d 947, 5 OSHC 1732 (8th Cir. 1977), and *Hartwell Excavating Co. v. Dunlop,* 537 F.2d 1071, 4 OSHC 1331 (9th Cir. 1976), the courts held that the failure of the inspector to present credentials to the employer before commencing the inspection and his failure to afford the employers an opportunity to accompany him on the walkaround were not sufficient to warrant suppression of the evidence obtained by the inspection, absent a showing of prejudice to the employer. See also *Able Contractors, Inc.*, 5 OSHC 1975 (1977). In *Marshall v. C. F. & I. Steel Corporation,* 576 F.2d 809, 6 OSHC 1543 (10th Cir. 1978), two employees of an independent contractor of C.F. & I. were killed in an accident on C.F. & I.'s worksite. C.F. & I. notified OSHA of the fatalities, and thereafter the inspector visited C.F. & I. At the conclusion of the inspection at the closing conference, C.F. & I. was informed that it might be cited for a violation under Section 5(a)(1) of the Act as well as the independent contractor. C.F. & I. alleged that the inspection was contrary to Sections 8(a), (b), and (e) of the Act in that the inspector failed to present his credentials and denied C.F. & I. the opportunity to have a representative accompany him during the inspection. The court, however, found that a "representative" of C.F. & I. had participated in the opening conference and that C.F. & I. was given the right to be present during a walkaround inspection. The court noted that

> . . . they were not given formal notice that they were the object of the inspection when the inspector entered the plant. As we view it, however, it was not possible to tell them this. The employees injured were those of . . . the contractor. The inspectors were investigating . . . the contractor first but there was no effort, conscious or otherwise to exclude any C. F. & I. representative from accompanying the inspectors. Furthermore, it is impossible to conclude that C. F. & I. was unaware of the possibility that the inspection would ultimately center on it. . . . We are of the opinion therefore that the inspectors substantially complied with the requirements of Section 8(e). . . . Moreover, C. F. & I. did not at any time make an objection that it was entitled to the presence of higher ranking representation. . . . Moreover, the record does not show that C. F. & I. suffered any prejudice as a result of not having had a more formal notice at the outset. *Citing Chicago Bridge and Iron Co. v. OSHRC,* 535 F.2d 371, 4 OSHC 1181 (7th Cir. 1976).

Thus, as a practical matter, if the general contractor in a multiemployer worksite situation consents to an inspection and is given the right to accompany the inspector on the walkaround and an attempt is made by the inspector to allow subcontractors to participate in portions of the walkaround concerning areas of the worksite where their employees are working, citations issued as a result of the inspection will not be vacated unless the employer who was denied a right to accompany the inspector somehow suffers prejudice in the preparation and presentation of its defense. See *Pullman Power Products, Inc.,* 8 OSHC 1930, 1932 (1980); *S & H. Riggers & Erectors, Inc.,* 8 OSHC 1173 (1980), *appeal filed* (5th Cir. 1980); *Titanium Metals Corp. of America,* 7 OSHC 2172 (1980), *appeal filed* (3d Cir. 1980). See also *Accu-Namics, Inc. v. OSHRC,* 515 F.2d 828, 3 OSHC 1299 (5th Cir. 1975), *cert. denied,* 425 U.S. 903 (1976); *Marshall v. Western Waterproofing Co.,* 560 F.2d 947, 5 OSHC 1732 (8th Cir. 1977); *Able Contractors, Inc.,* 5 OSHC 1975 (1977).

Normally in multiemployer situations, the employee representative should be an employee of the employer being inspected. In places of employment where groups of employees hav different representatives, a different employee representative for various phases of the inspection is acceptable to the extent that the arrangement does not interfere with the inspection. More than one employee representative may accompany the inspector during any phase of the inspection, if the inspector determines that such additional representatives will further aid the inspection. 29 CFR 1903.8(a). This situation could arise where there is a craft union and an industrial union representing employees in the same workplace. If the inspector is unable to ascertain with certainty the identity and the appropriateness of a particular employee representative or if employees are not represented, the inspector may conduct the inspection without an employee representative and consult with a reasonable number of employees concerning matters of safety and health in the workplace during the walkaround. 29 CFR 1903.8(b); FOM Ch. V.B–8. OSHA Instruction CPL 2.23A provides that the inspector may undertake additional private consultation with employees during working hours, if that is found necessary. Moreover, if good cause is shown why accompaniment by a third party may contribute to the conduct of an effective and thorough inspection, it is permissible per 29 CFR 1903.8(c). The nonemployee selected shall be cautioned not to discuss matters pertaining to operations of other employers during the inspection.

It should also be noted that the inspector is authorized to deny to any person whose conduct is interfering with the inspection the right to accompany him on the walkaround. 29 CFR 1903.7(d). If such disruptive behavior rises to the level of actual bodily harm or a threat of such harm to the inspector, there are criminal sanctions provided in the Act. Section 17(h)(1), (2); 29 U.S.C. §666(h)(1), (2); 18 U.S.C. §1111; 18 U.S.C. §1114.

2. Walkaround Pay

Because Section 8(e) of the Act gives an employee representative the right to accompany the inspector on the walkaround, employers are faced with the question of whether or not they are obligated to pay those employee representatives for the time they are involved in the OSHA inspection. This question is of no small consequence, since OSHA inspections of many worksites consume entire workdays. While collective bargaining agreements frequently provide compensation to employees for time spent accompanying the inspector, there is no right to compensation under the Fair Labor Standards Act. *Leone v. Mobil Oil Corp.*, 523 F.2d 1153, 1163–64, 3 OSHC 1715 (D.C. Cir. 1975). The Act itself does not specify whether an employer who accompanies an inspector on the inspection must be compensated for that time. However, OSHA has adopted a policy requiring all employers to pay employees for time spent accompanying an inspector on a walkaround; 29 CFR 1977.21 provides in pertinent part:

> [A]n employer's failure to pay employees for time during which they are engaged in walkaround inspection is discriminatory under Section 11(c).

Section 11(c) of the Act provides that "no person shall . . . in any manner discriminate against an employee . . . because of the exercise by such employee on behalf of himself or any others of any rights afforded by the Act." Section 8(e) of the Act provides in part that a representative authorized by his employees shall be given an opportunity to accompany the Secretary or his authorized representative during the physical inspection of the workplace. Under the referenced regulation, any employer failing to pay its employees for time spent on walkarounds will be charged with discrimination under Section 11(c) of the Act. The regulation was challenged in *Chamber of Commerce v. OSHA*, 636 F.2d 464, 8 OSHC 1648 (D.C. Cir. 1980), where the court held that 29 CFR 19771.21, which purported to be an "interpretative rule and general statement of policy," was actually a regulation invalidly issued without notice and comment proceedings. The court held that the "policy statement" did not clarify the meaning of the Act nor did it explain an existing duty under the Act; thus it was legislative rather than interpretive.

The court also held that no right to walkaround pay exists under the FLSA; an employer who refuses to compensate an employee for such time does not deprive him of any rights under the FLSA, and thus the employer's refusal to pay is not discriminatory under Section 11(c) of the Act. Whether OSHA will attempt to reissue the same regulations after satisfying the requirements of the Administrative Procedure Act, 5 U.S.C. §553, remains to be seen. Even if it chooses to do so, which is probable, it is still an open question whether or not ordering employee compensation for walkaround time is a "statutorily authorized, rational, nonarbitrary, and noncapricious method of supplementing the Act's provisions." *Marshall v. Ohio Bell Telephone Co.,*—F. Supp. —, 8 OSHC 1242 (N.D. Ohio 1980), which found the OSHA regulation to be invalid on the grounds that the legislative history of the Act is totally silent as to compensation for employee walkaround representatives; but see *Marshall v. Ohio Power Company,*—F. Supp.—, 8 OSHC 1323 (S.D. Ohio 1980), which held that the regulation is consistent with the promotion of the remedial purposes of the Act. The court in *Ohio Power Company* held that an employer must pay an employee for time spent participating in an opening conference held before the walkaround inspection but disallowed compensation for time spent by an employee in travel to and preparation for an opening conference, since the Secretary of Labor failed to prove that it was necessary to insure employee participation in the inspection.

3. Striking Employees

With regard to the walkaround rights of the employee representatives, a ticklish situation is presented during a strike. Although the inspector will try to avoid becoming embroiled in such a dispute, FOM, Ch. V.D.4.a, it seems to be OSHA's policy to allow the representative of the striking employees to accompany the inspector during the inspection. Indeed, OSHA may consider an employer's refusal to allow the authorized employee representative of striking workers from accompanying the inspector as a refusal of entry in violation of 29 CFR 1903.4. In response to OSHA statements that it would attempt to obtain a search warrant compelling entry of a union representative of the striking employees, one employer was initially unsuccessful in attempting to obtain an injunction barring OSHA from obtaining such a warrant. *Amoco Oil Co. v. Marshall,*—F. Supp.—, 8 OSHC 1077 (S.D. Tex. 1980). In *Amoco Oil Company v. Marshall,* 496 F. Supp. 1234, 8 OSHC 2030 (S.D. Tex. 1980), the court, which had denied Amoco's request for a preliminary injunction, above, finding that Amoco failed to establish the necessary element of irreparable harm, affirmed Amoco's refusal to comply with a warrant which compelled entry of a striking union employee representative into its workplace. The employer took this action even though it subjected itself to proceedings in civil contempt. In this case, the Secretary of Labor sought a warrant that ordered entry onto the premises of the employer not only of the inspector but also of employee representatives who were striking unionized employees. These employees were not present because of the strike, either at the time of the accident which gave rise to the inspection or at the time of the inspection itself. The employer had other employees working at that period of time which it would have permitted to accompany the inspector. The issue presented to the court was whether striking employees had the same walk-around right as they would normally have but for their strike status. The court, construing Section 8(e) of the Act and the regulations set forth in 29 CFR 1903.8(a) and (b) relative to employee representatives accompanying an inspector during inspections, and reviewing the legislative history of the Act relative to the importance of having an employee representative accompany an inspector during inspection to a worksite, concluded that an employee normally has an interest in the safety of its workplace and would have some level of expertise regarding his workplace's conditions and safety hazards, in the ordinary set of circumstances. The court stated that the instant case did not present these ordinary circumstances, since the unionized employees had not been working at the site for more than one month's time and other employees who were not on strike had the safety interest in the site, had an opportunity to observe the site, and knew the procedures surrounding the accident. Thus the striking employees would not have been knowledgeable persons of the conditions that precipitated the accident or of any hazards; thus their entry was denied.

4. Employee Interviews

The Act and its implementing regulations clearly authorize the inspector to conduct private employee interviews. Section 8(a)(2); 29 CFR 1903.3(a); 29 CFR 1903.7(t); 29 CFR 1903.10. Thus during the normal course of an inspection, the inspector will normally speak with a reasonable number of employees concerning matters of safety and health at the workplace. FOM, Ch. V.6.a(5); 29 CFR 1903.8(b). However, the inspector is obligated to avoid any "unreasonable disruption of the operations of the employer's establishment." 29 CFR 1903.7(d). See Section 8(d) of the Act.

The regulations also permit a third party who is not an employee of the employer, such as an industrial hygienist or a safety engineer, to accompany the CSHO on the inspection tour if it is reasonably necessary to the conduct of an effective and thorough inspection. 29 CFR 1903.8(c); FOM, Ch. V.B.4.b.

In the regulations, 29 CFR 1903.10 affords any employee an opportunity to bring any condition which he believes violates a standard or the general duty requirement to the attention of the inspector during an inspection.

Thus even though employees may be represented on the walkaround, this does not prevent the inspector from consulting with any employee who desires to discuss a possible violation, if it will not interfere with the conduct of the inspection. For example, in certain instances, the employer and/or employee walkaround representative cannot provide all the necessary information regarding an accident or possible violation. The inspector is therefore required to consult with employees while conducting his walkaround inspection and to schedule in-depth interviews with other employees to ascertain pertinent facts.

If an employee requests consultation at a time that would unduly hinder the production or work cycle of the employer's operation, the inspector should consult with the employee during break, or after working hours as appropriate. If these instructions cannot be met, the interview shall be held away from the establishment. Interviews may be held in the employee's home, in the Area Office, or at any other suitable place in the community.

Whenever an employee requests that an interview be held in private, the inspector should honor that request. In addition, even in the absence of such a request, the inspector should make every effort to conduct in-depth interviews with employees out of the view of employer representatives.

According to the FOM, written interview statements should be obtained whenever an employee is willing to sign such a statement. Such a statement may be crucial in subsequent legal proceedings if an employee changes his story due to faulty memory or fear of retaliation by his employer. A signed statement should be taken under the following circumstances:

(1) There is an actual or potential controversy between the employer and employee as to a material fact concerning a violation.
(2) There is a conflict or difference in the employee statements as to the facts, during willful or repeated violation investigations.
(3) In accident investigations, when attempting to determine if apparent violation(s) existed at the time of an accident, when advance notice has been given and there is reason to believe a violation would have existed at the time of inspection if advance notice had not been given.

If the employer attempts to deny the inspector or the employees the right to "interview," this refusal may be considered discrimination under Section 11(c) of the Act. Certainly, from the employer's point of view, a refusal to allow private employee interviews will be of little benefit since it will simply serve to raise the suspicions of the inspector and may cause a more extensive investigation than first intended, including contacting the employees away from the employer's workplace. Some commentators, relying on the legislative history of Section 8(a) of the Act, have argued that inspectors do not have the right to question employees privately. Specifically, they cite the following passage from the legislative history, which provides that:

> Consequently, in order to aid in inspection and provide an appropriate degree of involvement of employees themselves in the physical inspections of their own places of employment, the committee has concluded that an authorized representative of employees should be given an opportunity to accompany the person who is making the physical inspection of a place of employment under Section 9(a). Correspondingly, an employer should be entitled to accompany an inspector in his physical inspection, although the inspector should have an opportunity to question employees in private so that they will not be hesitant to point out hazardous conditions which they might otherwise be reluctant to discuss. [U.S. Code Congressional and Administrative News, Senate Rep. No. 91–1282, 91st Cong. 2d Sess. 5187 [1971]].

Reading this language as prohibiting an inspector from conducting private employee interviews is speculative at best. Moreover, such an approach, in addition to being possibly a violation of Section 11(c) of the Act, may produce more negative than positive results. However, where the inspection is being conducted pursuant to a search warrant, an employer may be able to refuse to allow private employee interviews if such interviews are not expressly included within the scope of the warrant itself. *Marshall v. Wollaston Alloys, Inc.*, 479 F. Supp. 1102, 7 OSHC 1944 (D. Mass. 1979).

In *Stephenson Enterprises, Inc. v. Marshall,* 578 F.2d 1021, 6 OSHC 1860 (5th Cir. 1978), the employer challenged the Secretary of Labor's failure to disclose the names of two employees with whom the inspector spoke during the walkaround tour of the plant, on the basis that he could not adequately challenge the evidentiary basis of the citations and determine whether the inspector had actually made the observations that backed up the citations without access to these statements. The court held that the Secretary of Labor properly refused to produce the names of the employees who were interviewed, in light of the interest in the efficient enforcement of the Act and the informer's right to be protected against the possibility of retaliation (as balanced against the employer's need to prepare for the hearing). Considering these three factors, the court found that the balancing analysis weighed in favor of protecting the employee's name from disclosure.[15] Names of persons who give statements to an inspector are also exempt from disclosure under the Freedom of Information Act exemption for investigating records which would disclose the identity of a confidential source. *T.V. Tower, Inc. v. Marshall,* 444 F. Supp. 1233, 6 OSHC 1321 (D.D.C. 1978); *Boca Rio Golf Club, Inc.*, 6 OSHC 1850 (1978).

5. The Walkaround

Once the opening conference is concluded and the employee and employer representatives are chosen, the walkaround will actually begin. The purpose of the walkaround is, of course, to determine if the employer is in compliance with the Act. The inspector will utilize the employees to assist him in fulfilling this purpose. The employer is considered the adversary.

OSHA issues several publications in its "Your Workplace Rights in Action" series for employees, informing them of their rights during the conduct of walkaround inspections (*e.g.*, OSHA Inspections: *How You Can Help, A Workbook & Guide;*[16] OSHA Health Inspections: *How You Can Help, A Workbook & Guide*). In these publications, several ways are suggested in which employees can help the inspector:

(1) Point out hazards. Ask the inspector to pay special attention to areas which you think are dangerous.
(2) Describe accidents or illnesses which have resulted from hazards. Unless you give the facts—who, when, where, how—the inspector may not be aware that a hazard has already caused harm. Tell the inspector if you have noticed an unusual number of illnesses or health problems of any type, even if you are not sure of their cause.

(3) Describe past worker complaints. Knowing about worker complaints may help the inspector identify hazards. In addition, in proposing fines against your employer, OSHA takes into account your employer's "good faith" in correcting hazards. OSHA needs to know whether workers have complained about a problem in the past, how long the hazard has existed, and whether it exists in other parts of the workplace as well.
(4) Tell the inspector if working conditions are not normal during the inspection. If machines are not producing as much noise or dust as usual, if certain operations have been shut down, or if work is being performed in a safer way than usual, the inspector should know.
(5) Help the inspector evaluate the records your employer must keep of deaths, injuries, illnesses, and employee exposure to dangerous substances.

OSHA Instruction CPL 2.23A, issued by OSHA, details the procedures for the participation of employee representatives in opening, closing, and informal conferences as well as in the walkaround, since neither the Field Operations Manual nor the rules promulgated under 29 CFR 1903 for inspections, citations, and penalties "provide sufficient guidance for establishing a uniform and consistent policy for insuring the involvement of affected employees or their representatives in the inspection and its attendant conferences." A brief outline of the OSHA Instruction follows:

(1) Area Directors are required to assure that employee representatives are afforded the opportunity, and are encouraged, to attend and express their views in discussions relating to workplace inspections and the issuance, amendment or withdrawal of citations. Joint opening and closing conferences should be conducted when practical with all parties represented. Where it is not practical to hold a joint conference, separate conferences shall be held for employee representatives and representatives of the employer. In those instances where separate conferences are held, a written summary of each conference shall be made and the summary made available on request to employee representatives and representatives of the employer.
(2) Where separate conferences are necessary, inspectors shall determine if their conduct will unacceptably delay observation or evaluation of workplace safety or health hazards. In such cases the conferences shall be brief and, if appropriate, reconvened after the inspector's inspection of the alleged hazards. During the course of the opening conference, pursuant to 29 CFR 1903.8, employer and employee representatives shall be informed of the opportunity to accompany the inspector during the physical inspection of the workplace.
(3) At the conclusion of the inspection, the inspector will conduct a joint closing conference with the employer and employee representatives. Where it is not practical to hold a joint closing conference, separate closing conferences shall be held. During the course of the closing conference, both the employer and employee representatives will be

advised of their right to participate in any subsequent conferences, meetings, or discussions as described herein.

(4) Employee representatives must receive copies of all citations and notifications of penalty issued pursuant to the Act. Where there is a collective bargaining agent at the workplace, a copy of each citation and notification of penalty shall be sent to the appropriate collective bargaining representative. If the workplace inspected does not have a collective bargaining representative, a copy of the Citation and Notification of Penalty shall be forwarded to the employee participating in the walkaround inspection. In those instances where the inspected workplace does not have a collective bargaining representative, and there was no employee participating in the walkaround inspection, the posting of the citation and notification of penalty shall be construed as compliance with this paragraph.

(5) Pursuant to 29 CFR 1903.19, either the representative of the employer or employee representative may request an informal conference. Whenever an informal conference is requested by either the employer or employee representative, both parties shall be afforded the opportunity and encouraged to participate fully. During the conduct of the informal conference, if matters of a delicate nature are brought up by either representative, separate or private discussions shall be permitted. In any event, the Regional Administrator or Area Director shall not amend or withdraw a citation or penalty without first obtaining the views of the employee representative. When the employee representative disagrees with the proposed amendment or withdrawal of a citation, the proposed disposition may be appealed to the Regional Administrator. However, as the fifteen working day period for filing a notice of contest may affect discussions regarding an amendment or withdrawal of a citation, telephonic communication shall be utilized in order to expedite the resolution of the matter under consideration. The authority of the Regional Administrator or Area Director to amend or withdraw citations and penalties is limited to the time before a notice of contest is received in the OSHA office.

The inspector will often take photographs at various locations in the workplace in order to record apparent violations or other hazardous conditions observed during the walkaround inspection which may change during the investigation or shortly thereafter to support apparent violations, taking care that the use of flashbulbs will not create a hazard. FOM, Ch. V.6.a.2(a). See 29 CFR 1903.7(b). If the employer objects to taking photographs, and the inspector determines they are essential, a warrant may be sought. FOM, Ch. V.D.2.e. Although the inspector will attempt properly to identify the photographs for evidentiary purposes, it is often wise for the employer also to take photographs of the same conditions from the same locations to prevent the possibility of mistaken identification and for purposes of self-analysis. It should be noted, however, that if the photographs taken by the inspector fail to develop properly, the employer's photographs may be subpoenaed. *Wheeling-Pittsburgh Steel Corp.*, 5 OSHC 1495 (1977).

In rare instances, the inspector may ask an employee to pose for a photograph. It is possible that no serious problem will arise when the employee is being used merely to indicate the scale of the items contained in the photographs; however, if the purpose of the photograph is to "restage" a violation by showing the employee exposed to a safety or health hazard, a citation based on the photograph may be vacated. *Leo J. Martone & Associates*, 5 OSHC 1228 (1977). Only on rare occasions will an inspector attempt to use a tape recorder. If the employer, employees, employer representative, or other witnesses object to recording their statements during any part of the investigation, the inspector should continue without the tape recorder. The inspector is instructed not to prejudice the conduct of the inspection or formulation of proposed penalties because of the refusal to permit recording.

Some apparent violations detected by the inspector may be corrected by the employer immediately. When these violations are corrected on the spot, the inspector will usually record such corrections in order to help in judging the employer's good faith in compliance. Even though corrected, however, the apparent violations will still serve as the basis for a citation. Nevertheless, it may be advisable for employers in certain situations to correct immediately any possible violations pointed out by the inspector, even where the employer is unsure whether the condition actually is a violation. This suggestion would not, of course, hold true if a greater hazard is caused by compliance, if the condition is clearly not a violation of the Act, or if abatement costs or procedures are prohibitive.

The inspector in an industrial hygiene inspection may conduct sampling and may also request that the employer instruct its employees to wear OSHA testing devices, such as dosimeters (noise-level) and air-contaminant samplers. However, OSHA's right to require the employer to have its employees wear such devices is seriously in question. Neither the Act nor the regulations seem to require an employer to have its employees wear various testing apparatuses in order to aid an OSHA investigation. Faced with an OSHA request for an injunction ordering an employer to rescind its policy of forbidding employees to wear measuring devices in aid of an OSHA inspection, the Ninth Circuit has held that the federal courts are without such power in the absence of specific legislative authorization. *Hutton v. Plum Creek Lumber Co.*, 608 F.2d 1283, 7 OSHC 1940 (9th Cir. (1979); but see *Establishment Inspection of Keokuk Steel Castings, Division of Cast Metals Corporation*, 493 F. Supp. 842, 8 OSHC 1730 (S.D. Iowa 1980), where the court rejected an employer challenge to a NIOSH inspection which requested that personal sampling devices be attached to employees, distinguishing *Plum Creek Lumber* since in that case the company had a policy of forbidding employees from wearing such devices. The company in Keokuk had not asserted that it had a similar company policy. See also *Marshall v. Miller Tube Corporation of America*, —F. Supp.—, 6 OSHC 2042 (E.D. N.Y. 1978).

When an inspector discovers what he believes to be an imminently dangerous situation during an inspection, he will immediately bring that situation to the attention of the employer and encourage the employer to attempt to eliminate the danger. FOM, Ch. IX.C.3. As with any other violation, the fact that the employer immediately eliminates the imminence of the danger and initiates steps to rectify the conditions which caused the danger will not protect the employer from a citation for a violation of the Act. 29 CFR 1903.13. If the employer does not voluntarily eliminate the danger or take steps to prevent employee exposure to the danger, the inspector will immediately call the OSHA Area Director and initiate enforcement proceedings, which may result in the issuance of a Temporary Restraining Order (TRO) by a United States District Court which will restrain the employer from allowing employee exposure to the danger. Section 13 of the Act; 29 CFR 1903.13; FOM, Ch. IX.C.3.c. Should OSHA "arbitrarily or capriciously" decline to bring a court action, the affected employees may bring a mandamus action against the Secretary of Labor to compel him to do so. Section 13(d) of the Act.

If the employer cannot immediately abate the danger but does assure the inspector that no employees will work in the area of the danger until the danger is eliminated, the inspector will determine whether the employer has, "through previous demonstrations of good faith, established a firm credibility." FOM, Ch. IX.C.3.b.2(c). If these assurances are satisfactory, no injunction will be sought. The inspector will not continue the inspection tour until the imminent-danger situation has been resolved. FOM, Ch. IX.C.1.b.

Where it has been necessary for OSHA to obtain a TRO in imminent-danger situations, a follow-up inspection will be conducted immediately after the court order has been issued, in order to determine if the employer has complied with the terms of that order. FOM, Ch. IX.E.1. If no court proceeding was necessary due to the employer's voluntary removal of exposed employees or by the issuance of personal protective equipment but the permanent correction of the imminent-danger condition was not achieved at the time of the inspection, a follow-up inspection will be conducted on the date set for abatement. FOM, Ch. IX.E.2. At OSHA's discretion, follow-up inspections may also be conducted, even in situations where the imminent-danger condition was immediately corrected. FOM, Ch. IX.E.3.

OSHA promulgated a regulation which permits an employee to refuse to perform work under conditions in which he reasonably believes he is in imminent and real danger of death or serious physical injury and alternative relief is unavailable:

> (1) On the other hand, review of the Act and examination of the legislative history discloses that, as a general matter, there is no right afforded by the Act which would entitle employees to walk off the job because of potential unsafe conditions at the workplace. Hazardous conditions which may be violative of the Act will ordinarily be corrected by the employer, once brought to his attention. If corrections are not accomplished, or if there is dispute about the existence of a hazard, the employee will normally have an opportunity to request inspection of the workplace pursuant to section 8(f) of the Act, or to seek the assistance of other public agencies which have responsibility in the field of safety and health. Under such circumstances, therefore, an employer would not ordinarily be in violation of section 11(c) by taking action to discipline an employee for refusing to perform normal job activities because of alleged safety or health hazards.
>
> (2) However, occasions might arise when an employee is confronted with a choice between not performing assigned tasks or subjecting himself to serious

> injury or death arising from a hazardous condition at the workplace. If the employee, with no reasonable alternative, refuses in good faith to expose himself to the dangerous condition, he would be protected against subsequent discrimination. The condition causing the employee's apprehension of death or injury must be of such a nature that a reasonable person, under the circumstances then confronting the employee, would conclude that there is a real danger of death or serious injury and that there is insufficient time, due to the urgency of the situation, to eliminate the danger through resort to regular statutory enforcement channels. In addition, in such circumstances, the employee, where possible, must also have sought from his employer, and been unable to obtain, a correction of the dangerous condition.

29 CFR 1977.12(b).

The United States Supreme Court upheld the validity of 29 CFR 1977.12(b) in the case of *Whirlpool Corp. v. Marshall*, 445 U.S. 1, 8 OSHC 1001 (1980). Thus, the Court held that the OSHA regulation authorizing employee "self-help" in some circumstances is permissible under the Act. The Court took into account the fact that "circumstances may sometimes exist in which the employee justifiably believes that the express statutory arrangement does not sufficiently protect them from death or serious injury." These situations may occur when the employee is ordered by his employer to work under conditions which the employee *reasonably* believes to pose an imminent risk of death or serious bodily injury and when the employee has good reason to believe that there is no sufficient time to seek effective redress from the employer or OSHA. The regulation simply provides that when an employee, in good faith, finds himself in this predicament, he may refuse to expose himself to the dangerous condition, without being subject to discrimination by the employer. This regulation, the Court concluded, does nothing more than effectuate the purposes of the Act:

> It would seem anomalous to construe an Act so directed and constructed [to prevent occupational death or injuries] as prohibiting an employee, with no other reasonable alternative, the freedom to withdraw from the workplace environment that he reasonably believes is highly dangerous.

Moreover, since OSHA inspectors obviously cannot be present at every workplace at all times, the regulation "insures that employees will in all circumstances enjoy the right afforded them by the 'general duty' clause."

The regulation, however, according to the Court, does not require employers to pay workers who refuse to perform their assigned tasks in the face of imminent danger, although an argument could be made that such a refusal to pay is discrimination against an employee. See *Marshall v. N. L. Industries, Inc.*, 618 F.2d 1220, 8 OSHC 1166 (7th Cir. 1980). Employees acting pursuant to the regulation, however, have no power to order an employer to correct the hazardous conditions or to clear the workplace of other employees. Finally, it is significant to note that any employee who acts in reliance on the regulation still "runs the risk of discharge or reprimand in the event a court subsequently finds that he acted unreasonably or in bad faith."

I. CLOSING CONFERENCE

At the completion of the walkaround inspection a conference will be held with both employer and employee representatives. The closing as well as the opening conference are considered to be a part of the inspection. *Marshall v. Able Contractors, Inc.*, 573 F.2d 1055, 6 OSHC 1317 (9th Cir. 1978). If, for some reason, it is impractical or undesirable to have a joint conference, separate closing conferences will be held. Representatives will be advised by the inspector of their right to participate in any subsequent conferences, meetings, or discussions. OSHA Instruction CPL 2.23a (1979). In construction site inspections upon completion of the walkaround, the inspector will confer with the general contractor(s) and all appropriate subcontractors or their representatives separately and advise each one of all apparent violations disclosed by their part of the inspection.

The inspector is required by regulation to confer with the employer and informally "advise him of any apparent safety or health violations disclosed by the inspection." 29 CFR 1903.7(e). A "Closing Conference Guide" will be given to employers informing them of their responsibilities and courses of action following the inspection. During the closing conference, the employer will be given an opportunity to bring to the attention of the inspector any information he deems pertinent regarding the conditions at the workplace. It is advisable for the employer, at this time, to remind the inspector of any information he may have obtained that contains trade secrets and that any photographs taken in those areas should be marked as confidential. FOM, Ch. V.6.b.1(d)(ii); 29 CFR 1903.9(c). In addition to informing the employer of any apparent violations that were discovered during the inspection, the inspector may also indicate the applicable standards involved and leave copies of those standards with the employer. The inspector will also inform the employees or their representatives of any apparent violations found during the inspection. FOM, Ch. V.7.a. Often-times the inspector will continue his discussion with the employees after the inspection; 29 CFR 1903.7(e) does not preclude the inspector from conducting such a meeting with employees and union representatives. For, as stated in *General Dynamics Corporation, Quincy Shipbuilding Division,* 6 OSHC 1753, *aff'd,* 599 F.2d 453, 7 OSHC 1373 (1st Cir. 1979):

> Indeed even assuming that Section 1903.7(e) precludes the compliance officer from continuing his inspection following a closing conference, that section does not govern the Secretary's powers of investigation, which continue even after the closing conference ends. Section 8(a)(2) of the Act authorizes the Secretary not only to inspect but also gives the Secretary the right "[to] investigate . . . any . . . place of employment . . . and . . . to question privately any . . . employee" [Emphasis supplied.] Additionally Section 8(b) of the Act gives the Secretary subpoena power to require the attendance and testimony of witnesses. It prescribes no time period within which such power shall be exercised and applies to investigations as well as inspections. Hence, even if we accept Quincy's argument that the Secretary's inspection powers end with a closing conference, Section 8(b) permits the questioning of witnesses pursuant to subpoena after a closing conference, at least with respect to his investigative powers. It follows therefore that the Secretary can

also question employees without a subpoena, at any time, so long as they consent. Thus, the Act provides the compliance officer with investigation as well as inspection authority and the power to speak with employees even after the formal inspection has been completed.

In any event, even assuming that the meeting with the union representatives following the closing conference was completely unauthorized and violated §1903.7(e), Quincy has not indicated how it was prejudiced as a result and therefore would not be entitled to any relief.

Usually, the employer will be informed as to any enforcement actions which may be taken and the possible penalties which may be assessed. The employer may take this opportunity to discuss with the inspector any practical problems involved in compliance and to estimate the time and cost of such compliance. Employers should not argue with the inspector at the closing conference or make frivolous or gratuitous statements which could possibly be used against them in a later hearing. Many employers do not take advantage of and refuse to participate in closing conferences, feeling that there is nothing to be gained from a closing conference and much can be lost. While this is certainly one approach to take and cannot be faulted, it must be noted that approaching a closing conference with the right attitude, that is, of obtaining information and giving none, can be quite helpful.

If the inspector decides to issue a citation at the closing conference, he will so inform the employer. FOM, Ch. V.7.a.3; FOM, Ch. V.7.e.2 (b). A copy of the citation will also be given to the employees or representatives of the employees should they so request. 29 CFR 1903.14(c). The employer is required, upon the receipt of any citation, immediately to post that citation, or a copy thereof, at or near the location of the violation referred to in the citation. 29 CFR 1903.16. Where it is not practical to post the citation near the place of the alleged violation, it must be posted in another conspicuous place where employees are likely to see it. *Note:* Notices of *de minimis* violations are not required to be posted. The employer must continue to post the citation for at least three working days. However, if the violation has not been abated during that period of time, the employer must continue to post the citation until it is abated. 29 CFR 1903.16(b). The fact that the employer has filed a notice of contest under 29 CFR 1903.17 does not effect this posting responsibility, unless the citation has been vacated by a Review Commission Order. Nevertheless, the employer may post his own notice at the same location, indicating that the citation is being contested and explaining the reasons for that contest or indicating the specific steps that the employer has taken to abate the violation. 29 CFR 1903.16(c).

Although postinspection closing conferences are not required by the Act, the Review Commission has determined that a failure to conduct closing conferences will be considered under the same rationale which has been applied to the statutory walkaround provisions. *Kast Metals Corp.*, 5 OSHC 1862 (1977). That is, although a closing conference is not mandatory, failure to afford an opportunity to have a closing conference will be grounds for vacation of a subsequently issued citation if the employer can demonstrate that it was "specifically prejudiced by the failure to hold a closing conference." See *Arthur Silva Lathing & Plastering Contractor, Inc.*, 5 OSHC 1753 (1977); *Henkels & McCoy, Inc.*, 7 OSHC 1674 (1979)—citation was dismissed on the basis of denial of walkaround rights, since the employer was prejudiced as a result of the failure to conduct an opening and closing conference, in addition to the fact that an employee representative only learned of the inspection by accident and the employee involved was not employed by the cited subcontractor. In *Edward Hines Lumber Co.,* 4 OSHC 1735 (1976), the Review Commission held that an inspector's failure to furnish the employer with a copy of cited standards at the closing conference did not warrant dismissal of the citation, since the provision in the FOM requiring such an action does not confer any substantive rights on the employer. In several cases, the Review Commission has stated that the advice and requirements set forth in the FOM are merely advisory and does not confer any substantive rights on an employer. *FMC Corporation,* 5 OSHC 1707 (1976). Similarly, the conduct of a closing conference by telephone after the inspection has been held to constitute substantial compliance with the closing conference requirement, even though this procedure may have prevented the employer from conveying to the inspector information that it was in fact in compliance with the cited standard. See *Westinghouse Elec. Corp.,* 4 OSHC 1952 (1976). The failure of an inspector to discuss or inform an employer of alleged violations is also found not to be a mandatory requirement of a closing conference. *Kisco Company, Inc.,* 1 OSHC 1200 (1975); *Moser Lumber Company,* 1 OSHC 3108 (1973).

At the conclusion of the closing conference the inspector will report back to the OSHA office, where the Area Director will determine what citation, if any, will be issued and what penalties, if any, will be proposed for any violations of the standards that the inspector may have uncovered. The citation will inform the employer and the employees of the regulations and standards which have been violated and of the time set for their abatement. The employer normally will receive citations and notices of proposed penalties by certified mail. 29 CFR 1903.14; 29 CFR 1903.15. If the inspector fails to inform the employer of possible violations and the employer has not received citations or other notifications of those possible violations, and a follow-up inspection is conducted, the employer will still not be excused for failure to comply with the Act. However, the fact that the employer probably would have corrected any violations had he been informed by the inspector of their existence will be taken into account in determining the proposed penalty. *Ideal of Idaho, Inc.,* 2 OSHC 3171 (1974); *Utah-Idaho Sugar Co.,* 1 OSHC 3227 (1973); Section 17(j) of the Act.

Once the employer has received citations, he will be faced with the decision of whether or not to contest them. The procedures and considerations that an employer should be aware of upon receipt of a citation are discussed in a later chapter of the book.

NOTES

1. The OSHA Field Operations Manual (FOM) defines a "workplace" as "the place . . . where work is performed by an employer of an employer," FOM, Ch. V.D.1.b, thus tracking the language contained in Section 8(a)(1) of the Act.

2. The Act and implementing regulations also authorize inspections by agents of the Secretary of Health and Human Services, most likely NIOSH. NIOSH inspectors, however, are not empowered to issue citations or proposed penalties for violations. Section 20(b) of the Act; 29 CFR 1903.1, 1903.3(a). In states which have state plans, state inspections may also be be conducted. See chapter III. For the purposes of this chapter, only federal OSHA inspections will be considered.

3. *Occupational Injuries and Illnesses in the United States, By Industry, 1974,* Bureau of Labor Statistics, U.S. Department of Labor, (1976), at 1.

4. Although employers under the OSH Act have an expectation of privacy, employers covered under the Mine Safety and Health Act do not and are not entitled to require search warrants because of the highly and pervasively regulated nature of the mining industry. *Marshall v. Stoudts Ferry Preparation Co.,* 602 F.2d 589, 1 MSHC 2097 (3d Cir. 1979). Thus probable cause need not be established prior to seeking entry in that industry. See chapter XIX.

5. Subsequent to the decision by a magistrate to issue a warrant, hearings have often been held regarding the proper scope and form of the warrant and in some cases have been turned into full-blown hearings, considering the issue of whether, in fact, the violation exists, assuming a function left to the Commission. *West-Point Pepperell, Inc. v. Marshall,* 496 F. Supp. 1178, 8 OSHC 1954 (N.D. Ga. 1980); *Marshall v. Chromalloy American Corp.,* 589 F.2d 1335, 6 OSHC 2151 (7th Cir.), *cert. denied,* 444 U.S. 884, 7 OSHC 2238 (1979). Moreover, even in cases where adversary hearings were held, resulting in the issuance of a warrant, further challenges have followed. *Marshall v. Milwaukee Boiler Mfg. Co.,* 626 F.2d 1339, 8 OSHC 1923 (7th Cir. 1980).

6. The Court recognized that "the Act . . . regulates a myriad of safety details that may be amenable to speedy alteration or disguise" and that there is a risk "that during the interval between an inspector's initial request to search a plant and his procuring a warrant following the owner's refusal of permission, violations of this . . . type could be corrected and thus escape the inspector's notice." *Id.* at 316. However, the Court was not convinced that this risk sufficiently justified warrantless inspections, noting that "the advantages of surprise would [not] be lost, if after being refused entry, procedures were available for the Secretary to seek an *ex parte* warrant and to reappear at the premises without further notice to the establishment being inspected." *Id.* at 320. Additionally, the Court noted that "[i]nsofar as the Secretary's statutory authority is concerned, a regulation expressly providing that the Secretary could proceed *ex parte* to seek a warrant or its equivalent would appear to be as much within the Secretary's power as the regulation currently in force and calling for 'compulsory process.' " *Id.* at 320, n.15. Finally, the Court recognized that *ex parte* warrants issued in advance of an inspection might become necessary in order to carry out the surprise inspections specifically contemplated by the Act. *Id.* at 316. See also *Marshall v. Gibsons Products, Inc. of Plano,* 584 F.2d 668, 6 OSHC 2092 (5th Cir. 1978)—the words "without delay" in Section 8(a) of the Act were inserted to preserve the element of surprise deemed essential to an inspection, which would be lost in the adversary proceedings.

7. The court stated that a rule is interpretive rather than legislative if it is not "issued pursuant to legislatively delegated power to make rules having the force of law or if the agency intends the rule to be no more than an expression of its construction of a statute or rule." Interpretive rules are statements as to what the administrative officer thinks the statute or regulation means. Such a rule only provides a clarification of statutory language.

8. The authority of NIOSH to obtain *ex parte* warrants has recently been affirmed by two courts. *In Re Establishment Inspection of Pfister and Vogel Tanning Company,* 493 F.

Supp. 351, 8 OSHC 1502 (E.D. Wis. 1980); *In Re Establishment Inspection of Keokuk Steel Castings,* 493 F. Supp. 842, 8 OSHC 1730 (S.D. Iowa 1980).

9. Section 3(8) of the Act provides: "The term 'occupational safety and health standard' means a standard which requires conditions, or the adoption or use of one or more practices, means, methods, operations, or processes, *reasonably necessary* or appropriate to provide safe or healthful employment and places of employment."

10. As commonly understood, a "standard" provides a means of determining what a thing should be. It is a measure against which a thing may be compared to make an immediate determination of whether the thing conforms to established criteria. As used in the definition of an occupational safety and health standard, the word "standard" most rationally denotes a similar meaning with respect to those components of the work environment which affect employee safety. Thus it may be concluded that Congress intended an occupational safety and health standard to establish a measure against which the conditions existing or the practices, means, methods, operations, or processes used in a given workplace may be compared for an immediate determination of whether the workplace is safe.

An occupational safety and health standard must also address particular hazards existing in a work environment. Before a workplace may be made safe by compliance with a standard, there must be a hazard which is reduced through conforming the conditions existing or the practices, means, methods, operations, or processes used in the workplace to those prescribed by the standard.

From this discussion the court concluded that Congress intended an occupational safety and health standard to possess two qualities: (1) it must address a particular hazard existing in a work environment; and (2) it must establish a measure against which the condition existing or the practices, means, methods, operations, or processes used in a workplace may be compared for an immediate determination of whether the workplace is safe with respect to the hazard addressed by the standard.

Having rejected the interpretation of a standard urged by the plaintiffs, it remains to be determined whether or not the rule in issue is an occupational health and safety standard within the interpretation of that term which this court has gleaned from the Occupational Safety and Health Act.

Insofar as whether the rule addresses a hazard, counsel for OSHA claimed at the hearing on plaintiffs' motion that the denial of access to employee medical records poses an occupational hazard. This conclusion is reflected in the preamble to the rule. See 45 Fed. Reg. 35213–14. "Whether or not such a denial actually does pose a hazard is not for this court to decide. For purposes of determining whether the court has jurisdiction it is necessary only to find that the rule was formulated with what could reasonably be considered a hazard in mind. The importance of medical records to the maintenance of good health cannot be denied. To find that the denial of access to medical records poses something of a hazard is therefore not unreasonable. The question of whether denial of access to the records covered by 1910.20 is actually a hazard is for the courts of appeals to decide. We simply find that for jurisdictional purposes the rule addresses a hazard and thus possesses one of the two essential qualities of a standard."

With regard to whether the rule establishes a measure against which the conditions existing or the practices, means, methods, operations, or processes used in a workplace may be compared for an immediate determination of whether the workplace is safe with respect to an identified occupational hazard, the court finds that it does. OSHA contends that the rule prescribes a practice designed to reduce the risk presented by a denial of access to employee medical records. We find that the denial of access to medical records may fairly be considered a practice.

11. The Third Circuit did not, however, require exhaustion when the Secretary of Labor attempted to seek an *ex parte* warrant. *Cerro Metal Products Division of Marman Group, Inc. v. Marshall,* 620 F.2d 964, 8 OSHC 1196 (3d Cir. 1980).

12. Although physical force is available for execution of a warrant, *Marshall v. Shellcast Corp.,* 592 F.2d 1369, 7 OSHC 1239 (5th Cir. 1979), the Secretary of Labor's policy is to utilize contempt proceedings.

13. OSHA has recently issued a directive which encourages the participation of employee representatives in all aspects of work-site inspections, including, in addition to the walkaround, the opening/closing conference as well as any other conference which may be held. See OSHA Instruction CPL 2.23A. This instruction will be discussed in detail in this part. An employee representative does not, however, include an employee who is interviewed during a walkaround tour.

It would behoove employers to make sure that the employee representative is designated prior to the inspector's arrival (*e.g,* member of a company safety committee, union representative). The employer does not have the right to select the employee representative.

14. This would not be an example of a situation where, under 29 CFR 1903.6, advance notice of an inspection will be given. OSHA will from time to time call pharmaceutical firms, medical research laboratories, and hospitals to ascertain mandatory immunization requirements but will not give the specific date of an inspection or indicate whether an inspection is actually planned. FOM, Ch. V.B.4.b.

15. The court also noted that the employee did not seek out the OSHA inspector; rather, the OSHA inspector chose the employee at random.

16. A number of other booklets concerning workers' rights under OSHA are available from any OSHA office. They include: (1) *Workers' Rights Under OSHA* (OSHA 3021); (2) *You Have a Right to Protect Your Life on the Job* (OSHA 3032); (3) *You Can't Be Punished for Insisting on Job Safety and Health. That's the Law.* (OSHA 3033); (4) *OSHA Health Inspections: How You Can Help* (OSHA 3024); (5) *Job Safety and Health: OSHA Inspections Are Only the Beginning* (OSHA 3029); (6) *Health and Safety Committees: A Good Way to Protect Workers* (OSHA 3035); and (7) *Job Safety and Health: Answers to Some Common Questions* (OSHA 3034). No similar publications are issued for employers.

CHAPTER X
Violations, Penalties, and Abatement

A. INTRODUCTION

Section 9(a) of the Act provides that if upon inspection the Secretary of Labor or his authorized representative believes that an employer has violated Sections 5(a)(1) or (2) of the Act, or any standard, rule, or order promulgated pursuant to Section 6 of the Act, he shall issue a citation describing the nature of the violation and fix a reasonable time for its abatement. Generally, three elements are required in order to establish a violation of the Act: (1) the existence of a hazard; (2) exposure or access of employees to the hazard; and (3) knowledge of the hazard. The Act can, of course, be violated even though no accident occurs. *REA Express, Inc. v. Brennan,* 495 F.2d 822, 1 OSHC 1651 (2d Cir. 1974).

B. CATEGORIES OF VIOLATIONS

The Act recognizes four basic substantive categories of violations: (1) serious, (2) other-than-serious, (3) willful, and (4) repeated.

1. Serious Violations

Under Section 17(k) of the Act, a serious violation exists "if there is a *substantial probability* that death or serious physical harm *could* result from a condition which exists, or from one or more practices, means, methods, operations, or processes which have been adopted or are in use, in [a] place of employment *unless* the employer did not, and could not with the exercise of reasonable diligence *know* of the presence of the violation." *Usery v. Hermitage Concrete Pipe Co.,* 584 F.2d 127, 6 OSHC 1887 (6th Cir. 1978); *Lloyd C. Lockrem, Inc. v. U.S.A.,* 609 F.2d 940, 7 OSHC 1999 (9th Cir. 1979). Despite the language of the statute suggesting that the employer has the burden of proving lack of knowledge to avoid a serious violation, the Commission has made knowledge an element to be proven by the Secretary of Labor before a violation can be found. *D. R. Johnson Lumber Co.,* 3 OSHC 1125 (1975).

Generally, in order for a serious violation to be found, the Secretary of Labor must show: (1) a substantial probability of death or serious physical harm attributable to the violation, and (2) the element of actual employer knowledge or inchoate knowledge had the employer been exercising reasonable diligence. *Environmental Utilities Corp.*, 5 OSHC 1195 (1977); *Martin-Tomlinson Roofing Co.*, 7 OSHC 2122 (1980). If these two elements are proven, a serious violation will be found even though no injury or accident has occurred. See, *e.g.*, *Titanium Metals Corp. v. Usery*, 579 F. 2d 536, 6 OSHC 1873 (9th Cir. 1978); *Brennan v. Butler Lime & Cement Co.*, 520 F.2d 1011, 3 OSHC 1461 (7th Cir. 1975); *Brennan v. OSHRC (Vy Lactos Laboratories)*, 494 F.2d 460, 1 OSHC 1623 (8th Cir. 1974). The occurrence of an injury, although it may be relevant to the finding of a violation, is not conclusive. See *Titanium Metals Corp.*, 579 F.2d at 542, 6 OSHC at 1877; *Cape & Vineyard Division v. OSHRC*, 512 F.2d 1148, 1150, 2 OSHC 1628, 1629 (1st Cir. 1975). This section will analyze this two-pronged test, as well as its refinements when applied to violations of the general duty clause and specific standards.

a. *The Substantial Probability Test*

The test for a serious violation is not whether the occurrence of an accident is probable; rather, a serious violation will be found when a substantial probability or likelihood of death or serious bodily harm exists *if* an accident were to occur. *Crescent Wharf and Warehouse Co.*, 1 OSHC 1219 (1973); *Austin Bridge Co.*, 7 OSHC 1761 (1979); *C. Kaufman, Inc.*, 6 OSHC 1295 (1978); *Shaw Construction, Inc. v. OSHRC*, 534 F.2d 1183, 4 OSHC 1427 (5th Cir. 1976). The accident itself need only be possible, not probable. *Bethlehem Steel Corp. v. OSHRC*, 607 F.2d 1069, 7 OSHC 1833 (3d Cir. 1979). Only the degree of the potential injury (*e.g.*, potential for serious injury) is relevant in determining whether a violation is serious; the probability of an accident occurring is relevant only to a determination of the gravity of the serious offense for penalty assessment purposes. *California Stevedore & Ballast Co. v. OSHRC*, 517 F.2d 986, 3 OSHC 1174 (9th Cir. 1975); *accord, Niagara Mohawk Power Corp.*, 7 OSHC 1477 (1979); *Coastal Pile Driving, Inc.*, 5 OSHC 1649 (1977); *RPM Erectors, Inc.*, 2 OSHC 1187 (1974). See also *Usery v. Hermitage Concrete Pipe Co.*, 584 F.2d 127, 6 OSHC 1886 (6th Cir. 1978). *Cf. Brennan v. OSHRC (Alsea Lumber Co.)*, 511 F.2d 1139, 2 OSHC 1646 (9th Cir. 1975).

Serious physical harm is defined in the FOM as a permanent, prolonged, or temporary impairment in which part of the body is made functionally useless or is substantially reduced in efficiency on or off the job. Illnesses that could shorten life or significantly reduce physical or mental efficiency are also included. FOM Ch. VIII–B(1). The Commission has not uniformly followed this definition. See *Lisbon Contractors, Inc.*, 5 OSHC 1741 (1977); *Pacaar, Inc.*, 3 OSHC 1133 (1977); *Hydrate Battery Corp.*, 2 OSHC 1719 (1975); *Fulton Instrument Co.*, 2 OSHC 1366 (1974); *Central Contracting Corp.*, 4 OSHC 2045 (1977).

Grouping together of two or more other-than-serious violations may be allowed in order to form a serious violation if such combination meets the substantial probability test of a serious violation. See FOM, Ch. VIII–B; *CTM, Inc.*, 4 OSHC 1468 (1976).

b. *Actual or Constructive Knowledge*

The element of knowledge is satisfied by knowledge of the physical conditions/hazards that constitute the violation; knowledge of the requirements of the law (standard) is not necessary. *Southwestern Acoustics & Specialty, Inc.*, 5 OSHC 1091 (1977); *Charles A. Gaetano Construction Corp.*, 6 OSHC 1465 (1978); *Westinghouse Broadcasting Co., Inc. d/b/a WBZ TV Group W Westinghouse Broadcasting*, 7 OSHC 2158 (1980). Moreover, knowledge of the probable consequences of the existence of a condition is not a necessary element. *Sun Outdoor Advertising, Inc.*, 5 OSHC 1159 (1977). The knowledge requirement is satisfied if the employer has actual knowledge of the condition that constitutes the violation. *Martin-Tomlinson Roofing Co.*, 7 OSHC 2122 (1980); *Diamond Roofing Co.*, 8 OSHC 1080 (1980); *Empire Boring Co.*, 4 OSHC 1259 (1976); *Magma Copper Company v. Marshall*, 608 F.2d 373, 7 OSHC 1793 (9th Cir. 1979); *Brennan v. OSHRC*, 494 F.2d 460, 1 OSHC 1623 (8th Cir. 1974). See also *H–30, Inc.*, 5 OSHC 1715 (1977) *rev'd*, 597 F.2d 234, 7 OSHC 1253 (10th Cir. 1979). Knowledge is imputed to the employer if he could have discovered or avoided the condition through the exercise of due diligence. See *Combustion Engineering, Inc.*, 5 OSHC 1943 (1977)—to avoid liability on the basis of lack of knowledge, the employer must effectively communicate to employees and enforce specific safety instructions and work rules addressing the peculiar hazards of the job: *Niagara Mohawk Power Corp.*, 7 OSHC 1447, 1449 (1979)—an exception to the rule exists if the employer has taken all necessary steps to comply with the Act, including adequate supervision of its supervisory personnel (communication and enforcement by discipline if necessary); *accord, Getty Oil Co. v. OSHRC*, 530 F.2d 1143, 4 OSHC 1121 (5th Cir. 1976)—knowledge of defective pressure vessel imputed to employer because of its failure to exercise due diligence in finding out whether vessel had been tested. See also *Austin Commercial v. OSHRC*, 610 F.2d 200, 7 OSHC 2119 (5th Cir. 1979); *National Industrial Constructors, Inc. v. OSHRC*, 583 F.2d 1048, 6 OSHC 1914 (8th Cir. 1978).

The Commission consistently has held that a supervisor's knowledge will be imputed to the employer unless the employer demonstrates that the supervisor was adequately supervised regarding safety matters and the supervisor's actions were unforeseeable. See *Wright & Lopez, Inc.*, 8 OSHC 1261 (1980); *Connecticut Natural Gas Corp.*, 6 OSHC 1796 (1978); *Kansas Power & Light Co.*, 5 OSHC 1202 (1977); *Western Waterproofing Co. v. Marshall*, 576 F.2d 139, 6 OSHC 1550 (8th Cir. 1978), *cert. denied*, 439 U.S. 965, 6 OSHC 2091 (1978). Similarly, when a supervisor's misconduct leads to a violation of the Act, knowledge of the violation generally will be imputed to the employer; the employer, however, is not *per se* liable for the supervisor's misconduct. See *Floyd S. Pike Electrical Contractor, Inc.*, 6 OSHC 1675 (1978); *Engineers Construction, Inc.*, 3 OSHC 1537 (1975). See also *Horne Plumbing & Heating Co. v. OSHRC*, 528 F.2d 564, 3 OSHC 2060 (5th Cir. 1976). In such cases, the Commission invariably has ruled that if the employer is to escape liability, it has the burden of demonstrating that it took all *feasible* precautions to prevent the occurrence of the violation. See, *e.g., F. H. Sparks of Maryland, Inc.*, 6 OSHC 1356 (1978); *Constructora Maza, Inc.*, 6 OSHC 1309 (1978). Recently, however, the Fourth and Tenth Circuits have held that the Secretary of Labor has the burden of proving that an employer's safety program which is designed to prevent violations was not adequate *vis–à–vis* supervisory actions. *Ocean Electric Corp. v. Secretary of Labor,* 594 F.2d 396, 7 OSHC 1149 (4th Cir. 1979); *The Mountain States Telephone & Telegraph Co. v. Marshall,* 623 F.2d 155, 8 OSHC 1557 (10th Cir. 1980). When violations are due to employee misconduct that could not reasonably have been detected at the moment of occurrence, the employer is considered to possess constructive knowledge of the activity if the misconduct might have been prevented through adequate safety training or other precautions. See *Butler Lime & Cement Co., supra,* 520 F.2d at 1017, 3 OSHC at 1465; *Mountain States Telephone & Telegraph Co.,* 6 OSHC 1504 (1978), *rev'd on other grounds,* 623 F.2d 155, 8 OSHC 1557 (10th Cir. 1980)—employer had constructive knowledge of employee misconduct because the employee had a consistent record of prior safety violations and the employer did not adequately discipline him; *Boise Cascade Corp.,* 4 OSHC 1205 (1976)—employer had constructive knowledge that employee, of his own volition, had removed chainguards in a sawmill; employer could have discovered violation through the exercise of due diligence. See also *Ames Crane & Rental Service, Inc. v. Dunlop,* 532 F.2d 123, 4 OSHC 1060 (8th Cir. 1976).[1]

c. *Violations of the General Duty Clause*

Section 5(a)(1) of the Act provides that an employer shall "furnish to each of his employees employment and a place of employment which are free from recognized hazards that are causing or are likely to cause death or serious physical harm to his employees." In the landmark decision of *National Realty & Construction Co. v. OSHRC,* 489 F.2d 1257, 1 OSHC 1422 (D.C. Cir. 1973), the court stated that in order to prove a violation of the general duty clause, "the Secretary must prove: (1) that the employer failed to render its workplace 'free' of a hazard which was; (2) 'recognized;' and (3) causing or likely to cause death or serious physical harm." *Accord, Empire Detroit Steel Division v. OSHRC,* 579 F.2d 378, 6 OSHC 1693 (6th Cir. 1978); *Usery v. Marquette Cement Manufacturing Co.,* 568 F.2d 902, 5 OSHC 1793 (2d Cir. 1977); *Georgia Electric Company v. Marshall,* 595 F.2d 309, 7 OSHC 1343 (5th Cir. 1979); *Magma Copper Company v. Marshall,* 608 F.2d 373, 7 OSHC 1893 (9th Cir. 1979); *General Dynamics Corp., Quincy Shipbuilding Div. v. OSHRC,* 599 F.2d 453, 7 OSHC 1373 (1st Cir. 1979); *Continental Oil Company v. Marshall,* 630 F.2d 446, 8 OSHC 1980 (6th Cir. 1980). The general duty clause, however, does not impose strict liability upon employers, but rather requires them to discover and exclude from the workplace "[a]ll preventable forms and instances of hazardous conduct. . . ."[2] *National Realty & Construction Co., supra,* 489 F.2d at 1266–67, 1 OSHC at 1427; *accord, Empire Detroit Steel Division, supra,* 579 F.2d at 384, 6 OSHC at 1697; *Brennan v. OSHRC (Hanovia Lamp Division),* 502 F.2d 946, 2 OSHC 1137 (3d Cir. 1974). The general duty clause is to be used only when no specific standards applicable to the particular hazard involved have been promulgated; compliance with a standard is deemed to be compliance with the general duty clause. *Cape & Vineyard Division, supra,* 512 F.2d at 1150 & n.2, 2 OSHC at 1629 & n.2 (1979); 29 CFR 1910.5(f); OSHA Field Operations Manual (hereinafter cited as

FOM), Ch. VIII, §A.2.c.(1). This clause, however, is still available when a specific standard is under review, but has not yet been adopted. *American Smelting & Refining Co.,* 501 F.2d 504, 8 OSHC 1041 (8th Cir. 1974). Because of the virtually identical language of Sections 17(k) and 5(a)(1) of the Act, it would appear that citations under the general duty clause are only appropriately issued for serious violations, including willful or repeated violations of a serious nature. FOM Chs. VIII, §A.2.d., and XI, §C.5.c.[3]

(1) *"Place of Employment."* In connection with the general duty clause, the Commission has stated that the term "place of employment" should be broadly defined to carry out the purposes of the Act. *Sugar Cane Growers Cooperative,* 4 OSHC 1320 (1976). Applying this rationale, the Commission found that a truck used by an employer to transport employees to a worksite was a "place of employment," even though the workers were not required to use such transportation. *Id.* Other cases addressing this issue have reached similar conclusions. See *Clarkson Construction Co. v. OSHRC,* 531 F.2d 451, 3 OSHC 1880 (10th Cir. 1976)—the shoulder of a public highway adjacent to a construction site was considered part of the place of employment; *REA Express v. Brennan,* 495 F.2d 822, 1 OSHC 1651 (2d Cir. 1974)—circuit-breaker room considered place of employment despite the fact that it was usually kept locked and was not available to all employees but only to authorized personnel; *Getter Trucking, Inc.,* 1 OSHC 1743 (1974)—railroad tracks owned by railroad considered place of employment of employer who had been hired to clear wreckage.

(2) *"Recognized" Hazard.* The test for a recognized hazard is an objective one; a hazard is deemed "recognized" if the standard of knowledge in the industry recognizes it as hazardous, even though every individual employer is not aware of its existence or potential for harm. *National Realty & Construction Co. v. OSHRC,* 489 F.2d at 1265, n.32, 1 OSHC at 1426 n.32. See also *Titanium Metals Corp. v. Usery,* 579 F.2d at 541, 6 OSHC at 1876; *R. L. Sanders Roofing Co.,* 7 OSHC 1566 (1979), *vacated,* 620 F.2d 97, 8 OSHC 1559 (5th Cir. 1980); *Schriber Sheet Metal & Roofers, Inc. v. OSHRC,* 597 F.2d 78, 7 OSHC 1246 (6th Cir. 1979). If an employer has actual knowledge[4] that a condition is hazardous, however, the hazard is deemed to be "recognized" regardless of recognition by the industry. See *Georgia Electric Co. v. Marshall,* 595 F.2d 309, 7 OSHC 1343 (5th Cir. 1979); *Brennan v. OSHRC (Vy Lactos Laboratories),* 494 F.2d at 464, 1 OSHC at 1625; *Copperweld Steel Co.,* 2 OSHC 1603 (1975). Recognized hazards are not limited to those detectable by the human senses. Thus, an employer may be required to use special instrumentation to monitor potential health-endangering conditions. *American Smelting & Refining Co. v. OSHRC,* 501 F.2d at 511, 2 OSHC at 1046–47.

(3) *"Likely to Cause" Death or Serious Physical Harm.* In analyzing the phrase "likely to cause," the Commission has held "that the proper question is not whether an accident is likely to occur, but whether, if an accident did occur, the result is likely to be death or serious physical harm."[5] *R. L. Sanders Roofing Co.,* 7 OSHC 1566 (1979). See also *Titanium Metals Corp. v. Usery,* 579

F.2d at 543, 6 OSHC at 1878; *National Realty & Construction Co.*, 489 F.2d at 1265 n.33, 1 OSHC at 1426 n.33.

d. *Violations of Specific Standards*

Section 5(a)(2) of the Act requires each employer to "comply with occupational safety and health standards promulgated under this chapter." Unlike the general duty clause, the Secretary need not prove that the hazard is "recognized," because he has already prohibited the activity by promulgating the standard.[6] *California Stevedore & Ballast Co. v. OSHRC*, 517 F.2d at 988 n.1, 3 OSHC at 1176 n.1. Furthermore, "[i]f the harm that the regulation was intended to prevent is death or serious physical injury, then its violation is serious *per se*." *Id.* (emphasis added).

e. *FOM Requirements*

In determining whether to cite employers for a serious violation, the FOM provides that the following factors must be considered: (1) the type of accident or health hazard exposure which the standard, alleged to have been violated, is designed to prevent in relation to the identified hazardous condition; (2) the injury or illness which could result from the type of accident or health exposure; (3) whether the types of injury or illness include death or serious physical harm; (4) whether the employer knew or with the exercise of reasonable diligence could have known of the presence of the hazardous condition. FOM, Ch. VIII. OSHA may issue a serious citation based upon a group of individual violations, which through standing alone are other-than-serious, but are considered serious in combination.

2. Other-than-Serious Violations

An other-than-(non)serious violation is not defined in the Act, but, by administrative and judicial interpretation, when a violation has a direct and immediate relationship to the job safety or health of the employees, *George A. Hormel & Co.*, 2 OSHC 1190 (1974); *Lee Way Motor Freight, Inc.*, 1 OSHC 1689 (1974), but no substantial probability of death or serious harm exists, *Hartwell Excavating Co.*, 2 OSHC 1236 (1974); *Texaco, Inc.*, 8 OSHC 1758 (1980); *Lisbon Contractors, Inc.*, 5 OSHC 1741 (1977), it is classified as other-than-serious.[7] See also *Usery v. Hermitage Concrete Pipe Co.*, 584 F.2d 127, 6 OSHC 1886 (6th Cir. 1978). Basically it is an intermediate step between a *de minimis* and a serious violation.

Initially, the Commission did not clearly articulate its position on whether actual or constructive employer knowledge of a violation was necessary to find an other-than-serious violation. See *CAM Industries, Inc.*, 1 OSHC 1564 (1974); *Mountain States Telephone & Telegraph Co.*, 1 OSHC 1077 (1973); See also *Brennan v. OSHRC (Interstate Glass Co.)*, 487 F.2d 438, 443 n. 19, 1 OSHC 1372 (8th Cir. 1973)—other-than-serious violations need not be supported by knowledge. Although it is still not entirely clear, more recent decisions by the Commission and the courts appear to have established actual or constructive knowledge as a prerequisite to the finding of such a violation. See

Dunlop v. Rockwell International, 540 F.2d 1283, 4 OSHC 1606 (6th Cir. 1976); *Horne Plumbing & Heating Co. v. OSHRC,* 528 F.2d 564, 3 OSHC 2060 (5th Cir. 1976); *Brennan v. OSHRC (Alsea Lumber Co.),* 511 F.2d 1139, 2 OSHC 1646 (9th Cir. 1975); *Scheel Construction, Inc.,* 4 OSHC 1825 (1976); *Welsh Farms Ice Cream, Inc.,* 5 OSHC 1755 (1977).

The Tenth Circuit has held that the Secretary need not show a hazard before an other-than-serious violation of a specific standard can be found; the standard presupposes that noncompliance presents a hazard. *Lee Way Motor Freight, Inc. v. Secretary of Labor,* 511 F.2d 864, 2 OSHC 1609 (10th Cir. 1975). The court emphasized, however, that a citation for a serious violation required a showing that the noncompliance had a direct and immediate relationship to the employees' safety and health. When no such showing is made, the noncompliance is only a *de minimis* violation. *Id.* at 869, 2 OSHC at 1612–13.

The denomination of a similar violation in one case as "other-than-serious" does not require a similar finding in other cases, since the Secretary of Labor has enforcement discretion. *I. J. Service Co.,* 6 OSHC 1509 (1978).

3. Willful Violations

Section 17(e) of the Act provides for the imposition of criminal sanctions against any employer who willfully violates any standards, rules, or regulations promulgated under the Act, if such a violation causes the death of an employee. In addition, civil penalties may be assessed for any willful violations of Section 5 of the Act, or of any standards, rules, and regulations under the Act. Because "willfully" is not defined in the Act, substantial litigation has dealt with the meaning of this term. The Commission defines as willful a violation that is intentional and knowing, as distinguished from accidental, and displays a careless or reckless disregard or plain indifference to the Act or its requirements. See *Dic-Underhill, A Joint Venture,* 5 OSHC 1251 (1977)—knowledge of law and plain indifference demonstrated by eleven outstanding final orders for violations of the same standard; *PAF Equipment Co.,* 7 OSHC 1209 (1979); *Georgia Electric Co.,* 5 OSHC 1112 (1977), *enf'd,* 595 F.2d 309, 7 OSHC 1343 (5th Cir. 1979)—violation willful because of evidence of indifference to employee safety and blatant ignoring of requirements of the Act; *Titanium Metals Corp. of America,* 7 OSHC 2172 (1980) —willfulness is established by a conscious, intentional, deliberate decision; *D. Federico Co.,* 3 OSHC 1970 (1976); *C. N. Flagg & Co.,* 2 OSHC 1195 (1974). *Cf. Graven Brothers & Co.,* 4 OSHC 1045 (1976)—knowledge of a standard and subsequent violation not a willful violation without demonstration of intentional disregard or plain indifference to the Act or its standards; *Environmental Utilities, Inc.,* 3 OSHC 1995 (1976).

There are three basic elements of a willful violation: (1) employer knowledge of a violative condition which violated the Act (standard); (2) subsequent thereto, noncompliance with the standard; and (3) an act, voluntarily committed with intentional disregard or plain indifference to the Act. While there is no requirement that a willful violation be serious, the strong knowledge requirement appears to rule out other-than-serious willful violations. Often a willful violation is shown by evidence of a prior violation of the same standard and/or prior OSHA warnings. See *Western Waterproofing Co.,* 5 OSHC 1064 (1977); *Intercounty Construction Co.,* 1 OSHC 1473 (1973), *aff'd* 522 F.2d 777, 3 OSHC 1337 (4th Cir. 1975), *cert. denied,* 423 U.S. 1072, 3 OSHC 1979 (1976). A civil willful violation can be set forth in a general duty clause citation; a criminal willful violation must allege the violation of a specific standard.

Although a willful violation requires that an employer have a particular state of mind, it does not require a showing of malicious intent or bad motive. *Kent Nowlin Construction Co.,* 5 OSHC 1051 (1977), *aff'd in part, rev'd in part,* 593 F.2d 368, 7 OSHC 1105 (10th Cir. 1979)—willfulness demonstrated by awareness of standard coupled with conscious and deliberate decision not to comply with requirements; *Stone & Webster Engineering Corp.,* 8 OSHC 1753 (1980)—proper test for detecting willfulness is employer's conscious and deliberate decision not to comply with standard or employer's careless disregard of or indifference to safety; *Tri-City Construction Corp.,* 8 OSHC 1567 (1980). A showing of lack of indifference or disregard will normally, however, defeat a willful characterization. See *Williams Enterprises, Inc.,* 4 OSHC 1663 (1973).

The circuit courts of appeal which have addressed this issue—with one notable exception—have agreed with the Commission's definition of willfulness as a conscious, intentional, deliberate, voluntary decision, even if there is no bad motive. See *National Steel & Shipbuilding Co. v. OSHRC,* 607 F.2d 311, 7 OSHC 1837 (9th Cir. 1979) and cases cited therein (adopting Commission's position and noting that First, Fourth, Fifth, Sixth, Eighth, and Tenth Circuits have all adopted similar definitions of willfulness).[8] The Third Circuit, however, in *Frank Irey, Jr., Inc. v. OSHRC,* 519 F.2d 1200, 2 OSHC 1283 (3d Cir. 1975), *aff'd on rehearing on other grounds,* 519 F.2d 1215, 3 OSHC 1329 (en banc), *aff'd sub nom. Atlas Roofing Co. v. OSHRC,* 430 U.S. 442, 5 OSHC 1105 (1977), took a different approach to the definition of willfulness. The court stated that a restrictive definition of willfulness was necessary to distinguish serious violations from willful ones and added:

> It is obvious from the size of the penalty that can be imposed for a "willful" infraction—ten times that of a "serious" one—that Congress meant to deal with a more flagrant type of conduct than that of a "serious" violation. Willfulness connotes defiance or such reckless disregard of consequences as to be equivalent to a knowing, conscious and deliberate *flaunting of the Act.* Willful means more than merely voluntary action or omission—it involves an element of *obstinate refusal to comply.*

Id. at 1207, 2 OSHC at 1289; see *Babcock & Wilcox Co. v. OSHRC,* 622 F.2d 1160, 8 OSHC 1317 (3d Cir. 1980). Other circuit courts interpreting the Third Circuit test have held it requires a showing of bad purpose.[9] *Georgia Electric, supra; Empire-Detroit Steel, supra.* The Third Circuit in *Frank Irey, supra,* remanded the case to the Commission, and, on remand, the Commission reluctantly applied the Third Circuit's definition, but emphasized that it would do so only in that case because the Third Circuit was unique in requiring a showing of malicious intent. *Frank Irey, Jr., Inc.,* 5 OSHC 2030, 2031, n.3. Whether there is in fact a difference between the approach taken by the Third Circuit and the other circuits is questionable. The D.C. Circuit concluded there is little, if any, difference between the approaches of the circuits; they are very likely to lead to the same results. *Cedar Construction Co. v. OSHRC,* 587 F.2d 1303, 6 OSHC 2010 (D.C. Cir. 1978). The Third Circuit itself has attempted to downplay the differences, equating its flaunting (bad motive) requirement with the intentional disregard of other circuits, in several cases,[10] including *Universal Auto Radiator Manufacturing Co. v. Marshall,* 631 F.2d 20, 8 OSHC 2026 (3d Cir. 1980). The court in that case cited the definition utilized by the Tenth Circuit in *United States v. Dye Construction Co.,* 510 F.2d 78, 2 OSHC 1510 (5th Cir. 1975):

> The failure to comply with a safety standard under the Occupational Safety and Health Act is willful if done knowingly and purposely by an employer who, having a free will or choice, either intentionally disregards the standard or is plainly indifferent to its requirement. An omission or failure to act is willfully done if done voluntarily and intentionally

and then stated that there is no material difference between the two "alleged" definitions.

a. *Willful Act by Supervisor*

If an employer's supervisor makes a conscious and deliberate decision to violate the Act or its standards, his conduct is willful and will be imputed to the employer for the purpose of finding the employer guilty of a willful violation. See, *e.g., F. X. Messina Construction Corp. v. OSHRC,* 505 F.2d 701, 2 OSHC 1325 (1st Cir. 1974); *Constructora Maza, Inc.,* 6 OSHC 1309, (1978); *C. N. Flagg & Co.,* 2 OSHC 1195 (1974). See also *United States v. Dye Construction Co.,* 510 F.2d 78, 2 OSHC 1510 (10th Cir. 1975)—corporation can be guilty of willfulness based on acts, conduct, and state of mind of authorized agents acting in the scope of their employment. An employer, however, can avoid such imputation if it demonstrates that its supervisors' acts were contrary to a consistently and adequately enforced company policy, that the supervisors were adequately trained in safety matters, and that it took reasonable steps to discover safety violations by supervisors. *Western Waterproofing Co.,* 5 OSHC 1064, *aff'd in part, rev'd in part,* 576 F.2d 139, 6 OSHC 1550 (8th Cir.), *cert. denied,* 439 U.S. 965, 6 OSHC 2091 (1978); see *Constructora Maza, Inc.,* 6 OSHC at 1312. *Cf. Dic-Underhill, A Joint Venture,* 5 OSHC 1037 (1977) —no willful violation by employer when violation was due to foremen's failure rigidly to enforce employer's safety rules and properly supervise employees. Frequent periodic safety meetings and reliance on "common sense" will not, however, relieve the employer of liability for a willful violation if no specific effort is made to acquaint supervisory personnel with the specific requirements of the Act. *Georgia Electric Co. v. Marshall,* 595 F.2d 309, 7 OSHC 1343 (5th Cir. 1979). See *Duane Meyer d/b/a D. T. Construction Co.,* 7 OSHC 1561 (1979)[11]—a finding of willful violation due to employer's failure to take positive steps to assure that its superintendant was informed of and complied with standards constituted careless disregard/plain indifference.

b. *Absence of Prior Citations; Good Faith*

The absence of any prior warnings or citations does not preclude the finding of a willful violation, although such a prior warning or citation may be a factor in finding the existence of willfulness. *National Steel & Shipbuilding Co. v. OSHRC,* 607 F.2d at 317, 7 OSHRC at 1841;[12] *Georgia Electric Co.,* 5 OSHC 1112 (1977), *aff'd* 595 F.2d 309, 7 OSHC 1343 (5th Cir. 1979). Because a willful violation does not, in most circuits, require malice or evil motive, an employer who knowingly chooses to violate an OSHA requirement will be found guilty of a willful violation regardless of any good faith belief that the workplace remains safe. See *Western Waterproofing Co. v. Marshall,* 576 F.2d 139, 6 OSHC 1550 (8th Cir.), *cert. denied,* 439 U.S. 965, 6 OSHC 2091 (1978); *Intercounty Construction Co. v. OSHRC,* 522 F.2d 777, 3 OSHC 1337 (4th Cir. 1975), *cert. denied,* 423 U.S. 1072, 3 OSHC 1879 (1976); *C. N. Flagg & Co.,* 2 OSHC at 1196–97. The Act, however, requires that the Commission give due consideration to good faith when it assesses civil penalties. *Dic-Underhill, A Joint Venture,* 5 OSHC at 1253; see *National Steel & Shipbuilding Co. v. OSHRC,* 6 OSHC 1680 (1978), *aff'd,* 607 F.2d 311, 7 OSHC 1837 (9th Cir. 1979); *D. Federico Co.,* 3 OSHC at 1975. Moreover, if the failure to comply arises out of a good faith dispute over a critical fact, the resolution of which determines whether a violation exists, no willful violation will be found. *Acme Fence & Iron Co.,* 7 OSHC 2228 (1980)—belief that compliance was impossible or greater hazard not a "good faith" defense, since employer failed to exercise reasonable diligence to find alternate protection; *General Electric Co.,* 5 OSHC 1448 (1977), *rev'd on other grounds,* 583 F.2d 61 (2d Cir. 1978)—employer's belief that certain area was not part of "platform" as defined by a specific standard was not so unreasonable as to show bad faith; no willful violation; *Northeastern Contracting Co.,* 2 OSHC 1539 (1975) —no willful violation because of employer's good faith belief that soil was a different composition than claimed by inspector. Finally, if the employer demonstrates a good faith effort to comply with the requirements of the Act, no violation will be found.

In *Wright & Lopez, Inc.,* 8 OSHC 1261 (1980), the employer was cited for a willful/serious violation of 29 CFR 1926.652(b), specific trenching requirements, because a trench in which one of its employees was working collapsed. Although the employer had a history of past trenching violations involving the same foremen as in the present violation, the Commission found that knowledge of a standard and a subsequent violation of that standard does not, in and of itself, prove a willful violation. See *Graven Brothers & Co.,* 4 OSHC 1045 (1976). The Commission, relying on its definition of willful violation set forth in *PAF Equipment Co., Inc.,* 7 OSHC 1209 (1979) as an "action taken knowledgeably by one subject to the statutory provisions of the Act in disregard of the action's legality", noted that, at the time of the cave-in, the employer was attempting to shore the trench so as to be in compliance with the standard. Thus it stated that "when an employer makes a good faith attempt at compliance the Commission has not found a willful violation." See *Williams Enterprises, Inc.,* 4 OSHC 1663 (1976).

Similarly in *Stone & Webster Engineering Corporation,* 8 OSHC 1753 (1980), the Commission held that:

> A willful characterization does not follow inevitably from evidence that, for whatever reason, conduct in compliance with the Act is discontinued. A contrary decision could undermine the purposes of the Act by discouraging employers self-enforcement and maintenance of safe working conditions. . . . Although the respondent used poor judgment in clearing the trench without protecting the employee who worked at the unguarded edge, the Secretary has failed to demonstrate that the conduct in question reached a level of intentional disregard, recklessness or plain indifference necessary to sustain a willful violation. In light of the respondent's belief that the guardrail would interfere with the installation of the waterproof membrane, its dismantling of that protective barrier was not devoid of reason.

Thus, the Commission again determined that a knowledge of the standard and subsequent violation do not in and of themselves prove a willful violation and that it is anomalous to find a respondent in willful violation solely because it previously had provided protective measures and then removed them. See also *Tri City Construction Company,* 8 OSHC 1567 (1980)—venal motive is not a necessary requirement to prove a willful violation; *Titanium Metals Corp. of America,* 7 OSHC 2172 (1980); *Morrison-Knudsen & Associates,* 8 OSHC 2231 (1980)—a recognized defense to a willful charge is an employer's good faith effort at compliance; *Bristol Steel and Ironworkers, Inc.,* 7 OSHC 2193 (1979); *Atlantic Gummed Paper Corporation,* 8 OSHC 1042 (1979); *Adrian Construction* Company, 7 OSHC 1172 (1979)—a violation was willful since the employer *ignored* an obvious and grave danger.

c. *Failure To Prove Willful Violations; Lesser Charge*

The Commission cannot find an employer guilty of a willful violation when the Secretary has not charged the employer with such a violation and the issue was not tried by the parties. *Wetmore & Parman, Inc.,* 1 OSHC 1099 (1973). Nevertheless, if a willful violation is charged but not proven and the evidence establishes a violation that is nonwillful, a nonserious citation may be found. *Dye Construction Co.,* 4 OSHC 1444 (1976). A serious violation, however, may not be found unless the issue of seriousness has been tried by express or implied consent of the parties. *Id.; Environmental Utilities, Inc.,* 3 OSHC 1995 (1976); *CPL Constructors,* 3 OSHC 1865 (1975).

d. *FOM Requirements*

The FOM provides that an inspector should examine the following factors prior to determining whether a willful violation should be charged: (i) the precautions taken, if any; (ii) whether the Act or a similar violation has been brought to the employer's attention; (iii) whether the nature and extent of the violations indicates a purposeful disregard of the employer's obligations under the Act; and (iv) the nature of the employer's business and the knowledge reasonably expected in that industry regarding safety and health.

e. *Criminal Willful Violations*

OSHA Instruction CPL 2.39 (October 9, 1979) sets forth guidelines for the investigation of potential criminal willful cases. See FOM Ch. X, D. In order to establish a criminal willful violation under Section 17(e) of the Act, and at the same time comply with due process safeguards, OSHA must show that the employer violated an OSHA standard (a criminal willful violation cannot be based upon the violation of Section 5(a)(1)); that the violation was willful in nature, that is, the employer had knowledge both of the hazardous working conditions *and* of the requirements of the applicable standards; and that the violation of the standard caused the death of an employee. See *United States v. Dye Construction Co.*, 510 F.2d 78, 2 OSHC 1510 (10th Cir. 1975). A criminal willful violation is distinguished from a civil willful violation *only* by the result of death of an employee and the requirement that the violation be of a specific standard rather than the general duty clause. With these exceptions, the elements of the two willful violations are identical. The guideline goes on to state that although it is not necessary to issue *Miranda* warnings to an employer, if a criminal, willful investigation is in progress, the Regional Solicitor should be involved in this decision. *United States v. Pinkston-Hollar, Inc.*,—F. Supp.—, 4 OSHC 1697 (D. Kan. 1976); *State of California v. Lockheed Shipbuilding and Construction Co.*, 1 OSHC 1450 (1973).

Since the Department of Justice will prosecute the criminal case, it is necessary to coordinate the issuance of civil citations with the referral of the criminal case to the Department of Justice. Civil citations should not be issued until the criminal case is referred to the Department of Justice, or a decision has been made not to refer the criminal case, or the six-month statute of limitations for the issuance of civil citations has nearly expired—whichever occurs first. A five-year statute of limitations exists for the issuance of a criminal, willful citation.

4. Repeated Violations

Section 17(a) of the Act prescribes a mandatory penalty of not more than $10,000 whenever an employer willfully or repeatedly violates the requirements of Section 5 of the Act. A repeated violation differs from a willful violation in that it may result from inadvertent, accidental, or negligent action on the part of the employer. FOM, Ch. VIII-B 20—an employer's state of mind is largely irrelevant, at least with respect to the Commission and most of the courts who have considered the matter. However, a repeated violation may also be a willful violation, and usually, if that is the case, OSHA will cite the employer for the willful violation and not the repeated violation. FOM, Chapter VIII. Unlike a willful violation, an employer must have been cited for a prior violation in order to be cited for a repeated violation, even though a prior citation may be a factor in determining if willfulness exists.

The first requirement, in order to sustain a repeated violation, is that a citation for repeated violation must be based on a prior violation which has become a *final* order. See *Northern Metal Co.*, 3 OSHC 1645 (1975); *General Steel Fabricators, Inc.*, 5 OSHC 1837 (1977). The OSHA Field Operations Manual, Chapter VIII, describes the methodology for citing a repeated violation in that "upon reinspection, another violation of . . . a previously cited section of a standard, regulation, rule, or order or the general duty clause is found and this repeated violation resulted from an inadvertent, accidental or ordinarily negligent act."

OSHA's position in the issuance of citations is that even though a prior citation has been contested and is not yet final, it is nevertheless to be cited as a repeated violation. If this is the case, OSHA usually states upon the citation that the prior violation upon which the repeated violation is based has been contested and that the repeated citation will be automatically rescinded, if the prior citation does not become a final order of the Review Commission. See FOM, Chapter VIII; *J. L. Foti Construction Co.,* 8 OSHC 1287 (1980).

A further distinction concerning the issuance of a repeated violation is whether or not the establishment involved is a fixed establishment. Citations which are issued to employers having fixed establishments will be based upon prior violations at the cited (same) establishment. FOM, Chapter VIII. For employers engaged in businesses having no fixed establishments (construction, painting, excavation), repeated violations will be issued based on prior violations occurring anywhere within the same state. Challenges have been raised against the fixed versus nonfixed establishments distinction, alleging that it violates the constitutional guarantee of equal protection of the law. However, the distinction has been upheld. For example, in *Desarrollos Metropolitanos, Inc. v. OSHRC,* 551 F.2d 874, 5 OSHC 1135 (1st Cir. 1977), the employer argued that, as a company with no fixed establishment, it could be found guilty of a repeated violation if violations occurred at two or more of its movable sites, whereas repeated violations could not be predicated upon actions at the separate locations for businesses with fixed establishments, such as factories, terminals, and stores. The court refused to overturn the distinction, saying that there is a reasonable basis for the distinction between permanent and transient work sites, since:

> A company with floating work sites will have little incentive to insure full compliance with safety standards at each new job site from the outset if it has one almost free bite of the apple at each site.

In *George Hyman Construction Co. v. OSHRC,* 582 F.2d 834, 6 OSHC 1855, *enforcing* 5 OSHC 1318 (4th Cir. 1978), the argument that the Act evidenced no legislative intent to discriminate between fixed and transient employers was again raised. Again, the court stated that it found nothing which militated against the Commission's adoption of this guideline as a reasonable interpretation of the Act, and, although the state boundary provision contained in the guidelines was not before it, the court, in the footnote, stated that there seems to be a rational basis for using state boundaries within the guidelines as found in *Desarrollos Metropolitanos.*[13] Separate corporate divisions at one geographical location do not constitute separate establishments for the purpose of issuance of a repeated violation.

a. *Early Review Commission Position*

Until recently, the Review Commission has been unable to reach a consensus in order to establish a definition of the word "repeatedly." In *George Hyman Construction Co.,* 5 OSHC 1318 (1977), the Commission members set forth in detail the basis of their differing interpretations. The Commission first noted that the Act does not define or mention the word "repeatedly," except for the admonition in Section 17(a) of the Act which ascribes a penalty of not more than $10,000 for each such violation. Nor was there any explicit indication in the legislative history of the Act concerning the intent of Congress in including the word "repeatedly" in Section 17(a).

Commissioner Barnako stated that a repeated violation need be based on only *one* prior violation, rejecting any significance of the fact that Section 17(a) uses the word "repeatedly" instead of "repeated." Commissioner Barnako would thus find a repeated violation when the circumstances show that an employer who has committed a prior violation has failed to take appropriate steps to prevent reoccurrences of a substantially similar violation. In determining what steps are appropriate to prevent violations from reoccurring, Commissioner Barnako felt that realities of corporate control and decision-making must be considered, *e.g.,* supervisory personnel. When a violation reoccurs in the area or responsibility of a particular supervisor, the inference arises that the employer has failed in his duty to prevent recurring violations. Thus the Secretary of Labor would establish a *prima facie* case that a violation is repeated if he showed that a violation has been cited and has become a final order, and thereafter a substantially similar violation occurs under the control of the supervisor who had responsibility for abating the first violation. Commissioner Cleary, rejecting the argument that the conduct necessary to support a repeated violation is similar to that necessary to support a willful violation, also rejected the argument that the repeated violation must be supported by a showing of two or more violations. Finally, Commissioner Moran, finding importance in the use of the term "repeatedly" as opposed to "repeated," found that "repeatedly" is descriptive of a rather *persistent course of conduct* and felt that it must be shown not only that an employer violated the same standard on more than one prior occasion, but that the subsequent violation constituted a violation of the requirements of the Act.

Subsequent to the *Hyman* case, the question of what constitutes a repeated violation came before the Commission on a number of occasions, but a majority was still unable to agree on the elements of a repeated violation and avoided taking action on citations involving such violations. See *FMC Corp.,* 5 OSHC 1707 (1977); *Seattle Stevedore Co.,* 5 OSHC 2041 (1977); *Brown & Root, Inc.,* 6 OSHC 1293 (1978). Although several plausible suggestions were made by individual commissioners, no consistent and authoritative answer emerged until the complexion of the Commission changed with the departure of Chairman Moran and the arrival of Member Cottine.

Reexamining the issue in light of the decisions of *George Hyman Construction Co.,* 582 F.2d 834, 6 OSHC 1885, (4th Cir. 1978), and *Todd Shipyards Corp. v. Secy. of Labor,* 566 F.2d 1327, 6 OSHC 1227 (9th Cir. 1977), the Commission announced an authoritative guideline on the issue of repeated violation in *Potlatch Corp.,* 7 OSHC 1061 (1979). The Commission stated that a violation is repeated under Section 17 of the Act if at the time of the alleged repeated violation there was *one* Commission *final order* against the same employer for a *substantially similar* violation. (A prior uncontested citation is a final order.) *Dun-Pat Engineered Form Co.,* 8 OSHC 1044 (1980); *cf. Otis Ele-*

vator Co., 8 OSHC 1019 (1980). The Commission rejected the "more than twice" concept of the Third Circuit in *Bethlehem Steel v. OSHRC,* 540 F.2d 157, 4 OSHC 1451 (3d Cir. 1976), discussed below. It also stated that the Secretary of Labor may establish substantial similarity in several ways. In cases arising under Section 5(a)(2) of the Act, which requires employers to comply with specific safety and health standards, the Commission stated that the Secretary of Labor may establish a *prima facie* case of similarity by showing that the prior and present violation are for failure to comply with the *same* standard. However, the Commission noted that standards vary from those that designate a specific means for preventing a hazard to those that do not designate specific means of preventing a hazard, *e.g.,* general versus specific standards. In those situations where the prior and present violations are for failure to comply with the same *specific* standard, the Commission stated the employer may have a difficult time rebutting the Secretary of Labor's case. But where the same *general* standard is involved, an employer may rebut the Secretary's *prima facie* case by showing that the prior violation was cited under dissimilar circumstances. The Commission stated:

> For example, 29 CFR §1926.28(a), one of the most commonly cited construction safety standards, is often alleged to be violated on the grounds that an employer failed to protect its employees from fall hazards by requiring the use of safety belts. On the other hand, the same standard has been cited under dissimilar circumstances, such as an employer's failure to require its employees to use seat belts in an earthmoving vehicle. A prima facie showing of similarity would be rebutted by evidence of the *disparate conditions and hazards* associated with these violations of the same standard. Of course, when the Secretary alleges a repeated violation of a general standard such as 1926.28(a), it is likely that he would introduce evidence of similarity other than that the prior violation and the alleged repeated violation are in contravention of the same standard.

In the absence of evidence that the antecedent and present violations concern noncompliance with the same standard, for example, where the violations involve the general duty clause, the Secretary must present other evidence that the violations are substantially similar in nature *(e.g.,* similar hazards or conditions). *Bethlehem Steel Corp.,* 8 OSHC 1309 (1980)—factual circumstances of present violation so disparate from those of the past violation that no reasonable relationship exists between them; *Belger Cartage Services, Inc.,* 7 OSHC 1233 (1979).

The Commission's decision thus left open the possibility of the issuance of repeated citation for a substantially similar violation, even if the *same standard was not involved (e.g.,* twenty-foot fall from an unguarded scaffold under 29 CFR 1926.451(a)(4) and a twenty-foot fall while using the same unguarded scaffold would be a violation of 29 CFR 1910.28(a)(3). Thus under *Potlatch* a repeated violation is not limited to subsequent violations of the same standard as long as the violations are substantially similar in nature. Similarly, a repeated violation may be found on the basis of a prior violation of Section 5(a)(1), the general duty clause, and a later Section 5(a)(2) violation, if they are found to be substantially similar.

Prior to *Potlatch,* questions concerning whether there was different equipment, different supervisory control, the same or different standard, different machinery, or different condition or practice or time between the violations may have been considered in terms of the effect of a repeated violation. In *Potlatch,* the Commission held that an employer's attitude, such as flaunting of the Act, the commonality of supervisory control over the violative condition, the geographical proximity of the violations, the time lapse between the violations, and the number of prior violations, do not bear on whether a particular violation is repeated, although they may be considered in determining the appropriateness of the penalty. *Becker Electric Co.,* 7 OSHC 1519 (1979). Moreover, the good faith of the employer is also relevant to the determination of the appropriateness of a penalty. Recent cases in which the Commission has followed the *Potlatch* reasoning include: *Triple "A" South, Inc.,* 7 OSHC 1352 (1979) and *FMC Corporation,* 7 OSHC 1419 (1979)—*prima facie* case of similarity shown by evidence that past and present violations were of same standard; *Stearns-Roger, Inc.,* 7 OSHC 1919 (1979) and *George Hyman Construction Co.,* 7 OSHC 2041 (1979)—employer's attitude, lack of commonality of supervisory control, lack of geographical proximity between work sites and length of time between violations go toward penalty determination only; *Mattson Construction Company,* 7 OSHC 2220 (1980)—an employer's state of mind is irrelevant in determining whether a violation is repeated; *J. L. Foti Construction Co., Inc.,* 8 OSHC 1281 (1980)—*prima facie* case not rebutted since employer failed to prove dissimilarity between cited violations; *Bethlehem Steel Corporation,* 8 OSHC 1309 (1980)—violations and standards, although not identical, were substantially similar, since a difference in the location of violations at the same work site is irrelevant to a determination of a repeated violation; *Austin Road Co.,* 8 OSHC 1916 (1980) —principal factor is whether the prior and instant violations resulted in similar hazards, not whether they both resulted in a likelihood of death or serious physical harm; that is, it makes no difference if the violations involved were serious or other-than-serious. But see *Williamette Iron & Steel Co.,* 6 OSHC 1304 (1978), *dismissing pet.,* 604 F.2d 1177, 7 OSHC 1641 (9th Cir. 1979)—a three-year interval between citations though repeated was not in bad faith and warranted a penalty reduction.

b. *The Courts' Position*

In *Bethlehem Steel Corp. v. OSHRC,* 540 F.2d 157, 4 OSHC 1451 (3d Cir. 1976), the court held that in order to establish a repeated violation of the Act, it must be shown that the employer violated the Act *more than twice* and that the employer's course of conduct amounted to a *flaunting* of the Act's requirements. The court's test, of course, would permit a finding of a repeated violation *whenever* the Act had been violated more than twice, no matter how factually unrelated the two infractions may have been or how nonserious those violations were. However, the violations must also constitute the flaunting necessary to be found before a penalty can be assessed. Thus if the general or specific duty clauses of the Act were repeatedly violated in such a way as to demonstrate a flaunting disregard of the requirements of the Act, it could be a repeated violation. See also *Frank Irey, Jr., Inc. v. OSHRC,* 519

F.2d 1200, 2 OSHC 1283 (3d Cir. 1975), *aff'd* 430 U.S. 442, 5 OSHC 1105 (1977).

The court stated that among the factors which should be considered in determining whether a course of conduct is flaunting are the number, proximity in time, the nature and extent of violations, their factual and legal relatedness, the degree of care of the employer in his efforts to prevent violations of the type involved, and the nature of the duties, standards, or regulations violated. The court stated that the acts themselves must flaunt the requirements of the Act and it is neither relevant nor necessary to determine whether the acts were performed with an intent to flaunt the requirements of the Act. Repeated in that respect is an objective test.

The Third Circuit's definition of the word "repeatedly" in a repeated violation is definitely at odds with those of the other circuits and has not been followed by the Review Commission, except in areas controlled by the Third Circuit. The other circuits which have considered the question of a repeated violation include *General Electric Co. v. OSHRC,* 540 F.2d 67, 4 OSHC 1512 (2d Cir. 1976), *vacating,* 3 OSHRC 1031; *George Hyman Construction Co. v. OSHRC,* 582 F.2d 834, 6 OSHC 1855 (4th Cir. 1978), in which the court stated that only a single prior infraction need be proven to invoke the repeated violation of sanction authorized by the Act. Nor, according to the Fourth Circuit, must the violation cited involve flagrant misconduct, as defined by the Third Circuit in *Bethlehem Steel.* The court stated that its interpretation of Section 17(a) includes a meaningful differentiation between willful and repeated violations which appears more consistent with the Act's enforcement scheme as well as its liberal terms. The court noted that it deliberately avoided setting forth an all-inclusive and rigid definition of "repeatedly" under the Act, but it did say that it is essential that an employer receive actual notice of the prior violation, and a reasonable time should elapse from receipt of the notice of the original citation in order that an employer can take corrective action. It rejected, however, the fact that the same supervisor must also be in charge of the work site at which a violation is cited as repeated. Finally, in *Todd Shipyards Corp. v. Secy. of Labor,* 566 F.2d 1327, 6 OSHC 1227 (9th Cir. 1977), and 586 F.2d 683, 6 OSHC 2122 (9th Cir. 1978), the court also rejected the *Bethlehem Steel* definition of a repeated violation, which it felt essentially equated willful with repeated, but did not address the general question of what is a repeated violation. It merely affirmed as a factual matter that violations were identical in character, occurred on the same ship, and took place within three months of each other. In *J. M. Martinac Shipbuilding Corporation v. Marshall,* 614 F.2d 776, 7 OSHC 2120 (9th Cir. 1980), it was held that evidence of flaunting, deliberate or reckless disregard of the Act's requirement, is not required to support a finding of a repeated violation; nor does a repeated violation require control of the same supervisor responsible for abatement. Finally, evidence that violations occurred over a five-year period, on different ships and in different locations, and presented different degrees of hazard, does not bar finding that violations were substantially similar for purposes of characterizing a violation as repeated. Since the decision in *Potlatch,* no courts have yet considered the Review Commission's new authoritative guidelines.

5. Failure To Abate Violations

The question still remains after *Potlatch* as to what happens if an employer is in the process of abating a first violation or is waiting for the abatement method to be approved and another inspection occurs. Can a citation for a repeated violation be sustained? *Concrete Technology Corp.*, 5 OSHC 1751 (1977); *Westinghouse Electric Corp.*, 6 OSHC 1095 (1977). Strictly speaking, the situation would probably be properly classified as a *failure to abate,* for if upon reinspection a violation of a previously cited standard is found on the *same* piece of equipment and at the *same* location and it appears that the original violation has *continued uncorrected* since the original inspection, it would be a failure to abate, rather than a repeated violation. Section 10(b) of the Act. *Glendale Mills, Inc.*, 6 OSHC 1361 (1978); *Braswell Motor Freight Lines, Inc.*, 5 OSHC 1469 (1977); FOM Ch. VIII. It may be a repeated violation, however, if it does not involve the same piece of equipment or same location or if the violation is not continuous. In *Kit Manufacturing Co.*, 2 OSHC 1672 (1975), a citation for failure to abate was vacated because the Secretary of Labor failed to prove that a hazard existed during the reinspection, despite the fact that the conditions which led to the first citation continued. Thus even though the conditions existing are the same and have continued to exist, the Secretary of Labor must prove the existence of a hazard both at the time of the original inspection and at the time of the reinspection in order to make out a *prima facie* case of a failure to abate. Moreover, the third element of a failure to abate, as set forth in Section 17(d) is that the original citation must have become a final order, *e.g.* no notice of contest was filed. *United States v. Shield Die Cutters and Embossers, Inc.*,—F.2d—, 6 OSHC 2071 (1978). An employer is subject to a penalty of up to $1,000 per day for *each* day a violation continues beyond the time allowed for abatement. Section 17(d) of the Act. Technically, a notice rather than a citation is issued for an employer's failure to correct a violation, but the fifteen-day notice of contest period is the same. 29 CFR 1903.18.

If the Secretary of Labor has made a *prima facie* case in proving a failure to abate citation *(e.g.*, a prior citation has become a final order, the same condition exists upon re-inspection and the condition is a violation of the Act), the burden then shifts to the employer to articulate a defense. Examples of defenses available include a lack of employee exposure, existence of a greater hazard, or there has been abatement and compliance. Challenges may also be raised, if the earlier, uncontested citation upon which the alleged failure to abate citation is based, was uncontested. *Res judicata* does not prevent such an action. See *York Metal Finishing Co.*, 1 OSHC 1655 (1974); *Kit Manufacturing Co.*, 2 OSHC 1672 (1975). The Secretary of Labor must, in those instances where a prior citation was not contested, prove that the alleged violative condition was in fact violative at the time of re-inspection. Thus, a prior uncontested citation can be reopened, since the failure of an employer to originally file a notice of contest does not mean there was a violation. However, where a notice of contest was filed and a violation was found, it is *res judicata* except as to the question of identical conditions. In *Marshall v. B. W. Harrison Lumber Co.*, 569 F.2d 1303, 6 OSHC 1446 (5th Cir.

1978) the court affirming the Commission held that in a failure to abate action, procedural defenses to an earlier uncontested citation which forms the basis for the failure to abate citation could be raised on the basis that the original citation failed to describe with particularity the nature of the violation. The court stated that the Act requires:

> that the description fairly characterize the violative condition so that the citation is adequate both to inform the employer of what must be changed and to allow the Commission in a subsequent failure to correct action to determine whether the condition was changed.

See also *General Electric Co. v. Secretary of Labor,* 576 F.2d 558, 6 OSHC 1541 (3d Cir. 1978)—failure to abate penalty may be stayed while an employer petitions for a variance.

Moreover, it is OSHA's policy to reinspect every employer who has been cited for a serious, willful, or repeated violation after the time set for abatement and correction of the violation. Thus it is conceivable, with respect to an abatement period of three days, that within the fifteen working-day period to contest a citation, an employer could be cited for failure to abate. Two courts have ruled that OSHA may specify an abatement date sooner than the contest period and may reinspect and issue a notice of failure to correct before the expiration of the contest date. *Brennan v. OSHRC (Kesler & Son Construction Co.),* 513 F.2d 553, 2 OSHC 1668 (10th Cir. 1975); *Dunlop v. Haybuster Mfg. Co.,* 524 F.2d 222, 3 OSHC 1594 (8th Cir. 1975). There will be occasions when a follow-up inspection is conducted to verify compliance on items of short abatement periods before an employer's fifteen working-day period allowed for filing a notice of contest has expired. When a follow-up inspection discloses that an employer has failed to correct the alleged violation in the period specified in the citation and has not yet exercised his contest rights, a notification or citation for failure to correct an alleged violation will not be issued until after the original fifteen working days allowed for the filing of the notice of contest has lapsed, if the employer has not contested during this period. It is OSHA's policy, however, that if an employer subsequently files a notice of contest, then a notification for failure to abate issued during the contest period will normally be vacated. FOM, Chapter XI.

6. De Minimis Violations

In addition to the four basic categories of violations explicitly set forth in the Act, the Act and the Review Commission have recognized an additional category, *de minimis* violation. A *de minimis* violation is one that is trifling, has no direct or immediate relationship to safety and health. Section 9(a) of the Act; *Turner Communications Corporation v. OSHRC,* 612 F.2d 941, 8 OSHC 1038 (5th Cir. 1980); *Southwestern Electric Power Co.,* 8 OSHC 1974 (1980); *Kenneth P. Thompson Co.,* 8 OSHC 1696 (1980); *Belger Cartage Service,* 7 OSHC 1233 (1979). A *de minimis* violation may also be found if there is evidence that no effective protection is available or that protective devices that are in use are as safe as those required by the standard. See *Roberts Sheet Metal Company,*

5 OSHC 1659 (1977); *Boring & Tunneling Company of America,* 5 OSHC 1867 (1977).

OSHA normally will not issue citations or penalty assessments or impose abatement requirements for *de minimis* violations (or for a failure to abate such a violation), even though they will be discussed at the closing conference following an inspection and noted in an employer's case file. *Combustion Engineering, Inc.,* 5 OSHC 1943 (1977). OSHA Program Directive 200–84, Sept. 1978. Rather, notices are issued. No notice of contest would be filed since no citation would have been issued.

Normally, a substantial record of an accident-free operation or that an injury, if it did occur, would be minor will support a finding of a *de minimis* violation. *Koller Craft Plastics Products, Inc.,* 7 OSHC 1409 (1979); *Fabricraft, Inc.,* 7 OSHC 1540 (1979). But the fact that no injuries have occurred is not controlling in the characterization of a violation as *de minimis;* a hazard may exist in the absence of injuries. *Alamo Store Fixtures Corp.,* 6 OSHC 1150 (1977). In *Nickylou Fashions, Inc.,* 6 OSHC 1824 (1978), an employer was cited for having sewing machines without point-of-operation guards as required by 29 CFR 1910.212(a)(3)(ii). However, an affidavit signed by the workers, stating that they would not use the guards because it slowed production and reduced income, and the fact that operators of 30, 38, 40, and 50 years' experience testified that they had never had an accident or injury while working with unguarded machines resulted in a finding that the violation had a negligible relationship to safety and was properly classified as *de minimis.* See also *Soule Steam Feed Works, Inc., Miner Edgar Division,* 6 OSHC 1852 (1978), where the fact that there was lack of evidence of past injuries in frequent use of a grinder and some protection afforded by a tool rest required a *de minimis* finding of violation of 1910.215(a)(2) for failure to provide safety guards to cover projections on the arbor bench grinder; compare *Pratt & Whitney Aircraft, Div. of United Technologies Corp.,* 8 OSHC 1329 (1980)—the fact that no employees had *yet* contracted illness from exposure to sodium silicate did not mean a violation was *de minimis.*

An employer's failure to provide separate compartments and doors on toilet facilities as required by the standard was ruled to be *de minimis* since it had no immediate relationship to safety and health. *Texaco, Inc.,* 5 OSHC 1193 (1977). But see *Sierra Construction Corp.,* 6 OSHC 1278 (1978)—failure to post illustration of hand signals is not *de minimis,* since hazard addressed by standard is not trifling; *T. J. Service Co.,* 6 OSHC 1509 (1978)—failure to provide single drinking cups is not considered *de minimis.*

The following are the criteria for issuance of *de minimis* violations set forth in the Field Operations Manual:

(1) An employer complies with the clear intent of the standard, but deviates from its particular requirements in a manner which has no direct or immediate relationship to safety and health. These deviations may involve distance requirements, material requirements, the use of color, specifications and sign wording, and slight variations in inspections, testing, recordkeeping and maintenance requirements. *Example:* 29 CFR 1910.27(b)(1)(ii) which requires 12 inches maximum

distance between the rungs of a ladder; if the rungs are 13 inches apart the violation would be *de minimis,* since climbing safety is not appreciably diminished. *Charles H. Thompkins,* 6 OSHC 1045 (1977).

(2) An employer complies with a proposed amended change to a standard, rather than with the standard presently in effect, and the proposed amendment provides equal or greater safety and health protection. *Example:* 29 CFR 1910.178(m)(6)—Powered Industrial Trucks, allowing trucks to open railroad freight car doors. See OSHA Program Directive #100–63.

(3) An employer's workplace is at the "state of the art," that is it is technically advanced beyond the requirements of the applicable standard or provides equivalent or better safety and health protection. *Alamo Store Fixtures Corp.,* 6 OSHC 1151 (1977).

7. Imminent Danger Violations

Section 8(f)(1) of the Act provides that if employees (or their representatives) believe that a violation of a standard exists which threatens physical harm or that an *imminent danger* exists, they may request an inspection. An imminent danger is danger that could reasonably be expected to cause death or serious physical injury immediately or before the danger can be corrected through normal enforcement proceeding. Section (13)(a) of the Act. Serious physical harm is defined as that type of harm which could cause permanent or prolonged impairment of the body or is the type of harm which, while not impairing the body on a prolonged basis, could cause such temporary disablement as would warrant incarceration at a hospital. FOM Chapter IX.

A request for an imminent-danger inspection must normally be in writing and set forth with reasonable particularity the basis therefor. If, upon receipt of such notification, OSHA determines that there are reasonable grounds to believe that an imminent danger exists, it will make a special inspection to determine if in fact such a violation or danger exists.

OSHA Instruction CPL 2.12A, issued September 1, 1979, provides guidelines for investigation of complaints concerning alleged unsafe and unhealthful conditions in the workplace. According to the guidelines, any complaint which constitutes an imminent danger must be investigated on the same day as it is received, not later than twenty-four hours after receipt. Other complaints will be prioritized depending on the gravity of the hazards as defined in the FOM.

Upon inspection of an imminent-danger complaint, a citation will be issued if the inspector believes or determines that there is a reasonable certainty that a danger exists to employees that can be expected to cause death or serious physical harm immediately or before the danger can be eliminated through normal enforcement procedures. If the inspector is unable to get the employer to immediately and voluntarily eliminate (abate) the danger, action will be initiated in the United States district courts pursuant to Section 13(a) of the Act to restrain the dangerous conditions or practices.

The required imminency under the Act is considered to be present where it is believed that death or serious physical harm could occur immediately or certainly within a short time. For a health hazard to constitute an imminent-danger situation, it must be concluded that there is a reasonable expectation that toxic substances or health hazards are present and exposure to them could cause irreversible harm to such a degree as to shorten life or cause reduction in physical or mental efficiency, even though the resulting irreversible harm may not manifest itself immediately. IHFOM, Chapter II; FOM, Chapter IX.A.2.c.

As stated, if an allegation of imminent danger appears to have merit, an inspector shall conduct an inspection within twenty-four hours, except in extraordinary situations. If it is determined that there are not reasonable grounds to believe that an imminent danger exists (after review of the complaint), OSHA will notify the complainant in writing of such determination. Any alleged imminent-danger situations brought to the attention of or discovered by the inspector during a regular inspection shall be inspected immediately, taking the place of additional inspection activity. (Regulations under 29 CFR 1903.6 authorize the advance notice of inspection for apparent imminent danger to enable the employer to eliminate the dangerous condition as quickly as possible.) If, as a result of the inspection, conditions or practices are revealed which could properly be characterized as imminently dangerous, the inspector must inform affected employees and notify them he will recommend that injunctive relief be sought. No imminent-danger proceeding will be brought if the employer voluntarily eliminates the imminent danger; the alleged violation will be treated as any other violation, *e.g.*, a citation for serious, repeated, or willful violation will usually be issued. Although there may be instances in which the employer will not be able to permanently eliminate the danger when it is brought to its attention, the inspector may consider voluntary elimination has been accomplished if the employer removes employees from the dangerous area or gives satisfactory assurance that it will eliminate the dangerous conditions before permitting the employees to work in the area. If the imminent danger is eliminated voluntarily, the inspector will inform affected employees that the imminent danger no longer exists and tell them of the steps taken to eliminate the danger.

If the employer cannot or does not voluntarily eliminate the hazard, the inspector shall contact the Area Director, who along with the Regional Solicitor will consider procedures to obtain a temporary restraining order and posting of the OSHA notice of alleged imminent danger. U.S. district courts are authorized to enter injunctive orders to avoid, correct, or remove (abate) such imminent danger. Section 13(a) of the Act. (A court has no authority to assess a penalty.) An *ex parte* temporary restraining order is effective for five days; injunctive relief is also provided pending the outcome of enforcement proceeding. In addition, the inspector will post a notice of alleged imminent danger at or near the area where the employees are working. The inspector himself has no authority to order the operation to be shut down or to direct employees to leave the area of the imminent danger or the workplace. The OSHA notice of alleged danger does not constitute a citation; it is only a notice that an imminent danger is believed to exist and that the Secretary of Labor will be seeking a court order to restrain the employer from permitting employees from working in an area until the imminent danger is eliminated. After an imminent danger has been found, an appropriate citation shall be completed and penalties proposed in accordance with the procedures applicable to those types of violations.

If a court has issued an injunction in an imminent-danger situation, a follow-up inspection will take place immediately after the court order has been issued to determine if the employer is in compliance with the terms of the order. If OSHA arbitrarily or capriciously fails to seek relief under Section 13(a) of the Act, an employee who might be injured as a result of such failure may bring a *mandamus* action against the Secretary of Labor to compel him to seek such relief, in one of several venues (where the danger exists, in the District of Columbia or where the employer has its principal office). Section 13(d); See also *Hodgson v. Greenfield and Associates,* 1 OSHC 1017 (1972)—a labor organization actively participated in an imminent danger action. The employer may contest the court action. After an imminent danger has been filed, an appropriate citation shall be completed in accordance with procedures contained in the FOM and shall be proposed in accordance with the procedures applicable to the violations involved along with the proposed penalties. No citation shall be issued, however, when court action is being or will be pursued without clearance from the Regional Administrator or Regional Solicitor.

The existence and handling of imminent-danger situations is inextricably intertwined with Section 11(c) discrimination complaints. In *Whirlpool Corporation v. Marshall,* 445 U.S. 1, 8 OSHC 1001 (1980), the court upheld the validity of the OSHA work refusal regulation, 29 CFR 1977.12(b), which protects an employee's right to refuse work because of a reasonable apprehension of death or serious injury. The regulation provides that hazardous conditions, which may be violative of the Act, ordinarily will be corrected by the employer when it is brought to his attention. If the correction is not accomplished, an employee will normally have an opportunity to request an inspection of the workplace, pursuant to Section 8(f) of the Act. Under such circumstances, an employer would not ordinarily be in violation of Section 11(c) of the Act by taking action to discipline employees who refuse to perform normal job duties because of alleged safety or health hazards. The regulation noted, however, that occasions may arise when an employee is confronted with a choice of not performing an assigned task or subjecting himself to serious injury or death arising from hazardous conditions in the workplace. If the employee has no reasonable alternative and refuses in good faith to expose himself to a dangerous condition, he is protected against subsequent discrimination. However, the condition causing the employee's apprehension of death or injury must be of such a nature that a reasonable person, under the circumstances then confronting the employee, would conclude that there is a real danger of death or serious injury and that there is insufficient time due to the urgency of the situation to eliminate the danger through resort to regular statutory enforcement channels. In addition, the employee, where possible, must have sought from his employer and have been unable to obtain a correction of the dangerous condition.

8. Other Violations

Sections 17(f)(g) and (h) of the Act set forth other violations, including: (1) falsifying records, reports, or applications, a penalty of up to $10,000 and or six months in jail; (2) posting requirements, up to a $1,000 penalty; (3) giving advance notice of inspections without authority, up to six months in jail and/or up to a $1,000 penalty; and (4) assaulting, resisting, opposing, intimidating, or interfering with a compliance officer, a $5,000 penalty and up to three years in jail.

C. PENALTIES

1. Civil Penalties

Section 17 of the Act authorizes the assessment of civil penalties for violations of the Act or of standards, regulations, and orders promulgated under the Act's authority. Penalties are proposed by OSHA upon the issuance of a citation, and are sent with the citation. If an employer does not contest a citation, the penalties assessed are final, unreviewable, and payable. If an employer does contest a citation, the penalties are subject to a determination of appropriateness by the Review Commission.

The Act sets forth the maximum penalties permissible for violations of the Act. Section 17(a) states that employers who *willfully* or *repeatedly* violate the Act may be assessed a penalty of *not more than $10,000* for each violation; Section 17(b) provides that an employer who receives a citation for a *serious* violation shall be assessed a penalty *of up to $1,000* for each violation;[14] Section 17(c) provides that an employer who receives a citation for a violation determined to be of an *other-than-serious* nature may be assessed a penalty of up to $1,000 for each violation; and under Section 17(d) any employer who fails to abate a violation for which a citation has been issued under Section 9, during the period permitted for its correction, may be assessed a civil penalty of not more than $1,000 for each day during which such failure continues. There is no penalty for a *de minimis* notice. Penalties determined to be owed under the Act must be paid to the Secretary of Labor for deposit in the Treasury of the United States, and such may be recovered in a civil action, under Section 17(1) of the Act, brought in the United States District Court for the district where the violation is alleged to have occurred or where the employer has its principal office.

OSHA's policy on the assessment of penalties is contained in the Field Operations Manual and is discussed in detail hereinafter. Suffice it to say at this point, however, that OSHA is not required to assess a penalty for every violation (except for serious). Examples: If an employer is cited for fewer than ten first-instance, nonserious violations, OSHA, pursuant to the Appropriations Act for fiscal year 1980, is prohibited from assessing a penalty. *Texaco, Inc.,* 8 OSHC 1758 (1980). Moreover, if a violation is *de minimis,* no penalty will be issued. If the penalty is less than $50, it will not be assessed. A penalty reduction of up to 40 percent is offered to small businesses, depending upon the number of employees,[15] up to 10 percent for a lack of prior violations, and up to 30 percent for good faith.

a. *Serious and Other-than-Serious Violations*

In assessing appropriate penalties for violations of the Act, the Commission has wide discretion, but four factors are given consideration: (1) the size of the employer's business, *Colonial Craft Reproductions, Inc.*, 1 OSHC 1063 (1972); *Allway Tools, Inc.*, 5 OSHC 1094 (1977); (2) the gravity of the violation, *State, Inc.*, 4 OSHC 1806 (1976)—high gravity; *T. J. Service Co.*, 6 OSHC 1509 (1978)—moderate gravity; *Niagara Mohawk Power Corp.*, 7 OSHC 1447 (1979)—low gravity; (3) the good faith of the employer (or lack thereof), *Williams Construction Co.*, 6 OSHC 1093 (1977); *Hullenkremer Construction Co., Inc.*, 6 OSHC 1469 (1978); *Scullin Steel Co.*, 6 OSHC 1764 (1978)—bad faith[16] demonstrated; and (4) the history of previous violations, *Westinghouse Electric Corp.*, 6 OSHC 1095 (1977)—no accident in 26 years; *Koller Craft Plastics Products, Inc.*, 7 OSHC 1409 (1979); *McGuire & Hester*, 2 OSHC 1504 (1975). *Wright and Lopez, Inc.*, 8 OSHC 1261 (1980); *Westburn Drilling, Inc.*, 5 OSHC 1457—a prior *de minimis* violation is not evidence of previous history, a prior violation must be a final order before it is considered in history; *Felton Construction Co.*, 4 OSHC 1817 (1976); *General Steel Fabricating, Inc.*, 5 OSHC 1837 (1977). See Section 17(j) of the Act; *Nacirema Operating Company, Inc.*, 1 OSHC 1001 (1972); *Truland Corp.*, 6 OSHC 1123 (1977); *Brennan v. OSHRC (Interstate Glass Co.)*, 487 F.2d 438, 1 OSHC 1372 (8th Cir. 1973); *Everhart Steel Construction Co., Inc.*, 3 OSHC 1068 (1975).

In *Dic-Underhill, A Joint Venture*, 5 OSHC 1251 (1977), the Review Commission affirmed a maximum permissible assessment of a $10,000 penalty for a willful violation upon consideration of the four factors used in penalty assessment. The Commission reached this decision since the element of good faith was not present, the employer had a history of previous violations, a great number of employees were affected, the violation was of long duration, and there was a high probability that employees could have been severely injured. Compare *House of Glass*, 5 OSHC 1171 (1977), where the gravity of the violation was moderate, the employer was small with no history of prior violations, all violations were abated prior to the hearing, and a $40 penalty was found to be appropriate.

The four criteria need not be given equal weight, nor must the computation percentages be followed. *J & H Livestock Co.*, 5 OSHC 1742 (1977). Of the four factors used in assessing penalties, the factor of good faith is critical in the imposition of the penalty. The presence of good faith will usually result in the reduction of a penalty, while the absence of good faith will normally result in the affirmance of the penalty at the prescribed or increased amount. Factors considered in determining good faith include the employer's overall safety and health program (utilization of safety meetings and committees), cooperation during the inspection, prompt abatement and a legitimate attempt to comply prior to the inspection. *Nacirema Operating Co.*, 1 OSHC 1001 (1972); *Continental Steel Corp.*, 3 OSHC 1410 (1975); *Hullenkemer Construction Co.*, 6 OSHC 1469 (1978); *Nip-Co Manufacturing, Inc.*, 5 OSHC 1632 (1977); *Allstate Trailer Sales, Inc.*, 3 OSHC 1183 (1975); *American Bag Company*,

4 OSHC 1105 (1970). See *Long Manufacturing Co. v. OSHRC,* 554 F.2d 903, 5 OSHC 1376 (1977), where the employer had acted in bad faith and with indifference to the safety of the employees; *Advance Pipeline, Inc.,* 6 OSHC 1065 (1977); *Western Waterproofing Co., Inc.,* 5 OSHC 1064 (1977); *Connecticut Natural Gas Corp.,* 6 OSHC 1796 (1978); *Newspaper Printing Corp.,* 5 OSHC 1810 (1977); *Hullenkramer Construction Co., Inc.,* 6 OSHC 1469 (1978). See also *Dell Cook Lumber Co.,* 6 OSHC 1362 (1978)—employer's reliance on an inspector's indication that guard rails not needed shows good faith in subsequent citation for failure to provide such protection. The fact that an employer demands a search warrant prior to inspection does not affect its good faith. *Walter H. Kessler Co., Inc.,* 7 OSHC 1401 (1979). But penalty deductions for good faith are not applicable in repeated or willful violations.

An employer's technical violation of the standard, which resulted from concern for safety, consisted of allowing employees to wear a type of foot protection which lessened the risk of foot injury but heightened the risk of more serious injury from falling or from electrical shock, causing the reduction of penalty from $810 to $1 for his good faith. *Republic Steel Corp.,* 6 OSHC 1658 (1978). A good-faith attitude toward safety is also considered in penalty assessments. See *B & L Excavating Co., Inc.,* 6 OSHC 1251 (1977); *Connecticut Natural Gas Co.,* 6 OSHC 1796 (1978). In one case, the employer hired a consultant to evaluate the situation and evaluate the records of abatement. *D. B. Drilling Corp.,* 6 OSHC 1806 (1978).

If an employer immediately corrects or initiates steps to correct, these actions are favorably considered in figuring a penalty adjustment factor for good faith. *Connecticut Natural Gas Corp.,* 6 OSHC 1796 (1978); *Williams Construction Co.,* 6 OSHC 1093 (1977); *Donald Harris, Inc.,* 6 OSHC 1267 (1978); *Hydraform Products Corp.,* 7 OSHC 1995 (1979). Prompt and aggressive initiation of abatement, absence of prior violations and injuries/illnesses, and/or thorough and effective safety and health efforts justifies a 30 percent reduction; reasonably prompt but slightly reluctant abatement, absence of serious repeated and willful violations of moderate or high gravity, and absence of most serious injuries/illnesses justifies a 20 percent reduction; barely acceptable abatement with some reluctance, absence of very grave violations, and absence of easily controlled injuries/illnesses justifies 10 percent reduction; and reluctance to undertake abatement, presence of grave violations, and presence of easily controlled injuries/illnesses justifies 0 percent reduction. FOM, Ch. XII.

Other factors which the Commission has considered in the assessment of penalties include whether the employer is financially able, is in dissolution, or is bankrupt. See *Pentrol Transportation Co.,* 4 OSHC 1746 (1976); *L. R. Ward Steel Products, Inc.,* 5 OSHC 1931 (1977); *Pittsburgh Brewing Company,* 7 OSHC 1099 (1978); *Price Potashnick, Codel Oman, A Joint Venture,* 6 OSHC 1909 (1978). However, the hardship factor does not relieve an employer of responsibility for violations of the Act. The brevity of employee exposure and the lack of injuries are also factors considered in determining appropriateness, *ALMCO Steel Products Corp.,* 6 OSHC 1532 (1978), as is the fact that the violation is a temporary aberration from past practice. *Hullenkremer Construction Co., Inc.,* 6 OSHC 1469 (1978).

According to the Field Operations Manual, the gravity of the violation is the primary factor in determining the basic penalty to be imposed. The size of the business, the good faith of the employer, and the history of previous violations, on the other hand, are primarily considered in determining whether the basic gravity-based penalty shall be *reduced.* Of importance in determining the gravity factor (high or low) of the penalty are the *severity or degree* of the injury or illness which could result from the alleged violation, the extent of exposure *and* the *likelihood or probability* that an injury or illness could occur as a result of the alleged violation. FOM, Chapter XI; *Truland Corp.*, 6 OSHC 1123 (1977); *Boltz Brothers Packing,* 1 OSHC 1118 (1973). An employer's injury and illness record as well as the actual occurrence of an injury or illness is important in the determination of these factors with respect to the violation. *Dreher Pickle Company,* 1 OSHC 1132 (1973); *Worcester Pressed Steel Co.,* 3 OSHC 1661 (1975); *Brennan v. Smokecraft, Inc.,* 530 F.2d 843, 3 OSHC 2000 (9th Cir. 1976); *George J. Ingall & Co.,* 6 OSHC 1642 (1975). The length of time that a condition exists is also an important factor in measuring the gravity of the violation. *Emerick Construction,* 5 OSHC 2048 (1977); *Forish Construction Co.,* 4 OSHC 1021 (1976). Precautions taken are also a consideration. *Broadview Construction Co.,* 1 OSHC 1083 (1973).

The classification of a violation as either serious or other-than-serious is the initial reflection on the *severity* of the injury or illness which could result from the violation and is the first step in determining the gravity of the violation. OSHA has published in the FOM separate penalty schedules for each violation, which reflect the difference in *severity.* Thereafter, the degree of probability that an injury or illness will occur as a result of a violation is estimated on the following scale:

Low	1–2
Moderate	3–6
High	7–8

Of significance in estimating probability vis-à-vis injuries are the number of employees exposed, the frequency and duration of exposure, the proximity of employees to the point of danger, and such factors as the speed of an operation which requires work under stress. The importance of each factor varies with each violation. FOM, Chapter XI.

With respect to illnesses which could result from exposure to a toxic substance or harmful physical agent, other procedures are used to estimate probability: for example, whether personal protective equipment is provided and, if provided, whether it is fitted, maintained, or worn as required. Once all of these factors are considered, a violation will be given a probability rating. A gravity-based penalty for each violation is then determined by consulting certain penalty tables, depending upon whether the violation has been classified as serious or other-than-serious. (See Attachment I.)

The gravity-based penalty may be adjusted downward as much as 80 percent, depending upon the employer's good faith, size of business, and history of previous violations. Up to a 40 percent reduction is permitted for size, a 30 percent reduction for good faith, and a 10 percent for history. The FOM contains detailed instructions for determining the extent of adjustments. As an example, for serious violations involving a high probability of injury or illness (*e.g.*, 7 or 8 as opposed to 1 or 2), the reduction permitted for good faith and history is severely limited (*e.g.*, 8 gets 0 percent; 7 gets 20 percent). When an adjusted penalty for a violation would amount to less than $50, no penalty shall be proposed.

b. *Imminent Danger Violations*

Penalties may be proposed in cases where citations are issued for imminent-danger situations, even though an employer immediately takes steps to abate such danger. An imminent-danger situation involves a serious, willful, or repeated violation, unless an employer's knowledge requirement is not satisfied—in that instance it would be cited as other-than-serious. The penalty policy for those violations applies to the imminent-danger situation.

c. *General Duty Clause Violations*

Citations under the general duty clause are restricted to serious violations. The procedure to be followed in calculating a proposed penalty is the same as for serious, willful, or repeated, whichever is applicable. Under the general duty clause, an activity or practice may be recognized even if the employer is ignorant of the activity's existence. Thus the fact that an employer lacked knowledge that certain conditions violated the Act and thus did not have an opportunity to correct violations prior to penalty assessment does not bar the Secretary of Labor from assessing penalties, because the absence of employer knowledge is a factor in deciding the appropriate penalty but not in determining whether a penalty can be assessed. *United States of America v. B & L Supply Co.*, 486 F. Supp. 26, 8 OSHC 1125 (N.D. Tex. 1980).

d. *Failure to Abate Violations*

A penalty will be applied only when an inspection discloses that an employer has not corrected an alleged violation for which a citation has been issued and a final order of the Commission has been entered. If the penalty is not contested within fifteen working days, it will be considered a final order. If the employer contests only the amount of proposed penalty, he must comply with the citation, which becomes a final order in the absence of a notice to contest. If an employer contests the abatement date in good faith, no citation for failure to correct or additional penalty should be issued. If the employer fails to correct or abate the violation on the day following the calendar correction date specified in the citation, a civil penalty of not more than $1,000 may be assessed for each day violation continues. For each day the violation has not been corrected, a gravity-based penalty shall be computed for the violation as either serious or other-than-serious, and the penalty shall be multiplied by the number of calendar days the citation continues uncorrected, considering the Section 17(j) penalty assessment factors. Normally, the total proposed penalty for a failure to correct a particular violation will not exceed ten times the amount of the daily additional penalty proposed. In unusual circumstances, such as the circumstances where the gravity of the violation is especially high or the employer has exhibited a high degree of negligence for failing to correct the violation, higher penalties may be proposed. An important factor in determining the amount of the penalty is the amount of the original penalty. *Empire Art Products Company,* 2 OSHC 1230 (1974).

e. *Repeated Violations*

The maximum penalty that may be assessed for repeated violations is not more than $10,000. (It is conceivable that other-than-serious repeated violations may carry no penalty, while on the other hand a serious, repeated violation must.) OSHA will determine the unadjusted penalty for repeated violations by using the same method used to determine gravity-based penalty for serious or other-than-serious violations. Absent special circumstances, once a gravity-based penalty is arrived at, it is doubled for the first repeated violation and quadrupled if the violation has been cited and repeated twice. If a third repetition of the previous violation occurs, it is increased by a factor of ten. If there was no initial proposed penalty assessed for the first violation, the minimum gravity-based penalty shall be $100 for repeated violations. An adjustment for good faith is normally not applicable, since OSHA feels that employers exhibit a lack of good faith by repeating a violation.

f. *Willful Violations*

The unadjusted penalty for a willful violation is determined by using the same method for evaluation as for serious or other-than-serious violations. A maximum penalty of $10,000 may be assessed for each violation that is willful. After determining if a violation is serious or not, an unadjusted penalty corresponding to the numerical gravity rating is determined. Adjustment factors of good faith, size, and history shall then be considered in arriving at an adjusted penalty. For each violation cited as willful, OSHA will take into account several factors in order to estimate the degree and gravity of willfulness. Before a violation can be classified as willful, the inspector must ascertain that the employer had specific knowledge of the standard or its requirements or the particular hazardous working conditions to which the employees were or are being exposed. The following examples aid the inspector in determining the degree of willfulness: (1) A *high degree* of willfulness means the employer intentionally disregards the necessary safety and health requirements, *e.g.,* consciously decides not to use guard rails in a situation where the need for such protection is obvious. (2) A *moderate degree* of willfulness is exhibited where the employer is indifferent to safety requirements, that is, there is a record of numerous prior violations not necessarily related. (3) A *low degree* of willfulness is shown when an employer neglects to provide protection which may be required, that is, an employer fails to implement safety and health standards where the exposure of employees was foreseeable but not specifically anticipated by the employer. There must be a voluntary choice on his part. OSHA will rate the degree of willfulness from one to ten, with ten being the highest degree. The amount of the unadjusted penalty for each willful violation is equal to the numerical rating. *(e.g.,* one to ten multiplied by the gravity-based penalty). The adjustment factors are then applied to arrive at the adjusted and proposed penalty.

g. *Other Violations*

The Field Operations Manual states that penalties shall be assessed and proposed for violations of recordkeeping requirements, 29 CFR 1903 and 1904, but only if the employer has received a copy of the recordkeeping requirements booklet or had knowledge of its requirements. Otherwise, citations without proposed penalties will be issued.

Any employer who violates posting requirements may be assessed a penalty of up to $1,000 for each violation, pursuant to Section 17(i) of the Act. If notices prescribed and furnished by OSHA *(e.g.,* 29 CFR 1903.2(a)) are not displayed, the penalty shall be $100. If citations are not posted, as required in 29 CFR 1903.16, $500 shall be the penalty for each citation not posted. If a copy of the Summary of Occupational Injuries and Illnesses is not *posted* as prescribed in 29 CFR 1904.5, a penalty of $200 shall be proposed. If an employer does not *maintain* the Log of Occupational Injuries and Illnesses or a Supplementary Record as prescribed in 29 CFR 1904, $50 shall be proposed for each form not maintained. If the Summary of Occupational Injuries and Illnesses is not *compiled and maintained* as prescribed in 29 CFR 1904.5, a $50 penalty shall be proposed. If an employer does not *report* the occurrence of an accident within 48 hours (fatality or requiring hospitalization), $400 shall be proposed. Violations of the posting and recordkeeping requirements which involve the same documents may be grouped for penalty purposes. Finally, an employer who has received advance notice of an inspection and fails to notify authorized employee representatives may be assessed a civil penalty of up to $1,000 for each violation pursuant to Section 17(c) of the Act; 29 CFR 1903.6.

h. *Appropriateness*

The determination of the amount of penalty is within the authority and discretion of the Commission and is normally made by the Commission rather than a court on review. Sections 12(i), 10(c) and 17(j) of the Act; see *Western Waterproofing Co. v. Marshall, Inc.,* 576 F.2d 139, 6 OSHC 1550 (8th Cir. 1978); *Brennan v. OSHRC (Interstate Glass Co.),* 487 F.2d 438, 1 OSHC 1372 (8th Cir. 1973)—the Commission is the final arbiter of penalties if the Secretary's proposals are contested. The Commission is not bound by the penalty proposed by the Secretary of Labor and may make its own determinations with respect to the penalty's appropriateness; that is, it can affirm the penalty as proposed, *e.g.,* to conform an amendment of a violation from serious to willful. It is not bound by the Secretary of Labor's computation formula. *Nacirema Operating Co.,* 1 OSHC 1001 (1972); *Brennan v. OSHRC (Interstate Glass),* 487 F.2d 438, 1 OSHC 1372 (8th Cir. 1973); *Clarkson Construction Co. v. OSHRC,* 531 F.2d 451, 3 OSHC 1880 (10th Cir. 1976); *Allied Structural Steel Co.,* 2 OSHC 1457 (1975); *Long Manufacturing Co. v. OSHRC,* 554 F.2d 903, 5 OSHC 1376 (8th Cir. 1977); *California Stevedore and Ballast Co. v. OSHRC,* 517 F.2d 986, 3 OSHC 1174 (9th Cir. 1975); *J. L. Foti Construction Co.,* 8 OSHC 1281 (1980); *Jensen Construction Co.,* 6 OSHC 1070 (1978) —judge assessed $100 even though the Secretary of Labor proposed no penalty. The Secretary of Labor of course has the burden of proving the appropriateness of a penalty. *C & C Plumbing Co. and M & M Contractors, Inc.,*

6 OSHC 1130 (1978). Evidence of harassment of an employer as a result of selective enforcement of the Act is not relevant to the appropriateness of penalty. *Turner Communications Corporation v. OSHRC,* 612 F.2d 941, 8 OSHC 1038 (5th Cir. 1980). However, the judge does not have the authority to assess no penalty for a serious violation, as it is inconsistent with the Act. Penalties must be assessed in conformity with the statutory criteria set forth in the Act. *Delaware and Hudson Railway Co.,* 8 OSHC 1253 (1980).

After a citation becomes final, employers cannot in a collateral attack contest the citation and penalties in a court action to recover penalties brought by the Secretary of Labor. See *U.S. v. Jan Hardware Manufacturing Company, Inc.,* 463 F. Supp. 732, 7 OSHC 1007 (E.D. N.Y. 1979); *U.S. v. J. M. Rosa Construction Co.,*—F. Supp.—, 1 OSHC 1188 (D. Conn. 1973); *contra, United States of America v. P & L Supply Co.,* 486 F. Supp. 26, 8 OSHC 1125 (N.D. Tex. 1980)—employer's allegation that misleading statements of OSHA inspector led it to believe that notice of contest need not be filed supports defense that time limit for filing notice should be equitably tolled.

i. *Constitutionality of Civil Penalties*

The United States Supreme Court, upholding the positions of the Third and Fifth Circuits, has ruled that civil penalties are constitutional and do not violate the Seventh Amendment guarantee of a jury trial. *Atlas Roofing Co.* and *Frank Irey, Jr., Inc. v. OSHRC,* 424 U.S. 965, 5 OSHC 1105 (1977); see also *Clarkson Construction Company v. OSHRC,* 531 F2d 451, 3 OSHC 1880 (10th Cir. 1976). The enforcement structure of the Act is not constitutionally infirm, because an aggrieved party has the right to appeal any Commission decision to the circuit courts. *Dan J. Sheehan Co. v. OSHRC,* 520 F.2d 1036, 3 OSHC 1573 (5th Cir. 1975), *cert. denied,* 424 U.S. 965, 4 OSHC 1022 (1976). In *Desarrollos Metropolitanos, Inc. v. OSAHRC,* 551 F.2d 874, 5 OSHC 1135 (1977), the First Circuit held that interpretations of the Act resulting in penalties of varying amounts based on the size of an employer's business do not violate the equal protection guarantee of the Fifth Amendment. Moreover, the fact that penalties differ in similar cases does not involve an unequal application of the Act.

2. Criminal Penalties

The Act provides in Section 17(e) for the imposition of criminal penalties against employers who willfully violate a standard, rule, or regulation under the Act, and which violation causes the death of an employee. Upon conviction, the employer may be punished by a fine of not more than $10,000 or imprisonment for not more than six months in jail or both. For a second conviction, punishment may be a fine of not more than $20,000 or imprisonment for not more than a year or both. Section 17(f) of the Act provides that any person who gives advance notice of any inspection conducted under the Act shall, upon conviction, be punished by a fine of not more than $1,000 or imprisonment for not more than six months or both (unless authorized by Secretary of Labor or his aid for the advance notice). Section 17(g) provides that any person who knowingly makes a false statement, representation, or certification in any application, report, or other document filed or required to be maintained under the Act shall, upon conviction, be punished by a fine of not more than $10,000 or imprisonment for not more than six months or both. Finally, Section 17(h) of the Act provides that any person who kills, assaults, or resists OSHA enforcement personnel engaged in the performance of their duties (other than by demanding a warrant) shall be subject to a fine of up to $10,000 and imprisonment for life.

In OSHA Instruction CPL 2.39, OSHA issued guidelines for the investigation of possible criminal violations under Section 17(e). In order to qualify for referral to the Justice Department, it must be shown that: (1) a specific OSHA standard was violated; (2) the violation was willful; and (3) the violation resulted in an employee's death.

A corporation charged with criminal violations under the Act has no Fifth Amendment privilege against self-incrimination. *United States v. Pinkston-Hollar, Inc.,*—F. Supp.—, 4 OSHC 1697 (D. Kansas 1976).

3. Collection of Penalties

Section 17(1) provides that civil penalties assessed and owed shall be paid to the Secretary of Labor. If not paid, civil proceedings may be instituted for their collection in the United States District Court where the violation occurred or where the employer has its principal office. The merits of the citation and penalty may not *normally* be contested in a penalty-collection suit. See *Arena Construction Co. v. Marshall,*—F. Supp.—, —OSHC—(S.D.N.Y. 1978); but see *United States v. Shield Dye Cutters and Embossers, Inc.,*—F. Supp. —, 6 OSHC 2071 (1978).

D. ABATEMENT

Each citation issued to an employer must fix a reasonable time for abatement of the cited violation. Section 9(a) of the Act; *Gilbert Manufacturing Co., Inc.,* 7 OSHC 1611 (1979). Although the determination of reasonableness must be made on a case-by-case basis, the Act seeks to achieve rapid abatement of conditions hazardous to employees. *Brennan v. OSHRC (Kesler & Sons Construction Co.),* 513 F.2d 553, 2 OSHC 1668 (10th Cir. 1975); *American Smelting & Refrig. Co. v. OSHRC,* 501 F.2d 504, 2 OSHC 1041 (8th Cir. 1974).

The citation will normally provide a specified date or period for abatement of the violation unless a short period of time, usually less than five working days, is involved, in which case abatement will be stated in terms of calendar days. Abatement normally does not exceed thirty days. However, a longer period of time may be necessary for correction of noise or air-contaminant violations; if so, a date will be specified in the citation. The abatement period should, in OSHA's perspective, be the shortest interval within which an employer can reasonably be expected to correct the violation. FOM, Chapter X. Normally, abatement periods exceeding thirty days are not considered necessary by OSHA, and abatement dates in excess thereof usually must be authorized by the Area Director's office. The granting of an abatement period is solely within the discretion of the Secretary of Labor, *Metalbestos Systems, Division Wallace Murray Corp.*, 7 OSHC 1656 (1979), but if abatement is challenged by an employer, the Secretary has to prove the reasonableness and feasibility of the proposed means of abatement by substantial evidence. *Voegele Co. v. OSHRC*, 625 F.2d 1075, 8 OSHC 1631 (3d Cir. 1980); *National Realty & Construction Co., Inc. v. OSHRC*, 489 F.2d 1257, 1 OSHC 1422 (D.C. Cir. 1973); *Noranda Aluminum, Inc. v. OSHRC*, 593 F.2d 811, 7 OSHC 1135 (8th Cir. 1979). The Commission then is responsible for determining the reasonableness of the abatement.

1. Contest to the Abatement Period

The reasonableness of an abatement period may be challenged either by filing an employer or employee notice of contest or by filing a petition for modification of abatement date (PMA). Section 10(c) of the Act. (See *Gilbert Manufacturing Co.*, 7 OSHC 1611 (1979), for a discussion of the differences between the two procedures.) An employer who files a notice of contest to the citation relative to abatement must do so within fifteen working days from receipt of the citation. When the reasonableness of abatement is in issue, the burden of proving reasonableness rests with the Secretary of Labor. *Druth Packaging Co.*, 8 OSHC 1999 (1980). The Commission stated in this regard:

> Judge Harris did not cite to any evidence of record in support of his finding that the abatement dates for each step in the citation as amended were reasonable. However, he did state in his decision that "[t]he respondent introduced no evidence which would indicate that the period for abatement of the alleged violations herein was in any way unreasonable. . . ." Accordingly, we conclude that the judge's finding was based not on the record evidence but rather on a misconception of the respective burdens of proof of the parties. When an employer contests a citation, it may place in issue the reasonableness of the abatement date specified in the citation, as the Respondent has clearly done in this case. When the reasonableness of the abatement date is in issue, the burden of proving reasonableness lies with the Secretary. *See Gilbert Manufacturing Co.*, 7 OSHC 1611 (1979) (lead, concurring, and dissenting opinions). Because the Secretary in this case failed to prove that the abatement dates in the citation as amended were reasonable, the judge erred in finding that they were reasonable.

See also *Boise Cascade Corp.*, 5 OSHC 1242 (1977); *Gilbert Manufacturing Co.*, 7 OSHC 1611 (1979). If the employer fails to contest a citation or contests only the proposed penalty, then the citation, including the abatement date it contains, becomes an irreversible final order.

The following considerations are used by the Commission in arriving at a decision concerning the reasonableness of an abatement period: the gravity of the alleged violation, the availability of needed equipment, material, and personnel, required training, and time required for delivery, installation, and modification or construction of controls.

In situations where an employer contests either the abatement period or the citation itself (and the contest is made in good faith and not solely for delay or avoidance of penalty), the abatement period is considered not to have begun until all administrative or court proceedings have been exhausted and have resulted in a Commission final order affirming the citation and abatement date. Section 10(b) of the Act. That is, it *stays* or *tolls* the abatement period. Under Section 11(a) of the Act, an appeal of a Commission order to a court of appeals does not automatically operate as a *stay* of the order. An employer's contest of only the abatement date does not change the abatement requirements of a citation (to accomplish that, the citation itself must be contested); in a contest to only the amount of penalty, the abatement period continues to run unaffected by that contest. If the employer does not contest, he must abide by the abatement date set forth in the citation, even if such date is within the fifteen-day notice of contest period, or he would be subject to a failure to abate the citation. The Secretary may require abatement within the fifteen-working-day notice of contest period. *Brennan v. OSHRC,* 513 F.2d 553, 2 OSHC 1668 (10th Cir. 1975); *Dunlop v. Haybuster Manufacturing Company,* 524 F.2d 222, 3 OSHC 1594 (8th Cir. 1975); *Brennan v. OSHRC (Kesler and Son Construction Co.),* 513 F.2d 553, 2 OSHC 1668 (10th Cir. 1975); *George T. Gerhardt Co., Inc.,* 4 OSHC 1351 (1976). A citation for failure to abate, however, according to the Field Operations Manual, will not normally be issued until after the contest period for the citation has expired and it has been determined that the employer has not filed a notice of contest.

2. Petition for Modification of Abatement

If an employer does not contest a citation but nevertheless finds it is unable to meet the abatement date, it may file a petition for modification of abatement with the OSHA Area Director. *K. L. Spring and Stamping Corp.,* 7 OSHC 1651 (1979). Section 10(c) of the Act; Commission Rule 34(c). An employer may also file a petition for modification of abatement date when it has made a good-faith effort to comply with the abatement requirements of a citation but such abatement has not been completed because of factors beyond the employer's reasonable control. Both are permitted since Section 10(c) employs the term abatement requirements when speaking of a petition for modification. The PMA shall be in writing and must include the following information, per Commission Rule 34(c):

(1) All steps taken by the employer, and the dates of such action, in an effort to achieve compliance during the prescribed abatement period;
(2) The specific additional abatement time necessary in order to achieve compliance;
(3) The reasons such additional time is necessary including the unavailability of professional or technical personnel or of materials and equipment, or because necessary construction or alteration of facilities cannot be completed by the original abatement date; and
(4) All available interim steps being taken to safeguard the employees against the cited hazard during the abatement period.

A petition for modification of abatement date must be filed with the Area Director who issued the citation, no later than the close of the next working day following the date on which abatement was originally required. If filed later, the PMA must be accompanied by the employer's statement of exceptional circumstances explaining the delay. A PMA, like a notice of contest, must be posted. *Aspro. Inc., Spun Steel Division,* 6 OSHC 1980 (1978).

Normally a PMA is filed because an employer who has made a good-faith attempt and acted with reasonable diligence cannot abate because of factors beyond his control. On the other hand, notice of contest to the abatement period is filed because the abatement period is thought to be unreasonable. Unlike the notice of contest, the PMA does not automatically toll the abatement period. *Brennan v. OSHRC (Kessler & Sons Construction Co.),* 513 F.2d 553, 2 OSHC 1668 (10th Cir. 1975); *Gilbert Manufacturing, supra.* The citation becomes a final order after the expiration of the fifteen-working-day period.

The Secretary of Labor has authority to grant the extension (at the end of a fifteen-day period following posting) requested in a PMA, unless the petition is objected to by affected employees. In such instance, the petition is forwarded to the Commission for hearing, which may then modify the abatement date. Commission Rule 34. Thus, unlike the notice of contest route, Commission proceedings are instituted *only* if a dispute exists or an objection is entered. Apart from the procedural differences between PMA's and notices of contest of the abatement period, an employer should be aware that the burden of establishing the reasonableness of the extension of the abatement period rests with the employer if he files a PMA. *Gilbert Manufacturing Company, supra;* Commission Rule 73(b); Rule 34(d)(3). The employer must prove good faith efforts, reasonable diligence, and impossibility of compliance. Moreover, during the pendency of a PMA proceeding an employer is subject to a citation for a failure to abate. FOM, Ch. X; *Gilbert Manufacturing Company, supra.*

The Review Commission considers employer requests for additional abatement time or extension of the abatement date as a petition for modification of abatement rather than a contest of the citation, even if filed within the fifteen day notice of contest period, since "the goal of expeditious adjudication suggests that proceedings before the Commission should not be required in the absence of a dispute requiring adjudication and that such consideration favors such treatment."[16] See *Gilbert Manufacturing Company, Inc.,* 7 OSHC 1611 (1979); *K. L. Spring & Stamping Corp.,* 7 OSHC 1651 (1979). The effect of these decisions is that under the petition for modification of abatement, the employer rather than OSHA has the burden of proving the reasonableness of the requested extension. *Corhart Refractories Co.,* 6 OSHC 1224 (1977)—good faith and technology unavailable; *Mueller Brass Co.,* 8 OSHC 1776 (1980). Moreover, intended PMA's should clearly so state since it appears that ambiguous communications relative to abatement will be construed as such rather than a notice of contest. See *Kimball Office Furniture, Inc.,* 4 OSHC 1276 (1976).

In *Amoco Chemicals Corporation,* 8 OSHC 1085 (1980), the Commission held that, in some cases, the Secretary will not be able to determine whether the request for extension of the abatement date is justified. It would be improper for the Secretary to approve such a petition, for granting an abatement extension that may be unjustified is inconsistent with the Act's purpose to secure rapid abatement of hazards. See *Gilbert Manufacturing Co.*, 7 OSHC 1611 (1979). Thus the Secretary's proper course is to transmit the petition to the Commission, noting that he does not have sufficient information to determine whether the petition is justified. Such a course of action does not indicate the Secretary's lack of objection, but merely that the Secretary seeks a hearing at which the employer must justify the extension. The requirement of Rule 34 that each objecting party file its objections to the petition, 29 CFR 2200.34(d)(4), does not require a different conclusion. It permits notice pleading.

3. Method of Abatement

The term abatement is not defined in the Act. Thus, its achievement is solely a factual matter to be resolved on a case-by-case basis—but it must be noted that there is a difference between the abatement date, time for abatement, and the actual abatement of the alleged hazardous condition.

For correction of most violations, short-term abatement is all that is required. In some instances, however, depending upon the nature of the violation, OSHA may require long term or multistep abatement. Those situations are usually ones in which ultimate abatement will require the implementation of feasible engineering controls, as distinguished from feasible administrative methods or use of personal protective equipment. Interim abatement may also be required. OSHA's policy on abatement is that there are usually sufficient engineering controls to abate the hazard and that engineering controls afford the best employee protection. In addition to feasible engineering controls, administrative controls should be required for full abatement, and only in the most unusual situations are there not feasible engineering or administrative controls, at which point abatement would be allowed by personal protective devices. Some standards require abatement by either engineering or administrative controls *(e.g.*, the noise standard, 29 CFR 1910.95); others designate engineering controls as the preferred method *(e.g.*, the air contaminant standard, 29 CFR 1910.134).

During the abatement process, an employer must attempt to meet the terms of the cited standard; where possible, afford through alternate means the same protection that could be afforded by strict compliance with the standard; and, therefore, attempt to needlessly avoid exposing employees to the hazard. *Floyd S. Pike Electrical Contractors, Inc.*, 5 OSHC 1088 (1977), *aff'd*, 576 F.2d 72, 6 OSHC 1781 (5th Cir. 1978). But see *H. S. Holtze Construction Co.*, 7 OSHC 1753 (1979), *aff'd in part and rev'd in part*,—F.2d—, 8 OSHC 1785 (8th Cir. 1980), where the court stated that a reasonable interpretation of the Act is necessary vis-à-vis the alternate means of protection. As the court stated:

> Additionally, a variance application is here inappropriate, as we are discussing a situation of employees working on a third floor level for the very purpose of

> eliminating unguarded edges. Petitioner does not maintain or desire that the edge should remain unguarded. The real question is only whether or not while the hazard is being eliminated, are the employees adequately protected? We believe they were more adequately protected than they would have been by erecting and then removing a separate guardrail.

Thus the requirement that alternative means of protection be provided must be approached realistically.

If a hazard is trifling and if equivalent protection is provided, a violation is usually considered to be *de minimis.* In such a situation, an employer will not be required to abate the violation. *Alamo Store Fixtures Corp.,* 6 OSHC 1150 (1977); *Alfred S. Austin Construction Co.,* 4 OSHC 1166 (1976).

In citations involving the general duty clause, the Secretary of Labor has an obligation to specify what constitutes an appropriate means of abatement. *Young Sales Corporation,* 7 OSHC 1297 (1979). An employer's duty to abate vis-à-vis a general duty clause citation is not, however, controlled by the custom and practice in its industry but is solely dependent upon a showing of feasibility. Although the level of hazard recognition required by a general duty clause citation is measured against industry employer knowledge, an abatement order under Section 5(a)(1) of the Act may require that work practices and safety precautions be upgraded to a feasible level which is above that considered customary or reasonable by an industry. The question is whether the precaution is recognized by safety experts as feasible, not whether the precaution's use has become customary. See *Continental Oil Company v. OSHRC,* 630 F.2d 446, 8 OSHC 1980 (6th Cir. 1980); *Voegele Co., Inc. v. OSHRC,* 625 F.2d 1075, 8 OSHC 1631 (3d Cir. 1980); *Ray Evers Welding Co. v. OSHRC,* 625 F.2d 726, 8 OSHC 1271 (6th Cir. 1980); *General Dynamics Corp. v. OSHRC,* 599 F.2d 453, 7 OSHC 1373 (1st Cir. 1979); *Ford, Bacon & Davis Construction Corp.,* 6 OSHC 1910 (1978). However, it is the Secretary of Labor's burden to show that any abatement which is ordered is a demonstrably feasible measure which would materially reduce the likelihood that such injury as that which resulted from the cited hazard would have occurred. *Champlin Petroleum Company v. OSHRC,* 593 F.2d 637, 7 OSHC 1241 (5th Cir. 1979); *Titanium Metals Corp. of America v. Usery,* 579 F.2d 536, 6 OSHC 1873 (9th Cir. 1978); *Getty Oil Co. v. OSHRC,* 530 F.2d 1143, 4 OSHC 1121 (5th Cir. 1976).

The prevailing view until recently has been that the Commission only has the authority to review the reasonableness of the time period for abatement. The merits of an abatement plan may only be considered if evidence is offered to show that abatement can be accomplished in a shorter time under an alternate plan. *Automobile Workers v. OSHRC,* 557 F.2d 607, 5 OSHC 1525 (7th Cir. 1977), but see *American Airlines, Inc. v. Secretary of Labor,* 578 F.2d 38, 6 OSHC 1691 (2d Cir. 1972) and *United Parcel Service of Ohio, Inc. v. OSHRC,* 570 F.2d 806, 6 OSHC 1347 (8th Cir. 1978), where the Commission was directed to determine the appropriateness and methods of abatement. In *American Cyanamid Company,* 8 OSHC 1346 (1980), employee representatives objected to a proposed settlement agreement between the Secretary of Labor and the employer on the basis that the conditions set forth in the citation had not been abated as set forth in the agreement. The Commission stated that "while the ordinary uncontested settlement will not provoke an inquiry such as this, an attack by a party which calls the occurrence of one of these conditions into question requires that we determine whether the condition at issue, in this case abatement, has occurred before approving the settlement agreement. . . . [w]e conclude that under the circumstances of this case we do have the jurisdiction and authority to determine whether abatement has occurred." In *Marshall v. Sun Petroleum Products Company,* 622 F.2d 1166, 8 OSHC 1422 (3d Cir. 1980), the court asserted that an employee or his representative has a limited right to challenge a settlement agreement or a citation on the ground that the time provided for abatement of the violation is unreasonably long. The court stated that the Act does not give the employees the right to be heard on any other matters relative to abatement. Section 10(c) of the Act. Thereafter, the Commission, in *Farmer's Export Company,* 8 OSHC 1655 (1980), rejected the Court of Appeals' assertion in *Sun Petroleum Products Co.* that the Commission's authority to review settlement agreements in contested cases is limited to those instances in which employees or their representatives express some interest in the case and that even in those cases the Commission's authority to review proposed settlement agreements is limited to entertaining objections from the employee or employee representative that the abatement period proposed by the settlement is unreasonable.

In *Deemer Steel Casting* Company, 5 OSHC 1157 (1977), the Commission implied that it would continue to issue a stay of abatement period where an application for a variance is pending. Moreover, once an employer has made a *prima facie* case for extension of an abatement date, the burden is on OSHA and others who object to the abatement and extension to rebut the justification offered by the employer with persuasive evidence. *Corhart Refractories Co.,* 6 OSHC 1224 (1977).

NOTES

1. The Commission and the courts often have stated that an employer is not guilty of a violation that results from an unforeseeable or isolated incident. The Fourth Circuit recently has placed on the Secretary the burden of proving foreseeability before a violation can be found. See *Ocean Elec. Corp. v. Secretary of Labor,* 594 F.2d 396, 7 OSHC 1149 (4th Cir. 1979). *Cf. General Dynamics Corp. v. OSHRC,* 599 F.2d 453, 7 OSHC 1373 (1st Cir. 1979); *National Realty & Constr. Co. v. OSHRC,* 489 F.2d 1257, 1 OSHC 1422 (D.C. Cir. 1973). The language used by the *Ocean* court suggests that foreseeability is an element of constructive knowledge. See 594 F.2d at 403, 7 OSHC at 1154. The Eighth Circuit appears to have taken the position that the employer has the burden of proving that the act of its employee was unforeseeable or isolated. See *Western Waterproofing Co. v. Marshall,* 576 F.2d 139, 6 OSHC 1550 (8th Cir.), *cert. denied,* 439 U.S. 965, 6 OSHC 2091 (1978). But see *Danco Constr. Co. v. OSHRC,* 586 F.2d 1243, 6 OSHC 2039 (8th Cir. 1978); *Arkansas–Best Freight System, Inc. v. OSHRC,* 529 F.2d 649, 3 OSHC 1910 (8th Cir. 1976). The Tenth Circuit has taken a middle ground. In *The Mountain States Telephone & Telegraph Co. v. OSHRC,* 623 F.2d 155, 8 OSHC 1557 (10th Cir. 1980), the court held that an employer had the burden of proving unforeseeability of a subordinate's conduct where a supervisor had knowledge of such conduct. Where the noncomplying conduct was the supervi-

sor's own, however, the burden of proving foreseeability rested with the Secretary. The Commission unequivocally has stated that the unforeseeability/isolated incident issue is an affirmative defense. See, *e.g., Everglades Sugar Ref., Inc.*, 7 OSHC 1410 (1979); *Ford Motor Co.*, 5 OSHC 1765 (1977); *B–G Maintenance M'gmt. Inc.*, 4 OSHC 1282 (1976); *Murphy Pac. Marine Salvage Co.*, 2 OSHC 1464 (1975). The language used in some of the cases, however, reveals that the Commission is also confused about whether foreseeability is an element of knowledge. See, *e.g., Briscoe/Arace/Conduit, A Joint Venture,* 5 OSHC 1167 (1977)—employer had burden of showing that employee's acts were a departure from enforced work rule and employer had neither actual nor constructive knowledge of conduct.

2. Under this test, the employer is required to take any safety precautions recognized by safety experts as "feasible," even though the precaution is not a customary one in the industry. *National Realty & Constr. Co. v. OSHRC,* 489 F.2d 1257, 1266 n.37, 1 OSHC 1422, 1427, n.37 (D.C. Cir. 1973); see *Southern Ry.*, 3 OSHC 1566, 1570 (1975); *Southern Ry.*, 3 OSHC 1657, 1658 (1975). The Secretary must specify the particular safety precautions to be taken and has the burden of proving their feasibility. See, *e.g., Bristol Steel & Iron Works, Inc. v. OSHRC,* 601 F.2d 717, 7 OSHC 1462 (4th Cir. 1979); *accord, Champlin Petroleum Co. v. OSHRC,* 593 F.2d 637, 640, 7 OSHC 1241, 1243 (5th Cir. 1979); *National Realty & Constr. Co. v. OSHRC,* 489 F.2d at 1268, 1 OSHC at 1428. But see *Royal Logging Co.*, 7 OSHC 1744, 1751 (1979)—stating that the burden of proof rests on the Secretary, but that this burden is met when he proves that an abatement method exists; the burden then shifts to the employer to show infeasibility.

3. No Commission decisions or court cases specifically have stated that an other-than-serious violation of the Act cannot be a violation of the general duty clause. Nevertheless, the most reasonable conclusion to be derived from the language in the statute and in *National Realty* is that this clause is breached only by a serious violation; a serious violation requires a "substantial probability" that death or serious physical harm will result, while a violation of the general duty clause requires that death or serious physical harm be a "likely" consequence of the hazard. See U.S.C. Section 654(a), 666(j), (1976). See also note 5 *infra* and accompanying text.

4. This use of the word "knowledge" should not be confused with the use of the same word in the context of the requirements of a serious violation. "Knowledge" as used here means the recognition of a condition as hazardous. When used as one of the requirements for a serious violation, it means only actual or constructive knowledge of the condition itself. If the employer has actual knowledge that an existing condition is hazardous, it would obviously also have actual knowledge of the condition itself. If the industry recognized the condition as hazardous but the employer had not, the employer apparently would have to have at least constructive knowledge of the condition itself before a serious violation could be found.

5. This analysis is the same one used by the courts and the Commission in interpreting the "substantial probability that death or serious physical harm could result" test necessary for the finding of a serious violation. See discussion *supra.*

6. Similarly, although the general duty clause places on the Secretary of Labor the burden of proving the feasibility of safety precautions, see note 2, *supra,* such a burden is not always placed on him when a violation of a specific standard is alleged. The employer has the burden of proving that compliance is not feasible if the regulation or standard specifies the precautions to be taken. *Ray Evers Welding Co. v. OSHRC,* 625 F.2d 726, 8 OSHC 1271 (6th Cir. 1980); *accord, A. E. Burgess Leather Co. v. OSHRC,* 576 F.2d 948, 6 OSHC 1661 (1st Cir. 1978); *Ace Sheeting & Repair Co. v. OSHRC,* 555 F.2d 439, 5 OSHC 1589 (5th Cir. 1977). *Cf. Southern Colo. Prestress Co. v.*

OSHRC, 586 F.2d 1342, 6 OSHC 2032 (10th Cir. 1978). If, however, the regulation does not specify the method of compliance, the burden of proving feasibility remains on the Secretary. *Diebold, Inc. v. Marshall,* 585 F.2d 1327, 6 OSHC 2002 (6th Cir. 1978); see *General Elec. Co. v. OSHRC,* 540 F.2d 67, 4 OSHC 1512 (2d Cir. 1976).

7. In *Lisbon Contractors, Inc.,* 5 OSHC 1741 (1977), the Commission found that a violation was nonserious because it was *unlikely* to cause death or serious physical harm.

8. *Kent Nowlin Constr. Co. v. OSHRC,* 593 F.2d 368, 7 OSHC 1105 (10th Cir. 1979); *Western Waterproofing Co. v. Marshall,* 576 F.2d 139, 6 OSHC 1550 (8th Cir. 1978); *Empire-Detroit Steel v. OSHRC,* 579 F.2d 378, 6 OSHC 1693 (6th Cir. 1978); *Intercounty Construction Co. v. OSHRC,* 522 F.2d 777, 3 OSHC 1337 (4th Cir. 1975); *F. X. Messina Construction Corp. v. OSHRC,* 505 F.2d 701, 2 OSHC 1325 (1st Cir. 1974); *Georgia Electric Co. v. Marshall,* 595 F.2d 309, 7 OSHC 1343 (5th Cir. 1979).

9. The Commission, and the circuit courts that follow the majority rule, view this language of the Third Circuit as imposing a requirement of malicious intent or evil motive to establish a willful violation. See, *e.g., National Steel & Shipbuilding Co. v. OSHRC,* 607 F.2d 311, 7 OSHC 1837 (9th Cir. 1979); *Intercounty Constr. Co. v. OSHRC,* 522 F.2d 777, 3 OSHC 1337 (4th Cir. 1975), *cert. denied,* 423 U.S. 1072, 3 OSHC 1879 (1976); *Frank Irey, Jr., Inc.,* 5 OSHC 2030 (1977). One circuit, however, suggested that, rather than creating a conflict among the circuits, the language used in *Frank Irey* merely denoted a difference in emphasis in the Third Circuit's approach. *Cedar Constr. Co. v. OSHRC,* 587 F.2d 1303, 6 OSHC 2010 (D.C. Cir. 1978). The Third Circuit recently has agreed with this interpretation of the *Irey* language. See *Babcock & Wilcox Co. v. OSHRC,* 622 F.2d 1160, 8 OSHC 1317 (3d Cir. 1980).

10. "To our way of thinking, an 'intentional disregard of OSHA requirements' differs little from an 'obstinate refusal to comply'; nor is there in context much to distinguish 'defiance' from 'intentional disregard.' 'Flaunting the act' or 'flouting it,' as some would say, again carries the same meaning." *Babcock & Wilcox v. OSHRC,* 622 F.2d 1160, 1167, 8 OSHC 1317, 1322 (3d Cir. 1980).

11. The language in all these cases suggests that the employer has the burden of proving that his safety program is adequate and that the employee or supervisor was adequately trained. Because a willful violation requires knowledge, however, an argument can be made that the Secretary has the burden of proving the inadequacy of the employer's program before a willful violation can be found.

12. The court noted that "to hold otherwise would be to obliterate . . . the distinction between 'repeat' and 'willful' violations." *Id.*

13. The OSHA FOM, Ch. VIII provides that "[f]or purposes of considering whether a violation is repeated, citations issued to employers having fixed establishments (e.g., factories, terminals, stores) will be limited to the cited establishment. For employers engaged in businesses having no fixed establishments (construction, painting, excavation) repeated violations will be alleged based on prior violations occurring anywhere within the same State." In *Desarrollos Metropolitanos, Inc. v. OSHRC,* 551 F.2d 874, 5 OSHC 1135 (1st Cir. 1977), it was argued that the standard is arbitrary, since, even for businesses with peripatetic sites, a repeat violation can be found if citations are assessed within a state but not if violations occur at sites in more than one state. It was also argued that as a company with no fixed establishments, it may be found guilty of a repeat violation if transgressions occur at two or more of its moveable sites, whereas repeat violations cannot be predicated upon actions at separate locations for businesses with fixed establishments, such as factories.

The court upheld the standard, finding a reasonable basis for the distinction between permanent and transient sites. As to the distinction between intrastate and interstate businesses, it was sustained, because it serves administrative convenience.

14. This "language" requires that an employer must be assessed a penalty (e.g. $1 to $1000) for each serious violation. *Continental Steel Corp.*, 3 OSHC 1410 (1975).

15.

Employees	*% Reduction*
10 or less	40
11 to 25	30
26 to 60	20
61 to 100	10
over 100	0

Differentiation in penalties based upon an employee's size has been held to be neither unreasonable nor unconstitutional. *Desarrollos Metropolitanos, Inc. v. OSHRC*, 551 F.2d 874, 5 OSHC 1135 (1st Cir. 1977).

16. Even though a PMA proceeding is considered expeditiously, pursuant to Commission Rules 34(d)(2) and 101 and processed as an expedited proceeding, practically speaking in many cases because of the delay in hearing the petition, the issue becomes moot unless a new PMA has been filed.

APPENDIX 1 to CHAPTER X

PENALTY TABLE A
Serious Violations

Probability Rating	Gravity-based Penalties	Percent Reduction							
		10	20	30	40	50	60	70	80
1	300	270	240	210	180	150	120	90	60
2	400	360	320	280	240	200	160	120	80
3	500	450	400	350	300	250	200	150	100
4	600	540	480	420	360	300	240	180	120
5	700	630	560	490	420	350	280	210	140
6	800	720	640	560	480	400	320	240	
7	900	810	720	630	540	450	360		
8	1000	900	800	700	60				

PENALTY TABLE B
Other-than-Serious Violations

Probability Rating	Gravity-based Penalties	Percent Reduction							
		10	20	30	40	50	60	70	80
1									
2									
3									
4									
5	100	90	80	70	60	50			
6	165	150	130	115	100	85	65	50	
7	230	205	185	160	140	115	90	70	45
8	300	270	240	210	180	150	120	90	60

CHAPTER XI
The Citation and Contest Process

A. INTRODUCTION

At the closing conference the inspector will normally tell the employer what conditions, practices, structures, devices, or equipment in the workplace he considers to be violations, indicating what standards, rules, or regulations have been violated. The inspector may also explain what possible abatement measures can be taken to correct the violative conditions and possible abatement dates that may be required. After the completion of the closing conference and hence the inspection, the inspector will return to his office and prepare his reports, recommending or not recommending the issuance of a citation. The OSHA Area Director will review the inspector's reports and recommendations and issue a citation, which is the primary method for enforcement of the Act.

B. CITATION AND PROPOSED PENALTY

Upon completion of an inspection at a workplace, the OSHA inspector will make an initial determination as to whether or not the employer has violated the Act. The inspector is also responsible for making the initial determination whether or not a citation should issue. The term "citation" is not defined in the Act: Commission Rule 1(i) states that it means a written communication issued by the Secretary of Labor to an employer pursuant to Section 9(a) of the Act. A citation is a unique document, because it has purposes other than simply to initiate a legal proceeding. It informs the employer of his right to file a notice of contest and sets forth information needed by employers to accomplish abatement where appropriate or to make the decision whether or not to contest. *Everhart Steel Construction Co.,* 3 OSHC 1068 (1975).

Section 9(a) of the Act provides that if, as a result of an inspection, the Area Director, in his review of the inspector's report, believes that an employer has violated a requirement of Section 5 of the Act, of any standard, rule, or order promulgated pursuant to Section 6 of the Act, or any regulation prescribed pursuant to the Act, he shall with reasonable promptness issue a citation (or notice) to the employer.[1] 29 CFR 1903.14(a). A citation or notice will normally be issued even though the employer has immediately abated an alleged violation.

The citation (which must be in written form) will set forth the nature of the alleged violation and the period of time OSHA deems appropriate for its correction, which is called the abatement period.[2] Every citation must also state that its issuance does not constitute a finding that a violation of the Act has occurred, unless there is a failure to contest as provided for in the Act, or, if contested, unless the citation is affirmed by the Commission. 29 CFR 1903.14(e). After or concurrent with the issuance of a citation, OSHA must also notify the employer in writing of the penalty, if any, proposed for the violation. The proposed penalty is normally set forth in the citation but it may be stated in a separate document.

1. Promptness

The Secretary of Labor, as the complainant or prosecutor under the Act, has the authority to decide what conditions will be cited; whether, in fact, a citation will issue; and if issued, what will be its scope. *Carnation Can Co.*, 6 OSHC 1730 (1978); *International Harvester Co.*, 7 OSHC 2197 (1980). However, Section 9(c) of the Act provides that no citation may be issued after the expiration of six months following the occurrence of the violation. See also 29 CFR 1903.14(a). The six-month time period is considered a limitations period and is not to be confused with the reasonable promptness requirement of Section 9(a). *Dravo Corp.*, 3 OSHC 1085 (1975). To comport with the *reasonable promptness* requirement every effort is made by OSHA to issue citations within seventy-two hours following the completion of an inspection or investigation. FOM, Chapter X. Most citations are not, however, issued within seventy-two hours, and the failure so to issue a citation has not been considered to be a fatal flaw. *Chicago Bridge & Iron Co.*, 1 OSHC 1485 (1974), *rev'd on other grounds, Brennan v. Chicago Bridge & Iron*, 514 F.2d 1082, 3 OSHC 1056 (7th Cir. 1979). Thus the Review Commission has adopted and continues to adhere to a rule set forth in *Coughlin Construction Co.*, 3 OSHC 1636 (1975), which permits the dismissal of a citation on reasonable promptness grounds *only* if the employer has suffered prejudice due to the delay, that is, it has affected its ability to prepare and present a defense to the allegations in contest. *Craig D. Lawrenz and Associates, Inc.*, 4 OSHC 1464 (1976). Delays from thirteen days to four months have been held not to be prejudicial. See *A. C. & S. Inc.*, 4 OSHC 1529 (1976); *Laclede Gas Company*, 7 OSHC 1874 (1979); *Louisiana Pacific Corp.*, 5 OSHC 1994 (1977); *Mayfair Construction Company*, 5 OSHC 1877 (1977); compare *Jack Conie & Sons Corp.*, 4 OSHC 1378 (1976), where a 125-day delay was found to be unjustified; *E. C. Ernest, Inc.*, 2 OSHC 1468 (1975)—a sixty-nine-day delay in the issuance of a citation resulted in the unavailability of a witness and was found to be unjustified and prejudicial.

It is the Commission's position that a defense based upon a lack of reasonable promptness under Section 9(a) of the Act is an affirmative defense and *must* be raised by an employer sufficiently in advance of the hearing so that the Secretary of Labor is not prejudiced in the trying of the case. *Kast Metals Corp.*, 5 OSHC 1862 (1977); *Chicago Bridge & Iron Co.*, 1 OSHC 1485 (1974) *rev'd,* 514 F.2d 1082, 3 OSHC 1056 (7th Cir., 1975); *Puterbaugh Enterprises, Inc.,* 2 OSHC 1030 (1974). Several court of appeals decisions have affirmed the Commission in holding that prejudice must be shown by an employer before a citation is dismissed due to the delay between inspection and citation. *United Parcel Service of Ohio, Inc. v. OSHRC,* 570 F.2d 806, 6 OSHC 1347 (8th Cir. 1978); *Todd Shipyards Corp. v. Secretary of Labor,* 566 F.2d. 1327, 6 OSHC 1227 (9th Cir. 1977); *Stephenson Enterprises, Inc. v. Marshall,* 578 F.2d 1021, 6 OSHC 1860 (5th Cir. 1978); *Noblecraft Industries, Inc. v. Secretary of Labor,* 614 F.2d 199, 7 OSHC 2059 (9th Cir. 1980); *Bethlehem Steel Corp. v. OSHRC,* 607 F.2d 871, 7 OSHC 1802 (3d Cir. 1979).

The Section 9(c) limitations requirement of the Act, that a citation be issued within *six* months following the occurrence of a violation, has been construed more particularly than the reasonable promptness requirement. A violation that has occurred outside the six-month limitations period can not be relied upon as a basis for citation. See *Bloomfield Mechanical Contracting, Inc. v. OSHRC,* 519 F.2d 1257, 3 OSHC 1403 (3d Cir. 1975); *General Electric Co. v. OSHRC,* 540 F.2d 67, 4 OSHC 1512 (2d Cir. 1976); *Wean United, Inc.,* 7 OSHC 2086 (1979). See also *Sun Petroleum Products Corp. v. Marshall,* 622 F.2d 1196, 8 OSHC 1422 (3d Cir. 1980)—an employer will not be required to defend a stale citation. An amendment to a citation under Rule 15(b) of the Federal Rules of Civil Procedure, arising out of conduct set forth in that citation, relates back to the date of that citation and would not be considered to constitute the issuance of a new citation, which would be barred by the six-month limitation period. *Kaiser Aluminum and Chemical Corp.,* 5 OSHC 1180 (1976); *Structural Painting Corp.,* 7 OSHC 1682 (1979); *Duane Smelser Roofing Co. v. Marshall,* 617 F.2d 448, 8 OSHC 1106 (6th Cir. 1980). (See discussion, *infra.)*

The six-month limitation period applies only to the civil citation procedure; it does not apply to criminal actions, which have a five-year limitation. *U.S. v. Dye Construction Co.,* 510 F.2d 78, 2 OSHC 1510 (10th Cir. 1975).

2. Particularity

Each citation is required to be in writing and must describe with particularity the nature of the violation, including the reference to the provision of the Act, standard, rule, regulation, or order alleged to have been violated. Section 9(a) of the Act; 29 CFR 1903.14(b). The citation shall also fix a reasonable time or times for abatement of the violation.

According to the FOM, in certain cases the same factual situation may present a possible violation of more than one standard. Thus where more than one standard is applicable to any given factual situation *and* it appears that compliance with any one of the applicable standards would effectively eliminate the hazard, OSHA deems it permissible and in compliance with the Act's particularity requirement to cite *alternative* standards, by means of an "and/or" designation. *Henkels & McCoy, Inc.,* 4 OSHC 1502 (1976); *Usery v. Marquette Cement Manufacturing Co.,* 568 F.2d 902, 5 OSHC 1793 (2d Cir. 1977)—employers may assert alternate affirmative defenses in the answer.

The FOM also addresses such questions as whether a violation should be cited under a specific (vertical) standard such as 29 CFR 1926 (construction) or under a general (horizontal) standard such as 29 CFR 1910 (general industry).

Separate violations of the same standard may be grouped into one violation; conditions which violate more than one standard may be cited as violations of both or cited in the alternative. FOM, Ch. X, C-1.

The FOM states that a citation which does not describe a violation with sufficient clarity and specificity may be subject to legal challenge on grounds of particularity. It similarly states that an ambiguous citation could result in problems relative to the employer's abatement efforts and to the precise legal basis for issuance of subsequent failure to abate or repeated citations. Ambiguities in the citation may also work to the detriment of the employees, who, upon posting of the citation pursuant to 29 CFR 1903.16, may not adequately be able to identify the hazard to which they are or may become exposed. Thus the FOM lists five essential elements which should be considered in the preparation of an alleged-violation description prior to the issuance of a citation. These elements are:

(1) Identification of the specific items of equipment or procedure which pose a hazard;
(2) A description of the location where a hazardous condition existed or hazardous practice took place;
(3) The description of the observed hazardous conditions or practices;
(4) The date on or about which the violation took place; and
(5) Identification of the fact that an employee is, has been, or potentially is exposed to the hazardous condition.[3]

Minute detail is not required to comply with the particularity requirement, only *fair* notice of the nature and general location of the violation and/or hazard. *Meadows Industries, Inc.*, 7 OSHC 1709 (1979); *H. S. Holtze Construction Co. v. Marshall,* 627 F.2d 149, 8 OSHC 1785 (8th Cir. 1980)—a typographical error in the citation giving an erroneous reference within the cited standard was not so ambiguous as to mislead or misinform the employer; *Savina Home Industries, Inc. v. Secretary of Labor,* 594 F.2d 1358, 7 OSHC 1154 (10th Cir. 1979); *Ringland-Johnson Inc. v. Dunlop,* 551 F.2d 1117, 5 OSHC 1137 (8th Cir. 1977)—failure to state the date of inspection in the citation does not prejudice employer. *Todd Shipyards Corp. v. Secretary of Labor,* 586 F.2d 683, 6 OSHC 2122 (9th Cir. 1978)—a citation for a repeated violation did not lack particularity, since the employer could determine which prior citation formed the basis for the repeated violation from information in the citation and by examination of its own files; *Schiavone Construction Co.*, 5 OSHC 1385 (1977)—particularity was established by reading two citations together, which had been issued at the same time; *Gold Kist, Inc.*, 7 OSHC 1855 (1979)—analysis of pleadings, motions, and inquiries of discovery rendered a citation sufficiently particular, so that an employer could decide whether or not to contest. Confusion about the penalty provisions of a citation does not cause a lack of particularity in a citation, since penalties do not affect an employer's ability to identify violative conditions. *FMC Corporation,* 5 OSHC 1707 (1977).

In *B. W. Harrison Lumber Co.,* 4 OSHC 1091 (1976), *aff'd sub nom.* 6 OSHC 1446, 569 F.2d 1303 (5th Cir. 1978), the Commission held that if a citation is so lacking in specificity that it cannot be determined what action is required of an employer, then the citation must be vacated. For an employer must be able to interpret a citation in order to make an intelligent decision whether or not to contest and, if it decides not to contest, in order to determine what hazards are involved and how to correct them. Specifically, the Commission held that:

> The *test of particularity* is whether the citation provided *fair notice* of the alleged violation. In determining whether fair notice has been afforded, consideration may be given to factors external to the citation, such as the nature of the alleged violation, the circumstances of the inspection, and the employer's knowledge of his own business. *REA Express v. Brennan,* 495 F.2d 822, 1 OSHC 1651 (2d Cir. 1974). We do not, however, believe that oral statements made by a compliance officer during the inspection and at a closing conference may be used to cure lack of particularity in a later issued citation. As the record herein shows, such statements can be interpreted differently by the compliance officer and the employer.
>
> Additionally, Section 9(a) of the Act requires that a citation be in writing. This would appear to *preclude incorporation* in a citation of allegations made *orally* during the inspection. Perhaps most importantly, however, is the fact that the compliance officers who conduct inspections for complainant are usually not the same officials authorized to issue citations. . . . Thus, statements made by a compliance officer during an inspection are at best an uncertain guide to the nature of the alleged violation that may thereafter appear in a citation. (emphasis added)

Upon review, the Fifth Circuit affirmed the Commission's decision in all respects. See also *National Realty & Construction Co. v. OSHRC,* 489 F.2d 1257, 1 OSHC 1422 (D.C. Cir. 1973); but see *REA Express, Inc. v. Brennan,* 495 F.2d 822, 1 OSHC 1651 (2d Cir. 1974).

The *B. W. Harrison Lumber* case stated that in a determination of fair notice, consideration may be given to certain external factors, such as the nature of the alleged violation, the circumstances of the inspection, and the employer's knowledge of his own business. However, these circumstances should relate *solely* to the question of whether the particularity requirement is satisfied and whether the employer has sufficient information to make a meaningful decision to contest, *at the time the citation was issued.* It would seem therefore that a defective citation could not be cured by information obtained by an employer through oral statements at an informal conference, through the discovery procedure, or during a hearing. See *Wheeling & Pittsburgh Steel Corp.,* 7 OSHC 1581, 1588 (1979) (concurring opinion of Commissioner Barnako); *Meadows Industries, Inc.,* 7 OSHC 1709, 1713 (1979) (concurring opinion of Commissioner Barnako); *Goldkist, Inc.,* 7 OSHC 1855, 1864 (1979) (dissenting opinion of Commissioner Barnako).

The Review Commission has held, however, that postcitation proceedings, such as pleadings, motions, discovery, and the hearing itself, may indeed provide the particularity required by Section 9(a) of the Act. See

Meadows Industries, Inc., supra; Del Monte Corp., 4 OSHC 1383 (1977); *Louisiana Pacific Corp.,* 5 OSHC 1994 (1977)—an employer's knowledge of his own workplace may cure alleged defects in particularity with which an alleged condition is described; *B. F. Goodrich Textile Products,* 5 OSHC 1458 (1977); *Todd Shipyards Corp.,* 5 OSHC 1012 (1977), *aff'd,* 566 F.2d 1327, 6 OSHC 1227 (9th Cir. 1977)—whether a citation gives fair notice depends upon factors other than the language of the citation and may also include circumstances of the compliance inspection and the employer's familiarity with his own business; *Republic Steel Corp.,* 6 OSHC 1204 (1977); *American Airlines, Inc.,* 6 OSHC 1252 (1977)—instructions given at a closing conference may be considered in determining whether a citation gave fair notice.

In *General Motors Corp., Central Foundry Div.,* 8 OSHC 1735 (1980), the Commission held that an employer's failure to inquire of the Secretary of Labor as to the precise areas in issue in the citation or to make discovery required rejection of its defense of a lack of particularity. It was also noted that the employer seemed to know his business, which would serve to remedy the lack of clarity. Furthermore, the Commission felt that in order to cure any possible defect in a citation and to avoid possible prejudice an employer believes he may suffer, he may be granted a continuance to prepare an adequate defense.

The key question in each case is whether the citation provides sufficient facts to put the employer on notice of the purported noncomplying condition, but also important in terms of prejudice is the employer's ability to prepare a response to and defense of the citation. Moreover, if a citation is lacking in particularity, it may also be questioned whether employees adequately understood the citation. If the citation was not sufficiently clear, employees may have been rendered unable to assist the employer in providing and preparing relevant facts and information concerning the alleged violation in a contest of the citation. The insight and experience of employees may be invaluable to an employer who must decide whether or not to contest a citation. The testimony by employees or their election of party status in support of the employer could present strong evidence that the violations were unjustified. Moreover, the citation should be of sufficient particularity so that an employee may be made aware of the hazardous conditions set forth therein. If a citation fails to describe the hazard sufficiently, employees are left without knowledge of the alleged dangers and remain unable to make a special effort to prevent possible injury. As a result, the employer could be exposed to increased liability for injuries which might otherwise have been prevented had the citation clearly described the hazards alleged. Thus if an employer posts a citation whose language is not sufficiently particular to inform the employees of the alleged hazardous condition, and the period of abatement specified therein has expired, and an employeee is injured, it is possible upon reinspection for the employer to be cited for a failure to abate. Moreover, the Act requires that an employer has an obligation, through posting, to keep his employees informed of their protections and obligations under the Act. The citation must be posted prominently at or near the place where the violation referred to in the citation took place. Section 9(b) of the Act. Employers may be cited under the Act for failure to post citations or notices. Section 17(i) of the Act. If lack of posting, which affects employees, may inure to the employer's detriment through a citation and penalty, then lack of particularity in a citation, which affects employees, should inure to the employer's benefit. The citation's lack of particularity also could be a reason for the employees' failure to challenge the reasonableness of the abatement period, for the Section 10(c) right of employees to participate in OSHA proceedings is well-established. *Eaton Corp.*, 6 OSHC 1906 (1978).

A challenge to the particularity of a citation is an affirmative defense, which must be raised in a pleading or motion or tried by the express or implied consent of the parties before the Review Commission, or a court of appeals will be precluded from considering the argument on review. *Todd Shipyards Corp. v. Secretary of Labor,* 566 F.2d 1327, 6 OSHC 1227 (9th Cir. 1977); *Wheeling-Pittsburgh Steel Corp.*, 7 OSHC 1581 (1979); *Gannett Corporation,* 4 OSHC 1383 (1976); *Union Camp Corp.*, 5 OSHC 1799 (1977)—in a citation for excess noise levels in screen room, of which the employer had three, the inspector could not identify in which room the alleged violation occurred, thus the citation was vacated.

3. Abatement and Penalties

The citation is required to fix a reasonable time for abatement of the violation. Section 9(a) of the Act. The FOM states that ordinarily a specific date should be set for abatement, but where a short abatement period is appropriate and it is not certain whether the citation will be received during the time period for abatement, the abatement period may be stated as a specific date in terms of calendar days. When immediate abatement is appropriate, the alleged violation must be corrected immediately upon receipt and the word immediately should be specified on the citation.

In conjunction with the issuance of a citation under Section 9(a) of the Act, OSHA must also notify the employer of the imposition of the penalty by certified mail or personal service, if any is proposed to be assessed under Section 17 of the Act, or that no penalty is being proposed. 29 CFR 1903.15(a). The notice must state that the proposed penalty shall be deemed to be a final Commission order and not subject to review unless within fifteen working days from the date of receipt the employer notifies the Area Director that it intends to contest the citation or penalty.

4. Notice to Employer

A citation may be served in any manner which is calculated to give notice to the employer, *i.e.*, personal service, certified mail. Commission Rule 4(d); *Donald K. Nelson Construction, Inc.*, 3 OSHC 1914 (1976); *Josten-Wilbert Vault Company,* 4 OSHC 1577 (1976). Questions relevant to the general validity of service may be determined by reference to Federal Rule of Civil Procedure 4.

The citation (and notification of penalty) must, according to one court, be given to an official of the employer "who has the authority to disburse corporate funds to abate the alleged violation, pay the penalty, or contest the citation and penalty." *Buckley & Co., Inc. v. Secretary of Labor,* 507 F.2d 78, 2 OSHC 1432 (3d Cir. 1975). The Commission, however, disagrees with this holding and has recently widened the scope of the appropriate official for purposes of receipt of notices. It states the test to be applied is whether "the service is reasonably calculated to provide an employer with knowledge of the citation and proposed penalty and an opportunity to determine whether to abate or contest." *B. J. Hughes, Inc.,* 7 OSHC 1471 (1979); *Capital City Excavation Co., Inc.,* 8 OSHC 2008 (1980). Thus service on an employee who will know to whom to forward the citation is considered acceptable. For construction employers, OSHA will send the citation and penalty to the work site with copies to the home office. FOM, Ch. VII. Employee representatives also receive copies of the citation and penalty especially if it was issued in response to an employee complaint. 29 CFR 1903.11 and 14(c). If the employees are not represented, notice will be sent to the employees who participated in the walkaround.

5. Amendments

At times, the details of a citation may be incorrect, misleading, or incomplete, and the citation may be considered appropriate for amendment. FOM, Ch. X. (Amendments to the complaint are discussed in Chapter XIII.) Examples of proper cause for amendment which have been upheld are the citation of an incorrect standard, the incorrect or incomplete description of an alleged violation, the incorrect calculation of a penalty, or the specification of an unreasonable time for abatement. *Prior to the contest* of a citation, an amendment or even a withdrawal of a citation by the Area Director may be undertaken and is considered permissible, assuming that the fifteen-working-day period for filing of the notice of contest has not expired or, where a notice of contest has been filed, that the contest is directed only to the abatement period or a petition for modification of abatement is received. In these circumstances, the OSHA Area Director may on his own initiative amend the citation. FOM, Chapter X.

Once an employer has contested a citation which the Area Director believes should be amended, it must be brought to the attention of the Regional Solicitor of Labor whether or not the fifteen-working-day period has expired.[4] The Solicitor's office can then take the necessary legal steps, *vis-à-vis* initiation of appropriate Review Commission proceedings, to make the necessary amendments. The Commission stated in *P. A. F. Equipment Co.,* 7 OSHC 1209 (1979), that the FOM provisions deal only with amendments prior to the filing of a notice of contest. They do not purport to deal with amendments occurring after notice of contest has been filed, presumably in recognition of the fact that this type of amendment is governed by the Commission's Rules and cannot be accomplished unilaterally by the Secretary. Moreover, the FOM has been held to be "merely directory" and serves only as an internal administrative guide for OSHA. It does not have the force of law and is thus not binding on the Secretary or on the Commission. *FMC*

Corporation, 5 OSHC 1707 (1977). *Note:* If the fifteen-working-day period has expired, the Secretary of Labor cannot amend the citation without Commission approval, even if other matters, such as the penalty, are in contest.

If a citation is amended, the employer will receive fifteen additional working days from receipt of the amended citation and notification of penalty to contest *those portions* of the original citation *which have been amended.* However, the amended citation does not extend the original contest period for those portions of the original citation which are unchanged by the amendment. Like the citation, each amendment to the citation must be prominently posted until the violation has been abated or for three working days, whichever is later.

In situations where the fifteen-working-day period for filing a notice of contest has expired or where a notice of contest has been filed, the amendment will or will not be allowed in accordance with the procedures of the Review Commission. Amendment must be made by motion. *Texaco, Inc.,* 8 OSHC 1677 (1980). An amendment (or withdrawal) of a citation may not be made to a citation which has become a final order. *Hartwell Excavation Co.,* 3 OSHC 1196 (1973).

As a general rule administrative pleadings are liberally construed and easily amended. *National Realty & Construction Co. v. OSHRC,* 489 F.2d 1257, 1267 n.40., 1 OSHC 1422 (D.C. Cir., 1973). The question is whether a citation and a notice of contest are "pleadings" subject to amendment. Clearly it appears from Commission Rule 33(a)(3), that the citation, even if it is not strictly speaking a pleading, can be amended through the complaint. With respect to the notice of contest, because of the fifteen-working-day limitation period set forth in Section 10 of the Act, it is arguable that the notice of contest is not a pleading subject to amendment (once the limitation period has expired).

Moreover, no equivalent provision is contained in Commission Rule 33(b) for amending a notice of contest through the answer and thus it appears that the failure to contest an item in a citation may not be cured by an amendment to the answer after the 15 day notice of contest period has expired. See, *Mississippi Valley Erection Co.,* 1 OSHC 1527 (1973). But see, *Goldkist, Inc.,* 7 OSHC 1390 (1979), which appeared to indicate that pleadings include notices of contest.

Amendments are usually allowed at any time to correct technical errors. In *John Hill, d/b/a Leisure Resources Corporation,* 7 OSHC 1485 (1979), as a result of an inspection which was conducted, two citations were issued to Leisure Resources Corporation. A timely notice of contest was filed by Leisure Resources Corporation and signed by J. E. Hill, President. Thereafter, the Secretary of Labor filed a complaint with the Commission naming John Hill d/b/a Leisure Resources Corporation as the respondent. The complaint alleged that the respondent was an individual proprietor doing business under the name and style of Leisure Resources Corporation. In its answer to the complaint, respondent John Hill denied that allegation. Following receipt of this information, the Secretary of Labor moved to amend the citation (and complaint) to name Joseph Hill d/b/a Leisure Resources Corporation as the respondent. The judge denied the motion. The Secretary then moved to amend the pleadings so as to name Leisure Resources Corporation as the respondent. The judge also denied this motion and entered an order dismissing the complaint against John Hill d/b/a Leisure Resources Corporation. Upon review the Commission reversed, stating:

> Business was being conducted in the name "Leisure Resources Corporation" and we find that the entity operating under that name is the employer, whether it was incorporated or not. Whether "Leisure Resources Corporation" was incorporated or unincorporated at the time of the inspection and citation does not affect its obligation as an employer to comply with the occupational safety and health regulations or its liability under the Act if it has failed to do so. The requested amendment to name that body as respondent should have been allowed. The fact that the complaint named an individual as respondent instead of the corporation, to which the citation was issued, does not require dismissal of the citation. The only question in this case seems to be the proper name or style of the case, and not whether the wrong company was cited. It is not contended that the wrong respondent was cited, nor that the respondent is unaware of the citation and of these proceedings. Leisure Resources contested the citation on Leisure Resources Corporation's stationery; and the notice of contest was signed by the company's president. Leisure Resources appeared at the hearing through its president. Leisure Resources obviously had notice of the proceedings and should have known that, but for the Secretary of Labor's erroneous belief that John Hill was the individual proprietor doing business as Leisure Resources Corporation, the action would have been maintained against Leisure Resources Corporation. The citation named the correct employer. We find that the Secretary's attempt to more accurately identify the employer in the complaint was a mere technical misnomer which did not affect the nature of the proceeding or the allegations against the employer.

a. *Pre-hearing Amendments*

Prior to the hearing, amendments to the citation are normally permitted, especially if there is no change in the cause of action, unless it would result in undue prejudice to the parties. Prejudice would generally be "defined" as a loss of evidence or the unavailability of witnesses. See *Usery v. Marquette Cement Manufacturing Company,* 568 F.2d 902, 5 OSHC 1793 (2d Cir. 1977). (Even in the absence of employer prejudice, a continuance may be requested to avoid surprise.) In *Noblecraft Industries, Inc. v. Secretary of Labor,* 614 F.2d 199, 7 OSHC 2059 (9th Cir. 1980), an amendment of a citation to increase the violation from an other-than-serious to serious characterization and to increase the penalty was upheld in spite of the employer's argument that such amendment constituted a withdrawal of the citation. Since the citation was the subject of dispute once it was contested, the Commission had the authority to affirm, modify, or vacate the citation or direct other appropriate relief. See also *Long Manufacturing Co. v. OSHRC,* 554 F.2d 903, 5 OSHC 1376 (8th Cir. 1977), which stated that:

> Petitioner was not surprised or prejudiced in any way by the change in the nature of the charge, and it was not prejudiced by the increase in the amount of the proposed penalty except in the sense that if petitioner had accepted the initial proposal it would probably not have been faced with a higher proposal later. That however, is not legal prejudice. 554 F.2d at 907–908.

The court also stated:

> An employer may accept the citation and proposed penalty without protest if he chooses to do so, but in our view if he chooses to contest the matter before the Commission, he does not have any vested right to go to trial on the specific charge mentioned in the citation or to be free from exposure to a penalty in excess of that originally proposed. 554 F.2d at 907.

See also *Usery v. Marquette Cement Mfg. Co.*, 568 F.2d 902, 5 OSHC 1793 (2d Cir. 1977); *Southern Colorado Prestrees Co. v. OSHRC*, 586 F.2d 1342, 6 OSHC 2032 (10th Cir. 1978)—amendment permitted on day of hearing since employer had prior notice due to filing of motion to amend, did not seek continuance, and did not demonstrate prejudice; *P. A. F. Equipment Co.*, 7 OSHC 1209 (1979); *Morton Buildings, Inc.*, 7 OSHC 1703 (1979)—there was no charge in the cause of action because the amended charge arose out of the same conduct as the original charge. But see *Cornell & Co. v. OSHRC*, 573 F.2d 820, 6 OSHC 1436 (3d Cir. 1978)—an amendment was denied nine days before the hearing, since it would have changed the factual and legal nature of the case and would have prejudiced the employer's ability to locate witnesses necessary to present its defense; *P & Z Company, Inc.*, 7 OSHC 1589 (1979)—motion to add a new party by means of amendment rather than through issuance of a new citation was denied. The failure of a party to object to an amendment may constitute a waiver.

b. *Amendments at and after the Hearing*

Pursuant to Rule 15(b) of the Federal Rules of Civil Procedure,[5] amendments of pleadings to conform to the evidence presented at the hearing or after the hearing are normally permitted if the issues involved in the amendment have been tried by the *implied* or *express* consent of the parties and there is no prejudice to the employer. *National Steel & Shipbuilding Co.*, 6 OSHC 1680 (1978), *aff'd sub nom. v. OSHRC*, 607 F.2d 311, 7 OSHC 1837 (9th Cir. 1979); *Kuhn Construction Co.*, 5 OSHC 1145 (1977). Implied consent to an amendment may be found by the failure of the employer to object to the introduction of evidence at the hearing. *Rogers Manufacturing Company*, 7 OSHC 1617 (1979). Proper subjects for amendment include technicalities such as the proper name and address of the employer, the standard cited, the penalty, the severity of the violation, and the abatement.

In *Asplundh Tree Expert Co.*, 6 OSHC 1951 (1978), the Commission permitted an amendment to a citation to allege a violation of a general industry standard rather than a construction industry standard, because the standards were identically worded and there would have been no change in the factual basis of the allegation. See also *Long Manufacturing Co. v. OSHRC*, 554 F.2d 903, 5 OSHC 1376 (8th Cir. 1977); *Western Waterproofing, Inc. v. Marshall*, 576 F.2d 139, 6 OSHC 1550 (8th Cir. 1978); *P & Z Company, Inc. and J. F. Shea Company*, 7 OSHC 1589 (1979)—motion granted to amend citation to add a party defendant; *Miller Brewing Company*, 7 OSHC 2155 (1980)—motion to amend charge from general duty clause to specific standard was approved, since it was based on the same facts and resulted in no prejudice.

Where there is no showing of prejudice to an employer in the preparation or presentation of a defense, a citation alleging a serious violation of the Act may be amended to allege a *willful* or *repeated* violation even after an employer files his notice of contest to the original citation. *Structural Painting Corporation,* 7 OSHC 1682 (1979); *P. A. F. Equipment Company,* 7 OSHC 1209 (1979). See also *National Realty and Construction Company, supra; Dan J. Brutger, Inc.,* 4 OSHC 1745 (1976); compare *McLean-Behm Steel Erectors, Inc. v. OSHRC,* 608 F.2d 580, 7 OSHC 2002 (5th Cir. 1979), where a posthearing amendment was rejected since an employer was found not to have impliedly consented and the evidence introduced at the hearing was relevant to both the original and amended citation.

Amendments to a citation which arise out of the conduct set forth in the original citation date back to that time and do not violate the six-month statute of limitations set forth in Section 9(c) of the Act. Federal Rule of Civil Procedure 15(c). In *Southern Colorado Prestress Company v. OSHRC,* 586 F.2d 1342, 6 OSHC 2032 (10th Cir. 1978), the court held that an amendment of a citation more than six months after the occurrence of the alleged violation was not barred on timeliness grounds, since the amended claim arose out of the same conduct, transaction, or occurrence set forth in the original pleading. In both *Higgins Erectors and Haulers, Inc.,* 7 OSHC 1736 (1979), and *Structural Painting Corporation,* 7 OSHC 1682 (1979), the Commission reaffirmed this position, stating that amendments to a standard relate back to the time of the original citation, need not be within six months of the time the violation occurred, and do not violate the six-month statute of limitations set forth in the Act. Compare *Bloomfield Mechanical Contracting, Inc. v. OSHRC,* 519 F.2d 1257, 3 OSHC 1403 (3d Cir. 1975)—the failure to issue an original or amended citation within six months was contrary to Section 9(c); *General Electric Co. v. OSHRC,* 540 F.2d 67, 4 OSHC 1512 (2d Cir. 1976)—the dismissal of the citation was upheld where the evidence failed to show a violation had occurred within six months of the issuance of the citation; in fact, ten months had elapsed from the occurrence of the alleged violation.

6. Posting

Citations are required to be prominently posted at or near the place of the alleged violation or, if that is not practicable, in a prominent place where they will be readily observable by affected employees. Section 9(b) of the Act; 29 CFR 1903.16(a); *Siggins Company,* 3 OSHC 1562 (1975). Citations must remain posted until the violation is abated or for three working days, whichever is longer. 29 CFR 1903.16(b). The employer may also post, in the same location, a notice that the citation is being contested and explain its reasons therefore or explain its reasons and methods of abatement. Violations of the posting requirement may bring a civil penalty up to $1,000 for each violation. Section 17(i) of the Act. Failure to post a citation *may* also result in the dismissal of the employer's notice of contest. *C & H Erection Co.,* 3 OSHC 1293 (1978); *Hersman Construction Co.,* 1 OSHC 1185 (1973).

Posting is normally considered to be for the benefit of affected employees, but as is discussed *infra,* it may also benefit the employer.

7. Withdrawal of Citations

A citation may be withdrawn by an Area Director at any time prior to the expiration of the fifteen-working-day period, provided that a notice of contest has not been filed. The letter of withdrawal must be posted for three working days in those locations where the citation was originally posted. FOM, Chapter X. If a citation was issued pursuant to an employee complaint, under Section 8 of the Act, employees must be afforded the right to an informal review prior to withdrawal, governed by the rules and procedures set forth in 29 CFR 1903.12(a) and 1903.14(d). Such procedures also apply with respect to amendments of those citations responding to employee complaints.

In *IMC Chemical Group, Inc.*, 6 OSHC 2075 (1978), the Commission held that Rule 41(a)(1) of the Federal Rules of Civil Procedure which permits the withdrawal of a complaint without leave of court before an answer has been filed does not permit the Secretary of Labor to withdraw a citation after an employer has filed a notice of contest, and Commission jurisdiction is triggered. Since the citation is analogous to the complaint, a notice of contest corresponds to the answer in federal court procedure. The Secretary of Labor's proper course of action, therefore, would be to file a motion to withdraw the citation which an employer has contested. Granting of such motion would be within the discretion of the Commission, since Rule 41(a)(2) of the Federal Rules of Civil Procedure is applicable to such situations. *Molinos de Puerto Rico, Inc.*, 1 OSHC 1402 (1973). But see *Farmers Export Company*, 8 OSHC 1655, 1660 n.6 (1980) (concurring opinion of Member Barnako), in whose view the Secretary of Labor has an absolute right to withdraw a citation prior to the filing of an answer or a motion for summary judgment by the employer, whichever occurs first. Nevertheless, such withdrawal must still be effectuated through a Commission order, and affected employees must be notified. In *Wean-Pori, Inc.*, 7 OSHC 1545 (1979), a motion to withdraw a citation was approved, since the employees received notice of and had an opportunity to be heard on the withdrawal motion.

The attempt to amend a citation which is the subject of dispute does not amount to a withdrawal of the citation. *Noblecraft Industries Inc. v. Secretary of Labor,* 614 F.2d 199, 7 OSHC 2057 (9th Cir. 1980).

8. Dismissal of Citations

Normally, motions requesting the dismissal of a citation will not be granted as a sanction for procedural violations by the Secretary of Labor such as: (1) failure to respond to interrogatories, *Circle T Drilling Co.*, 8 OSHC 1681 (1980); (2) failure to prosecute (unless in the face of a clear record of delay or contumacious conduct), *Ralston Purina Co.*, 7 OSHC 1730 (1979); (3) failure to timely file a prehearing conference report, *Duquesne Light Co.*, 8 OSHC 1218 (1980); or (4) failure to follow internal procedures or memoranda *(e.g.,* FOM), *Northwest Marine Iron Works,* 7 OSHC 1398 (1979). Several factors determine whether the dismissal sanction is applicable for failure to follow the Act or the Commission's rules and procedures and include: (1) the amount of time involved; (2) whether the party's claim lacks substance; (3) whether a party is prejudiced (the most important factor); (4) whether there is a clear record of delay; (5) whether the party exhibited contumacious conduct; (6) whether there is a showing of willful default on the part of the party; (7) the rights of the other party to be free from costly and harassing litigation; and (8) the time and energies of the court and the right of would-be litigants awaiting their turn to have other matters resolved. Lesser sanctions should be considered first, since there is a policy expressed in the law of deciding cases on their merits. See *National Roofing Corp.*, 8 OSHC 1916 (1980)—motion to dismiss notice of contest denied; *National Industrial Constructors, Inc. v. OSHRC*, 583 F.2d 1048, 6 OSHC 1914 (8th Cir. 1978); *Marshall v. C F & I Corp.*, 576 F.2d 809, 6 OSHC 1543 (10th Cir. 1978).

The assertion that a citation must be dismissed because an inspector does not personally observe the violation has been rejected, even though the FOM requires this, because the FOM is "merely a tool for guidance." *Combustion Engineering Inc.*, 5 OSHC 1943 (1977). Moreover, unless prejudice is shown, the failure to discuss an alleged violation at a closing conference is a technical and harmless violation of procedure. *Kast Metals Corp.*, 5 OSHC 1861 (1977).

C. INFORMAL CONFERENCE

Subsequent to the receipt of the citation and *prior* to the expiration of the fifteen-working-day period in which an employer can file a notice of contest to the citation, employers (or employee representatives) have a right to request an informal conference with the OSHA Area Director to discuss the propriety of the citation, penalty, or abatement, to receive an explanation of the cited violations and standards, discuss ways to abate violations, and obtain answers to other questions. 29 CFR 1903.19. Upon request, an informal conference will be held and, depending upon its outcome, the citation, period of abatement, and/or the penalty may be amended or settled. An informal conference may not be requested, however, once the matter is in contest, and it does not extend the fifteen-working-day time limit for contesting the citation. *Spartan Construction Co.*, 4 OSHC 1988 (1977). Moreover, holding of an informal conference is within the discretion of the Area Director. *Flemming Foods of Nebraska, Inc.*, 6 OSHC 1233 (1978).

The settlement of any issue at an informal conference is subject to the rules of procedure prescribed by the Review Commission. According to OSHA Instruction CPL 2.23A, whenever an informal conference is requested by either the employee or employer representative, *both* parties shall be afforded the opportunity to fully participate. If "delicate matters" are brought up, separate or private discussions may be held. The Area Director may not amend or withdraw a citation or penalty without first obtaining the views of the employee representative. When the employee representative disagrees with the proposed amendment or withdrawal of a citation, the proposed position may be appealed. However, as the fifteen-working-day period for filing a notice of contest may affect discussions regarding an amendment or withdrawal of a citation, telephonic communication shall be utilized in order to expedite the resolution of the matter under consideration. The authority of the Area Director to amend or withdraw citations and penalties is limited to the time before the notice of contest is received in the OSHA office. Any party may be represented by counsel at such conference or later in the actual contest of citations. *Yaffe Iron & Metal Co.,* 5 OSHC 1057 (1977). No such conference shall operate as a stay of the fifteen-working-day period for filing a notice of intention to contest as prescribed in 29 CFR 1903.17.

After a notice of contest is filed, the case will be referred to the Solicitor of Labor, and the Area Director will not be able to grant a request for an informal conference.

D. NOTICE OF CONTEST

The Secretary of Labor may issue a citation for an alleged violation of the Act under Section 9(a), a notice of failure to abate a prior violation under Section 10(b), a notice of proposed penalties under Section 10(a), and a *de minimis* notice if a violation is not directly or immediately related to safety or health under Section 9(a). An employer has fifteen *working* days from receipt of the citation or notice and notification of proposed penalty to file a *notice of contest* with the OSHA Area Director, in writing. Section 10(a) of the Act. This language has been construed to require that the notice of contest be postmarked within the time specified. 29 CFR 1903.17(a); *United States of America v. B & L Supply Co.,* 486 F. Supp. 26, 8 OSHC 1125 (N.D. Tex. 1980); *Electrical Contractor's Associates, Inc.,* 2 OSHC 1627 (1974).

If an employer does not timely contest the citation it becomes a final order of the Review Commission not subject to review. Section 10(a) of the Act; *Penn Central Transportation Co.,* 2 OSHC 1379 (1974), *aff'd,* 535 F.2d 1249, 3 OSHC 2059 (4th Cir. 1976); *Brennen v. Winters Battery Manufacturing Co.,* 531 F.2d 317, 3 OSHC 1775 (6th Cir. 1975). Moreover, as is discussed herein, although it is possible to amend an answer to include such matters as affirmative defenses which were not originally raised, it appears that a notice of contest is not a pleading subject to amendment, and that the failure to contest an item in a citation may not be cured by an amendment after the 15 working day notice of contest period has expired. See *Mississippi Valley Erection Co.,* 1 OSHC 1527 (1973).

The notice of contest should indicate that the employer intends to contest part or all of the citation, proposed penalty, or the abatement time allowed for correction of alleged violations. Section 10(a) of the Act. Citations, abatement, and penalties become enforceable, final orders of the Commission, not subject to review by any court or agency, if an employer fails to file a timely notice of contest. *Marshall v. Sun Petroleum Products Co.,* 622 F.2d 1176, 8 OSHC 1422 (3d Cir. 1980). Section 10(b) of the Act outlines an identical finality provision for notices alleging a failure to correct a violation within the prescribed abatement period. The filing by an employer of a notice of contest in good faith and not solely for purposes of delay or avoidance of penalties suspends the abatement period. In other words, the time given for correcting the violation alleged in the citation does not begin to run until the Commission finally determines any of the alleged violations that are contested.

1. Form of Notice

No special pleading is required for filing the notice of contest, but it must be properly filed, and it should be carefully *written* to make it clear what is being put into contest. 29 CFR 1903.17(a).

The requirement that the notice of contest be in writing has not been an absolute requirement in every case, depending upon the circumstances. In *Wood Products Co.,* 4 OSHC 1688 (1976), the employer met with the Area Director within the fifteen-day period and orally communicated his intent to contest the citation. The written notice was thereafter inadvertently not submitted. The Commission upheld the oral notice of contest, saying that the employer did not act in bad faith or a dilatory manner, but, rather, acted in a reasonable manner. See also *Florida Power & Light Co.,* 5 OSHC 1277 (1977).

In *Keppels, Inc.,* 7 OSHC 1442 (1979), the Commission overruled both *Wood Products* and *Florida Power & Light,* holding that:

> Having re-examined our prior decisions regarding oral notices of contest we hold that a notice of contest must be in writing.
>
> Section 10(a) of the Act, 29 U.S.C. §659(a), reads as follows:
>
> * * *
>
> Although the statute does not expressly require that a notice of contest to the Secretary be in writing, both the Secretary of Labor in a regulation published at 29 CFR 1903.17(a) shortly after the Act's effective date and Commission Rule 32, 29 CFR 2200.32, contemplate a writing. These rules interpret section 10(a), 29 U.S.C. §659(a), of the Act. The Act, specifically section 9(a), 29 U.S.C. §658(a), requires the Secretary to issue citations in writing. Section 10(a), 29 U.S.C. §659(a), requires the Secretary to notify respondent of the proposed penalty, if any, by certified mail. Similarly, under section 10(a) an affected employee must file a notice with the Secretary if he wishes to appear under section 10(c), 29 U.S.C. §659(c). Equal treatment would require that a notice of contest by an employer be in the form of a writing placed with the area director. Failure to file such a written notice of contest within the fifteen working day period is not merely a violation of the Secretary's rule, but operates to preclude Commission jurisdiction.

Although there are no specific procedural requirements for a notice of contest, the notice should be clear enough to indicate that the employer is contesting the citation *or* the proposed penalty or both and not just contesting the penalty or requesting an informal conference (unless that is what he desires). If only the penalty is in contest, the citation becomes a final order after the fifteen-working-day period has expired. *Brennan v. Bill Echols Trucking Co.,* 487 F.2d 230, 1 OSHC 1398 (5th Cir. 1973). The Secretary of Labor bears the burden of proving that a citation has become a final order by virtue of the employer's failure to contest. *J. A. McCarthy, Inc.,* 4 OSHC 1358 (1976). If the citation is contested, it serves to put in issue both the citation and penalty. *Danco Construction Co.,* 3 OSHC 1114 (1975).

Example: There may be two citations, and if the employer wishes to contest only one of them, the citation being contested should be clearly identified. Or there may be six different items alleged as violations in a citation, and if the employer only wishes to contest items 3, 4, and 6, they should be identified. If the employer wishes to contest the total amount of the proposed penalty or only the amount for one citation or for specific items on one citation or only the abatement period for some or all of the violations alleged, it should be stated. See 29 CFR 1903.17(a); *Tapco, Inc.*, 6 OSHC 1894 (1978); *Arena Construction Co. v. Marshall*, —F. Supp.—, —OSHC—(S.D. N.Y. 1978); *John W. McGrath Corp.*, 2 OSHC 1369 (1974)—uncontested item in citation results in item becoming final order despite claim of typographical error. See also *United States of America v. B & L Supply Co.*, 486 F. Supp. 26, 8 OSHC 1125 (N.D. Tex. 1980)—a statement that a penalty was unfounded, although it did not suggest an intent to contest the citation, did constitute a notice of contest to the penalty. The notice of contest need not, however, be in the form of a formal pleading. In *Tice Industries*, 2 OSHC 1490 (1975), a handwritten note on the citation itself was sufficient. If the employer does not contest any of the items listed in a citation or the proposed penalty, the abatement requirements must be fully met, and the employer must pay whatever penalty is proposed for the items in the citation. Payment should be made according to instructions issued by the Secretary of Labor.

2. Posting

The notice of contest must be posted at the same place the citation was posted pursuant to Commission Rule 7(g). Certification of posting must be submitted to the Commission pursuant to Commission Rule 7(d). Failure to do so *may* cause the notice to be subject to a motion to dismiss. *Carbitow Corp.*, 1 OSHC 1503 (1973); *Colonnade Cafeteria*, 7 OSHC 2234 (1980). In *Car & Truck Doctor, Inc.*, 8 OSHC 1767 (1980), evidence that an employer's failure to provide certification that a copy of notice of contest had been posted at the work site may have been due to the filing by the Secretary of a motion to withdraw the citation required the vacation of the judge's decision dismissing the notice of contest and a remand of case for purposes of ordering employer to post notice and submit proof thereof. Some commentators have stated that all parties *(e.g.*, employees and authorized employee representatives) must be notified by means other than posting, such as personal or mail service. This is not completely clear nor is it clear whether sanctions may be imposed for the failure to so notify such persons.

3. Time and Place for Filing

For purposes of determining the fifteen-working-day period for filing the notice of contest, Commission Rule 4 provides that in computing the period of time prescribed, the day from which the designated period begins to run shall not be included, *i.e.*, the day of receipt of the citation. The last day of the period so computed shall be included unless it is a Saturday, Sunday, or federal holiday, in which event the period runs until the end of the next day which is not a Saturday, Sunday, or federal holiday. "Working day" must be distinguished from "day," which means calendar day. Commission Rule 1(k). Working day is defined in Rule 1(1) of the Rules to mean "all days excepting Saturdays, Sundays or federal holidays." Employer holidays which vary from federal holidays *(e.g.,* Day after Christmas, Good Friday) are not excluded from the computation period. *B. J. Hughes, Inc.,* 7 OSHC 1471 (1979).

Example: The citation is received Monday, June 3. In computing fifteen *working* days, June 3 is not included. Begin counting days on June 4. Thus the period for filing a notice of contest ends on June 24. Saturdays, Sundays, and federals holidays are not included, since they are not working days.[6] An amendment to the citation will extend the time period for filing a notice of contest with respect to the portion of the citation so amended. An employer who does not wish to contest, but wishes to protect itself from the use of an uncontested citation in a later proceeding should consider including exculpatory language in a "notice of no contest" sent to the Area Director.

The fifteen-day period begins to run only when an appropriate corporate official has received the citation. See *Buckley & Co. v. Secretary of Labor,* 507 F.2d 78, 2 OSHC 1432 (3d Cir. 1975); *Service Specialty, Inc.,* 7 OSHC 1770 (1979). In the absence of receipt by a corporate official empowered to act on the citation and penalties, untimely filing of a notice of contest may be justified. *Norkin Plumbing Co., Inc.,* 5 OSHC 2062 (1977); *Otis Elevator Co.,* 6 OSHC 1515 (1978). This rule is based upon a concept of fairness, for if at the time of the receipt of the citation and notice of proposed penalty there is no official with authority to commit corporate funds to pay the penalty, abate the alleged violation, or contest the enforcement proceedings as set forth, then the fifteen-day period should not begin to run. The court in *Buckley* considered that notice should at the very least be sent to officials at the corporate headquarters and rejected as insufficient mailing to the superintendent in charge of the work site where the violation occurred. The Commission, however, through recent interpretation has rejected the concept of appropriate official first articulated in *Buckley* and now holds that service is proper if it is reasonably calculated to provide an employer with knowledge of the citation and proposed penalty and an opportunity to contest or abate. *B. J. Hughes, Inc.,* 7 OSHC 1471 (1979).

In *Capital City Excavating Co., Inc.,* 8 OSHC 2008 (1980), the citation was sent to the employer by certified mail and was delivered to a clerical employee. The employer's corporate secretary received it two days later and filed its notice of contest within fifteen working days thereafter. Relying on Rule 4(d) of the Federal Rules of Civil Procedure, which provides in pertinent part that

> (d) [s]ervice shall be made as follows: . . .
>
> (3) [u]pon a domestic or foreign corporation . . . by delivering a copy of the summons and of the complaint to an officer, a managing or general agent, or to any other agent authorized by appointment or by law to receive service of process . . .

and the Commission's holding in *B. J. Hughes, Inc.*, 7 OSHC 1471 (1979), that commencement of the period within which a notice of contest must be filed begins with service upon an employee who is situated so as to bring the citation to the attention of the appropriate corporate official, the administrative law judge concluded that the date of receipt by the corporate secretary was the date of effective service. The Commission held that the judge's reliance on Federal Rule 4(d) was misplaced and that *B. J. Hughes, Inc.* was misapplied:

> We have held that, when service of a citation and notice of proposed penalty is made by certified mail as provided by section 10(a) of the Act, Rule 4(d)(3) is not applicable. *Joseph Weinstein Electric Corp.*, 6 OSHC 1344 (1978). Section 10(a) of the Act provides:
>
> "If, after an inspection or investigation, the Secretary issues a citation under section 9(a), he shall, within a reasonable time after the termination of such inspection or investigation, notify the employer by certified mail of the penalty, if any, proposed to be assessed under section 17 and that the employer has fifteen working days within which to notify the Secretary that he wishes to contest the citation or proposed assessment of penalty. If, within fifteen working days from the receipt of the notice issued by the Secretary the employer fails to notify the Secretary that he intends to contest the citation or proposed assessment of penalty, and no notice is filed by an employee or representative of employees under subsection (c) within such time, the citation and the assessment, as proposed, shall be deemed a final order of the Commission and not subject to review by any court or agency."

Delivery of the citation and notice of proposed penalty to an employer's place of business and receipt by an employee who accepts delivery of certified mail constitutes service on the company as of the date of initial receipt. See *Henry C. Beck Co.*, 8 OSHC 1395 (1980). The date of the return certified receipt of a citation is normally controlling for determining the fifteen-day period, regardless of who signs for it.

In *Atlantic Marine, Inc. and Atlantic Dry Dock Corp. v. Dunlop*, 524 F.2d 476, 3 OSHC 1755 (5th Cir. 1975), the Fifth Circuit suggested that an employer should not be denied review for failure to file a notice of contest within the fifteen-day limit prescribed in the Act, if the Secretary's error, deception, or failure to follow proper procedures was responsible for the late filing. This approach has been accepted by the Commission in *Keppel's, Inc.*, 7 OSHC 1442 (1979); *B. J. Hughes, Inc.*, 7 OSHC 1471 (1979); *National Roofing Corp.*, 8 OSHC 1916 (1980); and *Seminole Distributors Inc.*, 6 OSHC 1894 (1978)—OSHA failed to inform the employer it had fifteen days to contest the citation. See also *United States of America v. B & L Supply Co.*, 486 F. Supp. 26, 8 OSHC 1125 (N.D. Tex 1980)—equitable tolling of the fifteen-day time period was permitted due to misleading statements of the inspector, as consistent with the purpose of the Act.

If, however, the employer is responsible for the failure to receive timely notice, *(e.g.*, refusal to accept service), the fifteen days would begin to run with the employer's action. *Service Specialty, Inc.*, 7 OSHC 1770 (1970); 29 CFR 1903.18(c).

In *Norkin Plumbing Co., Inc.*, 6 OSHC 1539 (1978), the Commission stated that the absence of a corporate official to receive the citation did not justify the late filing of a notice of contest, since there was implied authority on others to act. Similarly, in *Womack Construction Co., Inc.*, 6 OSHC 1125 (1978), the citation and notice of proposed penalty, which were received but lost by an employee who was not a corporate officer and had no authority to contest the citation, did not excuse the untimely filing, since there were corporate officers available when the citation was received.

Although minor instances of noncompliance with the Commission's Rules of Procedure will be permitted in exceptional circumstances pursuant to Rule 108, this rule cannot be used to waive or extend the fifteen-working-day period for filing a notice of contest, since it is a statutory not a procedural requirement. See *American Airlines, Inc.*, 2 OSHC 1326 (1974). When a notice of contest is untimely filed, the Commission is without jurisdiction to review the citation, which is final by operation of law pursuant to Section 10(a) of the Act. *Kerr-McGee Chemical Corp.*, 4 OSHC 1739 (1976); *City Mills Co.*, 5 OSHC 1129 (1977).

Good-faith mistakes do not warrant extension of the contest period, even if there is no prejudice to the Secretary of Labor. For example, a notice of contest must be filed with the Area Director of the OSHA office that mailed the citation. The Area Director's name and address will be listed on the citation. The notice of contest must *not* be sent to the Review Commission. In *Fitchburg Foundry, Inc.*, 7 OSHC 1516 (1979), a notice of contest was filed with the Review Commission rather than the Area Director and was not accepted.

In *Revco Steel Corporation,* 8 OSHC 1235 (1980), a petition to re-open a default judgment which occurred because corporate mismanagement led to a citation becoming a final order, due to the failure to file a notice of contest, was disallowed. The employer had filed the request pursuant to Rule 60(b) of the Federal Rules of Civil Procedure, which provides relief from mistake, inadvertence, surprise, visible neglect, or other good reason.[7] The Commission did not accept this excuse, saying that a petitioner must show why he was justified in failing to avoid mistake and inadvertence; mere carelessness does not justify relief. Neglect must be shown to be excusable. The Commission in other cases, such as *Monroe & Sons, Inc.*, 4 OSHC 2016 (1977) *aff'd,* 615 F.2d 1156, 8 OSHC 1034 (6th Cir. 1980), has permitted a case to be reinstated if the requirements of the Rule 60(b) were satisfied.

4. Scope of the Notice

It is possible, in the contest of a citation, for the employer to challenge the violation, the penalty, the abatement period, or a combination of all three. Uncontested items, however, will become final orders not subject to review. A notice of contest to either the citation or abatement period will extend the abatement period (and relieve an employer of his obligation to pay the penalty) until a final Commission order is issued. The abatement period will not be extended, however, if the notice of contest was filed only for delay or if only the penalty is contested. Moreover, if only the penalty is contested, the violation is not in contest and must be abated, since an employer is in effect waiving his right to contest the violation. It must be noted, however, that a notice of contest, for example, to an abatement period, is not a notice of contest to the penalty or to the violation and care should be taken to insure that the proper aspects of the citation have been put in contest. Moreover, a notice of contest to only part of a citation does not relieve the employer of its duty to comply with the uncontested aspects of the citation.

In order to avoid injustice, especially with respect to *pro se* employers, the Commission has attempted to liberally construe ambiguous employer communications. Specifically, notices of contest are to be liberally interpreted and construed in their most favorable light. *Data Electric Company, Inc.*, 5 OSHC 1077 (1977); *Seaboard Coast Line RR Co.*, 3 OSHC 1758 (1975). To the average employer, the penalty is what strikes home; it symbolizes the violation. Often a challenge to the proposed assessment of a penalty was intended also as a contest to the validity of the underlying violation. Similarly, many employers do not contest other-than-serious violations, even though they can form the basis for repeated or willful violations since no fine is imposed. Thus a knowing, informed contest of only the penalty is rare (e.g., a *limited* notice of contest). Therefore the Commission will look to an employer's intent. An employer should attempt to avoid this situation by adequately explaining in his notice what is actually being contested (the violation, the penalty, the abatement). It should date its communication and then sign it. Nothing more is required. This is important since it appears that a notice of contest is not generally subject to amendment as are other "pleadings." See *Tice Industries*, 2 OSHC 1489 (1975).

In *Turnbull Millwork Co.*, 3 OSHC 1781 (1975), an employer was not aware of the distinction between a contest to the penalty and a contest to the citation. The Commission, modifying its earlier decision in *Florida Eastcoast Properties, Inc.*, 1 OSHC 1532 (1974), held that a notice of contest to the penalty only would be construed to include a contest to the citation itself, *if* it was shown at the time the notice of contest was filed that the employer actually intended to contest the violation. The employer's true intent could, for example, be indicated by amendment in a later filed pleading. Such amendment should be made pursuant to Federal Rule of Civil Procedure 15(a) which provides that a party may amend his pleading once as a matter of course at any time before a responsive pleading is served, or by leave of the court or with the written consent of the adverse party and that such leave shall be given freely when justice so requires. *Philadelphia Coke Division, Eastern Associated Coal Corp.*, 2 OSHC 1171 (1974); *Eastern Knitting Mills, Inc.*, 1 OSHC 1677 (1974); *State Home Improvement Co.*, 6 OSHC 1249 (1977). *Marshall v. Haugan, d/b/a Gil Haugan Construction Co.*, 586 F.2 1263, 6 OSHC 2067 (8th Cir. 1978); *Penn-Dixie Steel Corporation v. OSHRC*, 553 F.2d 1078, 5 OSHC 1315 (7th Cir. 1977)—employer's intent was to contest both penalty and underlying citation. But see *F. H. Sparks of Md., Inc.*, 6 OSHC 1356 (1978)—the employer understood the distinction between contesting the citation and penalties and chose to contest only the penalties; *Juhr & Sons*, 3 OSHC 1871 (1876)—the employer's contest was limited solely to the penalties; *Brennan v. Bill Echols Trucking Co.*, 487 F.2d 230, 1 OSHC 1398 (5th Cir. 1973).[8] In order to determine what aspects of the citation were "in contest," the Commission would look at subsequent pleadings to determine the employer's intent; it made no difference whether these pleadings were filed *pro se* or by counsel. See *Collator Corp.*, 3 OSHC 2041 (1976); *Nielsen Smith Roofing & Sheet Metal Co.*, 4 OSHC 1765 (1976); *Acme Metal, Inc.*, 3 OSHC 1932 (1976); *Superior Boat Works, Inc.*, 4 OSHC 1764 (1976). Instances of OSHA deception or failure to provide adequate instructions, if it causes confusion would also be a basis to open up the contest. See *Haugan, d/b/a Gil Haugan Construction Co. v. Marshall*, 586 F.2d 1263, 6 OSHC 2067 (8th Cir. 1978).

In *Gilbert Manufacturing Co., Inc.*, 7 OSHC 1611 (1979), the Commission, in a partial reversal of precedent, held that an employer's letter that requests only an extension of the abatement date will be treated as a petition for modification of abatement[9] under Commission Rule 34. The Commission further concluded that "there should not be a distinction in the manner that employer's requests for extensions of abatement dates should be processed depending on whether the requests are filed within the fifteen day contest period." *Gilbert* overruled earlier decisions in *Philadelphia Coke Division, Eastern Associated Coal Corp.*, 2 OSHC 1171 (1974), and *Eastern Knitting Mills, Inc.*, 1 OSHC 1677 (1973), to the extent that they held that a request solely for an extension of the abatement date filed within the fifteen-day contest period, put in contest and gave the Commission jurisdiction over the entire citation.

The Commission recognized in *Gilbert*, however, that an initial request for relief filed by an employer may not always be clear as to whether the employer's intent was to seek modification of the abatement date alone or also to contest the citation. For example, in *Maxwell Wirebound Box Co., Inc.*, 8 OSHC 1995 (1980), the employer filed a document entitled "Notice of Intention to Contest/Petition for Modification of Abatement." While that notice appeared to contest only the abatement date, it was possible to infer, through the use of the terms "contest," "alleged violations," and "any noise deficiencies" in the accompanying letter, that the employer also intended to contest the merits of the citation. In its answer to the complaint, the employer clearly stated that it was "contesting the citation in total and in the alternative requesting a delay of the [abatement] date. . . ." It also specifically denied in its answer that it was in violation of the Act.

The Commission held that the employer's initial request for relief was an ambiguous one as envisioned by *Gilbert* and that it required that it decide whether the document was a petition for modification of abatement or a notice of contest of the entire citation. Continuing to follow that portion of *Philadelphia Coke Division* not overruled by *Gilbert,* that requests for relief filed within the fifteen-day contest period be given a liberal interpretation, the Commission held that the employer's initial request for relief must be read as a contest to the citation as well as a request for extension of time for abatement. See also *Bushwick Company, Inc.*, 8 OSHC 1654 (1980); *Gil Haugan, d/b/a Haugan Construction Co.*, 5 OSHC 1956 (1977), *aff'd*, 586 F.2d 1263, 6 OSHC 2067 (8th Cir. 1978). The Commission further stated in *Maxwell Wirebound Box, Inc., supra,* that:

> Under the *Turnbull* [*Millwork*] test, therefore, we consider the intent of the employer at the time of filing its original request for relief as clarified by that em-

> ployer's later pleadings. We now hold that the *Turnbull* test shall apply to initial requests for relief which appear to be limited to the time for abatement and would normally proceed under Commission Rule 34 pursuant to *Gilbert.* Thus, an initial request for relief solely directed to the time allowed for abatement and filed within the fifteen day contest period will be construed as a notice of contest to the citation where later 'pleadings' demonstrate that Respondent's intent at the time of filing the initial request was to contest the citation.
>
> In this case, Respondent indicated in its answer to the complaint that its intention at the time it filed its notice of contest was to contest the citation and in the alternative to request a modification of the abatement date. We conclude, therefore, that the merits of the citation, as well as the abatement dates, are in issue. Accordingly, this case will be remanded to the judge for a hearing on the merits.

Until that point in time when the Secretary of Labor adopts the suggestion of the Fifth Circuit that a citation be accompanied by a reply form on which an employer can check on appropriate box to indicate what aspect of the citation, penalty, or abatement he wishes to contest, it will be necessary to review pleadings filed subsequent to the notice of contest for evidence of intent in cases of ambiguous communications. See *Brennan v. Bill Echols Trucking Co.,* 487 F.2d 230, 1 OSHC 1398 (5th Cir. 1973).

5. Employee Notice of Contest

If an employer files a notice of contest, or even if the employer does not file a notice of contest, affected employees or their representatives may contest the reasonableness of the period of time specified in the citation for the abatement of the violation. Section 10(c) of the Act; 29 CFR 1903.17(b). Employees must file their notice of contest within fifteen working days from the date the citation was received (and posted) by the employer. An employee notice of contest need only state that the affected employee(s) and/or designated representatives wish to contest the reasonableness of the abatement period. Employees who file a notice of contest must provide a copy thereof to the employer for posting. Commission Rule 7(1).

Within ten days of receipt of the employee notice of contest, OSHA must file a statement of reasons explaining why the abatement period is reasonable. An employee response must thereafter be made within ten days stating why the period is unreasonable. Commission Rule 35. The Secretary of Labor, as the complainant in employee notices of contest, bears the burden of establishing that the abatement period allowed the employer to correct the cited condition is not unreasonable. *Kawecki-Berylco Industries, Inc.,* 1 OSHC 1210 (1973). The affected employees must for their part attempt to prove that the abatement period is unreasonable. Where a case involves only an employee contest to the reasonableness of the abatement period specified in the citation, the time allowed to the employer to correct the violation is not extended while the case is in progress. Employers may elect to participate as parties on employee contests.

Interpreting Section 10(c) of the Act relative to employee contests, the Review Commission stated, in *Automobile Workers, Local 588,* 4 OSHC 1243

(1976), that an employee's right to contest a citation is limited to contesting the reasonableness of the period of time which is fixed in the citation for abatement; it may not challenge the means or the adequacy of abatement. In *UAW Local 588 v. OSHRC,* 577 F.2d 607, 5 OSHC 1525 (7th Cir. 1977), the Court of Appeals affirmed the Review Commission's finding and stated that the Act gives an employee/union the right to challenge the Secretary of Labor's citation only on the basis that the abatement period provided therein is unreasonable. The Third Circuit, in *Marshall v. Sun Petroleum Products Co.,* 622 F.2d 1176, 8 OSHC 1922 (3d Cir. 1980), agreed that affected employees have no right to be heard on matters other than the reasonableness of the abatement period. See also *United States Steel Corp.,* 4 OSHC 2001 (1977); *Koppers Company, Inc.,* 5 OSHC 1306 (1977); *Southern Bell Telephone & Telegraph Co.,* 5 OSHC 1405 (1977); *Geuder Paeschke & Fry,* 5 OSHC 1417 (1977). *Contra, Farmers Export Co.,* 8 OSHC 1655 (1980). As is discussed, *infra,* the Commission has found a way around this restriction on employee challenges to matters other than the abatement period by granting wide latitude on employee participation in settlement agreements. *American Cyanamid Company,* 8 OSHC 1346 (1980).

Section 10(c) of the Act also provides that affected employees or their representatives, even though they have not filed a notice of contest, may participate as parties to hearings. Commission Rule 20 implements this provision by authorizing party status upon request. Party status must, however, be implemented upon the commencement of the hearing. Employee contests are treated as expedited proceedings under Commission Rule 101.

6. Docketing

Once a notice of contest is filed, the Secretary of Labor must advise the Commission, and the matter moves from the jurisdiction of OSHA to the jurisdiction of the Review Commission, and the rules, regulations, and procedures of the Commission are then applicable. Section 10(c) of the Act. The Secretary of Labor, within seven days of receipt of a notice of contest, must transmit the original to the Commission, together with copies of all relevant documents. Commission Rule 32. Normally, unless there is a showing of prejudice, a citation will not be vacated if the Secretary of Labor does not meet the seven-day requirement for transmittal of a notice of contest. *Rollins Outdoor Advertising Corp.,* 4 OSHC 1861 (1976); *Brennan v. OSHRC (Bill Echols Trucking Co.),* 487 F.2d 230, 1 OSHC 1398 (5th Cir. 1973). Once transmittal is received, the Review Commission takes jurisdiction of the case and will shortly thereafter assign it a docket number and send a notice of receipt of the matter and instructions to the employer. It will indicate that all future documents and communications relating to the case should include the docket and the region numbers, and be addressed to the Commission. A copy of the Commission's rules, a copy of the notice of contest, a notice notifying affected employees of the matter,[10] and a return postcard supplied by the Commission, which must be returned within forty-eight hours from receipt, are sent to the employer. The employer must thereafter post the notice of contest and certify that the notice to employees supplied by the Commission and the notice to contest have been posted at each place that the OSHA citation is required to be posted. Commission Rule 7; *Car & Truck Doctor, Inc.*, 8 OSHC 1767 (1980). The reason that affected employees must be notified is to apprise them of the case in the event they wish to elect party status. After a case is docketed with the Commission it will be assigned to an Administrative Law Judge who will be responsible for the case through hearing and decision. Commission Rule 66 sets forth the powers and duties of the judge.

7. Withdrawal/Dismissal

Normally, a party may withdraw its notice of contest or any portion thereof at any stage of a proceeding prior to final order, subject to the approval by the Commission, and certification that abatement has been or will be accomplished, assurance of compliance, payment of penalty and notification of affected employees. *Dawson Brothers, Inc.*, 1 OSHC 1024 (1972); Commission Rule 100a; *ITO Corp. of New England, John T. Clark & Son of Boston, and Terminal Services, Inc.*, 540 F.2d 543, 4 OSHC 1574 (1st Cir. 1976); *Hartwell Excavation Co.*, 3 OSHC 1196 (1975). Withdrawal is made by motion, which must include a statement that a promise of another party has not led to the withdrawal. Commission Rule 100a. If such a promise has led to the withdrawal, Rule 100, which establishes settlement procedure, must be followed. Withdrawal will be approved if notice is served on affected employees and no objections are received, payment of the proposed penalty has been made, and abatement has been or will be accomplished. Employees must be notified of the withdrawal by posting or by service of a copy thereof, depending upon whether or not they are represented, and they will have the right to object to the withdrawal, assuming they have elected party status. *IMC Chemical Group, Inc.*, 6 OSHC 2075 (1978); *R. P. Drywall, Inc.*, 3 OSHC 1186 (1975). After withdrawal, the citation becomes a final, nonrenewable order. *Matt J. Zaich Construction Co.*, 1 OSHC 1225 (1973).

The failure of an employer who has contested a citation to file an answer and appear at the hearing on the matter or comply with a discovery order may result in the dismissal of its notice of contest, unless in the interest of fairness and justice it would be inappropriate to do so and good cause is shown for his failure to appear. See *Dickherber (Kustom Kar Kare)*, 6 OSHC 1972 (1978); *Duquesne Electric & Mfg. Co.*, 5 OSHC 1843 (1977); *Capital Dredge & Dry Dock Corp.*, 1 OSHC 1066 (1972); Commission Rule 62.

In *Browar Wood Products Company, Inc.*, 7 OSHC 1165 (1979), a complaint and answer were duly filed, but the employer failed to file proof of its compliance with the Commission's requirement that its notice of contest be *posted* for the information of affected employees. Consequently, Browar's notice of contest was dismissed. Three days after the judge's order became final, Browar filed an undated motion asking the Commission to review the dismissal because it was "based on a technicality which, in view of the overall picture, is minor and has hurt no one." In explanation, Browar averred that it had posted its notice of contest but through "inadvertence and/or human error" no proof of the posting was ever sent, as required by Commission Rule 7.

Relying on the fact that there was a *technical* procedural requirement which was not complied with, that it failed to certify posting through inadvertance or human error, that the employer did submit, although late, a sufficient proof of proper posting of its notice of contest, that the employer appeared *pro se* throughout the proceedings, that the notice of contest was posted at or around the place where citation was posted, that the employer properly answered the complaint that employees were notified of their rights to participation and that the Secretary of Labor had not been prejudiced by the failure to file proof of posting, and that the employer was apparently a small business and may not have been aware of the technicalities with respect to process, the Commission vacated the judge's order and reinstated the case pursuant to Rule 60(b) of the Federal Rules of Civil Procedure. *Cf. Monroe & Sons, Inc.*, 4 OSHC 2016 (1977).

E. PETITION FOR MODIFICATION OF ABATEMENT

Challenges to abatement periods and methods may, of course, be raised during contests to the citation. In fact, an employer may file a notice of contest to the abatement itself, for example, if he disagrees with the abatement suggested or is not clear as to what would constitute acceptable abatement. *Gurney Industries, Inc.*, 1 OSHC 1218 (1973); *Ricardo's Mexican Enterprises*, 4 OSHC 1081 (1976). There may, however, be an employer who does not wish to enter a contest to the citation, the abatement, or the penalty, but nonetheless feels that he is unable to abate the alleged hazard within the period specified in the citation or feels that the period provided is unreasonable. In such a case a petition for modification of the abatement period (PMA) may be appropriate. An employer may also file a petition for modification of abatement date when he has made a good faith effort to comply with the abatement requirements of the citation and such abatement can not be completed because of factors beyond his reasonable control. 29 CFR 1903.14(a)(a). In filing a PMA, it should be remembered that the employer has the burden of proving he made a good-faith effort to comply with the abatement requirement of the citation but that he was not able to do so because of factors beyond his reasonable control. Rule 34(d)(3); *Midwest Steel Fabricators, Inc.*, 5 OSHC 2068 (1977); *Corhart Refractories Co.*, 6 OSHC 1224 (1977). An employer should be aware that, unlike the notice of contest, a PMA does *not* stay the abatement period. Thus in a PMA, since no contest is filed, the citation becomes a final order after the expiration of the fifteen-working-day period and the abatement period begins to run and is binding on the employer. It would appear, therefore, that a notice of contest affords greater relief and may be the preferred manner of proceeding.

A PMA is filed with the Area Director who issued the citation in the same manner as a notice of contest, must be filed *no later* than the end of the next working day following the date initially set for abatement, and should state why abatement cannot be completed within the given time. 29 CFR

1903.14(a)(c); Commission Rule 34(c). A later filed petition must be accompanied by a statement of exceptional circumstances explaining the delay. The Commission decision in *Newspaper Printing Corp.*, 5 OSHC 1810 (1977), states that the extension of the abatement period must be requested in writing and cannot be done orally.

In *Gilbert Manufacturing Co., Inc.*, 7 OSHC 1611 (1979), the Commission overruled established precedent, which had held that a request for an extension of an abatement date filed within the fifteen-day contest period gave the Commission jurisdiction over the entire citation, *Philadelphia Coke Division, Eastern Associated Coal Corp.*, 2 OSHC 1171 (1974); *Eastern Knitting Mills, Inc.*, 1 OSHC 1677 (1974). It articulated in *Gilbert* a new rule which holds that a request for an extension of an abatement date, even if filed within the fifteen-day period for contesting a citation or proposed penalty, should be treated as a petition for modification of abatement and processed under Commission Rule 34. See *KL Spring & Stamping Corp.*, 7 OSHC 1651 (1979). If, however, the employer's request is ambiguous such that there was a genuine question as to whether the employer intended to contest the citation or proposed penalty, or if the employer disputes the Secretary's interpretation of its request, the matter is to be referred to the Commission for appropriate action. See *Gil Haugan Construction Co.*, 5 OSHC (1977), *aff'd*, 586 F.2d 1263, 6 OSHC 2067 (8th Cir. 1978); *Bushwick Commission Co., Inc.*, 8 OSHC 1653 (1980).

The regulations, the FOM, and Commission Rule 34 specify the substantive aspects of a PMA, in that it must be in writing and the following information must be included and be proved by the employer: (1) a description of all steps taken by the employer to achieve compliance, the dates of such action, and a description of the "good-faith" effort to achieve compliance taken during the prescribed abatement period, *Mueller Brass Co.*, 8 OSHC 1776 (1980); (2) the specific additional abatement time necessary to achieve compliance; (3) the reasons such additional time is necessary, including, for example, the unavailability of professional or technical personnel, materials, and equipment, or because necessary construction or alteration of a facility cannot be completed by the original abatement date *(i.e.,* failure to comply is due to factors beyond employer's control), *Canton Drop Forging & Manufacturing Co.*, 7 OSHC 1513 (1979); *Marshfield Steel, Inc.*, 7 OSHC 1741 (1979); (4) all available interim steps have been or are being taken to safeguard employees against the cited hazards during the abatement period, that is, the employer has shown diligence, *Florida Power & Light Co.*, 5 OSHC 1277 (1977); and (5) a certification that a copy of the PMA has been posted and, if appropriate, served on the authorized representative of affected employees as well as the date on which service and/or posting is completed. (When an abatement period is challenged by means of notice of contest, a less focused showing is required; the employer must only establish the unreasonableness of period.) *Kawecki-Berylco Industries, Inc.*, 1 OSHC 1210 (1973).

A copy of the PMA must be posted in "a conspicuous place where all affected employees will have notice" thereof or near such location where the violation occurred and must remain posted for a period of ten working days. Rule 34(c)(1); 29 CFR 1903.14(a)(c)(1); *Auto Bolt & Nut Co.*, 7 OSHC 1203

(1979). In *Aspro, Inc., Spun Steel Div.*, 6 OSHC 1980 (1978), the Commission set aside a settlement agreement modifying an abatement date, since there was inadvertent noncompliance with the posting requirements which deprived affected employees of their rights of objection.

By virtue of the Commission's general rules on service and notice of pleadings and other documents, it seems clear that notification of the filing of a PMA must also be given to every other party or intervenor, such as "affected employees" and/or their "representatives." Commission Rule 7; 29 CFR 1903.14 (a)(c)(1). An "affected employee" is defined in the Commission Rules of Procedure as "an employee of a cited employer who is exposed to the alleged hazard described in the citation, as a result of his assigned duties. . . ." Commission Rule 1(e). "Representative" is defined as "any person, including an authorized employee representative, authorized by a party or intervenor to represent him in a proceeding. . . ." Commission Rule 1(h). "Authorized employee representative" means a labor organization which has a collective bargaining relationship with the cited employer and represents the affected employees. Commission Rule 1(g). If affected employees are represented, service of the PMA may be made on the representative by mail or personal delivery. Commission Rules 7(b), (c) and (f). If affected employees are not represented by an authorized employee representative, service may be made by posting the PMA at the place where the citation is required to be posted. Commission Rule 7(g). A notice informing affected employees of their rights to party status and of the availability of pleadings for inspection and copying shall also be posted. The notice, which will be provided by the Commission after the PMA is docketed, states *inter alia* that:

> The reasonableness of the period prescribed by the Secretary of Labor for abatement of the violation has been contested and will be the subject of a hearing before the Occupational Safety and Health Review Commission.

(Authorized employee representatives, if any, must be served with a copy of this notice).

Affected employees or their representatives may file an objection in writing to the PMA with the Area Director within ten working days from notification of the petition. 29 CFR 1903.14(a)(c)(2). The right of objection carries out the mandate of Section 10(c) of the Act, which permits employees to challenge the reasonableness of abatement in a PMA proceeding. If the employees or their representatives fail to file objections within ten working days, it shall constitute a waiver of any further rights to object to the PMA. 29 CFR 1903.14(a)(c)(2); Rule 34(c)(2).

The Secretary of Labor has the authority to approve "uncontested" petitions for modification of abatement, and uncontested petitions become final orders pursuant to Sections 10(a) and (c) of the Act. Commission Rule 34; 29 CFR 1903.14(a)(c)(3). However, the Secretary may not exercise his approval power until the expiration of fifteen working days from the date the petition was posted. *Gilbert Manufacturing Co.*, 7 OSHC 1611 (1979); *Noranda Aluminum v. OSHRC*, 593 F.2d 811, 7 OSHC 1135 (8th Cir. 1979). The primary factor in determining whether a PMA will be granted is whether the employer exhibited good faith in seeking to meet the initial abatement period.

If a petition is objected to by the Secretary of Labor or affected employees (or their representative), the petition, the citation, and the objections must be transferred to the Commission within three working days after the expiration of the fifteen-working-day period. 29 CFR 1903.14(a)(c)(4)(d); Commission Rule 34(c)(4). The matter will then be docketed. Within ten working days after receipt of notice of docketing, each objecting party must file a response, setting forth the reasons against the granting of an abatement date different from that set forth in the petition. The employer must establish that abatement cannot be completed for reasons beyond his control. Thus, the PMA in such situations is treated and processed as a notice of contest by the Review Commission. PMAs and employee contests to the reasonableness of the abatement period are handled as expedited proceedings under Commission Rule 101. See also FOM, Ch. X.

At times the issuance of a stay of abatement period may be intertwined with the filing of an application for a variance, for an employer may need extra time for filing of the variance application. See *Deemer Steel Casting Co.*, 5 OSHC 1157 (1977); *Max Durst, Inc.*, 6 OSHC 1194 (1977).

NOTES

1. Four types of notices (citations) may be issued to employers: (1) a Section 9(a) citation for violation of the Act, its standards or regulations; (2) a Section 10(b) notification for failure to abate a prior violation; (3) a Section 10(a) notification of proposed penalties; and (4) a Section 9(a) *de minimis* notice.

2. Generally, only violations actually observed will be cited, except in cases of accident investigation or in advance notice situations, where there is evidence that the alleged violation was corrected prior to inspection. FOM, Ch. X. However, an inspector's failure to comply with the FOM and his issuance of a citation not actually observed does not invalidate a citation for violations not actually observed. *H–30, Inc.*, 5 OSHC 1715 (1977), *rev'd on other grounds*, 597 F.2d 234, 7 OSHC 1253 (10th Cir. 1979); *Combustion Engineering, Inc.*, 5 OSHC 1943 (1977).

3. Any description of a violation of the general duty clause should be completed by stating the condition or practice which constitutes the alleged violation and describing how that condition or practice violates the particular element of the general duty clause. In situations where citations are issued as a result of an accident investigation, no reference should be made to fact that injury or death resulted from such violation. The operative aspect of the particularity requirement is to give an employer fair notice of a violation and inform him that corrective action may be taken. FOM, Ch. X.

4. In cases where the employer has only contested abatement and specifies a date it believes to be reasonable, OSHA may unilaterally amend the citation to include this abatement date. FOM, Ch. X.

5. The Commission's Rules provide that, in the absence of specific applicable provisions, its procedure shall be governed by the Federal Rules of Civil Procedure. Commission Rule 2; *Pukall Lumber Co.*, 2 OSHC 1675 (1975).

6. If abatement is required before the end of the fifteen-working-day period for filing a notice of contest, the employer may in fact have less than fifteen working days to contest, or else he may face a possible exposure for failure to abate the citation, for it has been held that OSHA does have the authority to cite an employer within the

fifteen-working-day period provided for filing the notice of contest. See *Brennan v. OSHRC (Kessler and Sons Construction Company),* 513 F.2d 553, 2 OSHC 1668 (10th Cir. 1975).

7. The pertinent portions of Rule 60(b) provide that "on the motion and upon such terms as are just the court may relieve a party . . . from a final judgment, order or proceeding for the following reasons: (1) mistake, inadvertence, surprise, or excusable neglect; . . . or (6) any other reason justifying relief from the operation of the judgment."

8. In *Brennan v. Occupational Safety and Health Review Commission,* 487 F.2d 230, 1 OSHC 1398 (5th Cir. 1973), the letter from the employer read as follows: "This is to inform you that the signaling device has been installed within the three day period after the citation. We request that the penalty be abated since corrective action has been taken well within the time allotted [sic]. . . ." The court held that this letter was insufficient as a notice of contest of the citation. The court found more difficult the question of whether the letter adequately contested the penalty but deferred to the Commission's decision that the letter was sufficient in this regard, noting that "the Commission properly gave a liberal interpretation to the letter."

In *Dan J. Sheehan Co. v. Occupational Safety and Health Review Commission,* 520 F.2d 1036, 3 OSHC 1573 (5th Cir. 1975), *cert. denied,* 424 U.S. 965, 4 OSHC 1022 (1976), an employer sent this letter to the Commission: "We are not contesting your August 3, 1973 citation. We admit that the scaffold was erected too close to the energized wire. However, we would like for you to look further into the matter and possibly with this new information, lift the proposed $500.00 penalty. . . ." The letter went on to request "release" of the penalty. The court rejected the contention that this letter merely admitted the facts relating to the violation and not the violation itself and held that the letter, construed in a light most favorable to its author, "failed to contest the violation and was at most a weak attempt to contest the proposed penalty." The Commission treated the letter as constituting a notice of contest of the penalty only, and this treatment was affirmed.

9. A side effect of this ruling is that construing a request for extension of abatement as a petition to modify abatement shifts the burden of proof regarding reasonableness from the Secretary of Labor to the employer. Commission Rule 34.

10. The notice to employees specifies: "Your employer has been cited by the Secretary of Labor for violation of the Occupational Safety and Health Act of 1970. The citation has been contested and will be the subject of a hearing before the Occupational Safety and Health Review Commission. Affected employees are entitled to participate in this hearing as parties under terms and conditions established by the Occupational Safety and Health Review Commission in its Rules of Procedure. Notice of intent to participate should be sent to the Occupational Safety and Health Review Commission, 1825 K Street, N.W., Washington, D.C. 20006. All papers relevant to this matter may be inspected at:"

CHAPTER XII
The Answer and Affirmative Defenses

A. INTRODUCTION

After a citation and notice of contest have been filed, the Secretary of Labor will then serve an employer with a Complaint. Commission Rules of Procedure require an employer who has been served wtih a complaint to file an answer within fifteen days. The answer must contain a short, plain statement denying those allegations which the employer intends to contest. Any allegation not denied shall be deemed admitted. Any affirmative defense which may be available to the employer should also be set forth in the answer. The Commission's Rules of Procedure do not specifically provide for pleading affirmative defenses; thus pursuant to Commission Rule 2(b), the Federal Rules of Civil Procedure would apply. Rule 8(c) thereof requires that affirmative defenses or matters of avoidance be specifically pleaded. Strictly speaking this would require raising only those matters outside the Secretary of Labor's *prima facie* case.

The failure to plead an affirmative defense during the issue formulation stage of the proceedings prior to the hearing, either in the answer or in an amendment to the answer, may prevent its assertion and proof at the hearing. *Chicago Bridge & Iron Company,* 1 OSHC 1485 (1974), *rev'd on other grounds* 514 F.2d 1082, 3 OSHC 1056 (7th Cir. 1975); but see *Connecticut Natural Gas Corporation,* 6 OSHC 1797 (1978). Leave to amend an answer to allege affirmative defenses should normally and freely be granted under Federal Rule of Civil Procedure 15 when justice so requires. *Continental Kitchens, Inc.,* 3 OSHC 1859 (1975); *Brown and Root, Inc., Power Plant Division,* 8 OSHC 1055 (1980). Substantive and procedural aspects of the complaint and answer are discussed in Chapter XIII.

A number of defenses may arise as a result of OSHA's failure to follow proper procedures during the promulgation of a standard, during the inspection of the employer's workplace, or in the actual preparation of the citation or complaint. These procedural defenses have been discussed previously in Chapters II, IV, IX, and XI, *supra,* but should be considered when the answer is drafted and included as defenses if applicable.[1] Besides these procedural defenses, there are also a number of important substantive defenses which employers may be able to raise in proceedings before the Commission and the courts. Affirmative defenses, however, are narrowly construed and the burden of their proof, by substantial evidence, lies with the employer.

Greyhound Lines West v. Marshall, 575 F.2d 759, 6 OSHC 1638 (9th Cir. 1978); *Southern Colorado Prestress Co. v. OSHRC,* 586 F.2d 1342, 6 OSHC 2032 (10th Cir. 1978); *United States Steel Corp. v. OSHRC,* 537 F.2d 780, 4 OSHC 1424 (3d Cir. 1976); *Buckeye Industries, Inc.,* 3 OSHC 1837 (1975), *aff'd,* 587 F.2d 231, 6 OSHC 2181 (5th Cir. 1979); *F. H. Lawson Co.,* 8 OSHC 1063 (1980). The defenses are set forth below.

B. THE ISOLATED INCIDENT

The isolated incident/occurrence and employee misconduct defenses derive from the concept that an employer should not be held responsible for the violation of a standard which is the result of an act by an employee of which the employer had no knowledge and which was contrary to the employer's instructions. *Brennan v. OSHRC and Raymond Hendrix, d/b/a Alsea Lumber Co.,* 511 F.2d 1139, 2 OSHC 1646 (9th Cir. 1975); *Jensen Construction Company,* 7 OSHC 1477 (1979). The Secretary of Labor has set forth several criteria which the employer must show to establish the defense. *Horne Plumbing & Heating v. OSHRC,* 528 F.2d 564, 3 OSHC 2060 (5th Cir. 1976):

(1) The violation must have been caused by or resulted from exclusively employee action.
(2) It must have been an isolated and unanticipated instance.
(3) It must have been of short duration.
(4) It must not have been participated in, observed by, or performed with the knowledge of supervisory personnel and must not have been foreseen by the employer in the exercise of reasonable diligence.
(5) The employee's action was in contravention of a well-established policy, program or rule which was used to instruct employees in safe work practices and which was enforced by disciplinary action when necessary.

See *Standard Glass Co., Inc.,* 1 OSHC 1045 (1972); *Floyd S. Pike Electrical Contractors, Inc.,* 5 OSHC 1088 (1977), *aff'd,* 576 F.2d 72, 6 OSHC 1781 (5th Cir. 1978); *Brennan v. Butler Lime & Cement Co.,* 520 F.2d 1011, 3 OSHC 1461 (7th Cir. 1975).

Where an employer is unable to prove that a work rule was enforced, that the employees received safety instructions, that the incident was a deviation from normal practice, and that the employees were not allowed to use their discretion in compliance with the instructions, the defense will not be sustained. *Marrano Leon & Sons, Inc.,* 3 OSHC 1117 (1975); *Maryland Shipbuilding & Drydock Co.,* 3 OSHC 1585 (1975); *Lease Constr. Co.,* 3 OSHC 1979 (1976); *American Bechtel Inc.,* 6 OSHC 1246 (1977); *American Gypsum,* 6 OSHC 1900 (1978). An employer must therefore take all reasonably necessary steps to insure that its employees understand the import of its safety rules and instructions. *General Dynamics Corp., Quincy Shipbuilding Div. v. OSHRC,* 599 F.2d 453, 7 OSHC 1373 (1st Cir. 1979).

Where a work practice has been observed at the job site, there will be a presumption that it is the employer's accepted general practice; the burden is on the employer to show otherwise. Specific safety instructions and work rules relative to hazards and the use of equipment and machinery (preferably reduced to written form), together with evidence that they have been uniformly enforced, are essential to prove the existence of a policy; it then need only be shown that there was deviation. *The Kansas Power & Light Company,* 5 OSHC 1202 (1977); *Sletten Construction Company,* 6 OSHC 1091 (1977); *Leo J. Martone and Associates, Inc.,* 5 OSHC 1228 (1977); *Southwestern Roofing & Sheet Metal Company,* 5 OSHC 1427 (1977). A good safety record will also help to support the defense. *Georgia Steel, Inc.,* 5 OSHC 1728 (1976). Further, the rules and regulations must be adequately explained to new employees before they commence work. *Maryland Shipbuilding and Drydock Company,* 5 OSHC 2019 (1977); *Teal Construction, Inc.,* 5 OSHC 1571 (1977). See also *Atlantic & Gulf Stevedores, Inc. v. OSHRC,* 534 F.2d 541, 4 OSHC 1061 (3d Cir. 1976)—employee resistance, including the threat of strike or boycott, does not relieve an employer from compliance with the Act; they must be forced to comply. *Weyerhaeuser Company,* 3 OSHC 1107 (1975).

The fact that supervisory personnel may have condoned or participated in the violation has been held to bar the defense, unless the employer can show it took all feasible steps/precautions to prevent the violation. *Floyd S. Pike Electrical Contractor, Inc.,* 6 OSHC 1675 (1978); *Leo J. Martone & Associates, Inc., supra.* Supervisory personnel must therefore be properly supervised to prevent an imputation to the employer of the foreman's knowledge of employees' violative conduct. *Jake Heaton Erecting Company, Inc.,* 5 OSHC 1908 (1977); *Interstate Roofing Co., Inc. of Georgia,* 6 OSHC 1710 (1978); *Southwestern Electrical Company, Inc.,* 5 OSHC 1920 (1977); *General Telephone of Michigan,* 6 OSHC 1555 (1978). In *Mountain States Telephone & Telegraph Co.,* 6 OSHC 1504 (1978), *rev'd,* 623 F.2d 155, 8 OSHC 1557 (10th Cir. 1980), the Commission reasoned that ordinarily the knowledge and actions of a supervisory employee are to be imputed to the employer and that the employer could defend by showing that the violation was unpreventable and therefore unforeseeable:

> The employer can establish this defense by showing that it had an effective safety program designed to prevent the violation, including adequate safety instructions effectively communicated to employees, means of discovering violations of these instructions and enforcement of safety rules when violations are discovered.

The Commission then held that *Mountain States* failed to show the supervisor's violation of the standard was unpreventable because it did not show the enforcement of its safety program was adequate.

The court, however, relying on Commission Rule 73(a), which provides that "[i]n all proceedings commenced by the filing of a notice of contest, the burden of proof shall rest with the Secretary," held that, reasonably construed, this rule requires the Secretary to prove the elements of a violation. The court also referred to a decision of the Fourth Circuit, which held that the Commission may not place the burden of proving that a violation was unpreventable on the employer. *Ocean Electric Corp. v. Secretary of Labor,* 594 F.2d 396, 7 OSHC 1149 (4th Cir. 1979). See also *Horne Plumbing & Heating*

Co. v. OSHRC, 528 F.2d 564, 3 OSHC 2060 (5th Cir. 1976); *Brennan v. OSHRC,* 511 F.2d 1139, 2 OSHC 1646 (9th Cir. 1975). But see *Danco Const. Co. v. OSHRC,* 586 F.2d 1243, 6 OSHC 2039 (8th Cir. 1978). The Tenth Circuit agreed with the result which was reached by the Fourth Circuit and stated:

> The Commission has consistently held that to establish a serious violation the Secretary must prove the employer knew or should have known of the likelihood of the noncomplying condition of conduct. *E.g., Harvey Workover, Inc.,* 7 OSHC 1687 (1979). . . . Here, however, the Commission seems to have determined that the Secretary's burden of showing the employer's knowledge was met by proof that Halverson had some supervisory responsibilities and that Halverson knew his own failure to wear rubber gloves was a violation. The premise is that because a corporate employer acts and acquires knowledge through its agents, ordinarily the actions and knowledge of its supervisory employees are imputed to the employer.
>
> When a corporate employer entrusts to a supervisory employee its duty to assure employee compliance with safety standards, it is reasonable to charge the employer with the supervisor's knowledge—actual or constructive—of noncomplying conduct of a subordinate. Upon a showing of the supervisor's knowledge, it is not unreasonable to require the employer to defend by showing the failure to prevent violations by subordinates was unforeseeable. But when the noncomplying behavior is the supervisor's own a different situation is presented. Halverson knew he personally violated the safety standards, of course; if we impute that knowledge to the employer—and declare that now the employer must show the noncomplying conduct was unforeseeable—we are shifting the burden of proof to the employer. All the Secretary would have to show is the violation; the employer then would carry the burden of nonpersuasion.
>
> The same flaw is present in the Secretary's argument that, even if it bears the burden of showing the violation was foreseeable, it need only present a prima facie case, after which the burden shifts to the employer to prove unpreventability. In the present context the prima facie case would be made by showing the supervisor's violation and imputing that knowledge, because he is a supervisor, to the employer. To interpret the rule as the Secretary requests would be to leave the Secretary only the initial burden of making out its case on a critical element and to impose on *Mountain States* the ultimate risk of nonpersuasion in violation of the Commission's own procedural rule 73(a).

If an employer is aware that a supervisor has not scrupulously observed its work rules, the employer should, however, be on notice that he must be watched and should thereafter make an effort to ensure that the rule is observed and enforced. *Floyd S. Pike Electrical Contractor Inc. v. OSHRC,* 576 F.2d 72, 6 OSHC 1781 (5th Cir. 1978). The fact that a foreman feels free to breach a company policy is considered strong evidence that the policy's implementation and enforcement was lax. *National Realty & Construction Co., Inc. v. OSHRC,* 489 F.2d 1257, 1 OSHC 1422 (D.C. Cir. 1973).

The isolated incident defense must be raised by the employer; there is no burden on the Secretary of Labor to show that an incident was *not* "isolated." *Mississippi Valley Erection Co.,* 1 OSHC 1527 (1973).

C. UNPREVENTABLE EMPLOYEE MISCONDUCT OR NEGLIGENCE

Often, employees resist compliance with OSHA standards; there are, however, no sanctions under the Act which would require employees to comply. The duty to comply with the Act rests solely with the employer.

An employer usually has been found to have complied with his general duties under the Act when he: (1) employs "safe" employees; (2) instructs them in the proper operation of machinery and the use of equipment; (3) adopts and enforces a safety program with rules and instructions; (4) adequately communicates the rules to the employees; (5) takes necessary steps and exercises diligence to discover hazards; (6) effectively enforces the rules when violations are discovered; and (7) permits employees to work at a workplace where their jobs can be performed safely with the exercise of reasonable care, even though a hazard may be present. *Jensen Construction Co.,* 7 OSHC 1477 (1979); *Maryland Shipbuilding & Dry Dock Co.,* 5 OSHC 2019 (1977); *Wonder Iron Works, Inc.,* 8 OSHC 1354, 1356 n.5 (1980); *Kansas Power & Light Co.,* 5 OSHC 1202 (1977); *Acchione & Canuso, Inc.,* 7 OSHC 2128 (1980); *General Dynamics Corp.,* 6 OSHC 1753 (1978); *B–G Maintenance Management, Inc.,* 4 OSHC 1282 (1976); *Danco Construction Co. v. OSHRC,* 586 F.2d 1243, 6 OSHC 2039 (8th Cir. 1978).

In Mountain States Telephone & Telegraph Co., 6 OSHC 1504 (1978), *rev'd on other grounds,* 623 F.2d 155, 8 OSHC 1557 (10th Cir. 1980), the Commission found that an employer did not take appropriate steps to assure compliance with its work rules. Evidence that the employer had not discharged or suspended any employees for violations of the work rules (and that only verbal reprimands were given) established that the rules were not effectively enforced by taking appropriate disciplinary measures. The court in reversing the Commission stated that a showing by the employer that it has an adequate and effectively enforced safety program gives rise to the inference that the employer's reliance on employees to comply with applicable safety rules is justifiable; violations then are not forseeable or preventable. But see *Asplundh Tree Expert Co.,* 6 OSHC 1951 (1978), where evidence that an employer had not discharged any employees for violations of a work rule requiring the wearing of body belts in aerial lifts did not bar a finding that the employer effectively enforced the work rule, in view of the existence of a three-step disciplinary procedure whereby discharge was the last recourse taken in view of *evidence that employee violations of the rule were infrequent, with no more than one violation by each employee in noncompliance.* See also *Sletten Construction Co.,* 6 OSHC 1091 (1977)—defense rejected for failure to show uniform and effective enforcement of work rule; *H. B. Zachary Co.,* 7 OSHC 2203 (1979)—defense rejected since employer failed to prove it effectively communicated and enforced work rules; *Wallace Roofing Co.,* 8 OSHC 1492 (1980)—defense rejected because a safety rule was not enforced beyond oral reprimands; *General Motors Corp.;* 8 OSHC 1412 (1980); *Tunnel Electric Construction Co.,* 8 OSHC 1961 (1980). Recently, in *Automatic Sprinkler Corp. of America,* 8 OSHC 1384 (1980), the Commission has made clear that the requirements to invoke the defense have become more stringent:

> We reject Automatic Sprinkler's contention that it could not have known that its employee would use the unguarded scaffold because the company's precautions against the occurrence of this type of violation were adequate. The company's precautions do not constitute a sufficient level of *continuing diligence* under the circumstances of this case.
>
> The Commission has held that an employer must make *a reasonable effort to anticipate* the particular hazards to which its employees may be exposed in the course of their scheduled work. Specifically, an employer must inspect the area to determine what hazards exist or may arise during the work before permitting employees to work in an area, and the employer must then give specific and appropriate instructions to prevent exposure to unsafe conditions. *Bulter Lime & Cement Co.*, 7 OSHC 1973 (1979); *Southwestern Bell Telephone Co.*, 7 OSHC 1058 (1979); *J. H. MacKay Electric Co.*, 6 OSHC 1947 (1978). A preliminary inspection must be made even where the employees are experienced. *J. H. MacKay Electric Co., supra.* Here, Automatic Sprinkler knew where its employee would have to fasten hangers. If the company had inspected the area, it could have observed that the plumbing pipes could immobilize the scaffold where hangers were required to be fastened. Since Automatic Sprinkler *could have anticipated* that the employee might resort to using the scaffold, the company should have taken specific measures against occurrence of the violation. See *Automatic Sprinkler Corp. of America*, 7 OSHC 1957 (1979).

An orderly and comprehensive safety training program is also essential to establish the defense, if the evidence shows it may have prevented the violation.[2] *Bechtel Power Co.*, 7 OSHC 1361 (1979). Specific safety instructions and work rules concerning hazards *peculiar* to the job being performed are essential to an adequate safety program. *Springfield Steel Erectors, Inc.*, 6 OSHC 1313 (1978); *Iowa Southern Utilities Co.*, 5 OSHC 1138 (1977).

An employer who fails to train his employees adequately cannot claim that he could not have known of his employees' preventable actions. *Danco Construction Company v. OSAHRC*, 586 F.2d 1243, 6 OSHC 2039 (8th Cir. 1978). Moreover, the need for comprehensive training is especially important if an employee must work without supervision. *Baroid Division of NL Industries, Inc.*, 7 OSHC 1466 (1979). More importantly, an employer may not ignore readily available opportunities to take simple precautionary measures that will protect an employee from exposure to life-threatening hazards simply because the employee is experienced. See *Getty Oil Company*, 530 F.2d 1143, 4 OSHC 1121 (5th Cir. 1976); *Sutter Lime and Cement Co.*, 7 OSHC 1973 (1979), *on remand from Brennan v. Butler Lime & Cement Co.*, 520 F.2d 1011, 3 OSHC 1461 (7th Cir. 1975)—which specifically noted that an employer may be responsible for a violation created by the negligent act of an employee if the act could have been prevented by proper training or instructions.

The Commission has refused to hold an employer in violation of the Act when an accident occurs that was the *direct* result of an employee's negligence which *could not have been foreseen* by the employer. *A & W Drill Rentals, Inc.*, 2 OSHC 1394 (1974). The courts, too, have rejected a construction of the Act which would make employers *strictly* liable for violations arising from employee misconduct. In *National Realty & Construction Co. v. OSHRC*, 489 F.2d 1257, 1 OSHC 1422 (D.C. Cir. 1973), the court stated:

> A hazard consisting of conduct by employees, such as equipment riding, cannot, however, be totally eliminated. A demented, suicidal, or willfully reckless employee may on occasion circumvent the best conceived and most vigorously enforced safety regime. . . . Congress intended to require elimination only of preventable hazards.

The court stated, however, that an employer's duty to prevent and suppress hazardous conduct by employees is not qualified by such common law doctrines as assumption of risk, contributory negligence, or comparative negligence. The employer's duty, it stated, is, however, qualified by the simple requirement that it be achievable and not be a mere vehicle for strict liability. The question, then, is whether a precaution is recognized as feasible. Conceivably, hazardous employee conduct could be precluded, suggested the court, through feasible precautions concerning hiring, training, and sanctioning of employees. If an employer is not reasonably diligent in taking measures to protect the safety of his employees, he will be in violation of the Act, regardless of an employee's negligence.

The standard of care to which an employer is to be held for a violation resulting from employee misconduct applies to specific standards to the same extent as it applies to the general duty clause, *Atlantic & Gulf Stevedores, Inc. v. OSHRC,* 534 F.2d 541, 4 OSHC 1061 (3d Cir. 1976).

An accident in which an employee was killed when windrowed railroad ties fell upon him after he cut the steel band holding them did not result in a citation against the employer, since the employee was told to stack ties only and had been ordered not to interfere with unloading operations. The court held that an employer is not required to train employee in preventive procedures for operations which he has been ordered not to undertake. *Brennan v. OSHRC (Republic Creosoting Co.),* 501 F.2d 1196, 2 OSHC 1109 (7th Cir. 1974). However, in another case where the Commission held an employer was not responsible for the death of an employee who had entered into an operation for which he was neither hired nor trained, the court reversed, stating that although the employer could not have detected the hazard at the moment of occurrence, the employer could have prevented the accident through better safety training and other precautions. *Brennan v. Butler Lime & Cement and OSHRC,* 520 F.2d 1011, 3 OSHC 1461 (7th Cir. 1975). While an employer is not an insurer and need not take steps to prevent hazards which are not generally foreseeable, including the idiosyncratic behavior of employees, it must do all it feasibly can to prevent foreseeable hazards, including dangerous, negligent conduct by employees. *General Dynamics Corp., Quincy Shipbuilding Div. v. OSHRC,* 599 F.2d 453, 7 OSHC 1373 (1st Cir. 1979). Consequently, an employer can defend a charge that it violated the Act on the basis of "employee misconduct" if it can show it took all necessary precautions/feasible steps *(e.g.,* training & supervision of employees) to prevent the occurrence of the violation.

In *Ocean Electric Corporation v. OSHRC,* 594 F.2d 396, 7 OSHC 1149 (4th Cir. 1979), the court observed that an employer may be held responsible if a violation by an employee is reasonably foreseeable. If, however, the employee's act is an isolated incident, unforeseeable, or idiosyncratic behavior, the purpose of the Act and common sense requires vacating the citation. The court also concluded that the *burden* of proving unforeseeability and unpreventability is on the Secretary of Labor and not on the employer. In *The Mountain States Telephone and Telegraph Co. v. OSHRC,* 623 F.2d 155, 8 OSHC 1557 (10th Cir. 1980), the court held that normally a supervisor's knowledge of the Act's violation can be imputed to an employer, and then the employer could defend by showing the violation was unpreventable. But, in cases where the Act was violated by the supervisor, no such imputation can be made, because to do so would shift the burden of proof to the employer, in violation of Commission Rule 73(a), which places that burden on the Secretary of Labor. See also *Cape & Vineyard Div. of Bedford Gas & Edison Light Co. v. OSHRC,* 512 F.2d 1148, 2 OSHC 1628 (1st Cir. 1975), which held that, simply because an accident could be prevented by using more protective equipment, this does not establish a violation when employee conduct is unforeseeable; *Ryder Truck Lines, Inc. v. Brennan,* 497 F.2d 230, 2 OSHC 1075 (5th Cir. 1974), where the court held that if an employee's conduct is willfully reckless, the employer cannot reasonably be expected to prevent the existence of the hazard which the reckless behavior creates; *Horne Plumbing & Heating Co. v. OSHRC,* 528 F.2d 564, 3 OSHC 2060 (5th Cir. 1976); *Anning-Johnson Co. v. United States & OSHRC,* 516 F.2d 1081, 3 OSHC 1166 (7th Cir. 1975).

D. "BORROWED SERVANT"; LACK OF CONTROL

In many instances, especially when construction work sites are involved, one employer may "loan" or "lease" its equipment and/or employees to another employer. Where the loaning employer has relinquished total control over the equipment and/or employees involved, he may be able to raise and establish what has been called the "borrowed servant" defense against a subsequent citation issued to him for exposure of the loaned employees to a citable hazard. This defense is also applicable jurisdictionally, for the person with no control over an employee is not considered to be the employer. The Commission, although recognizing the existence of the borrowed servant defense, has stated that it must be defined in terms of employment relationships, considering the realities of the situation and the remedial purposes intended by Congress in the Act. For "purposes of the Act," the Commission has considered the following factors to be relevant in identifying a person as an employer: (1) the person whom the employees consider to be their employer; (2) the person who pays the employees' wages; (3) the person who is responsible for controlling the employees' activities; (4) the person who has the power, as distinguished from the responsibility, to control the employee; and (5) the person who has the power to fire the employee or to modify his employment conditions. *Griffin and Brand of McAllen, Inc.,* 6 OSHC 1702 (1978); *Acchione & Canuso, Inc.,* 7 OSHC 2128 (1980); *Weicker Transfer & Storage Co.,* 2 OSHC 1493 (1975). The Commission has held that the control element, although important in determining whether an employment relationship exists under the Act, is by no means conclusive. *Brennan v. Gilles & Cotting, Inc.,* 503 F.2d 1255, 2 OSHC 1243 (4th Cir. 1974); *Evansville Material Inc.,* 6 OSHC 1706 (1978); *Acchione & Canuso, supra.*

This defense has been narrowly construed and, in many cases, the Commission and the courts have found the lessor-employer responsible for exposure of leased employees despite the existence of very little control. In *Frohlick Crane Service, Inc. v. OSHRC,* 521 F.2d 628, 3 OSHC 1432 (10th Cir. 1975), the court held that the lessor of the crane and its operator was properly cited as the employer, since the lessee relied upon the expertise of the crane operator supplied by the lessor and the lessee gave no particular instruction as to the performance of the operator's assigned tasks. Additionally, the lessor was not permitted to avoid a citation simply because of the existence of a "hold harmless clause" in the lease agreement, in which the lessee agreed that the equipment and operator were under the exclusive supervision and control of the lessee.

Even where a lessee admitted that he supervised the lessor's employees and was solely responsible for their safety, the Commission found the presence of the lessor's foreman, in conjunction with the absence of any lessee's supervisors at the work site, sufficient to show that there had been no actual redistribution of responsibility for the leased employees' safety. *Bayside Pipe Casters, Inc.,* 2 OSHC 1206 (1974). See also *Weicker Transfer & Storage Co.,* 2 OSHC 1493 (1975), in which the Commission found a lessor-employer responsible where the leased employees punched in at the lessor's office and reported trouble to the lessor's supervisors and where the lessor's supervisor had "some" authority to refuse to pick up a dangerous load and to stop the crane; *Gordon Construction Co.,* 4 OSHC 1581 (1976)—a contractor hired by a city to assist with sewer project remained the employer of backhoe operator which he assigned to the project, since he directed him to perform specific tasks incidental to the job and held himself out to the public as being in charge.

There have been a number of cases, however, in which lessors have been able to avoid responsibility for leased or loaned employees. In *Williams Crane & Rigging, Inc.,* 2 OSHC 3291 (1975), a lessor was found to have relinquished all control over the employees and was without right to come onto the work site. The Commission held he could not reasonably be said to be in a position to fulfill the objective of the Act and was not an employer within the meaning of the Act. See also *Savoy Construction,* 2 OSHC 3207 (1974); *Joseph Bucheit & Sons Co.,* 2 OSHC 1001 (1974). The key to avoiding responsibility seems to be proof that there has been no continued supervision on the part of the lessor-employer once his employees have been loaned. See *Williams Enterprises, Inc.,* 5 OSHC 1904 (1977); *Colorado Const. Ltd.,* 3 OSHC 1354 (1975).

In the case of *Lidstrom, Inc.,* 4 OSHC 1041 (1976), Commissioner Cleary and former Commissioner Moran, in their majority and dissenting opinions, provided a thorough discussion of the cases and various points of view surrounding the borrowed servant rule. According to Cleary, the important factors in his determination that the lessor of the crane remained the crane operator's employer were: (1) the lessee relied on the operator's expertise and did not supervise the crane's operation; (2) the lessor paid the operator; and (3) the crane operator had final authority to prevent the operation of the crane in an unsafe manner. Moran, however, felt that control over the employee is the paramount consideration, and that it is the right of control, not the exercise of it, that is governing. Subsequent to *Lidstrom,* however, the cases clearly indicate that Cleary's view has predominated. *Evansville Material, Inc.,* 6 OSHC 1706 (1978); *Acchione & Canuso, Inc.,* 7 OSHC 2128 (1980); *Sam Hall & Sons,* 8 OSHC 2176 (1980).

E. CREATION AND CONTROL OF HAZARDS (Multiemployer Work Site)

In situations where a multiemployer work site (that is, several employers working on a single work site) is involved, employees of different employers may be continuously and varyingly exposed to the same hazard; it is not a typical employment situation. The question arises as to which employer is responsible for abatement and to whom liability should attach for a violation of the Act. The answer appears to be that OSHA is not limited to citing just one of the employers—two or more employers may be liable for the same violative condition. This approach to enforcement naturally has important implications for both joint and multiemployer situations.

For example, on a construction site, several employers are involved with varying degrees of responsibility for completion of the work. The general contractor has overall responsibility for all work;[3] the various subcontractors have responsibility for various portions of the work *(e.g.,* electrical, steel erection, plumbing, concrete). Thus the multiemployer construction work site situation presents an additional consideration regarding employee safety and health for in this situation a hazard created and controlled by one employer can affect the safety of employees of other employers on the site. A multi (prime and sub) contractor situation exists when the creation/control of a hazard and the exposure of employees (usually those of the sub) to that hazard is divided between two or more employers; under certain conditions both employers can be cited, even though the employees of only one are affected. See *Anning-Johnson Co. and Workinger Electric, Inc. v. OSHRC,* 516 F.2d 1081, 3 OSHC 1243 (4th Cir. 1974); *Giles & Cotting, Inc. v. Brennan,* 504 F.2d 1255, 2 OSHC 1243 (4th Cir. 1974). A simultaneous or joint employment relationship exists when more than one employer controls the employment relationship of an employee, *e.g.,* one employer leases from another employer a piece of heavy equipment such as a crane and the employee who operates the crane. Employer responsibility in such situation generally devolves upon the employer who controls, supervises or directs the work of the loaned employee. The fact that the salary is paid by the other employer, is not determinative. The Commission and the courts have generally refused to apply the agency concept of borrowed servant in this respect. See *Frolich Crane Service, Inc. v. OSHRC,* 521 F.2d 628, 3 OSHC 1432 (10th Cir. 1975); *Weikord Transfer and Storage Co.,* 2 OSHC 1493 (1973); *Louisiana Paving Co.,* 4 OSHC 1023 (1975).

In order to provide guidance, OSHA initially issued instructions to its inspectors in the FOM for determining which employer(s) to cite in a multiemployer situation. According to the FOM, each employer was responsible for the working conditions of his own employees. Citations under the general duty clause and specific standards were to be issued only against employers whose *own* employees were exposed or potentially exposed to unsafe or unhealthful working conditions—even if he did not create the condition. An employer would not be cited if his employees were not exposed or potentially exposed to an unsafe or unhealthful working condition, even if that employer created the condition. FOM, Ch. XIV. Having taken this position, it was left to the Commission and courts to decide the propriety of its application.

1. Early Commission Position

Any employer whose employees were exposed to a hazard could be cited, regardless of the fact that another employer created and controlled the hazardous condition. An employer whose employees were not exposed, even though he created and controlled the hazard, was not citable. *R. H. Bishop Co.*, 1 OSHC 1767 (1974); *Gilles & Cotting, Inc.*, 1 OSHC 1388 (1973); *Otis Elevator Co.*, 2 OSHC 1239 (1974). OSHA construed the Commission's decisions to mean that on a multiemployer work site, several employers could be cited for the same hazard, as long as the employer's employees were exposed.

2. Court Decisions

In *Brennan v. Gilles & Cotting, Inc.*, 504 F.2d 1255, 2 OSHC 1243 (4th Cir. 1974), the court held, in deference to the Commission's interpretation, that a general contractor is not jointly responsible with a subcontractor for violations hazardous to the subcontractor's employees, even though the general contractor had primary safety responsibility on the job. Thereafter, in *Brennan v. OSHRC and Underhill Construction Corp.*, 513 F.2d 1032, 2 OSHC 1641 (2d Cir. 1975), a subcontractor was cited under Section 5(a)(2), because it had created several hazards. Employees of other subcontractors were exposed to these hazards. Finding that the general purpose of the Act is to prevent workplace hazards, the court upheld a citation for the employer who created and controlled the hazard, even if its employees were not exposed:

> In this regard it is not insignificant that it was *Dic-Underhill* that created the hazards and maintained the area in which they were located. It was an employer on a construction site, where there are generally a number of employers and employees. [footnote omitted] It had control over the areas in which the hazards were located and the duty to maintain those areas. Necessarily it must be responsible for creation of a hazard.

Subsequently, in *Anning-Johnson Co. and Workinger Electric, Inc. v. OSHRC*, 516 F.2d 1081, 3 OSHC 1166 (7th Cir. 1975), the issue was whether a sub on a multiemployer construction site could be cited for an other-than-serious violation of standards, to which its employees were *exposed* but which it did *not* create and for which it was not responsible for abatement. The court held that the sub was not responsible for hazards it neither created nor controlled, rejecting the position that the exposing employer is subject to liability merely because its employees were exposed to hazards, created and controlled by the general.

3. Commission Position Reconsidered

On remand from the Fourth Circuit, the Commission in *Giles & Cotting, Inc.*, 3 OSHC 2002 (1976), stated that an employer is liable for violations if its employees had access thereto and if the employer knew or with the exercise of reasonable diligence could have known of the presence of the violation. Access can be actual or potential and could be shown to exist when

> employees either while in the course of their assigned working duties, their personal comfort activities while on the job, or their normal means of ingress-egress to their assigned workplaces, will be, are, or have been in a *zone of danger*.

See also, *Stahr & Gregory Roofing Co., Inc.*, 7 OSHC 1012 (1979), which held that the brevity of exposure or access to a "zone of danger" does not negate the finding of a violation.

The Commission also adopted the *Underhill Construction, supra,* rationale as its official position on multiemployer liability to the extent that it imposed liability on an employer (either a sub or general) who creates/controls a hazard, even though only employees of *other* employers are exposed. Thus creating and controlling employers are responsible for the exposure of their own *and* other employers' employees. Employers who have no control over the operations of other employers in the work site, *e.g.*, subcontractor, still have a duty to exert reasonable efforts to protect their own employees from the violations of others. At the same time, OSHA issued guidelines for issuance of multiemployer work site citations, which basically comported with the new Commission position.

In *Anning-Johnson Co.*, 4 OSHC 1193 (1976), on review from Seventh Circuit, the Commission first held with respect to the subcontractor's liability that:

> even if a construction subcontractor neither created nor controlled a hazardous situation, the exposure of its employees to a condition that the employer knows or should have known to be hazardous, in light of the authority or "control" it retains over *its own* employees gives rise to a duty under Section 5(a)(2) of the Act. This duty requires that the construction subcontractor do what is "realistic" under the circumstances to protect its employees from the hazard to which a particular standard is addressed[18] even though literal compliance with the standard may be unrealistic.
>
> [18] Commissioner Cleary would find that a contractor has a virtually identical duty under section 5(a)(1) of the Act [29 U.S.C. Sec. 654(a)(1)], the general duty clause, if the specific standard and hence, the duty under section 5(a)(2) were found to apply only to those contractors capable of literal compliance. Commissioner Cleary would find a breach of the duty under section 5(a)(1) so long as (1) the hazard is "recognized" within the meaning of the section and (2) there is a likelihood of death or serious physical harm.

Then, with respect to the general contractor, the Commission held that it has overall responsibility for standard violations which it could reasonably have been expected to abate by reason of its supervisory capacity. Furthermore, the duty of a general contractor is not limited to the protection of its own employees from safety hazards but extends to the protection of all of the employees engaged at the work site. This holding, however, was contrary to the Fourth Circuit's decision in *Gilles & Cotting.* In addition, the Commission declined to make the distinction drawn by the Seventh Circuit in its *Anning-Johnson* decision between an employer's duty to abate violations depending on their classification as serious or other-than-serious. *Armor Elevator Co.,* 5 OSHC 1630 (1977).

In *Grossman Steel & Aluminum Corp.,* 4 OSHC 1185 (1976), the Commission reached the same result as it did in its *Anning-Johnson* decision, holding that a general contractor possesses sufficient control over the entire work site to give rise to a duty under Section 5(a)(2) to comply with the Act or issue compliance or abatement, and that a subcontractor is liable for a violation it neither created nor controlled unless it can show it made reasonable efforts to detect and correct violations to which its employees had access.

The Commission rules on multiemployer construction work site liability, which do not limit liability solely to the exposure of an employer's own employees and place the burden on employers to prove the affirmative defense, have been judicially approved as reasonable and consistent with the purposes of the Act. *New England Telephone & Telegraph Co. v. Secretary of Labor,* 589 F.2d 81, 6 OSHC 2142 (1st Cir. 1978); *Beatty Equipment Leasing, Inc. v. Secretary of Labor,* 577 F.2d 534, 6 OSHC 1699 (9th Cir. 1978); *Marshall v. Knutson Construction Co.,* 566 F.2d 596, 6 OSHC 1078 (8th Cir. 1977); *Central of Georgia R.R. Co. v. OSHRC,* 576 F.2d 620, 6 OSHC 1785 (5th Cir. 1978).

4. Multiemployer Guidelines

The OSHA guidelines, revised after its reconsideration of the Commission's *Anning-Johnson* position, provide that the Secretary of Labor may issue citations: (1) to an employer on a multiemployer work site who creates or causes a hazardous condition to which his employees or those of any other contractor or subcontractor engaged in activities on the multiemployer work site are exposed; (2) to an employer who has the ability to abate the hazardous condition (and fails to do so) regardless of whether he created the hazard; and (3) where violative conditions are readily apparent, not only to the employer who controls or is responsible for the violative conditions, but also to the employer who permits his employees to be exposed to the hazardous conditions, regardless of whether the latter employer created the hazard or has the power to abate it. FOM, Ch. X.

It should be noted, however, that the *Anning-Johnson* decision may not require what item (3) of the Secretary's guidelines state. Item (3) seems to have contemplated the situation where OSHA was not certain who was or should be responsible for the violation.

5. Requirements of the Defense

The *Anning-Johnson* affirmative defense relieves an employer from liability under the Act for a hazardous condition it neither created nor controlled but to which its employees are exposed, if it can show *either:*

(1) that its employees were protected by *realistic alternatives* to the extent possible, *Cornell & Co., Inc.*, 5 OSHC 1736 (1977); *Stahr & Gregory Roofing Co., Inc.*, 7 OSHC 1012 (1979); *Dutchess Mechanical Corp.*, 6 OSHC 1795 (1978); *or*

(2) that it neither had nor could have had, with the exercise of reasonable diligence, *notice* of the hazardous condition, *Data Electric Co., Inc.*, 5 OSHC 1077 (1976); *Bratton Corporation,* 6 OSHC 1327, *aff'd* 590 F.2d 273, 7 OSHC 1004 (8th Cir. 1979); *Fischback and Moore,* 8 OSHC 2125 (1980). [In this connection, the Commission held, in *Anning-Johnson,* that an employer's duty under Section 5(a)(2) should be no less than that of an employer under Section 5(a)(1) in preventing exposure of its employees to "recognized hazards."]

Before an employer may invoke this defense, it must first demonstrate that it *neither* created nor controlled the hazardous condition. *Central of Georgia Railroad Co. v. OSHRC,* 576 F.2d 620, 6 OSHC 1785 (5th Cir. 1978); *New England Telephone & Telegraph Co. v. Secretary of Labor,* 589 F.2d 81, 6 OSHC 2142 (1st Cir. 1978); *Marshall v. Knutson Construction Co.*, 566 F.2d 596, 6 OSHC 1077 (8th Cir. 1977)—general contractor on a multiemployer construction site; *Bratton Corporation v. OSHRC,* 590 F.2d 273, 7 OSHC 1006 (8th Cir. 1979)—subcontractor on a multiemployer work site; *Dun-Par Engineered Form Co.,* 8 OSHC 1044 (1980)—subcontractor had responsibility for constructing form work, but did not erect guardrails to protect employees, thus it created hazard, and defense is not applicable; *Masonry Contractors, Inc.*, 8 OSHC 1155 (1980)—subcontractor did not establish it was noncreating, noncontrolling; *Sun Rise Plastery Corp.*, 8 OSHC 1765 (1980). The term "controlled" has been defined as the ability to abate the violation within the literal terms of the standard. *Williams Enterprises of Georgia,* 7 OSHC 1901 (1979).

The two branches of the defense have received considerable interpretation. The realistic, reasonable alternative measures which must be taken to protect employees have been interpreted to include attempts to have the general contractor correct the condition, attempts to have the employer responsible for the condition correct it, instruction of employees to avoid the area where the condition exists, and if possible providing employees with alternative means of protection. *Bratton Corporation v. OSHRC,* 590 F.2d 273, 7 OSHC 1005 (8th Cir. 1979); *Western Waterproofing Co., Inc.* 5 OSHC 1496 (1977), *aff'd in part, rev'd in part,* 576 F.2d 139, 6 OSHC 1550 (8th Cir. 1978); *Derr Construction Co.,* 5 OSHC 1333 (1977). In some cases notice to the controlling employer has been held to be sufficient. *Mayfair Construction Co.*, 5 OSHC 1877 (1977); *Granger Construction Co., Inc.*, 6 OSHC 1099 (1977). But in other cases notice has found to be lacking. *Gotham Electric Co.*, 6 OSHC

1265 (1977)—requests to general contractor for abatement were not sufficiently definite and forceful to be a realistic alternative; *Automatic Sprinkler Corp. of America,* 8 OSHC 1384 (1980)—where no means are available by which employer can protect its employees against a hazard, it must nevertheless take the step of requesting the responsible contractor to abate it; *Mclean-Behm Steel Erectors, Inc.,* 6 OSHC 1712 (1978); *Wonder Iron Works, Inc.,* 8 OSHC 1354 (1980); *Forest Park Roofing Co.,* 8 OSHC 1181 (1980); *Williams Enterprises of Georgia, Inc.,* 7 OSHC 1900 (1979)—having failed to seek other means of protection or to show that other means were unavailable, the defense was not established. This branch of the defense *does not,* however, require that a noncreating/controlling employer remove his employees from the work site. *Williamette Iron & Steel Co.,* 5 OSHC 1479 (1977).

With respect to the second branch of the defense, OSHA has the initial burden of showing that the employer had actual/constructive knowledge of the violation; then it is incumbent upon the employer to show that it had no notice, that is, it did not/could not know the condition was hazardous. *Data Electric Co.,* 5 OSHC 1077 (1976); *Stahr & Gregory Roofing Co., Inc.,* 7 OSHC 1010 (1979)—subcontractor knew of hazard but failed to require general to correct condition; *Kenneth P. Thompson Co.,* 8 OSHC 1696 (1980)—failed to prove that employer had no knowledge and failed to show that realistic, protective action to correct the condition was undertaken; *Cotner & Cotner, Inc.,* 6 OSHC 1163 (1977).

While the defense is applicable to both serious and other-than-serious violations, *Armour Elevator Co., Inc.,* 5 OSHC 1630 (1977); *Anastasi Brothers Corp.,* 5 OSHC 1634 (1977), it applies *only* to employers on a multiemployer work site. *T. J. Service Co.,* 6 OSHC 1509 (1978).

The end result of this discussion is this: where the usual elements of employer liability are established, some but not all *subcontractors* may assert a limited affirmative defense as defined in the case law. The defense is available to those *subcontractors* who neither create nor control the violative condition and thus lack the ability to abate the condition within the literal terms of the standard. Accordingly, a subcontractor asserting the *Anning-Johnson/Grossman Steel* defense must make a *threshold* showing that it is a "noncreating," "noncontrolling" employer. If the subcontractor does not make this showing, the defense is not applicable to the circumstances, and the employer is liable for the violation on the basis of the exposure of its employees to and its creation and/or control of the violative conditions. See generally, *Masonry Contractors, Inc.,* 8 OSHC 1155 (1980). Moreover, more than one employer can still be cited by OSHA for the same violation, and contractual agreements between employers have no bearing on liability. *Armor Elevator Co.,* 1 OSHC 1409 (1973); *Electric Smith, Inc.,* 8 OSHC 1391 (1980); *Charter Builders, Inc.,* 8 OSHC 1232 (1980).

The upshot of *Anning-Johnson* vis-à-vis general contractors is that general contractors with overall supervisory responsibility for a multiemployer work site (including violations it could reasonably have been expected to prevent or abate by its supervisory capacity) have no recourse to the defense; they have the responsibility (and normally the means) to assure that other contractors fulfill their obligations with respect to employee safety, where those obligations affect the construction work site. *Geico Builders, Inc.*, 6 OSHC 1104 (1977); *Sierra Construction Corp.*, 6 OSHC 1278 (1978); *Smith & Mahoney*, 8 OSHC 2070 (1980). In order to determine if an employer is a general contractor for purposes of the application of *Anning-Johnson*, the appropriate test is to look at the construction documents/contracts, *vis-à-vis* its authority and obligations. *Jess Howard Electric Co.*, 5 OSHC 1939 (1977); *Truland Elliot*, 4 OSHC 1455 (1976); *Caldwell-Wingate Corp.*, 6 OSHC 1619 (1978); *Ray Boyd Plaster & Tile, Inc.*, 6 OSHC 1648 (1978); *Wander Iron Works, Inc.*, 8 OSHC 1354 (1980); *Automatic Sprinkler Corp. of America*, 8 OSHC 1384 (1980).

6. Recent Expansions

Prior to the decision in *Harvey Workover, Inc.* 7 OSHC 1687 (1979), employers would frequently argue that they were not engaged in construction work and thus not subject to the *Anning-Johnson* rules. In *Harvey Workover*, however, the Commission extended the rules to nonconstruction work sites, stating:

> While employers generally are held responsible for the exposure or access of only their employees, the Commission has developed an exception restricted to a multi-employer construction site at which an employer that creates or controls noncomplying conditions is held responsible if employees of other employers are exposed or have access to the conditions. Construction sites were distinguished from other worksites on the ground that the work of one employer's employees often requires those employees to work in or pass through areas where work has been performed by another employer's employees. Thus, one employer's employees easily may be exposed to hazards created or controlled by another employer.
>
> The number of employers at a construction site has not been shown to have a bearing on the movement of employees, and has not influenced application of the principles announced in the *Grossman Steel* and *Anning-Johnson* opinions. *See Northeast Marine Terminal Company*, 4 OSHC 1671 (1976) (two employers); *Cottner & Cottner, Inc.*, 6 OSHC 1163 (1977) (two employers); *4 G Plumbing & Heating, Inc.*, 6 OSHC 1528 (1978) (three employers). The movement of employees of one employer into areas where hazardous conditions are created or controlled by another employer is not restricted to construction sites, however, as the facts of this case demonstrate.
>
> We no longer find the distinction between construction sites and other worksites valid. The safety of all employees can best be achieved if each employer at multi-employer worksites has the duties to (1) abate hazardous conditions under its control and (2) prevent its employees from creating hazards.

The Commission has also affirmed its holding relative to construction sites, in that the general contractor is responsible for the violations of its subcontractors that the general contractor could reasonably be expected to prevent or detect and abate, by reasons of its supervisory capacity over the entire work site, even though none of its own employees is exposed to the hazard. *Sierra Construction Corp.*, 6 OSHC 1278 (1978); *Gil Haugan d/b/a Haugan Construction Co.*, 7 OSHC 2004 (1979). The holding in these cases was predicated on the presumption that, by virtue of its supervisory capacity over the entire work site, the general contractor on the site has sufficient control over its subcontractors to require them to comply with occupational safety and health standards and to abate violations. The burden of rebutting this presumption was placed on the general contractor. However, in *Red Lobster Inns of America Inc.*, 8 OSHC 1762 (1980), the Commission extended the application of this principle from a general contractor on a multiemployer construction site to *nonconstruction* worksites and to employers who are *not general contractors.*[4]

A significant decision which appears to limit the Commissioner's attempts to expand the applicability of the defense is *New England Telephone & Telegraph Co. v. Secretary of Labor,* 589 F.2d 81, 6 OSHC 2142 (1st Cir. 1978). In that case the company's supervisors, on July 24, 1974, inspected a construction site and found that the incompleted stairs were supplied with proper temporary safety fillers. On July 25, 1974, when one of its installers was, so far as appears, obliged to use the stairs, the general contractor had removed the fillers in anticipation of placing permanent treads. Finding that the fillers had been removed sooner than needed to be, the Secretary of Labor charged the company with an other-than-serious violation under 29 CFR 1926.501(f). Although the company sought to raise the *Anning-Johnson* defense, the Review Commission responded, without discussion, that the "record . . . lacks evidence." Upon review by the court, the Secretary of Labor contended that a company could be found liable *for not anticipating* that the general contractor might remove the fillers prematurely and *should have made an agreement with it.* The court rejected this argument, stating:

> Passing the fact that the Commission made no such, or any other finding, this is a burden of care we consider so unrealistic as to be unreasonable as a matter of law. If it is apparent from his having done so that a general contractor knows enough to install proper devices, we see no duty on a subcontractor to make an agreement with him, or, absent some affirmative reason for doubt, assume that he will improperly discontinue. If a passageway is properly lighted, a subcontractor does not have to stand around and see that the bulb does not burn out.
>
> On the undisputed evidence petitioner's supervisor inspected the worksite the day prior to the citation and found nothing amiss, nor anything that would create a reasonable apprehension that a hazard would develop. On this evidence we hold, as a matter of law, that petitioner has established an *Anning-Johnson, Grossman* defense.

Compare *Automatic Sprinkler Corp. of America,* 8 OSHC 1384 (1980). See generally *The Occupational Safety and Health Act of 1970 as Applied to the Construction Industry—The Multi-Employer Worksite Problem,* 25 Wash. & Lee L.R. 173 (1978).

F. FEASIBILITY

The only specific reference to feasibility in the Act is contained in Section 6(b)(5), which provides that the Secretary is required "in promulgating standards dealing with *toxic* materials or *harmful* physical agents . . . to set the standard which most adequately assures to the extent *feasible,* on the basis of the best available evidence. . . ." This section refers solely to the standard-setting process and only to health standards. There is no comparable requirement with respect to the Act's enforcement process or to the promulgation of safety standards, but the courts and the Commission have held that in order to prove an employer violated the Act, it must normally be shown that compliance was feasible. *Sampson Paper Bag Company, Inc.*, 8 OSHC 1519 (1980); *RMI Co. v. Secretary of Labor,* 594 F.2d 566, 7 OSHC 1119 (6th Cir. 1979), *on remand* 7 OSHC 1482 (1979).

Under both the general duty clause and specific standards, both safety and health, the lack of feasibility (according to a consensus of the courts and the legislative history of the Act), both economic and technological, has been held to constitute a defense to a citation for violation of the Act. See *General Electric Co. v. OSHRC,* 540 F.2d 67, 4 OSHC 1512 (2d Cir. 1976); *Diversified Industries Division, Independent Stave Co. and Independent Stave Co.*, 618 F.2d 30, 8 OSHC 1107 (8th Cir. 1980); *National Realty & Construction Co. v. OSHRC,* 489 F.2d 1257, 1 OSHC 1422 (D.C. Cir. 1973); *Turner Co. v. Secretary of Labor,* 561 F.2d 82, 5 OSHC 1790 (7th Cir. 1977); *RMI Co. v. Secretary of Labor,* 594 F.2d 566, 7 OSHC 1119 (6th Cir. 1979), *U. P. S. of Ohio, Inc. v. OSHRC,* 570 F.2d 806, 6 OSHC 1347 (8th Cir. 1978); *Marshall v. West Point Pepperell, Inc. v. OSHRC,* 588 F.2d 979, 7 OSHC 1034 (5th Cir. 1979); S. Rep. No. 91–1282, 91st Cong. 2d Sess.

Normally the burden of proof on the issue of feasibility (both technological and economic) lies with the employer, *except* in those situations where the standard provides otherwise. It has been held that feasibility may be assumed where the requirements for compliance are clear. *Ace Sheeting & Repair Co. v. OSHRC,* 555 F.2d 439, 5 OSHC 1589 (5th Cir. 1977). In situations where the requirements for compliance are not clear or feasibility is itself an element of the alleged violation, such as the general duty clause, the burden of proving feasibility rests with the Secretary of Labor. *National Realty & Construction Co. v. OSHRC,* 489 F.2d 1257, 1 OSHC 1422 (D.C. Cir. 1973). In this respect however, the Commission stated, in *S & H Riggers & Erectors, Inc.*, 7 OSHC 1265 (1979), that it considered the extension of the Section 5(a)(1) analysis to cases arising under a specific duty standard to be inappropriate:

> The *Frank Briscoe* [4 OSHC 1729 (1976)] burden of proof as to feasibility and utility was assigned by the majority to the Secretary because it agreed with the *National Realty* court that Congress intended only to impose achievable duties on employers. We do not dispute the court's interpretation. However, we disagree with the extension of that interpretation in *Frank Briscoe* that places the burden as to achievability with the Secretary [*re* compliance with a specific standard]. The Act is designed to protect employees from preventable workplace hazards. It is not intended to make employers insurers of safety or to hold employers to a standard of strict liability. As will be discussed later in this opinion in relation to the Respondent's affirmative defenses, the Commission has recognized that there are occasions when hazards cannot be completely eliminated from the workplace. However, the burden of proving unpreventability necessarily rests with the employer. It is the employer who has the greater knowledge of its workplace and work practices. The employer is best able to assess those factors which enable it to improve the safety and health of its employees and those factors which actually

> hinder its attempts. Furthermore, Congress clearly imposed the duty of achieving maximum safety in the workplace on the employer. *Atlantic & Gulf Stevedores, Inc.*, 3 OSHC 1003 (1975), *aff'd,* 534 F.2d 541, 4 OSHC 1061 (3d Cir. 1976). It is, therefore, properly the burden of the employer to show why this objective cannot be achieved at a particular worksite.

Another example where the burden of feasibility rests with the Secretary of Labor is 29 CFR 1910.95(b)(1), the noise standard, which provides that when employees are subjected to noise levels in excess of that permitted, feasible administrative or engineering controls shall be utilized (substantial evidence is the appropriate standard of review). Clearly, the Secretary of Labor has the burden of establishing feasibility according to the language of the standard. See *KLI, Inc.,* 6 OSHC 1097 (1977); *Louisiana Pacific Corp.,* 5 OSHC 1994 (1977); *Continental Can Co.,* 4 OSHC 1541 (1976); *Atlantic Steel Co.,* 6 OSHC 1289 (1978); *RMI Co., supra; West-Point Pepperell, Inc.,* 5 OSHC 1257, *aff'd,* 588 F.2d 979, 7 OSHC 1035 (5th Cir. 1979); *Diversified Industries Division, Independent Stave Co. and Independent Stave Co. v. OSHRC,* 618 F.2d 30, 8 OSHC 1107 (8th Cir. 1980).

With respect to compliance with other specific standards, however, the burden normally rests with the employer to show infeasibility, unless, as previously stated, feasibility is specifically referenced in the standard itself *(i.e.,* 29 CFR 1910.1000(e)). See *Easley Roofing & Sheet Metal Co., Inc.,* 8 OSHC 1410 (1980)—the Secretary of Labor does not have the burden of proving feasibility under 29 CFR 1926.28(a); *Republic Roofing Corp.,* 8 OSHC 1411 (1980)—the burden of proving the feasibility of use of personal protective equipment under 29 CFR 1926.28(a) is not on the Secretary of Labor; *Arkansas–Best Freight System, Inc. v. OSHRC,* 529 F.2d 649, 3 OSHC 1910 (8th Cir. 1976)—the burden of showing that an employer could not absorb the cost of compliance with the personal protective equipment standard, 29 CFR 1910.132(a), is on the employer; *S. H. Riggers and Erectors, Inc.,* 7 OSHC 1260 (1979)—feasibility of compliance with 29 CFR 1926.28(a) is on the employer; *Royal Logging Co.,* 7 OSHC 1744 (1979)—the Secretary of Labor need only prove that an abatement method exists that would provide protection against the cited hazard; the burden then shifts to the employer to produce evidence showing or tending to show that use of the method or methods established by the Secretary will cause consequences so adverse as to render their use infeasible. See also *I. T. O. Corp. v. OSHRC,* 540 F.2d 543, 4 OSHC 1574 (1st Cir. 1976)—the employer had the burden to show infeasibility due to concerted employee resistance. In *A. E. Burgess Leather Co. v. Secretary of Labor,* 576 F.2d 948, 6 OSHC 1661 (1st Cir. 1978), the court stated, with respect to a guarding citation under 29 CFR 1910.212(a)(1), that:

> Finally, Burgess argues that compliance with the standard would be infeasible, inasmuch as the operator must hold the die in order to use the beam dinker and the alternative technology, the "clicker," would unacceptably lower Burgess' productivity. But "where a specific duty standard contains the method by which the work hazard is to be abated, the burden of proof is on the employer to demonstrate that the remedy contained in the regulation is *infeasible* under the particular circumstances. . . ." [*Ace Sheeting & Repair Co., v. OSHRC,* 555 F.2d 439, 5 OSHC 1589 (5th Cir. 1977)]. Here Burgess has failed to carry its burden.

See also *Morton Buildings, Inc.*, 7 OSHC 1703 (1979); *Worley Brothers Granite Co.*, 4 OSHC 2042 (1977).

1. Technological Feasibility

The question of technological feasibility is often raised in the context of the enforcement of the noise standard, and, until recently, its application has generated very little controversy. A control capable of producing a *significant* reduction in exposure in an employer's workplace is considered to be technologically feasible. *Sampson Paper Bag Co., Inc.*, 8 OSHC 1515 (1980). To prove that an employer violated a standard, the Secretary of Labor must at the very least demonstrate that, at the time of the alleged violation, a technologically feasible method of abatement existed. *Love Box Co.*, 4 OSHC 1138 (1976). However, in *Sampson Paper Bag, supra,* the Commission, while holding that employers may not be found in violation of a standard if the technology for compliance does not exist, stated this does not mean that the Secretary of Labor must prove that "off the shelf" controls were available to the employer at the time of the violation. Rather it is sufficient to show the existence of technology that could be adapted to the employer's operations (feasible at the time the employer allegedly violated the standard). Adapting technology to a new use and the costs of doing so would then be properly considered in determining economic feasibility.

The apparent unanimity of the Commission on the issue of technological feasibility set forth in *Sampson Paper Bag* is belied by an analysis of the decision. Thus, the Secretary of Labor argued that the authority to cite employers for alleged violations of the noise standard is not limited to those employers which have failed to adapt presently existing technology to their operations. Making an analogy to OSHA's authority to force technology in the promulgation of standards, the Secretary argued that employers should be held in violation of the Act even if the technology for compliance does not presently exist. It asserted that the authority to enforce standards to the point of the technology-forcing stage is necessary "in order that the proper pressure be placed on employers to stimulate technological development."

Commissioner Barnako did not agree with the Secretary's conclusion. For him it goes beyond the requirement of the standard itself and demonstrates a fundamental misunderstanding of what is meant when a standard is labeled "technology forcing." An employer is only in violation of the standard and therefore subject to a penalty if it fails to implement controls that are feasible at the time it allegedly violated the standard. An employer cannot be penalized for failing to do something it could not possibly have done. *Cf. Central of Georgia Railroad Co. v. OSHRC,* 576 F.2d 620, 6 OSHC 1784 (5th Cir. 1978)—violation cannot be based upon unsupported generalizations that an employer's efforts have not gone far enough; *General Electric Co. v. OSHRC,* 540 F.2d 67, 4 OSHC 1512 (2d Cir. 1976). According to his reasoning the Secretary must prove the technology to reduce noise levels existed at the time of the alleged violation, not at some specified or unspecified future time. Of course he feels the (noise) standard places a continuing duty on an employer who is not already in full compliance. As technology advances, the employer must implement new controls as they become feasible, until full compliance is attained. An employer who fails to keep pace with technological advances may violate the standard in the future. But an employer cannot be found to have violated the standard in the past based on its failure to implement controls that were not feasible at the time of the alleged violation. Specifically addressing the Secretary of Labor's contention, Barnako further stated:

> The Secretary's argument that the authority to promulgate technology forcing standards carries with it the authority to enforce such standards during the technology forcing stage misses the point. In promulgating standards, the Secretary can certainly anticipate that advances in technology will occur and set objectives that cannot immediately be met. To conclude otherwise would mean that the Secretary would have to engage in a new rulemaking proceeding every time there is a technological advance that would permit a further reduction in a particular hazard. Such a conclusion would be inconsistent with the very nature of performance standards, which are preferred by the Act, and which assume that industry will itself develop technology in seeking the most efficient way to comply. See *Diebold, Inc.*, 3 OSHC 1897, (1976), *aff'd*, 585 F.2d 1327, 6 OSHC 2002 (6th Cir. 1978). But rulemaking is different from adjudication, and the fact that the Secretary can anticipate technological advances in rulemaking does not mean that employers can be found to violate a standard based on their failure to implement non-existent technology.
>
> . . . To go further is to impose a research requirement on all employers, and the Secretary's attempt to impose such a requirement in rulemaking has been held unauthorized by the Act. [*American Iron & Steel Institute v. OSHA*, 577 F.2d 825, 6 OSHC 1451 (3d Cir. 1978), *petitions for cert. filed*, 47 U.S.L.W. 3525 (U.S. 1978).]

Commissioner Cleary's opinion stated:

> Although the controls suggested by the Secretary were not the "off-the-shelf" type, their final form is not so much the development of new technology as it is the adaptation of existing technology to a new use. In addition, I agree that the record demonstrates that the reduction in noise upon implementation of the controls will be significant.

Commissioner Cottine basically agreed with the Secretary of Labor's interpretation, for he stated:

> The Secretary correctly observes that the authority to promulgate technology-forcing standards "carries with it the corresponding authority to enforce these standards during this technology forcing stage. . . . This can be accomplished by requiring either the development of new technology *or* the adaptation of already developed technology. 29 U.S.C. §§651(b)(1) and (b)(4). Sampson's argument that technological feasibility has not been established when effective controls are not shown to be immediately available must, therefore, be rejected. The evidence in this case clearly establishes that specific engineering methods are available, using conventional materials and existing technology. Moreover, these methods can be adapted to Sampson's machinery to achieve significant noise reduction. See *Castle & Cooke Foods*. . . .

See also *Wheeling-Pittsburgh Steel Co.*, 7 OSHC 1581 (1979)—controls are technologically feasible even though the noise reduction accomplished by their implementation does not comply with the permissible exposure levels required by the standard. The controls must, however, achieve a *significant* reduction in noise levels. *Continental Can Co.*, 4 OHSC 1541 (1976); *West Point Pepperell, Inc., supra; Cf. American Petroleum Institute v. OSHA*, 581 F.2d 493, 6 OSHC 1959 (5th Cir. 1978)—the Secretary may promulgate a standard requiring employers to implement technological methods not yet developed but looming on the horizon only if the Secretary can show by substantial evidence that the employment conditions imposed by the proposed standard are *reasonably necessary* to provide safe or healthful places of employment. In *Sampson Paper Bag Co., supra*, a 10-dba noise reduction was found to be *significant.* In *Druth Packaging Co.*, 8 OSHC 1999 (1980), a reduction in the noise exposure levels by 3 to 8 dba was considered significant. Requiring the use of protective methods or devices, even though complete protection cannot be afforded to employees, is consistent with the position the Commission has taken with respect to other standards. See, *e.g., Valley Roofing Corp. and J. B. Eurell Co.*, 6 OSHC 1513 (1978) (personal protective equipment); *Building Products Co.*, 5 OSHC (1977)) (saw guarding).

2. Economic Feasibility

Although the question of the proper role of economic feasibility continues to be unresolved at the Commission level, the balancing test of the cost of controls against the benefit achieved as set forth in *Continental Can Company*, 4 OSHC 1541 (1976), continues to be applied. Thus in determining whether controls are economically feasible, all of the relevant cost and benefit factors must be weighed, and it must be shown that the cost of the controls is justified by the benefits they produce. This interpretation of the noise standard has been approved by the two courts of appeals that have considered the question.

Even though the Commission continues to follow *Continental Can*'s precedent, it has by no means reached agreement on the issue of economic feasibility. Commissioner Cleary and Cottine, for example, feel that to the extent the decision in *Continental Can* requires the use of a *formal* cost-benefit test, it should be overruled. The amount of reduction that can be expected to be achieved by implementation of controls and the number of employees receiving the daily benefit of the reduction in the cited location are considered to be the principal aspects of benefit; thus where the cost *per* employee on a daily basis in the *cited location* is justified by the benefit, the controls are considered to be economically feasible. See *Carnation Co.*, 6 OSHC 1730 (1978); *Pabst Brewing Co.*, 4 OSHC 2003 (1978)—evidence pertaining to cost of controls throughout the plant is not relevant *vis-à-vis* reduction of noise levels in uncited locations; *Wheeling-Pittsburgh Steel Corp.*, 7 OSHC 1581 (1979)—in a decision written by Commissioner Cottine which did not actually involve the issue of economic feasibility, the Commission noted that a significant employee health *benefit* was derived from the reduction of noise levels in the workplace.

The division in the Review Commission has resulted in a bifurcated approach on the issue of economic feasibility, in that one faction now maintains that technological feasibility is the only issue which may be raised as a defense and that costs of compliance are irrelevant except in a determination of the appropriateness of the abatement period. See *Continental Can Company,* 4 OSHC 1541 (1976); *Atlantic Steel Co.,* 6 OSHC 1289 (1978); *Castle & Cook Foods, Inc.,* 5 OSHC 1435 (1977); *West End Brewing Co.,* 5 OSHC 1276 (1977); *Great Falls Tribune Co.,* 5 OSHC 1443 (1977). In *Price-Potashnick-Codell-Oman, A Joint Venture,* 6 OSHC 1179 (1977), the positions of the divided Commission on the proper role of the economic feasibility issue in enforcement proceedings were discussed. Commissioner Barnako felt that the term "feasible" includes an economic as well as a technological dimension; Commissioner Cleary believed that economic factors are relevant only with respect to abatement and should play no role in determining feasibility.

The Commission was unable to agree and continues to remain at odds over the appropriate test on economic feasibility, as is shown in *Sampson Paper Bag, supra,* the first opportunity for the presently constituted Commission to express its opinion concerning the noise standard. Thus *Continental Can* remains the dispositive Commission precedent on the issue. At the time of writing, the varying views on the issue were:

(1) *Secretary of Labor:* Costs only render controls infeasible if they threaten the financial viability of the employer, and the burden to establish economic feasibility is on the employer.
(2) *Barnako:* The Secretary of Labor has the burden of proving economic feasibility, a formal cost-benefit analysis is required, and it must be shown that costs are justified by the benefits they produce.[5]
(3) *Cottine:* Costs have no role in a determination of feasibility *(e.g.,* Cleary's old position); they only enter into establishing an appropriate abatement date (any employer can afford controls if their cost is spread out over a sufficient length of time); the burden of proof to support a lengthened abatement period is on the employer.
(4) *Cleary:* Economic considerations must play *some* role in feasibility determinations with respect to controls. But Cleary disagrees with Barnako as to the weight; for him, if the expected reduction through implementation of engineering controls is insignificant in terms of increased protection, and compliance is not expected to approach permissible exposure levels while the cost of implementing controls is great, then the controls are prohibitively expensive, that is, economically infeasible; Secretary of Labor must prove controls are technologically feasible and costs are not prohibitively expensive. As is evident, Cleary's view has changed somewhat from that expressed in *Price-Potashnick.*

The question of what is feasible in terms of economic considerations has also been addressed in several judicial cases. For example, in *Industrial Union Dept. v. Hodgson,* 499 F.2d 467, 1 OSHC 1631 (D.C. Cir. 1974), the court held, *vis-à-vis* challenges to the *promulgation* of a standard, that a standard which is prohibitively expensive is not feasible and that Congress did not intend to put employers out of business. However, the concept of economic feasibility does not necessarily guarantee the continued existence of individual employers. See also *Society of The Plastics Industry, Inc. v. OSHA,* 509 F.2d 1301, 2 OSHC 1496 (2d Cir. 1975).

Several of the courts considering the issue of economic feasibility in an enforcement context have advocated the use of a cost-benefit test *(e.g.,* weigh the estimated costs of compliance against the benefits reasonably expected therefrom). The Seventh Circuit, for example, in *Turner Co. v. Secretary of Labor,* 561 F.2d 82, 5 OSHC 1790 (7th Cir. 1977), in applying a cost-benefit test, stated that the term "feasibility" includes economic as well as technological feasibility and stated that the Commission must consider the relative cost of implementing engineering controls versus the effectiveness of an existing personal protective equipment program utilizing earplugs; the fact, however, that controls are expensive does not necessarily mean they are not economically feasible. (Of course, the reverse is also true: the fact that an employer can afford controls does not make them economically feasible.) The court approved the Commission's interpretation in *Continental Can* that feasibility has both technological and economic aspects. See *RMI Co. v. Secretary of Labor,* 594 F.2d 566, 7 OSHC 1119 (6th Cir. 1979)—in comparing costs and benefits, benefits to employees weigh more heavily in the calculus than costs to the employer; *International Harvester Co. v. OSHRC,* 628 F.2d 982, 8 OSHC 1780 (7th Cir. 1980). See also *AFL–CIO v. Marshall,* 617 F.2d 636, 7 OSHC 1775 (D.C. Cir. 1979)—although economic feasibility must be considered in the promulgation of standards, a formal cost-benefit analysis is not required. In *American Petroleum Institute v. OSHA,* 581 F.2d 493, 6 OSHC 1959 (5th Cir. 1978), a decision which held that a cost-benefit comparison is required in the promulgation of standards was affirmed by the Supreme Court in *Industrial Union Department, AFL–CIO v. American Petroleum Institute,* 448 U.S. 607, 8 OSHC 1587 (1980). However, the basis for the affirmance was not the cost-benefit question but was the requirements of Section 3(8) of the Act, that standards be reasonably necessary and appropriate. See also *American Iron & Steel Institute, v. OSHA,* 577 F.2d 825, 6 OSHC 1451 (3d Cir. 1978); *U. P. S. of Ohio Inc. v. OSHRC,* 570 F.2d 806, 6 OSHC 1347 (8th Cir. 1978).

In a different context the court in *Atlantic & Gulf Stevedores, Inc. v. OSHRC,* 534 F.2d 541 4 OSHC 1061 (3d Cir. 1976), stated that:

> To frame the issue in slightly different terms, can the Secretary insist that an employer in the collective bargaining process bargain to retain the right to discipline employees for violation of safety standards which are patently reasonable, and are economically feasible except for employee resistance?
>
> We hold that the Secretary has such power.
>
> * * *
>
> But as we observed in *AFL–CIO v. Brennan,* 530 F.2d 109, 3 OSHC 1820 (3d Cir. 1975), the task of weighing the economic feasibility of a regulation is conferred upon the Secretary. He has concluded that stevedores must take all available legal steps to secure compliance by the longshoremen with the hardhat standard.
>
> We can perceive several legal remedies which an employer in petitioners' shoes might find availing. An employer can bargain in good faith with the representa-

tives of its employees for the right to discharge or discipline any employee who disobeys an OSHA standard. Because occupational safety and health would seem to be subsumed within the subjects of mandatory collective bargaining—wages, hours and conditions of employment, *see* 29 U.S.C. Sec. 158 (d)—the employer can, consistent with its duty to bargain in good faith, insist to the point of impasse upon the right to discharge or discipline disobedient employees. See *NLRB v. American National Insurance Co.*, 343 U.S. 395 (1952).

* * *

In this case petitioners have produced no evidence demonstrating that they have bargained for a unilateral privilege of discharge or discipline, that they have actually discharged or disciplined, or threatened to discharge or discipline, any employee who defied the hardhat standard, or that they have petitioned the Secretary for a variance or an extension of the time within which compliance is to be achieved. We conclude that as a matter of law petitioners have failed to establish the infeasibility of the challenged regulation.

Remember, controls may be economically feasible even though they adversely affect profits. *KLI, Inc.*, 6 OSHC 2057 (1978); *Arkansas Best Freight System, Inc. v. OSHRC*, 529 F.2d 649, 3 OSHC 1910 (8th Cir. 1976).

G. LACK OF EMPLOYER KNOWLEDGE

The Act provides in Section 17(k) that an employer may be found in serious violation of the Act "unless the employer did not, and could not, with the exercise of reasonable diligence, *know* of the presence of the violation." No language comparable to that found in Section 17(k) is found in Section 17(c) of the Act, which deals with other-than-serious violations. However, the courts and the Review Commission have held that knowledge must be proved in other-than-serious violations as well as serious violations; otherwise, it would be tantamount to holding an employer strictly liable. See *Brennan v. OSHRC (Raymond Hendricks d/b/a Alsea Lumber Co.)*, 511 F.2d 1130, 2 OSHC 1646 (9th Cir. 1975); *Horne Plumbing & Heating Co. v. OSHRC*, 528 F.2d 564, 3 OSHC 2060 (5th Cir. 1976); *D. R. Johnson Lumber Co.*, 3 OSHC 1125 (1975); *Welsh Farms Ice Cream, Inc.*, 5 OSHC 1755 (1977); *Scheel Construction, Inc.*, 4 OSHC 1824 (1976). The knowledge element of the Act refers to knowledge of the physical conditions which constitute a violation, not to knowledge of the requirements of the law. *S. W. Acoustics & Specialty, Inc.*, 5 OSHC 1091 (1977); *Charles A. Gaetano Construction Corp.*, 6 OSHC 1463 (1978).

Strictly speaking, knowledge is an aspect of the Secretary of Labor's burden of proof in establishing a violation of the Act. However, as discussed previously, it would behoove an employer to both require the Secretary of Labor to prove this element and also raise knowledge as an affirmative defense. The Secretary of Labor bears the burden of proving, in the case of both serious and other-than-serious violations, that the employer knew or, in the exercise of reasonable diligence, should have known of the violative condition. *D & R Builders and M & M Ready Mix*, 5 OSHC 1069 (1977); *Acchione & Canuso, Inc.*, 7 OSHC 2128 (1980). The knowledge requirement applies in both specific standard and general duty clause citations and can be shown by either actual or constructive knowledge. Industry custom and practice will generally establish the conduct of the reasonably prudent employer for the purpose of interpolating specific duties from general OSHA regulations and establishing constructive knowledge. *Power Plant Division, Brown & Root, Inc. v. OSHRC,* 590 F.2d 1363, 7 OSHC 1139 (5th Cir. 1979); *B & B Insulation, Inc. v. OSHRC,* 583 F.2d 1364, 6 OSHC 2062 (5th Cir. 1978); *H–30 Inc. v. Marshall,* 597 F.2d 234, 7 OSHC 1253 (10th Cir. 1979). If, however, an employer has actual knowledge of a hazard, the problem of fair notice does not exist and evidence of industry custom and practice "may be" irrelevant. *Cape & Vineyard Division of the New Bedford Gas and Edison Light Co. v. OSHRC,* 512 F.2d 1148, 2 OSHC 1628 (1st Cir. 1975). However, there must be adequate warning to an employer and substantial evidence to support a finding of a "specific, confirmed" knowledge. *Cotter & Company v. Marshall,* 598 F.2d 911, 7 OSHC 1510 (5th Cir. 1979); *Magma Copper Co. v. Marshall,* 608 F.2d 373, 7 OSHC 1893 (9th Cir. 1979).

The acts and knowledge of supervisory personnel are normally imputed to employers in cases of certain violations of the Act, especially willful violations. *F. X. Messina Construction Corp. v. OSHRC,* 505 F.2d 701, 2 OSHC 1325 (1st Cir. 1974). *Cf. Intercounty Construction Co. v. OSHRC,* 522 F.2d 777, 3 OSHC 1337 (4th Cir. 1975); *Mountain States Telephone & Telegraph Co., supra; Minnotee Contracting & Erection Company,* 6 OSHC 1369 (1978); *Alder Electric Co.,* 5 OSHC 1303 (1977); *Iowa Southern Utilities Co.,* 5 OSHC 1138 (1977).

It has been recognized, however, that in some instances an employer will not be held absolutely liable for the *unpreventable* and *unforseeable* conduct of experienced supervisory employees. Thus while an employer has a duty to ensure that supervisory personnel are performing their work safely and are themselves adequately supervised with regard to safety matters, *Floyd S. Pike Electrical Contractor, Inc.,* 6 OSHC 1675 (1978); *Kansas Power & Light Company,* 5 OSHC 1202 (1977), if in spite of all necessary precautions taken to prevent the occurrence of violations by supervisory employees, accidents happen, the employer will ordinarily not be held liable. *Horne Plumbing and Heating Company v. OSHRC & Dunlop,* 528 F.2d 564, 3 OSHC 2061 (5th Cir. 1976); *Brennan v. OSHRC (Alsea Lumber Co.),* 511 F.2d 1159, 2 OSHC 1640 (9th Cir. 1975); *F. H. Sparks of Maryland, Inc.,* 6 OSHC 1356 (1978); *Coastal Pile Driving, Inc.,* 6 OSHC 1133 (1977); *Hogan Mechanical, Inc.,* 6 OSHC 1221 (1977). In *Ocean Electric Corp. v. Secretary of Labor,* 594 F.2d 396, 7 OSHC 1149 (4th Cir. 1977), the court held that the burden of proof is on the Secretary of Labor to show foreseeability and preventability of the supervisor's actions, not the employer to show the reverse. See also *The Mountain States Telephone and Telegraph Co. v. OSHRC,* 623 F.2d 155, 8 OSHC 1557 (10th Cir. 1980).

An employer must show it established work rules designed to prevent the violation, adequately communicated these rules to its employees, and adequately enforced these rules when violations were discovered to avoid liability for its supervisor's actions. *Wander Iron Works, Inc.,* 8 OSHC 1354 (1980); *Asplundh Tree Expert Co.,* 6 OSHC 1951 (1978); *Granite-Graves, A Joint Venture,* 6 OSHC 1909 (1978)—failure to show adequate safety training given to its foreman; *Bay Iron Works, Inc.,* 4 OSHC 1420 (1976); *Hammonds Construction, Inc.,* 3 OSHC 1260 (1975). In addition, supervisory employees must be adequately supervised and properly trained to prevent any imputation to the employer through constructive knowledge of a supervisory employee's actions. *Jake Heaton Erecting Co., Inc.,* 5 OSHC 1908 (1977); *Brisco/Arace Conduit, A Joint Venture,* 5 OSHC 1167 (1977).

The lack of employer knowledge defense is intimately related to the isolated occurrence and employee misconduct defenses, for in order to establish the latter two defenses, an employer must prove that the violation was a deviation from a well-known and well-enforced rule, policy, or instruction and that the deviation was unknown to and could not have been foreseen by the employer in the exercise of reasonable diligence. See *Remke Southern Div., Inc.,* 6 OSHC 1842 (1978); *Jobs Building Services, Inc.,* 6 OSHC 1535 (1978); *F. H. Sparks of Maryland, Inc.,* 6 OSHC 1356 (1978); and *Floyd S. Pike Electrical Contractor, Inc.,* 6 OSHC 1675 (1978).

H. IMPOSSIBILITY OF PERFORMANCE/COMPLIANCE

The defense of impossibility of performance, sometimes referred to as impossibility of compliance, is applicable whenever an employer can establish *either* that the specific nature of the work being performed would be functionally impossible/would preclude performance if compliance with the cited standard were required *or* if abatement action is required in order to comply with a standard that no feasible measures of compliance with the standard's requirements exist vis-à-vis the particular employer's specific operations. *M. J. Lee Construction Co.,* 7 OSHC 1140 (1979); *Masonry Contractors, Inc.,* 8 OSHC 1155 (1980); *F. H. Lawrence,* 8 OSHC 1063 (1980); *Diamond Roofing Co.,* 8 OSHC 1080 (1980); *Electrical Constructors of America;* 8 OSHC 2117 (1980). The employer must, however, demonstrate *impossibility,* not mere impracticability, loss of efficiency, or inconvenience. *International Paper Company,* 2 OSHC 3173 (1974); *Alberici-Koch-Laumand, A Joint Venture,* 5 OSHC 1895 (1977); *Otis Elevator Co.,* 5 OSHC 1429; *General Steel Fabricators, Inc.,* 5 OSHC 769 (1977); *Pass & Seymour, Inc.,* 7 OSHC 1961 (1979); *Turnbull Millwork Company,* 6 OSHC 1148 (1977); *United States Steel Corp. v. OSHRC,* 537 F.2d 780, 4 OSHC 1424 (3d Cir. 1976); *Taylor Building Associates,* 5 OSHC 1083 (1977). See *Robert W. Setterlin & Sons Co.,* 4 OSHC 1214 (1976)—requiring standard guardrails on open-sided tenth floor of building under construction would make impossible the installation of necessary cement blocks on the outside wall of building; *Worley Brothers Granite Company,* 4 OSHC 2043 (1977)—compliance would "significantly detract from the quality of the finished product"; *Sletten Construction Company,* 6 OSHC 1091 (1977); *Mushroom Transportation Company, Inc.,* 6 OSHC 1188 (1977); *M. J. Lee Construction Co.,* 7 OSHC 1140 (1979); *Geisler Ganz Corporation,* 8 OSHC 2083 (1980).

Performance under the standard must be truly impossible; it is not sufficient that compliance with the standard would be burdensome, result in added inconvenience, impracticability, or inefficiency, would involve additional time or manpower, or would cause a financial loss to the employer. *Havens Steel Company,* 6 OSHC 1564 (1978); *Underhill Construction Corp.,* 2 OSHC 1556 (1975); *Taylor Building Associates,* 5 OSHC 1083 (1977); *Turnbull*

Millwork Company, 6 OSHC 1148 (1977). Moreover, the employer who raises the defense is also required to establish: (1) that alternative means of protection are not available, *Dorey Electric Company v. OSHRC,* 553 F.2d 357, 5 OSHC 1285 (4th Cir. 1977); *Julius Nasso Concrete Co., S & A Concrete Co., Inc., A Joint Venture,* 6 OSHC 1171 (1977); and (2) that an application for a variance was made and rejected *or* would be inappropriate under the circumstances. *"AH" Metal Fabricators, Inc.,* 5 OSHC 1403 (1977); *Mikel Company, Inc.,* 5 OSHC 1190 (1977); *Warnel Corp.,* 4 OSHC 1034 (1976) *concurring opinion.*

Even though an employer may be able to "make his case" on the impossibility argument, it will not be sustained if protection was possible for part of the time. *Burk Construction Corporation,* 5 OSHC 1433 (1977); *Hurlock Roofing Co.,* 7 OSHC 1867 (1979); *A & S Millworks and Rentals,* 6 OSHC 1212 (1977) —saw guards could not be used during specialty type cuts, but could be used during other cutting; *Midsouth Glass Company, Inc.,* 4 OSHC 1958 (1977). Employee exposure must be shown, however, during a time when protection was possible. *Modern Fabrication & Equipment Company,* 5 OSHC 1890 (1977). The Commission in *F. H. Lawson Co.,* 8 OSHC 1063 (1980), recently seemed to require still another element of proof before an impossibility defense will be sustained:

> The flaw in respondent's proof of impossibility lies in its *failure to even consider* changes in its mode of production that might lead to compliance. *Hughes Brothers, Inc.* [6 OSHC 1630 (1978)]. In view of *Lawson's* failure to consider any change in its mode of operation, the Secretary's failure to rebut its evidence is without effect since respondent has not sustained the burden of proving impossibility.

I. GREATER HAZARD

If a hazard, which is cited as a violation of *either* the specific or general duty sections of the Act, would be *increased* by the abatement required in the citation, then an employer may properly argue that the citation must be vacated. *Parr, Inc.,* 4 OSHC 1449 (1976); *Price Cabinet Shop,* 4 OSHC 1024 (1976); *Kawneer Co., Inc.,* 4 OSHC 1120 (1976); but see *U.S. Steel Corporation v. OSHRC,* 537 F.2d 780, 4 OSHC 1424 (3d Cir. 1976), where the court rejected the "greater hazard defense" relative to the use of safety nets in the construction of the 109th floor of the Sears Tower, since the critical predicate that there was no safe and practical method of erection was not established by the employer. (A related defense which should be considered is that the means of protection provided is as safe or safer than that specified in standard. *Beach Boulevard Shell,* 5 OSHC 1816 (1977).)

The greater hazard defense is strictly applicable only where noncompliance is found to be safer than compliance vis-à-vis the particular cited hazard or work practice in issue. *National Steel & Shipbuilding Co.,* 6 OSHC 1680 (1978), *aff'd,* 607 F.2d 311, 7 OSHC 1837 (9th Cir. 1979); *Diamond Roofing Co.,* 8 OSHC 1080 (1980); *Marion Power Shovel Co.,* 8 OSHC 2244 (1980). The greater hazard defense is not applicable where compliance or abatement would create a new, different, separate, and basically unrelated hazard for the same employees. *Greyhound Lines West v. Marshall,* 575 F.2d 759, 6 OSHC 1636 (9th Cir. 1978). *Carpenter Rigging & Contracting Corp.,* 2 OSHC 1544 (1975); *Sierra Construction Corp.,* 6 OSHC 1278 (1978); *General Steel Fabricators, Inc.,* 5 OSHC 1768 (1977); *Technical Tubing Testers, Inc.,* 6 OSHC 1615 (1978). The fact that employees are more frequently exposed or exposed for a more prolonged period of time by compliance would, however, establish that the hazard is greater. *J. L. Foti Construction Company, Inc.,* 5 OSHC 2021 (1977); *C & W Roofing Company, Inc.,* 5 OSHC 1662 (1977). The hazard is also greater if compliance creates a greater possibility of a more severe injury occurring or a greater probability of an accident's occurring even though an injury would be less severe. *Otis Elevator Company,* 5 OSHC 1605 (1977).

An employer must establish that alternative means of protecting employees are unavailable or, if alternative means of protection are available, that they are used and are at least as safe as the methods required by compliance. *Meilman Food Industries,* 5 OSHC 2060 (1977); *Limbach Company,* 5 OSHC 1357 (1977); *Hughes Brothers, Inc.,* 6 OSHC 1830 (1978); *Russ Keller, Inc., t/a Surfa Shield,* 4 OSHC 1758 (1976); *National Steel, supra.* Finally, the employer must prove that a variance application under Section 6(d) of the Act was filed and rejected or would be inappropriate or futile. The court in *H. S. Holtze Construction Co. v. Marshall,* 627 F.2d 149, 8 OSHC 1785 (8th Cir. 1980), following the decision of other circuits, stated:

> On appeal, the Secretary argues that in order to plead the "greater hazard" defense, petitioner must show: (1) that the hazards of compliance are greater than the hazards of noncompliance; (2) that alternative means of protecting employees are unavailable; and (3) that a variance application under section 6(d) of the Act would be inappropriate.

See also *Noblecraft Industries, Inc. v. Secretary of Labor,* 614 F.2d 199, 7 OSHC 2059 (9th Cir. 1980); *J. F. Shea Company, Inc.,* 5 OSHC 1988 (1977); *Weyerhaeuser Company,* 4 OSHC 1972 (1977); *Giffen Industries of Tampa, Inc.,* 7 OSHC 1073 (1978); *M.J. Lee Construction Co.,* 7 OSHC 1140 (1979). The mere fact that enforcement action has been taken does not establish the futility of a variance. *General Electric Company v. Secretary of Labor,* 576 F.2d 558, 6 OSHC 1541 (3d Cir. 1978).

In *Fred Wilson Drilling Company, Inc.,* 6 OSHC 1942 (1978), the Commission (by means of a final order) rejected an employer's greater hazard defense, since it was not shown that the employer sought a variance from the standard. It was proved, however, that the use of a guard would have presented a greater danger to the employees than would have noncompliance with the standard. 1910.212(a)(1). On review, the Fifth Circuit, in *Fred Wilson Drilling Company, Inc. v. Marshall,* 624 F.2d 38, 8 OSHC 1921 (5th Cir. 1980), also found that the use of a guard would have created a greater hazard than the nonuse of a guard and, accordingly, vacated the citation for the violation of the standard. In doing so the court did not mention the fact that the employer had not applied for a variance. Thus, implicitly, the court rejected the requirement that variance be filed prior to establishment of the greater hazard defense. See also *Bristol Steel and Iron Works, Inc.,* 7 OSHC 2192 (1979)

—the failure to apply for a variance from compliance with 29 CFR 1926.105(a) did not bar the greater hazard defense.

J. ABSENCE OF RECORDABLE INJURIES

Quite often, an employer will raise the defense that since there was an absence or no evidence of "recordable" injuries, or that it had a favorable safety record via the cited condition, work area, or job, the cited condition should not be viewed as a violation of the Act or as hazardous to the employees. The position of the Review Commission appears to be that this evidence is not relevant to the characterization of the violation or the legitimacy of the citation. *F. H. Lawson Company,* 8 OSHC 1063 (1980). In *American Airlines, Inc.,* 4 OSHC 1630 (1976), the Commission stated in a footnote that the absence of recorded injuries, "while relevant, cannot be considered controlling," in that the frequency of injury does not alone negate the appropriateness of a citation. See also *Lombard Brothers, Inc.,* 5 OSHC 1716 (1977)—not a defense since the purpose of Act is to prevent all injuries; *Smead Manufacturing Company,* 7 OSHC 1515 (1979). But see *Chambers Engraving Company, Inc.,* 6 OSHC 1534 (1978)—during eighteen years of operation, only one injury occurred, a broken fingernail, which supported the fact that there was no violation of 29 CFR 1910.212(a)(3)(ii); *Johnson Steel and Wire Co., Inc.,* 6 OSHC 1308 (1977); *A. E. Burgess Leather Co. v. OSHRC,* 576 F.2d 948, 6 OSHC 1661 (1st Cir. 1978)—the fact that no substantial injury occurred on the machines despite years of use was entitled to great weight but is not conclusive.

Evidence of the absence of injuries or the existence of a low injury rate is, of course, relevant in determining: (1) whether a hazard exists, *Buckeye Industries, Inc.,* 3 OSHC 1837 (1975), *aff'd,* 587 F.2d 231, 6 OSHC 2181 (5th Cir. 1979); *Pass & Seymour, Inc.,* 7 OSHC 1961 (1979); *A. E. Burgess Leather Company, Inc.,* 5 OSHC 1096 (1977), *aff'd* 576 F.2d 948, 6 OSHC 1661 (1st Cir. 1978); *Dayton Tire & Rubber Co.,* 6 OSHC 2053 (1978); *Papertronics, Inc.,* 6 OSHC 1818 (1978); *General Electric Co.,* 7 OSHC 2187 (1980); *cf. Arkansas–Best Freight Systems, Inc. v. OSHRC,* 529 F.2d 649, 3 OSHC 1910 (8th Cir. 1976)—a hazard requiring abatement may exist even in the absence of recorded injuries; (2) whether a reasonably prudent man would believe that personal protective equipment is necessary, *Great Atlantic & Pacific Tea Co., Inc.,* 3 OSHC 1789 (1975); (3) whether the penalty assessment is correct, *Western Steel Manufacturing Co.,* 4 OSHC 1107 (1976); and (4) whether a citation is properly characterized as *"de minimis"* rather than serious or other-than-serious. *Nicky Lou Fashions,* 6 OSHC 1824 (1978); *Soule Steam Feedworks,* 6 OSHC 1852 (1978).

K. LACK OF EMPLOYEE EXPOSURE

An often presented defense is that there is a lack of employee exposure/access to the cited hazard. *Brennan v. Giles & Cotting Inc.,* 504 F.2d 1255, 2 OSHC 1243 (4th Cir. 1974). While the Secretary of Labor has the burden in this respect, *Arvin Millwork Co.*, 2 OSHC 1056 (1974); *Bechtel Corp.*, 2 OSHC 1336 (1974) he need not prove actual employee exposure to a hazard but must only show that employees will be, are, or have been exposed or that it is reasonably foreseeable or predictable that they will have access to a zone of danger or violative condition while in the course of performance of their normal work assignments,[6] during normal ingress or egress to the workplace, or during breaks taken during the workday. *General Electric Co.*, 7 OSHC 2183 (1980); *H. B. Zachary Co. (Int'l)*, 8 OSHC 1669 (1980); *Zwicker Electric Company, Inc.*, 5 OSHC 1338 (1977); *B. F. Goodrich Textile Products,* 5 OSHC 1458, (1977); *Frank C. Bibson,* 6 OSHC 1577 (1978); *Sierra Construction Corp.*, 6 OSHC 1279 (1978); *J. R. Simplot Co.*, 6 OSHC 1992 (1978); *Pullman Brick Co. Inc.*, 1 OSHC 3045 (1973). The risk of exposure, however, must be more than speculative. See *Weyerhaeuser Co.*, 4 OSHC 1972 (1977), *vacated and remanded sub nom., Noblecraft Industries, Inc., v. OSHRC,* 614 F.2d 199, 7 OSHC 2059 (9th Cir. 1980), relative to questions on applicability of the cited standard to the employers operations; *Stock Manufacturing & Design Co.*, 8 OSHC 2145 (1980); *cf. Roanoke Iron & Bridge Works, Inc. v. OSHRC,* 588 F.2d 1351, 7 OSHC 1014 (4th Cir. 1978). The proof must indicate not only the employees who could be expected to have access to the zone of danger but also must show by what means access could be achieved, in order to establish a reasonable or foreseeable predictability of access.

A defense of lack of employee exposure is not easily established, especially with respect to specific standards, for the Commission and the courts have held that an employer is in *prima facie* violation of Section 5(a)(2) of the Act if it fails to comply with a safety or health standard and its own employees or employees of another employer are exposed to the resulting hazard. *Brennan v. OSHRC (Underhill Constr. Corp.),* 513 F.2d 1032, 2 OSHC 1641 (2d Cir. 1975); *Marshall v. Knutson Constr. Co.*, 566 F.2d 596, 6 OSHC 1077 (8th Cir. 1977); *Beatty Equipment Leasing, Inc. v. Secretary,* 577 F.2d 534, 6 OSHC 1699 (9th Cir. 1978); *Harvey Workover, Inc.*, 7 OSHC 1687 (1979); *Gil Haugan,* 7 OSHC 2004 (1979); *Anning-Johnson Co.*, 4 OSHC 1193 (1976); *cf. Austin Building Company,* 8 OSHC 2151 (1980)—Secretary of Labor is not required to prove employee exposed to recognized hazard to establish violation of 1926.28(a) since extension of Section 5(a)(1) analysis to cases arising under a specific construction standard is inappropriate. In *Hydroform Products Corporation,* 7 OSHC 1995 (1979) the fact that an employer's president was exposed to a hazard did not relieve the employer of liability because the president as a working person was entitled to protection under the Act.

If an employer knows, or with exercise of reasonable diligence should know of the presence of a violation, the Commission has imposed a duty thereon to take action to prevent its employees from going into a zone of danger. *Stahr and Gregory Roofing Company, Inc.*, 7 OSHC 1010 (1979). Similarly, employers must insist that employees not only wear personal protection equipment but also use it, in order to avoid exposure. *Turner Communications Corporation v. OSHRC,* 612 F.2d 941, 8 OSHC 1038 (5th Cir. 1980).

An employer must show more than that the area where the violative condition is located is apart from normal work areas or barred to unauthorized employees to disprove exposure, since only access to the violative condition is necessary to establish exposure. *Brown-McKee, Inc.,* 8 OSHC 1248 (1980); *Otis Elevator Co.,* 6 OSHC 2048 (1978). The distance and conditions that are necessary to show lack of employee exposure are very important and may vary from "hazard to hazard," "machine to machine," and "zone to zone." For example, in *Fred Wilson Drilling Company,* 6 OSHC 1942 (1978), *rev'd on other grounds.* 624 F.2d 38, 8 OSHC 1921 (1980)—the proof that employees had to step over various pieces of machinery in order to come into a zone of danger created by unguarded belts, and that employees were not required to be in the danger zone, established no exposure and no hazard; *Circle Bar Drilling Co.,* 6 OSHC 2061 (1978)—the fact that employees had no duties that required their presence in danger zone and employer enforced work rule requiring that they stay clear of area required vacation of citation; *F. H. Lawson Co.,* 8 OSHC 1063 (1980)—operators' hands ten to twenty inches from point of operation during work cycle establishes exposure; *Roofing Systems Consultants, A Division of Bit U Tech. Inc.,* 8 OSHC 1446 (1980)—employees within four feet of an unguarded floor edge or opening are exposed; *Brown-McKee, Inc.,* 8 OSHC 1248 (1980)—access is established if it is shown that defective equipment is available for the use of employees;[7] *Cornell & Co., Inc.,* 5 OSHC 1736 (1975)—no employee exposure, since employees came no closer than fifteen feet to a fall hazard, but employees within ten feet were within the zone of danger, although employees working the same distance from a tripping hazard were not exposed; *Patterson & Wilder Construction Co., Inc.,* 5 OSHC 1153 (1977)—a citation for an unshored trench was vacated, since no employees were required to enter it, and a citation for having a gap in a railing was vacated, since it was necessary for repair work and was not likely that employees would be in the zone of danger; *Seaward Construction Co., Inc.,* 5 OSHC 1422 (1977).

When dealing with employee exposure to health hazards (*e.g.* 1910.1000–.1046), it must be established not only that an employee was exposed, but that the prescribed concentrations of the substance were exceeded for a certain period of time in excess of either the threshold limit value (TLV) or permissible exposure limit (PEL). *General Motors Corp., Delco Marine Division,* 8 OSHC 1718 (1980); *Asarco, Inc.,* 8 OSHC 2076 (1980); *Lutz, Daily & Brain—Consulting Engineers,* 8 OSHC 2241 (1980).

In *Witco Chemical Corp.,* 8 OSHC 1134 (1980), a citation based upon employee exposure to impermissible levels of "TDI" was changed from a serious to an other-than-serious violation, since there was only *brief exposure* of two employees in excess of the standard, and OSHA did not establish that brief exposure could result in serious bodily harm or death. See also *Stahr & Gregory Roofing Company, Inc.,* 7 OSHC 1010 (1979)—the brevity/duration of exposure is a factor to be considered in determining gravity for the purpose of penalty assessment; it does not negate the finding of a violation.

L. MACHINE NOT IN USE

It is possible to raise a defense in situations where equipment or machinery is cited for failure to have a guard, that said equipment or machinery is not in use, was inoperable, or was unguarded in order to effect repairs. In the cases of *Atlantic Coast Development Corp.*, 3 OSHC 1789 (1975); *Lembke Construction Co., Inc.*, 5 OSHC 1611 (1977); and *Iron Master, Inc.*, 5 OSHC 1776 (1976), the question of whether machinery or equipment was inoperable was an integral part of the citation.

This defense is also an aspect of employee exposure to a hazard. There can be no employee exposure if there is no machine in operation. See *General Electric Company v. OSHRC,* 540 F.2d 67, 4 OSHC 1512 (2d Cir. 1976), where the lack of evidence that a powered work platform was used without guardrails during the six months preceding the issuance of a citation required the vacation of citation for violation of 29 CFR 1910.23(c)(1), since the Section 9(c) limitations period was exceeded. In *Weyerhaeuser Co.*, 4 OSHC 1972, n.13 (1978), *vacated and remanded on other grounds,* 614 F.2d 199, 7 OSHC 2059 (9th Cir. 1980), the record showed that all of the saws were unguarded and were available for use by employees. Also, that some of the saws had been recently used was indicated by the presence of fresh sawdust. Some were plugged into electrical outlets, and one was seen in actual operation. This evidence was ample to satisfy the Secretary's burden of proof and discount the employer's "not in use" defense. *Thunderbird Coos Bay, Inc.*, 3 OSHC 1904 (1976); *Konkolville Lumber Company, Inc.*, 3 OSHC 1796 (1975); *Huber, Hunt & Nichols, Inc. and Blount Bros. Corp.*, 4 OSHC 1406 (1976).

M. INAPPLICABLE STANDARD

The defense of an inapplicable standard is often raised in conjunction with either (1) the application of the general duty clause in situations where a specific standard covering the same hazard is applicable or (2) the application of a general standard where a specific standard is applicable.

A general standard is inappropriately cited if there is a more specific standard applicable to the cited hazard. See discussion, Chapter IV, *supra.* 29 CFR 1910.5(c) provides in this respect that if a particular standard is specifically applicable to a particular operation or process, it shall prevail over any different general standard which might otherwise be applicable to the same condition, practice, means, method, operation, or process. See also FOM, Ch. X.

Some standards apply only to employers engaged in certain industries, such as construction, maritime, or agriculture. Thus citation under the general industry standards would be improper. *Chapman and Stephens Company,* 5 OSHC 1395 (1977); *Bechtel Power Corp.*, 4 OSHC 1005 (1976). For example, the use of the maritime personal protective equipment standard would properly apply to employers in the maritime industry rather than the general industry personal protective equipment standard.

Moreover, some standards are advisory and not mandatory. If a standard uses the word "should" instead of "shall" then it is directory. See *Wheeling-Pittsburgh Steel Corporation,* 5 OSHC 1495 (1977); *McHugh and McHugh,* 5 OSHC 1165 (1977); *Kennecott Copper Corp. v. Usery,* 577 F.2d 1113, 6 OSHC 1197 (10th Cir. 1977).

Similarly, a more specific general standard within the same CFR part may be more applicable to a condition than a less specific general standard *(e.g.,* guarding requirements of 29 CFR 1910.212 v. 29 CFR 1910.213). For example, a standard which requires that ladders be maintained in good condition is not applicable to an alleged violation for a ladder with a broken siderail, as a specific standard applies to broken siderails. *Allied Equipment Company,* 5 OSHC 1401 (1977). A standard which requires guarding of open floors or platforms (29 CFR 1926.500(d)(1) does not apply to flat roofs. See *Diamond Roofing Co. v. OSHRC,* 528 F.2d 645, 4 OSHC 1001 (5th Cir. 1976); *Langer Roofing and Sheet Metal Company v. Secretary of Labor,* 524 F.2d 1337, 3 OSHC 1685 (7th Cir. 1975). See generally *Matson Terminals, Inc.,* 3 OSHC 2071 (1976); *Saturn Construction Company, Inc.,* 5 OSHC 1686 (1977). Similarly, a citation under the general duty clause is not proper if a specific standard applies. *Mississippi Power & Light Co.,* 7 OSHC 2036 (1979); *Kaiser Aluminum Co.,* 4 OSHC 1162 (1977); *A. Prokosch & Sons Sheet Metal, Inc.,* 8 OSHC 2077 (1980).

N. REFUSAL TO COMPLY

Related to the employee misconduct/isolated occurrence defense is the defense that employees refused to obey an employer's reasonable direction relative to compliance with the Act. Of course, there are no sanctions available under the Act to make employees comply. A threat of a strike or refusal to comply has not been considered to be a defense. *Poston Bridge & Iron, Inc.,* 1 OSHC 3273 (1974).

In *Weyerhaeuser Co.,* 3 OSHC 1107 (1974) the employer argued that it had done everything reasonable to obtain compliance by its employees, but that the employees would strike if it enforced the wearing of personal protective equipment. The Commission affirmed the citation, stating that a fear of a strike does not relieve an employer of its duty to comply with the Act. See also *Wallace Roofing Co.,* 8 OSHC 1492 (1980)—argument that employer was not responsible for violation since employee refused to wear shield is rejected since employer failed to take any action beyond oral reprimands to enforce rule.

Moreover, the Act is not intended to prevent employees from purposely harming themselves. *Brennan v. OSHRC (Hanovia Lamp Div., Canrad Precision Industries),* 502 F.2d 946, 2 OSHC 1137 (3d Cir. 1974)—Section 5(a)(1) does not impose liability for suicidal self-exposure of employees; *National Realty, supra.*

Similar cases discussed previously which have followed the same viewpoint include *Atlantic & Gulf Stevedores, Inc. v. OSHRC,* 534 F.2d 541, 4 OSHC 1061 (3d Cir. 1976); *Asplundh Tree Expert Co.,* 7 OSHC 2074 (1979). See also *Nickylou Fashions, Inc.,* 6 OSHC 1824 (1978). It was not nearly enough to require adherence to strict safety measures. An employer has a duty to force its employees to comply, up to and including discharge. *Sheets Tree Expert Co.,* 4 OSHC 2042 (1977); *Iowa Southern Utilities Co.,* 5 OSHC 1138 (1977); *Howard P. Foley Co.,* 5 OSHC 1501 (1977).

O. LACK OF CAUSAL CONNECTION; NO MATERIAL REDUCTION

In *Champlin Petroleum Company v. OSHRC,* 593 F.2d 647, 7 OSHC 1241 (5th Cir. 1979), the employer was cited for violation of Section 5(a)(1) of the Act when three employees were injured, one fatally, by an oil fire which started when auto-ignitable hot oil escaped from a pipeline bleeder valve which had been opened with a wrench instead of a valve handle. The court vacated the citation because it found *no direct causal connection* between the missing valve handle and the fire. The record failed to show that an employee would be able to close the leaking valve before ignition of the hot oil, whether or not the valve had a handle. Moreover, wider communication of the employer's procedure on valve openings would have had no *material effect* on the likelihood of injuries occurring. See also *Brennan v. OSHRC,* 494 F.2d 460, 1 OSHC 1623 (8th Cir. 1974).

P. ADVERSE FINANCIAL CONDITION

The Commission has held that a business in reorganization is not exempt from compliance with the Act and that the federal bankruptcy courts lack jurisdiction to enjoin the Commission from enforcing the Act. *Chicago, Rock Island & Pacific Railroad Company,* 3 OSHC 1694 (1975). However, an employer's adverse financial condition or economic distress will usually support a general reduction in the penalties assessed for violation of the Act. *Perry Foundry, Inc.,* 6 OSHC 1690 (1978); *Willse Industries Inc.,* 6 OSHC 1424 (1978); *Pittsburgh Brewing Company,* 7 OSHC 1099 (1978).

Q. LACK OF HAZARD

It is the Secretary of Labor's burden to prove the existence of a hazard in order to establish a violation. If there is no hazard or exposure to a hazard, there is no basis upon which to base a citation. *Fred Wilson Drilling Company, Inc., supra; Seaward Construction Co., Inc.,* 5 OSHC 1422 (1977); *California Rotogravure Co.,* 2 OSHC 1515 (1975), *rev. denied,* 549 F.2d 807, 5 OSHC 1031 (9th Cir. 1977). Abatement efforts subsequent to the issuance of a citation neither excuse nor negate an employer's failure to comply with a standard. *Whirlpool Corporation,* 8 OSHC 2248 (1980).

Certain standards contain requirements or prohibitions that by their terms need only be observed when employees are exposed to the hazard described generally in the standard *(e.g.,* 29 CFR 1910.212(a)(3)(ii) and 1910.132(a)). *Parnon Construction, Inc.,* 5 OSHC 1232 (1977)—requirement that fire extinguishers be provided is inapplicable unless it is shown that a fire hazard exists.

The test for determining the existence of a hazardous condition requiring the use of personal protective equipment within the meaning of 29 CFR 1910.132(a) or 1926.28(a) is whether a reasonable person familiar with the circumstances surrounding an allegedly hazardous condition, including any facts unique to a particular industry, would recognize a hazard warranting the use of personal protective equipment. In *Brown-McKee, Inc.,* 8 OSHC 1248 (1980), the Commission stated:

> In order to prove noncompliance with section 1926.28(a), the Secretary must demonstrate either that there is exposure to hazardous conditions *or* that there is another standard in Part 1926 indicating the need for personal protective equipment. *S & H Riggers and Erectors, Inc.,* 7 OSHC 1260 (1979), *appeal filed,* No. 79–2358 (5th Cir. June 7, 1979). In order to make out his prima facie case under the former test, the Secretary must prove employee exposure to a hazardous condition requiring the use of personal protective equipment and must identify the appropriate form of personal protective equipment to elminate the hazard. *S & H Riggers and Erectors, Inc.,* 7 OSHC at 1266. The test of whether a hazardous condition exists within the meaning of section 1926.28(a) . . . is whether a reasonable person familiar with the factual circumstances surrounding the allegedly hazardous condition, including any facts unique to a particular industry, would recognize a hazard warranting the use of personal protective equipment.
>
> The Secretary sustained his burden of proof in this case. The record establishes employee exposure to an obvious fall hazard, thereby requiring the use of personal protective equipment. *See e.g., Hurlock Roofing Co.,* 7 OSHC 1867 (1979). The citation clearly refers to safety belts and lifelines as the appropriate means of protection. Moreover, the compliance officer testified without contradiction that a safety belt and lifeline would be the appropriate method of abatement.[8]

See *General Electric Co.,* 7 OSHC 2183 (1980); *Owens Corning Fiberglas Corp.,* 7 OSHC 1291 (1979); *Austin Bridge Co.,* 7 OSHC 1761 (1979). Using this test, the Commission in *General Electric* found that a reasonable person would not recognize the existence of a hazard from loads swinging and striking an employee. Furthermore, the record contained no evidence indicating any injuries to employees resulting from the allegedly violative condition. The absence of injuries supported the Commission's conclusion that a reasonable person would not recognize the existence of a hazard of objects falling from the crane or its load. See also *OSCO Industries, Inc.,* 8 OSHC 1799 (1980); *Continental Oil Co.,* 7 OSHC 1432 (1979)—the absence of injuries showed a lack of hazard and supported a finding of a *de minimis* violation.

Many standards, however, include requirements or prohibitions which by their terms must be observed whenever specified conditions, practices, or procedures are encountered or raise a presumption of a hazard which must be rebutted. *E.g.,* 29 CFR 1926.500(d)(1); *Williams Enterprises of Georgia, Inc.,* 7 OSHC 1900 (1979). These standards are predicated upon the existence of a hazard when their terms are met; accordingly, OSHA is not required to prove that noncompliance with these standards creates a hazard in order to establish a violation. *Vecco Concrete Construction, Inc.,* 5 OSHC 1960 (1977); *Austin Bridge Co.,* 7 OSHC 1761 (1979).

R. EQUIVALENT PROTECTION

In certain circumstances, the greater hazard defense is not met, but it is unreasonable to enforce the citation, since alternative equivalent protection was provided. *H. S. Holtze Co. v. Marshall,* 627 F.2d 149, 8 OSHC (8th Cir. 1980); *Diamond Roofing Co.,* 8 OSHC 1080 (1980); *cf. Brennan v. Pearl Steel Erection Co.,* 488 F.2d 337 1 OSHC 1429 (5th Cir. 1973); *Wander Iron Works, Inc.,* 8 OSHC 1354, 1355 n.5 (1980).

FAILURE TO FOLLOW "FOM"

The Commission has uniformly held that the issuance of citations in violation of the OSHA Field Operations Manual is not an affirmative defense, since the manual's guidelines do not have the force of law and are not binding on the Secretary of Labor or the Commission. The FOM is directory only, serving as a guide for implementing the Act, and does not afford procedural or substantive rights. *FMC Corporation,* 5 OSHC 1707 (1977); *Combustion Engineering, Inc.,* 5 OSHC 1943 (1977); *Laclede Gas Company,* 7 OSHC 1874 (1979); *P. A. F. Equipment Co.,* 7 OSHC 1209 (1979); *General Electric Co.,* 7 OSHC 2186 (1980)—the contention that a citation must be vacated because a compliance office did not personally observe a violation, contrary to FOM, was rejected.

This is not to say, however, that the contents of the FOM can never be accorded legal significance. *FMC Corporation, supra.*

HAZARD NOT RECOGNIZED

Section 5(a)(1) of the Act, the general duty clause, requires the employer to provide a place of employment which is free from recognized hazards that are causing or likely to cause death or serious physical harm to employees. Although general duty clause citations are discussed in another part of the book, one defense to an alleged violation of the general duty clause is that the Secretary of Labor has failed to prove that there is a recognized hazard. A recognized hazard can be shown either through the employer's actual knowledge of the hazard or through its constructive knowledge, relying upon the custom and practice of a reasonably prudent employer in the employer's industry. See *Schriber Sheet Metal and Roofers, Inc. v. OSHRC,* 597 F.2d 78, 7 OSHC 1246 (6th Cir. 1979); *H–30, Inc. v. Marshall,* 597 F.2d 234, 7 OSHC 1253 (10th Cir. 1979); *Magma Copper Co. v. Marshall,* 608 F.2d 373, 7 OSHC 1893 (9th Cir. 1979); *Bethlehem Steel Corp. v. OSHRC,* 607 F.2d 871, 7 OSHC 1802 (3d Cir. 1979); *Continental Oil Company v. OSHRC,* 630 F.2d 446, 8 OSHC 1980 (6th Cir. 1980).

Moreover, in addition to showing that a recognized hazard exists, the Secretary of Labor must also present substantial evidence that the implementation of the abatement required would *materially reduce* the hazard alleged. See *Champlin Petroleum Company v. OSHRC,* 593 F.2d 637, 7 OSHC 1241 (5th Cir. 1979); *National Realty, supra.*

U. IMPROPER INCORPORATION BY REFERENCE; REASONABLE AVAILABILITY

Many general industry standards, such as those issued by the American National Standards Institute (ANSI), have been incorporated by reference into OSHA standards. Publication is made in the Federal Register and not only constitutes constructive notice of the standard but also creates a rebuttable presumption that it has been validly published. Materials so incorporated by reference are presumed to be reasonably available to the class of persons affected by its incorporation. In order to allege a defense based upon invalid incorporation by reference, an employer must bear the burden of rebutting at least one of the presumptions. In *Charles A. Gaetano Construction Corp.*, 6 OSHC 1463 (1978), the employer challenged the reasonably available presumption. The Review Commission rejected its defense, holding:

> We agree with Judge Alfieri that, based upon the record presented at hearing, respondent has not proved the unavailability of the ANSI standard, and therefore the presumption of a valid incorporation by reference must stand.
>
> We note with approval Judge Alfieri's reliance upon respondent's failure to contact OSHA as evidence negating respondent's claim of unavailability and undermining its contention that the material was not reasonably available. However, we consider other facts more conclusive of the availability issue.
>
> Respondent's evidence concerning the availability of the standard relates solely to the period of time subsequent to citation. Respondent introduced no evidence of any attempt to obtain the ANSI standard in order to accomplish compliance with its requirements before the time of citation.
>
> More significant, however, is the fact that all of respondent's evidence relates to its own, geographically localized efforts to obtain the incorporated material. Respondent did not demonstrate the unavailability of the ANSI standards to employers in general, nor has it established the general unavailability of the standards. . . . For example, respondent did not show that other employers were unable to obtain copies of the ANSI standards, nor did respondent contact the national office of OSHA in its own attempt to obtain a copy of the standards.

See also *Empire Boring Co.*, 4 OSHC 1259 (1976); *Leader Evaporator Co., Inc.*, 4 OSHC 1292 (1976).

V. FAILURE OF PROOF

The Secretary of Labor is required to prove that an employer violated the Act by a preponderance of evidence. In a typical case arising under Section 5(a)(2) of the Act, the Secretary of Labor carries his burden of proof by establishing that a specific standard applies to the facts, that there was a failure on the employer's part to comply with the standard, and that employees of the employer had access to the hazard. If the Secretary of Labor fails to prove these three factors, the citation may be vacated. See *N. L. Industries, Inc.*, 5 OSHC 1524 (1977); *McHugh and McHugh,* 5 OSHC 1165 (1977); *Hillsdale Lumber Manufacturing, Inc.*, 5 OSHC 1281 (1977); *Republic Steel Corporation,* 5 OSHC 1248 (1977); *Cotter and Company, Inc. v. OSHRC,* 598 F.2d 911,

7 OSHC 1510 (5th Cir. 1979). Similarly, if the Secretary of Labor fails to prove all elements of a general duty clause citation, it will be vacated for failure of proof. *Schriber Sheet Metal & Roofers, Inc. v. OSHRC,* 597 F.2d 78, 7 OSHC 1246 (6th Cir. 1979); *Champlin Petroleum Co. v. OSHRC,* 593 F.2d 637, 7 OSHC 1241 (5th Cir. 1979).

W. KNOWLEDGE/FORESEEABILITY

The Commission clearly views unforeseeability as an affirmative defense, although the language in many of its decisions does not clearly separate this concept from constructive knowledge. The circuit courts apparently have split on the issue, and the language used in their decisions also demonstrates a great deal of confusion over these concepts. Research into violations of the Act has disclosed a great deal of confusion in the thinking of the Commission and the courts. Before a violation of the Act can be found, the Secretary of Labor generally has the burden of proving that the employer had either actual or constructive knowledge of the violation. See, *e.g., Dunlop v. Rockwell International,* 540 F.2d 1283, 4 OSHC 1606 (6th Cir. 1977); *Brennan v. OSHRC (Alsea Lumber Co.),* 511 F.2d 1139, 2 OSHC 1646 (9th Cir. 1975); *Environmental Utilities Corp.,* 5 OSHC 1195, 1199 (1977). Constructive knowledge exists when the employer could have discovered the condition through the exercise of due diligence. *Combustion Engineering, Inc.,* 5 OSHC 1943, (1977); see *Getty Oil Co. v. OSHRC,* 530 F.2d 1143, 4 OSHC 1121 (5th Cir. 1976).

Both the courts and the Commission, however, have encountered problems when they have attempted to separate the concepts of foreseeability and constructive knowledge. Logically, foreseeability should be viewed as another definition of constructive knowledge; if a condition is foreseeable, it is one that could have been anticipated or discovered through the use of due diligence. The Commission, however, consistently has viewed the isolated incident/foreseeability issue as an affirmative defense, and it has placed the burden of proof on the employer to show that the condition or accident was unforeseeable or isolated. See *Everglades Sugar Refinery, Inc.,* 7 OSHC 1410 (1979); *Ford Motor Co.,* 5 OSHC 1765, 1768 (1977); *B–G Maintenance Management, Inc.,* 4 OSHC 1282 (1976); *Murphy Pacific Marine Salvage Co.,* 2 OSHC 1464 (1975); *Steel Constructors, Inc.,* 8 OSHC 2146 (1980). Some of the language in these cases appears to be directly at odds with the concept that the Secretary has the burden of proving employer knowledge. Thus in *Briscoe/Arace/Conduit, A Joint Venture,* 5 OSHC 1167 (1977), the Commission held that the employer had the burden of showing that "the employee's action constituting non-compliance was a departure from a uniformly and effectively enforced workrule *and the employer had neither actual nor constructive knowledge of the employee's conduct.*" [Emphasis added.]. *Cf. Welsh Farms Ice Cream, Inc.,* 5 OSHC 1755 (1977); *Fisk-Oesco Joint Venture,* 5 OSHC 1350 (1977). The circuit courts which have addressed the question also have been unable to distinguish clearly the concept of knowledge and foreseeability.

The Eighth Circuit appears to have made unforeseeable idiosyncratic behavior an affirmative defense. See *Western Waterproofing Co. v. Marshall,* 576 F.2d 139, 6 OSHC 1550 (8th Cir. 1978). But see *Danco Construction Co. v. OSHRC,* 586 F.2d 1243, 6 OSHC 2039 (8th Cir. 1978); *Arkansas–Best Freight System, Inc. v. OSHRC,* 529 F.2d 649, 3 OSHC 1910 (8th Cir. 1976). Recently, however, the Fourth and Tenth Circuits have suggested that foreseeability and the adequacy of an employer's safety program are elements of constructive knowledge. See *Ocean Electric Corp. v. Secretary of Labor,* 594 F.2d 396, 7 OSHC 1149 (4th Cir. 1979); *The Mountain States Telephone & Telegraph Co.,* 6 OSHC 1504 (1978), *rev'd,* 623 F.2d 155, 8 OSHC 1557 (10th Cir. 1980). The courts clearly stated that the Secretary of Labor had the burden of proving these elements relative to a supervisor's knowledge of his own violations of the Act and could not impute such knowledge to the employer, requiring it thereafter to show the violation was unpreventable or unforeseeable. *Cf. General Dynamics Corp. v. OSHRC,* 599 F.2d 453, 7 OSHC 1373 (1st Cir. 1979); *New Bedford Gas & Electric Light Co., Cape & Vineyard Division v. OSHRC,* 512 F.2d 1148, 2 OSHC 1628 (1st Cir. 1975); *National Realty & Construction Co. v. OSHRC,* 489 F.2d 1257, 1 OSHC 1422 (D.C. Cir. 1973). A more careful analysis of these circuit court decisions might yield a basis for distinguishing them on the facts. The more plausible explanation, however, is that the courts and the Commission have been careless in their legal analysis and use of the English language.

The Commission clearly views the foreseeability/isolated incident issue as an affirmative defense. Language in its decisions, however, indicates that it has not adequately separated this concept from constructive knowledge. The circuit courts that have addressed the issue are equally confused, some suggesting that foreseeability is a part of knowledge, others that it is an affirmative defense.

NOTES

1. Examples of procedural defenses which may be raised to OSHA citations in the answer, and which have been discussed previously, include: (1) a statute of limitations defense based on Section 9(c) of the Act, which provides that no citation may be issued after the expiration of six months following the occurrence of any violation; (2) a lack of reasonable promptness defense, which is derived from Section 9(a) of the Act and which states that the Secretary of Labor shall, with reasonable promptness, issue a citation to the employer; (3) a defense based upon a failure timely to forward a notice of contest, as set forth in Rule 32 of the Commission's Rule of Procedure, which was intended to clarify Section 10(c) of the Act; (4) a defense based upon improper service under Section 10(a) of the Act, which requires the Secretary of Labor to notify the employer by certified mail of the penalty and to inform the employer that he has fifteen working days to contest a citation; (5) a defense based upon a lack of jurisdiction under Section 3(5) of the Act, on the basis that a person is not a covered employer, is not engaged in a business affecting commerce, or that there is a lack of an employer relationship; (6) a defense that the standard was improperly promulgated or that it is vague and does not give an employer fair notice of the conduct prohibited thereby and thus violates the employer's due process rights; (7) a defense of lack of particularity in that Section 9(a) of the Act requires that a citation shall be in writing and describe with particularity the violation; (8) an improper amendment defense, if an

amendment changes the citation or complaint in such a manner as to allege a new and different charge so as to prejudice the employer's defense; (9) a Section 4(b)(1) defense, in that jurisdiction is properly based within the control of another federal agency; (10) a defense based upon the failure to discuss an alleged violation at a closing conference in conjunction with a showing of prejudice; and(11) a defense based upon the failure to accord walkaround rights during an inspection.

2. The Commission has held that failure to provide proper training or instructions in circumstances where it could have prevented a violation by an employee constitutes constructive knowledge under the Act. See, *e.g.*, *Southwestern Bell Telephone Company*, 7 OSHC 1058 (1979); *United States Steel Corporation*, 8 OSHC 2147 (1980) *Enfield's Tree Service, Inc.*, 5 OSHC 1142 (1977). In *H. C. Nutting Co. v. OSHRC*, 615 F.2d 1360, 8 OSHC 1241 (6th Cir. 1980), the company was cited for violating 29 CFR 1926.21(b)(2), which reads as follows: "The employer shall instruct each employee in the recognition and avoidance of unsafe conditions and the regulations applicable to his work environment to control or eliminate any hazards or other exposure to illness or injury." The administrative law judge did not like Nutting's safety program and found it in violation of the standard. The problem, however, according to the court, was that the regulation did not outline any particular requirements for a safety program. Fairly read, the regulation requires that an employer inform employees of safety hazards which would be known to a reasonably prudent employer or which are addressed by specific OSHA regulations. See *Bristol Steel & Iron Works v. OSHRC*, 601 F.2d 717, 7 OSHC 1462 (4th Cir. 1979); *Cape & Vineyard Division v. OSHRC*, 512 F.2d 1148, 2 OSHC 1628 (1st. Cir. 1975). As so construed the court held that the regulation is not unenforceably vague and places reasonable duties on an employer. *National Industrial Contractors v. OSHRC*, 583 F.2d 1048, 6 OSHC 1914 (8th Cir. 1978); *General Dynamics v. OSHRC*, 599 F.2d 453, 7 OSHC 1373 (1st Cir. 1979); *Hoffman Construction Company v. OSHRC*, 546 F.2d 281 4 OSHC 1813 (9th Cir. 1976). Unfortunately, the record before the court did not contain any evidence whatsoever as to what standard of reasonable instruction was breached. The Secretary pointed to no evidence of industry practice or reasonable standards; thus there was no substantial evidence showing that the company violated the regulation, as construed, against a backdrop of reasonable industry practice.

Many safety and health standards require a specific training and instruction program. Examples of these were set forth previously in Chapter IV.

3. Under the Construction Safety Act, 40 U.S.C. §327–333, which was adapted into the Act as an established federal standard, the ultimate responsibility for compliance with safety and health standards rested with the general contractor.

4. Commissioner Barnako stated in his concurring opinion that "[t]he Commission has held that a general contractor is presumed to have supervisory authority over the entire worksite. *Gil Haugan, supra.* Contrary to my colleagues' assertion, however, we have not stated that '[n]ormally, the employer with supervisory control over a construction site is termed the "general contractor." ' *Cf. Bechtel Power Co.*, 4 OSHC 1005 (1976) (construction manager who was empowered to organize, plan and manage construction program as well as administer, inspect, approve and coordinate performance of prime contracts was held subject to construction standards even though it was neither general contractor nor subcontractor within the meaning of the *Grossman Steel* and *Anning-Johnson* decisions)."

5. According to Commissioner Barnako, the basic defect in the Secretary's argument equating feasibility with financial viability is that it assumes an employer's entire resources are available to correct an excessive noise problem without regard to the extent of the problem or the actual hazard presented.

6. A violation, however, cannot be based on the exposure to an employer's walk around representative during an inspection. *Brown-McKee, Inc.*, 8 OSHC 1248 (1980); *Bechtel Power Co.*, 7 OSHC 1361 (1979).

7. Commissioner Barnako did not agree that the Secretary proved that *Brown-McKee's* employees had access to the unsecured ladder. In his opinion, access is *only established* if the Secretary of Labor shows it is reasonably predictable that employees will be, are, or have been in a zone of danger. Absent an admission by an employer, the Secretary must prove access by evidentiary facts.

8. Commissioner Barnako disagrees with the elements of the Secretary of Labor's burden of proof assigned by the majority. Commissioner Barnako would require that the Secretary prove that: (1) Brown-McKee's employees were exposed to a hazard that a reasonable person familiar with the employer's industry would recognize as requiring the use of personal protective equipment; (2) a feasible means of protecting against the cited hazard exists; and (3) reference to other standards in Part 1926 indicates the need for using the personal protective equipment which the Secretary asserts Brown-McKee's employees should have used.

CHAPTER XIII
Hearing and Appeal Procedure

A. INTRODUCTION

The Occupational Safety and Health Review Commission (the Review Commission or the Commission) is an independent agency of the United States Government, not connected with either the Department of Labor or the Occupational Safety and Health Administration. The Commission's role is basically that of an adjudicator; it is not a prosecutor, policy-maker, or investigator. Enforcement of the Act is the Secretary's responsibility. *Dale M. Madden Construction, Inc. v. Hodgson,* 502 F.2d 278, 2 OSHC 1101 (9th Cir. 1974). He is the Act's prosecutor. *Atlas Roofing Co. v. OSHRC,* 430 U.S. 442, 5 OSHC 1105 (1975). *Marshall v. Sun Petroleum Products Co.,* 622 F.2d 1176, 8 OSHC 1422 (3d Cir. 1980). If a citation is issued as a result of an OSHA inspection and an employer decides to contest the citation, penalty, or abatement period, the Commission's jurisdiction is triggered, and the matter then moves from the Department of Labor to the Review Commission's tribunal for adjudication. The Secretary of Labor becomes the prosecutor in the case and through the Solicitor of Labor's office will bear the responsibility of enforcing the citation and proving that the proposed penalty and abatement are appropriate. *Atlas Roofing Co. v. OSHRC,* 430 U.S. 442, 5 OSHC 1105 (1975). The Commission's function is then (i) to issue orders, based on findings of fact that affirm, modify, or vacate the Secretary of Labor's citations and proposed penalties; (ii) to assess penalties; or (iii) to direct other appropriate relief.

The Commission is composed of three "members" appointed by the President (one of whom is designated by the President to serve as Chairman), with overlapping six-year terms. The members are appointed among persons who by reasons of training, education and experience are qualified to carry out the Commission's functions. The Chairman is responsible for the administration of the Commission and the appointment of more than forty-four administrative law judges (and a chief judge) who have life tenure. Section 12(a) of the Act. The judges hold hearings and decide matters arising under the Act; judges' decisions may thereafter be reviewed by the Commission. The following functions, powers, and duties are exercised by the Review Commission:

(1) Conduct adjudicatory hearings in accordance with Section 5 of the Administrative Procedure Act.

(2) Prescribe rules necessary for the orderly transaction of its proceedings.
(3) Administer oaths and affirmations; order and hear testimony; require persons to appear, to depose and to produce documentary evidence.
(4) Determine the appropriateness of proposed penalties and the reasonableness of the proposed periods of time fixed in citations for the abatement of violations.
(5) Provide that all hearings and records of hearings be open to the public.
(6) Issue orders, based on findings of fact, affirming, modifying, or vacating the Secretary of Labor's citations, proposed penalties, or proposed time for abating violations.
(7) Assess civil penalties provided in Section 17 of the Act, giving due consideration to the appropriateness of penalties with respect to the size of the business of the employer being charged, the gravity of the violation, the good faith of the employer, and the history of previous violations.
(8) File records of proceedings in United States Courts of Appeal when Commission cases are appealed to such courts.

For the purpose of any proceeding before the Commission, the provisions of Section 11 of the National Labor Relations Act, 29 U.S.C. Section 161, are made applicable to the jurisdiction and powers of the Commission. Section 12(i) of the Act.

B. THE COMMISSION AND ITS RULES OF PROCEDURE

The Review Commission was established in Section 12(a) of the Act, as an independent quasi-judicial administrative agency to adjudicate contested matters between OSHA and employers (or their employees) to whom OSHA has issued citations charging a violation of the Act.

Each of the three Commission members has a variety of functions to perform: (1) to examine decisions issued by administrative law judges within thirty days of their filing, in order to determine whether the decision should be ordered for review; (2) to issue decisions on all review-ordered cases; and (3) other review functions, such as interlocutory appeals and various motions before and after a case is assigned to a judge. Although any single member may order that a judge's decision be reviewed, the final disposition of any case must have the concurrence of at least two members. Each member has a staff to assist in performing these functions, consisting of a chief counsel, two attorneys, and two clericals.

In an effort to make appearances before the Commission and its judges more meaningful and less burdensome, the Commission has promulgated and published Rules of Procedure which govern all Commission proceedings. (Normally, a copy of these Rules is sent to an employer when a case is docketed with the Commission.) Congress empowered the Commission to promulgate such rules in Section 12(g) of the Act. In the absence of specific rules, procedure before the Commission is governed in accordance with the Federal Rules of Civil Procedure. Commission Rule 2; *Quality Stamping Products Company,* 7 OSHC 1285, 1287 n.5 (1979); *Ralston Purina Company,* 7 OSHC 1730 (1979).

The Commission Rules are codified in 29 CFR 2200.01–.211 and are similar in many respects to the rules of practice and procedure of other agencies dealing with labor relations statutes. In the Rules of Procedure, as published by the Commission, the codification prefix 2200 contained in the Code of Federal Regulations has been omitted. Throughout this discussion on contest, hearing, and appeal procedures, reference shall be made exclusively to the Commission Rules and not to the Code of Federal Regulations which contain those Rules.

The *conventional* Commission Rules cover such topics as parties and representatives, pleadings and motions, prehearing procedures and discovery, hearings, posthearing procedures, settlement, and other matters. An important, recent addition to the Commission Rules was the adoption of *simplified* proceedings, which are an experimental, alternative procedure to be used to resolve simple cases before the Review Commission. These simplified proceedings are contained in Commision Rules 200–211.

According to Commission statements, simplified proceedings are designed to: make the resolution of cases faster and easier for those appearing in Commission proceedings without an attorney; reduce paper work; and reduce the expense of litigation. Cases handled under simplified proceedings are not subject to many of the strictures generally applicable to conventional proceedings. Simplified proceedings are thus not appropriate for all cases, since certain legal tools used to litigate a case are not available in those proceedings. Much of the discussion in simplified proceedings takes place off the record, and the case may be more difficult for review at the Commission or court of appeals level should the case go that far.

Under simplified proceedings, the filing of pleadings or initial documents generally is not permitted. For example, the Secretary of Labor will not file a complaint; the employer will not respond by filing an answer. These functions will be served by the citation and notice of contest, functions which, by the way, are adopted by many states having state plans which do not utilize the complaint and answer. Discovery of factual material before the hearing is generally restricted and is possible only with the special permission of the administrative law judge. Interlocutory appeals of a judge's early rulings are also prohibited (thus there would be no immediate recourse to the denial of a discovery request). And, if the case actually goes to a hearing, the parties are not required to follow formal rules of evidence.

Prior to hearing, there are several opportunities provided for early settlement of the matters in dispute between the parties. The parties are required to engage in an informal discussion in order to settle the case or narrow the factual and legal issues to be decided by the judge. Such discussions are mandatory for all parties. At these meetings, issues will be discussed and narrowed, a statement of facts agreed upon by the parties will be required, specific defenses will be set forth, a list of the witnesses the parties intend to call should the case go to a hearing must be provided, and exhibits they intend to introduce to support positions must be specified. There will be few, if any, motions permitted. And, of course, discovery is greatly restricted and can only be accomplished with the express permission of the judge. If the case is not settled during the informal discussions between the parties, then a "prehearing" conference will be held which is presided over by the judge. At this conference the judge will again attempt to resolve or narrow the issues that are to be taken up at the hearing.

The hearing itself will be held in two parts. The first part of the proceeding will be a conference to resolve issues and to record agreements reached. At the conclusion of the conference, if there are still unresolved issues, the case will proceed to hearing. There will be the right to cross-examine witnesses, present evidence, present witnesses, and introduce exhibits, and testimony will be under oath at the hearing. A transcript will be taken and provided to the parties at the close of the hearing. A summation of the case is appropriate, and the right to submit briefs to the judge after hearing is also provided. The judge will then prepare and send to the parties a written decision, and, as in conventional cases, the parties will have the opportunity of accepting the judge's decision or asking the Review Commission to grant review.

Most cases are eligible for simplified proceedings; however, if a citation or any part of it involves an alleged violation of either Section 5(a)(1) of the Act (general duty clause) or certain health regulations,[1] then the case is ineligible for simplified proceedings and must go forward under conventional procedures. Merely because a case is *eligible* for simplified proceedings does not necessarily mean that the case is *appropriate* for such treatment.

The factual nature of the case and the legal issues involved should all enter into the decision of whether or not to use simplified proceedings. Basically the rule is that simplified proceedings are designed for simpler cases. (*E.g.*, you have only contested the citation because you believe the penalty is excessive.)

An election to proceed under simplified proceedings cannot be revoked at a later time. The decision to use simplified proceedings *must* be filed in writing with the Review Commission within ten days of an employer's receipt of a notice of docking (at which point in time the employer will be sent a memorandum entitled "Explanation of the Commission's Simplified Proceedings"). A simple statement is sufficient, *e.g.*, "Employer requests simplified proceedings." A copy of the request must be served upon the Secretary of Labor, employees or their representatives, and any other party. Service may and can be accomplished by first class mail or personal delivery. If employees are not represented, a request for simplified proceedings must be posted at the work site, at a location where it will be seen by affected employees. A brief statement indicating on whom, when, and how the request was served must accompany a request for simplified proceedings. If no party objects to the request for simplified proceedings, the Review Commission will approve it. If an objection is filed, the case *must* continue *per* conventional procedures. Objections must be filed within fifteen days after the request for simplified proceedings is filed. Objections need also only be a brief statement saying no more than "I object to simplified proceedings" and giving the docket number. The objection must be served on other parties in the same way as a request for simplified proceedings.

C. NOTICE, SERVICE, AND PARTIES

Except for the notice of contest and petition for modification of abatement, which are filed with the Area Director, subsequent pleadings or documents are filed with either the Executive Secretary of the Commission or the administrative law judge. Prior to assignment of a judge, all pleadings are filed with the Executive Secretary; thereafter pleadings are filed with the judge at the address given in the notice of the judge's assignment. Commission Rule 8. Service of the various pleadings filed with the Commission or judge is also required to be made on the *parties;* a statement that service has been made must be attached to all papers, pleadings, or documents which are filed. Commission Rule 7(d). Requests for extensions of time to file pleadings and other documents must be received in advance of the date on which filing is due. Commission Rule 5.

1. Computations of Time

The Commission Rules provide that in computing time, the day that starts the period shall not be included. The last day of the period is included, unless it is a Saturday, Sunday, or federal holiday. Day means a calendar day (and must be distinguished from the term "working day" used to compute the time period for filing a notice of contest). Commission Rule 1(k).

Example: The decision of the judge is filed Monday, June 3. In computing thirty days, June 3 is not included. Begin counting days on June 4. Thus, the review period ends on July 2. If July 2 is not a Saturday, Sunday, or federal holiday, it is included as the last day. However, if July 2 is a Saturday, Sunday or federal holiday, the last day of the review period would be the next working day.

Where the period of time allowed is less than seven days, intermediate Saturdays, Sundays, and federal holidays during that period are not included. Where service is made by mail upon a party, three days are added to the time period in which the party must serve his response. Commission Rule 4.

2. Notice to Secretary of Labor

Normally, the parties to the proceeding include only the employer and the Secretary of Labor; if others have requested party status, such as employees or their representative, the Review Commission or the judge will so notify the employer. *Reynolds Metals Co.,* 7 OSHC 1042 (1979); *Harshaw Chemical*

Co., 8 OSHC 1138 (1980). The Secretary of Labor is represented in all Commission proceedings by the Solicitor of Labor's office, and copies of all documents filed with the Commission should be sent to the Solicitor.

3. Notice to Employer

Section 10(a) of the Act requires the Secretary to "notify the employer by certified mail[2] of the *penalty*. . . ." Similarly, the Act in Section 9(a) provides that a citation shall issue to the employer. Technically the citation can be served separately and by ordinary mail. Commission Rule 7(c) provides that unless otherwise ordered, service of documents may be undertaken by first class mail (postage prepaid) or personal delivery and is effective at the time of mailing or personal delivery, respectively. Practically, the Secretary of Labor sends both the *citation* and notification of proposed *penalty* in one document to the employer, by certified mail.

The Act in its definition of "employer" does not actually specify the particular person to whom such service is to be made. Similarly, no assistance is provided in the Commission Rules of Procedure. In *Buckley & Co. v. Secretary of Labor,* 507 F.2d 78, 2 OSHC 1432 (3d Cir. 1975), the court held that the intent of Congress in enacting Section 10(a) of the Act was to provide notice to the person who has the authority to disburse funds to abate the violation, pay the penalty, or contest the citation or proposed penalty. The court considered that the notice should "at the very least" be sent to officials at the corporate headquarters and rejected as insufficient, mailing to the shop superintendent in charge of the job site where the alleged violation occurred. The court reasoned that because a worksite superintendent is primarily and normally responsible for the violation, he would be as likely to conceal the citation as to report the documents to his superiors.

Recent Commission cases have narrowed the construction and application of the *Buckley* rule. In *Joseph Weinstein Electric Corp.*, 6 OSHC 1344 (1978), a citation was sent to a construction work site rather than an employer's office, but since a corporate officer with requisite authority received *actual* notice from that service, it was upheld. In *Womack Construction Co., Inc.*, 6 OSHC 1125 (1977), the notice of proposed penalty was sent to the employer's corporate headquarters, where it was received and lost by a bookkeeper who was not management. When management became aware that the citation was received, it immediately filed a notice of contest. The Commission stated that the notice of contest was not timely, because it was sent to the proper location and there was a proper official with requisite authority present when it was received. See also *Otis Elevator Co.*, 6 OSHC 1515 (1978); *Norkin Plumbing Co., Inc.*, 6 OSHC 1539 (1978); *Truland Corp.*, 6 OSHC 1896 (1978)—citation not served by certified mail as required by Section 10(a) of the Act; however, since the employer received actual notice, it was not prejudiced by the manner of service.

In *B. J. Hughes, Inc.*, 7 OSHC 1471 (1979), the Secretary of Labor asked the Review Commission to reject the test set forth in *Buckley* for determining whether a citation and notification of proposed penalty had been properly served upon a corporate employer. The Secretary asserted that the test for determining sufficiency of service should be whether it was reasonably calculated to give notice of the citation and proposed penalty with an opportunity to contest or abate. The Secretary also argued that the service invoked in *B. J. Hughes* was reasonably calculated to give notice, because it was made upon an official who should know what to do with the citation and penalty, *i.e.*, know to whom in the corporate hierarchy to forward the documents. The Commission stated:

> We agree with the Secretary that the test to be applied in determining whether service is proper is whether the service is reasonably calculated to provide an employer with knowledge of the citation and notification of proposed penalty and an opportunity to determine whether to abate or contest. This approach is consistent with the test applied by courts in general when determining sufficiency of service in other areas of the law. E.g., *NLRB v. Clark,* 468 F.2d 459 (5th Cir. 1972); *American Football League v. National Football League,* 27 F.R.D. 264 (D. Md., 1964).
>
> Moreover we agree with the Secretary that service upon an employee who will know to whom in the corporate hierarchy to forward the documents will satisfy this test. Accordingly we accept as valid, service upon an employee at a local worksite who will know to whom the documents should be forwarded. To require the Secretary instead to determine who in the corporate hierarchy has the authority noted in *Buckley* places too great a burden upon him, particularly when the cited employer is a large corporation with offices throughout the United States. Since Congress intended that citations be issued with reasonable promptness, the Secretary's representatives should not have to spend time ferretting out the complexities of a corporate hierarchy, but should be able to rely instead upon the ability of individuals at a local worksite to direct the citation to the appropriate officials.

See also *P & Z Company, Inc.*, 7 OSHC 1589 (1979)—an employer who receives actual notice of a citation and proposed penalty has had receipt within the meaning of Section 10(a); *Henry C. Beck Co.,* 8 OSHC 1395 (1980)—mailing to a work site in the company name was proper (in *B. J. Hughes* the citation was addressed to a named employee, but the Commission held that this difference was not critical).

Although Section 10(a) requires that an employer who has been cited be served with notification of the proposed penalty by certified mail, an alternative method may be used if service by certified mail can not be affected, as long as the citation and notification of penalty are promptly received by a person responsible for taking appropriate action. In *Donald K. Nelson Construction, Inc.,* 3 OSHC 1914 (1976), the Commission stated that personal service could be approved if it conforms with Rule 4(d) of the Federal Rules of Civil Procedure and if it is made upon a responsible person. In *Nelson,* however, service was made upon the wrong person (corporation president's wife), who was not responsible for taking official action. See also *Truland Corporation,* 6 OSHC 1869 (1978)—employer received actual notice of citation and penalty.

4. Notice to Employees and Representatives

In the event that there are affected employees who are not represented by an authorized employee representative, upon receipt of notice that a case has been docketed by the Review Commission (notice sent to employer after notice of contest or PMA has been filed), the employer must post where citations are required to be posted: (1) a copy of the notice of contest or PMA; and (2) a notice informing affected employees of their right to participate in the case by electing party status and of the availability of pleadings for inspection and copying at reasonable times. Commission Rule 7(g). Both notices may be posted together. Proof of posting is required to be filed not later than the first working day following posting. Commission Rule 7(e).

In the case of employees who are represented by an authorized employee representative, the employer is also required to serve the two "notices," and said service may be accomplished by either mail or personal delivery. Proof of service shall be accomplished by a written statement of the same which sets forth the date and manner of service. Commission Rule 7(b), (c), (d), and (f).

If some affected employees are represented by a union and some are not, the employer must comply with all of the requirements listed. See generally *Weldship Corporation,* 8 OSHC 2044 n. 6 (1980).

Section 10(c) of the Act affords affected employees and their representatives the right to participate in Commission proceedings as parties, by either filing a notice of contest or specifically electing party status. Commission Rule 20; *American Airlines, Inc.,* 7 OSHC 1980 (1979); *Harshaw Chemical Co.,* 8 OSHC 1138 (1980). Affected employees are defined as employees who are exposed to hazards as a result of their assigned duties. Commission Rule 1(e); *Marshfield Steel Co.,* 7 OSHC 1741 (1979)—an employee who quit his job before filing an objection to his employer's PMA had no legal standing to object, thus a motion to strike his appearance as a party was granted; *IMC Chemical Group, Inc.,* 6 OSHC 2075 (1978). Any employee, including supervisors and officers, can be affected employees.

The right to elect party status may be exercised at any time *before* the commencement of the hearing. *Marshall v. Sun Petroleum Products Co.,* 622 F.2d 1176, 8 OSHC 1422 (3d Cir. 1980).[3] After that time, only for good cause will the Commission or the judge allow election of party status. To become a party, affected employees or their union must file with the Commission (or the judge) a statement of intent to participate or a statement of position (neither need be a formal pleading) which sets forth the affected employee's position about the issues. *General Electric Co.,* 7 OSHC 1277 (1979). A copy of the statement must be served on all other parties. Parties have the right to appear in person or through a representative; the representative need not be an attorney. Commission Rule 22(a)(d). If a party chooses to appear through a representative, the representative shall control all matters in the proceeding. Affected employees who are represented by an authorized employee representative may appear only through such representative. Commission Rule 22(c); *Babcock & Wilcox Company,* 8 OSHC 2102 (1980).

Neither the Act nor the Commission Rules specify the extent of the rights an employee has as a party; however, by Commission decision it appears that employee rights upon election of party status (in the context of a challenge to a settlement agreement) generally are much broader than the rights afforded normally in an employee contest under Commission Rule 35(a), in which they appear only to have the right to challenge the reasonableness of the abatement period. See *Automobile Workers v. OSHRC,* 557 F.2d 607, 5 OSHC 1525 (7th Cir. 1977), in which the court held that Section 10(c) of the Act only permits a consideration of whether the abatement period fixed by the citation is unreasonable. See also *Sun Petroleum Products Co. v. Marshall,* 622 F.2d 1176, 8 OSHC 1422 (3d Cir. 1980). The Commission, however, in two recent cases has expanded the rights of employees to include challenges to the means and method of abatement. *Farmers Export Co.,* 8 OSHC 1655 (1980); *American Cyanamid Co.,* 8 OSHC 1346 (1980). This "expansion" took place in the context of challenges to settlement agreements. See discussion below.

It would appear, however, that as parties, employees at least have the normal attendant rights including meaningful participation in Commission proceedings. *Gardner Inc.,* 7 OSHC 1738 (1979). For example, they have an unconditional right to a copy of all motions and pleadings filed. *Harshaw Chemical Co.,* 8 OSHC 1138 (1980). They would also have standing to object to such matters as the Secretary of Labor's withdrawal of a citation. *ITT Thompson Industries, Inc.,* 7 OSHC 1944 (1978). In such an instance, even though the Secretary may be granted permission to withdraw from a case, the proceedings may nevertheless continue based on the citation originally issued by the Secretary. In addition, the normal attendant rights of party status would include:

(1) The right to the service of all pleadings (papers filed with the judge or the Review Commission) that are filed by other parties;
(2) The right to present witnesses and evidence;
(3) The right to cross-examine witnesses called by other parties;
(4) The right to object to evidence sought to be introduced by other parties;
(5) The right to make oral argument before the administrative law judge;
(6) The right to present written arguments;
(7) The right to participate in any settlement negotiations;
(8) The right to receive a copy of the judge's decision;
(9) The right to petition the full Commission to review the judge's decision; and
(10) The right to appeal the Commission's decision to an appropriate court of appeals.

See *Harshaw Chemical Co.,* 8 OSHC 1138 (1980); *IMC Chemical Group, Inc.,* 6 OSHC 2075 (1978). Employees have not, however, been considered indispensable parties to Commission proceedings. *Interocean Stevedores Co.,* 5 OSHC 2009 (1977). See also *Trustees of Penn Coastal Transportation Co.,* 4 OSHC 2033 (1977).

In *International Harvester Co.,* 7 OSHC 2194 (1980), the Commission

(Members Cleary and Barnako) reached an impasse over the extent to which party status should be accorded employees and/or their authorized representatives relative to the consideration of objections to a settlement agreement. In view of the statutory purpose of expeditious adjudication, the members agreed to resolve their impasse by affirming the judge's decision but according it the precedential value of an unreviewed judge's decision. See, *e.g., Life Science Products Co.,* 6 OSHC 1053 (1977), *aff'd,* 591 F.2d 991, 7 OSHC 1031 (4th Cir. 1979). The facts in that case, however, are instructive. The employer timely contested the citation; the union thereafter notified the Commission that, as the authorized representative of certain employees who worked in an area covered by the citation, it was asserting party status in the case.

Following various pretrial motions and discovery, including an unsuccessful motion by the employer to have the union dismissed as a party (since the employees it represented were not exposed), the Secretary of Labor and the employer entered into a settlement agreement. The agreement provided that the employer would withdraw its notice of contest to the citation and proposed penalty and would abate the violation. The union objected to the settlement on the basis that the agreement would not protect the employees it represented. The judge convened a hearing to consider the objections to the settlement. At the hearing, the Secretary contended that the settlement should be approved because it provided for abatement of the hazard for which the employer was cited. The Secretary also contended that the citation did not reach the exposure of mechanics (the employees represented by the Union) to excessive noise because no mechanics were exposed during the inspection. Thus since the settlement would be fully effective in correcting the cited hazard, the Secretary urged that the settlement be approved. The employer also sought approval of the settlement, reiterating its position that the union was not a proper party because none of the employees it represented were encompassed within the hazard described in the citation. In briefs filed with the judge subsequent to the hearing, both the Secretary and the employer also argued that employees and/or representative of employees have no standing to object to a plan of abatement contained in a settlement agreement. They asserted that, under Commission precedent, an employee or employee representative can only object to the reasonableness of an abatement date contained in a settlement agreement, and that the union's objections did not concern the abatement date included in the agreement. In support of this position, the Secretary and the employer cited, among other cases, *Local 588, United Auto Workers (Ford Motor Co.),* 4 OSHC 1243 (1976), *aff'd,* 557 F.2d 607, 5 OSHC 1525 (7th Cir. 1977); *United States Steel Corp.,* 4 OSHC 2001 (1977). The union argued that its rights as a party go beyond the right to object to the time allowed for abatement in a settlement agreement and include the right to object to the plan of abatement itself.

The judge approved the settlement agreement and concluded that, since the alleged violation did not encompass the union's members, it had no standing to object to the agreement. On review before the Commission, Commissioner Barnako stated the union must demonstrate that its members *are exposed* to the cited condition before it would be granted party status, citing Commission Rules 20(a) and 1(e). Commissioner Cleary felt that where employees seek party status in a case that directly *involves* their working conditions, the Act requires the Commission to allow the employees to participate. Cleary felt the Commission must consider the nature of the hazard in determining whether employees are sufficiently affected to be entitled to party status. Thus:

> This is not a typical case involving a safety hazard, where it is relatively simple to determine whether particular employees are exposed to or have access to the hazard. Instead, it takes sophisticated instrumentation and careful measurements to establish that an employee is exposed to noise in excess of the limits provided in section 1910.95(b)(1). See, *e.g., Sun Shipbuilding and Drydock Co.,* 2 OSHC 1181 (1974).
>
> * * *
>
> The Secretary's assertion that his citation does not encompass the Union's members does not permit the Commission to ignore the union's concern. A citation involves a particular hazard, not particular employees. Indeed, the Commission has held that a violation can be predicated on exposure of an employee other than those mentioned in a citation. *R. Colwill Excavating Co.,* 5 OSHC 1984 (1977).
>
> * * *
>
> Where, as here, employees seek party status in a case that directly involves their working conditions, Chairman Cleary believes that the Act requires the Commission to allow the employees to participate. See *IMC Chemical Group, Inc.,* 6 OSHC 2075 (1978), appeal filed, No. 79–3041 (6th Cir. Jan. 16, 1979).

Accordingly, Chairman Cleary believed that the judge erred in rejecting the union as a party on the basis of the Secretary's assertion that he did not consider the citation to encompass the union's members. The union had shown a sufficient nexus between the cited hazard and the assigned work duties (employees worked in the general area where the alleged violation occurred) of its members to justify participation in the proceeding. Having concluded that the union was properly a party to the proceeding, Chairman Cleary would also have held that there was *nothing* limiting the Union's status as a party. In his dissent to the Commission's decision in *United States Steel Corp.,* 4 OSHC 2001 (1977), Chairman Cleary stated:

> There is nothing in either the Act or in the Commission Rules that limits the participation of an employee representative once party status has been elected. Neither the Act nor our rules envision that an employee representative be a party for limited purposes. I submit that an employee representative is a full party with all rights of any party, including the right to join or object to settlement.

The Commission recently has appeared to expand the role of affected employees or their representatives who have elected party status. Reemphasizing that they must be afforded an opportunity for meaningful participation in settlement negotiations, as required by *ITT Thompson Industries, Inc.,* 6 OSHC 1944 (1978), the Commission, in *American Cyanamid Co.,* 8 OSHC 1346 (1980), stated that they must first of all be served with a copy of the fully executed agreement when it is filed with the judge. See also *Babcock and Wilcox Co.,* 8 OSHC 2102 (1980). However, in an expansion of the role of employee parties, the Commission held that their participation is not limited to questioning the appropriateness of the abatement date set forth in the agreement; rather:

> When the authorized employee representative contends that a central requirement of the settlement, abatement, is missing, our approval of such an agreement without a judicial resolution of the factual controversy would be contrary to our congressional mandate.

Commissioner Barnako, relying on past Commission precedent, dissented and reaffirmed his view that employees have standing to raise objections only to the time fixed for abatement, citing *Reynolds Metals Co.*, 7 OSHC 1042 (1979) (concurring and dissenting opinion); *Kaiser Aluminum & Chemical Corp.*, 6 OSHC 2172 (1978) (dissenting opinion); *ITT Thompson Industries, Inc.*, 6 OSHC 1944 (1978) (concurring opinion).

See *Marshall v. Sun Petroleum Products Co.*, 622 F.2d 1176, 8 OSHC 1422 (3d Cir. 1980), where the court held that employees or their representatives may only raise objections that the abatement period proposed by the settlement agreement is unreasonable; they are not entitled to be heard on matters other than the abatement period. In *Farmers Export Co.*, 8 OSHC 1655 (1980), Chairman Cleary rejected the court's limited view of the employees' role as a party.

Even if an employee's interest was not sufficient to accord it party status, Chairman Cleary would conclude that the employee or its union had shown a sufficient interest to permit it to *intervene* pursuant to Commission Rule 21. As an intervener, the union could be afforded the opportunity for meaningful participation in the settlement or contest process.

In *Brown & Root, Inc.*, 7 OSHC 1526 (1979), the Commission held that a union which represented the employees of a subcontractor who were exposed to a hazard but did not represent the employees of the *cited* contractor failed to meet the definition of affected employees, as that term is defined in Commission Rule 1(e),[4] and thus was denied party status under Commission Rule 20. However, the Commission stated that the union, since it established an interest in the proceeding by virtue of its representation of the subcontractor's employees, and whose participation would aid the judge in reaching a decision because of the union's expertise in the steel erection industry, was allowed to intervene under Commission Rule 21(c). Interveners are given the right of meaningful participation in proceedings. *IMC Chemical Group, Inc.*, 6 OSHC 2075 (1978); *ITT Thompson Industries, Inc.*, 6 OSHC 1944 (1978). Interveners must file a petition for leave to intervene before commencement of the hearing. The petition must set forth the petitioner's interest in the proceeding and show that his participation will aid in the determination of the issues in question and will not unnecessarily delay the proceedings. Commission Rule 21(b); *Pennsylvania Truck Lines, Inc.*, 7 OSHC 1722 (1979).

In addition to the employer, other parties may be considered indispensable (as that term is defined in Federal Rule of Civil Procedure 19), in certain proceedings and must be joined in the action. *Penn Central Transportation Corp.*, 4 OSHC 2033 (1977); *Harry Pepper & Associates, Inc.*, 7 OSHC 1815 (1979).

5. Manner of Service

The Commission Rules are quite clear; all documents or pleadings filed with the Commission or judge must be served on all parties and interveners. Commission Rule 7. Service on parties who are represented is made on the representative. Service may be accomplished, unless the Commission or judge orders otherwise, by postage prepaid, first class mail or personal delivery, which is deemed effective at the time of mailing or delivery, respectively. Commission Rules 7 or 8. (Certified mail may also be used and may be a preferable method, since it provides evidence of mailing and delivery).

Service may also be accomplished by posting, and proof of such posting shall be filed not later than the first working day following the posting. If service is undertaken by mail, three days are added to the time period provided for a response. Commission Rule 4(b).

Proof of service must be made and can be accomplished by a written statement which sets forth the date and manner of service and which is filed with the pleading. Commission Rule 7(d). The statement must show the date and manner of service (mail or personal delivery) and the names of the persons served. This statement is called a Certificate of Service.

Ex parte communication with the Commission or a judge is prohibited; other parties must always be informed, except in the case of applications for subpoenas, which can be made *ex parte*. Commission Rule 103.

6. Formalities of Pleadings

All documents and pleadings filed should be typewritten on letter size (8½″ × 11″) paper, double spaced, should be signed by the party or its representative, should be properly captioned, and should include the case name and docket number. Commission Rules 30 and 31. Initial pleadings must also contain the filing party's name, address and telephone number. Commission Rule 6.

Prior to the assignment of a case to an administrative law judge, all papers required to be filed in the case must be filed with the Executive Secretary of the Occupational Safety & Health Review Commission, 1825 K Street, N.W., Washington, D.C. 20006. Subsequent to the assignment of the case to a judge and before the issuance of his decisions, all papers must be filed with the judge at the address given in the notice informing of such assignment. Commission Rule 8. Subsequent to the issuance of the decision of the judge, all papers must again be filed with the Executive Secretary.

D. PLEADINGS, DISCOVERY, AND MOTIONS

1. The Complaint

Within no later than twenty days from the date the Secretary of Labor receives a notice of contest, the Secretary of Labor must file a formal complaint with the Review Commission; copies must also be sent to all parties. Commission Rule 33. Day means calendar day, and must be distinguished from the "working day" used to calculate the time for filing a notice of contest. Rule 1(k). The Solicitor of Labor's office, which prepares the case for hearing before the Commissioner, may, and often does, seek and obtain an extension of time within which to file a complaint. A copy of the extension must, of course, be sent to the employer and any other parties to the case and must be requested in advance of the date on which the pleading is due. Commission Rule 5.

The courts and the Commission have rejected a strict application of the timeliness requirement in the filing of complaints, unless the delay has in some way prejudiced the employer. *Jensen Construction Co. of Oklahoma v. OSHRC,* 597 F.2d 246, 7 OSHC 1283 (10th Cir. 1979); *Austin Commercial, Inc.,* 7 OSHC 1100 (1979). Untimely complaints filed in excess of the twenty-day time period do not automatically warrant dismissal, unless the employer can show it suffered prejudice by the late filing of the complaint. See *National Industrial Constructors, Inc. v. OSHRC,* 583 F.2d 1048, 6 OSHC 1914 (8th Cir. 1978); *Builders Steel Co.,* 5 OSHC 1851 (1977); *Rollins Outdoor Advertising, Inc.,* 5 OSHC 1041 (1977). The use of alternative pleadings in the complaint (and answer) is permitted. *Henkles & McCoy, Inc.,* 4 OSHC 1502 (1976).

The Complaint must set forth all alleged violations and proposed penalties which are contested, stating with particularity:

(1) The basis for jurisdiction;
(2) The time, location, place, and circumstances of each such alleged violation; and
(3) The considerations upon which the period of abatement and the proposed penalty on each such alleged violation is based.

Commission Rule 33(a)(2). The citation may be attached to the complaint and incorporated by reference. *Sun Shipbuilding and Drydock Co.,* 3 OSHC 1413 (1979). In *Asarco, Inc., El Paso Division,* 8 OSHC 2156 (1980), the Secretary of Labor failed to file a complaint in response to the employers notice of contest, relying on *IMC Chemical Group, Inc.,* 6 OSHC 2075 (1978), asserting that the citation was the functional equivalent of and can stand as a complaint. The Commission rejected this argument, holding that the complaint and its twenty day time period are mandatory:

> The Secretary and the judge have in effect interpreted IMC as meaning that a citation may be treated as a complaint for the purpose of applying Commission Rule 33(a). For example, the Secretary argued, in opposing ASARCO's petition for interlocutory appeal, that "[t]he Commission's position, as stated in [IMC is] . . . that a citation and a notice of contest should be respectively treated like a complaint and answer in federal court litigation. . . ." We do not agree with this interpretation of IMC . . .
>
> In proceedings under the Act, a citation is not a complaint. A citation is a creature of statute. *See* Section 9 of the Act. Complaints owe their existence to the Commission's Rules of Procedure, specifically Rule 33(a). These rules clearly recognize that citations and complaints are separate documents. Thus, Rule 33(a)(1)

establishes a filing deadline for a complaint of twenty days after the Secretary has received a notice contesting his citation. Rule 33(a)(2) lists specific allegations that must be included in a complaint. Some of these allegations are neither customarily found in nor required to be included in citations. Compare *Cement Asbestos Products Co.*, 8 OSHC 1151 (1980) [complaints] with *Gold Kist, Inc.*, 7 OSHC 1855 (1979) [citations]. In addition, Rule 33(a)(3), 20 C.F.R. §2200.33(a)(3), established a means whereby the citation can be amended in the complaint.

We further conclude that the filing of a complaint by the Secretary in a proceeding initiated by an employer notice of contest is a mandatory requirement under the Commission rules. We base this conclusion on giving that term its ordinary and customary meaning.

The Commission further held that the judge exceeded his authority by waiving the requirement that the Secretary of Labor file a complaint, but held that:

It is well established under Commission precedent that a citation or notice of contest ordinarily should not be dismissed for failure of a party to comply with the Commission's Rules of Procedure or with other procedural requirements. See, *e.g., Circle T. Drilling Co.*, 8 OSHC 1681 (1980); *Duquesne Light Co.*, 8 OSHC 1218 (1980); *Rollins Outdoor Advertising, Inc.*, 5 OSHC 1041 (1977). Thus, the policy in the law in favor of deciding cases on their merits generally prevails unless the party's noncompliance results from its own contumacious conduct or results in prejudice to the opposing party.

The Commission found that the Secretary of Labor's conduct was not contumacious and the employer was not prejudiced and thus granted the Secretary's motion to file its pleading out of time.

Technically deficient complaints or complaints which lack specificity usually will not serve as a basis for dismissal, as the decisions of the courts and the Review Commission continue to indicate, as long as the complaint gives to the employer fair notice of the issues in controversy and does not result in prejudice. See *Savina Home Industries, Inc. v. Secretary of Labor*, 594 F.2d 358, 7 OSHC 1154 (10th Cir. 1979); *John Hill, d/b/a Leisure Resources Corporation*, 7 OSHC 1485 (1979). Technically deficient complaints are those in which the wrong employer is named, the wrong docket number is referenced, or the wrong inspection date is included. *Turner Communications Co. v. OSHRC*, 612 F.2d 941, 8 OSHC 1038 (5th Cir. 1980)—employer failed to demonstrate prejudice in preparing defense as a result of lack of descriptive detail in the complaint.

As a general rule, administrative pleadings are liberally construed and easily amended. *National Realty & Construction Co. v. OSHRC*, 489 F.2d 1257, 1267 n.40, 1 OSHC 1422 (D.C. Cir. 1973); *United States of America v. B&L Supply Company*, 486 F. Supp. 26, 8 OSHC 1125 (N.D. Tex. 1980); *Pukall Lumber Company, Inc.*, 2 OSHC 1675 (1975). Amendments to a complaint (or citation) can be made at either of three stages: prior to the hearing, at the hearing or subsequent to the hearing. Examples of amendments include: (1) one standard to another; (2) one standard to the general duty clause; (3) the general duty clause to a standard; (4) the severity or characterization of the violation; (5) an increase or decrease in penalty. See generally *Claude Neon Federal Company,* 5 OSHC 1546 (1977); *Southern Colorado Prestress Co.,* 4 OSHC 1638 (1976), *aff'd* 586 F.2d 1342, 6 OSHC 2032 (10th Cir. 1978); *Dye Construction Co.,* 7 OSHC 1617 (1979); *Structural Painting Corporation,* 7 OSHC 1682 (1979); *Asplundh Tree Expert Company,* 6 OSHC 1951 (1978); *Constructora Maza, Inc.,* 6 OSHC 1309 (1978); *PAF Equipment Co.,* 7 OSHC 1209 (1979); *D. Fortunato, Inc.,* 7 OSHC 1643 (1979).

The complaint may be amended prior to the hearing, pursuant to Commission Rule 33(a)(3) and Federal Rule of Civil Procedure 15. Since the answer is considered to be the employer's first responsive pleading, Federal Rule of Civil Procedure 15(a) would permit an amendment of right prior to its filing. (Literally read, Commission Rule 33 would only apply to a situation in which the Secretary seeks in the complaint to amend the citation and would not apply to an amendment to the complaint itself. Thus, if a complaint is filed which tracks the citation, the complaint arguably could later be amended without complying with Commission Rule 33(a)(3). *Pukall Lumber Co., Inc.,* 2 OSHC 1675 (1975). In *Long Manufacturing Company v. OSHRC,* 504 F.2d 903, 5 OSHC 1376 (8th Cir. 1977), the court stated:

> An employer may accept the citation and proposed penalty without protest if he chooses to do so, but in our view if he chooses to contest the matter before the Commission, he does not have any vested right to go to trial on the specific charge mentioned in the citation or to be free from exposure to a penalty in excess of that originally proposed. The Commission's rules recognized in the proper case both the citation and the proposed penalty may be amended.

The Secretary of Labor must, by motion, set forth the reasons for the amendment and state with particularity the changes sought. The Secretary has been allowed to amend his complaint in most instances and literal compliance with the requirements of a statement of reasons has not been required, unless there is a claim that the employer was misled or prejudiced by the Secretary's failure to explain the amendments. See *Roanoke Iron & Bridge Works, Inc.,* 5 OSHC 1391 (1977), *enf'd,* 588 F.2d 1351, 7 OSHC 1041 (4th Cir. 1978); *Southern Colorado Prestress Co. v. OSHRC,* 586 F.2d 1342, 6 OSHC 2032 (10th Cir. 1978); *Schiavone Construction Co.,* 5 OSHC 1385 (1977); *Lovell Clay Products, Inc.,* 2 OSHC 1121 (1974)—neither the employer nor the Commission raised the issue; *Mid-Plains Construction Co.,* 2 OSHC 1121 (1974)—dismissal of complaint is too severe a remedy for a peccadillo of this kind. The general rule is, then, that where fair notice is given and a party is not prejudiced, compliance with the procedural requirements of Commission Rule 33(a)(3) is not strictly required on the theory that administration proceedings are to be liberally construed and easily amended. As the court stated in *National Realty and Construction Co. v. OSHRC,* 489 F.2d 1257, 1264, 1 OSHC 1422 (D.C. Cir. 1973):

> So long as fair notice is afforded, an issue litigated at an administrative hearing may be decided by the hearing agency even though the formal pleadings did not squarely raise the issue. . . . Citations under the. . . . Act are drafted by nonlegal personnel acting with dispatch. Enforcement of the Act would be crippled if the

Secretary were inflexibly held to a narrow construction of citations issued by his inspectors.

In *Roanoke Iron and Bridge Works, Inc.*, 5 OSHC 1391 (1977), *aff'd.* 588 F.2d 1351, 7 OSHC 1014 (4th Cir. 1978), however, the Secretary of Labor, in his complaint, moved to amend a citation to allege an alternate violation of several standards. The employer opposed the Secretary's motion to amend and the Administrative Law Judge denied the motion. In support of its argument on review to the Commission, the Secretary of Labor noted that citations are issued by nonlegal personnel and that the Judge's reason for denying the amendment would lead to the result that the Secretary's legal theory in all OSHA proceedings would be irrevocably controlled by the opinion of a nonlegal staff. The Commission, however, upheld the refusal to permit the amendment, stating that in the particular case the amendment did not simply seek to change the legal theory, but also added new factual allegations:

> While it may be true the Secretary's nonlegal personnel who issue citations cannot be expected to always know the correct legal theory under which to proceed, they should be capable of identifying the appropriate means of abating safety hazards. Furthermore, in this case, the Secretary's legal staff seeks to alternatively allege violations of four separate standards . . . Thus, rather than attempting to determine the correct standard under which to proceed, the Secretary seeks to allege every standard which might conceivably apply . . . We do not however think it is justifiable for the Secretary to attempt to overcome these difficulties by means of the approach taken in this case. If the Secretary, who is presumably familiar with the requirements of the various standards, cannot determine the proper one under which to proceed, it is doubtful if an employer can be said to have had fair notice of the violations with which it is charged. We are therefore inclined to deny the Secretary's motion to amend.

Thus, preliminary amendments to the complaint will normally be permitted in the absence of a showing of prejudice to the employer in the preparation and presentation of its case. Typical amendments to the complaint will normally involve a change in the characterization of the violation, amount of the penalty or the cited standard, such as: (1) an amendment from the general duty clause to a specific standard, *Brisk Waterproofing Co.*, 3 OSHC 1132 (1973); *Claude Neon Federal Co.*, 5 OSHC 1546 (1977); *Higgins Erectors & Haulers, Inc.*, OSHC 1736 (1979); (2) an amendment from one standard to another, *Carr Erectors, Inc.*, 4 OSHC 2009 (1977); *Gerstner Electric Co.*, 2 OSHC 1130 (1974); *Southwestern Bell Telephone Co.*, 6 OSHC 2130 (1978); or (3) an amendment from a standard to the general duty clause, See *Dunlop v. Uriel G. Ashworth*, 538 F.2d 562, 3 OSHC 2065 (4th Cir. 1976); *Usery v. Marquette Cement Manufacturing Co.*, 568 F.2d 902, 5 OSHC 1793 (2d Cir. 1977); *B. Heckerman Ironworks, Inc.*, 1 OSHC 1352 (1973); *Royal Logging Co.*, 7 OSHC 1744 (1979).

Amendments to the complaint/citation are normally and liberally permitted if only the legal theory rather than the factual basis or cause of action of the charge is to be changed. *Henkels & McCoy*, 4 OSHC 1502 (1976); *Miller Brewing Co.*, 7 OSHC 2155. The general test for determining whether there has been a change in the cause of action is whether the original and the amended charges arise out of the same conduct, transaction or occurrence. See *Structural Painting Corp.*, 7 OSHC 1682 (1979); *D. Fortunato, Inc.*, 7 OSHC 1643 (1979); *J. L. Mabry Grading, Inc.*, 1 OSHC 1211 (1973); *P.A.F. Equipment Co.*, 7 OSHC 1209 (1979)—amendment of a complaint to allege a willful violation rather than a serious violation does not change the underlying cause of action; *Coastal Pile Driving, Inc.*, 7 OSHC 1133 (1977). In *Roanoke Iron & Bridge Works, Inc.*, 5 OSHC 1391 (1977), *enf'd*, 588 F.2d 1351, 7 OSHC 1014 (4th Cir. 1978), the Secretary of Labor moved in his complaint to amend the citation to allege a violation of a perimeter protection standard different from the one under which the respondent/employer was originally cited (1926.500(d)(1) v. 1926.750(b)(1)(iii). The Secretary also moved to amend in the alternative to charge that the employer's failure to use personal protective equipment violated 1926.28(a) and 1926.105(a). The Commission found that the proposed amendments did not simply seek to change the legal theory underlying the citation but also added new factual allegations. It recognized that while it may be true that the Secretary's non-legal personnel who issue citations cannot be expected to always know the correct legal theory under which to proceed, they should be capable of identifying the appropriate means of abating safety hazards. The Commission also noted that rather than attempting to determine the correct standard under which to proceed, the Secretary sought to allege every standard which might conceivably apply and therefore felt it was not justifiable for the Secretary to attempt to overcome the difficulties encountered in applying fall protection standards by the approach taken in this case and declined to permit the motion to amend.

In *Morrison-Knudsen & Associates*, 8 OSHC 2231 (1980), eleven weeks following the issuance of a citation, the Secretary of Labor filed a complaint which attempted to amend the citation. The judge, reasoning that the amendment would significantly change both the legal theory of the case and the legal burden of the parties, denied the motion to amend. The Commission overruled the judge's determination observing that an amendment well before a hearing rarely results in prejudice. It also rejected the contention that the amendment resulted in an impermissible change in the cause of action, since the violation alleged in the complaint was based on the same facts and on the same failure to comply with the same standards as the violation alleged in the citation. Even though the amendment created two alleged violations out of one, the legal theory of the case remained unchanged by the amendment. The Commission therefore stated that under its *P.A.F. Equipment Co.* theory, the amendment was permissible unless the record established prejudice to a respondent. In this case, the amendment came at the earliest involvement of counsel, the filing of the complaint, and was made five full months prior to the hearing. Thus, according to the Commission, the employer could not claim surprise. Compare *Miller Brewing Co.*, 7 OSHC 2155 (1980) and *Mississippi Power & Light Co.*, 7 OSHC 2036 (1979)—prejudice was found because the factual issues changed and the employer was not given a chance to present evidence to the new charge.

If an attempt is made to amend a complaint/citation, a respondent/employer must object and produce evidence and advance an argument in support of a claim of prejudice if it hopes to prevail. In the alternative, it should also request a continuance to prepare a new defense to the amendment, for the issue is whether the employer received fair notice. See *Structural Painting Corporation,* 7 OSHC 1682 (1979); *Brown & Root, Inc., Power Plant Division,* 8 OSHC 1055 (1980); *Turner Welding & Erection Co.,* 8 OSHC 1561 (1980); *Foster & Kleiser,* 8 OSHC 1639 (1980).

The principle of liberal granting of prehearing amendments normally extends to amendments which are made at the hearing itself (preferably at the beginning) if the non-moving party would not be prejudiced in the preparation or presentation of his case. If prejudice is claimed, the nonmoving party must produce evidence in support or request a continuance to prepare a new defense. Prehearing amendments basically follow Federal Rule of Civil Procedure 15(a), which provides that leave to amend shall freely be given when justice so requires. See *General Electric Co.,* 5 OSHC 1187 (1977), *aff'd.* 576 F.2d 558, 6 OSHC 1541 (3d Cir. 1978); *Asplundh Tree Expert Co.,* 6 OSHC 1951 (1978); *Structural Painting Corporation,* 7 OSHC 1682 (1979).

Amendments made well into the hearing and after the hearing (in order to conform the "pleadings" to the evidence presented) bear a more stringent standard and must comply with Rule 15(b) of the Federal Rules of Civil Procedure. See *Texaco, Inc.,* 8 OSHC 1677 (1980)—motion to amend made after judge's decision was filed with the Commission, Federal Rule of Civil Procedure 15(b) considerations are applicable. If the amendment of a complaint is requested after the hearing pursuant to Rule 15(b), the unpleaded charge or issue must have been tried with the express or implied consent of the parties.

Motions to amend pleadings after the hearing to conform to the evidence presented may be raised at the close of the hearing, in post hearing briefs and on review. A *sua sponte* recognition may also be raised by the judge or the Commission. Regardless of when or by whom a conforming amendment is sought, a prerequisite is the express or implied consent of the parties in trying the issues addressed—so that prejudice does not result. Prejudice is generally defined as the inability to present or defend a case on the merits.

In *Dwayne Smeltzer Roofing Company v. Marshall,* 617 F.2d 448, 8 OSHC 1106 (6th Cir. 1980), the court held that the Commission properly granted the Secretary of Labor's motion to amend the complaint to change the date of a violation to conform to the evidence since the employer had impliedly consented to the amendment by submitting evidence of work conditions on the amended date; *Brown & Root, Inc., Pipeline Div.,* 6 OSHC 1844 (1978); *First Colony Corp.,* 6 OSHC 1843 (1978); *D. Fortunato,* 7 OSHC 1643 (1979); *Southwestern Bell Telephone Co.,* 6 OSHC 2130 (1978). When a party opposing the amendment introduces evidence relevant to an unpleaded charge or fails to object to such evidence, then that partly impliedly consents to try the unpleaded charge. *Bill C. Caroll Co.,* 7 OSHC 1806 (1979). Moreover, if an amendment would only change the legal theory from that alleged in the pleadings, but not the facts, then consent to the amendment will normally be implied when the party opposing the amendment has not objected to the introduction of relevant evidence. See *McLean-Behm Steel Erectors, Inc.,* 6 OSHC 2081, *rev'd.* 608 F.2d 580, 7 OSHC 2002 (5th Cir. 1979); *Mississippi Power and Light Company,* 7 OSHC 2036 (1979).

The operative standard then for posthearing amendments under Rule 15(b) of the Federal Rules of Civil Procedure is that amendments made to conform to the evidence will be approved, if (1) there was *no objection* to the introduction of the evidence; (2) there was *no prejudice* that would result from amendment, in that the employer had an ample opportunity to present evidence; and (3) the issues related to the proposed amendment were tried by the *express or implied consent* of parties. See *Usery v. Marquette Manufacturing Co.,* 568 F.2d 902, 5 OSHC 1793 (2d Cir. 1977); *Builders Steel Co. v. Marshall,* 575 F.2d 663, 8 OSHC 1363 (8th Cir. 1978); *Southern Colorado Presstress Co. v. OSHRC,* 586 F.2d 1342, 6 OSHC 2032 (10th Cir. 1978). *John & Ray Carlstrom,* 6 OSHC 2101 (1978); *Rodney E. Fossett d/b/a Southern Lightweight Concrete Co.,* 7 OSHC 1915 (1979); *H. B. Zachary Co.,* 7 OSHC 2022 (1980). Compare *Mississippi Power & Light Co.,* 7 OSHC 2036 (1979); *Western Waterproofing Co. v. Marshall,* 576 F.2d 139, 6 OSHC 1550 (8th Cir. 1978); *McLean-Behm Steel Erectors, Inc. v. OSHRC,* 608 F.2d 580, 7 OSHC 2001 (5th Cir. 1979)—a *sua sponte* amendment of the citation/complaint after hearing, to allege a more specific standard than the one set forth in citation, was improper, since the employer objected to the amendment at the hearing, he was only prepared to defend against the cited standard, and the Solicitor of Labor declined to amend the citation at hearing; thus there was no consent to the amendment and prejudice was shown; *Cornell & Co., Inc. v. OSHRC,* 573 F.2d 820, 6 OSHC 1436 (3d Cir. 1978)—employer was prejudiced in presenting his defense by the amendment made nine days prior to the hearing, since relevant evidence became unavailable. Post-hearing amendments can be made at the close of the hearing, in post-hearing briefs or on review.

Such amendments, if they arise out of the same transaction, conduct, or occurrence set forth in the original pleadings, relate back to those pleadings and are not time barred by the six-month statute of limitations set forth in Section 9(c) of the Act. See Federal Rule of Civil Procedure 15(c); *Structural Painting Corp.,* 7 OSHC 1682 (1979); *Higgins Erectors & Haulers, Inc.,* 7 OSHC 1736 (1979).

2. The Answer

In response to the Secretary's complaint, the employer must file a timely answer with the Commission within fifteen days after service of the complaint. Commission Rule 33(b); *Cobia Boats, Inc.,* 1 OSHC 1151 (1973). The answer is considered to be the employer's first responsive pleading. *Mid-Plains Construction Co.,* 2 OSHC 1728 (1975). A copy of the answer must be sent to all parties. Failure to file an answer may result in the dismissal of a notice of contest pursuant to Commission Rule 38 which provides that the Commission may preclude further participation by party if it fails to file pleadings in a timely manner. *G&C Foundry Co.,* 1 OSHC 1076 (1972).

The answer must contain a short and plain statement denying those allegations in the complaint which the employer intends to contest. *Metals, Inc.,* 6 OSHC 1816 (1978). Thus the answer may either be a general denial of the complaint, such as "I deny all the charges in the complaint," or respond directly to those statements in the complaint that the employer admits, denies, or wishes to explain. Any allegation not denied shall be deemed admitted. If the employer fails to file an answer, a judge may grant a motion to dismiss the employer's notice of contest. *Daniel Rubin Painting Corp.*, 1 OSHC 1539 (1974). The failure to timely file an answer will usually result in the issuance of an Order to Show Cause why the notice of contest should not be dismissed. *Genco, Inc.*, 6 OSHC 2025 (1978). The Commission normally excuses an employer's failure and will reinstate an employer's notice of contest, if there are reasonably grounds on which to do so. Although Commission Rules fail to address this point, affirmative defenses should be set forth in the employer's answer. Federal Rule of Civil Procedure 8(c) would apply, by virtue of Commission Rule 2(b) and it requires that matters of avoidance or affirmative defense be specifically pleaded. Strictly construed however, Rule 8(c) only requires that matters not within the opposing party's prima facie case be pleaded. Thus affirmative defenses should only be asserted if they pertain to matters outside the Secretary of Labor's case. Unfortunately the Commission has applied its affirmative defense doctrine in a restrictive manner, requiring that *any and all* such defenses be raised, both procedural and substantive, and even if normally cognizable by motion pursuant to Federal Rule of Civil Procedure 12(b)(6). See *Concrete Construction Corp.*, 4 OSHC 1133 (1976)—vagueness; *Pasco Masonry Co.*, 5 OSHC 1864 (1977)—failure to comply with Section 8(a); *Amory Cotton Oil Co.*, 3 OSHC 1895 (1976)—failure to state a claim upon which relief can be granted. If an employer fails to include his affirmative defenses in his answer (or because of the 15-day time period in which to file a response, is unable to timely formulate them (five days fewer than under Federal Rule of Civil Procedure 12[a]), he should promptly move to amend his answer properly to include them during the issue formulation stage prior to hearing. *Chicago Bridge and Iron Company*, 1 OSHC 1485 (1974), *rev'd on other grounds*, 514 F.2d 1082, 3 OSHC 1056 (7th Cir. 1975). Leave to amend should be freely granted under Federal Rule of Civil Procedure 15(a) when justice so requires—especially since it appears that there is no provision for amending the notice of contest itself. In certain situations the Commission may refuse to allow amendments to an answer raising an affirmative defense, if prejudice results to the Secretary of Labor. *Signal Oil Field Service, Inc.*, 6 OSHC 1717 (1978); *Continental Kitchens, Inc.*, 3 OSHC 1859 (1975); Federal Rules of Civil Procedure 8(b) and (c). Moreover, unpleaded affirmative defenses may not normally be raised at the hearing. In *Connecticut Natural Gas Corporation*, 6 OSHC 1797 (1978) the employer moved at the hearing to amend the answer to include the affirmative defense of vagueness. The judge granted the motion to amend and interpreted the amendment as a motion to dismiss.

There is no specific form for the answer, and it need not conform to the technical aspects of pleading, as long as it is responsive to the allegations of the complaint. *Anesi Packing Co.*, 1 OSHC 1496 (1973); *Edward J. Huegel, Inc.*, 1 OSHC 1029 (1972)—a duplicate notice of contest was considered to be an answer.

3. Prehearing Conference

At any time prior to the hearing, the Commission or the judge, on its own motion or on the motion of a party, may direct the parties to exchange information or to participate in a prehearing conference to consider settlement or matters which will tend to simplify issues at the hearing. Commission Rule 51; *Williams Enterprises Inc.,* 4 OSHC 1663 (1976). A prehearing order will be issued by the judge or the Commission which will include the agreements reached by the parties, will be served on the parties, and will be part of the record. Rule 16 of the Federal Rules of Civil Procedure functions as a supplement to Commission Rule 51. The judge has considerable discretion in framing this order. Prior to the conclusion of the prehearing conference, the employer should seek the Solicitor of Labor's agreement that any exhibits and affidavits supplied may be included as a part of the Record in the case. Such an agreement eliminates the necessity of separately introducing each exhibit at the hearing and authenticating it prior to its admission into evidence. Also, at this time, any motions should be filed with the judge (*i.e.,* identification of expert witnesses).

If a prehearing order is issued, it may include the following matters: (1) the allowance and scope of depositions, interrogatories, or other discovery procedures; (2) disclosure of the identity of witnesses to be called at the hearing and a description of the "substance of their expected testimony;" (3) the identification of expert witnesses; and (4) the time for furnishing parties with additional exhibits.

Certain agreements or stipulations submitted for a prehearing order, such as a stipulation between the Secretary of Labor and an employer as to the appropriateness of a penalty, are not binding upon the Commission, but may be accepted in its discretion. *Bettendorf Terminal Co.,* 1 OSHC 1695 (1974). Failure to comply with a judge's prehearing order may result in the imposition of sanctions, or exclusion of testimony of witnesses if, for example they were not put on the witnesses' list. *Williams Enterprises, Inc.,* 4 OSHC 1663 (1976); *Hoerner Waldorf Corp.,* 4 OSHC 1836 (1976). The dismissal of a citation because of a party's failure to timely file a prehearing conference report is considered to be too harsh a sanction in the absence of a showing of prejudice by the delay. *Duquesne Light Company,* 8 OSHC 1218 (1980). A showing of actual prejudice is also required before vacating citations for failure to follow other procedural provisions of the Act or the Commission's rules or orders. *National Industrial Constructors, Inc. v. OSHRC,* 583 F.2d 1048, 6 OSHC 1914 (8th Cir. 1978); *Marshall v. CF&I Corp.,* 576 F.2d 809, 6 OSHC 1543 (10th Cir. 1978).

Parties may hold an informal prehearing conference which the judge has authorized but in which he does not participate or attend. Should the parties reach any agreement during the prehearing conference, the Commission or the judge may later issue a prehearing order which incorporates the agreements reached by the parties. There is some authority for the fact that matters set forth in a prehearing memorandum, if adopted as an order, should be binding on the parties at hearing. *Duquesne Light Co.,* 8 OSHC 1218, 1223 n.21 (1980); *W. C. Sivers Co.,* 1 OSHC 1733 (1974).

4. Discovery

During the period between the filing of the answer and the date the case is set down for hearing, various pleadings may be filed by the parties, and prehearing procedures and discovery may be conducted. The basic idea behind discovery is that a party to a hearing should be entitled to disclosure of all relevant information in the possession of any person, including parties, except what is privileged. Prehearing discovery affords an opportunity to acquire important and necessary information about the opposing party's case. There is, of course, no basic constitutional right to pretrial discovery in administrative proceedings; the Administrative Procedure Act (APA) does not provide for discovery, and the Federal Rules of Civil Procedure, in and of themselves, do not apply to agency proceedings. *Silverman v. Commodity Futures Trading Commission,* 549 F.2d 28 (7th Cir. 1977). The Review Commission's discovery procedures are similar to but more expansive than those of the National Labor Relations Board.

Section 12(g) of the Act gives the Commission authority to promulgate rules of procedure to conduct orderly hearings. These rules include discovery. To the extent the Commission Rules fail to address a specific discovery procedure, the Federal Rules of Civil Procedure will apply. See also Administrative Procedure Act, 5 U.S.C. Sections 554, 557, 559.

a. *Requests for Admissions*

Under Commission Rule 52, any party may request of the other party admissions under oath with respect to matters of fact. Each admission must be set forth separately. Specific written response is required within fifteen days (or a shorter or longer period if so directed), or the request will be deemed admitted at the hearing. See, *e.g., Armor Construction & Paving, Inc.,* 3 OSHC 1204 (1975); *McWilliams Forge Co.,* 8 OSHC 1792 (1980). Copies of all requests and responses must be served on all parties and filed with the Commission.

Requests for admission are the form of discovery most often used by Solicitor of Labor's Office. The employer may also wish to use this form of discovery to determine the scope of the Solicitor's case and the nature of the facts to be presented at the hearing. Employers may request admissions concerning, *inter alia,* dates of inspections, testing, sampling, photographs, observations of work processes or operations, discussions with the employer's officials during the opening or closing conferences, any circumstances justifying a delay in the issuance of citations, and the bases upon which citations were issued. A request for discovery may not be taken as an admission that a citation was improperly issued. *Thomas A. Galante & Sons, Inc.,* 6 OSHC 1945 (1978).

b. *Depositions and Interrogatories*

Under Commission Rule 53, depositions and interrogatories are not allowed as a matter of right; in fact, they are not allowed *unless* a judge or the Commission specifically allows, in its discretion, such a request by order.[5] *KLI, Inc.,* 6 OSHC 1097 (1977); *Cement Asbestos Products Co.,* 8 OSHC 1151

(1980). Such discovery must be undertaken prior to the commencement of the hearing. *Seattle Crescent Container Service,* 7 OSHC 1895 (1979). The order granting discovery will set forth appropriate time limits. Where such discovery is permitted, the Federal Rules of Civil Procedure, specifically Rules 26 through 37, would be the applicable provisions governing its use. The Federal Rules include therein provision for objections to the conduct of discovery, requirements for protective orders, provisions for the failure to make discovery, sanctions, motions for orders compelling discovery, and provisions for failure to comply with the orders. See, *e.g., Quality Stamping Products Company,* 7 OSHC 1285 (1979), where the Commission held that the informer's identity rule, within the meaning of Rule 26(b) of the Federal Rules of Civil Procedure, is relevant in Commission proceeding. The Commission stated, in *Quality Stamping,* that under Section 12(g) of the Act, the Federal Rules of Civil Procedure govern Commission proceedings in the absence of specifically applicable Commission rules. For example, the Commission has adopted a rule relating to interrogatories, Rule 53, but it only regulates the manner for serving interrogatories and specifies that appropriate time limits should be granted but does not specify what time limits are appropriate. In the event that the Commission or judge would set what appears to the employer to be an inappropriate time limit, it would be appropriate to cite Rule 33 of the Federal Rules of Civil Procedure, which provides that thirty days is normally considered to be the appropriate period of time for response to interrogatories.

Non-privileged information, and/or information which is not attorney work product must be disclosed if depositions and interrogatories are ordered. See *Gulf & Western Products Co.,* 4 OSHC 1436 (1976); *Hickman v. Taylor,* 329 U.S. 495 (1947); Federal Rule of Civil Procedure 26(b)(3).

The failure to permit discovery will not be grounds for reversal of the judge's decision, unless such refusal has been an abuse of discretion, resulting in substantial prejudice to the party. *Perini Corp.,* 5 OSHC 1596 (1977). Judges may issue appropriate orders and impose sanctions for a party's failure adequately to comply with a discovery order. Commission Rule 54; *Wheeling-Pittsburgh Steel Corp.,* 4 OSHC 1788 (1976). However, the Commission, in *Circle T Growing Company, Inc.,* 8 OSHC 1681 (1980), stated that an administrative law judge's dismissal of a citation because of the Secretary of Labor's failure to respond to an employer's interrogatories was too harsh a sanction where there was no indication of contumacious conduct on the Secretary's part or proof by the employer that it was prejudiced by the delay. Similarly, in *General Dynamics Corporation, Quincy Shipbuilding Division,* 6 OSHC 1753 (1978), *aff'd,* 599 F.2d 453, 7 OSHC 1373 (1st Cir. 1979), the Secretary of Labor's incomplete response to an employer's interrogatories did not prejudice the employer nor warrant dismissal of the citation, since a three-month delay in the hearing afforded the employer an ample opportunity to resolve the problem.

Discovery of an informant's name through interrogatories is protected, even though the informant was motivated by a wrongful purpose, such as harrassment of the employer, since the informer's privilege is applicable to any person furnishing information to the government. *Quality Stamping Products Company,* 7 OSHC 1285 (1979).

c. *Entry upon Land for Inspection and Other Purposes*

The Commission has held that a citation may issue on less evidence than the Secretary needs to prove a violation; thus post citation discovery inspections have been permitted for the purpose of obtaining additional relevant information; special permission is not needed—that requirement only applies to depositions and interrogatories. *Thomas A. Galante & Sons, Inc.,* 6 OSHC 1945 (1978). (The discovery provisions of the Federal Rules of Civil Procedure, *e.g.,* Rule 34, apply to those forms of discovery not specifically covered by Commission rules.) In *Bristol-Myers Company,* 7 OSHC 1039 (1979), the Commission held that the judge erred in denying a motion to conduct a postcitation discovery inspection for the purpose of obtaining additional information relevant to the allegation that feasible engineering or administrative controls exist to reduce excessive noise, since the Commission had held, in *Reynolds Metals Company,* 3 OSHC 1749 (1975), that a citation may issue on less evidence than the Secretary may need to prove a contested violation, and that discovery should be permitted under circumstances similar to the present case.

The Secretary of Labor will thus often use this discovery device to attain additional information from the employer through inspecting, measuring, sampling, surveying, photographing, or testing the property or any processes or operations on the employer's premises. *Contra, Pabst Brewing Co.,* 4 OSHC 2003 (1977)—discovery inspection is not permissible if the information sought to be obtained could have been discovered during the original inspection.

In *Reynolds Metals Co., I and II,* 3 OSHC 1749 (1975) and 6 OSHC 1667 (1978), the Secretary of Labor requested entry onto an employee's workplace by a nongovernmental noise expert to determine the feasibility of engineering and administrative controls. The employer argued that the Secretary of Labor could gain such information during his inspection and that a discovery inspection by a nongovernmental noise expert could endanger the confidentiality of its trade secrets. The Commission authorized *entry, but only by a government expert,* saying first of all that Rule 34 of the Federal Rules of Civil Procedure applied and that it permitted entry even though a party has had a prior opportunity to inspect the other party's premises. The Commission also stated that the entry should only be by a governmental expert, since the Secretary of Labor had not shown *good cause* why an outside individual from the private sector should be used.

Subsequently, a divided Commission, in *Owens-Illinois, Inc.,* 6 OSHC 2162 (1978), overruled *Reynolds I* to the extent that it required that certain discovery entries be limited only to federal employees and required a showing of good cause by the Secretary for use of an outside expert. The Commission held that, where a discovery entry might endanger an employer's trade secrets, the appropriate method of protecting those trade secrets was by protective order rather than by a federal employee limitation. The Commission also set forth certain minimum criteria that a protective order must satisfy. The four elements to be included in a protective order are:

> First, it should require that a reasonable time before the discovery inspection is to occur, the Secretary provide the employer with a resume of the expert who is to make the inspection. The resume should include the expert's name, address, employment history, clients, education and published works. The employer should have an opportunity to challenge the selection of the expert on the ground that the expert is closely aligned with a competitor of the employer. Any such challenge should be ruled on by the judge.
>
> Second, the expert should be ordered to sign and comply with a written oath stating that any information which he obtains during the discovery entry which the employer has established as a trade secret (or, in the event the Secretary consents to the entry of a protective order without a trade secret being established, any information obtained during the discovery entry which the employer designates to the expert in writing as a trade secret) shall not be disclosed except to the Secretary's representatives who are involved in litigation of the case and in proceedings before the Review Commission.
>
> Third, the order should require the Secretary to include in his contract with the expert a non-disclosure provision identical in content to the oath just described. The provision should be drafted to make clear that it is the express intent of the parties that the employer be a beneficiary of the provision and is granted by it a right to enforce the provision.
>
> Finally, any protective order entered by the judge shall be broad enough to include within its coverage the Secretary and his representatives and shall clearly indicate that the sanctions for a violation of the order apply to anyone acting in the Secretary's behalf.

See *Maxwell Wireboard Box Co., Inc.*, 8 OSHC 1995 (1980); Commission Rule 11, which provides that the judge shall issue such orders as are appropriate to protect the confidentiality of trade secrets and other confidential information. A motion for a protective order may also be filed pursuant to Rule 26 of the Federal Rules of Civil Procedure, alleging *inter alia,* annoyance, embarrassment, oppression, or undue burden or expense. *Ralston Purina Company,* 7 OSHC 1730 (1979); *American Can Co.,* 7 OSHC 1947 (1979). The judge may order that the information uncovered during such an entry and inspection be sealed and labeled "confidential" or "trade secrets" and thus will often grant a motion to permit entry since this protection is available.

If the Secretary agrees to the entry of such a protective order and agrees to be bound thereby, the employer need not prove its allegation of trade secrets. If the Secretary does not agree to be bound by the protective order, the order should be entered if the employer proves the existence of trade secrets that are likely to be revealed by the requested entry. See also *Reynolds Metals Company, Inc. (III),* 8 OSHC 1496 (1980), which overruled *Reynolds II* to the extent it was inconsistent with *Owens-Illinois,* and which held that the Secretary of Labor is not barred from using nonfederal experts to conduct a discovery inspection of an employer's workplace, despite an employer's claim that this type of discovery would compromise confidentiality of its trade secrets, as long as its trade secrets can be protected by use of an appropriate protective order that meets the requirements set out in *Owens-Illinois.* See also *National Manufacturing Co.,* 8 OSHC 1435 (1980).

A judge's order excluding evidence obtained by the Secretary of Labor during discovery as a sanction because of the Secretary's noncompliance with a protective order prohibiting disclosure of information to the employer's competitor does not constitute an abuse of discretion but the dismissal of the action was not permitted in absence of evidence of prejudice or that the Secretary acted willfully as opposed to negligently. *Metropak Containers Corporation,* 8 OSHC 2112 (1980).

The Commission no longer requires that the Secretary of Labor show good cause prior to securing an order for a nonfederal expert to enter an employer's premises for discovery purposes. *Kaiser Aluminum & Chemical Corp.,* 7 OSHC 1486 (1979); *World Color Press, Inc.,* 6 OSHC 1084 (1977).

d. *Request for Production*

Since discovery provisions of the Federal Rules of Civil Procedure apply to those forms of discovery not specifically covered by the Commission's Rules, Rule 34 may be used by employers in seeking access to or copies of OSHA's internal files contining notes of the inspector, reports, recommendations, or memoranda. In fact, in conjunction with the filing of the employer's answer to the complaint, a request for production of documents should be immediately filed and directed to the Solicitor of Labor's office, requesting production of such documents as the safety and health or accident inspection report (OSHA–1), narrative (OSHA–1A), worksheet (OSHA–1B), citation and notification of penalty (OSHA–2), notification of failure to correct alleged violation (OSHA–2B), complaint (OSHA–7), notice of alleged imminent danger (OSHA–8), and hazards not covered by standard (OSHA–9), depending upon their applicability. All of these reports are prepared by the inspector subsequent to the inspection and may provide informative detail which would be helpful to the employer in the contest.

Once an administrative law judge has been assigned, the request for production should be directed to the judge. The judge will then issue an order granting or denying the request for production of documents. See *Gulf & Western Food Products,* 4 OSHC 1436 (1976)—production of an area director's file required to extent it did not contain privileged matter; *Frazee Construction Co.,* 1 OSHC 1270, (1973)—citation vacated because Secretary of Labor failed to produce notes and memorandum of inspection prepared by compliance officer.

The Secretary of Labor may oppose a production request, arguing that such material is "privileged" and that the government's investigative files cannot be revealed prior to testimony being introduced during the hearing which is based on or relies upon such file documents. While most of these arguments are not well taken, it should be noted that *Frazee* has not been applied to require the disclosure of an employee-informant's name. In *Massman-Johnson (Luling),* 8 OSHC 1369 (1980), the Commission recently held that the "informer's privilege" applies to proceedings before the Commission and may be relied upon to establish the right of the government to withhold from disclosure the identity of persons furnishing information of violations of the law to law enforcement officers. See *Quality Stamping Products Co.,* 7 OSHC 1285 (1979); *Stephenson Enterprises, Inc.,* 2 OSHC 1080 (1974), 4 OSHC

1702 (1976), *aff'd* 578 F.2d 1021, 6 OSHC 1860 (5th Cir. 1978). The privilege is applicable to any person furnishing information to government officials concerning violations of the Act or its implementing standards and regulations, regardless of the informant's employment relationship to the cited employer. The essence of the privilege is the protection of the informer's identity. The privilege is qualified, however, when disclosure is essential to the fair determination of a case. In such an instance the privilege must yield or the case will be dismissed. In *Quality Stamping,* for example, the employer sought the name and relationship of a person who had given a statement to the Secretary of Labor concerning hazardous conditions at the employer's plant. The privilege applied to that information. In *Massman Johnson,* the *issue* was whether the employers are *entitled to the actual statements* given to the government by prospective witnesses in the case. The employer contended that the informer's privilege did not protect the content of the statements themselves. The Secretary of Labor argued that the privilege protects the contents to the extent that it would tend to reveal the identity of persons giving them. The Commission agreed and also stated that the information sought by an employer must be relevant to the subject matter of the case before the question of privilege is reached. The scope of discovery in Commission proceedings would be governed by Rules 26(b)(1) of the Federal Rules of Civil Procedure, which provides, in general, that parties may obtain discovery regarding any matter not privileged which is relevant to the subject matter of the pending action. The Commission found that the information contained in the witness statements was relevant to the violation and that therefore the burden of proving facts in support of the privilege rested with the Secretary of Labor. However, the Commission found that the Secretary of Labor had proved the existence and applicability of privilege in the case, stating:[6]

> Here, the Respondents seek statements given to the government by certain prospective witnesses concerning the case. Under our holding in *Quality Stamping,* the qualified privilege clearly applies to protect the identity of persons giving such statements. Thus, the fact that the individual companies, as opposed to the joint venture, did not employ the persons involved does not make the privilege inapplicable.
>
> Nor is the privilege applicable only to those persons who actually instigate investigations or act as confidential accusers in the criminal sense. It is the duty of every citizen to cooperate in the enforcement of the law and the privilege encourages the fulfillment of that obligation by preserving the anonymity of government sources generally. *Quality Stamping, supra.*
>
> The fact that two of the prospective witnesses are management personnel likewise does not render the privilege inapplicable. Supervisory and managerial personnel in some cases may be the only persons who could be aware of the existence of violations. They are entitled to the protection of the informer's privilege when communicating information concerning workplace hazards to the government.[5]
>
> ---
>
> 5. Of course, the nature of the employment relationship between the alleged informer and the respondent, or the lack of such a relationship, may be relevant in balancing the interests the respective parties have in the information. *Quality Stamping, supra,* at n.10. Also, we note that in this case the informer's privilege was waived as to the two managerial employees involved.

> The essence of the informer's privilege, however, is the protection of the informer's identity, and the confidential information in a statement is that which tends to indicate that the person giving it has cooperated with the government against the employer. As the Secretary points out, the identity of an informer might be revealed from the disclosure of even a basically factual statement relevant to an investigation, for example, if the tone and manner of the statement were accusatory or unfriendly to the respondent.

Since the Commission found that the informer's privilege applied to the statements in question and that there had been no general waiver,[7] it then addressed the question whether the employer had shown special circumstances which justified withdrawing the qualified privilege.

The Commission first stated that *even where* there is no queston of the informer's privilege involved, witness statements may be subject to the provisions of Federal Rule of Civil Procedure 26(b)(3), which provides that documents prepared by or for a party representative in anticipation of litigation or trial are not discoverable *except* upon a showing of substantial need in the preparation of its case and that it is unable, without undue hardship, to obtain the substantial equivalent of the materials by other means. The Commission found that the materials requested by the employer were, notwithstanding privilege arguments, not discoverable under Rule 26 of the Federal Rules of Civil Procedure, would not be subject to prehearing disclosure under the prehearing conference provisions of Federal Rule of Civil Procedure 16, and would not be subject to a subpoena under Federal Rule of Procedure 45(d)(1), since the Secretary took statements from persons in connection with the investigation of alleged violations in *anticipation of litigation.* See generally *Hickman v. Taylor,* 329 U.S. 495 (1947). The Commission thus held that the identities of persons who have given statements regarding alleged violations that are the subject of an ongoing investigation, along with the contents of the statements themselves, are not required to be disclosed in discovery *before the hearing, unless* the employer shows that the information is essential to prepare adequately for the hearing and that it is unable to obtain the information by any other means.

The Commission noted, however, that during the hearing itself different considerations come into play. Employers are entitled to full and effective cross-examination of each witness, which includes an opportunity to test the veracity and accuracy of the witness's testimony against prior statement by that witness on the same subject. The *Jencks* approach (discussed below) recognizes this need by permitting the employer access, upon request, to all of the witness's prior statements in the government's possession that relate to the subject matter of the witness's testimony after the witness has testified on direct examination.

One of the ways that an employer could show it is unable to obtain information from a prospective witness, *per* the Commission's *Massman-Johnson* decision, would be to show that it was unable to gain discovery from that witness. An employer is not entitled to relief on the grounds of inability to gain discovery unless a proper discovery request has first been attempted from the witness by one of the methods described in the Federal Rules of Civil Procedure or the Commission's Rule of Procedure. The Commission noted, in *Massman-Johnson,* however, that the employer did not raise the discovery issue until the hearing. A discovery matter, such as the inability to obtain discovery of a witness, should be raised at the discovery stage. There was no indication that the request could not have been timely raised; thus in *Massman-Johnson* the Commission found that the employer was not entitled to relief based upon the failure of a prospective witness to remember the facts when interviewed. See also *Okland Construction Co.*, 3 OSHC 2023 (1976), in which an employer requested an attachment of an employee's statement by means of an interrogatory. Upon receipt of the interrogatory, the statement was not attached. The employer's objection to the failure of the attachment of the statement to the interrogatories was overruled, since it had not been prejudiced and it had not, when it served its interrogatories, used a motion for production of documents.

e. *Subpoenas*

Another form of discovery provided for by both the Act and the Commission's Rules, specifically Rule 55, is the issuance of subpoenas. Section 8(b) of the Act. Information obtained by OSHA may be relevant to the conditions of an inspection. While the reports prepared by an OSHA inspector are usually discoverable, the testimony of OSHA personnel can usually not be obtained in the absence of a subpoena. Any judge or member of the Commission, upon the application of a party, may issue subpoenas requiring the attendance and testimony of witnesses (any person, not just a party) and the production of evidence, including relevant books, records, correspondence, or documents in his possession or under his control. In *Marshall v. Elward*, 8 OSHC 1244 (S. Ct. Ill. 1980), the court held that the doctrine of sovereign immunity does not prevent state court enforcement of a civil plaintiff's subpoena *duces tecum* served upon OSHA, since enforcement proceedings under the Act are not the same as a suit against the federal government.

Applications for subpoenas, if filed subsequent to the assignment of the case to a judge, must be filed with the judge and may be made *ex parte*. The subpoena shall show the person to whom it is directed and at whose request the subpoena was issued and shall specify the documents demanded and place and time that such a subpoena shall be complied with. Persons served with subpoenas, whether *ad testificadum* (which require the appearance and testimony of a witness) or *duces tecum* (which require the production of documents), must within five days after the date of service of the subpoena move in writing to revoke or modify it, if they do not intend to comply. A copy of such motion shall be served on the party at whose request the subpoena was issued, and the judge or the Commission, as the case may be, shall revoke or modify the subpoena, if, in its opinion, the evidence whose production is required does not relate to any matter under investigation in the proceedings, or if the subpoena does not describe with sufficient particularity the evidence whose production is required, or if it is insufficient in law or otherwise invalid. A statement of procedural or other grounds for the ruling on the motion to revoke or modify shall be made and the motion to revoke, its answer, and the ruling thereon shall become part of the record. Commission Rule 55.

Upon the failure of a person to comply with a subpoena, the Commission shall initiate proceedings in the appropriate district court for the enforcement, *if* in its judgment the enforcement of such subpoena would be consistent with law and with the policies of the Act. In *Marshall v. Milwaukee Boiler*

Manufacturing Co., Inc., 626 F.2d 1339, 8 OSHC 1923 (7th Cir. 1980), an employer failed to show prejudice as a result of the failure to enforce its subpoena requiring production of documents at a hearing.

f. *Discovery at Hearing*

The discovery procedures set forth above must be timely undertaken prior to the hearing. *Pabst Brewing Co.*, 4 OSHC 2003 (1977). Generally, the rules for discovery applicable to hearings are equivalent to those in other administrative proceedings. *Jencks v. United States*, 353 U.S. 657 (1957); *Edison Lamp Works*, 7 OSHC 1818 (1979). Notes and reports an inspector made during an inspection must be revealed when he appears as a witness. *Frazee Construction Co.*, 1 OSHC 1270 (1973); *Blakeslee-Midwest Prestressed Concrete Co.*, 5 OSHC 2036 (1977)—discovery of an employer's financial records is permitted but must be kept confidential; *Waterville Co., Inc.*, 1 OSHC 3124 (1972); *Stephenson Enterprises, Inc.*, 2 OSHC 1080 (1974); *Buckeye Industries, Inc.*, 3 OSHC 1837, *aff'd*, 587 F.2d 231, 6 OSHC 2181 (5th Cir. 1979). In *Massman-Johnson (Luling)*, 8 OSHC 1368 (1980)—the Commission gave a detailed account of the rights of parties at the hearing *vis-à-vis* witness statements. The Commission basically adopted the *Jencks* approach which permits an employer access, upon request, to a witness's prior statements in the government's possession that relate to the subject matter of the witness's testimony after the witness has testified on direct examination, noting that the NLRB has an essentially similar rule in 29 CFR 102.118. Thus when a witness has completed testifying for the Secretary of Labor on direct examination, the Secretary shall, *upon request or motion* by the employer, turn over to it all of the witness's prior statements that are in the government's possession and that relate to the subject matter of the witness's testimony. (If the employer fails to make a timely request, the right to receive the statement may be lost.) If the Secretary claims that a statement contains material that does not relate to the subject matter of the witness's testimony, the judge will order the Secretary to deliver the statement for the judge's inspection *in camera*. Thereafter (as stated by the Commission in *Massman-Johnson*):

> The Judge shall excise the portions of the statement that do relate to the subject matter of the witness's testimony with one exception: the Judge may in his discretion decline to excise any portion that is relevant to other matters raised by the pleadings. After making the appropriate excisions, the Judge shall direct delivery of the statement to the respondent and, if the respondent objects to any excisions the portions involved shall be preserved by the Secretary pending possible review by the Commission or appeal to the courts of the Judge's decision.
>
> The respondent shall be entitled to a recess for such reasonable time as is necessary to evaluate a statement and prepare to use it at the hearing. In the event that a statement discloses that the hearing contains material respondent could not have discovered previously and that bears on the issue of the case, the respondent shall be entitled upon request to a recess or continuance for such time as is reasonably necessary to meet or to take advantage of the new evidence.

g. *Orders and Sanctions*

Appropriate sanctions or orders may be issued pursuant to Commission Rule 54 where a discovery order has not been complied with by a party or where trade secrets or other confidential information is involved and may be divulged.

5. Freedom of Information Act Requests

A party has the choice of use of the discovery procedure or the Freedom of Information Act, 5 U.S.C. §552 (FOIA), to get records from an agency. FOIA, of course, has no effect on what is or is not privileged in a discovery context. *Association of Women in Science v. Califano,* 566 F.2d 339 (D.C. Cir. 1977).

FOIA has been implemented through procedures and regulations promulgated by the Review Commission in 29 CFR 2201 and by general regulations issued by the Department of Labor in 29 CFR Part 70. OSHA procedures for the disclosure of information under FOIA are set forth in Chapter XXIV of the FOM.

The several cases which have considered the applicability of requests for information under FOIA have honed in on the disclosure exemptions set forth therein.[8] For example, in *T. V. Tower, Inc. v. Marshall,* 444 F. Supp. 1233, 6 OSHC 1320 (D. D.C. 1978), the court held that informants' names are exempt from disclosure under FOIA, which excludes from its scope investigatory records compiled for law enforcement purposes. In *Moore-McCormack Lines, Inc. v. ITO Corp.,* 508 F.2d 945,—OSHC—(4th Cir. 1975), the court rejected the withholding of an OSHA inspector's accident report under the intra-agency communication exception, stating that the exemption only applies to materials reflecting deliberative or policy-making processes. Finally, in *Burrell v. Rodgers,* 438 F. Supp. 25,—OSHC—, (W.D. Okla. 1977), the court held that a party seeking disclosure of information from OSHA under FOIA must file a separate FOIA complaint before a federal court obtains jurisdiction to enjoin the withholding of information. See also *Lead Industries Association, Inc. v. OSHA,* 610 F.2d 70, 7 OSHC 1820 (2d Cir. 1979) —outside consultant's reports concerning the lead standard were protected from disclosure because they were inextricably intertwined in the decision-maker's deliberative process and thus exempted.

Both the Review Commission and the Department of Labor FOIA regulations set forth generally the categories of records accessible to the public and the type of records subject to prohibitions or restrictions on disclosure. Thus, upon the request of any person, submitted in accordance with the regulations, access will be provided to records reasonably described by the requester, and such records shall be made available for inspection and copying unless specifically exempt from disclosure. If a request is denied, an appeal can be filed within ten days of the decision.

Examples of records which are basically disclosable include papers and documents made a part of the official record in administrative proceedings in connection with the issuance, amendment, or revocation of standards, rules and regulations, evaluation reports on external programs, statements of policy and interpretations, staff manuals, instructions, procedures, and guidelines. Examples of the nine types of records which are normally exempted from disclosure are internal rules and practices, trade secrets, privileged or confidential information, interagency and intra-agency memoranda and letters, personnel, medical, and employment files, and investigatory records compiled for law enforcement purposes. See, *e.g.*, *Stokes, McNeill & Barbe v. Brennan,* 476 F.2d 699, 1 OSHC 1175 (5th Cir. 1973)—OSHA training manual must be disclosed; *Frazee Construction Co.,* 1 OSHC 1270 (1973); *Raza Ass'n v. Brennan,*—F. Supp.—, 2 OSHC 1263 (D. D.C. 1974). Agencies involved in FOIA litigation are required to file a Vaughn index (named for the case establishing the procedure, *Vaughn v. Rosen,* 484 F.2d 820 (D.C. Cir. 1973)) that lists all documents withheld and the exemption justifying the decision not to disclose.

In *Marshall v. Edward,* 399 N.E.2d 1329, 8 OSHC 1244 (Ill. S. Ct. 1980), the Illinois Supreme Court found that the FOIA does *not,* under the supremacy clause of the Constitution, preempt discovery of OSHA records under state law.

6. Motions

Commission Rule 30 sets forth the form pleadings and motions must take, but does not contain special requirements for filing. See *Smith's Transport Corp.,* 3 OSHC 1088 (1975)—the Commission's Rules do not provide for or require the filing of responsive pleadings. Commission Rule 37 provides that any party upon whom a motion is served shall have ten days from service of the motion to file a response. An additional three days is provided if service has been made by mail. Commission Rule 4(b); *American Cyanmid Co.,* 8 OSHC 1346 (1980). Parties are entitled to rely on the entire period for filing responses, and it is error for a judge to take action on a motion prior to the expiration thereof. *Reynolds Metals Co.,* 7 OSHC 1042 n.9 (1979). Failure to file any pleading, including motions when due, may, in the discretion of the Commission and/or judge, constitute a waiver of the right to further participation in the proceedings. Commission Rule 38.

Since no specific Commission Rule applies, motions which may be filed generally correspond to those types of motions permitted under the Federal Rules of Civil Procedure, *e.g.*, motions to compel production, motions to strike, motions to amend a citation, motions in opposition thereto, motions to file interrogatories, motions to remand, motions in opposition to remand, motions to vacate and leave to file brief, motions to stay, motions to consolidate or sever, motions to dismiss, motions to file amendment to complaint, motions to file amendment to answer, motions for summary judgment, and motions to quash and motions withdrawing citations. *National Roofing Corp.,* 8 OSHC 1916 (1980); *International Harvester Co.,* 7 OSHC 2194 (1980); *Marshfield Steel Co.,* 7 OSHC 1741 (1979); *Miller Ceramics, Inc.,* 7 OSHC 1740 (1979); *American Urethane Co., Inc.,* 5 OSHC 1543 (1977)—the Secretary of Labor's motion to strike certain allegations of complaint, which was granted, constitutes its abandonment and precludes it from later being raised before the Review Commission; *American Cyanmid Co.,* 8 OSHC 1346 (1980)—a settlement agreement is in the nature of a joint motion. Certain motions such as a motion for summary judgment or motion to dismiss are normally not granted; it is considered preferable to decide a case on its merits. *Harrington Construction Corp.*, 4 OSHC 1471 (1976).

In *Amory Cotton Oil Company,* 3 OSHC 1895 (1976), a motion to dismiss for failure to state a claim cognizable under Federal Rule of Civil Procedure 12(b)(6) was rejected, on the basis that the subject matter was required to be asserted in the answer (or by special motion) as an affirmative defense even though Rule 12(b)(6) provides that such defense is not waived even if not specifically pleaded in the answer.

In *Cement Asbestos Products Company,* 8 OSHC 1151 (1980), the Commission held that a complaint gave the employer fair notice of the nature of the claim and met the requirements of Commission Rule 33(a)(2); thus the employer's *motion for a more definite statement* should not have been granted. A motion for more definite statement generally will not be granted if the complaint states the jurisdictional grounds of the claim, identifies the sections of the Act and standards allegedly violated, and sets forth the time, location, place, and circumstances of each such alleged violation. A motion for more definite statement is designed under Rule 12(e) of the Federal Rules of Civil Procedure to strike at unintelligibility rather than want of detail. The Commission stated that such motions are not to be used as a substitute for discovery in order to obtain the facts in preparation for hearing. *Ralston Purina Company,* 7 OSHC 1730 (1979); *Gold Kist, Inc.*, 7 OSHC 1587 (1979)—a motion for more definite statement is not a substitute for discovery and, in any event, even when proper must be made *before* service of a responsive pleading.

A judge's order granting a motion for indefinite continuance, which lasted three months, in violation of Commission Rule 61's limitation of thirty days, was harmless error since the employer was not prejudiced thereby. *Maxwell Wireboard Box Co., Inc.*, 8 OSHC 1995 (1980).

The Review Commission has also held that a motion by the Secretary of Labor to withdraw a citation due to insufficient evidence to establish the violation alleged cannot be granted unless the authorized employee representative, if any, has had an opportunity to be heard. *Wean-Pori, Inc.,* 7 OSHC 1545 (1979).

7. Interlocutory Appeal

Interlocutory appeals, authorized by Commission Rule 75, are used to challenge an interlocutory ruling issued by an administrative law judge *(e.g.,* a grant or denial of a request for discovery), prior to the issuance of a final decision in the case. A judge's interlocutory ruling may only be appealed to the Review Commission in the manner prescribed by Commission Rule 75. A party desiring to appeal from an interlocutory ruling must file a written request for certification of the appeal with the judge within five days after receipt of the ruling. Responses to the request, if any, must be filed within five days of its service. A judge may certify an interlocutory appeal when the ruling involves an important question of law or policy about which there is substantial ground for difference of opinion and an immediate appeal of the ruling may materially expedite the proceedings. *Lawrence Brothers, Inc.*, 8 OSHC 1457 (1980); *Farmers Export Company*, 8 OSHC 1655 (1980); *National Manufacturing Co.*, 8 OSHC 1435 (1980).

Following certification, the judge will forward the matter to the Executive Secretary of the Commission along with supporting documents, responses filed by the parties, the ruling from which the appeal is taken, a copy of relevant portions of the record, and the judge's order certifying the appeal. The Commission may either accept or decline to accept a certification. In *Quality Stamping Products Company*, 7 OSHC 1285 (1979), the Commission held that the Secretary of Labor's request for an interlocutory appeal was premature and improvidently granted, since there was no order by the judge overruling the Secretary's objection to the employer's interrogatories, requiring the disclosure of an informer's name or his relationship to the employer. Nevertheless, the Commission, in the interest of judicial efficiency, considered the appeal.

If the judge denies certification, then within five days following receipt of the judge's order denying certification, a party may file with the Commission a petition for interlocutory appeal. Again, responses to the petition, if any, must be filed within five days following service of the petition. The Commission will grant a petition for interlocutory appeal only in exceptional circumstances where it finds that the appeal satisfies the criteria for certification *and* that there is a substantial probability of reversal.

If the Commission denies or declines to accept a certification or denies a petition for interlocutory appeal, it shall not preclude a party from raising an objection to the judge's interlocutory ruling in a later petition for discretionary review on the case as a whole. A party whose request for certification is denied by a judge and who elects not to file a petition for interlocutory appeal with the Commission is not precluded from raising, in a later petition for discretionary review, an objection to the ruling from which interlocutory appeal was sought.

Commission Rule 11 provides for protection of trade secrets. The judge accordingly shall issue orders as may be appropriate to protect the confidentiality of such trade secrets. If a request to certify an interlocutory appeal is filed with a judge concerning an alleged trade secret, that filing shall stay the effect of the judge's ruling either until the judge denies the request or, if the request is granted, until the Commission rules on the appeal or declines to accept certification. Commission Rule 75(e).

In all other cases, the filing or granting of a request to certify an interlocutory appeal, or the filing or granting of a petition for an interlocutory appeal, will not stay a proceeding or the affect of a ruling, unless otherwise ordered. The Commission may order briefs on issues raised by an interlocutory appeal, and, if so, the briefs would be regulated by Commission Rule 93.

E. SETTLEMENT AGREEMENTS

As the Supreme Court has noted, the Act creates public rights that are to be protected by the Secretary through government enforcement. Necessarily included within these prosecutorial powers is the discretion to settle citations issued against employers and to compromise, mitigate, or settle any penalty assessed under the Act. *Marshall v. Sun Petroleum Products Company and OSAHRC,* 622 F.2d 1176, 8 OSHC 1422 (3d Cir. 1980), *citing Atlas Roofing Co. v. OSHRC,* 430 U.S 442, 5 OSHC 1105 (1977). The large majority of the cases in which a notice of contest is filed by an employer is settled. Even if it is not totally settled, there is a great chance that the proposed penalty will be reduced or vacated entirely.

Commission Rule 100 provides that settlement is permitted at any stage of the proceedings, even after the judge's decision has been directed for review (at which time it shall be filed with the Executive Secretary). In a recently issued Department of Labor "Joint Memorandum on Settlement of Cases by Area Directors," effective October 1, 1980, area directors are authorized to enter into informal settlement agreements with employers prior to their filing of a notice of contest. Normally these precontest settlements will occur during the informal conference. Employers will be informed of the new procedure by means of a letter sent with each set of citations issued. The Area Director is authorized to amend the abatement dates, adjust the penalty, change the characterization of the violation, and withdraw items, if evidence is presented which will convince the Area Director the changes are justified. However, after a notice of contest has been filed, a settlement proposal will not become final until the Commission has approved it as consistent with the intent of the Act and the public interest. *Davies Can Co.,* 4 OSHC 1237 (1976); *Kaiser Aluminum & Chemical Corp.,* 6 OSHC 2172 (1978); *Empire Detroit Steel Division, Detroit Steel Corporation v. Marshall,* 579 F.2d 378, 6 OSHC 1693 (6th Cir. 1978)—an oral settlement agreement between attorneys has been held not to be enforceable; *Plum Creek Lumber Co.,* 8 OSHC 2185 (1980). Every settlement proposal submitted *must* include, where applicable: (1) a motion to amend or withdraw a citation, notification of proposed penalty, notice of contest, or petition for modification of abatement; (2) a statement that payment of the penalty has been tendered or a statement of the promise to pay; and (3) a statement that the cited condition has been abated or a statement of the date by which abatement will be accomplished. The settlement proposal *must* be served upon affected employees in the manner prescribed for notices of contest. Proof of service must accompany the settlement proposal. Finally, the settlement will not be approved until at least ten days following service.

In evaluating a proposed settlement agreement, the Secretary must determine whether it affectuates the purpose of the Act to "assure so far as possible every working man and woman in the Nation safe and healthful working conditions." At any time prior to final Commission action, the Secretary has the power to withdraw from a settlement if there is a reason to believe that the agreement does not effectuate the purpose of the Act. However, should the Secretary reject an otherwise agreed upon settlement, it could be argued that all parties should be placed in the position of *status quo ante* the issuance of the citation. Such was the case in *Sun Petroleum Products Co., supra,* where the parties entered into a settlement agreement from which the Secretary later withdrew, claiming that the settlement was not within the public interest. The court held that although the Secretary of Labor has the power to withdraw from a settlement agreement at any time prior to final review by Commission action, dictates of fairness and justice require that the employer will not be required to defend a stale citation. The parties must be placed in their respected positions prior to the issuance of a citation. A new inspection must be conducted, and, if circumstances warrant, a new citation must be issued, to prevent prejudice to the employer (and affected employees) in defending a stale citation.

Where settlement negotiations have been entered into and the Secretary forbears from preparing his case *(e.g.,* failure to request a discovery inspection), it is considered to be error to dismiss the case for failure to prosecute, if, upon breakdown of the settlement negotiations, the Secretary was not prepared to go forward. *Ralston Purina Co.,* 7 OSHC 1730 (1979).

The Commission has jurisdiction over and must approve any settlement. *PPG Industries, Inc.,* 3 OSHC 1852, *vacated*—F.2d—, 4 OSHC 1935 (D.C. Cir. 1976), *Dale M. Madden Construction, Inc. v. Hodgson,* 502 F.2d 278, 2 OSHC 1236 (9th Cir. 1974). The Commission has held that it is a necessary adjunct of its public interest and supervisory role that settlement agreements which do not comport with the purposes of the act be rejected. *American Airlines,* 2 OSHC 1391 (1974); *Farmers Export Co.,* 8 OSHC 1655 (1980). Thus, the jurisdiction of the Commission "to contest or disapprove" settlement agreements has been held essential to its statutory authority to review the actions of its administrative law judges, consistent with the purposes of the Act and with the public interest.

The Commission's Rules set forth specific requirements for notifying employees that a settlement agreement has been entered into by an employer and Secretary of Labor. Thus the agreement must be served upon both represented and unrepresented employees in the manner prescribed for notices of contest under Commission Rule 7. Rule 100(c). Proof of service must accompany the proposal, and it shall not be approved until at least ten days following its services. Absent this service, the Commission has held that affected employees may be deprived of their right to participate meaningfully in settlement adjudications. *Aspro, Inc., Spun Steel Division,* 6 OSHC 1980 (1978); *Snider Industries, Inc.,* 8 OSHC 2046 (1980); *Asarco, Inc.,* 8 OSHC 2201 (1980)—service by posting rather than mail or personal service is not sufficient for purpose of compliance with Commission Rule 7. Furthermore, where affected employees have indicated an intent to participate in settlement proceedings by electing party status, they must also be assured an opportunity for meaningful participation, that is, notified and afforded an opportunity to be heard. *ITT Thompson Industries, Inc.,* 6 OSHC 1944 (1978); *IMC Chemical Group, Inc.,* 6 OSHC 2075 (1978). Service is considered to be particularly important where party status has been elected. *Reynolds Metals Co.,* 7 OSHC 1042 (1979).

In *Kaiser Aluminum & Chemical Corp.,* 6 OSHC 2173 (1978), an employee representative who had elected party status was advised of a settlement agreement and affirmatively registered no objection. Thereafter, it filed a petition for discretionary review, claiming that the basis for the settlement was invalid. The Commission agreed, stating:

> This right includes a reasonable opportunity to review and comment upon settlements proposed by the other parties before they reach fruition, a right which may not have been accorded in this case. In this instance, there is considerable doubt whether the employee representatives were aware of the terms of the proposed settlement sufficiently in advance of the hearing to enable them to render a considered opinion. The length of time necessary for affected employees to adequately evaluate the merits of a settlement proposal will, of course, vary according to the complexity of the case. In no event, however, should affected employees be afforded less than 10 days in which to analyze a settlement proposal.

Employee objections to a settlement agreement, if any, should be filed with the judge who is considering the merit of the agreement. Ten days from service of the agreement is allowed to a party to file a response, and an additional three days is afforded if the settlement is served by mail. *Reynolds Metals Co.,* 7 OSHC 1042 (1979). No settlement proposal may be approved until at least ten days have elapsed following service of the proposal on affected employees. Commission Rule 100(c). The scope of the Commission's authority in this respect has, however, not been without challenge, as will be discussed below.

Before the Commission will approve a settlement of any case, the employer and the Secretary of Labor must agree to the settlement terms, and affected employees or their authorized representative must be afforded a meaningful opportunity to participate in the settlement. In *Gardinier, Inc. and ICW Local 439,* 7 OSHC 1738 (1979), the fact that the union was granted party status after the settlement negotiations had been concluded, was given a copy of the settlement agreement, had the terms explained to it, and orally stated that it had no objections and that the record was held open to allow comments established the fact that the union was given the opportunity for meaningful participation in settlement before the agreement was approved and thus justified Commission approval of the settlement agreement. In *Weldship Corporation,* 8 OSHC 2044 (1980), a settlement agreement was entered into between the Secretary of Labor and the employer, but it did not state when violations would be abated and whether the agreement was served on affected employees. The agreement also provided that the employer did not admit liability but "makes . . . payment [of the penalties] in order to avoid the expense of litigation." Upon review, the employer argued that the Secretary's agreement to any settlement "should be tantamount to a finding that it is within the purposes of the Act," because the Secretary is solely responsible for policy and enforcement under the Act. In the employer's view, the Commission's role was limited to the adjudication of disputes between the parties in contested cases. The employer further asserted that, because no violations of the Act were admitted, it was unnecessary and impossible to include in the settlement agreement assurances about abatement. Nevertheless, the employer also averred that all irregularities "alleged to exist in the company's workplace had been corrected and that the settlement was posted to give notice to affected employees as required by Commission Rule 100(c)." The Secretary argued that "[a]lthough we believe that [we] arrived at a settlement in this matter that adequately protected the employer's employees, it does not specifically provide for prospective enforceability of the Commission's order in accord with *Matt J. Zaich [Construction Co.],*" 1 OSHC 1225 (1973).

In response, the Commission reaffirmed its authority to review settlement agreements to assure they are consistent with the Act's purpose and to disprove a settlement agreement not consistent with the purpose of the Act:

> Further, we have consistently held that for a settlement agreement to receive favorable consideration, (1) the agreement must reflect, among other things, that abatement has been completed or the date by which it will be completed, and (2) proof that the proposed settlement agreement has been served upon represented and unrepresented affected employees must accompany the settlement proposal. . . . The settlement agreement in this case fails to meet these requirements.

In *International Harvester Company,* 7 OSHC 2194 (1980), the union argued that its right as a party goes beyond the right to object to the time period allowed for abatement in a settlement agreement. It asserted that party status should include the right to object to the plan of abatement itself and that the Commission should consider the adequacy of the abatement plan and not just the period of abatement. The administrative law judge disagreed, stating that, even if the union had standing to object to the agreement, it would be permitted only to challenge the reasonableness of the abatement date. The Commission, in a split decision by Commissioners Barnako and Cleary (Commissioner Cottine did not participate), affirmed the judge's order approving the agreement but accorded it the precedential value of an unreviewed judge's decision. In Cleary's opinion, however, the Commission has authority to disprove a settlement agreement as part of its function to act in a supervisory capacity over the Act's enforcement.

In a subsequent case, given another shot at the issue, with Commissioner Cottine participating, the Commission (through Commissioner Cleary) stated, in *American Cyanamid,* 8 OSHC 1346 (1980), that the right of meaningful employee participation is entirely independent of and severable from the Secretary of Labor's enforcement authority with respect to abatement. The union in the case had elected party status and had challenged the abatement aspects of the settlement agreement.[9] The Secretary of Labor had argued that whether or not abatement has occurred was not a concern of the Commission, for, in the event it had not taken place when settlement was approved, protection would still be afforded to employees by much more severe sanctions, such as issuance of willful or repeated violations, that could be brought against the employer by the Secretary. The employer also argued that the Commission had no authority to determine the reasonableness of abatement and had no authority to decide whether abatement had in fact occurred, citing *OCAW (Mobil Oil Corp.),* 1 OSHC 1104 (1973). It also argued that, since Section 10(c) of the Act limits participation by affected employees to challenges "to the reasonableness of the period of time fixed in the citation for abatement," and since abatement had already occurred, further participation by affected employees would only serve to impair settlement procedures. The employer also maintained that the employee representative was afforded "an opportunity for meaningful participation in settlement negotiations," as required by *ITT Thompson Industries, Inc.,* 6 OSHC 1944 (1978), since the employee representative was served with a copy of the settlement well in advance of the judge's approval, thus giving it an opportunity to participate in the proceedings. It failed, however, to do so before the judge approved the agreement.

The Commission rejected these arguments and held that authorized employee representatives must be given the opportunity to file objections or a response to the contents of a settlement agreement, within the time provided under Commission Rules 4(b) and 37. By electing party status, the employee representative is assured a role in settlement negotiations, which includes calling into question the truth of the assertions contained in the agreement (*e.g.*, that abatement has taken place). Even if a union has *not* elected party status, it must be served with a copy of the original and executed settlement agreement by mail or personal delivery as required by Commission Rules 7 and 100(c). *Babcock & Wilcox Company,* 8 OSHC 2107 (1980); *Asarco, Inc.*, 8 OSHC 2201 (1980).

The Commission also held that it has the authority and the right to approve settlement agreements *only* where the settlement is consistent with the provisions and objectives of the Act and the public interest, as expressed in Commission Rule 100 and its decision in *Dawson Brothers Mechanical Contractors,* 1 OSHC 1024 (1972), which requires that settlement agreements must include: (1) the date on which abatement has been or will be accomplished; (2) assurance by the employer of continuing compliance; (3) tender of payment of the penalty proposed by the Secretary of Labor; and (4) evidence that the affected employees or their authorized representatives were aware of the terms of the proposed agreement and have been afforded a meaningful opportunity to participate in the proceedings. The Commission thus stated that while the ordinary, uncontested settlement will not provoke inquiries and attack by a party questioning a condition at issue (in that case abatement), it has the authority to and must determine whether the condition has in fact occurred before approving a settlement agreement. Thus the Commission determined that it had the jurisdiction and authority to determine whether abatement in fact occurred; it was not limited to a determination of the reasonableness of the abatement period. The Commission thus remanded the case to the judge in order that he might hear the authorized employee's objections to the settlement agreement to determine if abatement has occurred. See also *National Steel and Shipbuilding Co.*, 8 OSHC 2023 (1980); *American Cyanamid Co.*, 8 OSHC 1346 (1980); *Kaiser Aluminum & Chemical Corp.*, 6 OSHC 2172 (1978); *Globe Industries, Inc.*, 7 OSHC 1312 (1979).

A contrary opinion relative to the scope of the Commission's authority was recently expressed by the court in *Sun Petroleum Products Co.,* 622 F.2d 1176, 8 OSHC 1422 (3d Cir. 1980), which stated that the Commission has the jurisdiction to review a settlement agreement, *but only* for the limited purpose of entertaining objections from affected employees or their representative that the abatement period proposed by the settlement is unreasonable.[10] The court decision came in direct conflict with the Commission's decision in *American Cyanamid Company, supra.* In *Farmers Export Co.*, 8 OSHC 1655 (1980), the Commission rejected the court's opinion in *Sun Petroleum Products Co., supra.*[11]

In approving a settlement agreement which contained exculpatory language which relieved an employer from liability or limited its liability from the consequences of a violation of the Act in future enforcement proceedings and/or precluded or limited the use of the settlement agreement in future proceedings, the Review Commission in *Farmers Export Company,* 8 OSHC 1655 (1980), stated that, as long as a proposed settlement agreement is consistent with the criteria set out in Rule 100 and its decision in *Dawson Brothers Mechanical Contractors,* 1 OSHC 1024 (1972), it would be approved. In so ruling, the Commission overruled the decisions in *Matt J. Zaich Construction Co.,* 1 OSHC 1225 (1973); *Greenfield and Associates,* 1 OSHC 1245 (1973) and *Blaisdel Manufacturing, Inc.,* 1 OSHC 1406 (1973). The Commission said it adds nothing to the safety and health of the employees to disapprove of settlements that contain exculpatory language or other language that does not admit the existence of violations. Settlement agreements may thus contain language that purports to exculpate employers from liability in causes of actions commenced in other forums or from liability for the purposes of future enforcement actions under the Act. See also *GAF Corporation,* 8 OSHC 2006 (1980)—the exculpatory language contained in the agreement explicitly precluded the use of the "affirmed" violations in any investigation or proceeding under the Act involving any facility of Respondent other than the Rensselaer, New York, facility involved in this proceeding; *Missouri Farmers Associates, Inc.,* 8 OSHC 2011 (1980)—the limiting language stated: "Nothing contained herein shall be deemed an admission by respondent of a violation of the Occupational Safety and Health Act or any regulation or standard issued pursuant thereto; Nor does this withdrawal [of the notice of contest] have any effect whatsoever in any action other than an action or proceeding specifically under the Occupational Safety and Health Act." *Independence Foundry & Mfg. Co., Inc.,* 8 OSHC 2020 (1980); *National Steel & Shipbuilding Co.,* 8 OSHC 2023 (1980); *cf. Western Electric Company, Inc.,* 7 OSHC 1021 (1978).

F. THE HEARING

If an employer observes certain procedural requirements, it is entitled to an administrative hearing before an administrative law judge and review of the citation issued to it. The administrative law judge is required to conduct a fair and impartial hearing. Commission Rule 66; *Hamilton Allied Corp.,* 6 OSHC 1946 (1978).

Prior to scheduling of the hearing, the Administrative Law Judge who has been assigned to hear the case will normally prepare a prehearing order requiring the parties to confer to explore and discuss the possibilities of settlement, to determine possible stipulations and admissions which will help identify and simplify contested items of the citation, and amicably to resolve discovery requirements. Thereafter, the parties will individually or jointly be required to file a conference report of the result of the prehearing conference, including therein the formulation of unresolved issues. The failure to file such a conference report or obtain, upon a timely written motion showing good cause, an extension of the time for filing may result in a prehearing conference being set to receive evidence bearing upon the failure to comply with the order. *Duquesne Light Co.*, 8 OSHC 1218 (1980). If there are unresolved issues remaining after submission of the conference report, the parties are normally required to exchange the following information in preparation for the hearing:

(1) The name and address of each witness and the summary of that witness's testimony;
(2) A list of the documents and other exhibits to be offered into evidence together with a copy of each such document; and
(3) An estimate of the time required to present the case.

In *Williams Enterprise, Inc.*, 4 OSHC 1663 (1976), a judge excluded a witness who was not listed in a prehearing exchange report.

After the filing of the pleadings, the conduct of the prehearing conference, and the completion of discovery, the judge who will preside at the hearing will issue a notice of hearing in accordance with Commission Rule 60. Without this notice, a hearing may be considered invalid. The time, place, and nature of a hearing will be given to the parties at least ten days in advance of the hearing, except in expedited proceedings governed under Commission Rule 101.[12] Information contained in a notice of hearing, which will often require the parties to file a prehearing statement, includes a statement of the issues and a list of lay and expert witnesses, exhibits, and stipulations. Hearings are normally open to the public, as required by Section 12(g) of the Act, although they may be closed to protect trade secrets. Section 15 of the Act. Hearings are normally held in or near the community where the alleged violation took place or at the nearest available facility and are to be set with due regard for convenience of the parties. *Bethlehem Steel Corp.*, 6 OSHC 1912 (1978).

In all proceedings commenced by the filing of a notice of contest, the burden of proof rests with the Secretary of Labor. Commission Rule 73(a). In proceedings commenced by the filing of a petition for modification of abatement period, the burden to establish the necessity for such modification rests with the employer. Commission Rule 73(b). Whichever party has the burden of going forward must do so with sufficient evidence of the facts to support a finding that the facts in fact exist. For example, it would be incumbent upon the Secretary of Labor to introduce sufficient factual evidence to establish a *prima facie* case that a violation exists as alleged.

At the commencement of the hearing, parties should be prepared to stipulate to the admission of all relevant facts and documents to which there is no reasonable dispute. In the event the parties agree as to all material facts and conclude that only issues of law are in dispute, the judge should be notified, and thereafter an appropriate joint stipulation as to said facts and a waiver of the right to a hearing must be submitted. Within thirty days thereafter, the parties must submit briefs on the contested legal issues, and a decision will be made based upon the stipulation of facts and the briefs.

Postponement of a hearing is ordinarily not allowed, and, except in cases of extreme emergency or unusual circumstances, no such request shall be considered unless received in writing at least three days in advance of the time set for the hearing. No postponement in excess of thirty days will be allowed without Commission approval. Commission Rule 61. In *Ralston Purina Co.,* 7 OSHC 1730 (1979), a one month continuance of the hearing was granted in order to permit conduct of additional discovery.

The failure of parties to appear at a hearing is deemed to be a default and a waiver of the parties' rights, except the right to be served with a copy of the judge's decision and to request Commission review. *United Terminals, Inc.,* 2 OSHC 3145 (1975). A party or intervenor may appear in person or through a representative, which need not be an attorney, at the hearing. Commission Rule 22(a). A request for reinstatement must be made, in the absence of extraordinary circumstances, within five days after the scheduled hearing date. Commission Rule 62. Upon a showing of good cause the Commission or the judge may excuse the failure to appear and reschedule the hearing. *Richard Rothbard, Inc.,* 8 OSHC 1408 (1980).

1. Evidence

The hearing shall be in accordance with the provisions of 5 U.S.C. §554 of the Administrative Procedure Act, the Commission's Rules of Procedure, and the Federal Rules of Civil Procedure. The use of exhibits, affidavits, depositions, and the oral examination of witnesses under oath is provided for in Commission Rules 68–71. Moreover, insofar as practicable the hearing shall be governed by the Federal Rules of Evidence. Commission Rule 72. The Commission, however, will normally not become involved in complicated evidentiary questions.

In all proceedings commenced by the filing of a notice of contest, the burden of proof is placed upon the Secretary of Labor. In all other proceedings, such as a petition for modification of abatement, the burden rests with the employer. Commission Rule 73(a).

There are basically four types of evidentiary questions which may be presented as problems of proof in establishing a *preponderance* of the evidence at the hearing.

a. *Types of Evidence*

Hearsay evidence is generally defined as a statement, other than one made by the declarant, while testifying at the trial or hearing, offered in evidence to prove the truth of the matter asserted. Federal Rules of Evidence, Rule 801(c). For example, a hearsay problem would arise if the employer or the Secretary of Labor placed a witness on the stand to testify about a condition which he did not personally observe, but about which he had been informed by other persons. The testimony would be hearsay because it would be based upon information conveyed to the witness by someone else and would not be based upon either the witness's own personal knowledge or observation. Generally, hearsay testimony is admissible evidence in administrative proceedings. But see *Hermitage Concrete Pipe Company,* 3 OSHC 1920 (1976)—article not admitted as hearsay exception; *American Airlines, Inc.,* 6 OSHC 1252 (1977), *rev'd and remanded,* 578 F.2d 38, 6 OSHC 1691 (2d Cir. 1978). The question, however, is what weight or probative effect is to be given to hearsay evidence and to what extent it may be used to establish a violation, either alone or in conjunction with other evidence. Section 7(c) of the Administrative Procedure Act permits the introduction of hearsay evidence but requires that a decision be based on a consideration of the whole record, supported by reliable, probative, and substantive evidence. *Tri-City Construction Co.,* 8 OSHC 1567 (1980). Normally, the Secretary of Labor's use of uncorroborative hearsay evidence would be insufficient in and of itself to establish a violation. *California Rotogravers Co.,* 2 OSHC 1515 (1975), *review denied,* 549 F.2d 807, 5 OSHC 1031 (9th Cir. 1977); *Milprint, Inc.,* 1 OSHC 1383 (1973); *National Engineering & Contracting Co.,* 7 OSHC 1207 (1979); *Champion Construction & Engineering Co.,* 6 OSHC 2116 (1978).

In *American Airlines, Inc.,* 6 OSHC 1252 (1977), a question concerning the business entry exception to the hearsay rule, as adopted in the Federal Rules of Evidence, was considered. The Commission decided that certain reports were admissible, pursuant to Section 7(c) of the Administrative Procedure Act rather than the Rules of Evidence, to show knowledge of a hazardous condition. This case seems to be unusual in its reference to the Administrative Procedure Act for evidentiary considerations, for it seems clear that the Commission will normally be guided by the Federal Rules of Evidence. But see *Hurlock Roofing Co.,* 7 OSHC 1867 (1979), where the Commission held that hearsay is admissible in administrative proceedings and may be used as probative evidence, consistent with the Administrative Procedure Act, which recognizes that an agency should not be bound by technical rules on the admissibility of evidence. The Commission further stated:

> Indeed the application of technical rules of evidence may well prevent the parties from obtaining the hearing they seek and the Commission from rendering an informed opinion on all relevant information. Therefore, as a matter of policy, the Commission's administrative law judges should not exclude evidence on the basis it is hearsay. Naturally, the weight to be assigned hearsay evidence will depend on its apparent reliability, and must take into account any possible bias or interest on the part of the person or persons who made the statements sought to be introduced. McCormick, *Handbook of the Law of Evidence,* §350 (2nd Ed. 1972). But the fact that certain evidence may ultimately be accorded little or no weight should not cause the judge to exclude the evidence from the record.
>
> Although we conclude that the judge erred in ruling the report inadmissible on hearsay grounds, the error was harmless.

See also *B & K Paving Co.,* 2 OSHC 1173 (1974), which held that hearsay is admissible and may have probative value but cannot in and of itself, uncorroborated, support a citation (in the absence of the employer's explicit adoption of the Secretary of Labor's hearsay presentation of the facts); *A. J. McNulty & Co. Inc.,* 4 OSHC 1097 (1976); *Stephenson Enterprises, Inc.,* 4 OSHC 1702 (1976), *aff'd,* 578 F.2d 1021, 6 OSHC 1860 (5th Cir. 1978).

The effect of *conflicting* evidence in a determination of whether or not one party has borne its burden of proof and/or persuasion and established its case by a *preponderance* of evidence has been discussed in cases such as *Billings Heights Construction Co.,* 3 OSHC 1581 (1975), in which it was stated:

> Where two witnesses before a court, equal in interest, in character and in opportunity to know the fact, have made irreconcilable contradictory statements and neither is corroborated, there is no preponderance, and the party relying on one of such witnesses, whose burden it is to go forward, has failed to sustain his burden.

See also *Flaherty-Sand Company,* 3 OSHC 1030 (1975); *Evansville Materials, Inc.,* 3 OSHC 1741 (1975).

Evidence must be *objective,* that is, of sufficient quality to support a determination and to prove the existence of a violation by a preponderance of evidence. Section 7(c) of the Administrative Procedure Act simply requires a rule or order to be issued in accordance with reliable, probative, and substantial evidence. It is left to the administrative law judge actually to determine the character thereof. Evidence has, however, been dismissed as insufficient in several cases, being described as vague, speculative, sloppy, or subjective. *H–E Lowdermilk Co.,* 1 OSHC 1663 (1974); *B. F. Goodrich Company,* 1 OSHC 3322 (1974); *Franklin Lumber Company, Inc.,* 2 OSHC 1077 (1974); *Nibco, Inc.,* 1 OSHC 1777 (1974). Thus there must be objective data to determine whether or not the conditions set forth in the citation are violative thereof. The evidence cannot be speculative, conjectural, obscure, or subjective. If it is found to be such, the citation may be dismissed for failure of proof.

Finally, although objective evidence is required to establish a violation, this does not mean that *circumstantial* evidence may not be utilized in situations where it is reasonable to do so. See *Sun Shipbuilding and Drydock Co.,* 2 OSHC 1181 (1974). Use of circumstantial evidence is often utilized in questions of the access or exposure of employees to a hazard. See *Stepter Brothers Lathing,* 2 OSHC 1213 (1974).

b. *Quantum of Evidence*

The quantum of evidence needed to sustain a party's burden of proof (*e.g.,* to support a violation or demonstrate the reasonableness of a PMA) may vary from case to case. However, as discussed below, the Commission has adopted a preponderance of evidence standard and rejected the use of both the substantial evidence test and the clear and convincing evidence test. *Olin Construction Company, Inc. v. OSHRC,* 525 F.2d 464, 3 OSHC 1526 (2d Cir. 1975).

The allegation that employees are exposed to excessive noise levels need not be supported by evidence of continuous monitoring or evidence establishing employee exposure throughout the course of a workday. The Commission has held that "grab samples" may be used to determine an employee's cumulative level of daily exposure. *Wheeling-Pittsburgh Steel Corp.,* 7 OSHC 1586 (1979); *Boise Cascade Corp.,* 5 OSHC 1242 (1977), *appeal filed,* No. 77–2201 (9th Cir. 1977). In *Sun Shipbuilding and Drydock Co.,* 2 OSHC 1181 (1974), the Commission held that the Secretary may use sample monitoring data to support a citation for excessive noise, provided that the sample data is supported by other evidence from which it may reasonably be inferred that employees were exposed to excessive sound levels as defined by Table G–16. See also *Weyerhaeuser Co.*, 4 OSHC 1972, *vacated and remanded on other grounds*, 614 F.2d 199, 7 OSHC 2059 (9th Cir. 1980).

c. *Admissibility of Evidence*

Questions of the admissibility of evidence may also be presented in Commission proceedings. See *Blakesley Midwest Prestressed Concrete Co.*, 5 OSHC 2036 (1977); *Modine Mfg. Co.*, 8 OSHC 1097 (1980). In *J. L. Foti Construction Co.*, 8 OSHC 1281 (1980), which contains a discussion of the best evidence or original document rule, which is applicable when the contents of a particular document or writing need to be established, the Commission stated that in order to trigger the application of the rule, an objection must be raised against a party against whom evidence is sought to be introduced. In *Noblecraft Industries, Inc.*, 3 OSHC 1727 (1975), *rev'd in part, aff'd in part*, 614 F.2d 199, 7 OSHC 2059 (9th Cir. 1980), a transcript of testimony from a prior Review Commission proceeding, to which the Secretary of Labor was party and exercised the right of cross-examination, was admissible under the prior testimony exception to hearsay rule. In *Weyerhaeuser Co.*, 4 OSHC 1972 (1977), *rev'd in part, aff'd in part*, 614 F.2d 199, 7 OSHC 2059 (9th Cir. 1980), statements made by employees or employer representatives to an inspector during an inspection were not hearsay but admissions. In *American Bechtel, Inc.*, 6 OSHC 1246 (1977) the Commission held that an attorney's statement during the course of trial does not conclusively bind his client, unless the statement can be properly characterized as a judicial admission. Formal judicial admissions are to be distinguished from mere evidentiary admissions.

The Secretary is not required to physically place an ANSI standard into evidence, if it has been validly incorporated by reference into the cited standard. *Charles A. Gaetano Construction Corp.*, 6 OSHC 1463 (1978). The Commission has noted, however, that in cases involving repeated violations, the Secretary of Labor should introduce the antecedent citation into evidence rather than merely rely on testimony of the inspector.

In *Minnotte Contracting & Erection Corp.*, 6 OSHC 1369 (1978), photographs taken by a compliance officer from a roadway, prior to his presentation of credentials, were properly admitted into evidence, and no violation of the employer's Fourth Amendment rights was found. The Commission held that a right to privacy under the Fourth Amendment exists only where there is a reasonable expectation of privacy. In *Environmental Utilities Corp.*, 5 OSHC 1195 (1977), the employer could not claim a reasonable expectation of privacy where a trench abutted into public view. Finally, in *Laclede Gas Company*, 7 OSHC 1874 (1979), the Commission held that the use of a telephoto lens, binoculars, and other sense-enhancing devices to take photographs of a work site prior to the presentation of an inspector's credentials did not violate an employer's Fourth Amendment rights and did not require the exclusion of the photographs from evidence, since the Fourth Amendment's right of privacy exists, and can be violated, only where there is a reasonable expectation of privacy. Where a worksite is visible from a public roadway, the worksite remains within the scope of the *plain view* doctrine, even though it is observed through telephoto devices.[13]

Moreover, the employer's Section 8(a) rights were not violated by the taking of photographs prior to the presentation of the credentials of the inspector. See also *Accu-Namics, Inc. v. OSHRC,* 515 F.2d 828, 3 OSHC 1299 (5th Cir. 1975), *cert. denied,* 425 U.S. 903, 4 OSHC 1090 (1976); *Able Contractors, Inc.,* 5 OSHC 1975 (1977), *aff'd,* 573 F2d.1055, 6 OSHC 1317 (9th Cir. 1978); *United States Steel Corporation,* 7 OSHC 1073 (1979)—employer's workplace open to public view, thus photograph of violative condition taken by inspector prior to beginning inspection was admissible in evidence; *Marshall v. Western Waterproofing Co., Inc.,* 560 F.2d 947, 5 OSHC 1732 (8th Cir. 1977); *Hartwell Excavating Company v. Dunlop,* 537 F.2d 1071, 4 OSHC 1331 (9th Cir. 1976). In *J. W. Conway, Inc.,* 7 OSHC 1718 (1979), a judge refused to admit an employer's photographs of the worksite into evidence since they were taken *after* the date of inspection.

Commission Rule 74 allows a party whose evidence is objected to or excluded to make an offer of proof which will be included in the record. *Cf. Williams Enterprises, Inc.,* 4 OSHC 1663 (1976). The offer of proof can be made by a summary of the testimony or through a question and answer format.

Official notice, the administrative counterpart to judicial notice, may be requested and taken as to matters of common knowledge not specifically related to the matter in dispute. *Franklin R. Lacy (Aqua View Apartments),* 4 OSHC 1115 (1976), *aff'd,* 628 F.2d 1226, 8 OSHC 2060 (9th Cir. 1980).

2. Witnesses and Testimony

Witnesses summoned before the Commission or a judge are to be paid the same fees and mileage that are paid to witnesses in the courts of the United States, and witnesses whose depositions are taken and the persons taking the same shall also be entitled to the same fees as are paid for like service in the courts of the United States. Witness fees and mileage shall be paid by the party at whose instance the witness appears, and the person taking the deposition shall be paid by the party at whose instance the deposition is taken. Commission Rule 63.

Witnesses may be examined orally and under oath, and there is a right of cross-examination. *Constructora Metalica, Inc.,* 1 OSHC 1058 (1972); Commission Rule 68. An affidavit or deposition may be submitted as evidence in lieu of oral testimony if the matters contained therein are otherwise admissible, the parties have agreed to its admission, and (in the case of depositions) good cause is shown in an application submitted to the judge. Commission Rules 69 and 70. Witnesses may, upon request of the parties, be sequestered pursuant to Rule 615 of the Federal Rules of Evidence.

The determination of a witness's credibility is a function of the administrative law judge, and his determination is normally entitled to deference by the Commission. Compare *Otis Elevator Company,* 8 OSHC 1019 (1980), to *Asplundh Tree Expert Co.,* 7 OSHC 2074 (1979). See also *Papertronics, Division of Hammermill Paper Co.,* 6 OSHC 1818 (1978)—employer's testimony credited, because, unlike the compliance officer, he had actual experience, not mere theoretical knowledge; *Connecticut Natural Gas Corp.*, 5 OSHC 1796 (1978)—inspector's testimony credited; *West Point Pepperell, Inc.*, 5 OSHC 1257, *aff'd*, 588 F.2d 979, 7 OSHC 1035 (5th Cir. 1979); *General Electric Co.*, 7 OSHC 2187 (1980)—the inconsistency between the compliance officer's testimony concerning the existence of a hazard and his contrary opinion formed at the time of the inspection cast doubt on the reliability of his testimony as a whole; *Empire Roofing Co.*, 8 OSHC 2195 (1980); *C. Kaufman, Inc.*, 6 OSHC 1295 (1978)—neither the Commission nor the Secretary is bound by the interpretation or opinion of a compliance officer; *Holman Erection Co., Inc.*, 5 OSHC 2079 (1977)—the testimony of the compliance officer in the case and his conduct during the inspection reveal that he did not believe that personal protective equipment was required under the cited conditions; *M. J. Lee Construction Co.*, 7 OSHC 1140 (1979)—OSHA compliance officer qualified as an expert by virtue of past trenching experience and conduct of fifty trench/excavation inspections.

The compliance officer's testimony is often the most important aspect of the Secretary of Labor's case; the employer's testimony is often necessary to rebut that testimony or establish a defense. *Stephenson Enterprises, Inc.*, 4 OSHC 1702 (1976), *aff'd*, 578 F.2d 1021, 6 OSHC 1860 (5th Cir. 1978); *Chemiquip Products Co.*, 8 OSHC 2202 (1980). In one case the alleged bias of an OSHA area director did not disqualify his testimony but was a factor going to the weight accorded. *Southern Colorado Prestress Co.*, 4 OSHC 1638 (1976), *aff'd*, 586 F.2d 1342, 6 OSHC 2032 (10th Cir. 1978).

Prior testimony from other proceedings has been admitted as an exception to the hearsay rule, in view of the fact that the party against whom it was offered was a party to the other proceedings and exercised his right of cross-examination. *Noblecraft Industries, Inc.*, 3 OSHC 1727 (1975), *rev'd in part, aff'd in part*, 614 F.2d 199, 7 OSHC 2059 (9th Cir. 1980). The testimony of recognized *expert* witnesses can also be crucial in many types of OSHA cases (*e.g.*, recognized hazard under the general duty clause, economic and technological feasibility, industry custom and practice relative to the use of personal protective equipment). An expert may testify about his opinion based on facts and data which are a result of his own observation or made known to him by others. *Anheuser Bush, Inc.*, 4 OSHC 1999 (1977); *Cape and Vineyard Division v. OSHRC*, 512 F.2d 1148, 2 OSHC 1628 (1st Cir. 1975). The testimony of an expert witness is governed by the Federal Rules of Evidence, *Hurlock Roofing Co.*, 7 OSHC 1867 (1979). The Secretary of Labor is more limited than the employer in his choice of experts. OSHA regional personnel are the first choice, state personnel are the second choice, national personnel are the third choice, and outside experts are the fourth choice. Commission members may not "serve as expert witnesses" even though they have been chosen by reasons of their training and experience. The court in *National Realty and Construction Co. v. OSHRC*, 489 F.2d 1257, 1 OSHC 1422 (D.C. Cir. 1973) stated that:

> In short the Commissioners attempted to serve as expert witnesses for the Secretary. This is not their role. The Secretary should have called his own expert or experts at the hearing. Only by requiring the Secretary, at the hearing, to formulate and defend *his own* theory of what a cited defendant should have done can the

> Commission and the courts assure even-handed enforcement of the general duty clause. . . .

Similarly, the Commission should not defend the case for the employer and raise nonjurisdictional affirmative defenses.

3. Exhibits and Objections

The use of exhibits in Commission proceedings is governed by Commission Rule 71. Photographs of the workplace and reports of compliance officers are the most common exhibits introduced at the hearing. See *J. W. Conway, Inc.*, 7 OSHC 1718 (1979). A copy of each exhibit shall be numbered, marked with a designation, given to the other parties and intervenors, introduced through the laying of a proper foundation, and admitted into evidence as part of the record, unless objected to or excluded by a judge. All exhibits offered into evidence but denied admission shall be identified and placed in a separate file designated for rejected exhibits. Exhibits should be relevant and probative to the issue. *Tri-City Construction Co.*, 8 OSHC 1569 (1980). Objections must be raised before the judge so that they may be preserved for review. In *Atlantic Steel Company*, 6 OSHC 1289 (1978), the Commission found as nonprejudicial error a judge's refusal to allow the introduction of relevant evidence, as there was no clear objection from that party and it withdrew its proffer of evidence.

Objections with respect to the conduct of the hearing (necessary for preservation of a case upon review), may be stated orally or in writing and may be accompanied by a short statement of the grounds therefor which shall be included for the record. Commission Rule 74; *J. L. Foti, supra*—failure to timely object to the admission of secondary evidence precludes application of best evidence rule; *Bill C. Carroll Co., Inc.*, 7 OSHC 1806 (1979)—failure to object to introduction of evidence constitutes implied consent to its trial. Objections are not deemed to be waived by further participation in the hearing, after their rejection. Offers of proof may be made whenever evidence is excluded from the record. Commission Rule 74.

4. Burden of Proof

In all proceedings commenced by the filing of a notice of contest, the burden of proof rests upon the Secretary of Labor, who is subject to the *preponderance of evidence* standard in the presentation of his case. Commission Rule 73; *Armor Elevator Co.*, 1 OSHC 1409 (1973); *Heath & Stitch, Inc.*, 8 OSHC 1641 (1980); *Charles H. Tompkins*, 6 OSHC 1045 (1977); *Arvin Millwork Co.*, 2 OSHC 1056 (1974); *United States Steel Corp.*, 7 OSHC 1073 (1978); *Diebold, Inc. v. Marshall*, 585 F.2d 1327, 6 OSHC 2002 (6th Cir. 1978). Preponderance of evidence means that, based upon the evidence, the facts asserted by the party having the burden of proof are more probably true than false. *Ceco Corporation*, 1 OSHC 1209 (1973); *Armor Elevator Co., Inc.*, 1 OSHC 1409 (1973); *McCormick on Evidence*, §339 (2d Ed. 1972). In *Usery v. Hermitage Concrete Pipe Company*, 584 F.2d 127, 6 OSHC 1886 (6th Cir. 1978), the court stated that in applying the substantial evidence test to the Commission's findings of fact, the courts will defer to the Commission's determination whether the burden of proof (preponderance of evidence) has been met:

> The quantum of proof which the Commission, as an independent body appointed by the President, may deem necessary to satisfy it of the existence of the "condition" within the meaning of the statute is a matter on which its expertise and experience is entitled to great deference.

The courts have generally held that the burden of proof rests with the Secretary of Labor to prove *all* elements of a violation of a general safety and health standard. *Voegele Co. v. OSHRC,* 625 F.2d 1075, 8 OSHC 1631 (3d Cir. 1980); *Power Plant Division v. OSHRC,* 590 F.2d 1363, 7 OSHC 1137 (5th Cir. 1979); *Bristol Iron & Steel Works v. OSHRC,* 601 F.2d 717, 7 OSHC 1462 (4th Cir. 1979)—These elements include *inter alia* employee exposure, employer knowledge, existence of a hazard, non-compliance with a standard, and in some instances feasibility. In this respect, the court in *Marshall v. Knutson Construction Company,* 566 F.2d 596, 6 OSHC 1077 (8th Cir. 1977) stated:

> An employer has a duty to comply with the safety standards promulgated by the Secretary under the Act arising from [Section 5](a)(2). In the typical case occurring under [Section 5](a)(2), the employer has either created or controlled the safety standard violation. The Commission has described the Secretary's burden of proof in this situation as follows:
>
> > In the typical case arising under section 5(a)(2) of the Act [a case involving an employer at a common construction site is atypical] the Secretary carries his burden of proving a violation by establishing (1) that a specific standard applies to the facts, (2) that there was a failure to comply with the specific standard, and (3) that employees of the cited employer had access to the hazard.
> >
> > An employer may rebut the Secretary's allegations by showing (1) that the standard cited by the Secretary as the basis for his charge is not applicable to the situation at issue, (2) that the situation at issue was in compliance with the cited standard, or (3) that its employees did not in fact have access to the hazard.
>
> *Anning-Johnson Co.,* 4 OSHC 1193 (1976).
>
> The multi-employer construction worksite situation presents an additional consideration regarding employee safety. In this situation, a hazard created and controlled by one employer can affect the safety of employees of other employers on the site. In light of this fact, the Commission has stated that in this situation an employer will have a duty under [Section 5](a)(2) regarding safety standard violations which it did not create or fully control. *Anning-Johnson Co., supra; Grossman Steel & Aluminum Corp.,* 4 OSHC 1185 (1976).

The "rule" set forth in *Knutson Construction* relative to specific standards is also the rule where the Secretary is seeking to prove a violation of the general duty clause. *Voegele Co. v. OSHRC,* 625 F.2d 1075, 8 OSHC 1631 (3d Cir. 1980). In situations involving violations of the general duty clause, and in order to establish a general duty clause violation, the Secretary of Labor must prove (1) that the employer failed to render its workplace "free" of a hazard which was (2) recognized and (3) causing or likely to cause death or serious physical harm. *National Realty & Construction Co. v. OSHRC,* 489 F.2d 1257, 1

OSHC 1422 (1973); *Getty Oil Co. v. OSHRC,* 530 F.2d 1143, 4 OSHC 1121 (5th Cir. 1976); *General Electric Co. v. OSHRC,* 540 F.2d 67, 4 OSHC 1512 (2d Cir. 1976); *Pratt and Whitney Aircraft,* 2 OSHC 1560 (1975). Proof of a recognized hazard requires evidence in addition to the inspector's testimony.

The general duty obligation is not designed to impose absolute liability or *respondeat superior* liability on an employer for an employee's negligence. Rather, it requires the employer to eliminate only "feasibly preventable" hazards. Thus it is also the Secretary of Labor's burden to show that demonstrably feasible measures existed which would materially reduce the likelihood that such injury as that which resulted from the cited hazard would have occurred. *Titanium Metals Corp. of America v. Usery,* 579 F.2d 536, 6 OSHC 1873 (9th Cir. 1978); *Champlin Petroleum Co. v. OSHRC,* 593 F.2d 637, 7 OSHC 1241 (5th Cir. 1979); *Voegele Co., supra.* The Secretary must also specify the particular steps the employer should have taken to avoid citation, and he must demonstrate the feasibility and likely utility of those measures. *National Realty,* 489 F.2d at 1268; *Bristol Iron & Steel Works, Inc. v. OSHRC,* 601 F.2d 717, 7 OSHC 1465 (4th Cir. 1979).

Once the Secretary of Labor makes out a *prima facie* case of a violation of the Act, the burden then shifts to the employer to articulate an affirmative defense and also meet the preponderance of evidence standard. *Geisler Gauge Corporation,* 8 OSHC 1539 (1980); *Murphy Pacific Marine Salvage Co.,* 2 OSHC 1464 (1975); *Mississippi Valley Erection Co.,* 1 OSHC 1527 (1973). If the Secretary of Labor fails to make out his *prima facie* case, a motion to dismiss is appropriately made by the employer. *AMP Construction Co.,* 1 OSHC 3054 (1973). This allocation of burdens is consistent with the traditional view that the burden of proving a particular issue should be assigned to the party which has knowledge of the relevant facts and access to relevant evidence. *Bratton Corp. v. OSHRC,* 590 F.2d 273, 7 OSHC 1004 (8th Cir. 1978); *Anning-Johnson Co., supra.*

The following quick checklist sets forth those evidentiary burdens which generally rest with the Secretary of Labor and the employer.

(1) The Commission has held that the burden of proving feasibility does not fall upon the Secretary, *unless* a particular standard requires it (*e.g.,* compare 29 CFR 1926.28(a) with 29 CFR 1910.95). *Easley Roofing and Sheet Metal Co.,* 8 OSHC 1410 (1980); *Republic Roofing Corp.,* 8 OSHC 1411 (1980); *Forest Park Roofing Co.,* 8 OSHC 1181 (1980); *Morton Buildings, Inc.,* 7 OSHC 1703 (1979); *Marion Power Shovel Co.,* 8 OSHC 2245 (1980); *Atlantic Steel Co.,* 6 OSHC 1289 (1978); *Great Falls Tribune Co.,* 5 OSHC 1443 (1977); *S&H Riggers, Inc.,* 7 OSHC 1260 (1979); *Ace Sheeting & Repair Co. v. OSHRC,* 555 F.2d 439, 5 OSHC 1589 (5th Cir. 1977) *aff'g* 3 OSHC 1868; *but see General Electric Co. v. OSHRC* 540 F.2d 67, 4 OSHC 1512 (2d Cir. 1976); *Diebold, Inc. v. Marshall,* 585 F.2d 1327, 6 OSHC 2002 (6th Cir. 1978); *Ocean Electric Corp. v. Secretary of Labor,* 594 F.2d 396, 7 OSHC 1149 (4th Cir. 1979). In general duty clause cases, however, feasibility of compliance must be shown by the Secretary. *National Realty & Construction Co. v. OSHRC,* 489 F.2d 1257, 1 OSHC 1422 (D.C. Cir. 1973).

(2) The Secretary of Labor must prove the existence of a hazardous condition and show that employees were exposed to those hazards in order to establish a *prima facie* case of a violation of the Act. *City Wide Transporting Service,* 1 OSHC 1232 (1973); *Hawkins Construction Co.,* 1 OSHC 1762 (1974); *Bechtel Corporation,* 2 OSHC 1336 (1974); *Stock Manufacturing & Design Co.,* 8 OSHC 2145 (1980); *Armor Elevator Co.,* 1 OSHC 1409 (1973); *General Electric Co.,* 7 OSHC 2183 (1980). Exposure has been defined as not just actual exposure to hazards but rather access to a zone of danger, based upon a reasonable predictability of exposure. *Usery v. Marquette Cement Manufacturing Co.,* 568 F.2d 902, 5 OSHC 1793 (2d Cir. 1977); *Havens Steel Co.,* 6 OSHC 1740 (1978); *Bechtel Power Corp.,* 7 OSHC 1361 (1979); *Winn Dixie Stores, Inc.,* 6 OSHC 1598 (1978). When dealing with exposure to health hazards, the time or duration of exposure is also important, since unlike safety standards, health standards such as noise, lead and cotton dust are violated *only* when there is exposure in excess of the threshold limit value (TLV) or permissible exposure limit (PEL). See *Wheeling Pittsburgh Steel Corp.,* 6 OSHC 1161 (1977); *Weyerhaeuser Co.,* 2 OSHC 1162 (1974); *Sun Shipbuilding & Drydock Co.,* 2 OSHC 1181 (1974); *Kropp Forge Co.,* 8 OSHC 2072 (1980).

When an employer is cited under a general standard, the Secretary has the added burden of establishing employee exposure to a *hazardous* condition. If the terms of the standard are precise or the hazard is apparent, the Secretary will not be required to meet this additional burden. *S&H Riggers and Erectors,* 7 OSHC 1260 (1979); *Connecticut Natural Gas Corp.,* 6 OSHC 1796 (1978); *Austin Building Co.,* 8 OSHC 2150 (1980)—Secretary of Labor is not required to prove exposure to recognized hazard in a 1926.28(a) case.

(3) The Secretary must show that an employer had knowledge of the existence of conditions violative of the Act. *National Realty, supra; Brennan v. OSHRC (Alsea Lumber Co).,* 511 F.2d 1139, 2 OSHC 1646 (9th Cir. 1975); *Charles A. Gaetano Constr. Corp.,* 6 OSHC 1463 (1978); *Harvey Workover, Inc.,* 7 OSHC 1687 (1979).

(4) The Secretary of Labor must prove that an employer is engaged in a business affecting commerce in order to establish coverage under the Act; he must do more than merely allege that coverage is obvious. *Marshall v. Able Contractors, Inc.,* 573 F.2d 1055, 6 OSHC 1317 (9th Cir. 1978); *Steven's, Inc.,* 8 OSHC 1269 (1980).

(5) The Secretary of Labor must prove all elements of a failure to abate. *Arvin Millwork Co.,* 2 OSHC 1056 (1974).

(6) In cases where a PMA is filed, the burden of establishing the necessity for and reasonableness of modification is on the employer. Commission Rule 73(b); *Gilbert Manufacturing Co., Inc.,* 7 OSHC 1611 (1979). But when the reasonableness of an abatement date is in issue in the contest of a citation, the burden of proving reasonableness rests with the Secretary. *Drath Packaging Co.,* 8 OSHC 1999 (1980).

(7) In situations involving a multiemployer worksite, an employer must prove it neither created nor controlled a hazard in order properly to

invoke the *Anning-Johnson* defense. *Central of Georgia R.R. Co. v. OSHRC,* 576 F.2d 620, 6 OSHC 1784 (5th Cir. 1978).

(8) An employer has the burden of proving affirmative defenses. *Star Mason Co., Inc.,* 1 OSHC 3223 (1974); *Mississippi Valley Erection Co.,* 1 OSHC 1527 (1973); *Garrison & Associates, Inc.,* 3 OSHC 1110 (1975).

5. Powers of the ALJ

Commission Rule 66 sets forth the duties and powers of the judge and states that his primary duties are to conduct a fair and impartial hearing, to assure that the facts are fully elicited, to adjudicate all issues, and to avoid delay. Subject to the rules and regulations of the Commission, the judge has authority to: (1) administer oaths and affirmations; (2) issue subpoenas; (3) rule upon petitions to revoke subpoenas; (4) rule upon offers of proof and receive relevant evidence; (5) take or cause depositions to be taken; (6) regulate the course of the hearing; (7) hold conferences for settlement or simplification of issues; (8) dispose of procedural requests, including motions; (9) make decisions in conformity with 5 U.S.C. Section 557; (10) call and examine witnesses and introduce into the record documentary or other evidence; (11) request parties to state positions; (12) adjourn the hearing as required; and (13) take any other necessary action authorized by the Commission's Rules. *E.g., A. Mazetti & Sons, Inc.,* 5 OSHC 1826 (1977). Commission Rule 67 sets forth the procedures for the disqualification of a judge.

The Commission and court decisions have basically added strength to the rules on the powers and authority of a judge in the conduct of a hearing. Thus it has been held that a judge: (1) can properly raise questions and engage in quasi-cross-examination of witnesses under Rule 66 but cannot raise procedural or substantive affirmative defenses *sua sponte; A. Mazetti & Sons, Inc.,* 5 OSHC 1826 (1977); *Concrete Construction Corp.,* 4 OSHC 1133 (1976); (2) has the power to request submissions from the parties and reopen a hearing; *D. Federico, Inc.,* 3 OSHC 1970 (1976); (3) may not raise an affirmative defense on behalf of a party and direct that supporting witnesses be summoned to testify; *Consolidated Pine, Inc.,* 3 OSHC 1178 (1975); (4) may not direct an employer to contest a citation or to raise defenses that he does not wish to raise; *Juhr & Sons,* 3 OSHC 1871 (1976); (5) has discretion to allow the appearance of an unscheduled witness or call witnesses and consider evidence not produced by the parties; *Southern Colorado Prestress Co. v. OSHRC,* 586 F.2d 1342, 6 OSHC 2032 (10th Cir. 1978); *Anheuser-Busch, Inc.,* 4 OSHC 1999 (1977); *Noblecraft Industries, Inc.,* 3 OSHC 1727 (1975), *aff'd,* 614 F.2d 199, 7 OSHC 2059 (9th Cir. 1980); (6) may permit the Secretary of Labor to amend a complaint at the close of his case due to a mistake; (7) may grant a continuance for presentation of additional expert testimony; *Fleming Foods of Nebraska, Inc.,* 6 OSHC 1233 (1977); *A. Mazetti & Sons, Inc.,* 5 OSHC 1826 (1977); (8) may not hold a party in contempt; *Owens-Illinois, Inc.,* 6 OSHC 2162 (1978); and (9) may not hear evidence on a citation that has become a final order; *Mississippi Valley Erection Co.,* 1 OSHC 1527 (1973); *Acme Metal, Inc.,* 3 OSHC 1932 (1976).

6. Conclusion of Hearing

At the close of the moving party's case in chief, the opposing party may move for a dismissal under Rule 41(b) of the Federal Rules of Civil Procedure, but the judge will usually reserve his ruling and wait until all of the evidence is in. *Stephenson Enterprises, Inc.*, 4 OSHC 1702 (1976), *aff'd*, 578 F.2d 1021, 6 OSHC 1860 (5th Cir. 1978); *Harrington Construction Corp.*, 4 OSHC 1471 (1976). A transcript of the proceedings may be obtained—hearings are transcribed verbatim. Commission Rule 65. In *Plum Creek Lumber Company*, 8 OSHC 2185 (1980), a motion to expunge the record was made and rejected.

At the close of the hearing, parties are entitled to and may, upon request, offer oral argument and submit timely briefs and/or proposed findings of fact and conclusions of law with the judge. See *Charter Builders Inc.*, 8 OSHC 1232 (1980). These submissions may normally not raise matters which have not been considered earlier at the hearing or are not contained in the record. *D. Federico Co., Inc.*, 3 OSHC 1970 (1976); *Williams Enterprises, Inc.*, 7 OSHC 1015 (1979). A reasonable period of time, initially not to exceed twenty days, is provided from receipt of the transcript of hearing for filing submissions. Commission Rule 76.

After consideration of the parties' submissions a decision will be issued which must be based upon *all* the relevant evidence before the judge. *C. Kaufman Inc.*, 6 OSHC 1295 (1978). In addition to making findings of fact and conclusions of law, the judge's decision should include the reasons for his findings. *P & Z Co., Inc.*, 6 OSHC 1189 (1977); *Asplundh Tree Expert Company, supra.* The judge's decision must follow Commission precedent and must comply with the Administrative Procedure Act. 5 U.S.C. Section 557. Commission Rule 90; *Gulf & Western Food Products, Inc.*, 4 OSHC 1437 (1974).

G. COMMISSION JURISDICTION

The general jurisdiction of the Commission is basically discretionary, a *certiorari*-type of review of both the legal and factual findings made by a judge. Rule 91(a); *Francisco Tower Service, Inc.*, 3 OSHC 1952 (1976). Review is normally limited to the issues raised by the parties before the administrative law judge, except in extraordinary circumstances, for, absent a question of compelling interest, the Commission will not usually address issues where a party has not raised the issue before the administrative law judge. *Huber, Hunt & Nichols, Inc. and Blount Brothers Corp.*, 4 OSHC 1406 (1976). Nor will it normally consider a request for review that does not challenge the judge's decision but rather only takes exception to certain findings of fact or conclusions of law. *Bethlehem Steel Corp.*, 7 OSHC 1053 (1979); *General Electric Co.*, 7 OSHC 2184 (1980).

The following is a guideline to the Commission's general jurisdictional authority.

(1) The Commission and its judges determine whether an employer is in violation of the Act. *Beall Construction Co. v. OSHRC,* 507 F.2d 1041, 2 OSHC 1398 (8th Cir. 1974).

(2) Review of the validity of standards is not foreclosed in enforcement proceedings. *Noblecraft Industries, Inc.,* 3 OSHC 1727 (1975), *aff'd,* 614 F.2d 199, 7 OSHC 2059 (9th Cir. 1980); *Marshall v. Union Oil Co. of California,* 616 F.2d 1113, 8 OSHC 1169 (9th Cir. 1980); *Atlantic & Gulf Stevedores, Inc. v. OSHRC,* 534 F.2d 541, 4 OSHC 1061 (3d Cir. 1976); *cf. National Industrial Constructors, Inc., v. OSHRC,* 583 F.2d 1048, 6 OSHC 1914 (8th Cir. 1978). The Commission has the authority to rule on the validity of OSHA standards. *Kennecott Copper Corp.,* 4 OSHC 1400 (1976), *aff'd,* 577 F.2d 1113, 6 OSHC 1197 (10th Cir. 1977); *Consolidated Freight Ways Corp.,* 5 OSHC 1481 (1977). A challenge to a standard's validity may involve substantive claims such as vagueness or procedural claims such as improper promulgation. *Cape and Vinyard Division v. OSHRC,* 512 F.2d 1148, 2 OSHC 1628 (1st Cir. 1975); *Brennan v. OSHRC (Santa Fe Trail Transport Co.),* 505 F.2d 869, 2 OSHC 1274 (10th Cir. 1974).

(3) Generally, the Commission has held that it is without authority to question the *wisdom* of a standard. *Fabricraft Inc.,* 7 OSHC 1540 (1979); *Hana Shoe Corp.,* 4 OSHC 1635 (1976); *MRS Printing Inc.,* 6 OSHC 2025 (1978); *Cornish Dress Mfg. Co.,* 3 OSHC 1850 (1975); *Van Raalte Co., Inc.,* 4 OSHC 1151 (1976); *Austin Bridge Co.,* 7 OSHC 1761 (1979); *Boise Cascade Corp.,* 3 OSHC 1804 (1976); *The Budd Co.,* 1 OSHC 1548 (1973), *aff'd,* 513 F.2d 20, 2 OSHC 1648 (3d Cir. 1975); *Tobacco River Lumber Company,* 3 OSHC 1059 (1975); *Daniel Construction Co.,* 5 OSHC 1713 (1977); *Charles A. Gaetano Constr. Co.,* 6 OSHC 1463 (1978).

The Commission has also decided vagueness and lack of fair notice challenges to standards, applying a standard of a reasonable man familiar with conditions in the industry. *Gold-Kist, Inc.,* 7 OSHC 1855 (1979); *Pratt & Whitney Aircraft,* 2 OSHC 1713 (1975); *American Airlines, Inc.,* 7 OSHC 1980 (1979).

(4) The Commission has generally held that it is without authority to rule on *fundamental* constitutional challenges to the Act. *Hydraulic Equipment Co.,* 5 OSHC 1892 (1977); *Lehr Construction Co.,* 6 OSHC 1352 (1978); *Pasco Masonry Co. Inc.,* 5 OSHC 1864 (1977); *Hurlock Roofing Co.,* 7 OSHC 1867 (1979). Constitutional issues are thus merely preserved for judicial review in Commission proceedings. *Robberson Steel Co.,* 6 OSHC 1430 (1978), *appeal dismissed,*—F.2d—, 7 OSHC 1052 (10th Cir. 1978); *Buckeye Industries, Inc. v. Secretary of Labor,* 587 F.2d 231, 6 OSHC 2181 (5th Cir. 1979).

Prior to the Supreme Court's decision in *Marshall v. Barlow's, Inc.,* 436 U.S. 307, 6 OSHC 1571 (1978), the Commission's decisions indicated that it had no authority to rule upon the constitutionality of Section 8(a) of the Act and declined to address arguments based upon Fourth Amendment challenges to the validity of inspection warrants. *E.g., Buckeye Industries, Inc.,* 3

OSHC 1838 (1975); *Electrocast Steel Foundry, Inc.*, 6 OSHC 1562 (1978).[14] Subsequent to *Barlow's,* however, the Commission felt that the Supreme Court, by deciding the issue involving the constitutionality of the Act, enabled the Commission to decide questions concerning the validity of an *individual* inspection warrant, thus putting the Commission in the posture where it is competent to address the inspection warrant issue. In *Chromalloy American Corporation,* 7 OSHC 1547 (1979), the Commission thus held that it is competent to address Fourth Amendment issues by applying constitutional principles to particular facts, vis-à-vis the validity of search warrants.

Most of the circuit courts of appeal presented with this question have held that the proper forum in which to raise the validity of the propriety of search warrants is initially with the Commission. See *Marshall v. Whittaker Corporation, Berwick Forge & Fabricating Company Division,* 610 F.2d 1141, 7 OSHC 1888 (3d Cir. 1979); *Marshall v. Central Mine Equipment Company,* 608 F.2d 719, 7 OSHC 1907 (8th Cir. 1979); *In Re Work Site Inspection of Quality Products,* 592 F.2d 611, 6 OSHC 1093 (1st Cir. 1979). The courts have also required that employers who challenge the constitutionality of inspections or inspection warrants must exhaust their administrative remedies before the Commission prior to petitioning the federal courts for relief.

This includes employer's motions to suppress evidence obtained during an inspection. See *Babcock and Wilcox Company v. Marshall,* 610 F.2d 1128, 7 OSHC 1880 (3d Cir. 1979). To date, only the Seventh Circuit, in *Weyerhaeuser Co. v. Marshall,* 592 F.2d 373, 7 OSHC 1090 (7th Cir. 1979), has denied that the Commission has authority to determine the constitutional questions in this area.

(5) The Commission has authority to determine if an employer is in violation of the Act and to vacate, affirm, or modify a citation at a higher or lesser degree than that alleged and may direct any other appropriate relief. *Edward Hines Lumber Co.,* 4 OSHC 1735 (1976); *Noblecraft Industries, Inc. v. Secretary of Labor,* 614 F.2d 199, 7 OSHC 2059 (9th Cir. 1980). This delegation of authority to the Commission has been held to be constitutional. *Beall Construction Co. v. OSHRC,* 507 F.2d 1041, 2 OSHC 1398 (8th Cir. 1974).

For example, where a willful violation is alleged and not proved, a serious or other-than-serious violation may be found by the Commission if the elements of such violation were tried and proved. *Toler Excavating Company,* 3 OSHC 1420 (1975); *Environmental Utilities Corp.,* 5 OSHC 1195 (1977). Generally, however, the Commission will not find a violation of a higher degree than that alleged. *Enfields Tree Service, Inc.,* 5 OSHC 1142 (1977); *Wetmore & Parman, Inc.,* 1 OSHC 1099 (1973).

(6) The Commission is not bound by the Secretary of Labor's penalty criteria and, may vacate or assess penalty amounts at a lower or higher level than those proposed, for the determination of the appropriateness of a penalty is *solely* within its discretion. *General Steel Fabricators, Inc.,* 5 OSHC 1837 (1977); *Long Manufacturing v. OSHRC,* 554 F.2d 903, 5 OSHC 1376 (8th Cir. 1977); *REA Express Inc. v. OSHRC,* 495 F.2d 822, 1 OSHC 1651 (2d Cir. 1974); *Dan J. Sheehan Co. v. OSHRC,* 520 F.2d 1036, 3 OSHC 1573 (5th Cir.

1975), *cert. denied,* 424 U.S. 965, 4 OSHC 1022 (1976); *Clarkson Construction Co. v. OSHRC,* 531 F.2d 451, 3 OSHC 1880 (10th Cir. 1976); *Noblecraft Industries, Inc. v. Secretary of Labor,* 614 F.2d 199, 7 OSHC 2059 (9th Cir. 1980); *Western Waterproofing Co., Inc. v. Marshall,* 576 F.2d 139, 6 OSHC 1550 (8th Cir.), *cert. denied,* 439 U.S. 965 (1978). See also, *Triple "A" South, Inc.* 7 OSHC 1364 (1979)—if all parties agree to the appropriateness of a proposed penalty, the Commission will normally accept the agreement, unless it is clearly repugnant to the purposes of the Act. In *Thorleif Larson & Sons,* 2 OSHC 1256 (1974), the Commission held that even though it is the final arbiter of penalties assessed under the Act, when the employer and the Secretary stipulate that the proposed penalties are acceptable, the Commission will abide by this agreement unless it is clearly repugnant to the Act. See *e.g. Shaffer Construction Engineering Co.,* 2 OSHC 1449 (1974); *George Hyman Construction Co.,* 7 OSHC 2041 (1979).

(7) The Commission may determine, affirm, reverse or modify the abatement period set by a citation which has been contested. *Boise Cascade Corp.,* 5 OSHC 1242 (1977); *Noranda Aluminum v. OSHRC,* 593 F.2d 811, 7 OSHC 1135 (8th Cir. 1979). Abatement includes not only the specific measures but also includes the element of feasibility. *American Airlines, Inc. v. Secretary of Labor,* 578 F.2d 38, 6 OSHC 1691 (2d Cir. 1978); *United Parcel Service of Ohio, Inc. v. OSHRC,* 570 F.2d 806, 6 OSHC 1347 (8th Cir. 1978). If an employer only challenges an abatement date, the challenge is treated as a PMA, not as a contest of the entire citation. *Gilbert Mfg. Co.,* 7 OSHC 1611 (1979). The effect of this ruling is to shift the burden of proof from the Secretary of Labor to the employer to show that the abatement period is unreasonable. The Commission may give expedited treatment to petitions for modification of abatement period (PMA). Under Section 10(c) it has authority to rule on (approve or disprove) all such petitions but actually considers only those contested by the Secretary of Labor or employees. Commission Rule 34.

(8) The Review Commission does not have authority to assess costs against a party, and a prevailing party does not have a right to attorney's fees. *John W. McGowan,* 5 OSHC 2028 (1977), *appeal declined, McGowan v. Marshall,* 604 F.2d 885, 7 OSHC 1842 (5th Cir. 1979). However, costs were awarded by the Second Circuit in *Northeast Marine Terminal Co.,* 573 F.2d 1300, 5 OSHC 2003 (2d Cir. 1977), and attorney's fees were awarded to the employer in *Marshall v. Nichols,*—F. Supp.—, 8 OSHC 1129 (E.D. Tex. 1980), because OSHA's conduct was marked by bad faith. In addition, a recently enacted statute, the Equal Access to Justice Act, 5 U.S.C. Section 504, provides for the awarding of attorney's fees, witness fees and costs to eligible persons if the Secretary of Labor's position in the matter was not substantially justified.

(9) The Commission originally held that it only has the power to make determinations as to the reasonableness of the abatement date and will not set forth specific abatement measures or grant a modification in the abatement plan, *United Automobile Workers, Local 588,* 4 OSHC 1243 (1976), but

recently has considered challenges to the abatement itself in the context of a settlement agreement. *American Cyanamid Co.,* 8 OSHC 1346 (1980).

(10) The Commission has no authority over citations that have become final orders. *S. J. Otinger Construction Co. v. OSHRC,* 502 F.2d 30, 2 OSHC 1215 (5th Cir. 1974); *Connecticut Aerosols, Inc.,* 8 OSHC 1052 (1980). Nor can it enlarge the scope of the citation the Secretary of Labor has issued. *International Harvester Co.,* 7 OSHC 2197 (1980).

(11) The Commission will generally decline to consider nonjurisdictional issues which have not been expressly or impliedly litigated prior to review. *Gulf Stevedore Corp.,* 5 OSHC 1625 (1977).

(12) With respect to settlement agreements, the Commission has recently held that it has the jurisdiction and authority to approve their terms (including the fact that abatement has occurred) only where the agreement is consistent with the provisions and objectives of the Act. *American Cyanamid Co.,* 8 OSHC 1346 (1980).

(13) The Commission has no authority to order an employer to alter working conditions which comply with the Act, or order abatement of a condition not charged in the citation. *International Harvester Co.,* 7 OSHC 2194 (1980). Nor, can the Commission order a non-employer to abate a hazard, pay a penalty or comply with the Act. See *Bloomfield Mechanical Contracting, Inc. v. OSHRC,* 519 F.2d 1257, 3 OSHC 1403 (3d Cir. 1975).

(14) In affirming a citation the Commission in effect has the authority to order an *employer* to cease and desist from noncompliance with the Act; it does not have this authority with respect to employees. See *Atlantic and Gulf Stevedores, Inc. v. OSHRC,* 534 F.2d 541, 4 OSHC 1061 (3d Cir. 1976).

(15) The Commission must base its decisions on the record and cannot serve as an expert witness for the Secretary of Labor. The court in *National Realty and Construction Co. v. OSHRC,* 489 F.2d 1257, 1276, n. 40, 1 OSHC 1422 (D.C. Cir. 1973) stated that:

> The Commission suggested that *National Realty* should have instructed drivers not to allow riders on their vehicles, but the Secretary introduced no evidence that *National Realty* had failed to issue instructions, or that *National Realty's* instructions were less frequent or forceful than experts regard as necessary to a sound safety program. The Commission suggested that riders and drivers involved in riding incidents should have been discharged, but the Secretary neither suggested the sanction nor offered testimony on its appropriateness or probable utility. In short, the Commissioners attempted to serve as expert witness for the Secretary. This is not their role. The Secretary should have called his own expert or experts at the hearing. . . . Thus, the Commission should be a neutral party reaching its decisions on the evidence presented facts and the applicable law.

Similarly, the Commission should not defend the case for the employer and raise nonjurisdictional affirmative defenses.

(16) The Commission has the authority to develop its own principles for determining the sufficiency of evidence and can overrule decisions of an administrative law judge not in conformity therewith. See *A.E. Burgess Leather Co. v. OSHRC,* 576 F.2d 948, 6 OSHC 1661 (1st Cir. 1978); *Usery v. Hermitage Concrete Pipe Co.,* 584 F.2d 127, 6 OSHC 1886 (6th Cir. 1978); *National Steel and Shipbuilding Co. v. OSHRC,* 607 F.2d 311, 7 OSHC 1837 (9th Cir. 1979).

(17) The Commission is authorized to reject an administrative law judge's decision that it considers contrary to the preponderance of the evidence. See *General Dynamics Corp.,* 6 OSHC 1753, *aff'd on other grounds,* 599 F.2d 453, 7 OSHC 1373 (1st Cir. 1979); *Champlin Petroleum Co.,* 5 OSHC 1601 (1977), *rev'd on other grounds,* 593 F.2d 637, 7 OSHC 1241 (5th Cir. 1979). The Commission gives deference to an administrative law judge only on the issue of a witness' credibility, but even in that area the Commission does not always adopt the judge's finding. In such an instance, the Commission may make its own findings of fact. See *Okland Construction Co.,* 3 OSHC 2023 (1976); *Kaufman, Inc.,* 6 OSHC 1295; *Hidden Valley Corp.,* 1 OSHC 1005 (1972).

(18) Disqualification of a Commission member is left to the discretion of the member involved. *National Manufacturing Co.,* 8 OSHC 1435 (1980).

H. POSTHEARING REVIEW

1. Review by the Commission

a. *Administrative Law Judge's Decision*

The Act provides that the administrative law judge who conducts the hearing shall make a "report" of any determination which constitutes his final disposition of the proceedings. Section 12(j) of the Act. The judge's decision (which is a "report" to the Commission) consists of findings of fact, conclusions of law, and an order affirming, modifying, or vacating the citation, penalty, and/or abatement. The decision must include the reasons for the decision and must follow Commission precedent and policy. *P & Z Co., Inc.,* 6 OSHC 1189 (1977); *Gulf & Western Food Products Co.,* 4 OSHC 1437 (1976).

Upon completion of a hearing and submission of posthearing briefs, proposed findings of fact, and conclusions of law by the parties, the judge shall prepare the decision and mail copies to all parties. Commission Rule 90. The parties can, within *twenty* days of the mailing, object to the judge's decision by filing a petition for discretionary review and statements in opposition to such a petition with the judge. Commission Rule 91(b)(1). Within the twenty-day period the judge can (if persuaded by the arguments set forth in the petition) either revise his decision or file it with the Commission as is. In any event, no later than twenty-one days after the decision is mailed to the parties, the judge must file with the Executive Secretary of the Commission a report consisting of the decision, the record and petitions for discretionary review (if any have been filed with him), and any statements in opposition to the petition. Commission Rule 90(a). Once the report is filed, the judge transfers jurisdiction of the case, unless he is subsequently reinvested with jurisdiction by the Commission on remand. *Murray Co. of Texas, Inc.*, 7 OSHC 1414 (1979); Commission Rules 90 and 91. Upon receipt of the judge's report, the Executive Secretary will docket the case and notify all parties of that fact. On or after the date of docketing all pleadings must be filed with the Executive Secretary. Issues must have first been raised at the hearing before a judge, or the Commission will not consider the issue in a subsequently filed petition for review. *Puterbaugh Enterprises, Inc.*, 2 OSHC 1030 (1974); *Edgewood Construction Co.*, 2 OSHC 1485 (1975); *Crescent Wharf and Warehouse Co.*, 2 OSHC 1623 (1975); *Huber, Hunt & Nichols, Inc.*, 4 OSHC 1406 (1976); *Gulf Stevedore Corp.*, 5 OSHC 1625 (1977); *Lombard Brothers, Inc.*, 5 OSHC 1716 (1977).

b. *Petition for Review*

Any part *aggrieved* by a judge's decision, who fails to file a timely petition for discretionary review, may be foreclosed from later court review of the judge's decision. *Keystone Roofing Co., Inc. v. Dunlop*, 539 F.2d 960, 4 OSHC 1481 (3d Cir. 1976); *McGowan v. Marshall*, 604 F.2d 885, 7 OSHC 1842 (5th Cir. 1979). In *PPG Industries (Caribe)*, 8 OSHC 2003 (1980), the Commission held that, because the judge found the employer's violation to be *de minimis*, the employer was not a party aggrieved by the decision.

As stated, a petition must be filed with the judge on or before the twentieth day after mailing a copy of the decision to the parties. Commission Rule 91(b)(1). The petition may also be filed with the Executive Secretary subsequent to the filing of the judge's report. Commission Rule 91(b)(3). Such (late) petition will be considered to the extent the Commission's time and resources permit, upon a showing that there was a valid reason for not filing on time within the twenty-day period. *Roadway Express, Inc.*, 5 OSHC 2039 (1977).

Once the judge's decision is filed with (received by) the Executive Secretary, the Commission members have thirty days from the date of docketing to direct review of the decision. *Gurney Industries, Inc.*, 1 OSHC 1376 (1973); *Robert W. Settrerlin & Sons, Inc.*, 4 OSHC 1214 (1976). (Any petition for discretionary review should be filed well within that period—as early as possible.) The direction of review by a Commission member must be timely made within thirty days from the time the judge's decision was docketed, or the decision of the judge will become a final order simply by operation of law. Section 12(j) of the Act; Commission Rule 90(b)(3); *H. S. Holtze Constr. Co. v. Marshall*, 627 F.2d 149, 8 OSHC 1785 (8th Cir. 1980); *Marshall v. West Point Pepperell, Inc.*, 588 F.2d 979, 7 OSHC 1034 (5th Cir. 1979). The Commission has no jurisdiction over final orders. *Connecticut Aerosals, Inc.*, 8 OSHC 1052 (1980). Both reviewed and unreviewed judge's decisions are styled "final orders," but both are distinguishable from those cases which receive full discretionary Commission review and are thus styled "Review Commission decisions."

Review may be directed on the basis of the petition for review *or* upon a

Commission member's own initiative *sua sponte,* upon any ground that could be raised by a party, but the issues are normally limited to novel questions of law or policy or questions involving a conflict in the administrative law judge's decisions. Section 12(j) of the Act; Commission Rule 92(d). This authority to review is not dependent upon the filing of a petition by one of the parties requesting discretionary review; such review is a matter of sound discretion of a Commission member. *GAF Corporation,* 8 OSHC 2006 (1980); *Francisco Tower Service, Inc.,* 4 OSHC 1459 (1976). Accordingly, it is for each Commission member to decide the novelty of a particular issue or the presence of "extraordinary circumstances" warranting a direction for review on an issue flowing from the judge's disposition of the case, regardless of whether the issue was raised by one of the parties. See Rule 92(d). See, however, *OEC Corporation,* 8 OSHC 1257 (1980)—*sua sponte* direction for review dismissed since there was a lack of either public or party interest. If review is not directed, the decision of the judge becomes a final order of the Commission (subject to judicial review) within thirty days after his report has been made. Section 12(j) of the Act.

A petition for discretionary review should contain a concise statement of each portion of the decision to which exception is taken and should be accompanied by a statement of points and authorities relied upon.[15] Commission Rule 91(c). Statements in opposition to a petition for discretionary review may be filed as specified in the Commission Rules. The Commission's failure to act upon a petition during the thirty-day review period is deemed to be a denial. Rule 92 provides that review is a matter of *sound discretion* of the Commission, and in the exercise of that discretion the following matters will be considered: (1) that a finding of material fact is not supported by a preponderance of the evidence; (2) that the decision is contrary to law or to the rules and decisions of the Commission; (3) that a substantial question of law, policy, or abuse of discretion is involved; or (4) that a prejudicial error was committed. In the absence of either party interest or compelling public interest, review is unlikely to be granted. *Weatherby Engineering Co.,* 8 OSHC 2013 (1980). When a petition for discretionary review is granted, review is limited to the issues specified in the petition, unless the order for review specifies differently. Commission Rule 92(c). In the absence of extraordinary circumstances, the Commission will not decide issues that were not raised before the judge. *A. Prokosch & Sons Sheetmetal, Inc.,* 8 OSHC 2077 (1980). Each issue in a petition should be separately numbered and simply stated. The issue should also make reference to specific pages of the transcript of the hearing, to the evidence, or to other authorities that the party thinks are supportive of its position. An original and three copies of the petition must be filed with the Commission.

When the Commission decides to review a judge's decision, the parties are notified and Rule 93 provides that the parties will ordinarily be given an opportunity to file a brief on the issues presented in the petition. Briefs, however, are no longer automatically required to be filed within a certain period after direction for review is issued. A briefing notice will be issued informing parties of the due date. Unless the briefing notice provides otherwise, the time for filing of briefs shall be as follows:

(1) *Appeal by one party.* A party whose petition for review or for interlocutory appeal is granted or whose interlocutory appeal is certified shall file a brief within forty days after the date of the briefing notice. All other parties shall file briefs within thirty days after the brief of the petitioning or appealing party is served.
(2) *Appeals by two or more parties.* When petitions of two or more parties are directed for review, each such party shall file an initial brief addressing the issues on which it appeals within forty days after the date of the briefing notice and may file a brief responding to the initial brief of the other party within thirty days after the initial brief of the other party or parties.
(3) *Direction for review on the motion of a Commission member.* When no petition for discretionary review is granted and a member directs review of a judge's decision on his own motion, all briefs shall be filed within forty days after the date of the briefing notice.

An extension of time to file briefs will ordinarily not be granted, but any motion for extension must be filed within the time prescribed for filing the briefs. The Commission may decline to accept untimely filed briefs. Except by permission of the Commission, briefs shall not exceed thirty-five pages; a table of contents must accompany briefs in excess of fifteen pages. Five copies of the brief must be filed.

The Commission considers the absence of briefs an indication of a party's lack of interest in review of the judge's decision. *Davison Wood Products,* 2 OSHC 1480 (1974); *General Electric Co.,* 7 OSHC 2184 (1980); *PPG Industries (Caribe),* 8 OSHC 2003 (1980); *Trans World Airlines, Inc.,* 7 OSHC 1047 (1979); *McGowan v. Marshall,* 604 F.2d 885, 7 OSHC 1842 (5th Cir. 1979). If briefs are not filed the direction for review is normally vacated and the administrative law judge's decision affirmed, even when review is ordered as a result of a party who had filed the petition for discretionary review. See *Prestressed Systems, Inc.,* 8 OSHC 1972 (1980); *Dic Underhill,* 4 OSHC 1146 (1976); *PPG Industries,* 8 OSHC 2003 (1980)—even though an employer filed a brief a finding of a *de minimis* violation made the employer not an aggrieved party. Similarly, an employer's argument at the hearing that its business did not affect commerce, which was omitted in its brief filed with the Commission in response to its direction of review, amounted to an abandonment of that contention. *S & S Diving Co.,* 8 OSHC 2041 (1980). Oral argument before the Commission is ordinarily not allowed, but in the event it desires to have argument, ten days notice will be provided. Commission Rule 95.

In cases where the Commission does not grant review of a judge's decision, the decision becomes a final order of the Commission thirty days after it has been filed. If a party is adversely affected or aggrieved with the result at that point, an appeal may be filed in a United States Court of Appeals. The party has the right of appeal in cases in which a judge's decision has become a final order of the Commission when the party has petitioned for review and the petition was not granted and also in cases in which the Commission has reviewed the case and rendered its own decision.

c. *Commission Decision*

Review by the Commission is basically *de novo,* except for the credibility evaluations of witnesses. *Accu-Namics, Inc. v. OSHRC,* 515 F.2d 838, 3 OSHC 1299 (5th Cir. 1975); *General Dynamics Corp. v. OSHRC,* 599 F.2d 453, 7 OSHC 1373 (1st Cir. 1979); *Okland Construction Co.,* 3 OSHC 2023 (1976); *Asplundh Tree Expert Co., supra; Able Contractors,* 5 OSHC 1975 (1977); *Titanium Metals Corp. of America,* 7 OSHC 2172 (1980). The Commission should and will ordinarily defer to the credibility findings of a judge where he has fairly considered the entire record and adequately explained his findings. *Deering Milliken, Inc.,* 6 OSHC 2143 (1978); *Otis Elevator Company,* 8 OSHC 1019 (1980); *Vampco Metal Products, Inc.,* 8 OSHC 2189 (1980); *Kent Nowlin Construction Co.,* 8 OSHC 1286 (1980)—the judge has lived with the case, heard the witnesses, and observed their demeanor. In *Accu-Namics, Inc. v. OSHRC,* 515 F.2d 828, 3 OSHC 1299 (5th Cir. 1975), *cert. denied,* 425 U.S. 903, 4 OSHC 1090 (1976), the Fifth Circuit discussed the relationship between the Review Commission and the judge, concluding:

> The judge's decision is merely a report, weighty of course, but not final until the Commission allows it to become so by inaction. Once the Commission decides to review, it may review the entire record, including the transcript, briefs, proposed findings, motions and the like.

Practically, however, the judge's decision weighs heavily with the Review Commission, and it reviews the judge's decision much like an appellate court would review the decision of a trial court.

If the judge fails to support his decision with factual findings or reasons or if additional evidence is required in the record, the Commission will either remand the case to the judge or, if the record on review is adequate, make its own findings of fact. *George J. Igel & Co.,* 6 OSHC 1642 (1978); *Weyerhaeuser Co.,* 8 OSHC 1091 (1980); *Evansville Materials, Inc.,* 3 OSHC 1741 (1975); *Gordon Construction Co.,* 4 OSHC 1581 (1976); *Asplundh Tree Expert Co., supra; J. L. Foti Construction Co., Inc.,* 8 OSHC 1666 (1980); *Spector Freight System, Inc.,* 3 OSHC 1233 (1979); *Butler Lime & Cement Co.,* 5 OSHC 1370 (1977); *National Industrial Constructors, Inc.,* 8 OSHC 1675 (1980). The Review Commission has the authority to make findings of fact on the entire record where the judge failed to do so. *Accu-Namics, Inc. v. OSHRC,* 515 F.2d 528, 3 OSHC 1299 (5th Cir. 1975); *American Package Co.,* 8 OSHC 2167 (1980). Commission decisions must also be supported by conclusions of law.

In reaching its decision the Commission will give deference to the views of the courts, but, "unless reversed by the Supreme Court," it feels it is obligated to establish its own precedent and is not bound to acquiesce to views of the courts that are in conflict with its uniform and national policy of adjudication. *Farmers Export Co.,* 8 OSHC 1655 (1980); *Triple "A" South, Inc.,* 7 OSHC 1352 (1979); *B. J. Hughes, Inc.,* 7 OSHC 1471 (1979); *Titantium Metals Corp., supra; Sampson Paper Bag Co., supra.* The Commission may thus decline to follow holdings of the courts of appeals with which it disagrees. See however *Jones & Laughlin Steel Corporation v. Marshall,* 636 F.2d 32, 8 OSHC 2217 (3d Cir. 1980) where the court said that an agency is not free to apply its own views in contravention of the courts precedent in dealing with matters within its jurisdiction.

In special circumstances the Commission has authority to reconsider and reinstate a case, pursuant to Rule 60(b) of the Federal Rules of Civil Procedure, even after a decision has become a final order of the Commission under Section 12(i) of the Act, as long as the case has not come within the jurisdiction of the courts. See *Monroe & Sons, Inc.*, 4 OSHC 2016 (1976) *aff'd,* 615 F.2d 1156, 8 OSHC 1034 (6th Cir. 1980); *Browar Wood Products Co., Inc.*, 7 OSHC 1165 (1979). The employer in *Monroe,* for example, failed to file an answer to the complaint, believing that its notice of contest was sufficient. The court, in upholding the Commission's decision to grant the motion for reinstatement, held:

> Fed. R. Civ. Pro. 60(b) provides for the granting of relief from a final judgment, order, or proceeding, on motion by one of the parties for a number of specified reasons including mistake, inadvertence, surprise and excusable neglect. . . . In order to be eligible for relief under 60(b)(1) the movant must demonstrate the following: (1) The existence of mistake, inadvertence, surprise or excusable neglect; (2) that he has a meritorious defense. . . .
>
> * * *
>
> Determinations made pursuant to Fed. R. Civ. Pro. 60(b) are within the sound discretion of the court and will not be disturbed on appeal unless the court has abused its discretion.

But see *Brennan v. OSHRC (S. J. Otinger, Jr. Constr. Co.)*, 502 F.2d 30, 2 OSHC 1218 (5th Cir. 1974); *Rebco Steel Corp.*, 8 OSHC 1235 (1980)—mere carelessness does not justify relief; neglect must be shown to be excusable.

In *Richard Rothbard, Inc.* 8 OSHC 1408 (1980), an employer failed to appear at the scheduled hearing, and the administrative law judge entered a default judgment. Thereafter the employer requested reinstatement of the case, pursuant to Commission Rule 62, which provides that requests for reinstatement must be made in the absence of extraordinary circumstances within *five* days after the scheduled hearing date. The Commission or the judge upon the showing of good cause may thereafter excuse such failure to appear. In this case, however, the Commission found that there was no request for reinstatement made within five days and that the employer failed to demonstrate extraordinary circumstances for waiver of the five-day rule. The Commission found that the employer's request did not set forth good cause for its failure to appear and was also an inadequate motion on its face.

An unreviewed judge's decision, as a Commission final order, is binding only upon the parties; it is not binding precedent upon other judges or the Commission. Similarly, issues decided by the judge but not addressed by the Commission are not considered to be precedent. *Northwest Airlines, Inc.*, 8 OSHC 1988 (1980); *O. E. C. Corporation,* 8 OSHC 1257 (1980); *Turner Co.*, 4 OSHC 1554 (1976); *RMI Co. v. Secretary of Labor,* 594 F.2d 566, 7 OSHC 1119, 1122 n.13 (6th Cir. 1979). At times a judge's decision will be reviewed and affirmed by the Commission, but because of certain procedural considerations (*e.g.*, Commission is equally divided or case raises no issue of compelling public interest), it is accorded the precedential value of an unreviewed decision. *Miller Ceramics, Inc.*, 7 OSHC 1740 (1979); *Leone Construction Co.*, 3 OSHC 1979 (1976); *Trojan Steel Co.*, 3 OSHC 1384 (1975); *Weatherby Engineering Co.*, 8 OSHC 2013 (1980).

Normally, where a party does not seek review (reversal) of the administrative law judge's decision, but only takes exception to his findings of fact or conclusions of law, the Commission will not address those issues unless they raise a question of compelling public interest. *Bethlehem Steel Corporation*, 7 OSHC 1053 (1979). *Cf. Union Camp Corporation*, 5 OSHC 1799 (1977); *Water Works Installation Corp.*, 4 OSHC 1339 (1976). Moreover, parties may not raise issues on review that were not addressed below or litigated without the express or implied consent of the parties. *Bechtel Power Corp.*, 4 OSHC 1005 (1976); *Northwest Airlines, Inc.*, 8 OSHC 1988 (1980); *Crescent Wharf & Warehouse Co.*, 2 OSHC 1623 (1975). A decision of the Commission constitutes the final administrative determination.

The doctrines of *res judicata* and collaterial estoppel are applicable to Commission proceedings.[16] *Continental Can Company, U.S.A. v. Marshall*, 603 F.2d 590, 7 OSHC 1521 (7th Cir. 1979);[17] *Kaiser Engineers, Inc.*, 6 OSHC 1845 (1978); *Cement Asbestos Products Co.*, 8 OSHC 1151 (1980). See also *International Harvester Co. v. OSHRC*, 628 F.2d 982, 8 OSHC 1780 (7th Cir. 1980), which stated that although *res judicata* is applicable in administrative proceedings, the action involved was not an appropriate application of it.

The Commission may affirm, modify, vacate, reverse, or remand the decision of the judge; it has jurisdiction until a final order is issued. *J. L. Foti Construction Co., Inc.*, 8 OSHC 1666 (1980); *National Industrial Constructors, Inc.*, 8 OSHC 1675 (1980); *Noranda Aluminum*, 8 OSHC 1450 (1980). Any party aggrieved by a final order of the Commission may, while the matter is still within the Commission's jurisdiction, file a motion for stay, setting forth the reasons a stay is sought and the length of the stay. Commission Rule 94. The Commission may thereafter order a stay for the period requested or for such longer or shorter period as it deems appropriate. After a petition for review has been filed with the court and it obtains jurisdiction, a stay of the Commission's decision pending the outcome of the appeal may be sought.

d. *Quorum*

Under the statutory provisions governing the Commission's discretionary review of determinations made by administrative law judges, no official action can be taken by the Commission without the affirmative vote of at least two of its members. Section 12(f) of the Act provides in this respect:

> For the purpose of carrying out its functions under this chapter, two members of the Commission shall constitute a quorum and official action can be taken only on the affirmative vote of at least two members.

In *Shaw Construction, Inc. v. OSHRC*, 534 F.2d 1183, 4 OSHC 1427 (5th Cir. 1976), the court held that an affirmance supported by only one vote (that is, an equally divided Commission) did not comply with the statutory requirement; thus it refused to conduct appellate review. See also *Dunlop v. Continental Oil Co.*,—F.2d —, 5 OSHC 2083 (6th Cir. 1977); *Thermo Tech, Inc. v. OSHRC*, 580 F.2d 1051, 6 OSHC 2012 (5th Cir. 1978). The Commission has attempted to circumvent the "quorum" requirements of the Act in cases where it is equally divided by making the judge's decision the final order of the Commission and according it the precedential value of an unreviewed judge's decision. *Bethlehem Steel Corp.*, 5 OSHC 1025 (1977); *Life Sciences, Inc.*, 6 OSHC 1053 (1977); *International Harvester Co.*, 7 OSHC 2194 (1980). But in *Cox Brothers, Inc. v. Secretary of Labor*, 574 F.2d 465, 6 OSHC 1484 (9th Cir. 1978), the court rejected this tactic, holding that official action requires the affirmative vote of two Commission members; in the absence of two such votes, the court lacked jurisdiction to review the Commission's order. In *George Hyman Construction Co. v. OSHRC*, 582 F.2d 834, 6 OSHC 1855 (4th Cir. 1978), however, the court held that a Commission order issued in a three-way split affirming a judge's decision was reviewable, since to hold otherwise would put the employer in a jurisdictional limbo. The court reasoned that inasmuch as a judge's decision becomes the final, unreviewable order of the Commission if no member directs review, then a divided Commission vote *a fortiori* would permit review. See also *Life Science Products Co. (Moore) v. OSHRC*, 591 F.2d 991, 7 OSHC 1031 (4th Cir. 1979)—the court did not really consider the Commission decision to be split, since although the members disagreed, they both voted to affirm the judge's decision to avoid an impasse; *Marshall v. L. E. Myers Co.*, 589 F.2d 270, 6 OSHC 2159 (7th Cir. 1978)—the court skirted the official action question and affirmed a divided Commission vote because it was not arbitrary, capricious, an abuse of discretion, or otherwise not in accordance with the law.

The Commission's formulation, in cases where its members are evenly divided, of agreeing to affirm the order of the judge but according his decision the precedential value of an unreviewed judge's decision was found not to constitute official action of the Commission by the Ninth Circuit in *Williamette Iron & Steel Co. v. Secretary of Labor*, 604 F.2d 1177, 7 OSHC 1641 (9th Cir. 1979), *cert. denied*, 100 S. Ct. 1337, 8 OSHC 1162 (1980), although subsequently it received the approval of the Third Circuit in *Marshall v. Sun Petroleum Products Co.*, 622 F.2d 1176, 8 OSHC 1422 (3d Cir. 1980), which considered it a final, reviewable order for purposes of appeal. Thereafter, the Commission stated in *Texaco, Inc.*, 8 OSHC 1758 (1980):

> We do not agree with the Ninth Circuit's conclusion that by affirming the judge's order but limiting its precedential value the Commission fails to take official action.
>
> However, the result that is attained by affirming the judge's order and according it the precedential value of an unreviewed judge's decision also can be achieved by vacating the direction for review. Since we believe that all courts of appeals are likely to find our vacation of a direction for review to constitute official action of the Commission, we adopt this mode of disposing of the present case. Therefore, the Commission members agree to vacate the direction for review, as they are unable to resolve the case on the merits. This action leaves any aggrieved party free to seek review in an appropriate court of appeals. Section 11(a) of the Act, 29 U.S.C. §660(a).

It is not necessary for *all* Commission members to participate for an agency to take official action; two members constitute a quorum. *Perini Corp.*, 6 OSHC 1611 (1978).

2. Judicial Review and Enforcement

Any person adversely affected or aggrieved by an order of the Review Commission may seek judicial review, assuming procedural prerequisites have been met. Section 11(a) of the Act. Provisions for review are necessary to ensure consistency, uniformity and compliance with both due process and the mandate of the Act. However, the Commission itself appears not entitled to party status in appellate proceedings. *Marshall v. Sun Petroleum Products Co.,* 622 F.2d 1176, 8 OSHC 1422 (3d Cir. 1980); *Dale M. Madden Construction Co. Inc. v. Hodgson,* 502 F.2d 278, 2 OSHC 1101 (9th Cir. 1974); but see *Diamond Roofing Co. v. OSHRC,* 528 F.2d 645, 4 OSHC 1001 (5th Cir. 1976); *Brennan v. Gilles & Cotting, Inc.,* 504 F.2d 1255, 2 OSHC 1243 (4th Cir. 1974).

Upon filing of a request for review of a Commission "final order," "the court shall have jurisdiction of the proceeding and of the question determined therein, and shall have power to grant such temporary relief or restraining order as it deems just and proper, and to make and enter upon the pleadings, testimony, and proceedings set forth in such record a decree affirming, modifying, or setting aside in whole or in part, the order of the Commission and enforcing the same to the extent that such order is affirmed or modified." Section 11(a) of the Act.[18] The filing of a petition for review does not stay the Commission's order; a stay should be sought from the court.

a. *Procedural Prerequisites*

A party who wishes to preserve its right to judicial review must first have submitted a petition for discretionary review with the Commission and followed its procedural rules, or else it may be foreclosed from court review of its objections to the judge's (or Commission's) decision. *Keystone Roofing Co., Inc. v. OSHRC,* 539 F.2d 960, 4 OSHC 1481 (3d Cir. 1976); *GAF Corp. v. OSHRC,* 561 F.2d 913, 5 OSHC 1555 (D.C. Cir. 1977); *Todd Shipyards Corp. v. Secretary of Labor,* 566 F.2d 1327, 6 OSHC 1227 (9th Cir. 1977); *Cf. Buckeye Industries Inc. v. Secretary of Labor,* 587 F.2d 231, 6 OSHC 2181 (5th Cir. 1979). Prior to that, however, a timely initial contest of the citation and penalty is a prerequisite to later judicial review (*e.g.,* if contest only of penalty, party cannot later appeal citation). *Brennan v. Winters Battery Mfg. Co.,* 531 F.2d 317, 4 OSHC 1240 (6th Cir. 1975); *Dan J. Sheehan Co. v. OSHRC,* 520 F.2d 1036, 3 OSHC 1573 (5th Cir. 1975).

Any person adversely affected or *aggrieved* by a Commission final order may seek judicial review in the court of appeals circuit in which the alleged violation occurred, in which the employer has its principal office, or in the District of Columbia Circuit, within sixty days following the issuance of such order. Section 11(a) of the Act; *United States v. Fornea Road Boring Co.,* 565 F.2d 1314, 6 OSHC 1232 (5th Cir. 1978). In *Savina Home Industries, Inc. v. Secretary of Labor,* 594 F.2d 1358, 7 OSHC 1154 (10th Cir. 1979), the Court refused to consider the merits of an employer's objections to the Commission's authority to increase penalties, holding the employer lacked standing, since there was no showing that it had been *aggrieved* or adversely affected by the Commission's decision which reduced the penalty. In *Midway Industrial Contractors, Inc. v. OSHRC,* 616 F.2d 346, 8 OSHC 1076 (7th Cir. 1980), an employer's failure to file a written petition for review of a Review Commission final order within sixty days from the date the order became final, as required by the Act, required the appellate court to dismiss the petition for lack of jurisdiction. See also *Hoerner Waldorf Pan American Bag Co. v. OSHRC,* 614 F.2d 795, 7 OSHC 2210 (1st Cir. 1980). A Review Commission interlocutory order upholding the validity of an inspection and remanding the case to the judge for further proceedings is not a final order. *Robberson Steel Co. v. Marshall,*—F.2d —, 7 OSHC 1052 (10th Cir. 1978). Neither is a Commission remand order to a judge. *Field Crest Mills, Inc. v. OSHRC,* 545 F.2d 1384, 4 OSHC 1845 (4th Cir. 1976); *Chicago Bridge & Iron Co. v. OSHRC,*—F.2d —, 1 OSHC 1244 (7th Cir. 1974).

Section 11(a) of the Act provides that matters not raised before the Commission cannot be raised by the employer for the first time and considered by the court on review, absent "extraordinary circumstances." In *Lloyd C. Lockrem, Inc., v. OSHRC,* 609 F.2d 940, 7 OSHC 1999 (9th Cir. 1979), extraordinary circumstances justifying a failure to exhaust administrative remedies was found to exist, since the Commission had twice previously considered the arguments. In *RMI Co. v. Secretary of Labor,* 594 F.2d 566, 7 OSHC 1119 (6th Cir. 1979), a subsequent Commission decision which altered its prior position on the question of economic feasibility was considered to be an extraordinary circumstance. See also *McGowan v. Marshall,* 604 F.2d 885, 7 OSHC 1842 (5th Cir. 1979)—the Commission's lack of authority over constitutional issues is an extraordinary circumstance.

Objections, issues, and arguments must be raised before the Commission, in order to vest the court with jurisdiction to consider the matter on review. *Stockwell Manufacturing Co. v. Usery,* 536 F.2d 1306, 4 OSHC 1332 (10th Cir. 1976). However, the Secretary of Labor, unlike employers, seems not to be limited to issues raised before the Commission. *Buckeye Industries, Inc. v. Secretary of Labor,* 587 F.2d 231, 6 OSHC 2181 (5th Cir. 1979); *Todd Shipyards Corp. v. Secretary of Labor,* 586 F.2d 683, 6 OSHC 2122 (9th Cir. 1977).

The Secretary of Labor also has the right to obtain review or enforcement of a final Commission order by filing a petition in the circuit in which the violation occurred or in which the employer has its principal office. Section 11(b) of the Act; *Atlantic & Gulf Stevedores, Inc. v. OSHRC,* 534 F.2d 541, 4 OSHC 1061 (3d Cir. 1976). Commission orders are not self-enforcing. Review must be based upon a "final order" and must be filed within sixty days after service of the Commission's order, but the Act does not provide a time limit for initiating enforcement action. Section 11(b) of the Act.

Filing of a petition for review does not operate as an automatic stay of the Commission's decision and order. An employer may request a stay from the Commission pursuant to Commission Rule 44, or it may submit a request to the appropriate court of appeals. Where an employer does not obtain a stay, the Secretary of Labor could collect the proposed penalties, and the employer could indeed be subject to further penalties and liability for a failure to abate the violations.

The Federal Rules of Appellate Procedure apply to the review of a Commission decision in the court of appeals. The court of appeals has the authority to reverse, modify, vacate, affirm, or remand the matter for evidentiary hearing and for additional findings of fact concerning the violation. See *Atlantic Drydock Corporation v. OSHRC*, 524 F.2d 476, 3 OSHC 1755 (5th Cir. 1975).

b. *Exhaustion of Administrative Remedies*

Under normal circumstances, prior to seeking judicial review, parties must exhaust administrative remedies under the Act. The section of the Act that provides for review of Commission decisions incorporates an exhaustion requirement and reads in pertinent part: "No objection that has not been urged before the Commission shall be considered by the court, unless the failure or neglect to urge such objection shall be excused because of extraordinary circumstances." Section 11(a) of the Act. See *Keystone Roofing Co., Inc. v. OSHRC*, 539 F.2d 960, 4 OSHC 1481 (3d Cir. 1976), where the court lacked jurisdiction to consider the merits of an employer's objection to a judge's penalty assessment in the absence of the employer's prior filing of a timely petition for discretionary review with the Commission. See also *McGowan v. Marshall*, 604 F.2d 885, 7 OSHC 1842 (5th Cir. 1979); *Stockwell Manufacturing Company v. Usery*, 536 F.2d 1306, 4 OSHC 1332 (10th Cir. 1976); *Felton Construction Co. v. OSHRC*, 518 F.2d 49, 3 OSHC 1269 (9th Cir. 1975); *Lloyd C. Lockrem, Inc., v. U.S.A.*, 609 F.2d 940, 7 OSHC 1999 (9th Cir. 1979)—the employer presented its objections to the Commission and gave it an opportunity to rule therein. In *Universal Auto Radiator Mfg. Co. v. Marshall*, 631 F.2d 20, 8 OSHC 2026 (3d Cir. 1980), a technical dispute over the filing of a petition for review arose:

> [T]he decision of the ALJ was mailed to the parties on August 8, 1979, 20 days in advance of its filing with the OSHRC on August 28. The decision was accompanied by a notice saying *inter alia* that such decision would become the final order of the Commission on September 27, 1979, 30 days after being filed with the OSHRC, unless within that time a member of the Commission directed that it be reviewed. This notice made clear that [the employer] might petition for review of the ALJ's decision by the OSHRC, presumably on or before September 27. On September 25, 1979, Universal mailed its petition for review to the OSHRC. This petition was not received until September 28. However, under OSHA's regulations, such mailing was deemed filed on September 25, 1979. 29 C.F.R. §2200.8 provides:
>
> (b) Unless otherwise ordered, all filing may be accompanied by first class mail.
> (c) Filing is deemed effected at the time of mailing.
>
> Thus, under OSHA's regulations, the petition was filed on September 25, 1979, which technically appears to comply with the requirements of 29 CFR §2200.91(a) and (b)(3). OSHA notes that in fact the petition was received on September 28, after the opinion became final, and argues that such filing should not be deemed to comply with the exhaustion requirement.

The Court did not have to resolve the technical dispute, because the Commission denied the petition for review on October 1, 1979; thus the case did not require dismissal.

The courts have also held that the exhaustion principle applies to Fourth Amendment challenges to the sufficiency and the authority of search warrants, unless exceptional circumstances exist. *Continental Can Co. v. Marshall*,

603 F.2d 590, 7 OSHC 1521 (7th Cir. 1978); *Marshall v. Northwest Orient Airlines,* 574 F.2d 119, 6 OSHC 1481 (2d Cir. 1978); *In Re Quality Products, Inc.,* 592 F.2d 611, 7 OSHC 1093 (1st Cir. 1979); *In Re Restland Memorial Park,* 540 F.2d 626, 4 OSHC 1485 (3d Cir. 1976); *Babcock & Wilcox Co. v. Marshall,* 610 F.2d 1128, 7 OSHC 1880 (3d Cir. 1979)—employer's motion to suppress evidence obtained during an inspection must be submitted to the Commission prior to seeking relief in the courts; *Chromalloy American Corp.,* 7 OSHC 1547 (1979). *Compare Weyerhaeuser Co. v. Marshall,* 592 F.2d 373, 7 OSHC 1090 (7th Cir. 1979)—exhaustion not required, since there was no benefit from a prior development of the factual record and no possibility of mooting the issue.

Interlocutory review of Commission orders to the courts has also not been permitted. *Fieldcrest Mills, Inc. v. OSHRC,* 545 F.2d 1384, 4 OSHC 1845 (4th Cir. 1976); *U.S. Steel Corp. v. OSHRC,* 517 F.2d 1400, 3 OSHC 1313 (2d Cir. 1975); *Robberson Steel Co. v. Marshall,*—F.2d —, 7 OSHC 1052 (10th Cir. 1978); but see *Continental Can Co. v. Marshall,* 603 F.2d 590, 7 OSHC 1521 (7th Cir. 1979)—Secretary of Labor enjoined from prosecuting citations because of *res judicata* principles.

c. *Standard of Review*

(1) *Factual Questions and Evidentiary Findings.* The *factual findings* of the Commission (*e.g.* existence of a violation, appropriateness of a penalty) are conclusive if supported by *substantial evidence* on the record considered as a whole. Section 11(a) of the Act. The courts have consistently applied this standard of review to the Commission's (or an administrative law judge's) factual determinations. The clearly erroneous test is not used.[19] See *Arkansas–Best Freight System, Inc. v. OSHRC,* 529 F.2d 649, 3 OSHC 1910 (8th Cir. 1976); *Marshall v. L. G. Meyers Co.,* 589 F.2d 270, 6 OSHC 2159 (7th Cir. 1978); *Madison Foods, Inc. v. Marshall,* 630 F.2d 628, 8 OSHC 2029 (8th Cir. 1980); *Dunlop v. Rockwell International,* 540 F.2d 1283, 4 OSHC 1606 (6th Cir. 1976); *D. Federico Co. v. OSHRC,* 558 F.2d 614, 5 OSHC 1528 (1st Cir. 1977); *Frank Irey, Jr., Inc. v. OSHRC,* 519 F.2d 1200, 2 OSHC 1283 (3d Cir. 1975); *Lee Way Motor Freight, Inc. v. Secretary of Labor,* 511 F.2d 864, 2 OSHC 1609 (10th Cir. 1975); *Brennan v. OSHRC (Republic Creosoting Co.),* 501 F.2d 1196, 2 OSHC 1109 (7th Cir. 1974); *Greyhound Lines–West v. Marshall,* 575 F.2d 759, 6 OSHC 1636 (9th Cir. 1978); *B & B Insulation, Inc. v. OSHRC,* 583 F.2d 1364, 6 OSHC 2062 (5th Cir. 1978); *Roanoke Iron & Bridge Works, Inc. v. OSHRC,* 588 F. 2d 1351, 7 OSHC 1041 (4th Cir. 1979); *National Steel & Shipbuilding Co. v. OSHRC,* 607 F.2d 311, 7 OSHC 1537 (9th Cir. 1979); *Schriber Steel Metal & Roofers, Inc. v. OSHRC,* 597 F.2d 78, 7 OSHC 1246 (6th Cir. 1979); *Austin Commercial v. OSHRC,* 610 F.2d 200, 7 OSHC 2119 (5th Cir. 1979); *Continental Oil Co. v. Marshall,* 630 F.2d 446, 8 OSHC 1982 (6th Cir. 1980); *H. S. Holtze Construction Co. v. Marshall,* 627 F.2d 149, 8 OSHC 1785 (8th Cir. 1980). Substantial evidence means more than a mere scintilla. It means such relevant evidence as a reasonable mind might accept as adequate to support a conclusion. *Marshall v. Knutson Construction Co.,* 566 F.2d 596, 6 OSHC 1077 (8th Cir. 1977); *Marshall v. West Point Pepperell, Inc.,* 588

F.2d 979, 7 OSHC 1034 (5th Cir. 1979). *Martin Painting and Coating Company v. Marshall,* 629 F.2d 437, 8 OSHC 2173 (6th Cir. 1980).

The actual requirements of the substantial evidence test have varied with the type of matter before the courts. For example, with respect to alleged violations of the general duty clause, "where evidence of the employer's *actual* knowledge is relied on as proof that a hazard is "recognized," the Secretary of Labor has the burden of demonstrating by substantial evidence that the employer's safety precautions were unacceptable in his industry or a relevant industry; . . . the standard is that of a reasonably conscientious safety expert familiar with the pertinent industry." *Magma Copper Co. v. Marshall,* 608 F.2d 373, 7 OSHC 1893 (9th Cir. 1979); *General Dynamics Corp. v. OSHRC,* 599 F.2d 453, 7 OSHC 1373 (1st Cir. 1979)—application of the harmless error rule. Where constructive knowledge is the basis for showing the existence of a recognized hazard, substantial evidence must include evidence of the standard of knowledge/custom and practice in the employer's industry relative to the hazard involved. *Usery v. Marquette Cement Manufacturing Co.,* 568 F.2d 902, 5 OSHC 1793 (2d Cir. 1977); *National Realty & Construction Co., Inc. v. OSHRC,* 489 F.2d 1257, 1 OSHC 1422 (D.C. Cir. 1973). Substantial evidence must be enough to warrant denial of a motion for directed verdict in a civil case tried by a jury. *Dunlop v. Rockwell International,* 540 F.2d 1283 (6th Cir. 1976).

The Second Circuit in *General Electric Co. v. OSHRC,* 540 F.2d 67, 4 OHSC 1512 (2d Cir. 1976), found that unsupported generalizations on the part of the Commission could not support a finding of violation of the general duty clause, and the court in *Schriber Sheet Metal & Roofers, Inc. v. OSHRC,* 597 F.2d 78, 7 OSHC 1246 (6th Cir. 1979), found that substantial evidence did not support a finding of feasibility under the general duty clause.

Construction and application of the substantive evidence requirements also vary when dealing with violations of specific standards. In *Bethlehem Steel Corporation v. OSHRC,* 607 F.2d 1069, 7 OSHC 1833 (3d Cir. 1979), the employer attacked the accuracy of an administrative law judge's finding regarding a serious violation of 29 CFR 1910.1000(a)(2), because the judge nowhere related the evidentiary basis upon which a violation might be found, arguing that the findings failed to meet minimum standards required by the Administrative Procedure Act. The Court agreed and stated:

> The Administrative Procedure Act provides in relevant part: all decisions, including initial . . . decisions are a part of the record and shall include a statement of (A) findings and conclusions, the reasons or basis therefor, on all material issues of fact, law, or discretion presented on the record . . . 5 U.S.C. §557E.
>
> We have held in *In Re: United Corp.,* 249 F.2d 168, 179 (3d Cir. 1957): the final order of the administrative agency must include the findings and conclusions upon all material issues presented on the record. The reasons or basis for the decision must also be clearly enunciated. OSHRC itself has held that "[the ALJ's] decision must show on its face what evidence has been considered in reaching his findings and conclusions." *P & Z Company, Inc.,* 6 OSHC 1189, 1191 (1977).
>
> The Commission has further indicated that findings of fact must include a determination of credibility where conflicting testimony is present. *Evansville Materials, Inc.,* 3 OSHC 1741, 1742 (1975).

> At a minimum, the ALJ's findings in this case should have indicated the evidentiary basis for his conclusion that the possible accumulation of excess fumes and smoke due to inadequate ventilation in Hancock's work space created a substantial probability of serious physical injury to him. . . . Yet, examination of the ALJ's decision reveals that absolutely no findings were made with respect to the alleged seriousness of the violation.
>
> We believe that the ALJ's findings do not comport with the minimum requirements of the Administrative Procedure Act, [and] we hold that the ALJ's findings of fact are inadequate to sustain the government's charges.

In *International Harvester Co. v. OSHRC,* 628 F.2d 982, 8 OSHC 1780 (7th Cir. 1980), the court placed a more stringent burden on itself in reviewing credibility factual determinations.

> [A]gency credibility resolutions are essentially nonreviewable unless contradicted by "incontrovertible documentary evidence or physical facts." *Olin Construction Co. v. OSHRC,* 525 F.2d 464, 467, 3 OSHC 1526, 1528 (2d Cir. 1975) [citation omitted]. We clearly may not substitute our judgment for that of the OSHRC. See *RMI Co., (v. Secretary of Labor),* 594 F.2d 566, 571, 7 OSHC at 1122 (6th Cir. 1979).

In *National Steel & Shipbuilding Co. v. OSHRC,* 607 F.2d 311, 7 OSHC 1837 (9th Cir. 1979), another court held in this respect that:

> Moreover, the Commission's special expertise calls for providing it with considerable discretion in drawing inferences from the evidence. *Greyhound Lines–West,* 575 F.2d at 762. Thus, "[i]f facts are open to conflicting inferences, we are not at liberty to draw an inference different from the one drawn by the [Commission], even though it may seem more plausible and reasonable to us."

The courts have also held that the Commission cannot fill in gaps in the Secretary of Labor's failure to meet his burden of proof by opinions and conjecture and cannot juggle and characterize the issues as findings of fact when they are actually conclusions of law. *Marshall v. CF&I Steel Corp.,* 576 F.2d 809, 6 OSHC 1543 (10th Cir. 1978). A case will be remanded if there are incomplete or incorrect findings. See *National Realty & Construction, Inc., supra.*

An important initial determination in the review of Commission decisions is whether the issue or finding in question is to be characterized as legal or factual. For example, the existence of technologically and economically feasible engineering controls has been considered a factual question and reviewed under the substantial evidence test. *Diversified Industries Division, Independent Stave Co. and Independent Stave Co. v. OSHRC,* 618 F.2d 30, 8 OSHC 1107 (8th Cir. 1980); *Marshall v. West Point Pepperell, Inc.,* 588 F.2d 979, 7 OSHC 1035 (5th Cir. 1979). In *RMI Co. v. Secretary of Labor, supra,* 594 F.2d at 569–70, 7 OSHC at 1121–22, the Commission had determined that engineering controls to reduce noise levels were feasible, even though such controls would not reduce noise to permissible levels. The court reviewed the Commission's determination under the Administrative Procedure Act's "arbitrary and capricious" standard. 5 U.S.C. Section 706(2)(A)(1976). Because the issue was one of policy or interpretation argued on uncontested facts, the court properly employed the narrower scope of review to test the Commission's conclusion. Predictably, therefore, even though the various circuits apply the same substantial evidence test, its application may result in different findings or may be accomplished with different accents. Compare *Mohawk Excavating, Inc., v. OSHRC,* 549 F.2d 859, 5 OSHC 1001 (2d Cir. 1977), with *Marshall v. Western Electric, Inc.,* 565 F.2d 240, 5 OSHC 2054 (2d Cir. 1977); see also *Duane Smelser Roofing Co. v. Marshall,* 617 F.2d 448, 8 OSHC 1106 (6th Cir. 1980)—Commission failed to make findings of fact as required; rather, it stated its holding on conclusory terms.

(2) *Questions of Law and Policy.* In reviewing questions of law or legislative policy (*e.g.* interpretation of standards) the courts have applied an arbitrary and capricious test rather than a substantial evidence test. *Arkansas Best Freight System, Inc. v. OSHRC,* 529 F.2d 649, 3 OSHC 1910 (8th Cir. 1976). The arbitrary and capricious test is taken from the Administrative Procedure Act, 5 U.S.C. Section 706(2)(A), which provides:

> To the extent necessary to decision and when presented, the reviewing court shall decide all relevant questions of law, interpret constitutional and statutory provisions, and determine the meaning or applicability of the terms of an agency action. The reviewing court shall . . . hold unlawful and set aside agency action, findings, and conclusions found to be arbitrary, capricious, an abuse of discretion, or otherwise not in accordance with law.

In the review of Commission determinations which involve questions of law, the courts have more or less uniformly applied this arbitrary and capricious test. The arbitrary, capricious or abuse of discretion test is highly deferential and presumes the action of the administrative agency to be valid. In its application, courts normally will not inquire into the wisdom of an agency's exercise of discretion and policy. See *Olin Construction Co. v. OSHRC,* 525 F.2d 464, 3 OSHC 1426 (2d Cir. 1975); *Long Manufacturing Co. v. OSHRC,* 554 F.2d 903, 5 OSHC 1376 (8th Cir. 1977); *Usery v. Hermitage Concrete Pipe Co.,* 584 F.2d 127, 6 OSHC 1886 (6th Cir. 1978)—the determination that a violation is not serious is a legal one of statutory interpretation, and therefore the substantial evidence test is inapposite. See also *Brennan v. OSHRC (Gerosa, Inc.),* 491 F.2d 1340, 1 OSHC 1523 (2d Cir. 1974); *Marshall v. Knutson Construction Co.,* 566 F.2d 596, 6 OSHC 1077 (8th Cir. 1977); *Dunlop v. Rockwell International,* 540 F.2d 1283, 4 OSHC 1606 (6th Cir. 1975); *Horne Plumbing & Heating Co. v. OSHRC,* 528 F.2d 564, 3 OSHC 2060 (5th Cir. 1976); *Marshall v. Cities Service Oil Co.,* 577 F.2d 126, 6 OSHC 1631 (10th Cir. 1978); *United Parcel Service of Ohio, Inc. v. OSHRC,* 570 F.2d 806, 6 OSHC 1347 (8th Cir. 1978)—there was substantial evidence to support the Commission's factual findings, but the legal remedy imposed was an abuse of discretion.

Under the "arbitrary and capricious" standard the scope of review is narrower than that employed under the substantial evidence test. "A reviewing court must consider whether the decision was based on a consideration of the relevant factors and whether there has been a clear error of judgment. Although this inquiry into the facts is to be searching and careful, the ultimate standard of review is a narrow one. The court is not empowered to substitute its judgment for that of the agency. Minor procedural violations are commonly overlooked in order to ensure that cases are decided on their merits. The agency must [however] articulate a rational connection between the facts found and the choice made." *Marshall v. Knutson Construction Company,* 566 F.2d 596, 6 OSHC 1077 (8th Cir. 1977), quoting from *Bowman Transportation Inc. v. Arkansas Best Freight System, Inc.,* 419 U.S. 281 (1974). In *Dunlop v. Haybuster Manufacturing Co.,* 524 F.2d 222, 3 OSHC 1594 (8th Cir. 1975), however, the court held as a general principle that "an agency determination of a question of law, unlike a finding of fact, does not have any presumption of correctness resulting from a substantial evidence rule and it must be closely scrutinized."

A primary question of course, for the court's initial consideration is whether an issue is a question of law or fact. *Marshall v. CFI Steel Corp.,* 576 F.2d 809, 6 OSHC 1543 (10th Cir. 1978). Only if a legal question is involved is the arbitrary and capricious test appropriate.

The courts have accorded varying degrees of deference to the legal interpretations of the Commission and the Secretary of Labor, with respect to the Act, the standards, and regulations.

(1) Where *both* the Secretary of Labor and the Commission have *congruent* views on the Act and its regulations and standards, those congruent determinations will carry substantive weight. *The Budd Co. v. OSHRC,* 513 F.2d 201, 2 OSHC 1698 (3d Cir. 1975); *Wisconsin Electric Power Co.,* 567 F.2d 733, 6 OSHC 1137 (7th Cir. 1977); *Clarkson Construction Co. v. OSHRC,* 531 F.2d 451, 3 OSHC 1880 (10th Cir. 1976); *Irvington Moore v. OSHRC,* 556 F.2d 431, 5 OSHC 1585 (9th Cir. 1977); *Lloyd C. Lockrem, Inc. v. U.S.A.,* 609 F.2d 940, 7 OSHC 1999 (9th Cir. 1979). Great deference is paid to the Secretary's interpretation of his own regulations, *particularly* when affirmed by the Commission. *UAW, Local 588 v. OSHRC,* 557 F.2d 607, 5 OSHC 1525 (7th Cir. 1977); *American Airlines, Inc. v. OSHRC,* 578 F.2d 38, 6 OSHC 1691 (2d Cir. 1978).

Some cases, however, state that the court should accord greater deference to the Commission *when it agrees* with the position of the Secretary. *Floyd S. Pike Electrical Contractor, Inc. v. OSHRC,* 576 F.2d 72, 6 OSHC 1781 (5th Cir. 1978). However, the Second Circuit and Sixth Circuit have refused to adopt this distinction and pay the same deference to Commission decisions, regardless of whether they dovetail with the Secretary of Labor's position. *General Electric Co. v. OSHRC,* 583 F.2d 61, 6 OSHC 1868 (2d Cir. 1978); *RMI Company v. Secretary of Labor,* 594 F.2d 566, 7 OSHC 1119 (6th Cir. 1979). These courts feel that varying the deference paid to a Commission decision, depending upon how it relates to the Secretary's position, is of questionable value analytically. But see *Marshall v. Anaconda Co., Montana Mining Division,* 596 F.2d 370, 7 OSHC 1382 (9th Cir. 1979)—when the Secretary's interpretation concurs with that of the Commission, it is entitled to "substantial weight;" when it does not, it "carries much less weight;" and when the regulation at issue is adopted from a consensus standard, the Secretary's interpretation is "less compelling;" *Brennan v. OSHRC (Ron M. Fiegen, Inc.),* 513 F.2d 713, 3 OSHC 1001 (8th Cir. 1975)—while the Secretary's interpretation was more consonant with the citation, the Commission's interpretation was reasonable and well within the plain meaning of its terms; thus its interpretation was accepted, since it is charged with the final adjudication of the Act.

(2) If the Commission and the Secretary of Labor *differ* on a legal interpretation, the courts are in disagreement as to the applicable standard of review and/or deference. Several courts give priority to views of the Secretary of Labor in this situation. See *GAF Corporation v. OSHRC,* 561 F.2d 913, 5 OSHC 1555 (D.C. Cir. 1977)—deference was given to the Secretary of Labor's interpretation, and no importance was attached to the Commission's differing view; *Marshall v. Western Electric, Inc.,* 565 F.2d 240, 5 OSHC 2054 (2d Cir. 1977)—the Commission's view was found to be unreasonable, and the Secretary's interpretation of the standard was found to be the only reasonable interpretation; *Clarkson Construction Co. v. OSHRC,* 531 F.2d 451, 3 OSHC 1880 (10th Cir. 1976)—the Secretary of Labor's interpretations are entitled to great weight and are controlling, even though there is another equally reasonable interpretation; *Brennan v. Southern Contractor Service,* 492 F.2d 498, 1 OSHC 1648 (5th Cir. 1974); *Brennan v. OSHRC (Kesler & Son Construction Co.,* 513 F.2d 553, 2 OSHC 1668 (10th Cir. 1975).

In *Marshall v. Anaconda Company, Montana Mining Division,* 596 F.2d 370, 7 OSHC 1382 (9th Cir., 1979), however, the court stated:

> Although we have recognized that the Secretary's interpretation of his own regulation, when affirmed by the Commission, must be accorded substantial weight, *Irvington Moore v. OSHRC,* 556 F.2d 431, 5 OSHC 1585 (9th Cir. 1977), it carries much less weight when at odds with the Commission's. Moreover, the Secretary's interpretation is less compelling when a regulation has been adopted from a consensus standard. *Bethlehem Steel Corp. v. OSHRC,* 573 F.2d 157, 160, 6 OSHC 1440 (3d Cir 1977). *See Skidmore v. Swift & Co.,* 323 U.S. 134, 140, 65 S. Ct. 161, 89 L. Ed. 124 (1944).
>
> Previously, we have held that although a section may be artlessly and ambiguously drafted, the Secretary's interpretation nevertheless is entitled to a certain deference. *California Stevedore & Ballast Co. v. OSHRC,* 517 F.2d 986, 3 OSHC 1174 (9th Cir. 1975). In *Irvington Moore v. OSHRC,* 556 F.2d 431, 5 OSHC 1585 (9th Cir. 1977), we upheld the Secretary's interpretation of a regulation as reasonable. But there the Commission had agreed with the interpretation of the Secretary. In this case, the Commission has repeatedly rejected the Secretary's interpretation.

Most courts accord a more significant degree of deference to the Commission's interpretation, when its position is different from the Secretary's. *The Budd Co. v. OSHRC,* 513 F.2d 201, 2 OSHC 1698 (3d Cir. 1975)—the interpretation is consistent with prior Commission rulings, is reasonable, and should be followed; *Diebold, Inc. v. Marshall,* 585 F.2d 1327, 6 OSHC 2002 (6th Cir. 1978)—where a regulation is ambiguous, great deference is given to the Commission's reasonable interpretation. These cases have held that the Commission's interpretations regarding the meaning of the Act should be given substantial deference by the courts, since it is the Commission, not the Secretary of Labor, which is charged with the final administrative adjudication of the Act. As stated by the Court in *Brennan v. OSHRC (Republic Creosoting Co.),* 501 F.2d 1196, 2 OSHC 1109 (7th Cir. 1974):

> We note at the outset that the Occupational Safety and Health Review Commission is presumed to have technical expertise and experience in the field of job safety. A court must, therefore, defer to the findings and analysis of the Commission unless such findings are without substantial basis in fact.

See also *Brennan v. OSHRC (Gerosa, Inc.),* 491 F.2d 1340, 1 OSHC 1523 (2d Cir. 1974); *Brennan v. OSHRC (Ron M. Fiegen, Inc.),* 513 F.2d 713, 3 OSHC 1001 (8th Cir. 1975); *Brennan v. Gilles & Cotting, Inc.,* 504 F.2d 1255, 2 OSHC 1243 (4th Cir. 1974); *Brennan v. OSHRC (J. W. Bounds),* 488 F.2d 337, 1 OSHC 1429 (5th Cir. 1973).

(3) In many cases, the courts have refused to defer to *either* the Commission's interpretation or the Secretary of Labor's interpretation. See *Ace Sheeting and Repair Co. v. OSHRC,* 555 F.2d 439, 5 OSHC 1589 (5th Cir. 1977); *Usery v. Marquette Cement Manufacturing Co.,* 568 F.2d 902, 5 OSHC 1793 (2nd Cir. 1977); *Marshall v. Western Waterproofing Co.,* 560 F.2d 947, 5 OSHC 1732 (8th Cir. 1977); *Amoco Oil Company v. OSHRC,* 549 F.2d 1, 4 OSHC 1791 (7th Cir. 1976). In *Kent Nowlin Construction Co. v. OSHRC,* 595 F.2d 368, 7 OSHC 1105 (10th Cir. 1979), the Secretary of Labor, the administrative law judge, and the Commission all disagreed about the interpretation of standard. Because of such ambiguity, the citation was vacated.

In *Westinghouse Electric Corp. v. OSHRC,* 617 F.2d 497, 8 OSHC 1110 (7th Cir. 1980), the court determined that the standards involved in the case were not easily intelligible to those subject to them, to those who administer them, or to those who try to determine what they mean, and that given the inartful drafting of the standard, neither interpretation could be branded as particularly unreasonable. Although the court was mindful of the "great deference" owed to the Commission's reasonable interpretation of the Secretary's regulations under ordinary circumstances, the court felt that deference would be dubiously bestowed when the Commission has shifting interpretations as well as likely divergent reviews arising in the various circuits.

In *General Electric Company v. OSHRC,* 583 F.2d 61, 6 OSHC 1868 (2d Cir. 1978), the court stated that it had the authority to decide whether the Commission's interpretation is unreasonable and inconsistent with its purpose and that a court need not defer to an interpretation that it finds unreasonable. The court stated that it has been suggested that the court of appeals should give greater deference to an interpretation of a regulation that the Secretary and Commission agree on, citing *Wisconsin Electric Power Co. v. OSHRC,* 567 F.2d 735, 6 OSHC 1137 (7th Cir. 1977), but the court did not agree, feeling neither bound by nor persuaded by the opinions of the Secretary or the Commission. It felt constrained to remind them of the responsibility to be reasonable, stating:

> The purpose of OSHA standards is to improve safety conditions in the working place, by telling employers just what they are required to do in order to prevent or minimize danger to employees. In an adjudicatory proceeding, the Commission should not strain the plain and natural meaning of words in a standard to alleviate an unlikely and uncontemplated hazard. The responsibility to promulgate clear and unambiguous standards is upon the Secretary. The test is not what

he might possibly have intended, but what he said. If the language is faulty, the Secretary has the means and the obligation to amend. *Bethlehem Steel Corp. v. OSHRC and Marshall,* 573 F.2d 187, 6 OSHC 1440 (3d Cir. 1978). Similarly, the Fifth Circuit has offered the following comments:

> If the regulation missed its mark, the fault lies in the wording of the regulation —a matter easily remedied under the flexible regulation promulgating structure . . . with no need to press limits by judicial construction in an industrial area presenting infinite operational situations.
>
> * * *
>
> To strain the plain and natural meaning of words for the purpose of alleviating a perceived safety hazard is to delay the day when the occupational safety and health regulations will be written in clear and concise language so that employers will be better able to understand and observe them.

Diamond Roofing Co. v. OSHRC and Usery, 528 F.2d 645, 648–50 [4 OSHC 1001] (5th Cir. 1976) [citations and footnote omitted]. Finally, the Seventh Circuit has noted:

> It may well be, as the Secretary urges, that employees working on a flat roof require perimeter protection. If this be so, the Secretary has the machinery available to amend the section in question or to promulgate an entirely new section that would lay the matter to rest. It appears to us that the time and effort —not to mention the expense—of the notice and hearing required for such regulation changes would be far less than the effort and expense of the instant case.

Langer Roof & Sheet Metal, Inc. v. Secretary of Labor and OSHRC, supra, 524 F.2d at 1339. We share these sentiments.

Simply put, to apply this standard to the surface here in question would go too far—it would be inconsistent with the wording of the standard and it would create considerable doubt that the standard provides to employers fair warning of the conduct which it prohibits or requires. *See Connally v. General Construction Co.,* 269 U.S. 385, 391 (1926); *Boyce Motor Lines, Inc. v. United States,* 342 U.S. 337, 340 (1952). *See also CTM, Inc. v. OSHRC and Marshall,* 572 F.2d 262, 264, 6 OSHC 1319 (10th Cir. 1978); *Wisconsin Electric Power Co. v. OSHRC and Secretary of Labor,* 567 F.2d 735, 740–41, 6 OSHC 1137 (7th Cir. 1977) (Pell, J., dissenting); *Diamond Roofing Co. v. OSHRC and Usery, supra,* 528 F. 2d at 649. *Compare McLean Trucking Co. v. OSHRC and Secretary of Labor,* 503 F.2d 8, 10–11 [2 OSHC 1165] (4th Cir. 1974); *Ryder Truck Lines, Inc. v. Brennan,* 497 F.2d 230, 233 [2 OSHC 1075] (5th Cir. (1975).

See also *D. Federico v. OSHRC,* 558 F.2d 614, 5 OSHC 1528 (1st Cir. 1977).

In *Diamond Roofing Co., Inc. v. OSHRC,* 528 F.2d 645, 4 OSHC 1001 (5th Cir. 1976), the court stated that a standard, rule, or regulation must be construed to give effect to the *natural and plain meaning* of its words and cannot be construed to mean what the agency intended but did not adequately express. Previously, in other cases, the court had accepted the Commission's interpretation of a standard, as long as it was one of several reasonable interpretations, even though it may not appear as reasonable as others. It declined to do so this time, noting that increasingly frequently the courts have been rejecting both the Commission's and the Secretary of Labor's interpretations of standards that are not reasonable. See *Brennan v.*

OSHRC, (Ron Fiegen, Inc.), 513 F.2d 713, 3 OSHC 1001 (8th Cir. 1975); *Brennan v. OSHRC (Interstate Glass Co.),* 487 F.2d 438, 1 OSHC 1372 (8th Cir. 1973). Deferral has been found to be inappropriate in those situations, even when applying an arbitrary and capricious test.

In *Lloyd C. Lockrem, Inc. v. U.S.A.,* 609 F.2d 940, 7 OSHC 1999 (9th Cir. 1979), the court, while recognizing that great deference must be paid to the Secretary of Labor's interpretation of his own regulations, particularly when affirmed by the Commission, held that an employer is not required to assume the burden of guessing what the Secretary intended his regulations to mean. The court held that an employer should not be held to standards, the application of which cannot be agreed upon by those charged with their enforcement. The question of the applicability of the regulation which was involved in the case (whether the excavation requirement applies to trenches) was so confusing that the court stated that the maximum *expressio unius est exclusio alterius* applies. See *Kent Nowlin Construction Co. v. OSHRC,* 593 F.2d 368, 7 OSHC 1105 (10th Cir. 1979)—"Nor can mere citation to OSHA's remedial purpose 'to assure so far as possible every man and woman in the Nation safe and healthful working conditions,' substitute for analysis of the problem at hand." *G.A.F. Corp. v. OSHRC,* 561 F.2d 913, 5 OSHC 1555 (D.C. Cir. 1977); *Southern Ry. Co. v. OSHRC,* 539 F.2d 335, 3 OSHC 1940 (4th Cir.), *cert. denied,* 429 U.S. 999, 4 OSHC 1936 (1979); *Diamond Roofing Co. v. OSHRC,* 528 F.2d 645, 4 OSHC 1001 (5th Cir. 1976); *Central of Georgia Railroad v. OSHRC,* 576 F.2d 620, 6 OSHC 1784 (5th Cir. 1978); *Usery v. Kennecott Copper Corp.,* 577 F.2d 1113, 6 OSHC 1197 (10th Cir. 1977); *Brennan v. OSHRC (Pearl Steel Erection Co.),* 488 F.2d 337, 1 OSHC 1429 (5th Cir. 1973); Davis., *Administrative Law Treatise* §30.12 (19).

(4) In several cases, the courts have criticized both OSHA's and the Review Commission's handling of its cases. See *Dravo Corp. v. OSHRC,* 584 F.2d 637, 6 OSHC 1933 (3d Cir. 1978); *Ringland-Johnson, Inc. v. Dunlop,* 551 F.2d 117, 5 OSHC 1137 (8th Cir. 1977).

The Commission frequently exhibits a lack of consistency in its decisions, a lack of unanimity by its members and a lack of clarity of its opinions. See *e.g., Keppels, Inc.* 7 OSHC 1442 (1979)—verbal notice of contest; *S.&H. Riggers and Erectors Inc.,* 7 OSHC 1260 (1979)—burden of proving feasibility; *Gilbert Manufacturing Co.,* 7 OSHC 1611 (1979)—contest of abatement date; *Potlach Corp.,* 7 OSHC 1061 (1979)—repeated violation; *IMC Chemical Group,* 6 OSHC 2075 (1978)—withdrawal of citation; *Reynolds Metals Co.,* 8 OSHC 1496 (1980)—entry of an inspector onto workplace for purposes of conducting discovery; *Farmers Export Co.,* 8 OSHC 1665 (1980)—settlement agreements. As the dissenting opinion in *Farmers Export Co.,* 8 OSHC 1655, 1665 (1980) noted:

> The failure to confront issues and to fully explain the rationale for a decision contributes to the subsequent issuance of inconsistent decisions by the Commissions Administrative Law judges and the need for a Commissioner's review of such decisions, thereby unnecessarily adding to the Commission's case load and the general confusion and delay that characterize the adjudication of OSHA cases.

Moreover, the Commission (and its judges) takes a long time to decide cases that proceed to hearing and are directed for review. (Since the notice of contest stays abatement, employers have little incentive to have cases decided promptly). Priority of review and assignment of cases for hearing should be given to health hazard violations, "durational" violations (*e.g.* those which continue for a limited period of time and then are terminated by a change in the working condition), safety violations of high gravity and actions for failure to abate. The lowest priority should be given to violations of low gravity and transitory violations which are short lived and usually self-abating.

(5) Questions concerning the relationship between the Commission and its judges have also been considered by the courts. In *Champlin Petroleum Company v. Marshall,* 593 F.2d 637, 7 OSHC 1241 (5th Cir. 1979), the court stated:

> [W]e observe that in reversing the decision of the Commission we need not and do not assess the sufficiency of support for the administrative law judge's conclusions. The relationship between the administrative law judge and the Commission differs from that of trial and appellate courts in that OSHA contemplates that the Commission be charged with fact-finding responsibility, 29 U.S.C.A. §659(c), and the ALJ is merely an arm of the Commission for that purpose. *Accu-Namics, Inc. v. OSHRC,* 515 F.2d 828, 3 OSHC 1299 (5th Cir. 1975), *cert. denied,* 425 U.S. 903, 96 S. Ct. 1492, 47 L.Ed.2d 752, 4 OSHC 1090 (1976). See 29 U.S.C.A. §661(i).

While the findings of an administrative law judge are to be considered on review and may weaken the contrary conclusion of the Commission, *D. Federico Co. v. OSHRC,* 558 F.2d 614, 5 OSHC 1528 (1st Cir. 1977), the Commission's decision will be overturned only because it lacks evidentiary support.

A court of appeals may, however, remand cases to the Review Commission for the taking of additional evidence or to supplement the record. Sometimes the courts may also remand cases to the Commission for reconsideration in light of subsequent Commission decisions or to conduct further factual findings or consider arguments raised below but not considered. *Floyd S. Pike Electric, Inc. v. OSHRC,* 557 F.2d 1045, 5 OSHC 1678 (4th Cir. 1977); *Westinghouse Electric Corp. v. OSHRC,* 617 F.2d 497, 8 OSHC 1110 (7th Cir. 1980); *Atlantic Marine Inc. v. OSHRC,* 524 F.2d 476, 3 OSHC 1755 (5th Cir. 1975). At other times the courts will apply the Commission's rules themselves. *Penn-Dixie Steel Corp. v. OSHRC,* 553 F.2d 1078, 5 OSHC 1315 (7th Cir. 1977); *RMI Company v. Secretary of Labor,* 594 F.2d 566, 7 OSHC 1119 (6th Cir. 1979).

(6) The Commission has held, in *United States Steel Corporation,* 5 OSHC 1289 (1977), that it is the administrative agency with expertise in the area of occupational safety and health. While recognizing that some courts afford deference to the interpretations of the Secretary of Labor, the majority of courts, in its opinion, have held that the Commission's interpretation of the Act and the standards promulgated thereunder will be accorded deference. Realizing that there is thus a conflict among the courts concerning whether the Commission's or Secretary of Labor's interpretations are entitled to deference, the Commission went on to state:

> [B]ut in any event that deference applies *only* on judicial review of an administrative decision. The criteria for judicial review of a decision by an agency are substantially different from those applied by the agency in reaching its decision. . . . In reaching our decisions we must apply the criteria applicable to administrative decision-making. Thus, the court decisions cited by Labor on the principle of judicial deference have no direct application to our proceedings. Furthermore, we think it would be inconsistent with the intent of Congress to limit our role to one of merely determining whether Labor's interpretation of a standard is reasonable.

Thus the Commission held that the principles of judicial deference to an expert administrative determination have no application to the Commission's review of the Department of Labor's interpretation of a standard.

I. REVIEW IN FEDERAL DISTRICT COURT

District courts have jurisdiction over penalty collection suits; however, the failure to contest the merits of a citation or penalty administratively will bar the right to contest in a later civil collection action brought by the Solicitor of Labor. *U.S. v. J. M. Rosa Construction Co. Inc.,*—F. Supp. —, 1 OSHC 1188 (D. Conn. 1973); *U.S. v. Fultz,* — F. Supp. —, 3 OSHC 1539 (N.D. Ind. 1975); *Marshall v. Painting by C.D.C., Inc.,* 497 F. Supp. 653, 8 OSHC 2085 (E.D. N.Y. 1980).

When the Secretary of Labor charges an employer with discrimination against an employee for exercising his rights under Section 11(c) of the Act or when the Secretary seeks an injunction against conditions or practices in any place of employment which present an imminent danger of death or serious physical harm under Section 13(g) of the Act, jurisdiction lies with the district courts. *Greenfield & Associates v. Hodgson,*—F. Supp. —, 1 OSHC 1015 (E.D. Mich. 1972); *Whirlpool v. Usery,* 416 F. Supp. 30, 4 OSHC 1391 (N.D. Ohio 1976); *Granite Groves, A Joint Venture v. Usery,*—F. Supp. —, 5 OSHC 1935 (D. D.C. 1977); *Pool Offshore Co. v. Marshall,* 467 F. Supp. 978, 7 OSHC 1179 (W.D. La. 1979). Although these sections specify district court jurisdiction in the matters referred to, they ought not be read to preclude district court jurisdiction over matters neither referred to nor directed to another forum. See *Stark v. Wickard,* 321 U.S. 288 (1944); *Abbott Laboratories v. Gardner,* 387 U.S. 136, 140 (1967); *Louisiana Chemical Associates v. Bingham,* 496 F. Supp. 1188, 8 OSHC 1950, 1951, n.3 (W.D. La. 1980).

NOTES

1. Health regulations involved are those contained in 29 CFR 1910.94, .95, .96, .97, and .1000 through .1045 and any future standard which might be published under Subpart z of Section 1910 involving toxic and hazardous substances and those contained in 29 CFR 1926.52, .53, .54, .55, .57 and .800(c).

2. In *Kerr-McGee Chemical Corp.,* 4 OSHC 1739 (1976), the Commission held that *prima facie* proof of receipt of a citation is shown by a certified mail receipt. The

burden then shifts to the employer to show that the person who signed the receipt had no authority to do so and is not a representative of the company.

3. If the employer files a notice of contest, as happened in this case, then the union is not required to file its notice within fifteen days of the citation because the employer's notice of contest has effectively triggered a hearing. Once the hearing mechanism is instituted, affected employees may elect to participate as parties at any time prior to the commencement of the hearing before the judge as provided by Commission Rule 20. The fifteen-day time limit for an employee filing under Section 10(c) of the Act is operative only when the employer has not contested the citation and a hearing is desired by the employee or his representative. Under these circumstances, the time limit operates as a fail-safe mechanism to insure employees the opportunity for a hearing.

4. An "affected employee" means an employee of the cited employer who is exposed to the alleged hazard described in the citation as a result of his assigned duties. Commission Rule 1(e).

5. Commission Rule 70 provides that an application to take the deposition of a witness in lieu of oral testimony may be permitted under special circumstances. The application shall be in writing and shall set forth the reasons why such deposition shall be taken along with other material. Such application shall be filed with the Commission or the judge, as the case may be, and served not less than seven days prior to the time when it is desired that the deposition be taken.

6. The Commission also held that the Secretary did not waive the confidentiality of the statement by submitting the names of prospective employees in his witness list, even though it is generally held that where the government voluntarily reveals the identity of informers, the privilege ceases to exist. The Commission stated that furnishing a list of witnesses and disclosing who among those witnesses has given the government statements is the distinction, and submission of the witness list in and of itself does not waive the informer's privilege. The informer's privilege is also waived if the informer voluntarily discloses his identity.

7. The Commission held that submission of a witness list or a summary of prospective witnesses' testimony does not constitute a waiver of the confidentiality of the identities of informers or their statements, thus making a distinction between furnishing a list of witnesses, on the one hand, and disclosing who among those witnesses had given the government statements, on the other. The informer's privilege belongs to the government, but it also is waived if the informer has voluntarily disclosed his identity.

8. 5 U.S.C. §552 provides in pertinent part that:

"(a) Each agency shall make available to the public information as follows:

> (1) Each agency shall separately state and currently publish in the Federal Register for the guidance of the public—
>
> > (A) description of its central and field organization and the established places at which, the employees . . . from whom, and the methods whereby, the public may obtain information,
> >
> > (B) statements of the general course and method by which its functions are channeled and determined, . . .;
> >
> > (C) rules of procedure, . . .

(D) substantive rules . . . and statements of general policy or interpretations of general applicability . . .; and

(E) each amendment, revision, or repeal of the foregoing.

* * *

(2) Each agency, in accordance with published rules, shall make available for public inspection and copying—

(A) final opinions, . . . as well as orders, made in the adjudication of cases;

(B) those statements of policy and interpretations which have been adopted by the agency and are not published in the Federal Register; and

(C) administrative staff manuals and instructions to staff that affect a member of the public;

unless the materials are promptly published and copies offered for sale. . . .

* * *

"(b) This section does not apply to matters that are—

(1) (A) specifically authorized under criteria established by an Executive order to be kept secret in the interest of national defense or foreign policy and (B) are in fact properly classified pursuant to such Executive order;

(2) related solely in the internal personnel rules and practices of an agency;

(3) specifically exempted from disclosure by statute (other than section 552b of this title), . . . ;

(4) trade secrets and commercial or financial information obtained from a person and privileged or confidential;

(5) inter-agency or intra-agency memorandums or letters . . .'

(6) personnel and medical files and similar files . . .'

(7) investigatory records compiled for law enforcement purposes, but only to the extent that the production of such records would (A) interfere with enforcement proceedings, (B) deprive a person of a right to a fair trial or an impartial adjudication, (C) constitute an unwarranted invasion of personal privacy, (D) disclose the identity of a confidential source . . . , (E) disclose investigative techniques and procedures, or (F) endanger the life or physical safety of law enforcement personnel;

(8) contained in or related to examination, operating, or condition reports . . . for the regulation or supervision of financial institutions; or

(9) geological and geophysical information and data . . . .

"Any reasonably segregable portion of a record shall be provided to any person requesting such record after deletion of the portions which are exempt under this subsection."

9. Affected employees need not expressly state that they wish to elect party status before it is conferred. *General Electric Company,* 7 OSHC 1277 (1979). Party status may

be conferred upon affected employees if they manifest an intent to be heard during the course of Commission proceedings. *IMC Chemical Group, Inc.,* 6 OSHC 2075 (1978), *petition for review docketed,* Nos. 79–3018 and 79–3041 (6th Cir. 1979). But see *Marshall v. Sun Petroleum Products Co.,* 622 F.2d 1176, 8 OSHC 1422 (3d Cir. 1980). Election of party status entitles the authorized employee representative to meaningful participation in settlements. *Reynolds Metals Co.,* 7 OSHC 1042 (1979); *Aspro, Inc., Spun Steel Division,* 6 OSHC 1980 (1978). A proposed settlement agreement should be served upon employee representatives, regardless of party status, since Commission Rules require that employee notice be provided by service upon represented and unrepresented affected employees. *Reynolds Metals Company,* 7 OSHC 1042 (1979). A settlement does not meet the criteria set forth in Commission Rule 100(c) where the record fails to demonstrate affirmatively that a copy of the settlement agreement was served upon represented and unrepresented affected employees in the manner prescribed for notices of contest. *National Steel & Shipbuilding Co.,* 8 OSHC 2023 (1980).

10. The court stated that "in the absence of a contest, neither the Review Commission nor the ALJ has jurisdiction to review a settlement agreement entered into between the Secretary and an employer. Moreover, we agree that even after the employer files a notice of contest, if no employee files a notice or has not acquired party status under 29 CFR 2200.20, then the Commission would lack jurisdiction to review any settlement entered into between the Secretary and the employer. If an employee formally expresses an interest in the proceedings, however, we believe that the Commission would have jurisidiction to review the settlement in order to protect that interest."

The Act provides that the Commission is required to afford an opportunity for a hearing upon the happening of either of two events: (1) if the employer notifies the Secretary that he intends to contest a citation, or (2) if "any employee or representative of employees files a notice with the Secretary alleging that the period of time fixed in the citation for the abatement of the violation is unreasonable." 29 U.S.C. §659(c). If settlement is reached after either of these events occurs but prior to the scheduled hearing, the ALJ would have jurisdiction to review the settlement, but only for the limited purpose of entertaining objections from the employee or employees' representative that the abatement period proposed by the settlement is unreasonable. (at 1428–1429).

11. The Commission stated: "On May 29, 1980, the United States Court of Appeals for the Third Circuit issued its opinion in *Marshall v. Sun Petroleum Products Co. and OSHRC,* Docket Nos. 79–1822 and 79–1828 [8 OSHC 1422] (3d Cir. May 29, 1980) ('Sun Petroleum'). In that opinion, the Court held, among other things, that the Commission's authority to review settlement agreements in contested cases is limited to those instances in which employees or their authorized representatives express some interest in the case either by filing a notice of contest under §10(c) of the Act, 29 U.S.C. §659(c), or by electing to participate as parties under Commission Rule 20, 29 CFR 2200.20. The Court went on to hold that even in those cases in which the Commission does have the authority to review the proposed settlement agreements, our authority is limited to 'entertaining objections from the employee or employees' representatives that the abatement period proposed by the Settlement is unreasonable.' *Sun Petroleum,* slip op. at 20 [8 OSHC 1429]. I respectfully disagree with the Court's overly narrow view of the Commission's role in the final administrative disposition of contested enforcement actions under the Act."

12. On the application of any party, a Commissioner may order an expedited proceeding, and the case shall be placed on a special docket. The judge assigned shall make necessary rulings with respect to time and filing pleadings and all other matters,

without reference to time set forth in the Commission Rules, consistent with fairness. Commission Rule 101.

13. The Commission also held, with respect to the employer's argument of prejudice in its Section 8(e) walkaround rights, that "[r]espondent next argues that the compliance officer violated Sections 8(a) and 8(e) of the Act by taking photographs of the worksite prior to his presentation of credentials, and that, therefore, the photographs should have been excluded from the evidence. Respondent asserts that it was prejudiced by the taking of photographs before being given an opportunity to exercise its Section 8(e) walkaround rights. According to respondent, it was prejudiced by not being present because it could not rely on the compliance officer, who was the only non-employee who witnessed the violation, to testify on its behalf in any grievance or arbitration proceeding concerning disciplinary actions against the employees involved. Respondent also argues that the allegation by an 'outsider' of employee misconduct based on an incident the employer did not witness would tend to stir up labor-management disputes. We find respondent's arguments unpersuasive.

"In order to warrant the exclusion of evidence obtained during an inspection conducted in violation of Section 8(e) of the Act, there must be a showing that the Secretary's failure to comply with Section 8(e) caused respondent to suffer prejudice in the preparation or presentation of its defense. *Marshall v. Western Waterproofing Co., Inc.*, 560 F.2d 947, 5 OSHC 1732 (8th Cir. 1977); *Able Contractors, Inc.*, 5 OSHC 1975 (1977).

"Respondent's claim that the compliance officer's alleged failure to comply with Section 8(e) could create labor-management problems is not the type of prejudice contemplated by the Act. The Commission has consistently stated that prejudice under Section 8(e) of the Act must be of the type that prejudices the employer in the preparation or presentation of its defense. *Accu-Namics, Inc. v. OSHRC, supra; Able Contractors, Inc., supra.* Thus we conclude that the circumstances of this inspection do not warrant the exclusion of any evidence obtained therefrom.

"The Commission has stated that the rights granted by Section 8(a) are coextensive with those granted by the Fourth Amendment, and that an employer's Section 8(a) rights therefore are not violated unless the circumstances also show a violation of the Fourth Amendment. *Western Waterproofing Co., Inc.*, (1976), *rev'd on other grounds*, 560 F.2d 947, 5 OSHC 1732 (8th Cir. 1977)."

14. The Commission had decided issues regarding the validity of inspections, but not in the context of challenges to the Act's constitutionality. Rather, the issues it considered related to whether an employer exercised control and consented to an inspection or whether an alleged violation was in plain or public view. See *Environmental Utilities Corp.*, 5 OSHC 1195 (1977); *Western Waterproofing Co., Inc.*, 4 OSHC 1301 (1976).

15. In the case of proposed settlements or other dispositions by consent, petitions for discretionary review are ordinarily not allowed except upon a showing of good cause. Commission Rule 91(b)(4).

16. The doctrine of *res judicata*, of which collateral estoppel is a part, is stated in §1 of the Restatement of Judgments (1942): "Where a reasonable opportunity has been afforded to the parties to litigate a claim before a court which has jurisdiction over the parties and the cause of action, and the court has finally decided the controversy, the interests of the State and of the parties require that the validity of the claim and any issue actually litigated in the action shall not be litigated again by them." *Accord, Brown v. United States*, 570 F.2d 1311, 1320 (7th Cir. 1978). The underlying policy of the doctrine was articulated by Lord Justice Blackburn over a century ago: "The object of the rule of *res judicata* is always put upon two grounds—the one public

policy, that is the interest of the State that there should be an end of litigation and the other, the hardship on the individual, that he should be vexed twice for the same cause." *Lockyer v. Ferryman,* 2 App. Cas. 519,530 (1877). Earlier this century the courts were unwilling to apply collateral estoppel to administrative determinations. *Pearson v. Williams,* 202 U.S. 281, 284–85 (1906). The courts slowly eroded this position so that by 1966 the Supreme Court could state that "[w]hen an administrative agency is acting in a judicial capacity and resolves disputed issues of fact properly before it which the parties have had an adequate opportunity to litigate, the courts have not hesitated to apply res judicata to enforce repose." *United States v. Utah Construction & Mining Co.,* 384 U.S. 394, 422 (1966). More recently, courts have expanded the application of collateral estoppel to serve the principle that one opportunity to litigate an issue fully and fairly is enough.

17. In the case, Continental argued that the only engineering controls that are technically feasible are machine enclosures and that it litigated the economic feasibility of these enclosures in a previous case, *Continental I.* The court agreed and stated: "Because the Commission held that they were not feasible, the Secretary should be estopped from relitigating the issue for each of Continental's plants. To analyze the cogency of this argument, we must determine whether the issue decided in *Continental I* is the same as that raised in the subsequent cases, whether the issue was actually litigated, whether the decision in *Continental I* depended on the resolution of the issue, and whether that decision was final. *IB Moore's Federal Practice* ¶0.441[2] (2d ed. 1974); *Restatement of Judgments* §68(1942)."

18. In *Noblecraft Industries, Inc. v. Secretary of Labor,* 614 F.2d 199, 7 OSHC 2059 (9th Cir. 1980), the court stated that validity of standards may be challenged in the review of enforcement orders under Section 11(a) of the Act, that the sixty-day time limitation of Section 6(f) of the Act applies only to preenforcement review. See also *Atlantic and Gulf Stevedores Inc. v. OSHRC,* 534 F.2d 541, 4 OSHC 1061 (3d Cir. 1976); but see *National Industrial Constructors, Inc. v. OSHRC,* 583 F.2d 1048, 6 OSHC 1914 (8th Cir. 1978).

19. Clearly erroneous is defined as: "although there is evidence to support it, the reviewing court on the entire evidence is left with a definite and firm conviction that a mistake has been committed." See *United States v. United States Gypsum Co.,* 333 U.S. 64 (1948).

APPENDIX 1 to CHAPTER XIII

Diagram of the Enforcement & Hearing Process

Promulgation of Standards

Consensus or Established Federal Standards

Permanent Standards, Emergency Temporary Standards

Variance Application (Optional)

OSHA Inspection—(Jurisdictional Questions)
General Schedule,
Follow-up,
Accident, Injury,
Employee Complaint
or
Fatality

Citation

Informal Conference (Optional)

Federal OSHA
Appeal to Federal District Court
e.g. Warrant

State OSHA
Appeal to State Court
e.g. Warrant

Citation Un-Contested (Final Order)

Notice of Contest/ Petition for Modification of Abatement (to Area Director)/

Notice of Contest/ Abatement Petition (to State Board of Appeals or State Commissioner of Labor/Industry)

Citation Failure to Abate

Prehearing Conference (Optional)

Hearing before Administrative Law Judge

Prehearing Conference (Optional)

Hearing before State Hearing Officer

Review by Review Commission	Review by a State Board of Appeals or State Commissioner of Labor/ Industry
Interlocutory Appeal	
Appeal to a United States Circuit Court of Appeals	Appeal to a State Court of Appeals
Appeal to the United States Supreme Court	Appeal to State Supreme Court

CHAPTER XIV
The Act and Civil Liability

A. INTRODUCTION

The use of OSHA standards, regulations, and citations as evidence in civil litigation, outside of the mechanisms specified in the Act itself, has manifested itself in a plethora of ways that can be broken down into two principal areas. First and foremost, the issue has arisen as to whether an OSHA violation constitutes proof of negligence or negligence *per se,* or indeed gives rise to an independent cause of action for damages implicitly based upon the Act itself. OSHA litigation has made inroads into other common law theories of civil liability as well. As for the second major area, a large number of states have worker's compensation statutes—either in the form of an exclusive remedy or one that is concurrent with common law actions sounding in tort. The effect of an OSHA violation in this important area has also not been insubstantial.

An employer who receives an OSHA citation has two basic courses open to him under the Act. He can pay the penalty and not contest the citation, or he can contest the citation and go to a hearing. In determining which of these two courses to follow, an employer will normally consider such things as the nature of both the violation and standard; the characterization of the violation (serious, other-than-serious, repeated, willful); the amount of the penalty assessed; monetary liability and possible injunctive relief; the monetary expense and practical consequences of attempting to abate the violation; and any available substantive or procedural defenses as well as the costs to be incurred in attempting to establish them. Particularly in view of the issues that will be developed in this chapter, however, it is suggested that another factor should be added to this equation. Namely, what will be the legal effect of the employer's decision to contest or not contest an OSHA citation with respect to potential common law civil liability?

B. COMMON LAW THEORIES OF TORT LIABILITY

1. Definition of Tort

Webster's New Collegiate Dictionary defines the term "tort" as "a wrongful act for which a civil action will lie except one involving a breach of contract."

Black's Law Dictionary defines tort as a "private or civil wrong or injury." Negligence is but one aspect of tort law, and has as its necessary elements to support a cause of action therefore: (1) a duty or obligation recognized by the law requiring one to conform to a standard of conduct; (2) a failure to conform to the standard required or breach of the duty; (3) a causal connection between the conduct and a resulting injury (proximate cause); and (4) actual loss or damage. Prosser, *The Law of Torts,* §30 (4th Ed., 1971).

At common law, the standard of conduct that is normally applied in negligence cases is that of the reasonable man—one of ordinary prudence, as measured by an objective rather than a subjective standard. *Reynolds v. City of Burlington,* 52 Va. 300, 308 (1880); *Hennessey v. Chicago & N.W. RR. Co.,* 99 W.S. 109 (1898). This mythical "reasonable man" has a duty or obligation to perform in such a manner as the ideal, prudent, and careful man would act in the same situation. An individual who fails to measure up to this standard, *i.e.,* who fails to act as a prudent and reasonable man, may be liable for injuries inflicted as a proximate result of his negligent acts. See 57 Am Jur. 2d *Negligence* §§66–89 (1971); *Restatement of The Law of Torts 2d,* §282 (1965); Prosser, *The Law of Torts,* §§32–33 (4th Ed. 1971); *Cf. B & B Insulation, Inc. v. OSHRC,* 583 F.2d 1364, 6 OSHC 2062 (5th Cir. 1978); *Diebold, Inc. v. Marshall,* 585 F.2d 1327, 6 OSHC 2002 (6th Cir. 1978); *Voegele Co. v. OSHRC,* 625 F.2d 1075, 8 OSHC 1631 (3d Cir. 1980); *General Dynamics v. OSHRC,* 599 F.2d 453, 7 OSHC 1373 (1st Cir. 1979); *Brown & Root, Inc., Power Plant Division v. OSHRC,* 590 F.2d 1363, 7 OSHC 1137 (5th Cir. 1979), which have applied the reasonable man concept in the context of industry custom and practice in OSHA cases involving vagueness challenges to standards.

In many cases, the reasonable man standard is established by statute or legislative enactment. In such cases, the existence of an OSHA standard, rule, or regulation may be introduced either as evidence of negligence or to irrefutably establish negligence *per se.*[1] This is true, regardless of whether the statute is civil or penal in character. In order to invoke the negligence *per se* doctrine, however, a plaintiff must show that the statute was intended to protect the class of persons of which the plaintiff is a member against the type of harm from which he has suffered. Furthermore, authority exists for the proposition that the violation of a statute that merely sets forth a general rule of conduct rather than a specific requirement either to act or not to act in a certain way cannot be considered as negligence *per se.* This proposition would be particularly important, for example, if the OSHA violation introduced as evidence in a civil lawsuit involved the "general duty clause." See, 57 Am. Jur. 2d *Negligence* §§234–251 (1971); Miller, *The Occupational Safety and Health Act of 1970 and the Law of Torts,* 38 Law and Contemporary Problems 174 (1974); Comment, *Occupational Safety and Health Act of 1970: Its Role in Civil Litigation,* 28 S.L.F. 999 (1974); Note, *Statutory Negligence,* 27 Drake L. Rev. 178 (1977); Morow, *Use of OSHA in Negligence Suits Against Those Responsible For Maintenance of Safe Work Sites,* 22 Trial Lawyers' Guide 167 (1978).

Administrative regulations have been admitted into evidence in order to prove the standard of conduct of a reasonable man, that is, the standard of care owed by one person to another. Prosser, *The Law of Torts,* §36 (4th Ed.

1971); *Restatement of the Law of Torts 2d,* §286 (1965); *Delta Airlines, Inc. v. United States,* 561 F.2d 381 (1st Cir. 1977). Normal requirements for the use of statutes, standards or regulations in negligence actions require that: (1) the regulation or statute was in effect at the time of the accident—*Bell v. Buddies Supermarket,* 516 S.W.2d 447, 448 (Tex. Civ. App. 1974); *Gubalki v. Estate of Anthese,* 189 Neb. 385, 202 N.W.2d 836 (1972); (2) that a causal connection existed between a violation of the regulation and the injury—*Vines v. Plantation Motor Lodge,* 336 Southern 2d 338 (Ala. 1976); and (3) the plaintiff was among the class of persons protected by the statute or regulation —*Betesh v. United States,* 400 F. Supp. 238 (D.D.C. 1974). It is also possible to introduce administrative standards and regulations to support a defendant's position in civil actions in order to show actions were reasonable or that a standard of care was met. *Spangler v. Kranco, Inc.,* 481 F.2d 373 (4th Cir. 1973); *Hagerman v. Signal L.P. Gas, Inc.,* 486 F.2d 479 (6th Cir. 1973).

The use of a statute or regulation even when used to support a finding of negligence *per se* does not automatically result in a finding for a plaintiff—it only results in establishing common law negligence. There would still be a requirement of proof of a causal connection and the defense would be entitled to raise defenses of contributory negligence or assumption of risk. See Prosser, *The Law of Torts,* §36 (4th Ed. 1971).

2. The Law of Negligence

In many cases decided since the enactment of the Act, both state and federal courts have held that the existence (and violation) of occupational safety and health standards, rules, or regulations may be material, relevant, and admissible in private litigation as evidence of negligence, in order to establish a standard of care (and its breach), or indeed to irrefutably establish negligence *per se* (even in cases where no employment relationship is involved). In many cases, where evidence of an OSHA standard, regulation, or rule is introduced, an OSHA violation or citation may also be introduced but would probably not be dispositive of the matter on either *res judicata* or collateral estoppel principles. It would also seem to be irrelevant, prejudicial, and immaterial. OSHA standards and regulations have, however, been admitted into evidence, and the courts have reached widely divergent results with respect to their effect. Some courts hold that OSHA violations show some negligence, some courts hold they prove negligence *per se,* and others say they are irrelevant or even inadmissible. *Note:* A cause of action has not been allowed based solely on the Act. See, *e.g., Schroeder v. C. F. Braun & Co.,* 502 F.2d 235 (7th Cir. 1974)—an Illinois safety rule on ladders (similar to the federal safety standard) was admissible as evidence of negligence, since it helped to establish the "standard of care" to be exercised by the employer; *Hanson & Orth, Inc., v. M/V Jalatarang,* 450 F. Supp. 528, 537 (S.D. Ga. 1978); *Lacaze v. Olendorff,* 526 F.2d 1213, 1220 (5th Cir. 1976)—OSHA provisions and regulations may be admissable to determine the standard of care that should have been followed; *Gallardo v. Westfal-Larsen & Co.,* 435 F. Supp. 484 (N.D. Cal. 1977); *Buhler v. Marriott Hotels, Inc.,* 390 F. Supp. 999, 1000 (E.D.

La. 1974); *Duncan v. Pennington County Housing Authority,* 283 N.W.2d 546 (S.D. 1979); *Kraus v. Alamo Nat'l Bank of San Antonio,* 586 S.W.2d 202 (Tex. Civ. App. 1979)—OSHA standards admissible as evidence even though plaintiff was not within the Act's protected class; *Ceco Corp. v. Maloney,* 404 A.2d 935 (D.C. App. 1979); *Knight v. Burns, Kirkley & Williams Constr. Co.,* 331 So. 2d 651 (Ala. 1976); *and Michel v. Valdastri, Ltd.,* 59 Hawaii 53, 575 P.2d 1299 (1978). *Accord, Scrimager v. Cabot Corp.,* 23 Ill. App.3d 193, 318 N.E.2d 521 (1974)—the failure of the owner of a warehouse to have a scaffold or catch platform erected at the edge of a roof, in violation of state safety and health regulations, was held admissible as evidence of negligence in an action by an employee of a painting subcontractor who was injured when he fell off the roof. See also *Barger v. Mayor & City Council of Baltimore,* 616 F.2d 730, 8 OSHC 1114, 1117 (4th Cir. 1980)—"the [OSHA] noise regulation was a codification of industry practices, practices by which the city could fairly be judged."; Note, *Use of OSHA in Products Liability Suits Against the Manufacturers of Industrial Machinery,* 11 Val. U. L. Rev. 37 (1979).

In *Arthur v. Flota Mercante Gran Centro Americana S.A.,* 487 F.2d 561 (5th Cir. 1973), the Fifth Circuit held that the violation of a federal safety standard was admissible to establish negligence *per se,* since the plaintiff had shown that the standards violated were designed to protect persons such as himself against the type of risk created. See also *Watwood v. R. R. Dawson Bridge Co.,* 293 Ala. 578, 307 So.2d 692 (1975)—violation of an OSHA standard concerning the guarding of a power tool supplied by an employer who was a subcontractor on a construction job constituted evidence of negligence in a negligence action against the general contractor; *Kelley v. Howard S. Wright Constr. Co.,* 582 P.2d 500, 508 (Wash. 1978)—violation of an OSHA safety net statute constituted negligence *per se; Davis v. Crook,* 261 N.W.2d 500 (Iowa 1978); *Bachner-Northwest v. Rich,* 554 P.2d 430 (Alaska 1976); *Dunn v. Brimer,* 259 Ark. 859, 537 S.W.2d 164, (1976); *Knight v. Burns, Kirkley & Williams, Constr. Co.,* 331 So.2d 651, 4 OSHC 1271 (Ala. 1976); *Hanson & Orth, Inc. v. M/V Jalatarang,* 450 F. Supp. 528 (S.D. Ga. 1978); *National Marine Service, Inc. v. Gulf Oil Co.,* 433 F. Supp. 913 (E.D. La. 1977), *aff'd,* 608 F.2d 522 (5th Cir. 1980); *DiSabatino Brothers v. Baio,* 366 A.2d 508, 4 OSHC 1855 (Del. 1976)—OSHA standards measure the conduct of the reasonable man; *Bowlus v. The North-South Supply Co.,* 1975–76 OSHD ¶20,409 (Ky. Sup. Ct. 1976); *Koll v. Manatt's Transp. Co.,* 253 N.W.2d 265, 5 OSHC 1398 (Iowa 1977); *Rosalee Taylor v. Moore-McCormack Lines, Inc.,* 621 F.2d 88, 90, 8 OSHC 1277 (4th Cir. 1980)—OSHA regulations have been repeatedly held to impose an affirmative duty on the stevedore, the violation of which would be negligence. See also *Barger and Iwanico v. Mayor & City Council of Baltimore,* 616 F.2d 730, 8 OSHC 1114 (4th Cir. 1980).

In *Northern Lights Motel v. Sweeney,* 561 P.2d 1176, 1183 (Alaska 1977), the Alaska Supreme Court reiterated four prerequisites to the use of general safety regulations to establish negligence *per se.* Thus to establish negligence *per se,* the purpose of the regulation must be:

(1) To protect a class of persons which includes the one whose interest is invaded;

(2) To protect the particular interest which is invaded;
(3) To protect the interest against the kind of harm which has resulted; and
(4) To protect that interest against the particular hazard from which the harm results.

This test is taken from the *Restatement of the Law of Torts 2d,* §268 (1965); See also *Bachner v. Rich,* 554 P.2d 430, 443 (Alaska 1976). Once these criteria are met, the court in its discretion may adopt the regulation as the applicable standard of conduct.

Recent decisions cast doubt, however, on the applicability of the negligence *per se* approach to the use of OSHA violations. See *Chavis v. Finnlines, Ltd., O/Y,* 576 F.2d 1072, 1082 (4th Cir. 1978), which indicates that *Flota Mercante, supra,* is no longer good law; *Dravo Corp. v. OSHRC,* 613 F.2d 1227, 1230 (3d Cir. 1980), which indicates that *Flota Mercante, supra,* is no longer good law; *Bell v. Buddies Super-Market,* 516 S.W.2d 447 (Tex. Civ. App. 1974)—evidence of a violation of an OSHA safety standard concerning ramps—promulgated twenty months after the accident—was not admissible as evidence of negligence in an action by a customer against a supermarket. See also *Pruett v. Precision Plumbing, Inc.,* 27 Ariz. App. 288, 554 P.2d 655 (1976)—evidence of the substance of the OSHA standard was excluded where the defendant was not the plaintiff's employer on the basis that no duty to the plaintiff exists and that any evidence of the standard of care was therefore immaterial; *Otto v. Specialties, Inc.,* 386 F. Supp. 1240, 2 OSHC 1424 (N.D. Miss. 1974); *Clary v. Ocean Drilling and Exploration Co.,* 429 F. Supp. 905 (W.D. La. 1977), *aff'd.,* 609 F.2d 1120 (5th Cir. 1980); *Macey v. United States,* 454 F. Supp. 684 (D. Alaska 1978)—OSHA regulations may not be used to establish liability where the injured party is not an employee nor on the premises of the defendant for a business purpose.

Finally several courts have not permitted the introduction of OSHA standards or violations as evidence, finding them irrelevant. *Jasper v. Skyhook Corp.,* 89 N.M. 98, 547 P.2d 1140 (1976), *rev'd on other grounds,* 90 N.M. 143, 560 P.2d 943 (1977)—they are irrelevant and inadmissable; *Cochran v. International Harvester Co.,* 408 F.Supp. 598 (W.D. Ky. 1975); *Otto v. Specialties, Inc.,* 386 F. Supp. 1240 (N.D. Miss. 1974); *Gonzales v. R. S. Novick Constr. Co.,* 70 Cal. App.3d 131, 139 Cal. Rptr. 113 (1977)—OSHA standards are inadmissible.

3. Union Liability

Employees, as private litigants, have often sought to sue their unions in negligence actions for allegedly failing to discover and correct safety hazards in the workplace. For example, in *Higley v. Disston, Inc.,* 1975–76 OSHD ¶20,689 (Super. Ct. Wash. 1976), a worker sued his union for alleged negligence in conducting safety inspection tours, as provided in the collective bargaining agreement. Holding that state law did not create a duty on the part of the union to ensure a safe workplace for its members, the court dismissed the employee's cause of action. In this respect, the court expressed particular concern that any such duty would discourage unions from participating in walkaround inspections or negotiating safety standards in collective bargaining agreements, and would thereby contravene national labor policy. See also *House v. Mine Safety Appliances Co.*, 417 F. Supp. 939 (D. Idaho 1976); *Wentz v. International Brotherhood of Electrical Workers*, 578 F.2d 1271 (8th Cir.), *cert. denied*, 439 U.S. 983 (1978)—failure to enforce safety provisions of contract; *Gerace v. Johns-Manville Corp.*, 95 LRRM 3282 (Pa. Ct. of Common Pleas 1977); *Brooks v. New Jersey Mfr's Ins. Co.*, 405 A.2d 466 (N.J. App. Div. 1979) —union's right to appoint members to a company safety committee did not give rise to damage action when members failed to report unsafe conditions; *contra, Hake v. Helton*, 564 S.W.2d 313, 98 LRRM 2905 (Mo. App. 1978), *cert. denied*, 439 U.S. 959 (1978)—union was held liable for negligently failing to enforce a safety rule set forth in a collective bargaining agreement; *Helton v. Hake*, 386 F. Supp. 1027 (W.D. Mo. 1974). A possible avenue of relief against unions in the event a negligence action proved fruitless would, however, be under Section 301 of the Labor Management Relations Act for breach of the duty of fair representation. *Powell v. Globe Industries, Inc.*, 431 F. Supp. 1096, 5 OSHC 1250 (N.D. Ohio 1977); *Bryant v. UMW*, 467 F.2d 1 (6th Cir. 1972). See also *Safety in the Workplace: Employee Remedies and Union Liability;* 13 Creighton L. Rev. 955 (1980).

4. Common Law Defenses

Defenses available against plaintiffs in a civil action for negligence may not be available in an OSHA case. Contributory negligence, assumption of risk, and comparative negligence are types of defenses that are not available in an OSHA case[2] but which can be used in a civil suit where an OSHA violation is alleged as the basis for or proof of negligence. As a general rule courts, in interpreting statutes that set specific standards, have barred defendants from asserting the defense of assumption of risk. They have, however, permitted defendants to assert contributory negligence defenses in situations where specific statutory mandates have been violated. *Hewitt v. Safeway Stores, Inc.*, 404 F.2d 1247 (D.C. Cir. 1968). See also *Bertholf v. Burlington Northern R.R.*, 402 F. Supp. 171 (E.D. Wash. 1975). The counterpart in an OSHA contest is employee misconduct or isolated incident. *National Realty & Construction Co. v. OSHRC*, 489 F.2d 1257, 1 OSHC 1422 (D.C. Cir. 1973); *Brennan v. OSHRC and Raymond Hendrix, d/b/a Alsea Lumber Co.*, 511 F.2d 1139, 2 OSHC 1646 (9th Cir. 1975).

5. Use of Exculpatory Language

Arguably, even uncontested citations which are final Review Commission orders could be admissible in evidence during the course of a negligence trial as either an admission or a statement against interest. Although the Act does not provide for a plea in the nature of a *nolo contendere*, the prudent employer who does not elect to contest a citation should attempt to obtain a nonadmission of liability statement from the Area Director. Other federal agencies in the labor area, such as the NLRB and the EEOC, have utilized the nonadmission of liability clause as an effective tool in reducing their litigation caseload and obtaining speedy compliance. Use of this device under OSHA would achieve this same goal from the standpoint of the agency and the public, yet would simultaneously protect the employer from inadvertently making damaging admissions that could be used against it in collateral proceedings. The Review Commission has in fact recently held that settlement agreements containing exculpatory language are not contrary to the purposes of the Act. *Farmers Export Co.*, 8 OSHC 1655 (1980); *Missouri Farmer's Association, Inc.*, 8 OSHC 2011 (1980).

Damaging admissions may take other forms as well. A good example of a situation where admissions made in OSHA proceedings *did* come back to haunt an employer in subsequent collateral proceedings is *Moore v. Allied Chemical Corp.*, 480 F. Supp. 377 (E.D. Va. 1979). *Moore* arose out of the kepone catastrophe that took place in Hopewell, Virginia, during the mid-1970s. Life Science Products Company and its corporate officers had been charged with willful violations of the Act in connection with that situation, and had admitted in pleadings before an administrative law judge that they were responsible for those willful violations. The administrative law judge incorporated these admissions into his findings of fact, which were not challenged by either Life Science or its corporate officers before either the Review Commission or the Court of Appeals. Subsequently, a Life Science corporate officer sued Allied Chemical Corporation, the company that supplied the raw material for processing the kepone to Life Science. This lawsuit was based, *inter alia,* upon the theory that Allied Chemical intentionally withheld information on the toxicity of kepone from Life Science—thereby preventing Life Science from providing a safe workplace for its employees. Accordingly, the Life Science corporate officer claimed that Allied Chemical had intentionally inflicted emotional distress upon him and was "liable for the emotional distress *which he suffered* from such injuries, suits, and penalties [as arose from the kepone catastrophe and the OSHA litigation]." *Moore, supra* at 385 [Emphasis in the original]. The trial court, however, dismissed this claim on the ground that the Life Science corporate officer was collaterally estopped from asserting that toxicity information was withheld from him by Allied Chemical by virtue of the admissions that he had made during the OSHA litigation. Thus by admitting in the OSHA litigation that he was responsible for willful violations of the Act, this corporate officer was held to be barred, as a matter of law, from claiming a lack of "knowledge of the toxicity of kepone" in the subsequent lawsuit. *Id.* at 386.

6. Miscellaneous

Conceivably, a violation of an OSHA standard could also be used as proof of a negligent or intentional nuisance, trespass, or unreasonable interference with the property interests of an individual in the use or enjoyment of his land. See 58 Am. Jur. 2d *Nuisance* §136 *et seq.* (1970). No reported decisions, however, have addressed this issue.

C. THE LAW OF CONTRACTS, PRODUCT LIABILITY, AND SALES

The effect of contract language which provided that a contractor shall comply with federal and state safety and health laws, provide safety equipment and devices, and be liable for damages, was found to be too general and insufficient to support a *negligence* action. *Macey v. United States,* 454 F. Supp. 684 (D. Alaska 1978). However, in *Horn v. C. L. Osborn Contracting Company,* 591 F.2d 318, 7 OSHC 1256 (5th Cir. 1979), contract language was found to have created additional obligations and duties which were actionable by their breach. The contract incorporated and required the contractor to comply with the Act's regulations, although not the Act itself. A duty was imposed by virtue of the express terms of contract; thus a private right of action based thereon was permitted.[3] The Commission in *Electric Smith, Inc.,* 8 OSHC 1388 (1980) held that evidence that an employer had entered into contract with a general contractor which relieved it of responsibility for correcting violative conditions and that it had warned the general contractor repeatedly of existence of hazardous conditions did not justify vacation of citations for violations of the Act, since an employer's obligation to provide safe working conditions could not be shifted by contract.

There are several actions which are conceivable against manufacturers of products, materials, machinery, or equipment which have failed or were defective and in which OSHA standards, rules, and requirements may be important. Potential liability could arise under the laws of contracts and of sales, for example, with respect to those companies that supply Material Data Safety Sheets to customers or provide them to their employees for a description of hazardous materials. See Chapter VII of the book for a description of Material Data Safety Sheets. Employees barred from suing their employers because of worker's compensation laws may be able to sue the manufacturer or seller of a product used in the workplace if it caused an injury or illness. To qualify for such relief the employee must show a "defect" in the product; Material Safety Data Sheets may provide evidence of the defect.

Material Data Safety Sheets are required to be prepared and utilized by employers in the ship repairing, shipbuilding, and ship-breaking industries (29 CFR 1915, 1916, and 1917). Employers in the referenced industries are prohibited from using any hazardous chemical product or process material unless and until the employer has ascertained the potential fire, toxic, and reactivity hazards which are likely to be encountered in the handling, application, or utilization of such material. In order to ascertain the hazards, employers in the above industries are required to obtain certain types of information applicable to the product or material being used. This information must be recorded on the Material Safety Data Sheet or on a substantially similar form which has been approved by OSHA.

The requirement to obtain a Material Safety Data Sheet is imposed only upon employers and not necessarily on the manufacturer of the product or material unless it also has employees exposed to the product. However, since the employers in the above-listed industries are required to obtain the information contained on the Material Safety Data Sheet, they will of necessity have to request and obtain such information from the manufacturer. As a practical matter, therefore, if the manufacturer wishes to sell the product or material to the employer, it will have to supply the required "hazard" information. Furthermore, at least one recently enacted New York statute, euphemistically known as a "right-to-know" statute on toxic materials, requires manufacturers of toxic substances to provide substantial information on those materials. This information includes the generic and trade name of the material, the level of exposure that is deemed hazardous, acute and chronic effects of exposure as well as its symptoms, recommended emergency treatment, and the recommended procedures for safe use and for the cleanup of spills and leaks. See 134 BNA Daily Labor Report at A-7 (July 10, 1980).

Apart from any state "right-to-know" statute, most employers request Material Safety Data Sheets, even though they are not required to do so. Employers are increasingly concerned about the potential hazards of the products and materials to which their employees are exposed. Manufacturers —if they wish to continue to do business with their customers—will normally supply such requested information. For even though Material Safety Data Sheets may not be required of the employer by OSHA, an employer still could be liable and possibly in violation of the Act if he exposed his employees to a product or substance which was hazardous. The situation would be quite serious if the employer knowingly failed to ascertain whether or not the product or substance was hazardous. Therefore it is encumbent upon an employer who is working with hazardous substances to find out their ingredients, their physical data, their reactivity data, their health hazard data, toxicity hazards, and any special protection information or precautions which are applicable. This information would of course be contained on the Material Safety Data Sheet.

The Toxic Substances Control Act, 15 U.S.C. §2601 *(et seq.)* (TSCA), does not regulate or require the use of the Material Safety Data Sheet. Depending upon the chemical substance or mixture, the EPA may, however, require other forms of reporting, recordkeeping, or control procedures under TSCA. None of these must be supplied to the customers of the manufacturer, processor, or distributor.

Under commercial law, contract codifications (sales), sellers of goods owe certain duties to buyers of goods. Express warranties by a seller may be created by an affirmation of fact or promise made by the seller to a buyer which relates to the goods and becomes a basis of the bargain. Any description of the goods which is made a part of the basis of the bargain creates an express warranty that the goods shall conform to their description. Thus the Material Safety Data Sheet or information provided by a manufacturer under a state "right-to-know" statute, by virtue of the fact that it contains a description of the goods and an affirmation of fact, may create an *express warranty* that the goods will conform to the information contained therein. Any deviation from an infirmity in that description could create a cause of action on the part of the buyer against the seller *(e.g.,* defectiveness in process, inspection, testing, labeling, or warning).

Even were the Material Safety Data Sheet or similar state-required material not sufficient to create an express warranty, it is possible that a *warranty of merchantability* may be implied if the seller is a merchant as to goods of that kind. Such a warranty may also arise in the course of dealing or usage of trade. For instance, goods, to be merchantable, must be adequately contained, packaged, and labeled and must conform to the promises or affirmation of facts made on the labels *(e.g.,* the Material Safety Data Sheet). An implied *warranty of fitness for particular use* may also arise where the seller at the time of contract had reason to know any particular purpose for which the goods were required and that the buyer was relying upon the seller's skill or judgment (as set forth on the Material Safety Data Sheet) to select or furnish suitable goods.

Breach of any of these warranties may give rise to a cause of action by a buyer or third party against the seller, assuming that an accident, injury, or illness resulted from the use of a product. See *Moore v. Allied Chemical Corp.,* 480 F. Supp. 364 (E.D. Va. 1979)—breach of warranty lawsuit against seller for failure to warn buyer of the dangers of kepone toxicity. Moreover, the seller's warranty is applicable and extends to persons affected by the goods, such as employees of the buyer, if it is reasonable to expect that such persons will use or be affected by the goods and if such persons are in fact injured by breach of the warranty. Thus there could possibly be a warranty cause of action against the manufacturer, processor, or distributor of goods by the employees who are exposed to the substance. *Cf. Borel v. Fibreboard Paper Products Corp.,* 493 F.2d 1076 (5th Cir. 1973), *cert. denied,* 419 U.S. 869 (1974) —industrial insulation worker contracted asbestosis, sued the manufacturer, and the court upheld liability on the basis it breached its duty to warn of the foreseeable dangers associated with using asbestos; *Scott v. Dreis & Krump Mfg. Co.,* 326 N.E.2d 74 (Ill. App. 1975)—employee of owner of press brake sued manufacturer on strict products liability theory and the manufacturer attempted to introduce the OSHA press-brakes standard, arguing that the employer should have installed safety devices, but the court held the manufacturer's duty to produce a safe product was not delegable; *Heinrich v. Goodyear Tire & Rubber Co.,* C.A. No. M 80–1956 (D. Md. 1980); *compare, Spangler v. Kranco, Inc.,* 481 F.2d 373 (4th Cir. 1973)—manufacturer of crane not liable for injuries sustained by an employee since OSHA did not require the warning devices he claimed were necessary; *Jasper v. Skyhook Corp.,* 89 N.M. 98, 547 P.2d 1140 (Ct. App. 1976), *rev'd on other grounds,* 90 N.M. 143, 560 P.2d 534 (1977)—OSHA standards irrelevant; *Bowlus v. North-South Supply Co.,* 1975–76 OSHD ¶20,409 (1976).

In order to be eligible for recovery in such an action, a plaintiff must show that there was an injury, a defective product, and a causal relationship between the two. *Restatement of the Law of Torts 2d,* §431 (1965). These same requirements apply regardless of whether a product liability action proceeds under a theory of breach of warranty, strict liability, or negligence. *Hemingsen v. Bloomfield Motors, Inc.,* 32 N.J. 358, 161 A.2d 69 (1960); *Greenman v. Yuba Power Prods., Inc.,* 59 Cal. 2d. 57, 377 P.2d 897 (1963).

Negligent use of the product by the purchaser (employer) could also be the basis of an indemnity action by the manufacturer, but at least one court has held that OSHA creates no cause of action against an employer by the manufacturer arising out of a products liability action brought by an employer. *Davis v. Niagara Machine Co.,* 1978 OSHD ¶22,965 (Wash. 1978).

In order to exclude or disclaim an implied warranty of merchantability, the seller must prepare and communicate language to that effect mentioning merchantability. The disclaimer should be in writing and it should be conspicuous. In order to exclude a warranty of fitness, the disclaimer must also be in writing and conspicuous. An example would be: "there are no warranties which extend beyond the description on the face hereof." It is more difficult to exclude or limit express warranties, and usually a disclaimer to this effect will be denied effect when inconsistent with the language of the express warranty.

Whether the act of selling or supplying a product amounts to negligence depends both upon the resulting injury and upon the danger or injury which might reasonably have been expected to result from the sale. The normal rule of negligence is that a manufacturer or seller of goods or products is liable for any injury sustained in connection with such sale as a consequence of a breach of duty which he owed to the injured party. A determination of whether a company may be liable under a negligence theory to either a buyer of its products or to an affected person such as an employee would depend upon an analysis of such familiar tort-based questions as: the existence of a duty of care, foreseeability, proximate cause, breach of the duty, damages, and injuries.

It must also be remembered that liability in a negligence sense is not necessarily dependent upon preexisting privity in a legal relationship between the person injured and the person causing the injury. Traditionally, there has been a rule that no cause of action in *tort* (negligence) can arise from the breach of a duty existing by virtue of a *contract* unless there is between the person causing the injury and the person injured what is termed "privity of contract."*McPherson v. Buick Motor Co.,* 217 N.Y. 382, 111 N.E. 1050 (1916). This means that a plaintiff in a negligence action who bases his suit upon the theory of a duty owed to him by a defendant as a result of a contract must be a party or privy to that contract. Otherwise, he fails to establish a duty toward himself on the part of the defendant and fails to show any wrong done to himself. Although the traditional rule requires privity of contract in order to recover in tort from the breach of duty existing by virtue of the contract, certain exceptions have been engrafted onto this rule, the most notable of which is the one specified under commercial law in the section on sales which was discussed previously. The restriction of privity of contract has also been eliminated in certain other situations, and thus the requirement of privity in negligence suits against manufacturers and sellers of products is gradually vanishing. For example, no privity of contract is required to support liability for negligence with respect to acts or circumstances which are imminently dangerous, *e.g.,* sales of hazardous substances. In some situations, therefore, one may be held liable for negligently supplying inaccurate information to another person, which caused an injury, despite the fact that there is no privity of contract between plaintiff and defendant. The liability in such instances is independent of privity of contract and depends merely upon the duty of every man to act so as not to injure the persons or property of others. Thus negligence in the preparation, supply, and distribution of the Material Safety Data Sheet or similarly required material *(e.g.,* inaccurate or incomplete information) could expose a company to liability not only from its purchasers but also from employees of its purchasers.

According to a recent article entitled *Compensating Victims of Occupational Disease,* 93 Harv. L. Rev. 916 (1980), to qualify for recovery, an "injured or ill" employee must show a defect, error, or omission in the product or its descriptive literature:

> [A] manufacturer's failure to give an adequate warning of the risks of a product makes it defective. The duty to warn extends to all risks known or reasonably foreseeable at the time of sale. While some courts have extended this duty to both buyers and users of the product, others have limited the manufacturer's duty to providing cautionary information only to the purchaser of the product (the employer). Under the latter view, "an employee who is injured because he was never warned about a hazard may be prevented from recovering against a manufacturer." This rule might make sense in industrial accident cases, since the employer has good reason to warn employees of the hazards of a product that poses the risk of an accident.

It must be remembered, however, that affirmative defenses—if applicable—(at least in a negligence context) may provide a relief from liability. Employees may be considered to have assumed the risk of working in a workplace exposed to hazard, or contributing to the situation by virtue of his own actions *(e.g.,* smoking and its synergistic reaction to certain toxic substances such as asbestos).

Disclaimers or exculpatory provisions exempting persons from liability for negligence have often been held to be invalid under both statutory restrictions and contrary to public policy, especially in the contract or product liability areas. It has also been stated that one cannot avoid liability for negligence by contract, especially with respect to non-parties to that contract. Therefore, it is unlikely that an exculpatory provision in a contract will be enforceable, but this would, of course, depend upon the circumstances of each particular case. Alternatives to a broad exculpatory provision may be to have an agreement requiring the buyer to assume the risk or to indemnify the seller if he incurs liability *(e.g.,* mishandling of the product, improper use, or use not in conformity with the use set forth by the seller).

A company which is a seller, distributor, or processor of a product may be held liable for injuries caused by the sale or use of its product under a strict liability theory, where the danger in its activity greatly threatens others. In such a situation a company may be held liable merely because, as a matter of social adjustment, the responsibility should be placed on the company. Hazardous enterprises, even though socially valuable, must pay their way and make good the damage inflicted. In short, they insure against the consequences of their own conduct because of the creation of an undue risk of harm. See generally *Restatement of the Law of Torts 2d,* § §519—524A (1965). Depending upon the company's product, it may be liable under a strict liability theory, and the information contained in the Material Safety Data Sheet or similar state-required material may serve as an affirmation that its activities and products are hazardous.

D. PRIVATE CAUSES OF ACTION

The federal courts have uniformly held to date that the Act in Section 4(b)(4) does not create a private cause of action or civil remedy for the recovery of damages that are proximately caused by an employer's OSHA violations.

In *Russell v. Bartley*, 494 F.2d 334, 1 OSHC 1589 (6th Cir. 1974), the Sixth Circuit held that Congress did not specifically indicate that the Act created a private cause of action for damages, but to the contrary disavowed the creation of a private civil remedy in Section 4(b)(4) of the Act. This is true whether an action is brought against an injured employee's employer or its officers, or against another employer who may have violated the Act and proximately caused the plaintiff's injuries or damages. In *Russell* the injured employee brought the action against a construction site's "supervision engineer," who was not his immediate employer, as well as against his employer. Similarly, in *Jeter v. St. Regis Paper Co.*, 507 F.2d 973, 2 OSHC 1591 (5th Cir. 1975), an employee of an independent contractor was injured when he fell off a swinging scaffold while painting a silo. He attempted to bring a lawsuit against the silo owner on the theory that there was a violation of the Act. The court found that there was no basis in the Act for the employee or an employee of an independent contractor to expand his rights or remedies based upon an OSHA violation. The case also held that there was no cause of action created by alleged OSHA violations, particularly given the legislative history and statutory declaration of public policy, against a company by an employee of an independent contractor. See also *Dravo Corp. v. OSHRC*, 613 F.2d 1227, 1230 n.2, 7 OSHC 2089 (3d Cir. 1980); *Jack M. Otto v. Specialties, Inc.*, 386 F. Supp. 1240, 2 OSHC 1424 (N.D. Miss. 1974); *Byrd v. Fieldcrest Mills, Inc.*, 496 F.2d 1323, 1 OSHC 1743 (4th Cir. 1974)—no private right of action exists under OSHA against employer, since state worker's compensation was the exclusive remedy for the injured party; *Skidmore v. Travelers Ins. Co.*, 356 F. Supp. 670, 1 OSHC 1173 (E.D. La. 1973), *aff'd. per curiam*, 483 F.2d 67, 1 OSHC 1294 (5th Cir. 1973)—no private right of action exists under the Act against executive officers of the employer; *Davis v. Niagara Machine Co.*, 1978 OSHD ¶22,965 (Wash. 1978); *Cochran v. Int'l Harvester Co.*, 408 F. Supp. 598, 4 OSHC 1385 (W.D. Ky. 1975); *Hare v. Federal Compress and Warehouse Co.*, 359 F. Supp. 214 (N.D. Miss. 1973).

Finally, in *Federal Employees for Non-Smokers' Rights v. United States*, 446 F. Supp. 181, 6 OSHC 1407 (D.D.C. 1978), *aff'd*, 598 F.2d 310, 7 OSHC 1634 (D.C. Cir.) *cert. denied*, 100 S. Ct. 265, 7 OSHC 2238 (1979), the court held that the Act does not create a private right of action against a federal agency. See also *Horn v. C. L. Osborn Contracting Co.*, 591 F.2d 318, 321, 7 OSHC 1256 (5th Cir. 1979)—private right of action created by contract which incorporated OSHA regulations (but not the Act); *Taylor v. Brighton Corp.*, 616 F.2d 256, 8 OSHC 1010 (6th Cir. 1980)—Section 11(c) of the Act does not create a private right of action for employees complaining of retaliatory discharge;

Buhler v. Marriott Hotels, Inc., 390 F. Supp. 999, 3 OSHC 1199 (E.D. La. 1974) —the Act does not create a private civil remedy for recovery for damages caused by a violation; see also *Laffey v. Northwest Airlines*, 567 F.2d 429 (D.C. Cir. 1976); *Davis v. United States*, 536 F.2d 758 (8th Cir. 1976). *Cf.*, *W. P. Moore v. OSHRC*, 591 F.2d 991, 7 OSHC 1031 (4th Cir. 1979)—under Virginia law, company's officers and directors incurred personal liability under the Act for violations that occurred during the period between dissolution and reinstatement of the company's corporate charter.

In this vein, most courts have also rejected the argument that a general contractor has an obligation under OSHA to provide a safe workplace for all employees on the premises *(e.g.*, employees of subcontractors); it is only responsible for its employees, *Koll v. Manatt's Transportation Co.*, 253 N.W. 2d 265, 5 OSHC 1398 (Sup. Ct. Ia. 1977)—evidence of an employer's violation of an OSHA standard is negligence *per se* to its employees and negligence to other persons exposed or likely to be exposed due to the violation; *Frith v. Harah South Shore Corp.*, 552 P.2d 337 (Nev. 1976); *Pruett v. Precision Plumbing, Inc.*, 27 Ariz. App. 288, 554 P.2d 655 (1976); *Gonzales v. R. J. Novick Construction Co.*, 70 Cal. App.3d 131, 13a Cal. Rpt. 113 (1977); *Horn v. C. L. Osborn Contracting Company*, 591 F.2d 318, 7 OSHC 1256 (5th Cir. 1979); *Dunn v. Grimer*, 259 Ark. 855, 537 S.W.2d 164 (1976)—injured employee of general contractor sued subcontractor claiming a defective ladder was provided, that the ladder failed to comply with OSHA standards, but the court held the Act was irrelevant since there was no employment relationship; but see *Knight v. Burns, Kirkley & Williams Constr. Co.*, 331 So. 2d 651 (Ala. 1976)—under state law a contractor is liable if its negligence injures a subcontractor's employee, but no action could be based on OSHA liability alone; *Bachner v. Rich*, 554 P.2d 430 (Alaska 1976); *Schroeder v. C. F. Braun & Co.*, 502 F.2d 235 (7th Cir. 1974).

At least two courts have decided that no cause of action exists against the federal government for the negligence or nonfeasance of OSHA officials under the Federal Tort Claims Act, 28 U.S.C. §1346(B) (FTCA). In one case, no cause of action existed under the FTCA for the wrongful death of an employee allegedly killed because an OSHA inspector failed to properly perform his duties by not conducting a follow-up inspection after a citation. *Davis v. United States*, 395 F. Supp. 793 (D. Neb. 1975), *aff'd*, 536 F.2d 758, 4 OSHC 1417 (8th Cir. 1976). In the second case, *Caldwell v. United States*,—F. Supp.—,6 OSHC 1410 (D.D.C. 1978), no action was held to exist under the FTCA for an alleged failure of OSHA to respond to an imminently dangerous condition and to perform its duties with due care. But see *Wallace v. Alaska*, 557 P.2d 1120 (Alaska 1976)—the negligent conduct of a safety inspector was not immune from civil suit. See also *Brock v. California*, 1978 OSHD ¶22,834 (Cal. Ct. App. 1978); *White v. Utah*, 1978 OSHD ¶22,906 (Utah 1978); and *Estate of Klee v. New York*, 1978 OSHD ¶22,888 (N.Y. Ct. C1. 1976); *Mudlo v. United States*, 423 F. Supp. 1373 (W.D. Pa. 1976)—the FTCA does not apply to enforcement activity of OSHA inspectors.

Several state courts have also held that the Act does not create a private cause of action by employees against employers. See *Deckle v. Todd*, 132 Ga. App. 156, 207 S.E.2d 654 (1974); *Childers v. Int'l Harvester Co.*, 569 S.W.2d

675 (Ky. App. 1978); *Knight v. Burns, Kirkley and Williams Constr. Co.*, 331 So.2d 651, 4 OSHC 1271 (Ala. 1976); *Frith v. Harrah South Shore Corp.*, 92 Nev. 447, 552 P.2d 337 (Sup. Ct. Nev. 1976); *Arvas v. Feather's Jewelers*, 92 N.M. 89, 582 P.2d 1302 (N.M. 1978). But see *Shimp v. New Jersey Bell Telephone Co.*, 145 N.J. Super. 516, 368 A.2d 408 (1976)—a common law duty to provide a safe and healthful workplace may arise regardless of the Act, thus an employer was ordered to restrict the smoking of employees to nonworking areas.

E. WORKER'S COMPENSATION

Prior to enactment of worker's compensation laws, employees injured on the job had to bring negligence actions against their employers, but because of the existence of common law defenses often times they could not recover. In order to remedy this situation, worker's compensation laws were enacted in every state plus the District of Columbia. In order to qualify for worker's compensation under these laws, an employee must suffer a personal injury by accident arising out of and in the course of employment or develop an occupational illness or disease as a result of exposure which was occasioned by employment.

With respect to the Act's relationship to worker's compensation laws, Congress stated in Section 4(b)(4) of the Act that:

> Nothing in this Act shall be construed to supersede or in any manner affect any workmen's compensation law or to enlarge or diminish or affect in any other manner the common law or statutory rights, duties, or liabilities of employers and employees under any law with respect to injuries, diseases, or death of employees arising out of, or in the course of employment.

The legislative history of this provision is extremely sparse. The plain language of the statute itself, however, bespeaks its intent to essentially leave untouched the present state of the law concerning civil actions and to have courts apply the common law tort principles discussed above. Notwithstanding either Section 4(b)(4) or its intent that OSHA has no bearing on worker's compensation claims (N. Ashford, *Crisis in the Workplace* 386–441, 1976), there is little question but that the Act has and will continue to have a significant impact on civil litigation and worker's compensation based upon the same facts that give rise to an OSHA violation. In *United Steelworkers of America, AFL-CIO-CLC v. Marshall*,—F.2d—, 8 OSHC 1810 (D.C. Cir. 1980), the employer petitioners charged that the medical removal provision (MRP) of the lead standard, 29 CFR 1910.102(k), violated the prohibition contained in Section 4(b)(4) of the Act. The court rejected the argument, concluding that although MRP may indeed have a great practical effect on worker's compensation claims, it leaves the state *schemes* wholly intact as a legal matter and thus does not violate the Act. The court continued:

> The question remains, then, what *does* Section 4(b)(4) mean, if it *does not* mean that OSHA is barred from creating medical removal protection? We see two plausible meanings. First, as courts have already held, Section 4(b)(4) bars workers from

> asserting a private cause of action against employers under OSHA standards, *Jeter v. St. Regis Paper Co.*, 507 F.2d 973 (5th Cir. 1975); *Byrd v. Fieldcrest Mills, Inc.*, 496 F.2d 1323 (4th Cir. 1974). Second, when a worker actually asserts a claim under workmen's compensation law or some other state law, Section 4(b)(4) intends that neither the worker nor the party against whom the claim is made can assert that any OSHA regulation or the OSH Act itself *preempts* any element of the state law. For example, where OSHA protects a worker against a form of disablement not compensable under state law, the worker cannot obtain state relief for that disablement. Conversely, where state law covers a wider range of disablements than OSHA aims to prevent, an employer cannot escape liability under state law for a disablement not covered by OSHA. In short, OSHA cannot *legally* preempt state compensation law, even if it *practically* preempts it in some situations.

Thus the Court found that there is a great difference between a regulation that has the effect of reducing the number of claims made with state law and one that actually alters the terms of such a law.

Evidence of OSHA violations has been held to be admissible in worker's compensation proceedings. *International Harvester Co. v. Childers,* 1978 OSHD ¶22,993 (Ky. 1978). In many states, however, the existence of worker's compensation is the exclusive remedy for an injured employee. In those states, an employee may not be able to sue his employer for negligence, and evidence of an OSHA violation would be of little use to him vis-à-vis his employer. See, *e.g., Byrd v. Fieldcrest Mills, Inc.*, 496 F.2d 1323 (4th Cir. 1974); *Wood v. Aetna Casualty & Surety Co.*, 260 Md. 651, 273 A.2d 125 (1971); *Russell v. Bartley*, 494 F.2d 334 (6th Cir. 1974)—OSHA did not create a private cause of action by an injured employee against his employer. To like effect see *Bertholf v. Burlington Northern Railroad,* 402 F. Supp. 171 (E.D. Wash. 1975) —OSHA has no applicability to claims under the Federal Employees Liability Act; *Groshaw v. Koninklijke Necloyd,* 398 F. Supp. 1224 (D. Ore. 1975)—the Longshoreman's and Harborworkers' Compensation Act was applicable and reference to OSHA was ineffective.

In situations where deceitful, gross, intentional, or willful acts of misconduct are involved, the employer may not be free from liability at law even if the injuries to his employees are compensable under worker's compensation laws. *Andrews v. Ins. Co. of North America,* 60 Mich. App. 190, 230 N.W.2d 371 (Mich. 1975); *Johns-Manville Products Corp. v. Contra Costa Superior Court,*— Cal.—, 612 P.2d 948 (Cal. Sup. Ct. 1980)—an employee alleged that his employer, an asbestos producer, fraudulently concealed the potential of occupational lung disease, thereby inducing further work and aggravating his untreated disease; exceptional circumstances existed and thus worker's compensation was not the sole and exclusive remedy against the asbestos producer; *McDaniel v. Johns-Manville Sales Corp.,* 487 F. Supp. 714 (N.D. Ill. 1978)—although worker's compensation is ordinarily the exclusive remedy for employees, it is not the sole remedy in cases of intentional torts under Illinois law.

Employee actions against nonemployers such as suppliers, co-workers, and other companies are generally not precluded by worker's compensation laws. See, *e.g., Hutzell v. Boyer,* 252 Md. 227, 249 A.2d 449 (1969); *Brocker Mfg. & Supply Co. v. Mashburn,* 17 Md. App. 327, 301 A.2d 501 (1973). An interesting variation upon that theme, which also involved the federal OSH Act as well, occurred in *Horn v. C. L. Osborn Contracting Co.*, 591 F.2d 318, 7 OSHC 1256 (5th Cir. 1979), where a damage action was brought against a contractor by an employee of its subcontractor. Osborn, the general contractor, was responsible for constructing portions of a sewer system, and subcontracted a portion of that work to Bama. Bama used only its own equipment and employees in carrying out its work. An employee of Bama, who was directly supervised by other employees of Bama, sustained personal injuries on two separate occasions. Because of these injuries, this employee received worker's compensation benefits from Bama's insurance carrier. The employee also sought to recover damages from the general contractor, Osborn. Osborn, under an indemnity provision contained in the subcontract, filed a third-party action against Bama, claiming a contingent right to recover against Bama in the event that Bama's employee recovered anything from Osborn. The court found that the general contractor was not liable for injuries to Bama's employee under a Georgia statute that imposes liability on the general contractor for wrongful acts in violation of a duty imposed by statute. While the employee relied upon the Act as imposing such a duty, it was held not to impose liability on the general contractor for the safety of employees of a subcontractor. As the Fifth Circuit observed, the key issue was "whether the term 'employer' as used in the Act should be interpreted to envelop general contractors as joint or statutory employers of an independent subcontractor's employees and thus impose upon them a duty to provide the employee with a safe working environment." *Horn, supra* at 321. Answering this question in the negative, the Fifth Circuit held "that an employer cannot be held in violation of the above subsection if his employees are not affected by the noncompliance with a standard." *Id.*

NOTES

1. Negligence is defined as "[t]he omission to do something which a reasonable man, guided by those ordinary conditions which ordinarily regulate human affairs, would do, or the doing of something which a reasonable and prudent man would not do." *Black's Law Dictionary.* Negligence *per se* is "[c]onduct, whether of action or omission, which may be declared and treated as negligence without any argument or proof as to the particular surrounding circumstances, either because it is in violation of a statute or valid municipal ordinance, or because it is so palpably opposed to the dictates of common prudence that it can be said without hesitation that no careful person would have been guilty of it." *Id.*

2. "In view of the clear purpose of the statute to set new standards of industrial safety, we cannot accept the proposition that common law defenses such as assumption of the risk or contributory negligence will exculpate the employer who is charged with violating the Act." *Rea Express, Inc. v. Brennan,* 495 F.2d 822, 1 OSHC 1651 (2d Cir. 1974).

3. Although Section 4(b)(4) of the Act provides that it does not in any manner affect, enlarge, or diminish the rights, duties, or liability of employers and employees, the contract at issue incorporated the regulation only, not the Act. Thus it was not limited by Section 4(b)(4).

CHAPTER XV
Discrimination under the Act

A. INTRODUCTION

Like many state and federal labor statutes, the Act prohibits discrimination against those employees who invoke its processes or otherwise engage in protected activity. Section 11(c)(1) of the Act specifically proscribes discrimination against an employee because he "has filed any complaint or instituted or caused to be instituted any proceeding under or related to" the Act; because he "has testified or is about to testify in any such proceeding;" or because he has exercised, either on his own behalf or on behalf of others, "any right afforded by this Act."

As with many other state and federal labor statutes, the Act also establishes an administrative mechanism for employees who feel they have been subject to discrimination. This mechanism provides for the filing of a complaint with the Secretary of Labor; a determination being reached by the Secretary, after an investigation, as to whether or not discrimination has occurred; and proceedings brought in federal district court by the Secretary, in order to obtain relief—including reinstatement and backpay—from any discrimination found to have occurred. See Section 11(c)(2) of the Act.

The Secretary has promulgated extensive regulations on the subject of discrimination—many of which expand greatly upon the literal language of Section 11(c) itself, a number of which have been addressed by various courts, including the Supreme Court.

B. PERSONS PROHIBITED FROM DISCRIMINATING

Section 11(c) of the Act prohibits *persons* from discriminating. That term is defined in Section 3(4) of the Act as "one or more individuals, partnerships, associations, corporations, business trusts, legal representatives, or any group of persons." According to the Secretary's regulations, an employment relationship is not essential to a finding of discrimination. Thus:

> A person may be chargeable with discriminatory action against an employee of another person. Section 11(c) would extend to such entities as organizations representing employees for collective bargaining purposes, employment agencies, or any other person in a position to discriminate against an employee. [29 CFR 1977.4.]

Notwithstanding this regulation, at least one district court and one court of appeals have held that only *employers* are "subject to the OSHA prohibitions on termination for exercising OSHA rights." *Marshall v. Lummus Co.,*—F. Supp. —, 8 OSHC 1358, 1360 (N.D. Ohio 1980); *Marshall v. Certified Welding Corp.,* —F.2d—, 7 OSHC 1069 (10th Cir. 1978), *disapproved on other grounds, Whirlpool Corp. v. Marshall,* 445 U.S. 1, 8 OSHC 1001 (1980). In *Lummus,* a petroleum refinery arranged for a construction contractor to build a "coker unit" at an existing facility. The alleged discriminatee was the project construction manager, who was ordered off the job by the refinery after complaining that an electrical substation used for building the coker unit was hazardous and unsafe under OSHA standards. The construction contractor subsequently terminated the project manager for failing to maintain good relations with the refinery. Under these circumstances, the court held that the refinery could not be sued by the Secretary under Section 11(c) of the Act, notwithstanding the Secretary's position to the contrary, because the refinery was not the project manager's employer. The facts in *Certified Welding,* as well as the result reached, were essentially the same as in *Lummus*—except that the alleged discriminatee was an employee rather than the project construction manager.

On the one hand, *Lummus* and *Certified Welding* might be criticized for their interpretation of the "no person shall" discriminate language of Section 11(c) to mean that "no employer shall discriminate against any of his employees." In that sense, *Lummus* and *Certified Welding* appear to contravene the explicit language of the Act. On the other hand, the regulations probably go too far, and the result reached in *Lummus* and *Certified Welding* may be closer to the correct interpretation and application of the Act. Thus 29 CFR 1977.4 indicates that the prohibition against discrimination extends to *any* "person in a position to discriminate against an employee." Carried to its logical conclusion, the regulation would forbid a mother from disciplining her son for missing dinner in order to testify at an OSHA proceeding. Obviously, such a result would be absurd. This hypothetical example does demonstrate, then, that the proper interpretation of exactly who is a "person prohibited from discriminating" lies somewhere in between the *Lummus/Certified Welding* rationale and the regulation. In line with the Act's purpose of assuring safe and healthful *working* conditions for *working* men and women in their *workplace,* it would seem that only persons *with a direct involvement in* the *employment* relationship should be held accountable under Section 11(c) for discrimination against an employee. Such an interpretation strikes a balance between the Act's goal of ensuring a safe workplace and the societal goal of discouraging the proliferation of governmental bureaucracies into marginally effective areas of regulation.

C. "EMPLOYEES" PROTECTED BY THE ACT

Section 11(c)(1) prohibits persons from discriminating against "any employee." That term is circuitously defined in Section 3(6) of the Act as any "employee of an employer who is employed in a business of his employer which affects commerce." According to 29 CFR 1977.5, Congress "clearly" intended that a determination of whether someone is "employed," within the meaning of Section 11(c), "is to be based upon economic realities rather than upon common law doctrines and concepts." The regulation does not go on to explain what it means by "economic realities"—except to note that applicants are employees and that someone need not be an employee of the discriminator in order to fall within the protection of Section 11(c).[1] Indeed, one court has held that Section 11(c) is violated when an employer refuses to allow a *union representative* to attend a monthly safety meeting in retaliation for the union's filing an OSHA complaint. *Marshall v. Kennedy Tubular Products,*—F. Supp. —, 5 OSHC 1467 (W.D. Pa. 1977).

The vagueries of the Secretary of Labor's "economic realities" test can perhaps best be demonstrated by a hypothetical. Assume that the owner-operator of a tractor-trailer has sufficient control over his operation to be regarded as an independent contractor under the National Labor Relations Act (NLRA). If a distributor stops using this owner-operator because he has complained to OSHA about an unsafe loading dock at one of the distributor's major clients, would the Secretary consider the owner-operator to be an employee under his "economic realities" test? While this precise issue has apparently not yet been decided in any reported decision, it is not unreasonable to assume that the Secretary *would* indeed consider the owner-operator to be an employee. While this issue was not raised by the employer, the Secretary took the position in *Marshall v. Lummus Co.,*—F. Supp.—, 8 OSHC 1358 (N.D. Ohio 1980) that a project manager on a construction site was an employee of the contractor. Just how far, then, the Secretary will go with his "economic realities" test as to who is an employee, let alone how far the courts will let him go, remains to be seen.

D. THE DETERMINATION OF CAUSATION

Fortunately, from an employer's standpoint, the Secretary of Labor has recognized that an employee engaged in protected activity under Section 11(c) is not automatically "immune from discharge or discipline for legitimate reasons, or from adverse action dictated by non-prohibited considerations." 29 CFR 1977.6(a). The regulations go on to provide in subsection (b) of Section 1977.6 that the following must be shown, in order "to establish a violation of section 11(c)":

> [T]he employee's engagement in protected activity need not be the sole consideration behind discharge or other adverse action. If protected activity was a *substantial reason* for the action, or if the discharge or other adverse action would not have taken place *"but for"* engagement in protected activity, section 11(c) has been violated. [Emphasis supplied.]

Section 11(c) is substantially similar to and, indeed, was modeled after Section 15(a)(3) of the Fair Labor Standards Act (FLSA). In recognition of that fact, both of the cases which are cited in subsection (b) of the regulations in support of the "substantial reason/but for" test are FLSA cases interpreting Section 15(a)(3). Both of those cases held that, in order to make out a violation of Section 15(a)(3), the protected activity must be a *substantial* reason for the adverse action[2] or the adverse action must be shown to have not taken place *but for* the employee's having engaged in protected activity. See *Mitchell v. Goodyear Tire & Rubber Co.,* 278 F.2d 562 (8th Cir. 1960) and *Goldberg v. Bama Mfg. Co.,* 302 F.2d 152 (5th Cir. 1962). Accordingly, while other federal labor statutes may use different tests for determining whether an adverse employment decision was caused by an illegal consideration, the sole and only appropriate test under OSHA, by explicit regulation, is the "substantial reason/but for" test.

This test had been utilized by courts on numerous occasions in Section 11(c) litigation. Given that the test invariably involves subtle issues of fact and widely ranging sets of circumstances, different courts have reached varied conclusions, and the methods of proof have been many, See, *e.g., Marshall v. Babcock and Wilcox Co.,*—F. Supp. —, 7 OSHC 2021, 2026 (E.D. Mich. 1979) —("[M]ere sequential proximity of calls to OSHA followed by discharge . . . [does not give] rise to a presumption that the discharge was retaliatory"); *Usery v. Granite-Groves, A Joint Venture,*—F. Supp. —, 5 OSHC 1935 (D.D.C. 1977); *Marshall v. Klug & Smith Co.,*—F. Supp. —, 7 OSHC 1162 (N.D. Ohio 1979); *Marshall v. Commonwealth Aquarium,* 469 F. Supp. 690, 1 OSHC 1387 (D. Mass.), *aff'd,* 611 F.2d 1, 1 OSHC 1970 (1st Cir. 1979); *Marshall v. P & Z Co., Inc.,*—F. Supp.—6 OSHC 1587 (D.D.C. 1978), *aff'd,* 600 F.2d 280, 7 OSHC 1633 (D.C. Cir. 1979); *Dunlop v. Trumbull Asphalt Co., Inc.,*—F. Supp. —, 4 OSHC 1847 (E.D. Mo. 1976); *Marshall v. Dairymen's Creamery Assoc., Inc.,* —F. Supp. —, 6 OSHC 2186 (D. Idaho 1978); and *Marshall v. Chapel Elec. Co.,*—F. Supp. —, 8 OSHC 1365 (S.D. Ohio 1980)—fact that employee, who filed complaint with state safety commission, was discharged the day after the state inspection, gives rise to a reasonable inference that the reason for discharge was engagement in protected activity. The following passage from *Dunlop v. Bechtel Power Corp.,*—F. Supp. —, 6 OSHC 1605 (M.D. La. 1977), however, is particularly instructive on the issue of causation:

> The *Secretary* bears the burden of proving that the employee's discharge was causally connected to the OSHA complaint filed by the employee. . . . Absent a showing that an employee was discharged for exercising a right protected by the Act, an employer may discharge an employee for good reason, a bad reason, or no reason at all.

See also *Marshall v. National Indus. Constructors, Inc.,*—F. Supp. —, 8 OSHC 1117 (D. Neb. 1980)—retaliatory discharge was not proved since there was no evidence that the employees made any complaints to the employer prior to the discharge; evidence showed that real reason for refusal to work on high crossbeams was desire for higher pay.

As in many other areas of labor and equal employment opportunity law, the discharge of someone "for no reason at all" may turn out to be indefensible. If that "someone" was fired the day after complaining to OSHA, had a fourteen-year spotless record, and was the first person ever discharged for purportedly doing something that other employees had been doing for years, the employer's stated reason would probably be deemed to be a pretext, and the discharge probably would not stand. In contrast, use of a progressive discipline system that includes oral and written warnings, suspension and discharge, together with a demonstrable history of uniform application of the standard that the alleged discriminatee was disciplined for violating, will go a long way toward successfully defending a Section 11(c) lawsuit, if not convincing the Secretary to keep from filing a complaint in the first place.

E. FOM DISCRIMINATION PROCEDURE

Chapter XV of the FOM sets forth guidelines for handling discrimination complaints under Section 11(c) of the Act. An employee or employee representative may file a discrimination complaint with any of several persons, including the Inspector, Area Director, Regional Director, or the OSHA National Office. Although such complaints may be accepted through oral or written filing, the latter method is preferable as it establishes the exact date a complaint was filed in order to insure timeliness. At a minimum, when an oral complaint is filed, a complainant's name, address, and telephone number must be obtained.

The OSHA Operations Review Office is responsible for the investigation of Section 11(c) discrimination complaints, and with the Office of the Solicitor of Labor, will seek appropriate relief for complainants who have "suffered" discrimination. At the time of filing the complaint, the complainant is encouraged by OSHA to retain all relevant documents, statements, and other materials supportive of his complaint for possible later use.

On September 1, 1979, OSHA adopted Instruction CPL 2.12 (A) which relates to safety and health complaint processing procedures. Although the instruction pertains primarily to and provides guidelines for investigation of complaints concerning alleged unsafe and unhealthful conditions at a workplace, it also references and updates certain procedures for the investigation of discrimination complaints.[3] All safety and health complaints are required to be carefully evaluated for discrimination implications and promptly forwarded to the appropriate office pursuant to OSHA's policy. Discrimination complaints will be processed in accordance with Chapter XV of the FOM and OSHA Instructions DIS .1 and DIS .2. In the investigation of safety and health-hazard complaints, the complainant shall be advised of the protection against discrimination afforded by Section 11(c) of the Act and informed of the procedure for filing an 11(c) complaint. In accordance with FOM instructions, the identity of all formal and nonformal complainants who wish to remain anonymous will be kept confidential, pursuant to Section 8(f)(1) of the Act and the informer's privilege.

F. PROTECTED ACTIVITIES

1. Filing a Complaint

Section 11(c)(1) forbids discrimination against employees who file "any complaint . . . under or related to this Act. . . ." In 29 CFR 1977.9(a), the Secretary lists employee inspection requests pursuant to Section 8(f) of the

Act as an example of a complaint "under" the Act. These regulations go on to provide in subsection (b) that complaints registered with other federal, state, or local agencies "regarding occupational safety and health conditions would be 'related to' the Act . . . [so long as they] relate to conditions at the workplace, as distinguished from complaints touching only upon general public safety and health." Finally, even complaints to *employers* "about occupational safety and health matters" are deemed to be related to the Act, and therefore protected, as long as those complaints are "made in good faith. . . ." 29 CFR 1977.9(c). This latter subsection of the regulation, relating to complaints made to employers, was upheld in *Marshall v. Springville Poultry Farm, Inc.*, 445 F. Supp. 2, 5 OSHC 1761 (M.D. Pa. 1977) and in *Marshall v. Klug & Smith Co.*,—F. Supp. —, 7 OSHC 1162 (D.N.D. 1979).

2. Institution of Proceedings

In addition to proscribing discrimination against employees who file a complaint, Section 11(c)(1) forbids discrimination against persons who institute or cause to be instituted "any *proceeding* under or related to this Act." The Secretary has given a number of examples of what he interprets to be proceedings:

> Examples of proceedings which could arise specifically under the Act would be inspections of worksites under section 8 of the Act, employee contest of abatement date under section 10(c) of the Act, employee initiation of proceedings for promulgation of an occupational safety and health standard under section 6(b) of the Act and Part 1911 of this chapter, employee application for modification or revocation of a variance under section 6(d) of the Act and Part 1905 of this chapter, employee judicial challenge to a standard under section 6(f) of the Act and employee appeal of an Occupational Safety and Health Law Review Commission order under section 11(a) of the Act. In determining whether a "proceeding" is "related to" the Act, the considerations discussed in §1977.9 [relating to "complaints"] would also be applicable.

29 CFR 1977.10(a). The regulations also provide that an employee need only set "into motion activities of others which result in proceedings under or related to the act," in order to be protected; *e.g.*, an employee who complains about a safety condition to his union. 29 CFR 1977.10(b).

3. Giving Testimony

Section 11(c)(1) prohibits the imposition of discipline on an employee because he "has testified or is about to testify in any such proceedings" under or related to the Act. The Secretary, by regulation, has indicated that this protection "would extend to any statements given in the course of judicial, quasi-judicial, and administrative proceedings, including inspections, investigations, and administrative rule-making or adjudicative functions." 29 CFR 1977.11. In an analogous setting, Section 8(a) 4 of the NLRA provides that an employer commits an unfair labor practice when he discharges or otherwise discriminates against an employee for filing charges or giving testimony under the NLRA. The United States Supreme Court interpreted this section as prohibiting an employer's discharge of his employees for their activity because such a construction comported with the section's objective of providing the National Labor Relations Board (NLRB), with sources of employee information essential to enforcing the NLRA. *NLRB v. Scrivener,* 405 U.S. 117 (1972).

4. Exercising Rights

Section 11(c)(1), as the foregoing discussion makes clear, protects three relatively specific types of activities that employees engage in under the Act —filing complaints, instituting proceedings, and testifying in those proceedings. Section 11(c)(1) also forbids discrimination, however, against employees "because of the exercise by such employee on behalf of himself or others of any right afforded by this Act." This catch-all phrase is quite open-ended, and has been utilized by the Secretary to carve out areas of protection that fall well outside the literal language of the Act itself.

Of course, as the Secretary points out in his regulations, a number of rights are explicitly provided in the Act—the right to participate in enforcement proceedings, for example. 29 CFR 1977.12(a). That same regulation lists other rights that are said to exist "by necessary implication":

> For example, employees may request information from the Occupational Safety and Health Administration; such requests would constitute the exercise of a right afforded by the Act. Likewise, employees interviewed by agents of the Secretary in the course of inspections or investigations could not subsequently be discriminated against because of their cooperation. [*Id.*]

Outside of these specific protective rights and rights that "exist by necessary implication," the Secretary has identified another right, which was a source of extensive litigation until the Supreme Court recently clarified the issue. That much-litigated right is embodied in 29 CFR 1977.12(b), which the Supreme Court recently upheld in *Whirlpool Corp. v. Marshall,* 445 U.S. 1, 8 OSHC 1001 (1980).

This regulation disarmingly begins with the statement that as a general proposition, employees have no right under the Act, "to walk off the job because of potential unsafe conditions at the workplace." 29 CFR 1977.12(b)(1). This regulation goes on to indicate that under normal circumstances, an employer will not find himself in violation of Section 11(c) of the Act "by taking action to discipline an employee for refusing to perform normal job activities because of alleged safety or health hazards." *Id.* The regulation then leads into subpart (b)(2), which, until recently, has been the source of controversy. That subpart reads in its entirety as follows:

> However, occasions might arise when an employee is confronted with a choice between not performing assigned tasks or subjecting himself to serious injury or death arising from a hazardous condition at the workplace. If the employee, with

> no reasonable alternative, refuses in good faith to expose himself to the dangerous conditions, he would be protected against subsequent discrimination. The condition causing the employee's apprehension of death or injury must be of such a nature that a reasonable person, under the circumstances then confronting the employee, would conclude that there is a real danger of death or serious injury and that there is insufficient time, due to the urgency of the situation, to eliminate the danger through resort to regular statutory enforcement channels. In addition, in such circumstances, the employee, where possible, must also have sought from his employer, and been unable to obtain, a correction of the dangerous condition.[4] [29 CFR 1977.12(b)(2).]

Until the Supreme Court definitively resolved the issue, the courts of appeal had split on the validity of this regulation. Two decisions—*Marshall v. Daniel Constr. Co.,* 563 F.2d 707, 6 OSHC 1031 (5th Cir. 1977) and *Marshall v. Certified Welding Corp.,*—F.2d —, 7 OSHC 1069 (10th Cir. 1978)—had held that this regulation was invalid.[5] The Sixth Circuit, in *Marshall v. Whirlpool Corp.,* 593 F.2d 715, 7 OSHC 1075 (6th Cir. 1979) held that the regulation was valid. The Supreme Court, affirming the Sixth Circuit's decision, upheld the regulation. *Whirlpool Corp. v. Marshall, supra.*

The Supreme Court in *Whirlpool* first examined the Act's remedial scheme, concluding that "nothing in the Act suggests that those few employees who have to face this dilemma [of being ordered to work under adverse conditions when there is insufficient time to apprise OSHA of the danger] must rely exclusively on the remedies expressly set forth in the Act at the risk of their own safety." *Id.* at 890. The Court then concluded that the congressional purpose underlying the Act was furthered by the Secretary's regulation on unsafe conditions. As the Court observed:

> The Act does not wait for an employee to die or become injured. It authorizes the promulgation of health and safety standards and the issuance of citations in the hope that these will act to prevent deaths or injuries from ever occurring. It would seem anomalous to construe an Act so directed and constructed as prohibiting an employee, with no other reasonable alternative, the freedom to withdraw from a workplace environment that he reasonably believes is highly dangerous. [*Id.* at 890–91.]

The Court further noted that "OSHA inspectors cannot be present around the clock in every workplace, [and, hence,] the Secretary's regulation ensures that employees will in all circumstances enjoy the rights afforded them by the 'general duty' clause." *Id.* at 891. Finally, the Court observed:

> The regulation accords no authority to government officials. It simply permits private employees of a private employer to avoid workplace conditions that they believe pose grave dangers to their own safety. The employees have no power under the regulation to order their employer to correct the hazardous conditions or to clear the dangerous workplace of others. Moreover, any employee who acts in reliance on the regulation runs the risk of discharge or reprimand in the event a court subsequently finds that he acted unreasonably or in bad faith.

Id. at 895. For these reasons, the Court held that 29 CFR 1977.12(b) which proscribes any discrimination in response to an employee's *good faith* refusal to expose himself to conditions he *reasonably* believes are dangerous "was promulgated by the Secretary in the valid exercise of his authority under the Act." *Id.*[6] See *Marshall v. Firestone Tire & Rubber Co.,*—F. Supp. —, 8 OSHC 1637 (C.D. Ill. 1980).[7]

In reaching its decision, the court in *Whirlpool* noted that the regulation at issue does not require employers to pay workers who refuse to perform their assigned tasks in the face of imminent danger. It simply provides that in such cases the employer may not "discriminate" against the employees involved. An employer "discriminates" against an employee only when he treats that employee less favorably than he treats other similarly situated employees. Moreover, the court noted that the regulation simply permits private employees of a private employer to avoid workplace conditions that they believe pose grave dangers to their own safety. The employees have no power under the regulation to order their employer to correct the hazardous condition or to clear the dangerous workplace of others. Moreover, any employee who acts in reliance on the regulation runs the risk of discharge or reprimand in the event a court subsequently finds that he acted unreasonably or in bad faith. The Court did not address the question of an employer's offer of alternate work and the options available to the employer if the employee refuses such work.

In *Marshall v. N.L. Industries, Inc.,* 618 F.2d 1220, 8 OSHC 1166 (7th Cir. 1980), the court stated that *Whirlpool* did not proscribe backpay as judicially ordered relief, and in fact left the question of what relief would be appropriate for the district court to decide on remand. Thus the court felt that backpay was an appropriate remedy for discriminatory discharge.

G. PROCEDURAL CONSIDERATIONS UNDER SECTION 11(c)

1. The Complaint

Section 11(c)(2) provides that any employee "who believes that he has been" discriminated against should file a complaint with the Secretary alleging such discrimination. By regulation, the Secretary has provided that a complaint "may be filed by the employee himself, or by a representative authorized to do so on his behalf." The complaint should be filed with the OSHA Area Director responsible for the geographical area "where the employee resides or was employed." 29 CFR 1977.15(a) and (c). See generally OSHA Instruction CPL 2.12(A) (September 1, 1979).

The complaint, by statute and regulation, is required to be filed within thirty days after the alleged violation occurs. Section 11(c)(2) of the Act. This thirty-day period has spawned and will continue to spawn a substantial amount of litigation, as the discussion below demonstrates.

Under 29 CFR 1977.15(d)(2), the Secretary asserts that "complaints not filed within thirty days of an alleged violation will ordinarily be presumed to be untimely." Notwithstanding this regulation, the Secretary argued in *Usery v. Northern Tank Line, Inc.,*—F. Supp. —, 4 OSHC 1964, 1965–66 (D. Mont. 1976), "that the use of the word 'may' in the Act is directory and is not mandatory," such that the Secretary has discretion to proceed upon complaints filed outside of the thirty-day period. The court in *Northern Tank Line,* however, rejected that argument—thereby achieving a result that fully accords with other federal labor statutes, most notably Title VII of the Civil Rights Act of 1964. See, *e.g., McDonnell Douglas Corp. v. Green,* 411 U.S. 792, 798 (1973); and *Alexander v. Gardner-Denver Co.,* 415 U.S. 36, 47 (1974). A similar result was reached in *Marshall v. Lummus Co.,*—F. Supp. —, 8 OSHC 1358 (N.D. Ohio 1980), where the court dismissed the Secretary of Labor's action because the employee's complaint was untimely, filed fifty days after he was discharged.

Other decisions have recognized that, while this thirty-day period must be complied with by complainants, the Secretary is not required to file a lawsuit within any specific time frame. Most notably in this respect, even though the Act itself is silent as to how long the Secretary has to file a lawsuit under Section 11(c), state statute of limitations cannot be invoked by defendants to bar a Section 11(c) lawsuit.[8] *Marshall v. American Atomics, Inc.,*—F. Supp. —, 8 OSHC 1243 (D. Ariz. 1980); *Marshall v. Intermountain Elec. Co.,* 614 F.2d 260, 7 OSHC 2149 (10th Cir. 1980). This result is again consistent with Title VII case law. Thus lawsuits brought by the EEOC under Title VII cannot be barred by a state statute of limitations. *Occidental Life Ins. Co. v. EEOC,* 432 U.S. 355 (1977). While a state statute of limitations cannot be invoked, however, a laches defense may be available in Title VII lawsuits and under Section 11(c) as well. *Contra, Dunlop v. Bechtel Power Corp.,*—F. Supp. —, 6 OSHC 1604, 1606 (M.D. La. 1977)—"The Secretary is not subject to the doctrine of laches in enforcing a public right". Under a laches defense, a lawsuit will be dismissed if the Secretary waits an inordinate amount of time before filing a Section 11(c) lawsuit, and the defendant is thereby prejudiced in some material respect. See, *e.g., Occidental Life, supra* (EEOC); *EEOC v. Alioto Fish Co.,* 623 F.2d 86 (9th Cir. 1980) (EEOC); and *Intermountain Elec. Co., supra* (OSHA).

In 29 CFR 1977.15(d)(3), the Secretary of Labor asserts that even when complaints are not timely filed "there may be circumstances which would justify tolling of the thirty day period on recognized equitable principals or because of strongly extenuating circumstances. . . ." Whether this thirty-day period can *ever* be tolled, however, is an open question. Indeed, courts have split on whether analogous time periods in Title VII and the Age Discrimination in Employment Act (ADEA) can be tolled. See, *e.g., Wright v. State of Tennessee,*—F.2d —, 21 FEP Cases 1347 (6th Cir. 1980)—ADEA time limits cannot be tolled; *Dartt v. Shell Oil Co.,* 539 F.2d 1256 (10th Cir. 1976), *aff'd by equally divided court,* 434 U.S. 99 (1977) (*per curiam*)—ADEA time limitations can be tolled; *Laffey v. Northwest Airlines, Inc.,* 567 F.2d 429 (D.C. Cir. 1976), *cert. denied,* 434 U.S. 1086 (1978)—tolling is available under Title VII; and *Wong v. The Bon Marche,* 508 F.2d 1249 (9th Cir. 1975)—Title VII time periods are jurisdictional prerequisites. Undoubtedly, this same issue will be litigated within the context of Section 11(c) lawsuits.

The Secretary of Labor also gives examples of *when* the thirty-day period contained in Section 11(c) can be tolled. One example is "where the employer has concealed, or misled the employee regarding the grounds for discharge or other adverse action. . . ." 29 CFR 1977.15(d)(3). The Secretary is probably on safe ground with respect to this example.

The Secretary's position will probably be rejected with respect to another situation where he says tolling might be justified—*i.e.*, "where the employee has, within the thirty day period, resorted in good faith to grievance-arbitration proceedings under a collective bargaining agreement. . . ." *Id.* Courts will probably hold that Section 11(c) lawsuits and arbitrations are separate and independent, such that compliance with procedural prerequisites of one does not ensure compliance with procedural prerequisites of the other. At least one court of appeals, relying upon Title VII precedents, has held that Section 11(c) remedies and arbitration remedies are separate and independent, such that an arbitrator's decision under a collective bargaining agreement does not bar a later federal lawsuit.[9] *Marshall v. N. L. Indus., Inc.*, 618 F.2d 1220, 8 OSHC 1166 (7th Cir. 1980), *relying upon Alexander v. Gardner-Denver Co.*, 415 U.S. 36 (1974). *Accord, Marshall v. Firestone Tire & Rubber Co.*, —F. Supp. —, 8 OSHC 1637 (C.D. Ill. 1980). Furthermore, in the area of racial discrimination, the Supreme Court has indicated that even two federal statutes—Title VII and 42 U.S.C. 1981—are separate and independent, such that filing of an EEOC charge does not toll the statute of limitations under Section 1981. *Johnson v. Railway Express Agency, Inc.*, 421 U.S. 454 (1975). The decision that will be most likely to invalidate the Secretary's reliance upon resorting to grievance-arbitration proceedings in order to toll the thirty-day period, however, is *Electrical Workers v. Robbins & Myers, Inc.*, 429 U.S. 229 (1976). In that case the Supreme Court held that the filing of a grievance by a discharged employee under a collective bargaining agreement did not toll the period within which the employee was required to file a charge under Title VII with the EEOC.

As another example of a situation that might justify tolling the thirty-day period, the Secretary identifies an employee filing "a complaint regarding the same general subject with another agency. . . ." 29 CFR 1977.15(d)(3). Employers will undoubtedly argue that filing a complaint with a state or local agency cannot toll the thirty-day period under Section 11(c), under much the same rationale as in the case of filing a grievance under a collective bargaining agreement. One case that did construe this regulation did not go quite that far, but did construe the "regarding the same general subject with another agency" language very narrowly. In that case, *Marshall v. Certified Welding Corp.*,—F.2d —, 7 OSHC 1069 (10th Cir. 1978), *disapproved on other grounds* in *Whirlpool Corp. v. Marshall*, 445 U.S. 1, 8 OSHC 1001 (1980), an employee filed a state charge complaining that he was fired for refusing to work in an unsafe area. This state charge was held not to encompass "the same general subject" as the employee's Section 11(c) complaint, in which he alleged that he was fired for filing safety complaints, but not for refusing to work in an unsafe area. Accordingly, the Section 11(c) lawsuit was held to have been properly dismissed.

As a final example of a situation where circumstances would justify tolling, the Secretary identifies discrimination that is "in the nature of a continuing violation." 29 CFR 1977.15(d)(3). An obvious example of a continuing violation is where someone is paid a lower wage for having complained about safety conditions. Every time that this person receives a paycheck, the prohibition against discrimination contained in Section 11(c) would presumably be violated. Of course the continuing-violation theory is quite susceptible to being stretched beyond the breaking point. For example, a discharge is *not* a continuing violation, although some might argue that every day that the discharged employee is not brought back to work constitutes a separate violation of Section 11(c).

Few cases have addressed this continuing violation theory under the Act. But see *Usery v. Northern Tank Line, Inc.,*—F. Supp. —, 4 OSHC 1964, 1968 (D. Mont. 1976). Fortunately, however, there is a large body of law concerning what constitutes a continuing violation under a number of other federal labor statutes, including Title VII, the LMRA, and the ADEA. Perhaps the most important continuing-violation decision is the Supreme Court's decision under Title VII in *United Air Lines, Inc. v. Evans,* 431 U.S. 553 (1977). The *Evans* case involved the validity of an airline's policy of refusing to allow its stewardesses to be married. The plaintiff, a stewardess, had been forced to resign when she married, but did not challenge the airline's policy at that time. Instead, she waited four years and sued the airline for refusing to credit her with the seniority between her resignation and her rehiring. Since her forced resignation was concededly time-barred, the stewardess alleged that the airline had engaged in a "continuing violation" of Title VII, in an effort to state a timely claim. The Supreme Court rejected the stewardess's continuing violation theory, however, and held "that the Complaint was properly dismissed." *Id.* at 557. As the Court further observed:

> A discriminatory act which has not made the basis for a timely charge is the legal equivalent of a discriminatory act which occurred before the statute was passed. It may constitute relevant background evidence in a proceeding in which the status of a current practice is at issue, but separately considered, it is merely an unfortunate event in history which has no present legal consequences.

Id. at 558. See also, *e.g., Mobley v. Acme Markets, Inc.,* 473 F. Supp. 851 (D. Md. 1979).

2. Investigation of the Complaint, and Court Action

Section 11(c)(2) provides that the Secretary "shall cause such investigation to be made [of the discrimination complaint] as he deems appropriate. . . ." While there are no regulations on this point, it is reasonable to presume that the Secretary is bound by the same general considerations in conducting his investigation as the EEOC is under Title VII. See generally, Schlei & Grossman, *Employment Discrimination Law,* at 778–815. One court has indicated, however, that the Secretary of Labor is not *required* to conduct an investigation before he files a lawsuit. *Dunlop v. Hanover Shoe Farms, Inc.,* 441 F. Supp. 395, 4 OSHC 1241 (M.D. Pa. 1976). Nor does the failure to afford the employer the opportunity to a hearing, prior to filing suit in federal district court deprive him of due process, since judicial proceedings do so. *Id.*

If the Secretary determines that discrimination has occurred, on the basis of his investigation, Section 11(c) indicates that "he shall bring an action in any appropriate United States district court against" the alleged discriminator. Decisions that have addressed the point indicate that aggrieved employees cannot themselves sue directly in federal court but rather that only the

Secretary is authorized to bring a federal lawsuit under Section 11(c). *Taylor v. Brighton Corp.*, 616 F.2d 256, 8 OSHC 1010 (6th Cir. 1980). Section 11(c) does not create a private right of action. *Powell v. Globe Indus., Inc.*, 431 F. Supp. 1096, 5 OSHC 1250 (N.D. Ohio 1977).

While there is no specific statutory language or regulations on this point, it is reasonable to expect that, as under Title VII, only such claims of discrimination that are stated in the Section 11(c) complaint or developed during the course of a reasonable investigation by the Secretary of that complaint may form the basis for a subsequent civil action in federal court under Section 11(c). See, *e.g.*, *EEOC v. Chesapeake & Ohio Ry. Co.*, 577 F.2d 229 (4th Cir. 1978); *EEOC v. General Elec. Co.*, 532 F.2d 359 (4th Cir. 1976). Section 11(c) lawsuits have been held not to give rise to a right of trial by jury. *Dunlop v. Hanover Shoe Farms, Inc.*, 441 F. Supp. 385, 4 OSHC 1241 (M.D. Pa. 1976). This accords with Title VII precedents. See, *e.g.*, *Slack v. Havens*, 522 F.2d 1091 (9th Cir. 1975), and *EEOC v. Detroit Edison Co.*, 515 F.2d 301 (6th Cir. 1975), *vacated on other grounds*, 431 U.S. 951 (1977).

In addition to giving jurisdiction to restrain persons from committing acts of discrimination, Section 11(c)(2) authorizes federal courts to "order all appropriate relief including rehiring or reinstatement of the employee to his former position with back pay (plus interest)." See *Marshall v. N.L. Indus., Inc.*, 618 F.2d 1220, 8 OSHC 1166 (7th Cir. 1980)—upholding power to award backpay. See also *Marshall v. Wallace Bros. Mfg. Co.*,—F. Supp. —, 7 OSHC 1022 (M.D. Pa. 1979)—awarding reinstatement with backpay and retroactive seniority, expungement of incident from personnel records, and posting of court order in a public place on the company's premises. The award of backpay has been held to be an integral part of the Act's equitable remedy for discrimination and thus is not a claim for damages affording an employer a right to a jury trial. *Hanover Shoe Farms, supra.*

3. Notification of Determination

Section 11(c)(3) of the Act provides that the Secretary is required to notify persons who file a discrimination complaint of his determination "[w]ithin ninety days of the receipt of a complaint. . . ." A determination, however, need not be made before the Secretary files an action in federal court. Rather, filing the lawsuit can itself constitute the determination under the Act. *Marshall v. S. K. Williams Co.*, 462 F. Supp. 722, 6 OSHC 2193 (E.D. Wis. 1978). Furthermore, the Secretary's regulations provide that the ninety-day provision in Section 11(c)(3) is merely "directory in nature," such that failure to comply with the ninety-day notification will not affect the validity of any subsequent federal lawsuit. 29 CFR 1977.16. This regulation was upheld in *Marshall v. N.L. Indus., Inc.*, 618 F.2d 1220, 8 OSHC 1166 (7th Cir. 1980):

> [S]ince there has been no showing that this delay impeded any settlement negotiations, as defendant has alleged it could, or that defendant was in any other way prejudiced by the Secretary's failure to comply with Section 11(c)(3), the Secretary's failure to meet the 90-day limit should not bar the action here. A similar rule has been adopted for another provision of the Act. See *Marshall v. Western Waterproofing Co.*, 560 F.2d 947, 951–952, 5 OSHC 1732, 1739–1740 (8th Cir.

1977); *Chicago Bridge & Iron Co. v. Occupational Safety & Health Review Comm'n,* 535 F.2d 371, 375–377, 4 OSHC 1181, 1185 (7th Cir. 1976); *Accu-Namics, Inc. v. Occupational Safety & Health Review Comm'n,* 515 F.2d 828, 833–834, 3 OSHC 1299, 1301–1302 (5th Cir. 1975), certiorari denied, 425 U.S. 903, 4 OSHC 1090 (all construing Section 8(e) of the Act).

4. Withdrawal of Complaint

While the Act is silent on the point, the Secretary has provided by regulation that an investigation will not necessarily be terminated simply because an employee attempts to withdraw a discrimination complaint. 29 CFR 1977.17. The reason given by the Secretary for this policy is that enforcing Section 11(c) "is not only a matter of protecting rights of individual employees, but also of public interest." *Id.* The regulation also notes, however, that a "voluntary and uncoerced request" to withdraw a complaint "will be given careful consideration and substantial weight . . ." by the Secretary. *Id.*

5. Interface of Section 11(c) Remedies, Arbitration, and Other Administrative Remedies

Depending on the circumstances, an employee may be entitled to simultaneously pursue Section 11(c) relief, relief before other agencies, or remedies under the grievance and arbitration provisions of a collective bargaining agreement. In those situations, the Secretary has established the following policy:

> Where a complainant is in fact pursuing remedies other than those provided by section 11(c), postponement of the Secretary's determination and deferral to the result of such proceedings may be in order.

29 CFR 1977.18(a)(3). The regulations go on to provide just when the Secretary will postpone his determination under Section 11(c):

> Postponement of determination would be justified where the rights asserted in other proceedings are substantially the same as rights under section 11(c) and those proceedings are not likely to violate the rights guaranteed by section 11(c). The factual issues in such proceedings must be substantially the same as those raised by section 11(c) complaint, and the forum hearing the matter must have the power to determine the ultimate issue of discrimination.

29 CFR 1977.18(b).

It would appear that the regulation is at variance with the procedures adopted under other statutory schemes. For example, the National Labor Relations Board (NLRB), whose deferral rule inspired OSHA's rule, will not defer to arbitration in matters of individual discrimination under Section 8(a)(3) of the NLRA or in cases arising under Section 8(a)(4) for alleged discrimination resulting from filing of charges with or giving testimony before the NLRB. See, *e.g., Potter Electric Signal Co.,* 237 NLRB 209, 99 LRRM 1248 (1978); *Virginia Carolina Freight Lines, Inc.,* 155 NLRB 52, 60 LRRM 1331

(1965); see also *Alexander v. Gardner–Denver,* 415 U.S. 40 (1974). In the OSHA context, the courts have differed on the legitimacy of pre-arbitral and post-arbitral deference. For example, the postponement regulation was addressed recently in *Newport News Shipbuilding and Dock Co. v. Marshall,*—F. Supp. —, 8 OSHC 1393 (E.D. Va. 1980). In *Newport News,* seven employees charged their employer under Section 11(c) with retaliating against them for complaining about unsafe work conditions. OSHA completed its investigation of these charges, but postponed reaching a determination. OSHA cited a pending NLRB complaint, instigated by the union of the seven employees, as the reason why it was postponing its determination. The employer challenged this postponement decision in federal court, alleging that irrespective of the Secretary's regulations, OSHA was under a duty to reach a determination. The court agreed with the employer, holding as follows:

> Nothing in the Act says or indicates the Secretary may defer to another agency or body. The exclusive remedy to redress the alleged wrong is in an action by the Secretary. . . . The duty to act has been recognized as a mandatory duty. . . . Though defendant says he had followed the procedure of deferral in other cases, and has a regulation providing for deferral, nothing in the Act provides the Secretary may defer action. The Act says he shall bring the action. While he has leeway as to when he will file the action, nothing in the Act authorizes him to permit another agency to bring the action, or make a determination of liability under the Act. The Act says the action shall be brought in the appropriate district court. Such a court is the body authorized to determine the question." [*Newport News, supra* at 1395.]

In addition to the referenced regulation on postponement of a Section 11(c) determination, the Secretary has also established a correlative regulation outlining the circumstances under which he will defer to the outcome of other proceedings. Relying upon NLRB precedent, the Secretary has established the following as a precondition to his deferring to the results of other proceedings:

> [I]t must be clear that those proceedings dealt adequately with all factual issues, that the proceedings were fair, regular, and free of procedural infirmities, and that the outcome of the proceedings was not repugnant to the purpose and policy of the Act.

29 CFR 1977.18(c). This regulation was upheld in *Brennan v. Alan Wood Steel Co.,*—F. Supp. —, 3 OSHC 1654 (E.D. Pa. 1975), where the court held that the Secretary of Labor was entitled, in his discretion, to defer to an arbitration award, but was not required to do so. An example of a situation where the Secretary will *not* defer is where proceedings before a separate forum "are dismissed without adjudicatory hearing thereof. . . ." 29 CFR 1977.18(c). However, the court in *Marshall v. Firestone Tire & Rubber Co.,*—F. Supp. —, 8 OSHC 1637 (C.D. Ill. 1980) stated that the existence of an arbitrator's award does not in any way preclude the Secretary of Labor from seeking relief under Section 11(c). See also *Marshall v. N.L. Industries,* 618 F.2d 1220, 8 OSHC 1166 (7th Cir. 1980); *Marshall v. General Motors Corp.,*—F. Supp. —, 6 OSHC 1200 (N.D. Ohio 1977).

Finally in a Memorandum of Understanding between OSHA and the

NLRB a procedure was established for coordinating Section 11(c) litigation and litigation arising under Section 8 of the National Labor Relations Act. 40 Fed. Reg. 26083, June 20, 1975. The memorandum provides that since an employee's right to engage in safety and health activity is specifically protected by the OSH Act and is only generally included in the broader right to engage in concerted activities under the NLRA, it is appropriate that enforcement actions to protect such safety and health activities should primarily be taken under the OSH Act rather than the NLRA.

H. MISCELLANEOUS CONSIDERATIONS

1. Walkaround Pay

The Secretary has promulgated regulations establishing a *per se* rule that "an employer's failure to pay employees for time during which they are engaged in walkaround inspections, is discriminatory under Section 11(c)." This regulation further provides that employers must pay employees for time engaged in "other inspection related activities, such as responding to questions of compliance officers, or participating in the opening and closing conferences. . . ." 29 CFR 1977.21.

Courts have split on the validity of this regulation, which was promulgated in 1977. *Marshall v. The Ohio Bell Tel. Co.,*—F. Supp —, 8 OSHC 1242 (N.D. Ohio 1980)[10] (Valid); *Marshall v. Ohio Power Co.,*—F. Supp. —, 8 OSHC 1322 (S.D. Ohio 1980) (Invalid); and *Chamber of Commerce of the United States of America v. OSHA,* 636 F.2d 464, 8 OSHC 1648 (D.C. Cir. 1980) (Invalid). The most definitive rejection of this regulation, of course, is that of the D.C. Circuit in the *Chamber of Commerce* decision. There the court held that a company which "refuses to compensate an employee such [walkaround] time does not deprive the employee of rights conferred by the FLSA, and thus the employer's refusal to pay is not discriminatory under section 11(c)(1) of the Act." *Id.* at 1650. Furthermore, the court held that the walkaround pay regulation had to be vacated because, in promulgating that regulation, OSHA failed to comply with the notice and comment provisions of the Administrative Procedure Act. See also *Leone v. Mobil Oil Corp.,* 523 F.2d 1153, 3 OSHC 1715 (D.C. Cir. 1975), which invalidated the previous OSHA regulation on walkaround pay). Pursuant to the *Chamber of Commerce* decision, OSHA has withdrawn the regulation; it is uncertain whether it will be reissued.

2. Employee Refusal to Comply with Safety Rules

According to the Secretary's regulations, employees are not engaged in protected activity under Section 11(c) when they refuse to comply with safety rules implemented by an employer. Accordingly, an employer can take disciplinary action against employees who refuse to comply with safety rules without violating Section 11(c). Presumably, however, an employer that uses a safety rule as a pretext for discrimination would still violate the Act. Furthermore, according to the Secretary, refusal to comply with safety rules is different from refusing to work because of safety hazards, the latter being protected activity under certain circumstances. 29 CFR 1977.22. Disciplining (or not disciplining) an employee for refusing to comply with safety regulations can raise other issues, most notably in the arbitral forum or in the pleading and proof of certain affirmative defenses.

3. State Plans

According to the Secretary of Labor, state occupational safety and health programs must have provisions protecting employees from discrimination that are at least as effective as Section 11(c). Furthermore, while such provisions in a state plan do not divest either the Secretary or federal courts of jurisdiction under Section 11(c), the Secretary takes the position that he "may refer complaints of employees adequately protected by state plans' provisions to the appropriate state agency." 29 CFR 1977.23. OSHA Instruction CPL 2.21(A) (September 1, 1979), discussed *infra,* sets forth requirements that states adopt the procedures set forth therein for processing safety and health complaints or explain the reaons why it is not necessary to do so.

NOTES

1. The regulation also states that "employees of a state or political subdivision thereof would not ordinarily be within the contemplated coverage of Section 11(c)." 29 CFR 1977. 5(c).

2. Adverse action includes not only discharge, but suspension, reprimand, warning, transfer, reduction in pay, failure to promote, etc. *Marshall v. Firestone Tire and Rubber Company,*—F. Supp. —, 8 OSHC 1637 (C.D. Ill. 1980); *Usery v. Northern Tank Line, Inc.,*—F. Supp. —, 4 OSHC 1964 (D. Mont. 1976); *Whirlpool Corp. v. Marshall,* 445 U.S. 1, 8 OSHC 1001 (1980).

3. The ability of an inspector to enter onto an employer's workplace to investigate a complaint of discrimination filed under Section 11(c), appears to have not yet been questioned. Although Section 11(c) provides that upon receipt of a complaint of discrimination, the Secretary of Labor shall cause such investigation to be made as he deems appropriate, unlike the cognate OSHA inspection provisions set forth in Section 8(a), Section 11(c) authority for an inspector to enter a workplace is absent. Moreover, although there are regulations in 29 CFR 1903 with respect to entry into a workplace to conduct an inspection, there are no similar provisions in 29 CFR 1977, with respect to entry into a workplace for purposes of conducting an "11(c)" investigation. In addition, the Act in Section 11(c) contains no authorization for privately questioning employees on an employer's premises during discrimination investigations as does Section 8(a)(2).

4. Section 8(f)(1) entitles employees who believe that an "imminent danger" exists at the workplace to notify the Secretary of Labor of the danger. If the Secretary determines an inspection is unnecessary, he must notify the employee of his decision. If the Secretary is satisfied, however, that the notice provides him with reasonable

grounds to believe that an imminent danger exists, an OSHA inspector may enter the workplace to inspect and investigate the premises.

Section 3 of the Act, which sets forth Congress's definitions of principal words and phrases in the Act, does not provide a definition of "imminent danger." Under Section 13(a), however, the Secretary has the authority to enjoin "any conditions or practices in any place of employment which are such that a danger exists which could reasonably be expected to cause death or serious physical harm immediately or before the imminence of such danger can be eliminated through the enforcement procedures otherwise provided by [the Act]." The legislative history supports the conclusion that this is to serve as Congress's definition of the phrase "imminent danger." *E.g.*, S. Rep. No. 1282, 91st Cong., 2d Sess. 12, 35 [hereinafter cited as S. Rep.], *reprinted in* Subcommittee on Labor of Senate Committee on Labor and Public Welfare, 92d Cong., 1st Sess., *Legislative History of the Occupational Safety and Health Act of 1970,* at 152 (Comm. Print 1971).

5. The court in *Daniel* held that "at no point does the Act permit workers to make a determination that a dangerous condition exists in fact and that their employment or their employer's business may be halted by their refusal to work; moreover, neither the legislative history nor analogous case law implies the existence of this right."

6. Compare *Gateway Coal Co. v. Mine Workers,* 414 U.S. 368 (1974). The National Labor Relations Board has taken the position, with the concurrence of at least one court, that employees who refuse to work out of a good faith belief that they would subject themselves to an "abnormally dangerous" workplace condition are engaged in protected concerted activity within the meaning of the National Labor Relations Act, Section 502, *irrespective* of whether their sincere beliefs are reasonable. *NLRB v. Modern Carpet Indus.,* 611 F.2d 811, 103 LRRM 2167 (10th Cir. 1979). *Contra, Wheeling-Pittsburgh Steel Corp. v. NLRB,* 618 F.2d 1009, 104 LRRM 2054 (3d. Cir. 1980)—employee's belief as to danger of a workplace hazard must be sincere, made in good faith, *and* be reasonable. Moreover, under Section 7 of the National Labor Relations Act, employees have a protected right to strike over safety issues. But see *NLRB v. Washington Aluminum Co.,* 370 U.S. (1964).

7. Several courts subsequent to *Whirlpool* have held that OSHA must prove each of the following conditions in order to show discrimination under Section 11(c), following adverse employer action due to refusal to work in conditions of "imminent danger": (i.) That the conditions of the employment were, by objective test, in fact, dangerous; (ii.) That the employee's refusal was in good faith; (iii.) That the employee had no reasonable alternative to refusing; (iv.) That the employee's apprehension of death or serious injury was based on circumstances then facing him which would cause a reasonable person to reach the same conclusion; (v.) That the urgency of the situation provided insufficient time to eliminate the danger by resort to regular statutory enforcement channels; and (vi.) That the employee, where possible, sought from his employer, and was unable to obtain, correction of the dangerous condition. *Marshall v. National Industrial Constructors, Inc.,*—F. Supp. —, 8 OSHC 1117 (D. Neb. 1980); *Marshall v. N. L. Industries, Inc.,* 618 F.2d 1220, 8 OSHC 1166 (7th Cir. 1980).

8. Section 4(b)(1) of the Act has also been held not to bar a Section 11(c) action, since the phrase "working conditions" of employees under the statutory authority of other agencies does not include allegations of discrimination. *Marshall v. American Atomics, Inc.,*—F. Supp. —, 8 OSHC 1243 (D. Ariz. 1980).

9. An arbitration *decision* may not bar a subsequent lawsuit under Section 11(c), but settling a grievance *may* have this effect. *Marshall v. General Motors Corp.,*—F. Supp.—, 6 OSHC 1200 (N.D. Ohio 1977). This is consistent with Title VII precedent. *Alexander v. Gardner-Denver Co.,* 415 U.S. 36, 52 (1974) ("[A]n employee may waive his

cause of action under Title VII as part of a voluntary settlement. . . ."). See also, *e.g.*, *Strozier v. General Motors Corp.*, 442 F. Supp. 475, 481 (N.D. Ga. 1977); and *Lyght v. Ford Motor Co.*, 458 F. Supp. 137 (E.D. Mich. 1978).

10. The court's decision actually upheld the regulation's validity vis-à-vis pay for employees' time spent in inspection-related activities, such as the opening conference, rather than walkaround time. The court disallowed pay for travel to and preparation for the opening conference, in *absence* of proof it was necessary.

CHAPTER XVI
National Institute for Occupational Safety and Health

A. INTRODUCTION

The National Institute for Occupational Safety and Health (NIOSH) is the primary federal agency under the auspices of the Secretary of Health, Education, and Welfare (Health and Human Services) engaged in research, education, and training to eliminate on-the-job hazards potentially detrimental to the health and safety of employees. NIOSH was established within the Department of Health, Education, and Welfare (Health and Human Services) under the provisions of Section 22 of the Act in order to carry out the policies set forth in Section 2 of the Act and to perform the functions of the Secretary of Health, Education, and Welfare (Health and Human Services) under Sections 20 and 21 of the Act. NIOSH, under its own initiative or upon request of OHSA, is authorized to conduct research and experimental programs which are necessary for the development of criteria for new and improved safety and health standards and to recommend such standards to the Secretary of Labor. NIOSH has additional research responsibilities relative to the promulgation of standards under Section 101 of the Mine Safety and Health Act.

Administratively, NIOSH is located within the Department of Health, Education, and Welfare's (Health and Human Services) Center for Disease Control of the Public Health Service, headquartered in Rockville, Maryland. Its main research laboratories are located in Cincinnati, Ohio, where the studies that are conducted include not only those related to the physical effects of exposure to toxic or hazardous substances used in the workplace, but also the psychological, motivational, and behavioral factors involved in occupational safety and health. At NIOSH's laboratory in Morgantown, West Virginia, studies in respiratory disease, agricultural and noncoal mining health, and energy research are carried out. Additionally, all of NIOSH's safety research efforts, including the evaluation and certification of workers' personal safety equipment, are conducted at Morgantown.

B. STANDARDS DEVELOPMENT

NIOSH is required under Section 20(a) (1) and (2) of the Act to conduct research, experiments, and demonstrations relating to occupational safety and health, including studies of technological factors involved and relating to innovative methods, techniques, and approaches for dealing with occupational safety and health problems. In conjunction with this responsibility, NIOSH is required to consult with the Secretary of Labor in order to develop as necessary, criteria identifying toxic and hazardous substances which will enable OSHA to meet its responsibility for the formulation of safety and health standards under the Act. The purpose of the criteria documentation process, like most of NIOSH's research activities, is to evaluate thoroughly the problems in the workplace as the first step in protecting the health and safety of workers exposed to an ever increasing number of potential hazards in the workplace. Unless the health and safety hazards of a material, agent, substance, or process are known and documented, appropriate control measures by labor, industry, and government may prove difficult, if not impossible.

The National Institute for Occupational Safety and Health (NIOSH) evaluates all available research data and criteria and recommends standards for safe work practices and occupational exposure to toxic substances. The Secretary of Labor will weigh these recommendations along with other considerations, such as feasibility and means of implementation, in promulgating regulatory standards. NIOSH will also periodically review the recommended standards to ensure continuing protection of workers and will make successive reports as new research and epidemiologic studies are completed and as engineering controls for the workers' safety are developed.

Prior to publication of criteria documents recommending adoption, revocation, or modification of exposure standards, NIOSH will review and analyze existing (and available) literature on the subject and NIOSH will of course prepare and add its own. NIOSH criteria documents, for a given "hazard," contain recommended environmental exposure limits, medical examination requirements, labeling and environmental methods, engineering controls, workplace practices, personal protective equipment and clothing, and recordkeeping requirements, and, as such, have an impact on workplace conditions even prior to promulgation as standards. Labor and industry use criteria documents as guides for the control of hazards, even though not effective as law. [An attempt to utilize a published NIOSH criteria document on the permissible limits of phosgene exposure, to establish the existence of a recognized hazard under Section 5(a)(1) of the Act was not permitted in *Toms River Chemical Corporation,* 6 OSHC 2192 (1978)]. Once the criteria document has been drafted, it is reviewed by professionals in the particular area including union and industry safety and health experts. Their comments and criticisms are submitted to NIOSH for final review.

Upon completion, NIOSH's recommended criteria are transmitted to OSHA which has the responsibility for the actual development, promulgation and enforcement of toxicity standards. NIOSH has no direct authority to promulgate or enforce standards. It has been argued, however, that the

Secretary of Labor is bound by the NIOSH criteria in promulgating standards to the extent that a proposed standard is otherwise feasible. The District of Columbia Circuit has rejected this argument holding that NIOSH criteria are advisory only, meant to be an "aid" to OSHA, not a conclusive determination:

> It is the Secretary, rather than NIOSH who conducts hearings and receives the comments of interested persons.
>
> The Secretary may also appoint a special advisory committee to assist him in his standard-setting functions, and receive recommendations from it, as it did here.
>
> The Act, or so it seems to us, must be taken as contemplating that the Secretary may consider all of this information as well as that received from NIOSH. [*Industrial Union Dep't., AFL-CIO v. Hodgson,* 499 F.2d 467, 1 OSHC 1631 (D.C. Cir. 1974).]

The Commission has also rejected similar arguments. *GAF Corp.*, 3 OSHC 1686 (1975).

NIOSH has issued criteria documents recommending the regulation of some ninety separate toxic or hazardous substances. 45 Fed. Reg. 45212 (May 23, 1980). In the future, however, NIOSH will adopt a new approach of documenting hazards by *groups of similar substances.* Such criteria documents will consider the interaction between substances and the combined effects they might have on the health of employees who are seldom exposed to only one hazard in their working lifetime. Moreover, new critiera documents will concentrate on methods of controling exposure rather than establishing permissible exposure levels. Documents on single substances will be prepared only for special reasons (*e.g.*, reproductive hazards).

C. RESEARCH AND SURVEILLANCE STUDIES

In order to comply with its responsibilities under Section 20(a)(2) of the Act and in order to develop information regarding potentially toxic substances or harmful physical agents, NIOSH is authorized to prescribe regulations requiring employers to measure, record, and report on employees' exposure to those substances and agents which it reasonably believes may endanger the health and safety of employees. NIOSH is also authorized to establish programs of medical examinations and tests which are necessary to determine the incidences of occupational illnesses and susceptibility of employees to such illnesses. Nothing, however, requires or authorizes medical examinations or treatment for those employees who object on religious grounds except where it is necessary for the protection of the health or safety of others. Section 20(a)(5) of the Act. Upon the request of any employer who is required to measure and record exposure of employees to substances or physical agents, NIOSH is required to furnish financial and other assistance.

NIOSH relies on an extensive research program, conducted in its own laboratories and under contract with universities, public agencies, and private research institutes, for the development of information on which to base recommendations for new health and safety standards. Section 20(c) of the Act. NIOSH-sponsored research covers a wide spectrum, ranging from how a chemical acts within the body to the psychological effects of a stressful job. Much of the research falls within the following categories: animal toxicology, techniques for measurement and analysis of workplace contaminants, physical hazards (noise, radiation, heat, vibration), and psychological and social stresses (*e.g.*, studies of job stress indicate that the daily psychological stress imposed on some workers makes those people more susceptible to physical and mental illness).

Preparation of industrial hygiene, medical and epidemiological studies of occupational exposure to workplace hazards is also specifically authorized under the Act. To this end, NIOSH conducts field research studies in selected industries, evaluates its findings, reports on these findings, and recommends "safe" conditions, practices, and standards for personnel, materials, substances, and personal protective equipment found in the workplace. Section 20(a)(7) of the Act also requires that NIOSH conduct and publish *industry-wide studies* of the long-term effects of chronic or low-level exposure to industrial materials, processes, and stresses on the potential for illness, disease, or loss of functional capacity in aging adults. NIOSH conducts some 40 industry-wide studies each year in a wide range of occupational groups. Recent investigations have explored the effects of anesthetic gases on hospital operating room employees; of grain dust on grain elevator workers; of yeast and flour dust on bakery and confectionery workers; of the effect of asbestos, silica, talc, and solvents on painters; and of the effect of nonionizing radiation on VDT users. Current industry-wide studies are focusing on two of NIOSH's major areas of concern—occupational cancer and hazards to the reproductive systems of workers. NIOSH uses the results of these industry-wide studies and its Occupational Hazards Survey to develop relevant information toward improvement of employee occupational safety and health.

In addition to industry-wide studies, health hazard evaluations (HHE) serve as an early detection system for new and emerging occupational health problems. HHE's are used to collect and provide data on human exposure to toxic substances and thereby assist in providing support for validating, reassessing, and establishing criteria on which health standards can be based. Pursuant to Section 20(a)(6) of the Act, if employers, employees, or authorized employee representaitves request an on-site investigation of a health hazard in the workplace, NIOSH is required to conduct an HHE. A health hazard evaluation means an investigation of the potentially toxic or hazardous effects of any *substance*[1] normally found or used in any place of employment to which the Act is applicable.

NIOSH, in 42 CFR 85, has promulgated regulations applicable to requests submitted pursuant to Section 20(a)(6) of the Act for health hazard evaluations. A request for a health hazard evaluation should be submitted to NIOSH and should (1) be in writing and be signed by the employer or the authorized representative of employees; (2) set forth the requester's name, address, and telephone number; address of the place of employment where the substance is found; the specific workplace or workplaces involved; and the specific process or type of work which is the source of the substance or in which the substance is used; (3) specify with reasonable particularity the nature, conditions, circumstances, or other grounds on which the request was made; (4) state, where the requester is other than the employer, that he is an authorized representative for purposes of collective bargaining or that he is an employee of employer and is authorized by two or more employees employed in the workplace where the substance is found to represent them for purposes of the Act, or that he is one of three or less employees employed in the workplace where the substance is found; and (5) indicate whether the requester desires that NIOSH not reveal his name. NIOSH has developed a form entitled "Request for Health Hazard Evaluation" to assist persons in requesting evaluations, which is available upon request from NIOSH. (Attachment I).

When a request has been submitted, and NIOSH has concluded there is *reasonable cause* to believe that an HHE is warranted, NIOSH will arrange to inspect the place of employment, collect samples, and perform tests as are necessary.[2] If there is no reasonable cause to conclude that an investigation is warranted, the requester will be notified. 29 CFR 95.4(a) and (b). Normally, HHE's will be conducted without prior notification to the employer, but, in certain instances, advance notice of the visit will be given in order to expedite a thorough and effective investigation. Like inspections in workplaces, HHE's begin with the presentation of credentials and an explanation of the nature, purpose, and scope of the investigation. Where the investigation is the result of a request submitted by employees or an authorized representative of employees, a copy of the request is provided to the employer, unless the requester has asked that his name not be revealed. 29 CFR 85.7(a). As in workplace inspections, employers are permitted to identify trade secret information that might be obtained during the inspection. NIOSH is authorized to collect samples, photographs, employ other investigatory techniques, including medical examinations of employees with their consent, and question privately employers and employees. Again, conduct of HHE investigatons shall be such as to preclude unreasonable disruption of the operations of the employer.

Where the request for a health hazard evaluation has been made by an authorized representative of employees, both the representative of the employer and the authorized representative of the employees shall be given an opportunity to accompany NIOSH in the investigation of the workplace. Additional employer and employee representatives may accompany the NIOSH officer if they do not interfere with the conduct of the inspection. Procedures for disclosure of imminent dangers discovered during an HHE are similar to those provided in the regulations on investigations of places of employment. Pursuant to the OSHA/NIOSH Interagency Agreement, discussed *infra,* OSHA is also informed of serious safety and health hazards.

At the conclusion of an HHE, a determination is made pursuant to Section 20(a)(6) of the Act in order to identify and set forth the concentrations of the substances found in the place of employment, its conditions of use, and whether such substances have potentially toxic effects in such concentrations as used or found. A copy of the determination will be given to the employer and authorized represenative of affected employees. If NIOSH determines that the substance is potentially toxic in concentrations used or found and such substance is not covered by an occupational safety and health standard, it will submit such determination to OSHA, together with pertinent criteria. The employer must post a copy of the determination for a period of thirty calendar days at or near the workplaces of affected employees. 42 CFR 85.11(c).

The findings of HHE investigations are generally included and disseminated as part of NIOSH criteria documents and NIOSH technical reports. The information is normally presented in a manner that does not identify any specific place of employment. It should be noted, howevever, that specific reports of investigations of each place of employment are subject to mandatory disclosure, upon request, under the provisions of the Freedom of Information Act.

Under Section 101 of the Mine Safety and Health Act, NIOSH was given responsibility for conducting HHE's in covered places of employment. In order to include coverage activities involving occupational health in mining, NIOSH recently amended its regulations regarding requests for health hazard evaluations.

Located within the same NIOSH "branch" as health hazard evaluations, technical service assistance provides technical information to employers and employees concerning health or safety conditions at workplaces, such as the possible hazards involved in working with specific solvents and the proper use of personal protective equipment. Technical service assistance is rendered pursuant to NIOSH's authority to conduct workplace inspections. See discussion, *infra*. Unlike a health hazard evaluation study, the employer is entitled to have input in the formulation of the technical service assistance report. Other services provided by NIOSH include: (1) accident prevention, which provides technical assistance for controlling on-the-job injuries including the evaluation of special problems and recommendations for corrective action; (2) industrial hygiene, which provides technical assistance in the areas of engineering and industrial hygiene, including the evaluation of special health-related control measures problems in the workplace; and (3) medical service, which provides assistance in solving occupational medical and nursing problems in the workplace, including the assessment of existing medically-related needs and the development of recommended means for meeting such needs.

D. ACCESS TO INFORMATION

In order to carry out its research responsibilities, Section 20(b) of the Act sets forth NIOSH's right to enter places of employment and request and obtain information:

> The Secretary of Health, Education, and Welfare is authorized to make inspections and question employers and employees as provided in Section 8 of this Act in order to carry out his functions and responsibilities under this Section.

Although NIOSH is given the same authority to enter and inspect workplaces as is OSHA, unlike OSHA, it is not concerned with enforcement of the

Act; rather, it uses its authority to conduct research. NIOSH regulations implementing the relevant provisions of the Act for investigations of places of employment are contained in 42 CFR 85(a) and refer to NIOSH's authority, set forth in Sections 20 and 8 of the Act, to enter places of employment for purposes of obtaining information necessary to its research, experiments, studies, and other activities. 42 CFR 85(a)3. It is sufficient to observe that as to the authority to investigate, the regulation repeats in substantial part Section 8(a) of the Act.

By far the most significant judicial interpretation of the inspection portion of the Act is the Supreme Court's decision in *Marshall v. Barlow's Inc.*, 436 U.S. 307, 6 OSHC 1571 (1978), where it was held that "the Act [in Section 8(a)] is unconstitutional insofar as it purports to authorize inspections without a warrant or its equivalent." The court in *Barlow's* dealt with the level of probable cause necessary to obtain an administrative inspection warrant, stating that probable cause in a criminal law sense is not required. For purposes of an administrative search, probable cause justifying the issuance of a warrant may be based not only on evidence of a violation, but also on a showing that reasonable legislative or administrative standards for conducting an inspection are satisfied with respect to the particular establishment. Although there are few reported cases *specifically* extending *Barlow's* and the inspection requirements it imposed to NIOSH investigations, the decisions that have issued seem to make it clear that since the NIOSH inspection authority is also based upon Section 8(a) and an inspection must be based upon some showing of administrative probable cause. See *Establishment Inspection of Pfister & Vogel Tanning Company,* 493 F. Supp. 351, 8 OSHC 1693 (E.D. Wis. 1980); *Establishment Inspection of Keokuk Steel Castings, Division of Kast Metals Corp,* 493 F. Supp. 842, 8 OSHC 1730 (S.D. Iowa 1980). The court in *Keokuk Steel Castings* found that reasonable legislative standards for conducting a health hazard evaluation are set forth in Section 28(a)(6) of the Act. Reasonable administrative standards are set forth in 42 CFR 85. In particular, 42 CFR 85.3 requires the requester to include pertinent general information as well as to specify with reasonable particularity the nature of the conditions, circumstances, or other grounds on which the request is made. 42 CFR 85.4 requires that NIOSH conclude that there is reasonable cause to believe that the requested investigation is necessary before it proceeds with the investigation.

It should be noted, however, that the argument could still be made post-*Barlow's,* that a NIOSH inspection does not require a warrant. The basis for this argument is the language in *Barlow's* which suggests that the Court decided as it did was because of the well-established Fourth Amendment prohibition on warrantless searches during criminal investigations or other investigations which could expose persons to sanctions. The Court stated:

> If the government intrudes on a person's property, the privacy interest suffers whether the government's motivation is to investigate violations of criminal laws or breaches of other statutory or regulatory standards. [98 S. Ct. at 1820.]

This seems to make relevant the major difference between an OSHA investigation and a NIOSH investigation: NIOSH inspections do not search for breaches of regulatory standards and do not result in the issuance of citations or penalties; rather they are experimental and research-oriented. However, due to the fact that NIOSH will submit reports of its investigation to OSHA and report imminent-danger situations, there is a possibility of the initiation of regulatory sanctions following a NIOSH inspection. The better likelihood, then, is that a court will continue to extend the *Barlow's* holding to NIOSH inspections, largely on the grounds that *Barlow's* seemed to narrow the areas which are exempt from the warrant requirement beyond the facts presented by a NIOSH inspection.

It probably does not make much of a practical difference, however, whether a NIOSH inspection requires a warrant, since the probable cause standard set out in *Barlow's* for getting an administrative warrant is quite easily satisfied. This probable cause requirement has two dimensions: it "may be based not only on specific evidence or an existing violation but also on a showing that "reasonable legislative or administrative standards for conducting an inspection are satisfied with respect to a particular establishment.' " *Camara v. Municipal Court.* 98 S. Ct. at 1824 (fn. omitted). The only reported decision to specifically discuss NIOSH's authority to inspect (pre-*Barlow's*) was a consent decree issued in *U.S. v. American Annodizing Co.,* 1973–74 OSHD ¶16,728 (D. Ill. 1973), which orderd an employer to admit NIOSH inspectors as part of a "National Occupational Hazards Survey." Several post-*Barlow's* cases have, however, specifically discussed NIOSH's authority to inspect, and are discussed later. An important post-*Barlow's* OSHA case, however, which construed probable cause in the context of the conduct of inspections predicated on the "existence of health hazards," is *In the Matter of Establishment Inspection of Gilbert & Bennett Manufacturing Company,* 589 F.2d 1335, 6 OSHC 2151 (7th Cir.) *cert. denied,* 444 U.S. 884, 7 OSHC 2238 (1979). In *Gilbert & Bennett,* it was decided that "detailing the employee's complaint and indicating the bases for concluding that potentially significant health hazards to workers were alleged by it" are enough to satisfy the "specific evidence of an existing violation" language of *Barlow's.* The fact that the employer was selected for inspection on the basis of a national-local plan to reduce a high incidence of occupational injuries and illnesses in its industry also constitutes a sufficient showing of probable cause.

Assuming probable cause requirements are met, investigations must be conducted in a reasonable manner, during regular working hours or at other reasonable times and within reasonable limits. Normally, NIOSH will contact an employer's representative prior to the visit and explain the details of why an investigation of the place of employment is being conducted, except that advance notice will not be given when it would adversely affect the validity and effectiveness of an investigation. 42 CFR 85.a(4). Prior to beginning a visit, NIOSH will present credentials, explain the nature, purpose, and scope of the investigation and the records to be reviewed. 42 CFR 85.a(5). At this point in time, it is encumbent upon the employer to precisely identify that information which is a trade secret and to designate it as such.

A representative of the employer is permitted to accompany the NIOSH representative during the inspection of the place of employment. 42 CFR 85.a(5)(d)(1) and .a(3)(a) provide that NIOSH representatives may collect environmental samples, measure environmental conditions, and employee exposure; take or obtain photographs; employ other reasonable techniques, including medical examinations, anthropometric measurements, and standardized and experimental functional tests of employment with the informed consent of employees; review, abstract, and duplicate personnel records as are pertinent to mortality, morbidity, injuries, safety, and other studies; question and interview privately employers and employees and review employment records, medical records, and records required to be maintained by the Act and its regulations. The employer, under 42 CFR 85.a(6), is required to provide suitable space to NIOSH at the place of employment to conduct private interviews, medical examinations, and other such tests.

The results of the investigation will be made available by NIOSH to the employer, with copies to OSHA and appropriate state agencies. (If NIOSH has reasonable cause to believe an imminent-danger situation exists, NIOSH must advise the employer and OSHA or the appropriate state agency). 42 CFR 85(a)(7). Prior to release of the results, however, a preliminary report will be sent by NIOSH to the employer for review of trade secret information and technical inaccuracies, at which time the employer may review and make adjustments. The findings of the investigation generally will be made available to the public and may be disseminated as part of NIOSH's criteria documents, technical reports, informational packets, and other journals or publications. It should be noted that while the findings will be presented in a manner which does not identify the specific place of employment, specific reports of investigations of each place of employment are subject to mandatory disclosure upon request under the provisions of the Freedom of Information Act. 5 U.S.C. §552; 42 CFR 85.a(8)(c).

In those few instances where NIOSH has been refused a right entry to carry out its Section 20 research functions, NIOSH's right to enter has been judicially enforced.

1. Inspections

In Establishment Inspection of Inland Steel Company, 492 F. Supp. 310, 8 OSHC 1725 (N.D. Ind. 1980), NIOSH sought a warrant authorizing a physical inspection of the company's premises, including the taking of air, material, and surface wipe samples; medical examinations and private interviews of employees; and examination, copying, or abstracting of medical and job history records of employees. The court first stated that such an *inspection* was authorized by Section 8(a) of the Act and was justified by a showing of administrative probable clause [an affidavit of a NIOSH industrial hygienist stating that there were employee respiratory problems due to exposure to trichlorothane was found to be supportive of probable cause]. Stating thereafter that although Section 8(a) of the Act did not on its face authorize medical examinations of employees (although it did permit private interviews), the court found it to be within a reasonable construction of the Act. Moreover, NIOSH regulations provide for medical examinations (with employee consent). With respect to these examinations, NIOSH must conduct and bear the costs and they must be conducted so as to preclude unreasonable disruptions of the employer's operations. 42 CFR 85.7(c), .7(e) and .8. Thus the court felt that medical examinations would be only slightly more disruptive than private interviews which are authorized by the Act. In addition, the court stated that NIOSH's mandate under Section 20(a)(6) of the Act is to determine whether a substance found in a place of employment has potentially toxic effects. The determination of toxicity will almost necessarily require an assessment of the health of those persons exposed by means of a medical examination. 42 CFR 85.4(a), 85.7(c), and 85.8.

With respect to the actual inspection, copying, or abstracting of medical and employment records, however, the court stated it was not within the scope of Sections 8(a) and 20(a)(6) of the Act, even though authorized by 42 CFR 85.5(a). The Act's legislative history, in the court's view, supported the interpretation that Section 8(a) limits inspections to interviewing, physical observation, and measurement. Allowing the inspection of records and other documentary evidence by means other than an administrative subpoena was not within the scope of the provision, an expansive reading of Section 8(a) to cover documents was unnecessary in the court's opinion, for Section 8(b) provided for the use of subpoenas and required the testimony of witnesses.[3] The existence of a subpoena power was a clear indication that access to documents was not granted under the basic inspection authority of Section 8(a) of the Act. *Citing E. I. du Pont de Nemours & Co. v. Finklea,* 442 F. Supp. 821, 6 OSHC 1167 (S.D. W. Va. 1977).

In another case entitled *Establishment Inspection of Keokuk Steel Castings, Division of Kast Metals Corp.,* 493 F. Supp. 842, 8 OSHC 1730 (S.D. Iowa 1980), the court permitted NIOSH entry to a plant in order to conduct a health hazard evaluation requested by an authorized employee representative. The grounds for the HHE, which alleged exposure to coremaking chemicals including isocyanates and aliphatic amines, were set forth with reasonable particularity and the employee representative alleged information which could support probable cause for the warrant. In upholding the warrant, the court specifically rejected the employer's attempt to obtain an agreement from NIOSH to three specified conditions prior to allowing entry into its premises: (1) no private interviews with employees during working hours on the plant premises would be permitted; (2) the only records to be reviewed were those required to be kept by OSHA regulations; and (3) there would be no attachment of personal sampling devices to employees. The court stated that conducting private interviews with employees during regular working hours was consistent with 42 CFR 85.5(a); records other than those required by OSHA Regulations may be examined; and with respect to attaching personal sampling devices, the court stated that the employer, in *Plum Creek Lumber Co. v. Hutton,* 608 F.2d 1283, 7 OSHC 1940 (9th Cir. 1979), which was relied upon in the instant case to support the alleged right to condition entry upon an agreement not to use personal sampling devices, had a prior policy of forbidding employees from wearing devices; that the employer in this case had not asserted that it had a similar policy. The court, however, stated that the employees should be specifically advised by NIOSH that they could voluntarily decide whether or not to wear sampling devices.

An *ex parte* warrant was also upheld by the court in *Keokuk,* even in the absence of a NIOSH regulation expressly permitting such action. See also *Inland Steel, supra,* which stated that 42 CFR 85.6 does not forbid *ex parte* warrants.

Finally, in *Establishment Inspection of Pfister and Vogel Tanning Co.,* 493 F. Supp. 351, 8 OSHC 1693 (E.D. Wis. 1980), the court upheld the issuance of an inspection warrant to NIOSH in furtherance of its study into the health effects of occupational exposure in the leather tanning and finishing industry. The employer refused to consent to a walk-through inspection by NIOSH, at which point NIOSH obtained a warrant and conducted the inspection under protest. The purpose of the investigation, according to NIOSH, was to determine the suitability of including the employer in a subsequent in-depth industry-wide study. The employer, however, wished to avoid its inclusion in the further study and filed a motion to suppress the evidence obtained during the initial NIOSH walk-through inspection.[4] The court held that NIOSH had shown probable cause for the inspection warrant's issuance, since it was supported by an affidavit from a NIOSH epidemiologist which stated that NIOSH was investigating latent health effects in tanning workers due to their exposure to carcinogens found at other tanneries. The court felt that the affidavit set forth the necessary neutral criteria to meet the probable cause standard in the Supreme Court's *Barlow's* decision.[5]

The employer also claimed that NIOSH should have been required to rely on the results of a previous OSHA inspection which showed only trace amounts of carcinogens present in the employer's workplace. This OSHA inspection was not made known by NIOSH to the magistrate and according to the employer would have supported a basis for finding a lack of reasonableness in NIOSH's request to inspect the same premises. The court, however, rejected this argument, stating that the overall scope of the NIOSH inspection request appeared reasonable, despite the overlap with the OSHA study ("inefficiency is not a proper basis for invalidating a warrant").

2. Subpoenas

In *E.I. du Pont de Nemours & Co. v. Finklea,* 442 F. Supp. 821, 6 OSHC 1167 (S.D.W. Va. 1977), NIOSH, acting pursuant to Section 8(b) of the Act, issued an administrative subpoena *duces tecum* requiring duPont to produce, in connection with a research investigation, medical and personnel records of certain employees which were necessary to determine whether a statistical basis was present to conclude that the rate of cancer among the employees was job-related. DuPont requested its employees to indicate whether they would consent to such disclosure; for those who refused, duPont declined to comply with the subpoena. The court stated that Section 20(a)(6) of the Act, read in *pari materia* with Section 8(b), gave NIOSH the authority to issue subpoenas and that enforcement might be had if the inquiry was relevant and not too indefinite. While thereafter finding that a right of privacy existed with respect to the requested information, the court found that it did not appear that the right would be abridged, that is, used improperly, since NIOSH gave assurances of confidentiality. See also *United States v. McGee Industries,* 439 F. Supp. 296, 5 OSHC 1562 (E.D. Pa. 1977).

Recently, another court restricted NIOSH access to employee medical records during a health hazard investigation that would have identified the individual worker. The court ruled in *General Motors Corp. v. Finklea, Director of NIOSH,* 459 F. Supp. 235, 6 OSHC 1976 (S.D. Ohio 1978) that an employer "is not required to submit to NIOSH, medical records identified by name and address without the specific consent of the employee involved." General Motors agreed to turn over employee records provided that the employees involved signed a release. With respect to those workers who refused consent, General Motors refused to give NIOSH the medical records. The court in ruling on the refusal stated:

> If NIOSH is seeking only to accumulate statistical data regarding health conditions present in a group of employees who performed a certain process and not in others, [the court stated], it is clear that such information is pertinent, significant, and within its mandate of inquiry. Such information does not require the disclosure of identifying names or addresses.

Thus, while the court found that issuance of subpoenas was within NIOSH's authority, it also found, relying on Ohio law, that even though the requested information was relevant and pertinent to the investigation, it was privileged (patient/physician) and could not be disclosed without the patient's consent.

More recently, in *United States of America v. Westinghouse Electric Corporation,* 483 F. Supp. 1265, 7 OSHC 2179 (W.D. Pa.), *aff'd,*—F.2d—, 8 OSHC 2131 (3d Cir. 1980), the district court distinguished *General Motors* since there was no assertion of privilege under Pennsylvania law similar to that asserted under Ohio law. Even if there had been such an assertion, the court felt it would be irrelevant, since in federal question cases, reference must be to federal law on the existence and scope of privilege. No such privilege exists under federal law. In enforcing the subpoena, the court found that the evidence indicated there were assurances of nondisclosure in the general NIOSH procedure of safekeeping of records similar to those obtained from NIOSH in *du Pont,* even though the court did not specifically require it. See also, *In Establishment Inspection of Inland Steel Company,* 492 F. Supp. 310, 8 OSHC 1725 (N.D. Ind. 1980).

The *Westinghouse* court on appeal held that NIOSH, in conducting a health hazard evaluation, is not limited in a request for access to only those records which must be maintained under the Act or its regulations; other "reasonably relevant" records related to the purpose of the investigation may be inspected. With respect to the privacy question it stated:

> Although the full measure of the constitutional protection of the right to privacy has not yet been delineated, we know that it extends to two types of privacy interests: "One is the individual interest in avoiding disclosure of personal matters, and another is the interest in independence in making certain kinds of important decisions." *Whalen v. Roe,* 429 U.S. 589, 599–600 (1977).
>
> There can be no question that an employee's medical records, which may contain intimate facts of a personal nature, are well within the ambit of materials entitled to privacy protection . . . It has been recognized in various contexts that medical records and information stand on a different plain than other relevant material. For example, the Federal Rules of Civil Procedure impose a higher

burden for discovery of reports of the physical and mental condition of a party of other person than for discovery generally. *Compare* Fed. R. Civ. P. 35 with Fed. R. Civ. P. 26(b). *See also* 8 Wright and Miller, Federal Practice and Procedure: Civil, §§2237, 2238 (1970). Medical files are the subject of a specific exemption under the Freedom of Information Act. 5 U.S.C. §552(b)(6) (1976). This difference in treatment reflects a recogniztion that information concerning one's body has a special character.

In the cases in which a court has allowed some intrusion into the zone of privacy surrounding medical records, it has usually done so only after finding that the societal interest in disclosure outweighs the privacy interest on the specific facts of the case. In *Detroit Edison Co. v. NLRB*, 440 U.S. 301, 313-17 (1979), the interests of the NLRB and the labor union in giving the union access to employee's scores in psychological aptitude tests to assist the union in processing a grievance were weighed against the strong interest of the Company and its employees in maintaining confidentiality.

Thus, the interest in occupational safety and health to the employees in the particular plant, employees in other plants, future employees and the public at large is substantial. It ranks with the other public interests which have been found to justify intrusion into records and information normally considered private.

Turning next to NIOSH's need for access to the requested information, the court found that it showed a reasonable need for the employee's medical file. It then addressed the question of security provisions to prevent its unauthorized disclosure. Westinghouse argued that the security of the information was inadequate. It stressed that the statute authorized use of outside contractors for data processing and analysis, and contained neither means to police compliance with nondisclosure nor adequate sanctions for unwarranted disclosure. It also argued that removal by NIOSH of individual identifiers before disclosure to those parties is made will be inadequate, since only obvious identifiers, such as names and addresses, are deleted, but that identification may still be possible because more idiosyncratic identifiers will be included in the disclosed materials. For these reasons, Westinghouse conditioned its compliance with NIOSH's request for the records in issue on the provision by the government of written assurance that the contents of the employees' medical records would not be disclosed to third parties, a condition the government refused to accept.

The district court concluded that the "evidence indicates that NIOSH's procedures of safekeeping the records and of removing the names and addresses of the individuals in its compilation of published data represents sufficiently adequately assurance of non-disclosure by the petitioner [NIOSH]." The appellate court saw no reason to disturb this conclusion.

Westinghouse also suggested that the employees' private rights would not be adequately preserved unless their written consents were obtained. The court responded:

One district court has already conditioned the disclosure of personal identifiers on conducting a due process hearing. *General Motors Corp. v. Director of NIOSH*, 459 F.Supp. 235, 6 OSHC 1976 (S.D. Ohio 1978). We believe that the requirement of securing written consent may impose too great an impediment to NIOSH's ability to carry out its statutory mandate. Under the circumstances, we believe the most appropriate procedure is to require NIOSH to give prior notice to the employees whose medical records it seeks to examine and to permit the employees to raise a personal claim of privacy, if they desire. The form of notice may vary in each case but it should contain information as to the fact and purpose of the investigation and the documents NIOSH seeks to examine, and should advise the employees that if they do not object in writing by a date certain, specifying the type of material they seek to protect, their consent to disclosure will be assumed. The mechanics of the required notice can be arranged by the district court in accordance with the circumstances of the particular case.

3. Regulations on Access

On May 23, 1980, OSHA published an amendment to 29 CFR 1910.20, containing the regulations dealing with access to employee exposure and medical records for those employees exposed to toxic substances or harmful physical agents.[6] 45 Fed. Reg. 35212. The original purpose of the proposal for access to exposure and medical records was to make the data contained therein available to NIOSH, OSHA, and employees[7] in order to promote the recognition of workplace hazards and the subsequent reduction of occupational disease. 43 Fed. Reg. 31371 (July 21, 1978). The final standard differs from the proposal in that NIOSH is not covered, since, according to the regulation, NIOSH has independent legal authority under Section 20 of the Act to seek access to employee exposure and medical records. It was felt that administrative regulations may be ill-suited to NIOSH operations and thus OSHA decided to exclude NIOSH from the final standard's access provisions. See discussion of the court decisions upholding NIOSH's right to seek access to employee medical records, even though the right of privacy has been recognized as constitutionally protected. *Roe v. Wade,* 410 U.S. 113 (1973); *Griswold v. Connecticut,* 381 U.S. 479 (1975); *Whalen v. Roe,* 429 U.S. 589 n. 23 (1977).[8] In *United Steelworkers of America, AFL-CIO-CLC v. Marshall,* —F.2d—, 8 OSHC 1810 (D.C. Cir. 1980), the court, in upholding the validity of the lead standard also approved the provision providing for unrestricted access to "identifiable" employee medical records by OSHA and NIOSH, noting that the Act itself permits and even requires such access:

> [C]ontrary to LIA's contention, the Supreme Court's decision in *Whalen v. Roe,* 429 U.S. 589 (1977), does not invalidate the government's access to medical records as a violation of employees' constitutional rights. In holding that the New York system for recording the identities of people taking prescribed addictive drugs did not violate these people's "zone of privacy," the *Whalen* Court did rely in part on the detailed protections against unwarranted disclosure contained in the statute creating the system. *Id* at 601, 605. But the Court stated explicitly that it was expressing no opinion on the constitutionality of any system of government collection of confidential information that did not contain similar protection.
>
> * * *
>
> Moreover, to obtain express-permission for disclosure from each of millions of workers would create an unthinkable administrative burden and risk the health of workers who for unfortunate but understandable reasons might fear any disclosure of their health records. The rule on government access thus meets the demand of *Whalen* that such required disclosure be a reasonable exercise of

government responsibility over public welfare, *id.* at 597–598, and the requirement under the OSH Act that any OSHA program be reasonably related to the general goal of preventing occupational lead disease, 29 U.S.C. §652(8) (1976).

OSHA's own right to subpoena employer-generated exposure and medical records was upheld in *Marshall v. American Olean Tile Co.*, 489 F. Supp. 32, 8 OSHC 1136 (E.D. Pa. 1980). In that case, the court stated that although OSHA may obtain such information through recordkeeping requirements or investigations by NIOSH, the remedial purposes of the Act can also be furthered by inspections through subpoenas designed to explore the causes and prevention of occupational illnesses. Thus the court declined to find that the authorization of inspections to carry out the purposes of the Act extends only to enforcement. Rather, the inspection authority is also relevant to the Secretary's legislative functions, and therefore it enforced the subpoena.

In the amended 29 CFR 1913 concerning OSHA access to employee medical records, there is now a provision for OSHA access to medical records which contain personally identifiable employee medical information, whether or not pursuant to the access provisions of 29 CFR 1910.20(e). According to the standard, OSHA access to these medical records under certain circumstances is important to the agency's performance of statutory functions. However, since those records contain personal details concerning the lives of employees, and due to the substantial personal privacy interest involved, OSHA authority to gain access to personally identifiable employee medical information will be exercised only after the agency has made a careful determination of the need for this information and only with appropriate safeguards to protect individual privacy. Once the information is obtained, OSHA's examination and use of it is limited to that information needed to accomplish the purpose of access and will be retained only for so long as to accomplish that purpose and will not be disclosed to other agencies, except in narrowly defined circumstances. 29 CFR 1913.10(a). One of the agencies to which personally identifiable employee medical information may be transferred is NIOSH. 29 CFR 1913.10(m). In fact, the preamble to the rule states that its purpose was to make the data available to OSHA, NIOSH, and employees, for it is felt that access to such records is important to the detection, treatment, and prevention of occupational diseases, would facilitate occupational health research, and would assist employees to make complaints to OSHA and NIOSH.

E. EDUCATION AND TRAINING

Under Section 21 of the Act, the Secretary of Health, Education, and Welfare (Health and Human Resources), through NIOSH, is required to conduct educational programs to provide an adequate supply of personnel to serve the purposes of the Act and informational programs on the importance and proper use of adequate safety and health equipment. NIOSH is also authorized to conduct short-term training of personnel engaged in work related to its responsibilities under the Act and advise employers, employees, and organizations representing employers and employees as to effective means of preventing occupational injuries and illnesses. Sections 21(a), (b), and (c) of the Act. The long-term approach to an adequate supply of training personnel in occupational safety and health is found in the colleges and universities and other institutions in the private sector. NIOSH encourages such institutions, by contracts and training grants, to expand their curricula in occupational medicine, occupational health nursing, industrial hygiene, and occupational safety engineering. For those schools that already have funding, NIOSH offers consultation. NIOSH also extends its training reach by cooperating with such outside groups as the American Industrial Hygiene Association and the American Society of Safety Engineers. NIOSH develops courses, sometimes complete with equipment, and the private groups provide the lecturers who present the training at locations around the country. Every year NIOSH offers these courses, lasting from three days to two weeks, which are attended by persons from federal, state, and local governments; colleges and universities; insurance companies; labor unions; and private industry.

F. NIOSH PUBLICATIONS

NIOSH encourages employers and employees to ascertain if there is or potentially is exposure to occupational safety and health hazards or if employees have been in contact with toxic substances or hazardous materials. In addition to technical publications such as the criteria documents and the annual registry of toxic substances intended for specialized audiences, NIOSH also publishes booklets presenting good practices for persons who work in specific industries or with specific materials. These publications discuss problems faced by many kinds of workers—electroplaters, painters, foundry workers, battery workers, insulation installers, and printers.

NIOSH also publishes health and safety booklets that report on specific industries. Each booklet contains a checklist which can be used to perform informal inspections of the workplace and may be helpful to health and safety committees. Some of the industries covered are sawmills, battery manufacturers, users of carcinogens, pesticide formulators, and tanneries.

Two of the more important technical publications issued by NIOSH are:

(1) *Current Intelligence Bulletins.* NIOSH's research efforts result in the publication of current intelligence bulletins on recognized health hazards, health and safety guides for specific industries, and technical guides for industry regarding environmental monitoring and control technology for workplace hazards. This rapid-alert system is an effective way for NIOSH to inform health professionals in government, industry, organized labor, and universities about a new health hazard or about new data on an old one. Usually only a few pages long, these reports capsulize background information about the hazard and outline recommended action for controlling exposure to it.

(2) *Registry of Toxic Effects of Chemical Substances (RTECS).* Formerly titled the *Toxic Substances List,* the RTECS issuance is mandated by Section 20(a)(6)

of the Act. It is intended to provide basic information on the known toxic and biological effects of chemical substances for anyone concerned with the handling of chemicals. Presently, the (1978) RTCES contains 124,247 listings of chemical substances: 33,929 are names of different chemicals with their associated toxicity data and 90,318 are synonyms. The 1978 edition includes approximately 7500 new chemical compounds that had not appeared in the 1977 registry.[9] Approximately 1500 new toxic substances are added to the registry each quarter.

The RTECS is a single source document for basic toxicity information and for other data, such as chemical identifiers and information necessary for the preparation of safety directives and hazard evaluations for chemical substances. Moreover, the RTECS furnishes valuable information for those persons responsible for preparing material safety data sheets for chemical substances in the workplace. Entry of a substance in the RTECS does not automatically mean that it must be avoided. A listing does mean, however, that there is a potential for the substance to be harmful if misused.

G. OSHA/NIOSH COOPERATION

The Secretary of Labor is required under Section 20(c) of the Act to cooperate with the Secretary of Health, Education, and Welfare in order to avoid duplication of efforts on their safety and health responsibilities. To this end, an Interagency Agreement was entered into between OSHA and NIOSH effective April 5, 1979, to set forth the understanding between the two agencies for consultation, coordination, and cooperation in carrying out their respective safety and health functions under the Act. 44 Fed. Reg. 22834 (April 17, 1979). The Agreement between the two agencies covers the following issues.

(1) *Criteria.* NIOSH and OSHA jointly will establish priorities for criteria development and will develop parameters to be considered in criteria documents.
(2) *Standards.* OSHA will advise NIOSH of the schedule of rulemaking that establishes or revises various health and safety standards. NIOSH will set up a procedure for providing technical assistance to OSHA during preparation and review of standards and will provide expert technical witnesses for OSHA's public rulemaking hearings.
(3) *Health-Hazard Evaluations.* NIOSH will advise OSHA of all health-hazard evaluation requests that it receives and the Institute will contact OSHA and employer/employee representatives to obtain information relevant to the evaluation. If NIOSH discovers a situation during an HHE that may present a *serious safety or health hazard,* OSHA will be notified immediately "in order to take appropriate action." In addition, the two agencies will coordinate their field activities in each facility where a health-hazard evaluation has been requested. Copies of final reports on the HHE will be sent to OSHA.
(4) *Compliance Assistance.* NIOSH will provide technical assistance and

supportive field investigations to OSHA and the Institute will provide expert witnesses in support of OSHA for court actions, administrative proceedings, and other legal actions.

(5) *Training and Education.* OSHA and NIOSH will coordinate training and education activities so that there is minimal duplication of effort.

(6) *Testing and Certification.* NIOSH testing activities will "be responsive" to OSHA's needs regarding safety and health standards. OSHA agreed to use its regulations and educational programs to encourage use of items certified under the joint NIOSH/OSHA testing and certification program whenever the devices are appropriate.

(7) *Technical Information Exchange.* OSHA and NIOSH will provide one another with cooperative data, services, products, and technical information.

Other sections of the agreement indicate that both agencies will enter into "subagreements" to accomplish more specific goals and that the two groups will meet regularly for policy and technical discussions.

NOTES

1. "Substance" means any chemical or biological agent which has the potential to produce toxic effects. 42 CFR 85.2(h). Toxic effects are those which result in short or long-term disease, bodily injury, affect health adversely, or endanger the life of man. 42 CFR 85.2(i).

2. The requirement for a determination of reasonableness is *not* present in the general NIOSH regulations pertaining to workplace inspections and may suggest that a higher standard of probable cause is required to be met prior to entering the workplace.

3. The case law on access to records takes two forms. The first is a mysterious, lengthy footnote in *Barlow's,* which reads, in relevant part, as follows: "Delineating the scope of a search with some care is particularly important where documents are involved. . . . In describing the scope of the warrantless inspection authorized by the statute, Section 8(a) does not expressly include any *records* among those items or things that may be examined, and Section 8(c) merely provides that the employer is 'to make available' his pertinent records and to make periodic reports.

"OSHA's regulation, 29 CFR 1903.3, however, expressly includes among the inspector's powers the authority 'to review, records required by the Act and regulations published in this chapter, and other records which are directly related to the purpose of the inspection.' . . . It is the Secretary's position, which we reject, that an inspection of documents of this scope may be affected without a warrant." 98 S. Ct. at 1826 n.22.

Coming, as it does, as a footnote to a sentence indicating that the warrant has important functions to perform, "functions which underlie the court's prior decisions that the warrant calls for compliance with regulatory statutes," 98 S.Ct at 1826, this passage clearly suggests a sensitivity on the part of the majority to the issue of limiting the scope of the warrants for access to records. What it also seems to suggest, in passing, is that the inspector's powers under the regulation exceed those granted by the statute. If this were the case, of course, the regulations would be, to this extent, invalid. Likewise, NIOSH's regulations, which provide authority to review, abstract, or duplicate employment records, medical records, and other required and related records would also seem to be invalid in an inspection context. It is "questionable" whether the *Barlow's* decision which restricted OSHA access to workplaces without a search warrant would apply to NIOSH activity under the Mine Safety and Health Act, since warrantless inspections under the Mine Act have been upheld in the courts.

4. The employer challenged the basis of NIOSH criteria which led to its selection and its inclusion in the NIOSH study. The court, however, stated that these considerations were more appropriately left to the expertise of the agency, but that in any event there appeared to be a reasonable basis for the employer's inclusion. Another problem, however, was that the requested inspection (and warrant) included a request to review medical and personnel records. The court approved the request, stating that the records did not have to be produced in a personally identifiable form and that access thereto would pose no perceivable threat to confidential business and trade-secret information.

5. In holding as it did, the court recognized that there are some cases that could be read to support the view that an inspection may include review of medical and personnel records—*e.g.*, in the context of OSHA inspections, courts have upheld Section 8(a) inspection warrants which permit, among other things, inspection of records, files, and papers related to the safety of the workplace. *Marshall v. Chromalloy American Corp.*, 589 F.2d 1335, 6 OSHC 2151 (7th Cir. 1979). But the court in *Inland Steel* stated that even in that case the court did not specifically address the issue of what types of items could be examined pursuant to Section 8(a).

6. Toxic substances or harmful physical agents are defined as any chemical substances, biological agents, or physical stress which are regulated by any federal law or rule due to a hazard to health, are listed in the latest printed edition of the NIOSH Registry of Toxic Effects of Chemical Substances (RTECS), have yielded positive evidence of an acute or chronic health hazard in human, animal, or other biological testing conducted or known to the employer, or have a material-safety data sheet, indicating that the material may pose a hazard to human health.

7. The standard states that substantial occupational health benefits will result from improving the ability of employees to make beneficial use of six explicit statutory rights established by the Act. The fifth of these rights, enhanced by worker access to exposure and medical records, is the Section 20(a)(6) right of employees to request a NIOSH health-hazard evaluation.

8. *Whalen v. Roe,* which is the leading constitutional case with respect to the privacy of medical information, was specifically and explicitly applied to a NIOSH health-hazard evaluation in *E. I. du Pont de Nemours Co. v. Finklea,* 442 F. Supp. 821, 6 OSHC 1167 (S.D. W.Va. 1977).

9. The Director of the Federal Register has approved for incorporation by reference into 29 CFR 1910 the 1978 edition of the RTECS. See 29 CFR 1910.20(c)(11)(ii).

APPENDIX 1 to CHAPTER XVI

U.S. DEPARTMENT OF HEALTH, EDUCATION, AND WELFARE
NATIONAL INSTITUTE FOR OCCUPATIONAL SAFETY AND HEALTH

REQUEST FOR HEALTH HAZARD EVALUATION

This form is provided to assist in registering a request for a health hazard evaluation with the U.S. Department of Health, Education, and Welfare as provided in Section 20(a)(6) of the Occupational Safety and Health Act of 1970 and 42 CFR Part 85. (See Statement of Authority on Reverse Side).

Name of Establishment Where Alleged Hazard(s) Exist

____

Company Street_____________________ Telephone_______________

Address City_________________ State_____________ Zip Code__________

---

1. Principal Company Activity_________________________________
(manufacturing, construction, transportation, services, etc.)

2. Specify the particular building or worksite where the alleged hazard is located, including address.

____

3. Specify the name and phone number of employer's agent(s) in charge.

____

4. Describe briefly the hazard(s) which exists by completing the following information:

Identification of Hazard or Toxic Substance(s)

____

Trade Name (If Applicable)________________________________

Chemical Name________________________________

Manufacturer________________________________

Does the material have a warning label?____Yes ____No

If Yes, attach copy of label or a copy of the information contained on the label.

Physical Form: Dust ☐ Gas ☐ Liquid ☐ Mist ☐ Other ☐

Type of Exposure? Breathing ☐ Swallowing ☐ Skin Contact ☐

Number of People Exposed________________________________

Length of Exposure (Hours/Day)________________________________

Occupations of Exposed Employees________________________________

____

____

5. Using the space below describe further the nature of the conditions or circumstances which prompted this request and other relevant aspects which you may consider important, such as the nature of the illness or symptoms of exposure, the concern for the potentially toxic effects of a new chemical substance introduced into the workplace, etc.

____

____

____

____

____

____

6. (a) To your knowledge has this hazard been considered previously by any Government agency?

____

(b) If so, give the name and address of each,

____

____

(c) and the approximate date it was so considered.

____

7. (a) Is this request, or a request alleging a similar hazard, being filed with any other Government agency?__________ (b) If so, give the name and address of each.

____

____

____

____

The undersigned (check one)

☐ Employer

☐ Authorized representative of employees*

i ii iii (circle one)

believes that a substance (or substances) normally found at the following place of employment may have potentially toxic effects in the concentration used or found.

____

* "Authorized representative of employees" means any person or organization meeting the conditions specified in 42 CFR Part 85.3(b)(4(i), (ii) or (iii): (i)—that he is an authorized representative of, or an officer of the organization representing, the employees for purposes of collective bargaining; or (ii)—that he is an employee of the employer and is authorized by two or more employees employed in the workplace where the substance is normally found, to represent them for purposes of the Act. Each such authorization shall be in writing and included in the request; or (iii)—that he is one of three or less employees employed in the workplace where the substance is normally found.

____

Signature______________________________ Date__________

Typed or Printed Name______________________________

Telephone: Home-______________________

(Street______________________

(Business Address______________________

(City______________ State__________ Zip Code__________

____

If you are a representative of employees, state the name and address of your organization.

____

____

Please indicate your desire:

☐ I do not want my name revealed to the employer.

☐ My name may be revealed to the employer.

Authority:

Section 20(a)(6) of the Occupational Safety and Health Act, (29 U.S.C. 669(a)(6)) provides as follows: The Secretary of Health, Education, and Welfare shall . . . determine following a written request by any employer or authorized representative of employees, specifying with reasonable particularity the grounds on which the request is made, whether any substance normally found in the place of employment has potentially toxic effects in such concentrations as used or found; and shall submit such determination both to employers and affected employees as soon as possible. If the Secretary of Health, Education, and Welfare determines that any substance is potentially toxic at the concentrations in which it is used or found in a place of employment, and such substance is not covered by an occupational safety or health standard promulgated under section 6, the Secretary of Health, Education, and Welfare shall immediately submit such determination to the Secretary of Labor, together with all pertinent criteria.

Send the completed form to:
National Institute for Occupational Safety and Health
Hazard Evaluation Services Branch
U.S. Department of Health, Education, and Welfare
Cincinnati, Ohio 45202

CHAPTER XVII
Labor Relations, Collective Bargaining, and Title VII

A. INTRODUCTION

Although it was not intended that the Act would directly influence or affect the employment relationship between employees and their employers, it appears that the contrary is true. Safety and health provisions have made their way into collective bargaining agreements in increasing numbers, and grievance and arbitration disputes more frequently include the issues of safety and health. Section 502 of the National Labor Relations Act, as amended by the Labor Management Relations Act, overlaps with both the discrimination provisions of Section 11(c) of the [OSH] Act and the imminent-danger provisions of Section 13(a) of the [OSH] Act with respect to the refusal of an employee to work because he considers a condition to be imminently or abnormally dangerous. Moreover, in the past two to three years a seeming conflict between the [OSH] Act and Title VII of the Civil Rights Act of 1964 has also arisen when a worker is removed from or denied entrance into a workplace because of his or her particular health or safety susceptibilities.

B. GRIEVANCE AND ARBITRATION

Many collective bargaining agreements contain provisions for the resolution of "workplace" disputes through the avenue of grievance and arbitration before an "impartial" arbitrator. Depending upon their scope, the following issues may be considered within the framework of these procedures.

1. Discrimination Issues

Depending upon the circumstances, employees may be able to pursue several avenues of relief relative to safety and health discrimination. Section 11(c) of the Act provides that no one shall discharge or otherwise discriminate against an employee because he has filed a complaint, instituted a proceeding, or exercised any right under the [OSH] Act on behalf of himself or others. See Chapter XV. The discharge of an employee for filing a complaint with OSHA may also be an independent violation of the National Labor Relations Act, 29 U.S.C. Sections 151–168 (NLRA), since the NLRA gives employees the right to engage in certain protected activities for their mutual aid and protection. In *Alleluia Cushion Co.*, 221 NLRB 999, 91 LRRM 1131 (1975), for example, an employee complained to a state safety and health agency concerning safety and health conditions in the workplace. An inspection was conducted. The employee was discharged the day after the inspection at the company's plant, and thereafter filed a complaint with the National Labor Relations Board (NLRB). The NLRB stated that the employee's action was concerted activity; even though the employee's company was not organized, there was no collective bargaining agreement and the employee acted alone in challenging the lack of safety precautions. In the absence of evidence that other employees did not share his interest in safety or did not support his complaint, the matters of safe working conditions were considered to be matters of concern for all persons. The NLRB concluded:

> Where an employee speaks up and seeks to enforce statutory provisions relating to occupational safety designed for the benefit of all employees in the absence of any evidence that fellow employees disavow such representation, we will find an implied consent thereto and deem such activity to be concerted.

According to a Memorandum of Understanding between OSHA and the NLRB, where a charge involving issues covered under Section 11(c) has been filed under both the NLRA and the [OSH] Act with respect to the same factual matters of alleged discrimination, the General Counsel of the NLRB will normally defer the charge to OSHA or dismiss the charge. If no complaint has been filed with OSHA, the General Counsel of the NLRB will notify the employee of his right to do so within thirty days. If the employee does not file a complaint with OSHA, the General Counsel will then process under the NLRA those issues normally covered by Section 11(c) of the Act. The June 16, 1975, Memorandum of Understanding between the NLRB and OSHA states in pertinent part:[1]

> There may be some safety and health activities which may be protected under both Acts. However, since an employee's right to engage in safety and health activity is specifically protected by the OSH Act and is only generally included in the broader right to engage in concerted activities under the NLRA, it is appropriate that enforcement actions to protect such safety and health activities should primarily be taken under the OSH Act rather than the NLRA. [40 Fed. Reg. 26033.]

Although the Memorandum of Understanding provides that OSHA shall be primarily responsible for handling safety and health discrimination issues, the OSHA regulations at 29 CFR 1977.18 specifically recognize the national labor policy favoring resolution of disputes under collective bargaining procedures. This policy has been established in the following cases: *Boys Markets, Inc. v. Retail Clerks*, 398 U.S. 235 (1970); *Republic Steel Corp. v. Maddox*, 379 U.S. 650 (1965); *Carey v. Westinghouse Electric Co.*, 375 U.S. 261 (1964); *Steelworkers v. American Mfg. Co.*, 363 U.S. 564 (1960); *Steelworkers v. Warrior & Gulf Navigation Co.*, 363 U.S. 574 (1960); *Steelworkers v. Enterprise Wheel & Car*

Corp., 363 U.S. 593 (1960); *Textile Workers Union v. Lincoln Mills*, 353 U.S. 448 (1957). Therefore, in certain instances the OSHA regulations support the deferral of Section 11(c) discrimination complaints to grievance and arbitration procedures of collective bargaining agreements or to the jurisdiction of other forums such as the NLRB established to resolve disputes which may be related to Section 11(c) complaints. In those instances, the Secretary of Labor has established the following policy:

> Where a complainant is in fact pursuing remedies other than those provided in Section 11(c), *postponement* of the Secretary's determination and *deferral* to the result of such proceedings may be in order. *See Burlington Truck Lines, Inc. v. U.S.*, 371 U.S. 156 (1962). [29 CFR 1977.18(a)(3).]

Under the NLRA, a similar doctrine of judicial deference to grievance and arbitration proceedings has been invoked with respect to safety disputes. In *Gateway Coal Co. v. Mine Workers*, 414 U.S. 368, 1 OSHC 1461 (1974), for example, the Supreme Court held that it could enjoin a strike in breach of a no-strike clause and order the parties to arbitrate the underlying dispute, even where that dispute was over a safety matter.

The determination through grievance and arbitration procedures of the issue whether a Section 11(c) violation has occurred is considered to be justified if the rights (and factual issues) asserted in such proceedings are substantially the same as the rights protected under Section 11(c), those proceedings are not likely to violate the rights guaranteed by Section 11(c), *and* the forum hearing the matter has the power to determine the ultimate issue of discrimination. See *Rios v. Reynolds Metals Co.*, 467 F.2d 54 (5th Cir. 1972); *Newman v. Avco Corp.*, 451 F.2d 743 (6th Cir. 1971). Thus, according to the OSHA regulations, postponement and due deferral should, in such an instance, be paid to the jurisdiction of other forums to resolve disputes related to Section 11(c) complaints. 29 CFR 1977.18(b)(2). This policy is based upon the NLRB decisions in *Collyer Insulated Wire*, 192 NLRB 837, 77 LRRM 1931 (1971) and *Spielberg Mfg. Co.*, 112 NLRB 1080, 36 LRRM 1152 (1955). Deferral, although it must be made on a case-by-case basis after careful scrutiny of the available information, is normally permitted, under the OSHA regulations if (1) the proceedings deal adequately with all factual issues; (2) the proceedings are fair, regular and free of procedural infirmities; and (3) the outcome is not repugnant to the purpose of Act. 29 CFR 1977.18(c). An example of a situation in which the Secretary of Labor will not defer is where the proceedings before the other forum are dismissed without an adjudicatory hearing thereof.

The NLRB deferral to arbitration policy of course, has changed over the years. See generally Rosenzweig, *Spielberg and the Deferral to Arbitration: Recent Decisions*, 6 Emp. Rel. L.J. 173 (1980). For example, in *General American Transportation Corp.*, 228 NLRB 808, 94 LRRM 1483 (1977), the NLRB narrowed *its* deferral to arbitration doctrine by holding that it would not defer in Section 8(a)(1) and 8(a)(3) cases.[2] Furthermore, the NLRB does not defer in Section 8(a)(4)[3] cases which are the functional equivalent of Section 11(c) cases. *United Parcel Service*, 228 NLRB 1060, 94 LRRM 1731 (1977); *Potter Electric Signal Co.*, 237 NLRB 209, 99 LRRM 1248 (1978); *Virginia-Carolina*

Freight Lines, Inc., 155 NLRB 52, 60 LRRM 1331 (1965); *Kansas City Star,* 236 NLRB No. 119, 98 LRRM 1320 (1978); *Atlantic Steel Company,* 245 NLRB No. 107, 102 LRRM 1247 (1979); *Suburban Motor Freight,* 247 NLRB No. 2, 103 LRRM 1113 (1979); *Bay Shipbuilding Corporation,* 251 NLRB No. 114, 105 LRRM 1376 (1980); *Chemical Layman Tanklines,* 251 NLRB No. 146, 105 LRRM 1276 (1980).

In *Pincus Brothers, Inc.–Maxwell v. NLRB,* 620 F.2d 367, 104 LRRM 2001 (3d Cir. 1980), the Third Circuit found an employee's conduct "arguably unprotected" when he handed out leaflets critical of his employer's policies. In doing so, it rejected an NLRB decision which upheld an arbitrator's award that had found no just cause for the employee's discharge. In *NLRB v. Max Factor & Co.,*—F.2d—, 105 LRRM 2765 (9th Cir. 1980), the Ninth Circuit rejected the Third Circuit's approach in *Pincus Brothers* and declined to apply a determination of the *Spielberg* criteria which differed from the historic approach. See also *NLRB v. Washington Aluminum Co.,* 370 U.S. 9 (1962)—where no contract exists, no arbitration is required. The OSHA deferral regulation, at least as written, appears to be at variance with the statutory scheme and policy under the NLRA.

OSHA's deferral policy has however recently undergone a change, at least as interpreted and applied by the courts. For example, the regulation was initially upheld in *Brennan v. Allen Wood Steel Co.,*—F. Supp.—, 3 OSHC 1654, 4 OSHC 1598 (E.D. Pa. 1975, 1976), where the court stated that the Secretary of Labor was entitled in his discretion to defer to an arbitration award but was not required to do so. The postponement regulation was again recently addressed in *Newport News Shipbuilding and Drydock Co. v. Marshall,* —F. Supp.—, 8 OSHC 1393 (E.D. Va. 1980), however, with a different result. In that case, several employees charged their employer under Section 11(c) of the Act with retaliating against them for complaining about unsafe working conditions. The union also filed a complaint with the NLRB on the same matter. Upon completion of its investigation, OSHA postponed its determination, citing the pending NLRB complaint. The employer challenged the postponement decision in the court, alleging that OSHA was under a duty to reach a determination. The court agreed with the employer, holding:

> The right to bring an action because of the discharge of an employee in retaliation for reporting a violation of the Act, is in the Secretary. If he finds such a violation, the statute says that he shall bring an action. The Sixth Circuit in *Taylor v. Brighton Corporation,*—F.2d—, 8 OSHC 1010 (6th Cir. 1980) held that . . . Congress intends suits by the Secretary of Labor to be the exclusive means of redressing violations. . . . Nothing in the Act says or indicates the Secretary may defer to another agency or body. The exclusive remedy to redress the alleged wrong is in an action by the Secretary. The duty to act has been recognized as a mandatory duty. *Dunlop v. Hanover Shoe Farms,* 441 F. Supp. 385, 4 OSHC 1241 (M.D. Pa. 1976). Though defendant says he had followed the procedure of deferral in other cases, and has a regulation providing for referral, nothing in the Act provides the Secretary may defer action. The Act says he shall bring the action.
>
> The Secretary having made a determination that the provisions of [Section 11(c)] had been violated prior to the time the complaint was filed with the NLRB, and this action having been instituted prior to the time the complaint was filed

> with the NLRB, the Secretary may not escape his duty to act as required by the statute. He should act, rather than defer to the NLRB, particularly in view of the fact that the Secretary says he will not be bound by the decision of the NLRB, and there probably will be considerable delay in a decision from that body.

Thus if an employee pursues an avenue of relief under the NLRA or submits a grievance to arbitration and accepts the arbitrator's decision, this apparently will not automatically waive the employee's rights to obtain relief under Section 11(c). In *Brennan v. Allen Wood Steel Co.,*—F. Supp.—, 3 OSHC 1654, 4 OSHC 1598 (E.D. Pa. 1975, 1976) the court refused to dismiss a complaint brought by the Secretary of Labor, even though an arbitrator had previously ruled that the employee was wrong in refusing to work by claiming a hazardous condition. A later discrimination suit may be instituted, *provided* that the employee can comply with the procedural requirements of filing the complaint within thirty days of the alleged discrimination; filing of a grievance does not toll the thirty-day filing requirement. For example, in *Marshall v. NL Industries, Inc.,* 618 F.2d 1220, 8 OSHC 1166 (7th Cir. 1980), an employee filed a complaint with OSHA and a grievance under an arbitration procedure of a collective bargaining agreement relative to his refusal to use a payloader to load lead scrap into a melting kettle. The arbitrator determined the employee should be reinstated but without backpay. Later, in the Section 11(c) action, the Secretary of Labor requested backpay. The court, relying on the decision in *Alexander v. Gardiner-Denver Company,* 415 U.S. 36 (1974) which held that an arbitration decision does not bar a later employment discrimination suit under Title VII of the Civil Rights Act of 1964, stated that the *Gardiner-Denver* holding applies equally well to the [OSH] Act unless the employee voluntarily waives his right to judicial relief in effecting a settlement. The court also stated that backpay may be an appropriate remedy for discriminatory discharge even though the Supreme Court in *Whirlpool Corp. v. Marshall* 445 U.S. 1, 8 OSHC 1001 (1980) stated that the OSHA regulation, 29 CFR 1977. 12(b), which protects an employee's right to refuse to work because of a reasonable apprehension of death or serious physical injury, does not require employers to pay employees who refuse to perform their tasks in the face of such imminent danger. See also *Marshall v. Firestone Tire & Rubber Co.,*—F. Supp.—, 8 OSHC 1637 (C.D. Ill. 1980).

The OSHA regulation requiring employers to pay employees for time spent on walkaround inspections has recently been ruled invalid by the United States Court of Appeals in *Chamber of Commerce v. OSHA,* 465 F.2d 10, 8 OSHC 1648 (D.C. Cir. 1980). The court stated that the principles of the Fair Labor Standards Act, 29 U.S.C. Sections 201–19, do not require that employers pay employees for time spent during a walkaround accompanying inspectors. Moreover, the failure to pay for walkaround time is not discriminatory under Section 11(c). However, the decision does not prohibit employers from voluntarily compensating workers for walkarounds and related inspection activities. Employers may, by agreement, contract to pay employees for time spent in safety and health committee meetings, walkarounds, opening and closing conferences, and so forth. See *Ashland Chemical Co.,* 67 LA 52 (Belshaw, 1976). It is possible that OSHA, which has now repealed the regulation pursuant to court directive, may reissue it after it has complied with the notice and comment requirements of the Act (which it did not originally do).

Employers may be required to pay employees for other safety and health related activities, such as serving on safety and health committees or responding to a subpoena as a witness at an OSHA hearing even if they testify against the employer, depending upon the language of the collective bargaining agreement. In *Butler Lime & Cement Co.*, 60 LA 611 (Knudson, 1973), for example, the contract provisions stated that employees would be compensated for all time lost due to delays and court appearances as a result of violations involving federal, state, or city regulations. The arbitrator held that a violation could include an employer's OSHA violation.

2. Safety Rules and Regulations

An employer has the right and/or prerogative to establish reasonable safety and health rules, even if his employees are represented by a collective bargaining representative; if a collective bargaining agreement is silent on the subject, it has nevertheless normally been held that an employer has the right to promulgate and enforce such rules to insure the safety and health of its employees and in certain instances to comply with its contractual (and statutory) obligation to provide a safe and healthful place of employment. *Nu-Ply Corporations,* 50 LA 985 (Hnatiuk, 1968); *Ogden Air Logistics Center,* 73 LA 1100 (Hayes, 1980); *Lone Star Steel Co.,* 48 LA 1094 (Jenkins, 1967); *Cit-Con Oil Corp.,* 30 LA 252 (Logan, 1958). The operative and determinative factor in the promulgation of such rules is one of reasonableness. However, if the rules are ambiguous or not uniformly applied, their enforcement may be difficult, for inconsistency can defeat the application of even a reasonable rule. See *Chemetco,* 71 LA 457 (Gibson, 1978). Normally an employer also has the unilateral right to change its rules and regulations as part of its duty to provide safe working conditions. *General Telephone of Pennsylvania,* 71 LA 488 (Ipavec, 1978).

The treatise by Elkouri and Elkouri, *How Arbitration Works,* (BNA, 3d Ed. 1973), gives exhaustive treatment to safety and health considerations in the context of collective bargaining agreements (at pp. 667–679). Elkouri deals with such safety and health issues as crew sizes, the physical and mental condition of employees, refusal to obey an order, and the requirements for special protective equipment and clothing. According to the treatise, correlative to the right of employers to establish reasonable safety and health rules is that employers also have certain obligations in safety and health matters. As Elkouri states:

> Many agreements do contain safety and health provisions, and the typical clause may state that management will exert a reasonable effort to protect the safety and health of its employees. Under such a clause, arbitrators have [generally] ruled that while management must make a reasonable effort towards safety, it is not required to eliminate all possible hazards or to utilize all possible safety precau-

> tions. [For example], a contract clause requiring the company to make every reasonable effort to provide necessary and practical safety measures was construed as constituting the minimum obligation which the company had assumed for the safety and health of its employees; but the company could do more toward that end so long as it acted reasonably.

See generally *Chromallory American Corp.*, 61 LA 824 (Rothman, 1973); *Lone Stone Steel Co.*, 48 LA 1094 (Jenkins, 1967).

It must also be noted briefly that unions and employees may also have obligations with respect to safety and health matters. Such duties may be imposed by contract usually in a form of an agreement to cooperate in promoting safety, *e.g.*, through safety and health committees.[4] *Westmoreland Country Housing Authority*, 73 LA 32 (Bolte, 1979). This duty may also, of course, be imposed upon the employee and even the union, to a certain extent, by safety and health rules promulgated under the management's prerogative and obligation to provide a safe and healthful workplace.

As stated, however, the primary obligation lies with the employer. While it is generally accepted that an employer has the authority to adopt safety rules, *Lozier Corp.*, 72 LA 164 (Ferguson, 1979); *A. O. Smith Corporation*, 58 LA 784 (Volz, 1972), the authority to do so is subject to a limitation of reasonableness in application as well as in purpose and content. In *A. O. Smith, supra,* the arbitrator held that if an alternative safety procedure could be found, a reasonable administration of the safety rule would require that the alternative be incorporated. The arbitrator in *Lozier, supra,* on the other hand, held that the arbitrator's task was not to determine whether it was the best regulation, but rather to determine whether the company's regulation was reasonable. If it is supportive of an employer's position, he may wish to reference any OSHA standards justifying the management action that prompted the promulgation of the safety and health rules and regulations. For example, the requirement of wearing ear protection in areas of high noise levels for purposes of compliance with 29 CFR 1910.95 would justify a management rule to that effect. *National Brush Co.*, 66 LA 869 (Grant, 1976). The existence of an OSHA citation may also serve as a basis for a company to promulgate a rule or to take other action such as a refusal to hold to jurisdictional lines in the assignment of work, if safety considerations are involved.[5] *Champion International Corp.*, 61 LA 400 (Chalfie, 1973). Similarly, with respect to conducting tests relative to efficiency quotients. *Fiat-Allise Construction Machinery, Inc.*, 64 LA 365 (Sembower, 1975).

As is true with other workplace rules, rules related to safety and health must be promulgated in conformity with applicable provisions of any collective bargaining agreement and must be adequately communicated and even-handedly applied to employees. For the most reasonble rule may be set aside if it conflicts with the clear terms of a labor contract or if it is not uniformly applied. *Lynchburg Foundry Co.*, 64 LA 1059 (Coburn, 1975). In addition, the principle of progressive discipline may be appropriate, for the final arbiter of the reasonableness of the application of such rules will in most cases rest with an arbitrator. In this respect, it must be noted that some collective bargaining agreements limit the authority of an arbitrator to interpret only the terms of the contract, and restrict outside reference sources. *Checker Motors Corp.*, 61 LA 33 (Daniel, 1973).

a. *Refusal to Obey Rules*

Elkouri and Elkouri state that it is a well-established principle that employees must obey an employer's work rules and carry out their job assignments especially where safe and healthful working conditions are involved, even if they are believed to be unreasonable or violative of the [OSH] Act or a collective bargaining agreement. *Eberle Tanning Co.*, 71 LA 302 (Sloane, 1978); *Waterloo Industries, Inc.*, 62 LA 605 (Gilroy, 1974); *FMC Corp.*, 45 LA 293 (McCoy, 1965); *Hegler Zink Company*, 8 LA 826 (Elson, 1947). They may later turn to the grievance-arbitration procedure for relief. An exception to this "obey-now-grieve-later" doctrine exists where obedience would involve an unusual, imminent, or abnormal safety or health hazard. If this exception is invoked, however, it must be shown that a safety or health hazard was the real reason for the employee's action. See *Sperry Rand Corp.*, 51 LA 709 (Rohman, 1968); *Chesapeake Corp. of Virginia*, 50 LA 14 (Jaffee, 1967); *Thrikol Corp.*, 63 LA 633 (Rimer, 1974); *Chrysler Corp.*, 63 LA 677 (Alexander, 1974); *Hercules Inc.*, 48 LA 788 (Hopson, Jr., 1967); *FMC Corp.*, 45 LA 293 (McCoy, 1965); *City of Detroit*, 61 LA 645 (McCormick, 1973); *Artco-Bell Corp.*, 61 LA 773 (Springfield, 1973); *United States Steel Corp.*, 66 LA 489 (Caghan, 1976); *Pittston Co.*, 62 LA 1303 (Sergent, 1974); *Haveg Industries*, 68 LA 841 (McKone, 1977).

The exception has been held applicable where an employee asserts that if he were to perform as ordered, the safety or health of others would also be in jeopardy. This right of refusal, however, applies only to employees assigned to or affected by the work. It does not permit other employees not assigned to or affected by the work to justify their resort to self-help instead of using the grievance procedures. [It must be noted, too, that safety rules may not only require observance of certain procedures and practices of employees but may also impose a duty upon the employee to refrain from working under potentially unsafe or unhealthful conditions.]

Many contract clauses may expressly deal with the right of employees to refuse to do work in abnormally or imminently dangerous or hazardous conditions. Such clauses may, according to Elkouri, give employees the right to file a grievance at an advanced step for preferred handling or to be relieved of the job and be assigned to another available position. See *Bethlehem Steel Corp.*, 49 LA 1203 (Seward, 1968); *Lone Star Steel Corp.*, 48 LA 1094 (Jenkins, 1967).

Arbitration decisions relative to a refusal to obey an order because of alleged safety and health hazards are usually couched in such terms as an employee's reasonable belief or an employee's sincere and genuine fear for his life or an employee's good-faith belief. *Hercules, Inc.*, 48 LA 788 (Hopson, 1967); *Lone Star Steel Co.*, 48 LA 1094 (Jenkins, 1967); *Wilcolator Co.*, 44 LA 847 (Altieri, 1964); *Devaul Corp.*, 43 LA 102 (Myers, 1964); *Bunker Hill Co.*, 65 LA 182 (Kleinsorge, 1975); *Ri-Ja Machinery Co.*, 66 LA 474 (Waters, 1976). According to Elkouri and Elkouri, the greatest number of arbitrators appear to take some form of the reasonable man approach in this circumstance; that is, whether the facts and circumstances known to the employee at the time of the incident would cause a reasonable man to fear for his personal safety and health. See *Jefferson City Cabinet Co.,* 50 LA 213 (Williams, 1978); *Laclede Gas Co.,* 39 LA 833 (Bothwell, 1962); *Consolidated Edison of New York, Inc.,* 71 LA 238 (Kelly, 1978).

b. *Hard Hats and Other Personal Protective Equipment*

Section 5(b) of the Act provides that employees are required to comply therewith but there are no sanctions imposed for their failure to do so and there are no injunctive orders available to employers under the [OSH] Act to force resistant employees to wear protective equipment. See *Atlantic & Gulf Stevedores, Inc. v. OSHRC,* 534 F.2d 541, 4 OSHC 1061 (3d Cir. 1976); *Weyerhauser Co.,* 2 OSHC 1152 (1974). Moreover, the failure of an employee to wear personal protective equipment can involve the employer in a violation of the Act; it has a duty to ensure compliance.

Rules related to the wearing of employee personal protection can and should be promulgated and enforced by employers; if reasonable and related to the requirements of the workplace, they will most often be upheld. *General Telephone Co. of Florida,* 62 LA 997 (Turkus, 1974); *Penn-Walt Corp.,* 65 LA 751 (Conn. Bd., 1974); *Emerey Industries, Inc.,* 65 LA 1309 (Donnelly, 1974). Moreover, OSHA regulations set forth minimum standards; there is "nothing" to prevent an employer from instituting a more stringent standard than those required in the OSHA standards if required by the employer's operations. *Mrs. Baird's Bakery, Inc.,* 68 LA 773 (Fox, 1977); *Hugo Neu-Proler Co.,* 69 LA 751 (Richman, 1978). The gravamen is whether the rules are reasonable. For example, a rule requiring an employee to wear a hard hat at *all* times may be regarded as unreasonable. *York Water Co.,* 60 LA 90 (Tripp, 1972). But an employee complaint that a hard hat is too hot, too uncomfortable, or interferes with his hairstyle will not usually be sustained. See *Castleberry's Food Co.,* 62 LA 1267 (King, 1974); *Paul S. Yoney, Inc.,* 64 LA 60 (Purall, 1975). In one case a rule prohibiting an employee from wearing a hard hat at all was upheld for safety reasons. *National Park Service,* 72 LA 314 (Pritzker, 1978).

c. *Grooming*

If a safety policy interferes with an employee's personal grooming habit, such as his beard, moustache, sideburns or his/her hair, the arbitrator will first determine if the rule is reasonable and that all affected employees are treated the same. Then it must be shown that there is a hazard which requires employee protection. See *Challenge-Cook Brothers, Inc.,* 55 LA 517 (Roberts, 1970); *Western Kentucky Gas Co.,* 62 LA 175 (Dolson, 1974); *Sealright Co.,* 59 LA 305 (Goetz, 1972). For the rule must normally be demonstrably related to safety or health. *United Parcel Service of Pa.,* 58 LA 201 (Duff, 1971). For example, the trimming of a beard if it presents a potential safety hazard or interferes with the proper fit of a respirator would be reasonable and upheld. See *Springday Co.,* 53 LA 627 (Bothwell, 1969); *Phoenix Forging Co.,* 71 LA 879 (Dash, 1978); *Bethlehem Steel Corp.,* 66 LA 266 (Reel, 1976). Arbitrators have generally upheld the right of employers to promulgate rules which prohibit the wearing of facial hair when the use of respirators is required (and a "seal" is prevented), so long as the rule does not conflict with a collective bargaining agreement. See *J. R. Simplot Co.,* 68 LA 912 (Conant, 1977); *American Smelting & Refining Co.,* 69 LA 824 (Hutcheson, 1977); *Niagara Mohawk Power Corp.,* 74 LA 58 (Markowitz, 1980). If the wearing of beards does not constitute a safety hazard, however, a company cannot automatically discharge employees for violating such a rule. *Giles and Ransome Corp.,* 70–1 ARB ¶8319 (Loucks, 1969); *City of Waterbury, Connecticut,* 70–1 ARB ¶8395 (Stutz, 1970); *Allied Chemical Corp.,* 74 LA 412 (Eischen, 1980).

In *Phillips Petroleum Company,* 74 LA 400 (Carter, 1980), the arbitrator ruled that an employer did not have the right to require an employee to shave his beard on the basis of a presumption that the beard prevented a positive seal on his respirator face mask. The employer's visual observation that the beard would prevent a positive seal was arbitrary, in light of clear evidence that facial hair *does not always* prevent a positive seal and that the only manner in which to determine that the employee can obtain a positive seal is by an actual test. Also important was the fact that the union and the company agreed that the employees were not required to remove facial hair that did not interfere with the positive seal. Although the arbitrator agreed that the employer had the right to establish reasonable plant rules which could include a rule pertaining to the wearing of beards, the real issue in the grievance was whether the company had the right to determine solely by visual observation that an employee's beard interfered with the seal.

In *Randall Foods No. 2,* 74 LA 729 (Sembower, 1980), it was held that a grocery chain grooming code forbidding beards, was not an unreasonable rule, even though the standard had nothing to do with health requirements or safety problems. The arbitrator stated that it cannot be said that the rule was unreasonable as the company set its sights for maintenance of a reasonâble image. The rule was, therefore, sustained on the basis of customer approval. But see *Food Employer's Council, Inc.,* 66 LA 439 (Wyckoff, 1975).

d. *Sampling Devices*

The wearing of safety apparel and personal protective equipment can normally be required as can the wearing of sampling and monitoring equipment. In *Mobile Chemical Company,* 71 LA 535 (May, 1978), an employee repeatedly refused to wear sampling devices. The union argued that it was the company's responsibility, not the employees', to take air samples and that the question was not specifically covered in the contract. The arbitrator saw no difference between wearing protective equipment and a monitor and denied the grievance. Reasonable provisions for safety and health go beyond the wearing safety apparel. See, however, *Plum Creek Lumber v. Hutton,* 608 F.2d 1283, 7 OSHC 1940 (9th Cir. 1979)—the employer had a policy prohibiting its employees from wearing OSHA inspection sampling devices.

e. *Ear Protection*

Employers have the right and the obligation to require that employees exposed to high noise levels wear ear protection and to enforce such rules.

These rules must be adequately communicated to the employees and the employer must inform them which type of ear protection is proper. The rule must, as with every other "work" rule, however, be uniformly enforced. See *National Brush Co.,* 66 LA 869 (Grant, 1976). However, even if an employer has provided that a certain type of ear protection must be worn, discharge may be improper if an employee chooses to wear another type, based upon the reasonableness argument. See *Manistee Drop Forge Corp.,* 62 LA 1165 (Brooks, 1974).

f. *Eye Protection*

Employers have the right to require that employees wear whatever eye protection is necessary in the workplace, even though not specifically required under OSHA regulations. However, in this area the complaints of employees of discomfort may be considered more favorably by arbitrators. See *Marathon Electric Manufacturing Corp.,* 61 LA 1249 (Rauch, 1973); *United States Steel Co.,* 64 LA 948 (Witt, 1975). Again, the penalty must fit the crime; discharge may be too severe for first-instance failure. See *Hussman Refrigerator Co.,* 68 LA 565 (Mansfield, 1977); *Chayes Virginia, Inc.,* 71 LA 993 (Davis, 1978); *Reynolds Metals,* 59 LA 421 (Volz, 1972); *Northwest Engineering Co.,* 72 LA 486 (Grabb, 1978); *Gra-Iron Foundry Corp.,* 71 LA 702 (Carter, 1978); *Honey Well, Inc.,* 63 LA 151 (Sembower, 1974). In this light, however, a ban on the wearing of contact lenses may and usually will be considered reasonable. See *Peterbilt Motors Co. (Paccar, Inc.),* 61 LA 1107 (Karasick, 1973).

In *Bethlehem Steel Corporation,* 74 LA 921 (Fishgold, 1980) the arbitrator held that an employer, who refused to allow an employee to wear tinted safety glasses as protection from welding flashes from ultraviolet rays, did not violate its contract with the employee's union. The employer maintained that the employee's work area was not a designated area where employees were required to wear goggles and stated that the tinted glasses reduced the grievant's visibility to the point of jeopardizing the safety of other employees, contending that clear glasses were sufficient protection from the welding done in the grievant's work area.

g. *Hand Protection*

The wearing of work gloves usually gives rise to few problems, disciplinary or otherwise, except with respect to whom is obligated to pay the cost therefor. See *Virginia Electric Power Co.,* 68 LA 410 (Harkless, undated); *United States Steel Corp.,* 66 LA 487 (Garrett, 1976); *Big Rivers Electric Corp.,* 80–2 ARB ¶8487 (Erbs, 1980); *Bethlehem Steel Corp.,* 62 LA 708 (Strongin, 1974); *Southern Steel and Wire Co.,* 67 LA 435 (Goodstein, 1976). Recently in *FMC Corpóration,* (Loven, 1978) (unreported), an employee who believed it was safer to wear gloves while operating a rotating machine was found not to be justified in disobeying a company rule which prohibited wearing gloves while operating the machine. The failure of OSHA, which investigated a complaint that employees were required to work on the machine without gloves, to cite the company for a violation reinforced the company position.

h. *Foot Protection*

Similar questions may come up with respect to the providing of foot protection. Thus when an employee refuses to wear prescribed shoes or reports to work wearing shoes other than those considered safe, discipline may be in order, if there is a rule against it and the rule is consistently enforced. *McEvoy Oil Field Equipment Co.*, 68 LA 738 (Bailey, 1977); *Pennwalt Corp.*, 65 LA 751 (Blumrosen, 1975); *Bethlehem Steel Corp.*, 66 LA 1095 (Strongin, 1976); *Ingalls Iron Works*, 61 LA 1154 (Griffen, 1973). There are situations too where wearing certain types of socks may be a valid safety requirement, *New England Telephone Company*, 60 LA 155 (MacLeod, 1972), but again the rule must be applied fairly and evenly. *P & H Harnischfeger Corp.*, 65 LA 1276 (Gibson, 1976); *Lynchburg Foundry Co.*, 64 LA 1059 (Coburn, 1975). As in hand protection, the main question which normally arises in the area of foot protection is whether the employer or the employee is required to pay for such protection. *John Morell & Co.*, 65 LA 933 (Davis, 1975). This issue should be covered in the collective bargaining agreement, for in the absence of such language, the employer is normally not obligated to provide or pay for safety shoes. See *NL Industries, Inc.*, 66 LA 590 (Caraway, 1976)—work shoes were not considered to be protective devices within the meaning of the contract.

i. *Clothing*

Many recent decisions have addressed the issue of whether employers can require employees to wear or not to wear certain types of clothing (and/or jewelry) to protect their arms, legs, and torsos despite an employee's personal preference/complaint. *United States Steel Corp.*, 65 LA 360 (Friedman, 1975). As a general matter, such rules should be specific, so that the choice of wearing certain types of clothing is not left to the employees' discretion, *e.g.*, wearing *suitable, necessary, appropriate,* or *adequate* clothing. See *Pacific Telephone and Telegraph Co.*, 65 LA 96 (Boner, 1975)—wearing of tank tops was not considered appropriate; *Babcock & Wilcox*, 73 LA 443 (Strasshofer, 1979)—prohibition on the wearing of beach clothes while walking to and from plant gates was unreasonable; *United States Steel Corp.*, 69 LA 1215 (Garrett, 1978). A dress code, which banned the wearing of tank tops by a telephone company's maintenance and construction employees was found to be unreasonable, since it was not shown that the employees' safety would be adversely affected. *Cincinnati Gas and Electric Company*, 80–2 ARB ¶8592 (Dolson, 1980).

In several decisions the requirement that women wear pants rather than skirts has brought charges of discrimination and in certain instances arbitrators have looked to the reasonable interpretation of the rule to resolve the discrimination issue. In *A. O. Smith Corp.*, 58 LA 784 (Volz, 1972), the arbitrator held that under a contract providing that the employer and the union recognize the desirability of reducing industrial injuries and that the employees will observe the safety rules of the employer, the employer had a right to establish a rule requiring male employees to wear trousers and female employees to wear slacks, since the [OSH] Act requires employees to be safety-conscious. However, for those female employees who objected to dressing as a man because of their religion, their discharge for disobeying the rule was overturned provided that they presented themselves for work with suitable, full-length leg protection, even though they may also be wearing a dress or skirt. In *Lozier Corp.,* 72 LA 164 (Ferguson, 1979), the arbitrator called "reasonable" an unwritten rule requiring all employees to wear slacks in order to reduce the number of leg injuries that workers received from sharp-edged metal pieces, flying sparks, and dermatological solvents in the firm's textile manufacturing plant. A female employee refused to wear pants, saying that to do so would conflict with her religious beliefs. The company informed the employee that she had to abide by the rule or be fired. In answering the arbitrator's questions, the grievant changed the basis of her objection to one of "personal conviction." Arbitrator Ferguson disagreed with the union's contention that the employee should be allowed to wear alternative protective clothing in the form of a straight skirt and leather boots to the hemline. He observed that federal law grants companies a right and in certain cases a duty to adopt safety regulations, including dress requirements, and that this right would be negated if each employee could make his or her own personal decision about the validity of a rule.

It may be possible to reconcile the above two holdings. In *A. O. Smith,* the grievant based her objection to wearing trousers on religious principles, while in *Lozier,* the demurral to this requirement was grounded on a personal conviction. See also *Marhoefer Packing Co.,* 51 LA 983 (Bradley, 1968); compare *Colt Industries,* 71 LA 22 (Rutherford, 1978), in which although the complaint was based upon religious prohibitions, the rule was upheld since past accident reports of leg injuries rendered the rule reasonable. In another case the arbitrator ordered the transfer of an employee to a different area which did not require the wearing of slacks. See *Hurley Hospital,* 71 LA 1013 (Roumell, 1978).

Sometimes employees will demand that protective clothing be provided by the company or that a particular type of protective clothing be provided. Most often, the resolution of this question will depend upon reference to the applicable collective bargaining agreement. See *Housing Authority of Baltimore City,* 60 LA 167 (Bode, 1972). This matter is discussed in a later section of this chapter.

j. *Smoking*

Many companies have rules against smoking or eating in work areas or in other specified areas for safety and health reasons. *Illinois Fruit and Produce Corp.,* 66 LA 498 (Novak, 1976); *Aro Corp.,* 66 LA 928 (Lubow, 1976); *Schien Body & Equipment Co., Inc.,* 69 LA 930 (Roberts, 1977); *Gladieux Food Services, Inc.,* 70 LA 544 (Lewis, 1978). If the rule is based upon safety or health considerations related to fire hazards, cleanliness, and housekeeping considerations, it will normally be upheld. *Sherwood Medical Industries,* 72 LA 258 (Yarkowsky, 1979).

In *Johns-Manville Sales Corporation v. International Association of Machinists, Local Lodge 1609,*—F. 2d—, 8 OSHC 1751 (5th Cir. 1980), the doctrine that an employer has the duty to provide a safe workplace for its employees was invoked to induce the court to set aside an arbitration award. The company had adopted a rule prohibiting all smoking on its property (an asbestos plant) and providing escalating disciplinary sanctions for violation of the rule. The company asserted that damage to employees is greatest when both smoking and exposure to asbestos occur at the same time and therefore smoking on plant premises could add significantly to a carcinogenic risk. Medical evidence indicated that the risk of contracting asbestosis increases when combined with smoking, since asbestos reacts synergistically with smoke. The union protested the rule arguing that it violated the collective bargaining agreement and submitted the dispute to arbitration. The arbitrator declared that the rule was invalid but held that the company could adopt a rule permitting smoking only during work breaks, in specially designated work areas, and in separate portions of the cafeteria during meal breaks. The company asked the court to overturn the arbitration decision.

The court first noted that national labor policy encourages the settlement of labor-management disputes by arbitration and stated that an arbitrator's award is not subject to judicial review except in a few circumstances. *Hines v. Anchor Motor Freight, Inc.*, 424 U.S 554 (1976). An award must be enforced without judicial review of the evidence that draws its essence from the collective bargaining agreement. Citing *United Steelworkers v. Enterprise Wheel and Car Corp.*, 363 U.S. 593 (1960); *Piggly Wiggly Operators Warehousing v. Independent Truckdrivers Local 1,* 611 F.2d 580 (5th Cir. 1980). The court found that the award drew its substance from the collective bargaining agreement and that the employer did not dispute that fact, but contested the basis of the award as contrary to public policy that should not be enforced no matter what the agreement as interpreted by the arbitrator provides. The court stated, however, that like a contract, an arbitration award that is contrary to law will not be enforced, but since the company cited no statute or regulation that forbids smoking in an asbestos plant and no decision that denies the enforcement of a collective bargaining agreement entered into as a result of the union's proper action in representing the members of the bargaining unit that increases work danger inadvertently and without design, it upheld the award.

The Third Circuit, in a case entitled *Pincus Brothers, Inc.-Maxwell v. NLRB*, 620 F.2d 367, 104 LRRM 2001 (3d Cir. 1980), held that if an arbitration award is subject to alternate interpretations, one of which is not clearly repugnant to national labor relations policy, it must be honored by the NLRB under the *Spielberg* doctrine. The court stated that an arbitration award should be sustained if it is arguably correct and that the Board must defer to the decision of the arbitrator where there are two arguable interpretations of an arbitration award, one permissible and one impermissible.

k. *Job Evaluation and Complement*

Where safety and health are involved, employees who normally do not have a say in the performance of a job may be entitled to a say regarding such normal management prerogatives as the size and assignment of a crew. See *Bethlehem Steel Corp.*, 69 LA 162 (Sharnoff, 1977); *Western Airlines, Inc.*, 67 LA 486 (Christopher, 1976). But see *Champion International Corp.*, 61 LA 400 (Chalfre, 1973); *Robert Shaw Controls Co.*, 62 LA 571 (Gentile, 1974); *Fluor Engineers & Constructor's, Inc.*, 63 LA 389 (Unterberger, 1974). In *Corning*

Glass Works and Flint Glassworkers, 80–1 Arb. ¶8304 (High, 1980), the arbitrator found in a review of certain employee's jobs that given the number of injuries and major accidents suffered by the employees, the company's evaluation of the job as entailing only first-degree hazards was inaccurate. Thus the arbitrator ruled the job was to be reevaluated at a degree for frequent to severe to moderate injuries and hazards. See also *Portsmouth Naval Shipyard,* 69 LA 1097 (Whitman, 1977); *United States Air Force,* 74 LA 820 (Gersenfeld, 1980); *G. E. Wilson Co.,* 74 LA 1183 (Goldberg, 1980)—contract required certain number of employees (no working alone) to work in areas where hazardous materials were handled.

1. *Unsafe/Uncomfortable Conditions*

An employer may have an obligation to respond to employee complaints if a workplace is unsafe or unhealthful (too hot or too cold). If it fails to do so, discipline of employees for working in uncomfortable conditions may be reversed or modified. See *Parker Co.,* 60 LA 473 (Edwards, 1973); *United States Steel Corp.,* 65 LA 856 (Garrett, 1975); *Bethlehem Steel Corp.,* 65 LA 902 (Harkless, 1975); *Conalco, Inc.,* 65 LA 776 (Seinsheimer, 1975).

In *Bureau of Retirement and Survivors Ins., of Social Security Admin.,* 73 LA 267 (Eaton, 1979), an unpleasant working environment caused by poor ventilation and uncomfortable temperatures was ruled not to constitute a health risk under a collective bargaining agreement that guaranteed a "safe and healthful working environment" to employees. The employer, the Social Security Administration, acknowledged that its heating and cooling system was not completely satisfactory. However, it did not attempt independently to correct system inadequacies because the General Services Administration (GSA), was trying to get the building contractor to take responsibility for correcting the system's shortcomings. Employees filed numerous complaints with the union claiming that work areas were either too hot or too cold, with workplace temperatures fluctuating widely within any given day. Testimony showed that employees reported "headaches, lack of oxygen, allergies aggravated by dust, bursitis, contracting colds more frequently than in prior work locations and similar symptoms."

Article 13(a) of the applicable collective bargaining agreement stated as follows:

> The Bureau agrees to provide a safe and healthful working environment for all employees and will comply with applicable Federal, State, and local laws and regulations. Each supervisor will take prompt and appropriate action to correct any unsafe condition which is reported to or observed by him or her.

According to the union, "[M]anagement's response was to shrug its shoulders and to blame the air quality problems on the building contractor" or GSA, a "totally unacceptable" response violating Article 13(a) of the agreement. The union further charged that management did not "completely and accurately" describe working conditions created by heating and cooling system malfunctions or the effect of those conditions on worker health and safety. The union requested that the employer immediately correct system deficiencies and "promptly transfer" any union member working in work areas where "it is determined adverse environmental conditions prevail, to areas free of such conditions."

The employer's position was that while the working environment may be uncomfortable at times, such "discomfort or displeasure" as was claimed by its employees must be distinguished from health and safety risks.

The arbitrator first observed that worker complaints might be "subjective" or "minor" but "their number and persistence indicate beyond doubt that they are sincere." However, the issue was whether working conditions had risked worker health and safety. While faulting management's evidence concerning the workplace environment, "the union's case . . . (fell) short" of proving the existence of an unhealthful or unsafe workplace, according to the arbitrator. Accordingly, while mutual efforts to alleviate remaining problems should continue, evidence was not sufficient to form a basis for a finding of contract violation or to support the remedies requested.

In *Naval Air Rework Facility,* 73 LA 201 (Livengood, 1979), employees were not entitled to "environmental" pay for being subjected to a considerable amount of dust, noise, and heat. See also *Naval Air Rework Facility,* 73 LA 644 (Flannagan, 1979); *Bethlehem Steel Corp.,* 73 LA 825 (Sharnoff, 1979); *United States Steel Corp.,* 65 LA 856 (Witt, 1975); *Conalco Inc.,* 65 LA 776 (Seinsheimer, 1975).

m. *Unfit Employees*

In *Grigoleit Co.,* 79–2 Arb. ¶8612 (Lesar, 1979), an arbitrator overturned the suspension of an employee stating that he should not have been sent home when he refused to perform roof repairs with a deaf-mute employee. It was considered fair for the employee to insist on working with a partner with whom he could communicate in a case of an emergency. See also *United States Steel Corp.,* 61 LA 1202 (Miller, 1973)—employee was properly sent home because he suffered recurrent headaches and dizziness. In *Koppers, Inc.,* 79–2 Arb. ¶8613 (Lewis, 1979), the company's discharge of an employee who reported off work leaving two propane torches burning was upheld since if the torches had come into contact with pitch an explosion could have occurred. Arguments that the employee was not responsible since he suffered from alcoholism were not persuasive. According to the arbitrator, when a worker who abuses drugs or alcohol violates company work rules that apply to safety and productivity, regardless of whether the drugs (or alcohol) are prescribed (mental illness or epilepsy) or are illegal, he may be disciplined. Being under the influence of drugs or alcohol renders an employee incapable of properly performing his or her work and in an industrial environment it also renders an employee a hazard to himself, herself, or others. Similar reasoning applies when employee negligence or ill-advised action causes the unsafe or unhealthful condition. *Ogden Air Logistic Center,* 72 LA 445 (Richardson, 1978); *Bonded Scale & Machine Co.,* 72 LA 520 (Modjeska, 1978); *F. G. Olds & Son, Inc.,* 64 LA 726 (Jones, 1975).

The termination of employees whose physical condition renders them unfit to perform a job and the disqualification of employees because of mental unfitness for work has been upheld in many cases, even if the employee has medical evidence of his ability to work. In *Ormet Corp.,* 80–1 ARB ¶8034

(Seinsheimer, 1980), the facts were these: the employee had been examined by a number of doctors who concluded that the best way the employee could avoid skin irritations would be by refraining from being exposed to tar pitch at work. The employee himself admitted that he was affected by pitch. Only one physician—the worker's personal physician, not a skin specialist—stated that the worker might be returned to unlimited duty in areas involving pitch. With respect to this latter examination, however, it was noted that it took place only after the employee had been protected from tar pitch for approximately six months. Thus even though the employee himself claimed that he did not want to be protected from exposure to tar pitch, it was concluded that the employer had the right to determine whether the employee should be permitted to work in areas where he would be exposed to irritants. After all, the company had the responsibility not only to protect workers from recognized hazards, but it was also liable for the payments (either directly or indirectly) of the costs of treating any ailments incurred as a result of working in the plant. See *Matthiessen Hegeler Zinc Co.*, 51 LA 568 (Stouffer, 1968); *United States Steel Corp.*, 46 LA 545 (Rock, 1966); *Fisher Scientific Company*, 66–1 Arb. ¶8011 (Turkus, 1965); *Arandell Corp.*, 56 LA 832 (Hazelwood, 1971); *Westinghouse Electric Corp.*, 44 LA 634 (Blair, 1965).

There are many other cases which have upheld termination or disqualification because an employee's physical or mental condition rendered his continued employment unduly hazardous to himself as well as to others. *United Steel Corp.*, 48 LA 1271 (Garrett, 1967); *American Airlines Inc.*, 68 LA 527 (Turkus, 1977); *Sterling Pulp & Paper Co.*, 54 LA 213 (Klescewski, 1970)—obese employee; *E. W. Bliss Co.*, 42 LA 1042 (Cahni, 1964); *Sealtest Foods*, 39 LA 918 (Donnelly, 1961)—loss of an eye; *Chemetco*, 71 LA 457 (Gibson, 1978)—employee properly suspended for failure to release medical information to employer which had a material relationship to health and safety in workplace. But see *Colgate-Palmolive, Co.*, 64 LA 293 (Traynor, 1975); *Bethlehem Steel Corp.*, 64 LA 447 (Reel, 1975) and 64 LA 194 (Strongin, 1975); *Samuel Bingham Co.*, 67 LA 706 (Cohen, 1976).

C. CONFLICT WITH NLRA AND/OR TITLE VII

The [OSH] Act requires that employers provide a safe and healthful place of employment and as a corollary thereto, that they hire and continue to employ safe employees. However, in meeting this obligation the employer may run into problems with other employment laws. At times the removal of or the failure to place certain employees in the workplace or their disqualification from employment because of their sex, age, national origin, race, or handicap may be felt to be required by an employer from an OSHA perspective due to safety and health considerations. The recently enacted OSHA lead and cotton dust standards, for example, require the removal of certain classifications of employees from the workplace under certain conditions. Such actions may at times appear to conflict with both equal employment opportunity and labor laws. For example, during the past several years substantial changes have been made in the employment of women in industry. A majority of the women who are entering the workplace are of childbearing age. Many have found employment in industries where there is actual or potential exposure to hazardous or toxic materials which may have or has had unintentional chronic effects on their bodily systems. Women may, therefore, run into exclusionary policies which some commentators state perpetuate past sex discrimination based on outmoded protective laws and attitudes. The most complex situation arises when a woman who is pregnant or of childbearing age and capacity seeks a job, transfer, or promotion to a situation in which she will be exposed to hazardous or toxic substances at levels which are unhealthy to her and/or to her embryo, fetus, and possible future offspring.

The concerns raised in this issue with respect to sex may apply equally well to other susceptible classifications: handicapped, age, race, national origin, and—less well—religion. Individuals in these protected groups may possess particular characteristics or traits that would render them particularly susceptible to health threats.[6] Thus there are potential conflicts between the [OSH] Act and the discrimination laws and even the labor laws when the conditions of employment in a particular workplace may require exposure to an agent which is either toxic to one sex (or susceptible group) and not to the other, toxic to the human reproductive process in either sex, or toxic to the fetus but not the nonpregnant adult. As a result of this problem, employers have begun testing workers to determine first of all if their genes make them especially vulnerable to certain chemicals in the workplace (genetic testing), or by excluding certain hypersusceptible employees from the workplace. The problem will occur at any time when exposure to equal environments causes unequal physical and legal consequences.

1. OSHA Approach

The [OSH] Act has as its purpose, the assurance, *so far as possible,* to every working man and woman in the nation, safe and healthful working conditions in order to preserve our human resources. It also contains provision for the promulgation of certain standards which may require that no employee dealing with toxic materials or harmful physical agents will suffer a material impairment of health or functional capacity even if such employee has regular exposure to the hazard dealt with by such standard for the period of his working life. Section 6(b)(5) of the Act. To effectuate the purpose of the Act, the employer has a specific duty to comply with all OSHA standards and a general duty to furnish each employee employment and a place of employment free from recognized hazards causing harm or likely to cause death or serious physical harm. When OSHA promulgates standards to regulate the level of exposure to toxic or hazardous substances in the occupational workplace, it normally does not distinguish between the needs of men and women workers. It is reported, however, that until recently all OSHA standards were established to preserve the health of the white, Anglo-Saxon, protestant male, approximately 26–28 years old. This approach is now changing.

Although OSHA's purpose in the enforcement of the Act and promulgation of standards is to insure the existence of a safe and healthful workplace and equal employment for all persons, in several recently promulgated standards, *e.g.*, for occupational exposure to lead and cotton dust, OSHA specifically recognized that there are certain classes of susceptible persons who may be at a greater risk of occupational exposure to certain substances, materials, and agents than are other "persons," *e.g.*, pregnant females and fetuses. A special exposure limit, called an "action level," lower than the general exposure limit, was established in the lead standard for such persons.[7] Thus even though the lead standard establishes a general permissible exposure limit (PEL) of 50 ug/m^3 averaged over an eight-hour period, the lowest level for which there is evidence of feasibility of compliance, and OSHA states that compliance will insure equal employment for both men and women, in truth such a PEL will only assure equal carcinogenic exposure, for in order to provide necessary overall protection against the effects of lead exposure, the blood lead level of workers must be set below 40 ug/100g. Moreover, the blood lead level of men and women who wish to plan pregnancy (or women who are pregnant) should be maintained at less than 30 ug/100g during this period, which is also necessary to protect the fetus. The only way, therefore, in which the standard proposes to protect the recognized categories of susceptible persons, such as women and fetuses, is to establish a 30 ug/100g action level and *remove* workers (with rate retention and job protection) who wish to plan pregnancies or are pregnant when exposures reach that level. By mandating removal, OSHA specifically recognizes that the general PEL of the lead standard cannot feasibly protect either workers who are pregnant or workers who plan pregnancies. Removal under these OSHA standards is allegedly not a synonym for exclusion. It is only to be temporary, pending resolution of the situation in the workplace.

In an enforcement context, however, it is OSHA's approach that employers should make the workplace safe for *all* workers by reducing or eliminating dangerous exposure to toxic substances, and *not by* excluding workers. Women who are transferred or excluded from the workplace because they are of childbearing age, and may suffer harm from exposure to substances or conditions in the workplace, have been encouraged to file discrimination complaints. OSHA's antidiscrimination policy has recently culminated in the issuance of willful citations against employers under the Act's *general duty clause* in several cases, including *American Cyanamid Company,* C.A. No. 79–2438 (1980), for allegedly developing policies requiring that female employees be sterilized in order to continue working in certain areas of the companies' plants. The case was dismissed at the administrative law judge level, but review was granted by the Commission.

2. EEO Approach

Title VII of the Civil Rights Act of 1964, 42 U.S.C. Sec. 2000e *et seq.* (Title VII), like the NLRA, has an underlying legislative concern with the employment conditions of employees. Title VII, the most significant federal proscription against discrimination, which is administered by the Equal Employment Opportunity Commission (EEOC), has as its mandate the elimination of unlawful employment practices based on race, color, religion, sex, or national origin which discriminate against a person in compensation or in the terms, conditions, and privileges of employment.[8] It also prohibits any employment practice which limits, classifies, or excludes a person within the protected classes from equal opportunities with other employees or which adversely affects their classification status. Differentiation on the basis of pregnancy or reproductive capacity has evolved rapidly since the enactment of Title VII, and such condition is now specifically included within its protections as sex discrimination. 42 U.S.C. Sec. 2000e(k). See also *General Electric Co. v. Gilbert,* 429 U.S. 125 (1976); *Nashville Gas Co. v. Satty,* 434 U.S. 136 (1977); *Geduldig v. Aiello,* 417 U.S. 484 (1974); *Griggs v. Duke Power Co.,* 401 U.S. 424 (1971).

Congress made only one real exception to the applicability of Title VII and it is known as the "bona fide occupational qualification" (BFOQ), which permits a form of sex discrimination when the employee's genuineness is essential to and inherent in the job to be performed in a particular business enterprise. The "business-necessity" defense, on the other hand, is entirely a judge-made doctrine. The BFOQ defense technically arises only after a "plaintiff" in a civil rights case has made a showing of overt discrimination; the business necessity defense only comes into play when a plaintiff has shown that a particular employment practice, otherwise neutral, has a disparate impact upon a group protected under Title VII.

Both or either of these defenses could conceivably come into play if an employer excluded women generally, or even those women with reproductive (childbearing) capacities from the workplace for safety and health reasons. The BFOQ and business necessity defenses have been addressed by the Supreme Court in at least two cases. *Phillips v. Martin Marietta Corp.,* 400 U.S. 542 (1971); *Dothard v. Rawlinson,* 433 U.S. 321 (1977). See also Sirota, *Sex Discrimination: Title VII and the Bona Fide Occupational Qualification,* 55 Tex. L. Rev. 1025 (1977).

The legality of employment discrimination and differentiation under federal laws such as Title VII, in the absence of safety and health considerations, is a relatively straightforward proposition; it is not permitted. See generally Schlei and Grossman, *Employment Discrimination Law,* (BNA, 1976).

In an attempt to resolve the problem of fair and equitable treatment of women's rights of employment where there is exposure to toxic and/or hazardous substances in the workplace, the EEOC and the Office of Federal Contract Compliance Programs (OFCCP) issued proposed interpretive guidelines on employment discrimination and reproductive hazards. 45 Fed. Reg. 7514 (February 1, 1980); 41 CFR Part 60–20. In the opinion of the EEOC and OFCCP, exclusionary policies directed at women of childbearing capacity have been developed without apparent regard to whether exposure to the father can also result in harm to himself or the unborn child. Exclusions in and of themselves are *per se* unlawful, since by their terms they are based on membership in a sex-based class, thus they must be closely scrutinized. The basic premise of the guidelines is that an exclusionary policy based on reproductive hazards will *only* be justified where there is clear and specific scientific evidence that a threat to female employees from continued exposure exists and that no equal danger exists to male employees. Even where such scientific evidence exists, any exclusionary policy must be tailored to the narrowest possible class.

The guidelines also state that employers may establish neutral policies to protect *all* employees from workplace hazards. If such policies have an adverse impact on one sex, however, they would have to be justified in accordance with relevant legal principles. Thus if an employer seeks to determine hazardous reproductive effects on pregnant women from their exposure to a substance or condition in the workplace, the employer must also determine the effects on males and nonpregnant females from that exposure. In other words, if the data indicates danger only to pregnant females, an exclusionary policy should only encompass this group and should not extend more broadly to all fertile female employees. The EEOC also envisioned in its regulations that before adopting exclusionary policies, an employer should thoroughly investigate all alternatives. Among the alternatives suggested by the EEOC are reduction of the level of exposure, use of respirators or other protective devices, and transfer of affected employees without loss of pay, benefits, or accrual of seniority. The regulations, like OSHA's approach, contemplate a temporary emergency exclusion where reputable, scientific evidence exists that a hazard is present which is likely to cause significant harm to the reproductive health of employees of one sex only, or of pregnant employees, and also where there is insufficient reputable scientific evidence concerning the reproductive harm to other employees. The policy, if adopted, must be narrowly tailored to those persons to whom harm is indicated; a temporary exclusion can then be justified, provided job and rate retention are guaranteed. In addition, the guidelines would permit an employer to temporarily remove employees of both sexes from work areas containing reproductive hazards if they declare an intention to parent children and voluntarily request such exclusion, without providing job or rate retention. The guidelines were withdrawn upon review for it was concluded (partially because OSHA did not join in the issuance) that the most appropriate method of eliminating employment discrimination in the workplace where there is a potential exposure to reproductive hazards was through investigation and enforcement on a case-by-case basis. (46 Fed. Reg. 3916).

3. NLRA Approach[9]

Elkouri and Elkouri state that the legitimacy of policies disqualifying or excluding employees from employment for physical and mental reasons, such as epilepsy, heart condition, obesity, visual problems, alcoholism, pregnancy, diabetes, and physical incapacity on safety and health grounds will depend mainly upon the existence of medical evidence as it relates to the employee's condition in the performance of his work. The prevailing view of arbitrators is that management has not only the right but also the responsibility to take corrective action (transfer, demotion, layoff, or termination) when an employee has a physical or mental disability which endangers his safety and health or that of others. This assertion by Elkouri and Elkouri should, however, be carefully reviewed, especially in light of Title VII and federal and state handicap laws. See generally Guy, J., *The Developing Law on Equal Employment Opportunity for the Handicapped: An Overview and Analysis of the Major Issues,* 7 U. Balt. L. Rev. 183 (1978); Nothstein, G. and Ayres, J., *Sex Based Considerations of Differentiation in the Workplace: Exploring the Biomedical Interface between OSHA and Title VII,* 26 Vill. L. Rev. 2 (1980). These concerns aside for the moment, the equities of transferring, discharging, or disqualifying workers because of certain handicaps or susceptibilities may and should also be analyzed in an NLRA context.

In cases where a safety or health rule, policy, or regulation is challenged in arbitrations as discriminatory, the consensus suggests that it will be sustained as long as it is reasonably related to the object sought to be accomplished and there is evidence concerning the safety and health aspects of the grievant's condition as it relates to his work. *A. O. Smith Corporation,* 58 LA 784 (Volz, 1972); *Seaboard Coast Line R.R. Co.,* 53 LA 578 (Morris, 1969); *Reynolds Metals Co.,* 43 LA 734 (Boles, 1964); *Mutual Plastics Mold Corp.,* 48 LA 2 (Block, 1967); *Hughes Aircraft Co.,* 41 LA 535 (Block, 1963); *Great Atlantic and Pacific Tea Corp.,* 41 LA 278 (Scheiber, 1963); *E. W. Bliss Co.,* 43 LA 217 (Klein, 1964); *Southern Airways, Inc.,* 47 LA 1135 (Wallen, 1967); *St. Joe Paper Co.,* 68 LA 124 (Klein, 1977); *Hoerner Waldorf Corp.,* 70 LA 335 (Talent, 1978).[10]

In *Olin Corp.,* 73 LA 291 (Knudson, 1979), the arbitrator ruled that the company's decision to exclude female employees capable of childbearing, aged 18 to 50, from a plant area of potentially high lead exposure was made in good faith to protect employee health. Therefore, the decision did not violate the applicable collective bargaining agreement. In *Olin,* after NIOSH had identified lead as teratogenic *(i.e.,* causing fetal deformity) the company began a review of the chemical and physical agents used at its plant that could affect female employees capable of childbearing. This review led to the conclusion that the potential lead exposure in certain areas was significantly high to warrant the exclusion of fertile females from departments located in those areas.

While the union argued that an entire class of employees could not be excluded from certain departments, the company countered that the restriction applied only to a certain category of female employees and not to all female workers as a class. The company added that it has the sole authority, duty, and responsibility for the health and safety of employees, and that its decision was based on competent recommendations and medical knowledge.

In contesting the company's restriction, the union introduced a page from the Federal Register in which OSHA discussed its belief that women of childbearing age should not be excluded from the lead industry. The arbitrator said that the "equivocal" nature of the OSHA statements, however, left open the possibility that other medical evidence could justify more stringent standards than the current OSHA regulations.

The arbitrator also noted that the company's conclusion was the result of its *own* medical studies, that the contract specifically provided that the company was responsible for employee health, and that the company bore financial responsibility for occupational damage to employee health. "In light of OSHA's admission that even under its standard the removal of pregnant employees from areas of lead exposure might be advisable, the company's restriction appeared reasonable," the arbitrator concluded. Accordingly, because the company's action was founded on reasonable medical advice, it was ruled to be not discriminatory.

In another decision an obligation to report the first signs of pregnancy and a decision to discharge or transfer a woman employee from a job which was dangerous to an unborn child was upheld. *Amoco Oil Company,* 64 LA 511 (Brown, 1975). The arbitrator said there is no violation of the antidiscrimination provisions of the contract:

> The obligations that the company places upon pregnant women are not obligations based solely on sex; for pregnancy, although a condition associated with one sex, is not identified with or reducible to sex. An obligation that affects pregnant women is not an obligation that affects all women.

See also *Linde Company,* 37 LA 1040 (Wyckoff, 1962); *National Steel Corp.,* 60 LA 613 (McDermott, 1973); *General Portland Cement Co.,* 58 LA 1299 (Marshall, 1972); *Pet. Inc.,* 73 LA 181 (Erbs, 1979); *Eaton Corp.,* 73 LA 729 (Porter, 1979).

In *Cole (Division of Litton Business Systems, Inc.),* 79–2 Arb. ¶8610 (Stone, 1979) the arbitrator stated that an employer's refusal to install clean-up and shower facilities for its female employees similar to facilities installed and maintained for male employees was in violation of the employer's antidiscrimination pledge, since there were enough women employed at the company's plant to warrant the installation of such facilities even though showers are not required by OSHA regulations. In another case, an arbitrator held that the rigid application of a safety rule requiring employees who use respirators to be clean-shaven may constitute unnecessary discrimination against black males. One such employee grew a beard because he suffered from a skin condition common to many black males, which resulted in small hair tumors or shaving bumps when he shaved with a razor. The arbitrator in *Niagara Mohawk Power Co.,* 74 LA 58 (Markowitz, 1980), found the term "clean shaven" was therefore vague. In *Joy Manufacturing,* 73 LA 1269 (Abrams, 1980), the arbitrator held that management had the right to refuse to place an employee in a job that endangered the employee's health. In that case the employee had letters from his doctor which stated that oils and lubricants used in his work aggravated his dermatitis. He was then transferred to another position for medical reasons and was not allowed to return to his old job, even though he later presented a release letter from another specialist which stated that the dermatitis would not reoccur if the employee was later exposed to the substances. The arbitrator observed that the company was obligated to maintain safe work conditions.

Other medical conditions may also affect work performance. For example, the mere fact of being diabetic and the possibility that the employee may go into shock may not be sufficient reason not to hire or to transfer, but if the employee actually has a seizure, that may be a reason. *Bethlehem Steel Co.,*

64 LA 447 (Reel, 1975); *Kimberly Clark Co.,* 64 LA 544 (Cocalis, 1975). Similarly, with respect to an epileptic employee, see *Samuel Bingham Co.,* 67 LA 706 (Cohen, 1976).

Along this line, one can bar employees who fear heights from working at high elevations. *Union Carbide Co.,* 60 LA 605 (Duff, 1973); *Admiral Paint Co.,* 60 LA 418 (Carmichael, 1973). Moreover, if employers have rules against possession of alcoholic beverages or drugs on the plant, an employee can be discharged for violating such rules, for working under the influence of intoxicants can be a grave safety hazard. See *Tennessee River Pulp and Paper Co.,* 68 LA 421 (Simon, 1976). Although an employee can normally not be discharged on the basis of being an alcoholic or a drug addict, he may be discharged if his use thereof interferes with health and safety in the workplace. See *Greenley Brothers and Co.,* 67 LA 847 (Wolff, 1976); *Bord & Citrus Products Corp.,* 67 LA 1145 (Naehring, 1977). However, such action must relate to and affect his ability to perform his job. *C. F. Ormet Corp.,* 80–1 CCH ARB. 8034 (Seinsheimer, 1980)—employer could restrict the overtime of an employee who suffered irritation from prolonged exposure to tar pitch, even though employee wanted to work.

D. RELATIONSHIP TO COLLECTIVE BARGAINING

Section 8(a)(5) of the NLRA establishes an employer's duty to bargain collectively with the representative of his employees. Section 9 provides that representatives of employees selected or designated for purposes of collective bargaining shall be the exclusive representative with respect to rates of pay, wages, hours of work, and other conditions of employment.

1. Subjects of Bargaining

The fact that an employer complies with regulations, orders, and rules promulgated under the [OSH] Act does not mean that he automatically has satisfied his obligations with respect to safety and health hazards. Safety and health issues may also be raised in the context of an employer's labor relationship with his employees. As will be discussed below, an employer will often have an obligation to inform employees or discuss safety and health issues as they affect his workplace. There are, of course, limits to this obligation. If, for example, the union requests the employer to cease manufacturing one product and begin producing another on the ground that the first product is hazardous to employee health, the employer would probably be entitled to refuse to bargain. Matters involving the scope and ultimate direction of the company, even if impacting upon employee health, lie at the core of managerial prerogatives. *Kingwood Mining Co.,* 210 NLRB 844, 86 LRRM 1203 (1974), *enf'd,* 515 F.2d 1018, 90 LRRM 2844 (D.C. Cir. 1975); *General Motors Corp.,* 191 NLRB 951, 77 LRRM 1537 (1971).

The NLRA contains no simple "litmus test" for determining whether a particular subject is within the recognized philosophy of bargaining, thus the

NLRB assumed the role of deciding what are and what are not mandatory subjects of bargaining. Morris, *The Developing Labor Law,* BNA (1971 Rev. Ed.).

Safety and health rules, procedures, and practices have been held to be mandatory subjects of bargaining under the NLRA, and thus their resolution can be bargained to impasse. In *NLRB v. Gulf Power Co.,* 384 F.2d 822, 66 LRRM 2501 (5th Cir. 1967), *enf'g* 156 NLRB 622, 61 LRRM 1073 (1966), the court upheld the NLRB's ruling that an employer violated Section 8(a)(5) of the NLRA by refusing to bargain over safety rules and practices. *In Fibreboard Paper Products Corp. v. NLRB,* 379 U.S. 203, 57 LRRM 2609 (1964), safety practices were mentioned as a condition of employment for purposes of determining the bargaining duty of an employer. See also *NLRB v. Borg-Warner Co.,* 356 U.S. 342, 42 LRRM 2034 (1958). Unions have recently stated that in order to reinforce occupational safety and health laws and regulations, the bargaining process should be used to the hilt. However, because of the increasing volume of safety and health actions against unions, they have been instructed to exercise caution as they approach the bargaining table. (See Chapter XIV). These suits are in part the result of OSHA's efforts to better educate workers in safety and health matters. The actions which are and have been filed against unions include allegations of failure to: insure safe work-sites; warn employees of hazardous substances; not refer an incompetent employee for a job; and negotiate "sufficient" safety and health protection. In the case of *Dunbar v. United Steelworkers,*—Ia.—(S. Ct. 1978), (unreported), for example, survivors of employees who died in an employment accident sued the union for damages because of a contract clause which stated that the union shall inspect a mine, which it did not.

As of July, 1976, safety provisions were found in 897 (57 percent) of 1,570 contracts, each covering 1,000 or more workers, analyzed by the Bureau of Labor Statistics; environmental and health provisions were found in 62 contracts. *Characteristics of Major Collective Bargaining Agreements.* As of September 1978, occupational safety and health provisions appeared in 82 percent of the sample contracts analyzed, according to a BNA Report on "Safety and Health Patterns in Union Contracts." *(Daily Labor Report,* No. 178, September 13, 1978). [Guarantees against discrimination either by a union, company, or both on the basis of race, color, sex, national origin . . . appeared in 91 percent of all contracts analyzed.] Safety and health clauses were included in 87 percent of manufacturing agreements and 73 percent of nonmanufacturing contracts.

Of contracts with safety and health provisions, 36 percent included a statement that the company will comply with federal, state, and/or local laws, and 61 percent included a general statement of responsibility for employees' safety and health. More than two fifths (43 percent) of the general statements apply only to management and more than half (57 percent) to both management and union. Employee responsibilities and obligations in maintaining safety and health standards are stated in 37 percent of the agreements included in the survey, compared to 31 percent of contracts in a 1975 study and to 21 percent in a 1970 study.

Examples of clauses in collective bargaining agreements include employer payment of full-time wages to health and safety enforcement or training personnel, provision of first aid supplies, physical examinations, rate retention, workmen's compensation, provision of safety equipment,[11] employee rights to refuse to perform hazardous work, safety and health committees, instant arbitration of safety and health disputes by experts rather than traditional arbitrators, and vacation time or extra time for employees who perform particularly hazardous or unhealthful jobs. The single most important safety and health clause [from an employee perspective] in a contract today is one in which the employer agrees to provide a safe and healthful workplace patterned after Section 5(a)(1) of the [OSH] Act. This clause is more protective than merely requiring employers to abide by all federal, state, and local laws.

The most common safety provisions, in order of frequency, are joint pledges of cooperation in safety programs, the right to refuse unsafe work, the right to discipline employees for violating safety and health rules, the right to grieve unsafe work, the right of inspection by a joint or union safety committee, control over regulation of crew size, and the posting of safety and health rules. Other provisions may require physical or mental examination upon hiring, after breaks of service, upon termination, or upon an annual or semi-annual basis, such as in the case of audiometric testing. Many of these provisions are prompted by the need for both worker's compensation reports and claims and insurance reports and claims. Contracts may often provide for specification of certain eating places, sanitary drinking procedures, washrooms, showers, lockers, protective clothing, and other similar matters relative to working conditions and safety. Many contracts provide for the establishment of joint employer-union committees as an instrument for implementing safety and health programs. The functions and compositions of safety and health committees may vary but they usually address such matters as inspection tours, handling complaints, educating employees, and making recommendations relative to the elimination of unsafe and unhealthful conditions.

Several types of collective bargaining clauses relative to employee safety and health are attached as an appendum to this chapter.

Even when safety and health conditions and procedures are included in collective bargaining contracts, employers still basically retain the responsibility for determining policies and procedures. See *General Telephone Co. of Pennsylvania,* 71 LA 488 (Ipavec, 1978). However, in one instance an arbitrator ruled that all such policies and procedures must first be submitted to a safety and health committee prior to promulgation, since one had been established. See *FWD Corp.,* 71 LA 929 (Lynch, 1978). See also *General Electric Co.,* 3 OSHC 1031 (1973).

It should be noted that when union officials enter a plant, pursuant to their representation rights set forth in a collective bargaining agreement, they too are required to abide by safety and health rules. See *Honeywell, Inc.,* 63 LA 151 (Sembower, 1974); *Allegheny Ludlum Steel Corp.,* 66 LA 1306 (Yagoda, 1976). An employee's status as a union official should not save him from discipline in this regard. *Hoerner Waldorf Corp.,* 70 LA 335 (Talent, 1978).

2. Employee Discipline

Employees, although they have an obligation to comply with the (OSH) Act pursuant to Section 5(b), cannot be cited or fined by OSHA for violations of any standards or for their refusal to comply with the Act. *Atlantic & Gulf Stevedores, Inc. v. OSHRC,* 534 F.2d 541, 4 OSHC 1061 (3d Cir. 1976); *Nacirema Operating Co., Inc. v. OSHRC,* 538 F.2d 73, 4 OSHC 1061 (3d Cir. 1976). As stated by the court in those cases:

> To frame the issue in slightly different terms, can the Secretary insist that an employer in the collective bargaining process bargain to retain the right to discipline employees for violation of safety standards which are patently reasonable, and are economically feasible except for employee resistance?
>
> We hold that the Secretary has such power . . . the entire thrust of the Act is to place primary responsibility for safety in the work place upon the employer. That, certainly, is a decision within the legislative competence of Congress. In some cases, undoubtedly, such a policy will result in work stoppages. But as we observed in *AFL-CIO v. Brennan, supra,* the task of weighing the economic feasibility of a regulation is conferred upon the Secretary. He has concluded that stevedores must take all available legal steps to secure compliance by the longshoremen with the hard hat standard.
>
> We can perceive several legal remedies which an employer in petitioners' shoes might find availing. An employer can bargain in good faith with the representatives of its employees for the right to discharge or discipline any employee who disobeys an OSHA standard. Because occupational safety and health would seem to be subsumed within the subjects of mandatory collective bargaining—wages, hours and conditions of employment, see 29 U.S.C. Sec. 158(d)—the employer can, consistent with its duty to bargain in good faith, insist to the point of impasse upon the right to discharge or discipline disobedient employees. *See NLRB v. American National Insurance Co.,* 343 U.S. 395 (1952). Where the employer's prerogative in such matters is established, that right can be enforced under Sec. 301. Should discipline or discharge nevertheless provide a work stoppage, *Boys Markets* injunctive relief would be available if the parties have agreed upon a no-strike or grievance and arbitration provision. And even in those cases in which an injunction cannot be obtained, or where arbitration fails to vindicate the employer's action, the employer can still apply to the Secretary pursuant to Sec. 6(d) of the Act, 29 U.S.C. Sec. 655(d), for a variance from a promulgated standard, on a showing that alternative methods for protecting employees would be equally effective. *See Brennan v. OSHRC (Underhill Construction Corp.),* 513 F.2d 1032, 1036.

Thus refusal by employees to comply with a safety or health standard or labor relations problems generally does not absolve an employer from liability under the Act. In *Independent Pier Co.,* 3 OSHC 1674 (1975) and *Rollins Construction Co.,* 3 OSHC 1298 (1975), the Commission found the employer in violation of a protective equipment standard even though its employees refused to wear their hard hats and even though the employer believed he would suffer a strike if he enforced the rule. See also *I.T.O. Corp. v. OSHRC,* 540 F.2d 543, 4 OSHC 1574 (1st Cir. 1976); *Detroit Printing Pressman Local No. 13,* 1 OSHC 1071 (1972); *Alpha Poster Service, Inc.,* 4 OSHC 1883 (1976).

Under certain circumstances, however, the employee can and should be disciplined or discharged by the employer for noncompliance with OSHA standards, a position acknowledged by the Secretary of Labor, the Review Commission, the courts, and even labor organizations. See *McCord Corp.*, 64 LA 1291 (Simon, 1975), *Corbin Lavoy (Empire Boring Co.)*, 4 OSHC 1259 (1976). 29 CFR 1977.22 states in this respect that:

> Employees who refuse to comply with occupational safety and health standards or valid safety rules implemented by the employer in furtherance of the Act are not exercising any rights afforded by the Act. Disciplinary measures taken by employers solely in response to employee refusal to comply with appropriate safety rules and regulations, will not ordinarily be regarded as discriminatory action prohibited by Section 11(c). This situation should be distinguished from refusals to work, as discussed in [29 CRF] 1977.12.

In *Gateway Coal, supra,* arbitration was found to be the primary forum for resolving health and safety disputes. Consistent with that proposition, more and more safety and health issues are being determined by arbitrators under collective bargaining agreements. See *e.g., Bethlehem Steel Corp.*, 73 LA 825 (Strongin, 1979); *Naval Air Rework Facility,* 73 LA 644 (Flannagan, 1979); *Ogden Air Logistics Center,* 73 LA 1100 (Hayes, 1979).

One case, however, has pointed up the hurdles confronting an employee in his efforts to use either the NLRA or arbitration approach for the resolution of safety and health disputes. In *United Parcel Service, Inc.*, 232 NLRB 1114, 96 LRRM 1288 (1977), the NLRB deferred to an arbitrator's award upholding the discharge of an employee who refused to work despite three separate grievance committee determinations that his working conditions were not unsafe. But recent NLRB decisions have taken an opposite approach and upheld an employee's right to refuse to work in hazardous conditions. *Anheuser-Busch, Inc.*, 239 NLRB No. 23, 99 LRRM 1548 (1978); *Pink Moody, Inc.*, 237 NLRB No. 7, 98 LRRM 1463 (1978); *Modern Carpet Industries, Inc.*, 236 NLRB No. 132, 98 LRRM 1426 (1978), *enf'd,* 103 LRRM 2167 (10th Cir. 1979); *Brown Mfg. Corp.*, 235 NRLB No. 189, 98 LRRM 1347 (1978).

In order to establish and prove the employee misconduct and isolated incident defense, it is encumbent upon an employer to show that its safety rules and regulations were adequately communicated and enforced, usually by disciplinary measures. (See Chapter XII). Thus even though the Act does not require that an employer take disciplinary measures against employees when they fail to comply with OSHA standards, in order to prove it comes within the scope of certain affirmative defenses, an employer may in many instances be required to show that it did in fact take disciplinary measures to secure compliance. See *Leo J. Martone & Associates, Inc.*, 5 OSHC 1229 (1977); *B-G Maintenance Management, Inc.*, 4 OHSC 1282 (1976); *Combustion Engineering, Inc.*, 5 OSHC 1943 (1977). Moreover, it is usually held that an employer is responsible for violations caused by unsafe employees and must, even though not specifically required by a standard, adequately train and instruct them in safety and health procedures. See *Brennan v. Butler Lime & Cement Co.*, 520 F.2d 1011, 3 OSHC 1461 (7th Cir. 1975); *Rea Express, Inc. v. Brennan,* 495 F.2d 822, 1 OSHC 1651 (2d Cir. 1974); *National Industrial Constructors, Inc. v. OSHRC,* 583 F.2d 1048, 6 OSHC 1914 (8th Cir. 1978); *Getty Oil Co. v. Secretary of Labor,* 530 F.2d 1143, 4 OSHC 1121 (5th Cir. 1976); *Nickie Lou*

Fashions, Inc., 6 OSHC 1824 (1978); *Danco Construction Company v. OSHRC,* 586 F.2d 1243, 6 OSHC 1939 (8th Cir. 1979); *R. A. Pohl Construction Co.,* 7 OSHC 1483 (1978); *Horween Leather Company,* 7 OSHC 1664 (1979).

3. Cost of Personal Protective Equipment

Although it has been held that the [OSH] Act does not require that an employer pay for employee personal protective equipment and clothing (or their laundering and replacement), see *The Budd Co. v. OSHRC,* 513 F.2d 201, 2 OSHC 1698 (3d Cir. 1975); *Ryder Truck Lines, Inc.,* 1 OSHC 1290 (1973) *aff'd,* 497 F.2d 230, 2 OSHC 1075 (5th Cir. 1974); *Amoco Oil Co.,* 3 OSHC 1745 (1975), an employer may by agreement in a collective bargaining contract commit to pay the cost of personal protective equipment and may be required to bargain over such expenditures if they come under the heading of wages and working conditions. See *Metalcraft Products Co.,* 58 LA 925 (Levin, 1972); *Mrs. Baird's Bakeries, Inc.,* 68 LA 773 (Fox, 1977). Any allocation for the cost of safety shoes, safety glasses and other protective clothing, when the wearing of such items is mandated by OSHA regulations or safety and health considerations generally, is a matter for arbitration or collective bargaining; an employer is not automatically required to pay for such items. *Intalco Aluminum Co.,* 70 LA 1229 (Peterschmidt, 1978); *Alabama By-Products Corp.,* 61 LA 1286 (Eyraud, 1973); *Bagget Transportation Co.,* 68 LA 82 (Crane, 1977); *Virginia Electric Power Co.,* 68 LA 418 (Harkless, 1977); *United Telephone Co. of Florida,* 69 LA 87 (Carson, 1977); *Dart Industries,* 65 LA 1263 (Carter, 1975); *Ingram Mfg. Co.,* 70 LA 1269 (Johannes, 1978); *Mobil Oil Corp.,* 69 LA 828 (Ray, 1977); *Potash Co. of America,* 70 LA 194 (Harr, 1978). But see *Bendix Forest Products Corp. v. Div. of Occupational Safety and Health,* 600 P.2d 1339 (Cal. 1979)—the California Supreme Court upheld, *en banc,* a California OSHA requirement that employers absorb the cost of providing employees with personal safety equipment.

In *Hater Industries, Inc.,* 73 LA 1025 (Ipavec, 1980) an arbitrator held that an employer was under no obligation to furnish or to pay for safety shoes for workers who handled heavy materials and which were required under OSHA regulations. The OSHA regulation only required that the company make sure that shoes were used and worn. Thus it had no obligation to purchase protective footwear, either by agreement or past practice. Although safety and health provisions had been made part of the agreement, the company's purchase of safety equipment was not mentioned. The arbitrator also stated that although in the past the company had provided workers with some types of safety equipment, such as glasses, aprons, and gloves, those items were not personal in nature, such as shoes, but were items the company could reuse.

In another case the union argued that an employer had an obligation to bargain over whether in fact to provide protective equipment, but most arbitrators hold that management is free to take unilateral action during a contract term unless expressly prohibited by the contract from doing so. In the absence of a "zipper clause" expressly stating that a contract constitutes the entire agreement of parties or that the right to bargain on matters not covered by the agreement is waived, the company may be required to discuss proposed changes. See *Exxon Chemical Co.*, 68 LA 362 (Bailey, 1977).

The requirement that an employer pay for personal protective equipment under a collective bargaining agreement may depend upon whether the employees can put that equipment to other uses, such as work gloves or protective shoes. See *General Telephone Co. of Florida*, 62 LA 997 (Turkus, 1974). For example, if the equipment is suitable for personal wear off the job and comparable in price to regular shoes, the employer may not have to pay for such equipment. *Baggett Transportation Co.*, 68 LA 82 (Crane, 1977); *Dart Industries, Inc.*, 65 LA 1263 (Carter, 1975); *Manistee Drop Forge Co.*, 66 LA 65 (Rayl, Jr., 1976)—shoes with metatarsal guards may not be able to be used off-premises, therefore metatarsal subsidy was appropriate; *United Telephone Co. of Florida*, 69 LA 87 (Carson, 1977)—safety glasses; *Monongahela Power Co.*, 64 LA 73 (Hunter, 1975)—no obligation to reimburse employee if he breaks or loses equipment that must be replaced; *Marmon Group Inc.*, 73 LA 1279 (Bode, 1979)—no requirement to pay for ordinary glasses for employee who broke safety glasses; *Nichols Construction Corp.*, 74 LA 964 (Britton, 1980)—contract provided for refund for certain losses of work clothes, but did not include glasses; *Exxon Chemical Co.*, 68 LA 362 (Bailey, 1977)—provisions relative for the cost of laundering; *White Mfg. Co.*, 74 LA 1191 (LeBaron, 1980)—company which is required under collective bargaining agreement to purchase safety shoes for employees is unilaterally entitled to determine that it will pay no more than a certain amount for such shoes. In *Ingram Manufacturing Co.*, 70 LA 1269 (Johannes, 1978), the company received an OSHA citation, posted a notice requiring the purchase and use of safety shoes, and cited a contract provision stating that employees could purchase such items through payroll deduction. The union objected, stating it was the company's obligation to maintain safe and healthful working conditions in accordance with the Act. The arbitrator rejected the argument, stating that the purpose of the section cited by the union was to provide for the safety of the employees by assuring safety, not by buying personal protective equipment. Regardless of who pays the bill, the employer may be able to deny pay and time for an employee who is denied permission to work because he does not wear proper personal protective equipment. See *McEvoy Oil Field Equipment Co.*, 68 LA 738 (Bailey, 1977); *CVI Corp.*, 64 LA 50 (Dyke, 1975).

4. Rate Retention

A recent labor relations question which has arisen is the issue of rate retention. When an employee is transferred to a lower-paying job because of the occurrence or threat of occupational injury or illness, the question becomes, does his old rate of pay remain in effect at the job to which he is transferred? Normally such questions are handled under whatever rules are established regarding exercise of seniority or transfer. See *Anaconda Aluminum Co.*, 64 LA 24 (Larkin, 1974); *United Steel Corp.*, 67 LA 1192 (Garrett, undated).

Several recently promulgated OSHA standards, moreover, have contained medical removal and wage retention provisions and employers have argued that they conflict with and interfere with national labor policy. Examples of these are contained in the recently issued cotton dust standard, 29 CFR 1910.1043 and the recently issued standard for occupational exposure to lead, 29 CFR 1910.1025. In *United Steelworkers of America, AFL-CIO, CLC v. Marshall,*—F.2d—, 8 OSHC 1810 (D.C. Cir. 1980), the court, in reviewing the standard limiting occupational exposure to lead, considered the employer's argument that the medical removal and wage reduction provisions interfered with a traditional and mandatory subject of collective bargaining and thus violated the congressional principle that the substantive provisions of labor-management relations be left to the bargaining process. The court stated that it did not doubt that the provisions will have a noticeable effect in collective bargaining and stated that earnings protection is no doubt a mandatory subject, but so is any issue directly related to safety. The court concluded that there is nothing in the Act which would suggest that OSHA should not have the power to create such a program instrumental to achieving worker safety simply because it could otherwise be created through collective bargaining and is a reasonable means toward that end. The court also noted in a footnote that the question was raised relative to the validity of the provision under Title VII. It was argued by OSHA that without medical removal protection, employers will discriminate against fertile women, to whom lead exposure poses even a greater hazard than it does to other workers, by excluding them from all lead-exposed jobs at the outset. The court, while not addressing this hypothetical question, stated that it thought fertile women can find statutory protection from such discrimination in the Act's own requirement that OSHAstandards insure that no employee will suffer material impairment of health. In *AFL-CIO v. Marshall,* 617 F.2d 636, 7 OSHC 1775 (D.C. Cir. 1979) the court upheld similar but less stringent medical removal and wage retention provisions in the cotton dust standard.

5. Duty to Furnish Information

An employer has a duty under Section 8(a)(5) of the NLRA to furnish information to a union relevant to the effectuation of its collective bargaining obligations.[12] Since health and safety matters are Section 8(d) subjects, it follows that an employer has an obligation to furnish information to the union concerning these matters, absent a waiver of the right to information. Similar rights to furnish information have been held to exist with respect to other subjects of bargaining, such as discrimination.

In *Westinghouse Electric Corp.,* 239 NLRB No. 19, 99 LRRM 1482 (1978), for example, the union was found entitled to statistical data on minority group and female employees, charges, and complaints involving unit employees that had been filed against the employer under federal and state fair-employment practices laws, and copies of work force analyses contained in affirmative action programs that the employer was required to develop pursuant to Executive Order 11246. In *East Dayton Tool & Dye Co.,* 239 NLRB

No. 20, 99 LRRM 1499 (1978), the NLRB held that a union may obtain information as to the race and sex of job applicants, since this information is necessary and relevant to the union's policing of the contract. See also *Handy Andy, Inc.*, 228 NLRB 59, 94 LRRM 1354 (1977); *J. P. Stevens & Co.*, 239 NLRB No. 95, 100 LRRM 1052 (1978); *Bendix Corp.*, 242 NLRB No. 170, 101 LRRM 1459 (1979); *White Motor Corp. (White Farm Equip. Co.)*, 242 NLRB No. 201, 101 LRRM 1470 (1979); *Emporium Capwell Co. v. WACO*, 420 U.S. 50, 88 LRRM 2660 (1975).

The basis for the proposition that a union representing a unit of employees is entitled to information from an employer which may be relevant and reasonably necessary for the proper performance of its obligations to negotiate and administer a collective bargaining agreement is contained in *F. W. Woolworth Co. v. NLRB*, 352 U.S. 938, 39 LRRM 2151 (1956). The test of the union's need for such information is a showing that the desired information is *relevant* and that it would be of use to the union in carrying out its statutory duties and responsibilities. *NLRB v. Acme Industrial Company*, 385 U.S. 432, 64 LRRM 2069 (1967); *NLRB v. Yawman and Erbe Company*, 187 F.2d 947, 27 LRRM 2524 (2d Cir. 1951).

Recently, in several pending cases it has been argued that the duty to furnish information should be extended to the occupational health as well as safety area. In *Thatcher Plastic Packaging, Division of Dart Industries, Inc.* and *Premier Thermoplastic Company, Division of Rikkold Chemicals, Inc.*, the union requested that the employer provide it with information concerning the plant's chemical substances, alleged hazards, and environmental and medical monitoring. In addition, the union sought and was denied access for one of its experts to conduct a safety and health inspection of the plant. The collective bargaining agreement provided that the employer would take precautions relative to safety and health but contained no express provision relating to plant acess for union experts. An administrative law judge of the NLRB was urged to conclude that the employer must not only provide access to the records requested, but also was urged to state that the employer violated Sections 8(a)(5) and 8(a)(1) of NLRA by denying access to union experts for the purposes of conducting health and safety inspections of the plant. According to the arguments, which relied on the seminal NLRB case in this area, *Gulf Power Company*, 156 NLRB 622, 61 LRRM 1073 (1966), *enf'd*, 384 F.2d 822 (5th Cir. 1967), it can be argued that *unhealthy* conditions in a plant can be as harmful to the well-being of the employees as *unsafe* conditions at a plant. Thus health considerations, no less than safety conditions, should be considered Section 8(d) subjects. Given the proposition that health matters are Section 8(d) subjects, it follows that the employer has the obligation to furnish information to the union concerning this subject, absent a waiver of the right to the information. Moreover, it was argued, the existence of safety and health provisions in a collective bargaining agreement was not conclusive evidence that the parties bargained over the entire subject of safety and health, including rights to information and access. Finally, it was noted that these cases were not proper for deferral to OSHA since the Memorandum of Understanding between the NLRB and OSHA only deals with Section 11(c) cases and the Memorandum has no bearing on a labor organization's statutory right to information necessary for the conduct of its collective bargaining duties under the NLRA.

In a decision entitled *Winona Industries, Inc.,* No. 18-CA-6509 (1980), an NLRB administrative law judge determined that an employer refused to bargain in good faith by its failure to grant a union industrial hygienist access to its plant for the purpose of evaluating health and safety conditions. The judge said the conditions affecting the health and safety of employees are terms and conditions of employment, citing *Gulf Power Company, supra,* and *Winn-Dixie Stores,* 224 NLRB 1418, 92 LRRM 1625, modified in other respects, 567 F.2d 1343 (5th Cir. 1978).

A recent question which has also come to the forefront relates to the requests filed by unions with respect to unrestricted access to employee medical and exposure records. As set forth in 29 CFR 1910.20, access to employee medical records is accorded to union representatives *only* when an employee gives specific written authorization. Several unions have recently filed requests without such written authorization asserting their right under the NLRA to obtain information relative to their collective bargaining representative status. The unions argue that the [OSH] Act, like the FLSA's minimum wage laws and a variety of other governmental regulations, merely establish minimum requirements in their respective fields and are not intended to preempt the field of regulations to such an extent to exclude therefrom the concept of collective bargaining. This argument has some merit (except for the aspect of confidentiality), for OSHA itself states in 29 CFR 1910.20 that the regulation is not intended to preclude achievement of wider access to employee medical and exposure records through *collective bargaining* than that provided for therein.[13] This matter is now tangentially before the Board in three cases entitled *Minnesota Mining and Manufacturing Company,* No. 18-CA-5710 and 5711, *Colgate Palmolive Company,* No. 17-CA-8331, and *Borden Chemical* No. 32-CA-551. In those cases, the Oil, Chemical, and Atomic Workers Union requested information on chemicals used in the companies' facilities and company health and safety records on employees. The companies argued that the information requested need not be supplied in the absence of a specific agreement, while the union argued it was entitled to the requested information pursuant to their duty to bargain intelligently about health and safety matters.

E. CONCERTED ACTIVITY

Section 7 of the NLRA gives employees the right to engage in *concerted* activities for their mutual aid or protection, for the purposes of collective bargaining, with or without union participation. *NLRB v. Washington Aluminum Co.,* 370 U.S. 9, 50 LRRM 2235 (1962). Individual activities are normally not protected. Sections 8(a)(1) and 8(b)(1) of the NLRA prohibit employer and/or union unfair labor practices respectively, which restrain or coerce employees in the exercise of their rights guaranteed under Section 7. The right to engage in concerted activity (e.g., a strike) can however be bargained away by a union and prohibited under a contract as a *quid pro quo* for arbitration.

(Section 301 also provides an avenue to employees for redress of violations of contracts between an employer and a union). In the specific area of safety and health, Section 502 states that the withholding or quitting of work because of an *abnormally dangerous* condition is not considered to be a strike. (*e.g.*, concerted activity). Conceivably, then, employees would have recourse to Section 502 (and possibly Sections 301 and 8(a)(1) and/or 8(b)(1)) for protection and/or redress from abnormally dangerous conditions of safety and health in the workplace. In an early case involving Section 502, *Knight Morley Corp.*, 116 NLRB 140, 38 LRRM 1194 (1956) *enf'd.*, 251 F.2d 753, 41 LRRM 2242 (6th Cir. 1957), *cert. denied,* 357 U.S. 927, 42 LRRM 2307 (1958), the court said that Section 502 limits the right of an employer to require a continuance of work under conditions which an employee in good faith believes to be abnormally dangerous. (Subsequent decisions have restricted the applicability of Section 502 by imposing the requirement of objective proof. See *e.g., Redwing Carriers, Inc. v. NLRB,* 586 F.2d 1066, 100 LRRM 2171 (5th Cir. 1978)).

The Supreme Court first addressed the utilization of Section 502 in *Gateway Coal Company v. United Mine Workers,* 414 U.S. 368, 85 LRRM 2049 (1974). In that case, the Court was faced with the question of whether a collective bargaining agreement imposed a compulsory duty on the parties thereto to submit safety disputes to arbitration. Assuming that the answer was in the affirmative, a secondary question was did that duty to arbitrate give rise to an implied no-strike obligation supporting the issuance of an injunction pursuant to the guidelines set forth in *Boys Markets, Inc. v. Retail Clerks Union,* 398 U.S. 235, 74 LRRM 2257 (1970). The court first stated that no obligation to arbitrate a labor dispute arises solely by operation of law. The law compels a party to submit his grievance to arbitration only if he has contracted to do so. The court found however in the case that the parties had through negotiation agreed upon an arbitration clause to settle disputes. The court then concluded that the presumption of arbitrability announced in the *Steelworkers' Triology* applied to safety disputes, and that the dispute in the instant matter was covered by the arbitration clause in the parties' collective bargaining agreement.

The second question addressed by the court was assuming the above, whether it had the authority to enjoin the work stoppage. The court stated that the answer depends on whether the union was under a contractual duty not to strike. Although the collective bargaining agreement in *Boys Market* contained an express no-strike clause, it stated that injunctive relief may also be granted on the basis of an *implied* undertaking not to strike. Absent an explicit expression negating no-strike obligation, the agreement to arbitrate and the duty not to strike should be construed as having co-terminous application.

The court then addressed the question of the effect of Section 502 on the denial of injunctive relief to prevent a safety strike. The court first said that Section 502 provides a limited exception to an express or an implied no-strike obligation. The court also stated that a work stoppage solely to protect employees from immediate danger would be authorized by Section 502 and could not be the basis for either a damage award or a *Boys Market* injunction.

It disagreed with the decision of the court of appeals below which had stated that "An honest belief no matter how unjustified in the existence of abnormally dangerous conditions for work invokes protection of 502." The Court stated that the courts must require objective evidence that such conditions actually obtain or they face a wholly speculative inquiry into the motives of a worker. A union seeking to justify a contractually prohibited work stoppage under Section 502 must, therefore, present ascertainable, objective evidence supporting its conclusion that an abnormally dangerous condition does exist.[14] (Citations alleging imminent danger and serious violations may help to objectively establish the fact of abnormally dangerous conditions for the purposes of Section 502).

The limitation on an employee's right to refuse to work as set forth under Section 502 and discussed in the *Gateway Coal* decision, emphasizes the importance of the role of comprehensive collective bargaining negotiations with respect to safety and health matters. It also emphasizes the importance of alternative means of protection of employees. For example, under the Supreme Court's decision in *Whirlpool Corp. v. Marshall,* 445 U.S. 1, 8 OSHC (1980), discussed in detail in Chapter XV, refusals to perform work based upon an imminently dangerous condition are deemed protected under Section 11(c) of the Act. Employees, individually or through their unions, may attempt to gain contractual protection for refusing to perform hazardous work under abnormally or imminently dangerous conditions by keying into Section 11(c) language in addition to or in lieu of Section 502. Unions may also seek to negotiate a provision in collective bargaining agreements which would incorporate by reference the imminent danger provisions of Section 13(a) of the [OSH] Act. Thus, in addition to directly invoking the protections afforded by Sections 11(c) and 13(a) of the [OSH] Act, employees may indirectly be able to obtain their protections and refuse to perform hazardous or abnormally dangerous work pursuant to his or her own collective bargaining agreement. The propriety of a work refusal would then be determined through arbitration, expedited or otherwise. Possibly, such a refusal to perform work may also be non-enjoinable even in the absence of reasonable objective evidence, if based on Section 11(c).

It remains to be seen what will be the interplay between OSHA and the NLRB on the question of an employee's right to stop working when he believes he will be exposed to imminently or abnormally dangerous conditions. In order to avoid inconsistency and confusion[15] there must be development and definition with respect to the questions of the relationship between the two Acts, the relationship between a no-strike agreement and arbitration, the relationship between reasonable danger, imminent danger and abnormal danger and the relationship between a reasonable belief and objective evidence. See *I.T.O. Corp. v. OSHRC,* 540 F.2d 543, 4 OSHC 1574 (1st Cir. 1976); *Myers Industrial Electric,* 177 NLRB 817, 71 LRRM 1425 (1969).[16]

Outside the context of abnormal or imminently dangerous conditions, other issues and concerns in the areas of safety and health may arise. These matters must also be handled within the concept of concerted activity.

In *Alleluia Cushion Co.,* 221 NLRB No. 999, 91 LRRM 1131 (1975), an employee complained to his employer about what he considered to be unsafe working conditions. He also questioned the employer's failure to post a previous OSHA citation. When the employer failed to correct these conditions, the employee complained to California OSHA and requested that it conduct a plant inspection. The employee did not attempt to enlist any other employees in support of his efforts. Even though the employee acted alone and had no outward manifestation of support from his fellow employees, the NLRB stated that Section 7 of the Act protected the employee's safety-related activities, enlarging the definition of concertedness to include an employee's efforts to secure enforcement of federal, state, and local legislation on occupationally-related subjects:

> Accordingly, where an employee speaks up and seeks to enforce statutory provisions relating to occupational safety designed for the safety of all employees, in the absence of any evidence that fellow employees disavow such representation, we will find an implied consent thereto and deem such activity to be concerted.

Relying on *Bunney Brothers,* 139 NLRB 1516, 51 LRRM 1532 (1962), the NLRB stated that even a single employee's individual activity would be considered concerted if it amounted to enforcement of the terms and condition of a collective bargaining agreement, *unless* the other employees explicitly disavowed such activity. The NLRB also held that an employee's individual resort to a government agency constituted an extension of the concerted employee activity which gave rise to the bargaining agreement, and that such concertedness "emanated from the mere assertion of such statutory rights." The NLRB reached this decision because it felt it had an obligation to construe the NLRA in a manner supportive of the policies and purposes of other employment legislation.

Hence, where a solitary employee speaks up or seeks to enforce statutory provisions relating to occupational safety and health designed for the benefit of all employees, it will be presumed that his fellow employees concur in his efforts and his activity will be deemed concerted, even though he does not consult with or represent other employees. *P & L Cedar Products,* 224 NLRB 244, 93 LRRM 1341 (1976); *Wray Electric Contracting, Inc.,* 210 NLRB 757, 86 LRRM 1589 (1974); *C & I Air Conditioning, Inc.,* 193 NLRB 911, 78 LRRM 1417 (1971), *enforcement denied,* 486 F.2d 977 (9th Cir. 1973). This is known as the "constructive concerted activity" doctrine. For cases dealing with this doctrine in situations where employees are represented by a union under a collective bargaining agreement, see *Roadway Express, Inc.,* 217 NLRB 278, 88 LRRM 1503, *aff'd.* 532 F.2d 751 (4th Cir. 1976); *NLRB v. Interboro Contractors, Inc.,* 388 F.2d 495, 67 LRRM 2083 (2d Cir. 1967), *enf'g.* 157 NLRB 1295, 61 LRRM 1537 (1966); *Pelton Casteel, Inc. v. NLRB,* 627 F.2d 23, 105 LRRM 2124 (7th Cir. 1980); *NLRB v. Dawson Cabinet Co., Inc.,* 566 F.2d 1079, 97 LRRM 2075 (8th Cir. 1977). Moreover, in such cases, the NLRB will not defer a charge to arbitration since it alleges a violation of Sections 8(a)(1), 8(a)(2), 8(a)(3), or 8(a)(4) of the NLRA, even if the claim involves an arguably arbitrable issue. *General American Transp. Corp.,* 228 NLRB 808, 94 LRRM 1483 (1977). The Court of Appeals for the District of Columbia has also refused to allow dismissal of unfair labor practice charges where an arbitration award would have permitted an employer to require employees to violate state safety laws or to create safety hazards for themselves or others. *Banyard v. NLRB,* 202 NLRB 710, 82 LRRM 1652 (1973), *enforcement denied,* 505 F.2d 342, 87 LRRM 2001 (D.C. Cir. 1974).

The *Alleluia Cushion/Interboro* approach on "presumed" concerted activity, however, has not been uniformly accepted by the courts. See *NLRB v. C&I Air Conditioning, Inc.,* 486 F.2d 977, 84 LRRM 2625 (9th Cir. 1975)—the court recognized the *Interboro* doctrine but failed to see any evidence in that case showing that the purpose of the employee's conduct was for the mutual aid and protection of other employees or that it was an attempt to enforce the provisions of a collective bargaining agreement. See also, *ARO, Inc. v. NLRB,* 596 F.2d 713, 101 LRRM 2153 (6th Cir. 1979); *NLRB v. Bighorn Beverage,* 614 F.2d 1238, 103 LRRM 3008 (9th Cir. 1980); *Jim Causley Pontiac v. NLRB,* 620 F.2d 122, 104 LRRM 2190 (6th Cir. 1980); *NLRB v. Buddies Supermarkets, Inc.,* 481 F.2d 714, 83 LRRM 2625 (5th Cir. 1973); *Ontario Knife Company v. NLRB,*—F.2d —, 106 LRRM 2053 (2d Cir. 1980); *NLRB v. Northern Metals Co.,* 444 F.2d 881, 76 LRRM 2958 (3d Cir. 1971); *Kohls v. NLRB,* 629 F.2d 173, 104 LRRM 3049 (D.C. Cir. 1980); *Krispy Kreame Donut Corp. v. NLRB,* 635 F.2d 304, 105 LRRM 3407 (4th Cir. 1980). The courts have also split on the issue of whether employees must have a good faith, sincere *and* reasonable belief that a working condition is unsafe or unhealthful before an action, such as a work stoppage, constitutes protective concerted activity. See, *e.g., Wheeling-Pittsburgh Steel Corp. v. NLRB,* 618 F.2d 1009, 104 LRRM 2054 (3d Cir. 1980)—must be a reasonable basis in fact for the belief; *NLRB v. Modern Carpet Ind., Inc.,* 611 F.2d 811, 103 LRRM 2167 (10th Cir. 1979)—a good faith belief need only be sincere, not reasonable. In a case entitled *Allied Aviation Service Company of New Jersey, Inc.,* 248 NLRB No. 26, 103 LRRM 1454 (1980), the NLRB stated that a mechanic at the Newark, New Jersey, airport had a right under the NLRA to complain to his employer's (an airline) customers who landed at the airport concerning alleged safety hazards to which he was exposed. Letters which were written referred to numerous alleged hazards and mechanical operations at the airport. The employer, after receiving questions from one of the customers, first suspended and then discharged the employee. The NLRB concluded that the safety aspects of the letter were related to an ongoing labor dispute and as such were protected activity under the NLRA and ordered the employee reinstated with backpay.

In an arbitration context, if employees engage in concerted activity, for example, walk out over what they consider to be dangerous conditions, they must be able to prove that such condition in fact existed. They have the burden of proof to show that the conditions are unhealthy or unsafe; if not, their action may be considered an unlawful strike. See *Bunker Hill Company,* 65 LA 182 (Kleinsorge, 1975); *Consolidated Edison Co.,* 61 LA 607 (Turkus, 1973); *Affinity Mining Co.,* 64 LA 369 (Sherman, 1975); *Quaker Oats Co.,* 69 LA 727 (Hunter, 1977). But many arbitrators say that if the employee is sincere in his belief that danger exists, his activity should be protected. See *Western Airlines, Inc.,* 67 LA 486 (Christopher, 1976); *McLung-Logan Equipment Co., Inc.,* 71 LA 513 (Wahl, 1978). In many instances, if an employee's fear is legitimate and he has an honest belief, even though it will not *per se* immunize an employee from discipline, it would be a factor weighing against the severity of discipline. *Consolidated Edison Co.*, 61 LA 607 (Turkus, 1973). An employer may be within his right in firing a union steward who instigates a work refusal if his belief is mistaken that conditions are unsafe. See *Thiokol Corp.*, 63 LA 633 (Rimer, Jr., 1974); *R.I.-J.A. Machinery Co.*, 66 LA 474 (Waters, 1976). A strike by employees would be illegal if the agreement contains a no-strike clause. *ITT Thompsen Industries, Inc.*, 70 LA 970 (Seifer, 1978).

Some collective bargaining agreements have imminent-danger provisions or emergency provisions. An employee's failure to invoke them or improper invocation of such procedures may also be a basis for discharge. See *Hess Oil Virgin Islands Corp.*, 72 LA 81 (Berkman, 1978); *Agrico Chemical Corp.*, 70 LA 20 (Hebert, 1976). Moreoever, an employer should be wary of lack or inconsistent enforcement of a safety rule and must keep in mind the progressive discipline concept or any penalties imposed for a violation may be mitigated. See *Tumwater Valley Development Company*, 64 LA 981 (Peck, 1975). One arbitrator, however, has stated that the standards for review in ordinary discipline cases must be distinguished from cases involving safety issues, since safety rules and regulations are promulgated in order to preserve safety and health and not outwardly to punish. See *J. R. Simplot Company*, 68 LA 912 (Conant, 1975).

It must be remembered that action must not be taken under the guise of safety and health enforcement, for an employee is protected from discrimination under Title VII, under Section 8(a)(3) of the NLRA and under Section 11(c) of the [OSH] Act. Arbitrators have stated that even though an employer has more than one reason for discharging an employee, if one of those reasons is that the employee is engaged in, for example, protected concerted activity, the discharge could be in violation of the NLRA. See *Dura-Containers, Inc.*, 67 LA 82 (Maslanka, 1976).

F. WORKER'S COMPENSATION

Although considered in Chapter XIV, it must be noted at this point that a worker who is injured or made ill on the job is normally entitled to receive worker's compensation. Section 4(b)(4) of the [OSH] Act precludes a private remedy for damages suffered by employees as result of a violation of the Act by an employer who is covered under a state worker's compensation Act. *Byrd v. Field Crest Mills, Inc.*, 496 F.2d 1323, 1 OSHC 1743 (4th Cir. 1974). Suffice it to say, however, at this point that many contractually provided benefits, if awarded to an injured worker who receives worker's compensation, may provide the worker with double payment, as interpreted by an arbitrator under a collective bargaining agreement. *Mobil Oil Corp.*, 63 LA 986 (Nicholas, 1974); *United Steel Corp.*, 68 LA 1094 (Rimer, Jr., 1977). Other questions that arise in this context relate to whether an employee is entitled to worker's compensation payments during a period in which he receives full benefits pay from his employer. See *Toledo Blade Company*, 52 LA 728 (Kelliher, 1969). In one case an employee was found to be contractually entitled to such payments. *Gulf State Asphalt Co.*, 57 LA 837 (Schedler, 1971). Benefits may also be covered by insurance policies. *MIF Industries, Inc.*, 52 LA 1 (Summers, 1969). Certain contractual benefits, such as holiday and vacation pay, may, however, be withheld from employees who receive worker's compensation benefits during the time they are off due to injuries and illnesses. See *Roper Corp.*, 61 LA 342 (Bradley, 1973); *Walworth County*, 71 LA 1118 (Gundermann, 1978); *Oldberg Manufacturing Co.*, 51 LA 509 (Marshall, 1968); *Southern California Rapid Transit District*, 80-2 ARB ¶8625 (Tamoush, 1980).

NOTES

1. A similar memorandum of understanding has been entered into between the NLRB and MSHA—Fed. Reg.—See also Braid, *OSHA and the NLRA: New Wrinkles on old Issues*, 29 Lab. L.J. 755 (1978); Ashford, *Unsafe Working Conditions: Employee Rights under the NLRA and OSHA*, 52 Notre Dame L. Rev. 802 (1977); Tobin, *OSHA, Section 301 and the NLRB: Conflicts of Jurisdiction and Rights*, 23 Am. U. L.R. (1974).

2. Section 8(a)(1) and (3) provide that it shall be an unfair labor practice for an employer "(1) to interfere with, restrain, or coerce employees in the exercise of the rights guaranteed in Section 7; . . . (3) by discrimination in regard to hire or tenure of employment or any term or condition of employment to encourage or discourage membership in any labor organization."

3. Section 8(a)(4) provides that it shall be an unfair labor practice for an employer to discharge or otherwise discriminate against an employer because he has filed charges or given testimony under the NLRA.

4. In *General Electric Co.*, 3 OSHC 1031 (1975), the Commission agreed to the issuance of an order directing the employer to "cease and desist" from its policy of "unilaterally establishing health and safety programs without the advice, consent, and active participation of the 'Authorized Representative of Employees'." In *Marshall v. Kennedy Tubular Products*,—F. Supp. —, 5 OSHC 1467 (W.D. Pa. 1977), a district court enjoined an employer from excluding an employee representative from its premises in retaliation for filing a complaint with OSHA. The representative regularly had met with the employer's supervisors at arranged meetings during nonworking periods to discuss safety matters. Although the meetings were not specifically provided for in the collective bargaining agreement, the court held that the meetings were a part of the "industrial common law" which was part of the collective bargaining agreement.

5. This may run contrary to the employer's objective, for if it attempts to enforce harsh discipline for an other-than-serious violation as characterized by OSHA, an arbitrator may not agree with the appropriateness of the penalty. See *Phoenix Dye Works Co.*, 59 LA 912 (Dyke, 1972); *Precision Extrusions, Inc.*, 61 LA 572 (Epstein, 1973).

6. A recent series of articles on genetic screening (psychogenetics) noted the existence of gene variations associated with sex, race, and ethnic background. See Severo, "The Genetic Barrier: Job Benefit or Job Bias," *The New York Times*, February 3–4, 1980. Some members of these groups have genetically caused blood disorders which may place them at special risk when working with many chemicals. Moreover, alcoholics and persons with breathing difficulties, such as asthma, are very susceptible to exposure to chemicals in the workplace and are arguably protected as handicapped individuals under the Vocational Rehabilitation Act of 1973, 29 U.S.C. Sections 701–779. A handicapped individual is one who has a physical or mental impairment which substantially limits one or more of such person's major life activities, has a record of such impairment, or is regarded as having such impairment.

Hearing levels may differ significantly due to race and sex. In general, the black female population has the lowest or most sensitive hearing threshold, while the white male population shows the least sensitive hearing thresholds.

Many cancers are strongly age dependent, since cancer is a disease of old age, because the natural selection would be expected to have led to the evolution of defense mechanisms that operate early in life. OSHA Final Rule on Identification, Classification and Regulation of Potential Occupational Carcinogens. 45 Fed. Reg. 5001, 5026 (1980).

Black men have higher rates of cancer of the esophagus, lung, colon, rectum, and prostate. Stomach cancers are common to the Japanese, and white women have more breast and endometrial cancers than do blacks, but black women have a significantly higher rate of uterine cervical cancer.

7. 29 CFR 1910.1025. See also 29 CFR 1910.1043, which contains significant medical removal and wage retention provisions relative to the cotton dust standard; and 29 CFR 1910.1000(d)(2)(iv)(c), which provides that workers exposed to asbestos, who require a respirator but can not function with one, will be rotated to another job without loss of pay rate.

8. In an OSHA context, the most likely conflict will involve questions of sex or handicap discrimination. The Vocational Rehabilitation Act, 29 U.S.C. Sections 701–779, prohibits discrimination on the basis of handicap.

9. The typical discrimination case which arises under the NLRA (or LMRA) involves Section 8(b)(3) allegations of discrimination on the basis of union membership. See *Wright Line,* 251 NLRB No. 150, 105 LRRM 1169 (1980).

10. In the *St. Joe Paper Co.* and *Hoerner Waldorf Corp.* arbitration decisions, the arbitrators upheld the discharge of employees who made safety complaints, justifying the decisions on the basis of insubordination.

11. Safety equipment, such as guards and shields around machines and safety boots and goggles to be worn by employees, must be provided by the company in 42% of the contracts. Of these provisions, 46% specify that the company will furnish all safety equipment at no cost to the employee, and 21% state that employees will share some of the cost (often for replacements only) of wearing apparel.

12. See also *Ciba-Geigy Corp.,* 61 LA 438 (Jaffee, 1973); *Kawecki-Beryleco,* 1 OSHC 3097 (1973)—the union was entitled to data from monitoring an abatement program.

13. However, it is also stated therein that OSHA also does not accept arguments based on Section 4(b)(1) of the Act that OSHA must yield jurisdiction to the NLRB on all matters relating to employee or designated representative access to information pertinent to occupational safety and health. First, the Act and rules issued under it apply equally to employees who are represented by unions with collective bargaining status as well as employees who are not. The latter constitute approximately 80 percent of the work force. Second, Section 4(b)(1) preemption arguments are clearly overbroad, since any safety or health problem which OSHA could address would also be an appropriate subject for collective bargaining. The failure to bargain in good faith on any health or safety problem would then be the subject of NLRB's adjudicatory and sanctioning powers. It may also be an appropriate matter for arbitration under a collective bargaining argeement. But when Congress added the OSH Act to federal labor policy, OSHA was given rule-making authority, intended not only to fill gaps in existing labor-management relations and the collective bargaining process but the place a solid foundation under that process. Worker safety and health was thought by Congress to be too important to be resolved solely by the interplay of private economic forces and the rules established for economic warfare. OSHA, therefore, may appropriately set standards and provide remedies which are independent from, and in addition to, any rights afforded by the NLRA. Thus as an NLRB administrative

law judge recently stated in *Colgate-Palmolive Co.*, Case No. 17–CA–8331 (March 27, 1979): "Accommodation between the [NLRA] and OSHA is to be undertaken in a careful manner so as to preserve the objectives of each. See *Southern Steamship Company v. NLRB*, 316 U.S. 31, 47 (1942); *Western Addition Community Organization v. NLRB*, 485 F.2d 917, 927–28 (D.C. Cir. 1973); *Alleluia Cushion Co.*, 221 NLRB 999; Memorandum of Understanding Between Department of Labor, Occupational Safety and Health Administration and National Labor Relations Board, 40 Fed. Reg. 26033 (June, 1975)." OSHA agrees. *Cf. Brennen v. Western U. Tel Co.*, 561 F.2d 477 (3d Cir. 1977). Third, under the terms of Section 4(b)(1), OSHA's authority for this standard is not preempted, because the NLRB has not exercised statutory authority "to prescribe or enforce standards or regulations affecting or enforce standards or regulations affecting occupational safety or health" with respect to employees' working conditions. To the extent the NLRB has acted in this area, it has proceeded solely through adjudication, not through standard-setting or rule-making of general applicability. The object of its adjudications has only been to assure a fair bargaining process and not to assure safe and healthful working conditions for the workers affected. The triggering requirements of Section 4(b)(1) have therefore not been satisfied."

14. More recently, the Third Circuit ruled that specific contract language protecting a union's right to shut down an unsafe operation will insulate a work stoppage from an injunction. *Jones and Laughlin Steel Corp. v. United Mine Workers*, 519 F.2d 1155, 3 OSHC 1373 (3d Cir. 1975). The court vacated an injunction which had issued against a partial work stoppage at mine areas determined by the union's safety and health committee to pose an "imminent danger."

15. See also *Banyard v. NLRB*, 202 NLRB 710, 82 LRRM 1652 (1973) *denied enforcement*, 505 F.2d 342, 87 LRRM 2001 (D.C. Cir. 1974) in which a court refused to enforce an arbitrator's decision which would have forced a truck driver to violate a safety law. Section 11(c) of the OSH Act does not, on the other hand, require concertedness as a prerequisite to employee protection. *Dunlop v. Hanover Shoe Farms*, 441 F. Supp. 385, 4 OSHC 1241 (M.D. Pa. 1976).

16. In *Jones & Laughlin Steel Corp. v. United Mine Workers*, 519 F.2d 1155, 3 OSHC 1373 (3d Cir. 1975), an employer who was required by contract to accept a union safety committee's recommendation that employees be removed from an area in which the committee believed that an "imminent danger" existed was not entitled to an injunction against the work stoppage in the affected mine area.

APPENDIX 1 to CHAPTER XVII
Safety and Health Article (Aluminum Company)

A. The Company and the Union will continue to cooperate toward eliminating accident and health hazards and will continue to encourage employees to use the procedures stated herein in reaching this objective.

B. The Company, in accordance with applicable federal and state laws shall furnish to each employee, employment free from hazards that are causing or are likely to cause death or serious physical harm. Further, the Company shall comply with occupational safety and health standards promulgated under such laws.

C. It is intended that the International Union, local Unions, Union Safety Committees and its officers, employees, and agents shall not be liable for any work-connected injuries, disabilities, or diseases which may be incurred by employees. In this Article, the Union, through its various representatives, committees, officers, employees, and agents, has been accorded participation relating to employee safety and health; however, it is not the intention of the parties that these provisions shall in any way diminish the Company's exclusive responsibility. In this regard, the local Union, through the Chairman of the Union Safety Committee, will provide the Company with complete Union Safety Committee assignments by department and shift and will keep same up to date by notifying the Company of any and all changes as they occur.

D. Further, it is clearly understood by the parties that this Article shall not be used for the purpose of concerted activity or to avoid unpleasant working conditions and nothing contained in this Article shall abridge the rights of the Company under Article X and Article XI of this Agreement.

E. Joint Safety and Health Committees will function at each plant, meet monthly, and each shall concern themselves with the items outlined below:

1. Participate with the Plant Safety Department in considering practices and rules relating to safety and health.

2. Suggest appropriate changes to existing practices and rules and recommend adoption of such changes.

3. The Company will make available accident data (reports) and safe job procedures to review and develop data which will be useful in accident sources and injury trends.

4. A Company member of the Committee shall maintain minutes of its activities. Minutes will be mailed to appropriate levels of Company management, the International Union Staff Representative, local Union President, Union Safety Committee Chairman, and each Committee member within fifteen (15) days of the meeting.

5. Establish and support an Inspection Subcommittee. The purpose of this Inspection Subcommittee is to assist in locating potential accident causes and to help determine what safeguarding is necessary to protect against hazards before accident and personal injuries occur. It is not the purpose of this Subcommittee to replace or diminish existing inspection efforts being made by departments or individuals.

The Inspection Subcommittee will be composed of two members (one labor/one management) of the Joint Safety and Health Committee; the Chairman (Union) of the Joint Committee or his designated representative; and one member of the Safety Department (Management). One inspection will be conducted each month. The Safety Department will prearrange the date of inspection and shift to be inspected with the three other Committee members. Also, the Safety Department will arrange to rotate the members of the inspection team so that each member will have an equal opportunity to serve. The Union Chairman will receive prior notification of inspections.

6. All designated Committee members, whether Company or Union, shall have full authority and responsibility to bring to the attention of appropriate supervision and/or individual Union members unsafe conditions or practices which are considered detrimental to safe job procedures and/or good health. In all-cases where active enforcement of such unsafe practices and procedures is required by an individual Committee member, such action shall be immediately reported to the appropriate supervisor and to the departmental (shift) Safety Committeeman.

7. The Joint Safety and Health Committee will be composed of one designated Company representative from each production and maintenance department; one designated Union representative from each production and maintenance department; and the Company Safety Supervisor. However, additional members from any one department may be designated if deemed necessary by the departmental superintendent and agreed to by the Committee Chairmen. Each party shall designate a Co-Chairman. It shall be the joint responsibility of the Committee Chairman and area Safety Committeemen to:

   a. Investigate serious accidents or other accidents mutually agreed to be investigated by the Committee Chairman, and, at the next Committee meeting, report the results of such investigation to the full Committee;

   b. Jointly inspect work areas they deem appropriate for inspection and to report the results of such monthly inspection to the full Committee at its next meeting as outlined in subparagraph 5, above.

At every third monthly meeting the Committee shall review its activities conducted during the previous periods. The International Union Staff Representative servicing the location may attend and participate in this meeting.

F. The Company shall provide, at its expense, monitoring of health hazards as required by the Occupational Safety and Health Act, and shall provide protective devices, wearing apparel (excluding safety shoes), and other equipment necessary to protect employees from industrial illness and/or injury; and further, the Company will continue to maintain adequate first-aid coverage at all of its facilities.

G. Without detracting from the existing rights and obligations of the parties recognized in the other provisions of this Agreement, the Company and the Union will cooperate in encouraging employees afflicted with alcoholism or drug addiction to undergo a coordinated program directed to the objective of their rehabilitation.

H. An employee who believes that there exists an unsafe condition, beyond the normal hazards inherent in the operation, which involves a clear and immediate danger of injury to his person, may request an immediate meeting with his foreman. The foreman shall investigate to determine whether or not such conditions exist.

1. If the foreman determines that such imminently hazardous conditions do not exist and the employee still believes that such conditions do exist, the employee may then request a meeting between his foreman and the appropriate Union Safety Committeeman. Upon conferring, the parties must first decide if an immediate danger of physical injury exists. If the parties agree that no such danger of immediate physical injury exists, the employee shall perform the work.

2. If upon conferring the foreman and Union and Safety Committeeman cannot agree if an immediate danger of physical injury exists, the work shall be performed by the employee. If the employee believes that there still exists an immediate danger of physical injury to his person he will be offered other work in his classification or another classification as appropriate, substitute work, or he will be sent home. Nothing contained in this Article shall preclude the Company from assigning another employee to the job in question and no employee, other than communicating the facts relating to the safety of the job, shall take any steps to prevent the assigned employee from working on the job. The Company Safety Supervisor and the Chairman of the Union Safety Committee shall then confer immediately to decide if any immediate danger of physical injury does in fact exist. If unable to be resolved at this level, then the Company Industrial Relations Director and the Representative of the International Union will be asked to confer.

3. If the Plant Industrial Relations Director and the Representative of the International Union are in dispute, the matter shall be submitted immediately to the State Safety Representative, or such other person as may have been agreed upon in advance, whose decision shall be final and binding, and whose expenses shall be borne by the Company and the Union.

I. If the employee is sent home and it is later decided in accordance with the above procedure that such conditions did not exist, the Company may exercise its right under Articles X and XI of this Agreement. If, on the other hand, the employee is sent home and it is later decided in accordance with the above procedure that such conditions did exist, the employee shall be entitled to recover the pay for the time lost.

J. In providing the procedures outlined above, the parties agree to guard against the misapplication of this procedure by those seeking to use it for purposes not related to safety and/or by the misapplication of the immediate-danger provisions described above. It is therefore the continuing responsibility of all employees as well as all Company and Union representatives to make certain that the provisions of this Article XIV are applied solely for and in the manner intended herein. If the Company believes that this Article XIV procedure is being abused, use of this Section may be suspended until the Union and Company Negotiating Chairmen have an opportunity to review such abuse and take necessary corrective action.

K. An employee alleging an unsafe working condition beyond the hazards inherent in this particular operation, which does not pose an immediate danger of injury to his person, should file a grievance in Step 2 of the Grievance Procedure rather than use the procedure in paragraph H above. It is understood and agreed that if the grievance is filed on an unsafe working condition, such grievance shall receive preferred handling and shall be expedited through the Grievance Procedure by the Company and Chairman of the Union Safety Committee.

L. The Chairman of the Union Safety Committee shall have access to all places in or about the plant or property of the Company at reasonable times to investigate safety and health conditions, provided reasonable notice is given to his immediate foreman and the foreman (supervisor) of the department to be visited. Each Union member of the Safety Committee shall, upon reasonable notice to his immediate supervisor, be afforded such time off without pay as may be required for the purposes of making inspections and performing other duties provided for in the Article. The Director of the International Union Safety and Health Department or his designee may arrange to make a plant visitation by making arrangements with the Union Chairman of the Negotiations Committee, who will contact the Company Industrial Relations Director.

M. When an employee is temporarily reassigned from his job because of a Company medical determination establishing that exposure to a toxic substance involved in such job has adversely and temporarily affected his health, he shall retain his regular rate of pay for a period of not more than thirty (30) calendar days while working on any other Company-assigned job. The local parties may mutually agree to an extension of the rate retention period.

N. On-the-job training will be provided at the departmental level for employees in new jobs or assignments. Monthly crew safety meetings will be held to discuss safe job methods and other departmental safety issues.

O. Employees who must work alone on jobs should be checked periodically by another employee or prearrange periodic telephone or radio checks.

P. The Company agrees to provide complete toilet facilities, shower facilities with soap, and fresh, clean drinking water throughout the plant as needed.

Q. All persons entering the plant will be informed of and are required to abide by all Company safety rules and regulations. Exceptions can be mutually agreed to by both parties.

APPENDIX 2 to CHAPTER XVII
Safety and Health Article (Chemical Company)

By and between ("Company") and ("Union"), collective bargaining agent for those Company employees covered by the "Agreement" between Company and Union as of [date], (herein called "Agreement").

In recognition of the mutual belief that a useful purpose would be served by instituting industrial health research of the work environment by recognized, independent authorities in the field of industrial health, it is mutually agreed, for the term of the currently effective agreement, that:

1. There shall be established a Joint Labor-Management Health and Safety Committee, consisting of equal Union and Company representatives, and not less than 2 or more than 4 each.

2. The Company will from time to time retain at its expense qualified independent industrial health consultants, mutually acceptable to the International Union President or his designee and the Company to undertake industrial health research surveys, as decided upon by the Committee, to determine if any health hazards exist in the workplace.

3. Such research surveys shall include measurements of exposures in the workplace, the results of which shall be submitted in writing to the Company, the International Union President, and the joint Committee by the research consultant, and the results will also relate the findings to existing recognized standards.

4. The Company agrees to pay for appropriate physical examinations and medical tests at a frequency and extent necessary in light of findings set forth in the industrial consultant's report as may be determined by the Joint Committee.

5. The Union agrees that each research report shall be treated as privileged and confidential and will be screened by the Company to prevent disclosure of proprietary information or any other disclosure not permitted by legal or contractual obligations.

6. At a mutually established time, subsequent to the receipt of such research reports, the Joint Committee will meet for the purpose of reviewing such reports, to determine whether corrective measures are necessary in light of the industrial

consultant's findings, and to determine the means of implementing such corrective measures.

7. Within sixty (60) days following the execution of this Agreement and on each successive October 1 thereafter, the Company will furnish to the Union all available information on the morbidity and mortality experience of its employees.

8. The Joint Committee shall meet as often as necessary, but not less than once each month at a regularly scheduled time and place, for the purpose of considering, inspecting, investigating, and reviewing health and safety conditions and practices. Union Committeemen shall have the right to investigate accidents under procedures developed by the Joint Committee. The Joint Committee shall make constructive recommendations with respect thereto, including but not limited to the implementation of corrective measures to eliminate unhealthy and unsafe conditions and practices and to improve existing health and safety conditions and practices. All matters considered and handled by the Committee shall be reduced to writing, and joint minutes of all meetings of the Committee shall be made and maintained, and copies thereof shall be furnished to the International Union President. Time spent in connection with the work of the Committee by Union Representatives, including walkaround time spent in relation to inspections and investigations, shall be considered and compensated for as their regularly assigned work.

9. The Company will at its expense provide for the training of union members of the Joint Committee one (1) time during the term of this agreement. Such training will be limited to a maximum of five (5) days per committee member. The subject matter and the timing of such training will be determined by the Joint Committee. Such training will be conducted by qualified individuals, institutions, or organizations recognized in the field of health and safety.

10. In addition to the foregoing, the Company intends to continue its existing industrial hygiene program as administered by Company personnel.

11. Any dispute arising with respect to the interpretation or application of the provisions hereof shall be subject to the grievance and arbitration procedures set forth in the Agreement.

APPENDIX 3 to CHAPTER XVII
Safety and Health Article (Oil Company)

1. It shall be the policy of the Company to make every reasonable effort to provide safe working conditions and to provide safe working practices. Inspection of all equipment throughout the plant or place of employment shall be made by the Plant Manager or other persons designated by the employer from time to time. Complaints or suggestions by workmen employed on plant equipment regarding the safety of same shall be promptly investigated by the Company. The Workmen's Committee may make written suggestions to the Plant Manager or his representative as to the elimination of hazards in order to prevent accidents. The Union agrees to encourage its members to work in a safe manner and to cooperate with the safety program.

2. No employee shall be required to perform services that seriously endanger his physical safety other than the normal hazards and dangers of normal refinery operations, and his refusal to do such abnormally dangerous work shall not warrant or justify discharge. In all such cases, an immediate conference between the foreman, safety man, and Workmen's Committeeman or Committeemen shall be held to settle the issue.

3. Any employee whose clothes are destroyed through no fault of his own as a result of fire, or because of the spilling of chemicals, shall be provided with suitable clothes or be given a cash replacement allowance provided the damaged clothing is turned in together with a report to the employee's foreman.

4. Safety equipment and clothing will be furnished by the Company in whatever place and quantity it is warranted or needed. Should there be any complaint on the part of any employee as to the foregoing, it shall be handled by the regular grievance procedure as set up elsewhere herein.

5. An employee shall not be discharged on account of any accident if physically and mentally capable of continuing his duties, unless the accident was caused by the gross negligence, extreme carelessness, or malicious intent of the employee.

6. When an employee is injured while on duty and is required by a Company representative to take medical treatment, the employee shall not suffer any loss of pay on the first day he receives medical treatment. Pay under this section shall be limited to the end of the shift on the first day such employee receives medical treatment. Should subsequent treatment be required by a Company representative for the same injury after the employee has returned to work, such employee will not suffer loss of pay on such workday while en route to and from the place where the employee receives medical treatment and while treatment is being received.

7. When necessary, the Company will provide transportation to and from the place where medical treatment is given under the provisions of this section. When necessary, the Company will provide transportation for an employee to his home, who is too sick, as determined by the Company, to continue his work.

8. Nothing in this section shall modify the terms and benefits provided in the Accident and Sickness Benefit plan.

APPENDIX 4 to CHAPTER XVII
Safety and Health Article (Steel Company)

Section 1. The Company shall exert every reasonable effort to provide and maintain safe and healthy working conditions. The Union shall exert its best effort to have the employees work in a safe manner, observing all reasonable safety-health rules and regulations, and cooperate with the Company in maintaining the Company's reasonable policies and practices pertaining to safety and health.

Section 2. Joint Safety Committee: The Union will appoint three (3) members of the Union and the Company will appoint two (2) supervisors, the Safety Supervisor and/or his representative to serve on a joint Company-Union Safety Committee:

The Joint Safety Committee will meet monthly or more frequently depending on circumstances to discuss safety and health problems and make recommendations for the maintenance of proper safety standards. They will also receive and investigate, at an appropriate time, complaints regarding unsafe or hazardous conditions of work, equipment, or practices. Records of these meetings will be kept and furnished to the local Union Committee.

The entire plant will be inspected every six (6) to eight (8) weeks by weekly safety inspection tours of selected areas of the plant conducted by one member of the Union Safety Committee, the supervisor of the area, the unit supervisor, the maintenance supervisor for the area, and the safety director or his representative. If the Company Safety Committeeman cannot be present during the inspection, the Company will appoint another management employee to participate in the inspection. The Union will be furnished a copy of the inspection results.

The Union President or his designate and whatever representatives of management are required will accompany Federal and State Safety Inspectors when such inspections of the Plant are necessary.

The Joint Safety Committee will be informed of all recordable injuries. Union committee members on the plant at the time of the initial investigation of the injury will be given the opportunity to participate in the investigation. The Safety Committee will inspect any new operation or building prior to startup of same.

Union members of the Safety Committee will be paid by the Company for the time spent on safety and health activities during their regular scheduled working hours and this will be considered as time worked for the purpose of overtime computations.

Section 3. The Company agrees to maintain the following:

(a) Heated locker rooms with individual lockers.

(b) Adequate first-aid service on the Company premises.

(c) Adequate shower baths, washbasins, toilet facilities, and sanitary drinking fountains.

Section 4. The Company agrees to furnish all protective equipment necessary to protect the clothing and/or health of its employees without cost to the employees. The employees will be responsible for the care, cleanliness, and return of such equipment, subject to reasonable depreciation incident to its use.

Section 5. An employee may be required to submit to a physical examination by a physician designated by the Company whenever the Company deems the health of such employee or the general safety of other employees requires it.

Section 6. When an employee is removed from his work area or assigned to work not normal to his job content while his equipment is running, supervisory instructions to this effect will be entered in writing at the time of removal or assignment in the operating log book.

APPENDIX 5 to CHAPTER XVII
Safety and Health Article (Glass Company)

1. The Company and the Union recognize the importance of an effective safety program.

2. Each local Union President may appoint not less than three members of the local Union to function on the Company's Safety Committee.

3. The suggestions and recommendations set forth in the Safety Program proposed by the National Glass Container Labor-Management Committee are a guide which the Company will follow in developing a safety program applicable to its own operations.

4. The Company will provide first-aid facilities, and designate on each shift an individual or individuals who are trained in and capable of performing first aid to the extent necessary to provide adequate first aid for all employees. Such individual or individuals, who may or may not be members of the bargaining unit, will give first aid to injured employees, and, in cases of severe injury, will stay with the injured employee until relieved by a medical attendant. The Company will provide the necessary training of such individuals.

5. The industry standards shall be the minimum guidelines for first aid and medical facilities and personnel.

6. All employees will be furnished new gloves satisfactory to the local Union as needed. Old gloves are to be returned when replacements are issued.

7. The Company shall provide employees with special tools deemed necessary by the Company to operate machinery and equipment. The employees will be responsible for such tools when provided to them.

8. The Company will make available work uniforms, through a vendor, to the employee.

9. The Company agrees to continue its best efforts to provide adequate heat, light, and ventilation to employees, and to devise systems to control drafts, noise, fumes, dust, dirt, grease, and job hazards.

CHAPTER XVIII
Constitutional Developments

A. INTRODUCTION

Questions concerning the constitutionality of the Act continue to be raised in both standards promulgation and enforcement proceedings, despite the resolution of many of these issues during the Act's formative years. See *Industrial Union Dept., AFL-CIO v. American Petroleum Institute,* 448 U.S. —, 8 OSHC 1586 (1980) (Rehnquist, J., concurring). To date there have been few successful constitutional challenges to the Act except in the areas of warrantless inspections and due process/fair notice of standards.

B. SIXTH AND SEVENTH AMENDMENTS: DUE PROCESS AND RIGHT TO JURY TRIAL

Section 17 of the Act provides that employers who violate the Act or the rules, standards, regulations, or orders issued pursuant to the Act may be assessed civil penalties for the violations. Employers have argued that the imposition of penalties amounts to a punishment for past deeds rather than corrective action. Moreover, they have asserted that the assessments are criminal rather than regulatory in nature. Since the Act does not include any provisions either for a jury trial or for a hearing prior to the Secretary of Labor's assessment of a proposed penalty, employers have argued that both the Sixth and Seventh Amendments are violated.[1] Every court of appeals that has been presented with this argument, however, has held that the penalties authorized and imposed by the Act are regulatory rather than punitive, given the clear intention on the part of Congress to create a civil sanction. The Supreme Court has affirmed these holdings by ruling that the Act is not subject to constitutional challenges under the Seventh Amendment on the grounds that it permits the imposition of penalties. Moreover, parties to a proceeding before an administrative agency are not entitled to a jury trial on the issue of propriety of violations; with *de novo* review by the Commission, the denial of a jury trial is not unconstitutional. See, *e.g., Beall Construction Company v. OSHRC,* 507 F.2d 1041, 2 OSHC 1398 (8th Cir. 1974); *Clarkson Construction Co. v. OSHRC,* 531 F.2d 451, 3 OSHC 1880 (10th Cir. 1976); *Lake*

Butler Apparel Co. v. Secretary of Labor, 519 F.2d 84, 3 OSHC 1522 (5th Cir. 1975); *Brennan v. Winters Battery Mfg. Co.,* 531 F.2d 317, 3 OSHC 1775 (6th Cir. 1975), *cert. den.,* 425 U.S. 99, 4 OSHC 1240; *Mohawk Excavating, Inc. v. OSHRC,* 549 F.2d 859, 5 OSHC 1001 (2d Cir. 1977); *American Smelting and Refining Company v. OSHRC,* 501 F.2d 504, 2 OSHC 1041 (8th Cir. 1974); *Bloomfield Mechanical Contracting, Inc.,* 519 F.2d 1257, 3 OSHC 1403 (3d Cir. 1975); *Savina Home Industries, Inc. v. Secretary of Labor,* 594 F.2d 1358, 7 OSHC 1155 (10th Cir. 1979); *Atlas Roofing Co. v. OSHRC,* 518 F.2d 990, 3 OSHC 1490 (5th Cir. 1975), *aff'd,* 430 U.S. 442, 5 OSHC 1105 (1977); *Frank Irey, Jr., Inc. v. OSHRC,* 519 F.2d 1200, 2 OSHC 1283 (3d Cir. 1974), *aff'd en banc,* 519 F.2d 1215, 3 OSHC 1329 (3d Cir. 1975), *aff'd,* 430 U.S. 442, 5 OSHC 1105 (1977).

The Supreme Court has also held in other contexts that the Seventh Amendment does not guarantee the right to a jury trial in civil administrative proceedings. See, *e.g., Pernell v. Southall Realty,* 416 U.S. 363 (1974); *Curtis v. Loether,* 415 U.S. 189 (1974).

C. FIFTH AMENDMENT: DUE PROCESS AND SELF-INCRIMINATION

The courts have held that Section 10(a) of the Act which states that a citation may be issued and proposed penalties may be assessed prior to a hearing and that they become final unless a notice of contest is filed within fifteen working days after receipt of a citation is not an unconstitutional denial of procedural due process. As long as a procedure affords affected persons a reasonable opportunity to be heard and present evidence, it does not offend due process requirements of the Fifth Amendment.[2] An employer may contest a citation and proposed penalties and receive an administrative hearing thereon. *McLean Trucking Co. v. OSHRC,* 503 F.2d 8, 2 OSHC 1165 (4th Cir. 1974); *Winters Battery, supra; Mohawk Excavating, Inc., supra; Beall Construction Company, supra.* If the employer is given timely notice of the penalties, allowed to cross-examine witnesses and present evidence of its own, it is not denied due process; *Lake Butler Apparel Company, supra.* The courts have reasoned that there is no risk of an erroneous deprivation of an employer's interest because a penalty is only proposed in the citation and does not have to be paid unless and until administrative procedures are exhausted.

The court in *Winters Battery, supra,* also rejected the argument that the citation and penalty scheme violates the right against self-incrimination. While no Supreme Court case has addressed this question in the OSHA context, another case involving an interrogation by IRS agents could have significant ramifications on OSHA inspections. *Beckwith v. United States,* 425 U.S. 341 (1976). The Court held that individuals interrogated by federal inspectors will not be entitled to *Miranda*-type warnings as long as they are not in custody. Voluntary statements made to OSHA inspectors during an inspection may therefore probably be used against an employer without violating constitutional guarantees.

D. FOURTH AMENDMENT: UNREASONABLE SEARCH AND SEIZURE

The Act provides in Section 8(a) that the Secretary of Labor is authorized to enter without delay and at reasonable times any factory, plant, worksite, establishment, construction site, or other place of employment to inspect and investigate pertinent conditions, equipment, materials, devices, etc. Employers have often argued that OSHA inspections are in violation of the Fourth Amendment[3] because they constitute illegal searches. Where the employer has consented to the inspection, however, courts have not reached the constitutional issue. *Marshall v. Western Waterproofing Co., Inc.,* 560 F.2d 947, 5 OSHC 1732 (8th Cir. 1977); *Dorey Electric Co. v. OSHRC,* 553 F.2d 357, 5 OSHC 1285 (4th Cir. 1977); *Lake Butler Apparel Co., supra; Frank Irey, Jr., supra; Milliken & Co. v. OSHRC,* 605 F.2d 1201, 7 OSHC 1700 (4th Cir. 1979) —a compliance officer's announcement that he is at the workplace for purpose of conducting an inspection is not coercive and the official's acquiesence in the inspection constituted consent; *Stephenson Enterprises, Inc. v. Marshall,* 578 F.2d 1202, 6 OSHC 1860 (5th Cir. 1978). The Eighth Circuit, in *Western Waterproofing, supra,* added that where OSHA compliance officers obtained permission to enter upon the relevant parts of a place of employment to conduct their inspections from persons capable of giving such permission it was binding on the employer.

Where the employer has refused to permit entry of an OSHA inspector onto a worksite, however, the Supreme Court has stated that inspections must be conducted with a search warrant or they would be in violation of the Fourth Amendment. *Marshall v. Barlows, Inc.,* 436 U.S. 307, 6 OSHC 1571 (1978). The impact of the Court's ruling on OSHA was seemingly softened by the standard it adopted for the issuance of an inspection warrant. Thus the Secretary's authority to inspect will *not* depend on his demonstrating "probable cause to believe that conditions in violation of OSHA exist on the premises. Probable cause in the criminal law sense is not required. . . ." *Barlow's, Inc.,* 436 U.S. at 320. Rather, the court held:

> For purposes of an administrative search such as this, probable cause justifying the issuance of a warrant may be based not only on specific evidence of an existing violation but also on a showing that "reasonable legislative or administrative standards for conducting an . . . inspection are satisfied with respect to a particular [establishment]. *Camara v. Municipal Court* [387 U.S. 523 (at 538) 1967].

See also *Usery v. Rupp Forge Company,* 582 F.2d 1281, 6 OSHC 1788 (6th Cir. 1978).

For example, a warrant showing that a specific business has been chosen for an OSHA inspection on the basis of a general administrative plan for the enforcement of the Act derived from neutral sources such as the dispersion of employees in various types of industries across a given area and the desired frequency of searches in any lesser division of the area would protect an employer's Fourth Amendment rights.

Application of the *Barlow's* probable cause standards by the varying district courts and courts of appeal, however, have at times resulted in a much more restrictive standard of probable cause, necessary to support an inspection warrant. *Central Mine Equipment Co.,*—F. Supp. —, 7 OSHC 1185 (E.D. Mo. 1979) *rev'd,* 608 F.2d 719, 7 OSHC 1907 (8th Cir. 1979); *Urick Foundry,* 472 F. Supp. 1193, 7 OSHC 1497 (W.D. Pa. 1979); *Federal Clearing Die Cast Co.,*—F. Supp. —, 8 OSHC 1635 (N.D. Ill. 1980). In *Establishment Inspection of Northwest Airlines, Inc.,* 587 F.2d 12, 6 OSHC 2070 (7th Cir. 1978), the court reviewed the administrative probable cause standard and agreed that traditional criminal probable cause is not required for issuance of a warrant for an OSHA inspection, but found that insufficient information was presented therein to justify issuance of a warrant. See also, *Marshall v. W. and W. Steel Co.,* 604 F.2d 1322, 6 OSHC 1670 (10th Cir. 1979); *Plum Creek Lumber Co. v. Hutton,* 608 F.2d 1283, 7 OSHC 1940 (9th Cir. 1979).

One court held that a warrant application which stated only that an employee complaint had been received and that the Secretary of Labor had determined there was reasonable cause to believe a violation existed was insufficient to estabish probable cause where the nature of the alleged violation was not described with specificity. *Weyerhaeuser v. Marshall,* 592 F.2d 373, 7 OSHC 1090 (7th Cir. 1979). However, in a more recent case the same court held that

> Where probable cause to conduct an OSHA inspection is established on the basis of employee complaints, the inspection need not be limited in scope to the substance of those complaints . . . It will generally be reasonable in such a case to conduct an OSHA inspection of the entire workplace identified in the complaints. *Burkart Randall Division of Textron, Inc. v. Marshall,*—F.2d —, 8 OSHC 1467, 1476 (7th Cir. 1980).

Despite this decision, which would seem to favor expansion of the scope of an OSHA inspection, the Third Circuit labeled as sound judicial policy the Supreme Court's dictum in *Marshall v. Barlow's Inc., supra,* which "appeared to prohibit" OHSA compliance officers from seeking *ex parte* inspection warrants. *Cerro Metal Products v. Marshall,* 620 F.2d 964, 8 OSHC 1196 (3d Cir. 1980). See, however, *Marshall v. W. & W. Steel Company, Inc.,* 604 F.2d 1322, 7 OSHC 1670 (10th Cir. 1979), which held that the Secretary of Labor is authorized to apply for an inspection warrant *ex parte* and that OSHA need not give an employer prior legal notice since a regulation, 29 CFR 1903.4(d), which defines compulsory process to include *ex parte* warrants was properly promulgated. Bowing to the Third Circuit's reasoning, however, OSHA has repromulgated the regulation to cure procedural problems.

The Tenth Circuit held that where the state of the law at the time of an inspection was not such as to charge the OSHA inspector with knowledge of the unconstitutionality of a warrantless search, the exclusionary rule was not applicable. *Savina Home Industries v. Sec'y. of Labor,* 594 F.2d 1358, 7 OSHC 1154 (10th Cir. 1979); see also *Todd Shipyards, Inc. v. Sec'y. of Labor,* 566 F.2d 1327, 6 OSHC 2122 (9th Cir. 1978). Hence, evidence obtained from a warrantless search which took place prior to the rendering of the *Barlow's* decision in 1978 would not be excluded as a violation of the Fourth Amendment. See also *Gibsons Products, Inc., of Plano v. Marshall,* 584 F.2d 668, 6 OSHC 2092 (5th Cir. 1978), for an interesting discussion which asserts that district courts do not have jurisdiction to consider an action brought by the Secretary of Labor to compel an inspection under Section 8(a) of the Act.

Presumably, the Fourth Amendment's prohibition against unreasonable searches and seizures would also be applicable to and limit the permissible scope of an administrative subpoena issued under Section 8(b) of the Act. *Marshall v. American Olean Tile Company, Inc.*, 489 F. Supp. 32, 8 OSHC 1136 (E.D. Pa. 1980).

E. FIFTH AMENDMENT: CHILLING EFFECT

Employers have argued unsuccessfully that the potential for an increase in penalties by the Review Commission imposed after a citation has been contested, as provided under Section 17(j) of the Act, deters employers from contesting citations and penalties and consequently denies their rights to a hearing. The courts, however, have held that the opportunity for employers to petition a court of appeals for review under an abuse of discretion standard is deemed a sufficient safeguard against possible retributive penalty increases. *Savina Home Industries, supra; Dan J. Sheehan Co. v. OSHRC,* 520 F.2d 1036, 3 OSHC 1573 (5th Cir. 1975), *cert. den.*, 424 U.S. 965, 4 OSHC 1022 (1976); *Clarkson Construction, supra; Cf. Lake Butler Apparel Co., supra.*

The courts have also held that there is no chilling effect by reason of the fact that OSHA may propose abatement requirements, since the employer can contest a citation before the requirements become final. *Clarkson Construction, supra.*

F. FIRST AMENDMENT: FREEDOM OF SPEECH

29 C.F.R. 1903.2(a) requires that employers post notices furnished by OSHA informing employees of the protections and obligations provided for in the Act. The Fifth Circuit has held that the requirement that an informational sign or notice be posted at an employer's place of business does not violate the First Amendment,[4] and that the argument to the contrary is "seemingly nonsensical." The First Amendment, which gives employers the full and complete right to contest validity, cannot justify a refusal to post a notice. *Lake Butler Apparel Company, supra.* See also *Ryder Truck Lines, Inc. v. Brennan,* 497 F.2d 230, 2 OSHC 1075 (5th Cir. 1974)—the First Amendment does not apply to the personal protective equipment standard.

G. ARTICLE III, SECTION 1: DELEGATION OF POWER

Sections 9 and 10 of the Act, which provide for the conduct of inspections and investigations and the issuance of citations and proposed penalties, have been challenged as an unconstitutional delegation of judicial power to the Secretary of Labor and Review Commission in violation of the separation of powers principal found in Article III, Section 1[5] of the Constitution. The courts in *McLean Trucking Co., supra* and *Frank Irey, Inc., supra,* have rejected this argument, holding that the delegation of adjudicative functions to an administrative agency with special expertise in the subject matter, with the right of judicial review, does not violate the separation of powers principle. See also *Clarkson Construction, supra; Beall Construction Company, supra;* and *Bloomfield Mechanical Contracting, supra,* which upheld the delegation to the Secretary of Labor of the power to promulgate rules, regulations, and standards with respect to safe and healthful working conditions. See also *Blocksom & Company,* 582 F.2d 1122, 6 OSHC 1865 (7th Cir. 1978); *Savina Home Industries, supra.*

However, in a situation where an employer refuses to comply with a Commission order, OSHA is empowered to file a petition for enforcement in a United States Court of Appeals. In *Winters Battery, supra,* the Secretary of Labor argued that the Clerk of Court could enter an enforcement order, pursuant to Section 11(b) of the Act, since this was only a summary/ministerial rather than a judicial action. The Sixth Circuit rejected this argument, holding that since the entry of an enforcement order forms the basis of a contempt citation, it is a judicial action, and express court authorization and review is required. Otherwise, enforcement procedures would constitute an improper delegation of power in violation of the Constitution.

In an important Supreme Court test of OSHA's authority, OSHA's promulgation of a regulation interpreting Section 11(c)(1) of the Act, which prohibits an employer from discharging or discriminating against an employee who exercises "any right afforded by" the Act, was challenged as inconsistent with the objectives of the Act and the intent of Congress. *Whirlpool Corp. v. Marshall,* 445 U.S. 1, 8 OSHC 1001 (1980). Among these rights, according to the regulation, is the ability to choose not to perform an assigned task because of a reasonable apprehension of death or serious injury, coupled with a reasonable belief that no less drastic alternative is available.

Although the employer argued that this regulation was inconsistent with the Act, the Supreme Court held that:

> The regulation in question was promulgated by respondent in the valid exercise of his authority under the Act and clearly conforms to the Act's fundamental objective of preventing occupational deaths and serious injuries. *Id.* at 4189.

An even more recent decision by the Supreme Court affirmed a ruling of the Fifth Circuit Court of Appeals invalidating the two-year-old OSHA benzene regulation and in contrast to *Whirlpool* may have diminished OSHA's power. In *Industrial Union Department, AFL-CIO, et al. v. American Petroleum Institute, et al.,* 448 U.S. —, 8 OSHC 1586 (1980) (see esp. concurring opinion of Rehnquist, J.), the Court held that the Secretary of Labor exceeded his standard-setting authority under the Act by promulgating standards which limited permissible employee exposure to airborne concentrations of benzene to one part per million and which prohibited dermal contact with solutions containing benzene. The standard was found to be invalidly promulgated since OSHA failed to make a determination that it was "reasonably necessary or appropriate to provide a safe and healthful place of employment" under Section 3(8) of the Act. See also *United States Chamber of Commerce v. OSHA,*

636 F.2d 464, 8 OSHC 1648 (D.C. Cir. 1980), holding OSHA's walkaround pay regulation invalid.

H. FOURTEENTH AMENDMENT: EQUAL PROTECTION

In *Desarrollos Metropolitanos, Inc. v. OSHRC,* 551 F.2d 874, 5 OSHC 1135 (1st Cir. 1977) and *George Hyman Construction Co. v. OSHRC,* 582 F.2d 834, 6 OSHC 1855 (4th Cir. 1978), the courts rejected Fourteenth-Amendment[6] equal protection challenges by an employer who argued that an OSHA regulation which distinguished between businesses with fixed establishments (factories, stores) and businesses with nonfixed establishments (construction sites) relative to the issuance of repeated violations was arbitrary, capricious, and discriminatory. The courts stated that when a classification in the area of economic or social welfare legislation is called into question, if any state of facts can be reasonably construed that would sustain it, it will be upheld. A reasonable basis was found for the distinction between transient and permanent work sites—thus the classification was found to be rational and passed constitutional muster.

The *Desarrollos Metropolitanos* court also held that consideration of the size of an employer's business for penalty assessment purposes under the Act does not violate equal protection considerations.

The Act also does not violate equal protection guarantees even though the Secretary of Labor has enforcement authority through the Courts of Appeals, while the employer must first exhaust administrative remedies. *Winters Battery Mfg., supra.*

I. FIFTH AND FOURTEENTH AMENDMENTS: DUE PROCESS (VAGUENESS)

Within the myriad applications of the due process clauses, it is a fundamental principle that statutes and regulations which purport to govern a person's conduct must give an adequate warning of what they command or forbid. Even a regulation which governs purely economic or commercial activity, if its violation can engender penalties, must be so framed as to provide a constitutionally adequate warning to those whose activities are governed. Several standards have been challenged on the basis of a lack of notice or vagueness, alleging that enforcement would violate the due process clause of the Fifth and/or Fourteenth Amendment. In *Diebold, Inc. v. Marshall,* 585 F.2d 1327, 6 OSHC 2002 (6th Cir. 1978), the court held that the question of whether a regulation or standard gives an employer sufficient warning, in that its machinery, equipment, or operations are within the scope of its requirements, is to be answered in the light of the conduct to which the regulation is to be applied. The constitutional adequacy or inadequacy of the warning given must be measured by common understanding and commercial practice.

An employer is not required to assume the burden of guessing what the

Secretary of Labor intended a standard or regulation to mean. Laws regulating conduct must give the person of ordinary intelligence a reasonable opportunity to know what is prohibited, so that he may act accordingly. A regulation's validity must thus be considered in light of the conduct to which it is applied. *Diebold, Inc., supra; R. L. Sanders Roofing Company v. OSHRC,* 620 F.2d 97, 8 OSHC 1559 (5th Cir. 1980); *Schriber Sheet Metal & Roofers, Inc. v. OSHRC,* 597 F.2d 78, 7 OSHC 1246 (6th Cir. 1979); *General Dynamics v. OSHRC,* 599 F.2d 453, 7 OSHC 1373 (1st Cir. 1979); *Diamond Roofing Co. v. OSHRC,* 528 F.2d 645, 4 OSHC 1001 (5th Cir. 1976); *Kent Nowlin Construction Co. v. OSHRC,* 593 F.2d 368, 7 OSHC 1105 (10th Cir. 1979).

In determining whether an administrative regulation provides adequate notice, courts have inquired whether an employer familiar with the circumstances of the industry could reasonably be expected to have had an adequate warning of the conduct required by the regulation; if so, due process does not forbid its application. *National Industrial Constructors, Inc. v. OSHRC,* 583 F.2d 1048, 6 OSHC 1914 (8th Cir. 1978); *Ray Evers Welding Company v. OSHRC,* 625 F.2d 726, 8 OSHC 1271 (6th Cir. 1980). This "reasonableness" test is not entirely dependent upon industry standards and customs because conceivably an entire industry practice may be negligent. Due process would thus be satisfied by using the standard of a reasonably prudent employer under similar circumstances familiar with the industry and its hazards in assessing whether a regulation meets notice requirements. *Ray Evers Welding Company, supra; Bristol Steel & Iron Works v. OSHRC,* 601 F.2d 717, 7 OSHC 1462 (4th Cir. 1979); *American Airlines, Inc. v. Secretary of Labor,* 578 F.2d 38, 6 OSHC 1691 (2d Cir. 1978); *Brennan v. Smoke-Craft, Inc.,* 530 F.2d 843, 3 OSHC 2000 (9th Cir. 1976); *Jensen Construction Co. v. OSHRC,* 597 F.2d 246, 7 OSHC 1283 (10th Cir. 1979); *Diebold Inc., supra.*

An employer not charged under the general duty clause of the Act lacks *standing* to raise a claim that such a provision is unconstitutionally vague. *Savina Home Industries, Inc., supra.*

J. EXHAUSTION OF ADMINISTRATIVE REMEDIES

An employer, according to some authorities, must exhaust its administrative remedies before the Review Commission prior to seeking equitable relief in the federal courts, even though such action is based on alleged constitutional issues. The basic purpose of the exhaustion doctrine is to allow an administrative agency to perform functions within its special competence, make a factual record, and apply its expertise. *Blocksom Co. v. Marshall,* 582 F.2d 1122, 6 OSHC 1865 (7th Cir. 1978); *Babcock and Wilcox Co. v. Marshall,* 610 F.2d 1128, 7 OSHC 1886 (3d Cir. 1979). The Commission's final orders are reviewable as of right in the Courts of Appeals. However, the Fifth Circuit held that because the Review Commission does not have authority to rule on the constitutionality of the Act, such a situation is an "extraordinary circumstance" which excuses an employer's failure to first seek administrative review. The appellate court thus has jurisdiction to hear the issue of constitutionality on appeal even though an employer failed to exhaust its administrative remedies. *McGowan v. Marshall,* 604 F.2d 885, 7 OSHC 1842 (5th Cir. 1979); *Buckeye Industries, Inc. v. Secretary of Labor,* 587 F.2d 231, 6 OSHC 2181 (5th Cir. 1979).

The Seventh Circuit has also held in other circumstances that exhaustion of administrative remedies is not required if the established administrative procedures would prove of no benefit, be unavailing, or futile (*e.g.,* special circumstances). *Weyerhauser Company v. Marshall,* 592 F.2d 373, 7 OSHC 1090 (7th Cir. 1979). See also *Cerro Metal Products, Div. Marmon Group Inc. v. Marshall,* 620 F.2d 964, 8 OSHC 1196 (3d Cir. 1980), which held that, unless required by statute, exhaustion of administrative remedies is a matter of sound judicial discretion. The court also stated that the exhaustion doctrine must be guided by the purpose of (1) promoting administrative efficiency; (2) respecting executive autonomy by allowing an agency the opportunity to correct its own errors; (3) facilitating judicial review by affording courts the benefit of the agency's experience and expertise; and (4) serving judicial economy by having the agency rather than the court compile a factual record. When a party presses a constitutional claim, exhaustion serves an additional purpose of furthering parsimony in judicial decision-making. An agency or tribunal decision favorable to the moving party and based on facts, regulations, or law prevents the creation of unnecessary constitutional precedent. The court in this case felt, however, that no legitimate purpose would be served by requiring exhaustion of administrative remedies in the context of the process involved. See also *Continental Can Co. U.S.A. v. Marshall,* 603 F.2d 590, 7 OSHC 1521 (7th Cir. 1979). But see *Whittaker Corporation, Berwick Forge-Fabricating Company,* 610 F.2d 1141, 7 OSHC 1888 (3d Cir. 1979)—once an inspection pursuant to a warrant has been completed, the aggrieved employer must exhaust administrative remedies prior to judicial review.

The First Circuit also held that the federal courts should not review a search warrant issued under the Act unless all administrative remedies have been exhausted. *In re Worksite Inspection of Quality Products, Inc.,* 592 F.2d 611, 7 OSHC 1093 (1st Cir. 1979), the court stated that in the majority of OSHA enforcement cases, challenges to a warrant can be adequately considered in statutory enforcement proceedings by the Commission. The court also stated that the only circumstances that it could foresee where equitable considerations might allow a court to review the validity of an OSHA inspection warrant in the absence of prior administrative review would be in cases where an employer states a substantial challenge to the validity of a warrant application. Under the guidelines of *Franks v. Delaware,* 438 U.S. 154 (1978), an effective consideration of such a Fourth Amendment claim would require an evidentiary hearing. The court felt it would not be competent to conduct and the Commission could not conduct such a hearing, so that the claim would not be fully and fairly adjudicated unless the District Court stepped in. See also *In re Restland Memorial Park,* 540 F.2d 626, 4 OSHC 1485 (3d Cir. 1976); *In the Matter of the Inspection of Central Mine Equipment Co.,* 608 F. 2d 719, 7 OSHC 1907 (8th Cir. 1979).

K. ARTICLE VI: SUPREMACY CLAUSE

In *P and Z Co., Inc. v. District of Columbia,* 408 A.2d 1249, 8 OSHC 1078 (D.C. 1979), an employer sought review of a misdemeanor conviction for failure to report employees' injuries as required by the District of Columbia Industrial Safety Act. The employer argued that the D.C. statute was inoperative because it was preempted by congressional passage of the Occupational Safety Health Act. This argument, which was derived from Article VI, Clause 2, the Supremacy Clause[7] of the Constitution was, according to the court, a mistake. The court said that the Supremacy Clause arises only when federal-state regulatory efforts come in conflict but that it was not applicable in this case since both statutes were acts of the same legislative body, the United States Congress. It stated that when two statutes are capable of coexistence it is the court's duty, absent a clearly expressed congressional intent to the contrary, to regard each as effective. Moreover, there is no provision in the Act, which even arguably indicates that the Congress intended to repeal by implication the D. C. statute.

NOTES

1. The Sixth Amendment provides that "[i]n all criminal prosecutions, the accused shall enjoy the right to a speedy and public trial, by an impartial jury of the State and district wherein the crime shall have been committed, which district shall have been previously ascertained by law, and to be informed of the nature and cause of the accusation; to be confronted with the witnesses against him; to have compulsory process for obtaining witness in his favor, and to have the Assistance of Counsel for his defense."

The Seventh Amendment provides that "[i]n suits at common law, where the value in controversy shall exceed twenty dollars, the right of trial by jury shall be preserved, and no fact tried by a jury, shall be otherwise re-examined in any Court of the United States, than according to the rules of the common law."

2. The Fifth Amendment provides that "[n]o person shall be held to answer for a capital, or otherwise infamous crime, unless on a presentment or indictment of a Grand Jury, except in cases arising in the land or naval forces, or in the Militia, when in actual service in time of War or public danger; nor shall any person be subject for the same offense to be twice put in jeopardy of life or limb; nor shall be compelled in any criminal case to be a witness against himself, nor be deprived of life, liberty, or property, without due process of law; nor shall private property be taken for public use, without just compensation.

3. The Fourth Amendment provides that "[t]he right of the people to be secure in their persons, houses, papers, and effects, against unreasonable searches and seizures, shall not be violated, and no Warrants shall issue, but upon probable cause, supported by Oath or affirmation, and particularly describing the place to be searched, and the persons or things to be seized."

4. The First Amendment provides that "Congress shall make no law respecting an establishment of religion, or prohibitng the free exercise thereof; or abridging the freedom of speech, or of the press; or the right of the people peaceably to assemble, and to petition the Government for a redress of grievances."

5. Article III, Section 1 provides that "the judicial Power of the United States, shall be vested in one supreme Court, and in such inferior Courts as the Congress may from time to time ordain and establish. . . ."

6. The Fourteenth Amendment provides that "[a]ll persons born or naturalized in the United States, and subject to the jurisdiction thereof, are citizens of the United States and of the State wherein they reside. No State shall make or enforce any law which shall abridge the privileges or immunities of citizens of the United States; nor shall any State deprive any person of life, liberty, or property, without due process of law; nor deny to any person within its jurisdiction the equal protection of the laws."

7. Article VI, Clause 2, provides that "[t]his Constitution, and the Laws of the United States which shall be made in Pursuance thereof; and all Treaties made, or which shall be made, under the Authority of the United States, shall be the supreme Law of the Land; and the Judges in every State shall be bound thereby, any Thing in the Constitution or Laws of any State to the Contrary notwithstanding."

CHAPTER XIX
Mine Safety and Health

A. INTRODUCTION

The (Federal) Mine Safety and Health Act of 1977, 30 U.S.C. Section 801 *et seq.* (the Mine Act), declares that the first priority of "all in the coal or other mining industry must be the health and safety of its most precious resource, the miner." At the time of the Mine Act's enactment, there was an urgent need to provide more effective means and measures for improving the working conditions and practices in the nation's mines in order to prevent death and serious physical harm. Utilizing established precedent to accomplish this goal, many of the Mine Act's provisions were modeled upon those of the OSH Act. Many of the controversies that arise under it are therefore similar to those which arise under the OSH Act. There are also, however, several significant differences between the two statutes. The thrust of this chapter, although it will not be comprehensive, will be to provide an overview of the Mine Act and attempt to highlight those areas of differences and similarity in the Mine Act and OSH Act.

B. OVERVIEW

The (Federal) Mine Safety and Health Amendments Act, P.L. 95–164, 91 Stat. 1290, which resulted in the Mine Act, consolidated the regulation of mine safety and health under one statute. Safety and health in *coal* mines had previously been regulated by the Department of the Interior's Mining Enforcement and Safety Administration (MESA) under the Federal Coal Mine Health and Safety Act of 1969, P.L. 91–173, 83 Stat. 742, 30 U.S.C. Section 801 *et seq.* (the 1969 Coal Act). Safety and health in *metal* and *nonmetal* mines had been regulated by MESA under the Federal Metal and Nonmetallic Mine Safety Act, P.L. 89–577, 80 Stat. 772, 30 U.S.C. Section 721 *et seq.* (the 1966 Metal Act).

The Mine Act amendments: (1) substantially amended and renamed the 1969 Coal Act, the (Federal) Mine Safety and Health Act of 1977; (2) repealed the 1966 Metal Act and placed all miners under the safeguards of the Mine Act; (3) transferred enforcement responsibilities from the Interior Department to the Labor, Department as of March 9, 1978; (4) established in the Department of Labor, the Mine Safety and Health Administration (MSHA); (5) established, as an independent adjudicative body, the (Federal)

Mine Safety and Health Review Commission (MSHRC or the Mine Commission); (6) provided new procedures to expedite the rule-making process; (7) strengthened the mechanisms for enforcing compliance; (8) gave increased emphasis to protecting miners' health; (9) allowed for greater involvement of miners in processes affecting their health and safety; (10) provided procedures for assessing and collecting (precontest) civil penalties resulting from violations of mine health and safety regulations; and (11) provided for broad mandatory training for miners.

C. COVERAGE

Each coal or other mine, the products of which enter commerce, or the operations or products of which affect commerce, and each operator of such mine, and every miner in such mine is subject to the provisions of the Mine Act. Section 4 of the Mine Act. A mine is defined in Section 3(h)(1) as:

> (A) an area of land from which minerals are extracted in nonliquid form or, if in liquid form, are extracted with workers underground, (B) private ways and roads appurtenant to such area, and (C) lands, excavations, underground passageways, shafts, slopes, tunnels and workings, structures, facilities, equipment, machines, tools, or other property including impoundments, retention dams, and tailings ponds, on the surface or underground, used in, or to be used in, or resulting from, the work of extracting such minerals from their natural deposits in nonliquid form, or if in liquid form, with workers underground, or used in, or to be used in, the milling of such minerals, or the work of preparing coal or other minerals, and includes custom coal preparation facilities.

Marshall v. Wallach Concrete Products, Inc.,—F. Supp. —, 1 MSHC 2337 (D. Mex. 1980), *appeal filed* No. 80–1334 (10th Cir. March 26, 1980)—a sand and gravel operation is a mine within the meaning of Section 3(h)(1); *Marshall v. Kniseley Coal Company,* 487 F. Supp. 1376, 1 MSHC 2353 (W.D. Pa. 1980)—a family-owned, family-operated mine is covered by the Act and subject to safety inspections.

"Operator" is defined in Section 3(d) of the Mine Act to mean any owner, lessee, or other person who operates, controls, or supervises a mine *or* any independent contractor performing services or construction at such mine. A miner means any individual working in a coal or other mine. Section 3(g). *Marshall v. William R. Kraynak,* 604 F.2d 231, 1 MSHC 2131 (3d Cir. 1979)—four brothers who own and operate a mine and are its only miners are both *operators and miners* within the meaning of the Mine Act and must comply with its mandatory standards; *Old Ben Coal Company,* 1 MSHC 2177 (1979), *aff'd* — F.2d —, 2 MSHC 1065 (2d Cir. 1980)—an owner or operator of a mine can be held responsible for violations committed by an independent contractor engaged by the owner to perform work on the owner's premises; *Cowin & Company, Inc.,* 1 MSHC 2010 (1979)—an independent construction contractor who performed work on the mine premises under a contract with the mine owner is considered to be an operator within the meaning of the Mine Act.

D. STANDARDS

MSHA administers the provisions of the Mine Act so as to achieve a safe and healthful environment in the Nation's coal and other mining operations. It oversees the development and enforcement of safety and health standards and programs throughout the mining industry in order to achieve this environment. Enforcement of these standards is assured through regular and special inspection efforts at the Nation's 22,000 coal, metal, and nonmetal mining establishments. Civil monetary penalties may be assessed for any violations of the Mine Act's standards.

Standards which were in force under Title 30 of the Code of Federal Regulations on the date of the Mine Act's enactment remained in effect and were not immediately affected by the new Act, although some of these subsequently have been or in the future may be revoked or modified. These existing standards were both mandatory and advisory. An advisory committee was designated to review all of the advisory standards and recommend which, if any, should be made mandatory.

In addition to *permanent* mandatory standards, discussed below, the Mine Act authorizes the issuance of *emergency temporary* mandatory standards without going through the formal rule-making process in situations that pose grave dangers to miners. Section 101(b)(1). A nine-month limit is imposed on such standards in order to insure due process protection to mine operators.

Section 101(a) of the Act authorizes the Secretary of Labor to develop, promulgate, and revise "improved *permanent* mandatory health or safety standards." Whenever the Secretary of Labor, upon the basis of information submitted to him by an interested person, a representative of any organization of employers or employees, a nationally recognized standards-producing organization, NIOSH, or a state or political subdivision, or on the basis of information developed or otherwise available to him, determines that mandatory standards should be promulgated, the Secretary may request the recommendation of an advisory committee appointed under Section 102(c). Reliance on an advisory committee is discretionary, but if it is utilized certain procedures must be followed. When the Secretary receives recommendations from NIOSH with appropriate criteria that a rule be promulgated, he *must* within sixty days after receipt refer the recommendation to the advisory committee or publish such as a proposed rule or publish his reasons for not doing so. If the standard is new in effect or application, an advisory committee must also be utilized.

Whether or not an advisory committee is utilized, a proposed rule promulgating, revoking, or modifying a permanent mandatory standard must be published in the Federal Register (within sixty days of the advisory committee's recommendation if one is received). Section 101(a)(2). Interested persons are then afforded a thirty-day comment period for filing comments and objections and requesting a hearing. Thereafter a hearing notice will be published and a hearing held within sixty days. Within ninety days after certification of the record of the hearing (or within ninety days after the period for filing objections has expired) the Secretary shall promulgate, modify, or revoke the standard and publish the same in the Federal Register or publish reasons for not doing so. Section 101(a)(6) and (7). Mandatory standards shall be effective upon publication, unless another date is specified.

When promulgated, safety and health standards provide a point of reference for compliance efforts of employees and employers, a legal basis for subsequent enforcement actions, and a benchmark for voluntary compliance on the part of mine operators. A copy of every proposed mandatory standard published in the Federal Register is sent to all miners and miners' representatives and must be posted. Section 101(e).

Anyone claiming to be adversely affected by the issuance of a mandatory standard has sixty days after the standard is issued to petition the appropriate U.S. court of appeals for relief. Section 101(d) of the Mine Act. This section is intended to be the *sole means* of challenging the validity of a health or safety standard. The standard will remain in effect during litigation unless the court rules otherwise and grants a stay. In *National Industrial Sand Association v. Marshall,* 601 F.2d 289, 1 MSHC 2033 (3d Cir. 1979), the court stated that in considering the scope of statutory authority of the Secretary of Labor to issue standards under the Mine Act, a reviewing court will defer to the Secretary's interpretation of the scope of such authority so long as the interpretation is reasonable. The court further stated that under Section 10(e) of the Administrative Procedure Act, challenged regulations are to be reviewed to insure that they are not *arbitrary, capricious, and an abuse of discretion* or otherwise not in accordance with the law. In sum, MSHA, if its rule-making is to be sustained, must demonstrate that it has considered the relevant factors brought to its attention by interested parties during the course of rule-making and that it has made a reasoned choice among the various alternatives presented.

While some of the standards-setting provisions of the Mine Act are similar to those of the OSH Act, there are many important differences. The standards-setting provisions, for example, appear more favorable to the Department of Labor than the cognate OSHA provisions. For example, Section 6(b)(5) of the OSH Act requires the Secretary to "set the [toxic material] standard which most adequately assures, to the extent feasible, . . . that no employee will suffer material impairment of health," and also makes, in its next sentence, "the feasibility of the standards" a consideration in standard-setting. The comparable provision in the Mine Act, Section 101 (a)(6)(A), omits the phrase "to the extent feasible," but retains "the feasibility of the standards" as a criterion. The definition of "mandatory health or safety standard" in Section 3(1) of the Mine Act does not use the phrase "reasonably necessary or appropriate" found in Section 3(8) of the OSH Act, that was so important in the benzene decision, *Industrial Union Department, AFL-CIO v. American Petroleum Institute,* 448 U.S. —, 8 OSHC 1586 (1980). In addition, as stated previously, the pre-enforcement challenge permitted by Section 101(d) of the Mine Act is exclusive. Compare *Atlantic Gulf & Stevedores v. OSHRC,* 534 F.2d 541, 4 OSHC 1061 (3d Cir. 1976). See also *Southern Clay, Inc. v. Department of Labor,* 612 F.2d 574, 1 MSHC 2225 (3d Cir. 1979)—the court lacks jurisdiction to consider an operator's petition pursuant to Section 101(d) of the Mine Act for review of rules relative to the training of miners, since the contention that the rules cannot be applied to the operator's milling operations is cognizable in the enforcement proceeding but not in a Section 101(d) petition for review by the court of appeals; *Oracle Ridge Mining Partners,* 1 MSHC 2401 (1980).

The Secretary, after petition by an operator or miner representative, may modify the application of any mandatory safety—but not health—standard to a given mine, if he finds that another method will protect miners as well as the standard, or if application of a standard will make a mine more hazardous Section 101(c) of the Act. This is analogous to OSHA's variance procedure.

E. ROLE OF NIOSH

NIOSH conducts research, develops criteria for mine safety and health standards, and provides technical services to government, labor, and industry, including training in the recognition, avoidance, and prevention of unsafe or unhealthful working conditions and the proper use of adequate safety and health equipment. The most important product of NIOSH's research efforts in the mine safety and health area is the publication of criteria documents which, for a given hazard, contain recommended environmental exposure limits, medical examination requirements, labeling and warning, environmental monitoring methods, engineering controls, workplace practices, personal protective equipment and clothing, and recordkeeping requirements. Criteria documents may have an impact on workplace conditions even before being promulgated as standards, for labor and industry use them as guides for control of hazards, even though the documents do not have the force of law. In addition, NIOSH's research efforts result in a publication of current intelligent bulletins on previously recognized health hazards, health and safety guides for specific industries on avoidance of workplace hazards, and technical guides for industry regarding environmental monitoring and control of technology for workplace hazards. NIOSH's various functions are designed to reduce the high economic and social costs of occupational illnesses and injuries through the prevention and control of occupational diseases and hazards.

NIOSH and MSHA entered into a memorandum of understanding dated May 4, 1978, setting forth the agreement between the agencies for the consultation, coordination, and cooperation in effectively and efficiently carrying out their respective safety and health functions under the Mine Act. Topics covered in the memorandum of understanding include agreements relative to the identification of toxic materials and agents found in mines, development of mine health criteria and mine health standards, devices for the measurement, testing and certification of personal protective equipment and measurement and monitoring equipment, and an agreement to respond to health-hazard evaluation requests under the authority of Section 501 of the Mine Act. NIOSH is also required to coordinate its activities with MSHA in each mine for which a NIOSH health-hazard evaluation has been requested. NIOSH has also agreed to provide technical assistance and supportive field investigations to MSHA and will provide assistance in performing training and education responsibilities under the Mine Act. NIOSH, of course, was also required within eighteen months after the effective date of the Mine Act to determine whether substances are potentially toxic at the concentrations used or found in mines and make recommendations to the Secretary of Labor arising from these findings.

F. INSPECTIONS, INVESTIGATIONS, AND RECORDKEEPING

Inspections are required to be conducted for the purposes of: (1) obtaining and utilizing information relating to safety and health conditions and causes of accidents, diseases, and physical impairments originating in mines; (2) gathering information for mandatory health and safety standards; (3) determining whether an imminent danger exists; and (4) determining whether there is compliance with MSHA standards. Section 103(a) of the Act. Provisions for the issuance of subpoenas for the purpose of making investigations of accidents or other occurrences are also provided. Advance notice is prohibited. The Secretary of Labor is required to make inspections of underground mines at least four times a year and of surface mines at least two times a year. Moreover, inspection requirements are tightened in order to deal with especially gassy mines. Spot inspections must be made at irregular intervals every five days, every ten days, and every fifteen days, at mines found to liberate more than one million cubic feet, more than five hundred thousand cubic feet, and more than two hundred thousand cubic feet of methane or other explosive gas, respectively, over a twenty-four hour period. Section 103(i) of the Act.

Section 103(f) of the Act accords to representatives of miners both the right to accompany the inspector during an inspection and a limited right to walkaround pay. See generally *Magma Copper Corp.*, 1 MSHC 2227 (1979), *pet. for rev. filed*, No. 79–7535 (9th Cir. Oct. 15, 1979); *The Helen Mining Co.*, 1 MSHC 2193 (1979), *pets. for rev. filed*, Nos. 79–2518, 79–2537 (D.C. Cir. Dec. 19, 21, 1979)—walkaround pay right does not apply to "spot" inspections; *Kentland-Elkhorn Coal Corp.*, 1 MSHC 2230 (1979), *pets. for rev. filed*, Nos. 79–2503, 79–2536 (D.C. Cir. Dec. 17, 21, 1979)—walkaround pay limited to regular, complete inspections; *Island Creek Coal Company*, 1 MSHC 2521 (1980)—walkaround pay provision only applies to annual regular inspections and not spot inspections.

Section 103(a) of the Mine Act states that the Secretary of Labor "shall have a right of entry to, upon, or through any coal or other mine." The legislative history of the Mine Act indicates that the Congress intended to authorize such searches without warrants. S. Rep. No. 95–181, 95th Cong., 1st Sess. 27 (1977) ("absolute right of entry without need to obtain a warrant"), reprinted in 1977 U.S. Code Cong. & Ad. News 3401, 3427. The courts have not agreed on the constitutionality of this scheme. In fact, the United States Supreme Court agreed to review the constitutionality of Section 103(a) of the Mine Act, which authorizes warrantless inspections, in a direct appeal from a federal district court decision which had struck down the section as unconstitutional under the Fourth Amendment, at least with respect to inspections of stone quarries. *Marshall v. Dewey, Douglas and Waukesha Lime & Stone Company*, 493 F. Supp. 963, 1 MSHC 2433 (E.D. Wis. 1980), *cert.*

granted,—U.S. —, 2 MSHC—(1981). Four federal courts of appeals had previously distinguished the Supreme Court's 1978 decision in *Marshall v. Barlow's, Inc.,* 436 U.S. 307, 6 OSHC 1571 (1978), on the basis that warrantless inspections have traditionally been allowed with respect to closely regulated businesses which have a long history of regulation. *Marshall v. Stoudt's Ferry Preparation Company,* 602 F.2d 589, 1 MSHC 2097 (3d Cir. 1979), *cert. denied,* 444 U.S. 1815, 1 MSHC 2259 (1980); *Marshall v. Nolichukey Sand Company,* 606 F.2d 693, 1 MSHC 2161 (6th Cir. 1979), *cert. denied,* 446 U.S. 908, 1 MSHC 2384 (1980); *Marshall v. Texoline Co.,* 612 F.2d 935, 1 MSHC 2289 (5th Cir. 1980); and *Marshall v. Sink,* 614 F.2d 37, 1 MSHC 2273 (4th Cir. 1980). The district court in *Dewey* noted that stone quarries have been pervasively regulated only since 1966.

In the most recent appellate court decision, the Ninth Circuit held that the Secretary of Labor was not entitled to a court order directing a mine operator to permit a warrantless entry, because the mining operation there was not so pervasively regulated as to destroy the mine operator's expectation of privacy or justify a finding of implied consent to warrantless entries. *Marshall v. Wait,* 628 F.2d 1255, 1 MSHC 2529 (9th Cir. 1980). The court's decision emphasized that the mining operation involved—the quarrying of decorative rock—sat "on the fringe of traditional mining operations," the operator was not subject to extensive federal oversight, the operation was owned and run by a husband and wife without employees, and the operation had begun "many decades" before it was subjected to federal regulation. The court did not strike down the Mine Act on its face, however, but only as applied to the facts before it.

G. ENFORCEMENT

The Mine Act is enforced by the Mine Safety and Health Administration (MSHA) which is headed by the Assistant Secretary of Labor for Mine Safety and Health. If, upon inspection or investigation, the Secretary of Labor (through the person of an MSHA inspector) believes that an operator has violated the Mine Act or any mandatory standard issued thereunder, he will issue a citation with reasonable promptness. Section 104(a). The citation must be in writing, describe the violation with particularity, and fix a reasonable time for abatement.

If, upon a follow-up inspection, it is found that a violation described in a citation issued pursuant to 104(a) has not been totally abated within the time period prescribed, the inspector will order all persons (with certain enumerated exceptions) to be withdrawn from the mine. Section 104(b). The types of withdrawal orders that can be issued under Section 104 are: failure-to-abate withdrawal orders (Section 104(b)); unwarrantable failure withdrawal orders (Section 104(d)); and pattern of violation withdrawal orders (Section 104(e)). Imminent danger withdrawal orders are issued under Section 107(a). Accident, rescue, and recovery orders can also be issued under Section 103(j), (k). See *Pittsburgh and Midway Coal Mining Company,* 1 MSHC 2354 (1980)—conditions and practices which constitute an imminent danger and justify withdrawal; *CF&I Steel Corporation,* 2 MSHC 1057 (1980)—the prerequisite to the issuance of a withdrawal order under Section 104(c)(2) is the absence of an intervening "clean" inspection.

If a violation is found and it is also found that, even though the conditions created by the violation do not cause imminent danger, such violation is of a nature which could significantly or substantially contribute to the cause and effect of a mine safety or health hazard and which was due to the operator's *unwarrantable failure* to comply with the law, the inspector shall include such finding in his citation. Section 104(d)(1). "Unwarrantable failure" by an operator to comply means the failure of an operator to abate a violation he knew or should have known existed or the failure to abate a violation because of a lack of due diligence or because of indifference or lack of reasonable care. From an inspector's standpoint, this means ordinary negligence on the part of the operator or his agent. The effect of such a finding is that if, during the same or subsequent inspection (within ninety days) after issuance of the citation, *another* violation is found to be caused by an unwarrantable failure, a withdrawal order shall be issued.

Like the Coal Act, the Mine Act also provides for closure of all or part of a mine or withdrawal of miners from affected areas when an inspector finds a condition exists which could cause *imminent danger* to miners before the condition could be abated. Section 107(a). "Imminent danger" is defined as the existence of any condition or practice in a coal or other mine which could reasonably be expected to cause death or serious physical harm before such condition or practice can be abated. Section 3(j). In such instance withdrawal is required. Thus along with the unwarrantable failure mine closure order, the imminent danger provision is another enforcement tool which will be applied to all mining, both metal and nonmetal. Moreover, when a miner or employee without the required training set forth in Section 115 is found by an inspector to be working in a mine, the miner shall be declared to be a hazard to himself and others and an order (of withdrawal) may be issued for the "untrained" employee. Section 104(g)(1). No employee ordered withdrawn because of lack of training may be discharged or otherwise discriminated against because of the order, or suffer a loss of compensation while receiving the required training. No withdrawn miner may reenter the mine until he has received the required training.

Another important provision requires that an operator be given written notice of a *pattern of violations* which exist in his mine, if the violations could significantly and substantially contribute to the cause and effect of a health or safety condition. Section 104(d) of the Act. If any such violation is found within ninety days after an operator is issued a pattern of violations notice (citation), MSHA must issue an order to withdraw mine personnel from the hazardous area or areas until the condition is abated. Once such an order has been issued, each successive violation which could significantly and substantially contribute to the cause and effect of a health or safety hazard will result in the issuance of a withdrawal order until such time as an inspection of the entire mine reveals no such violations and the notice of a pattern of such violations is terminated. If the pattern is reestablished, however, the sequence for issuing withdrawal orders will begin again.

Another enforcement tool authorizes the Secretary of Labor to seek an injunction in federal court against operators who engage in a pattern of violations constituting a continuing hazard to miners when withdrawal orders have not been effective in achieving lasting compliance. Section 108 of the Mine Act. The Mine Act also broadens the effect of MSHA in the event of metal/nonmetal accidents (such authority already existed in the Coal Act). At an accident scene in which rescue and recovery work is necessary, MSHA may take whatever action it considers appropriate to protect lives, including supervision and direction of rescue and recovery activities. Section 103(j) of the Act. An operator must obtain MSHA approval in consultation with state representatives, when feasible, of any recovery plan.

Citations, orders, and notices issued under the Mine Act can, of course, be contested and later reviewed by the Mine Commission. Statutory provisions for contest are set forth in Sections 105(a) and (d). In *Sunbeam Coal Corporation,* 1 MSHC 2314 (1980), it was held that an operator is entitled to review of a Section 104(a) violation and the inspector's additional findings of "significant and substantial," even where the citations were abated, because the operator was still vulnerable under Section 104(e) to a withdrawal order resulting from a finding of a pattern of significant and substantial violations. Moreover, it was noted that all violations of mandatory standards under the Mine Act are significant and substantial except those violations which pose no risk of injury at all or those violations which pose a source of injury that has only a speculative chance of happening. In *Helvetia Coal Company,* 1 MSHC 2024 (1979), the Commission stated it has jurisdiction under the Mine Act to review, prior to proposal of penalties by the Secretary of Labor, a citation issued under Section 104(a) of the Act, even though an alleged violation has been abated by the mine operator.

Section 105(a) *mandates* that MSHA must *propose* civil penalties under Section 110(a) if a citation or order is issued under Section 104. Unlike OSHA practice, MSHA citations are usually issued on the spot, and penalties are proposed later. The Mine Commission has assessment authority. Section 110(i).

Penalties (civil) are established at not more than $10,000 for each occurrence of violation of a mandatory health and safety standard. Section 110(a). Operators who fail to correct a violation for which a citation has been issued under Section 104(a) within the prescribed period may also be assessed a penalty of any amount up to $1,000 a day for every day beyond the prescribed abatement period during which the condition or practice cited is continued. Provision is also made for extending such liability to the directors, officers, or agents of the corporate operator.

Violations are not classified according to the OSHA framework, *e.g.,* there is no distinction between serious and other-than-serious violations, but there is a distinction between knowing and willful violations. Whenever an operator *willfully* violates a standard or *knowingly* violates or fails to comply with an order, he shall be subject upon conviction to a penalty of up to $25,000 or up to a year in jail or both. Criminal penalties of up to $1,000 and/or six months in jail are also sanctioned for persons who give advance notice of mine inspections, except that NIOSH may give advance notice in carrying out its health responsibilities under the Mine Act. Section 110(e). Miners who willfully violate the mandatory smoking standards are subject to a civil penalty of $250. Section 110(g).

After the issuance of a citation or order under Section 104, the Secretary of Labor will, within a reasonable time, notify the operator by certified mail of the civil penalty proposed to be assessed under Section 110(a). An operator has thirty days after notification of assessment of a proposed civil penalty to notify the Secretary that he intends to contest the citation or proposed assessment of penalty. Section 105(a). If he does not contest the citation or proposed assessment, it becomes a final order of the Mine Commission and is not subject to review by any court or agency. Section 105(a). Once a proposed penalty is contested before the MSHRC or a federal court, it cannot be compromised, mitigated, or settled without approval of the Mine Commission. Section 110(k).

There are six statutory criteria which the Secretary of Labor and the Commission must consider prior to proposing and assessing a penalty: history, size of business, negligence, viability, gravity, and good faith. Section 105(b)(1)(B); *Peabody Coal Company,* 1 MSHC 2422 (1980); *Shamrock Coal Company,* 1 MSHC 2069 (1979)—under Section 110(i) a *de novo* assessment of penalties is within the authority of the Mine Commission and its judges, and the assessment must be supported by substantial evidence; *Co-Op Mining Company,* 1 MSHC 2356 (1980). Unlike OSHA, MSHA has developed a formal, internal procedure for determining the amount of a proposed penalty in which the mine operator is allowed to participate. See 30 CFR Part 100.

Another difference from OSHA citations is that the abatement and penalty payment requirements of MSHA citations and withdrawal orders are not stayed by the filing of a notice of contest. Compare Sections 10(b) and 17(d) of the OSH Act with Sections 104(b) and (h), 105(b)(1)(A) and (b)(2) (penultimate sentence) of the Mine Act. See generally *Energy Fuels Corp.,* 1 MSHC 2013, 2018 n.9 (1979). Moreover, a mine operator who disobeys an MSHA citation or withdrawal order may be issued a failure-to-abate withdrawal order (Section 104 (b)), and an additional proposed penalty (Sections 105 (b)(1)(A) and 110 (b)). See *Marshall v. C. S. & S. Coal Co. (Conway),* 491 F. Supp. 1123, 1 MSHC 2417 (E.D. Pa. 1980)—upholding constitutionality.

Mine operators have two thirty-day periods to contest an MSHA citation: thirty days after its issuance and thirty days after the associated penalty is proposed. Sections 105(a) and (d); 29 CFR 2700.22 and 2700.22; *Energy Fuels Corp.,* 1 MSHC 2013, 2015 (1979). The same cannot be said for an MSHA withdrawal order, however, which apparently must only be contested within thirty days after its issuance. A notification of proposed assessment of penalty must be contested within thirty days after its issuance.

The Mine Act also authorizes the Secretary of Labor to issue regulations requiring operators to maintain records of employee exposure to potentially toxic materials or harmful physical agents which are required to be measured or monitored under any standard. Miners or their representatives must be given access to those records as well as given the opportunity to observe the monitoring and measuring. Section 103(c) of the Act. Operators are also required to make reports required by the Secretary of Labor. Section 103(h) of the Act.

In *Collins v. United States,* 621 F.2d 832, 1 MSHC 2385 (6th Cir. 1980), the court held that the United States is not liable under the Federal Tort Claims Act, 28 U.S.C. Section 1346 (b) and 2671 *et seq.,* for injuries and deaths of miners from an explosion because of an inspector's failure to inspect a mine. In *Bernitsky v. United States,* 620 F.2d 948, 1 MSHC 2321 (3d Cir. 1980); *cert. denied,* — U.S. —, 1 MSHC 2532 (1980), the court held an action for damages brought by the mine operator under the aegis of the Federal Tort Claims Act, alleging that an inspector's issuance of a withdrawal order pursuant to Section 104(b) of the Mine Act caused a mine collapse, was properly dismissed, because the inspector's decision falls within the discretionary function exception to the Federal Tort Claims Act.

General mine safety and health regulations are contained in 30 CFR 1–50. Civil penalties regulations for violation of the Mine Act are contained in 30 CFR 100. Metallic and nonmetallic mine mandatory standards are contained in 30 CFR 55–57. Coal mine mandatory standards are contained in 30 CFR 70–90. Variance provisions are contained in 30 CFR 44. Mine safety and Health Review Commission rules of procedure are contained in 29 CFR 2700. Freedom of Information Act regulations are contained in 29 CFR 2702.

H. DISCRIMINATION

Miners' rights and entitlements are significantly expanded under the Mine Act. Section 103(f), (g). For example, a representative of miners may accompany (walkaround) an MSHA inspector during a mine inspection and may attend any preinspection or postinspection conference at full pay for the time spent in such activities. Miners' rights to file complaints and request inspections for violations of the Mine Act and imminent danger situations have also been expanded. A miner's name will not appear on the complaint which is provided to MSHA and the operator.

Full compensation of up to one week is provided for miners where closure orders are given for the failure of an operator to comply with any health or safety standard, as distinguished from closure due to unwarrantable failure of an operator, which was compensated under the Coal Act. If an inspection reveals that a miner who is employed at a mine has not received the safety training required under Section 115 of the Act, the Secretary shall issue an order declaring him to be a hazard and withdrawn from the mine until he has received the required training. No miner so withdrawn may be discharged and must receive compensation for the period of his training.

Provisions to prevent discrimination against miners or their representatives in exercising their statutory rights have also been made more specific in the Mine Act. Section 105(c)(1), (2), (3), and (4). The Mine Act defines a miner as any person working in a mine and includes both supervisory and nonsupervisory employees working in a mine. Job applicants as well as miners are covered in the antidiscrimination provisions. The following activity of miners under the Mine Act is protected:

(1) Filing or making a complaint concerning an alleged danger or safety/health violation;
(2) Instituting a proceeding, *e.g.*, a claim for compensation or discrimination complaint;
(3) Testifying in an application for review, civil penalty, or other proceeding;
(4) Being the subject of medical evaluation and potential transfer, as provided for in regulations promulgated under Section 101(a);
(5) Being withdrawn from mine, as a hazard, because of lack of required training;
(6) Exercising rights afforded by the Act on behalf of the miner himself or other miners, *e.g.*, walkaround or request for inspection.

A miner or job applicant who believes that he has been discriminated against has sixty days after the violation occurs to file a complaint with the Secretary of Labor. Section 105(c)(2). If the Secretary of Labor, upon investigation, finds that the complaint of a miner is not frivolous, he shall immediately order the miner reinstated, pending final action on the complaint by the MSHRC. Thereafter he must file a complaint with the MSHRC, which, after hearing, will issue an order based on the finding. Mine Commission authority to redress the discrimination includes rehiring or reinstatement with back pay and interest.

If the Secretary finds upon investigation that there is no violation, the miner may within thirty days, file his own complaint with the Commission for final disposition. Whenever an order is issued sustaining such a charge the person found to have committed the violation must pay the fees and expenses (including attorneys fees) reasonably incurred by the miner in connection with the proceedings.

MSHA discrimination cases are not litigated before the federal district courts; Section 105(c) of the Mine Act gives jurisdiction to the MSHRC. The MSHRC has construed the Mine Act to afford miners a right to refuse to work in conditions believed to be unsafe. In *Pasula v. Consolidation Coal Co.*, 2 MSHC 1001 (1980), *pet. for rev. filed,* No. 80 2600 (3d Cir. Nov. 12, 1980), the Mine Commission held that the right to refuse work in some situations is an activity protected by Section 105(c)(1), even though this right is not set forth in the plain language of the Act. The Commission also noted that the statutory right to refuse work under the discrimination clause was broader than the miners' right set forth in their union contract which permitted refusal to work only in abnormal imminent danger:

> Pasula's contractual right to refuse work, however, is limited to a narrow class of hazards: those that are "abnormally and immediately dangerous . . . beyond the normal hazards inherent in the operation which could reasonably be expected to cause death or serious physical harm before such condition or practice can be abated." As the arbitrator's and the safety committeeman's views indicate, this language does not appear to encompass the condition here. In our view, the

statutory right to refuse work under the 1977 Mine Act is broader and does apply to the condition here. The contractual language permits refusals to work in only what might be called an "abnormal imminent danger." We do not construe the 1977 Mine Act to limit a miner's refusal to work only to such conditions.

Adopting the Supreme Court's approach in *Alexander v. Gardner Denver Co.*, 415 U.S. 36(1974), relative to Title VII claims of discrimination, the Mine Commission held that the arbitrator's findings, even those addressing issues perfectly congruent with those before a judge, were not controlling on the judge. The employee's claim may be and was considered *de novo.* Finally, the Commission said where a miner's discharge was motivated by both an unlawful reason and by a lawful reason, he is still entitled to a remedy under the Act. Reviewing the various tests applied in "discrimination" cases, the Commission stated:

Various approaches have been suggested, the most common being the "in any part" test and the "but for" test. The "in any part" test can be simply stated as follows: If any part of the motivation for an employer's adverse action against an employee has been that employee's protected activity, the adverse action is unlawful. It matters not that the employee's unprotected activities were outrageous, would have alone justified adverse action, and did in fact partially motivate the adverse action. The partial illegality of the employer's motive irretrievably taints the adverse action. The "but for" test can be simply stated as follows: it is not enough to support relief to find that the protected activity played a part, however great, in the adverse action; the evidence must also show that the employer would not have acted against the employee but for the protected activity, *i.e.,* that in the absence of the protected activity, no adverse action would have been taken.

The Commission, after discussing the two types of discrimination tests, in effect adopted both:

We hold that the complainant has established a *prima facie* case of a violation of section 105(c)(1) if a preponderance of the evidence proves (1) that he engaged in a protected activity, and (2) that the adverse action was motivated *in any* part by the protected activity. On these issues, the complainant must bear the ultimate burden of persuasion. The employer may affirmatively defend, however, by proving by a preponderance of all the evidence that, although part of his motive was unlawful, (1) he was also motivated by the miner's unprotected activities, and (2) that he would have taken adverse action against the miner in any event for the unprotected activities alone. On these issues, the employer must bear the ultimate burden of persuasion. It is not sufficient for the employer to show that the miner deserved to have been fired for engaging in the unprotected activity; if the unprotected conduct did not originally concern the employer enough to have resulted in the same adverse action, we will not consider it. The employer must show that he did in fact consider the employee deserving of discipline for engaging in the unprotected activity alone and that he would have disciplined him in any event.

By adopting this approach, we have adopted both the "in any part" and "but for" tests, but we have allocated differing burdens of persuasion to each party.

See also *Mine Workers, UMW, Local 9800 v. Dupree,* 2 MSHC 1077 (1980)—MSHA is a person under Section 105(c) of the Act, subject to liability with respect to discrimination.

I. THE MINE COMMISSION AND JUDICIAL REVIEW

The Mine Act creates an independent five-member Mine Safety and Health Review Commission (MSHRC or the Mine Commission) appointed by the President to act as an administrative adjudicatory body for enforcement. Section 113 of the Act. The primary function of the MSHRC is the review of administrative law judges' decisions concerning citations, withdrawal orders, civil penalty assessments, discrimination complaints, and miner compensation claims based on lost wages resulting from mine closure orders. Proceedings before the Commission begin with a hearing before an administrative law judge, who is assigned to hear a case and who then issues a decision deciding the matters in dispute. Section 113(d)(1). The decision of the judge becomes the final decision of the Commission within forty days after issuance, unless a petition for discretionary review is filed by a person affected or aggrieved by the decision, and review is ordered by the Commission within that time period. A request that the Commission exercise such discretion may be made by any person adversely affected or aggrieved by the judge's decision within thirty days of its issuance and must be made on any one or more of the following grounds:

(1) A finding or conclusion of material fact is not supported by substantial evidence;
(2) A necessary legal conclusion is erroneous;
(3) The decision is contrary to law or to the duly promulgated rules and decisions of the MSHRC;
(4) That substantial question of law, policy or discretion is involved; and
(5) A prejudicial error of procedure was committed.

See *Scotia Coal Company v. Marshall,*—F.2d —, 2 MSHC 1051 (6th Cir. 1980) —a final order is required before a party is adversely affected and aggrieved; *Council of Southern Mountains, Inc. v. Martin County Coal Corporation,* 2 MSHC 1058 (1980); *Pontiki Coal Corporation,* 1 MSHC 2208 (1979)—the validity of a withdrawal order may not be considered by the Commission in a civil penalty proceeding; *Marshall v. C.S.&S Coal Co. (Conway),* 491 F. Supp. 1123, 1 MSHC 2417 (E.D. Pa. 1980)—the due process rights of mine owners are not offended by policy of the Mine Act which does not allow an owner a hearing prior to issuance of citations and orders for violation of the Act, because the Mine Act provides for elaborate review procedures which afford the mine owners a full opportunity to be heard; *New York State Department of Transportation,* 1 MSHC 2481 (1980)—the Mine Commission does not have jurisdiction to pass on the constitutionality of the Mine Act but does have jurisdiction to determine whether the Act may be constitutionally applied to the facts of a particular case; thus the Tenth Amendment to the Constitution does not prevent the enforcement of the Act against state-owned sand and gravel mining operations; *Grundy Mining Company, Inc. v. Secretary of Labor,*—F.2d —, 2 MSHC 1049 (6th Cir. 1980)—parties appearing before the MSHRC and administrative law judges are entitled to have disputes arising under the Mine Act judged fairly and on their own merits.

The MSHRC has no jurisdiction to hear matters concerning the promulgation of mandatory health or safety standards, applications for modifications or variances with respect to existing standards, or mine accidents and other such occurrences or disasters under Section 103(b). These matters are within the jurisdiction of the Assistant Secretary of Labor for Mine Safety and Health.

There are two basic judicial review provisions provided in the Mine Act. Section 101(d), discussed previously, provides for review of newly issued mandatory permanent health and safety standards. There is also a right to judicial review of the Commission's (enforcement) decisions by aggrieved persons in the appropriate United States court of appeals within thirty days of the issuance of a Mine Commission order. Section 106 of the Act. The administrative adjudication scheme under the Mine Act is very similar to that under the OSH Act.

No objection not urged before the Commission shall be considered by a court, unless extraordinary circumstances exist. The standard of review is substantial evidence on the record. *Shamrock Coal Company,* 1 MSHC 2069 (1979), *appeal filed* No. 79–3393 (6th Cir. 1979)—assessment of penalties by a judge must be supported by substantial evidence; *Westmoreland Coal Company v. MSHRC,* 606 F.2d 417, 1 MSHC 2129 (4th Cir. 1979)—substantial evidence must support an administrative law judge's ruling; *Clinchfield Coal Company,* 620 F.2d 292, 1 MSHC 2337 (4th Cir. 1980)—substantial evidence test is applicable; *Swope Coal Company v. FMSHRC,* 626 F.2d 863, 1 MSHC 2420 (4th Cir. 1980)—a holding that a lessor coal company was an operator of a coal mine within the meaning of the Act is supported by substantial evidence; *Magma Copper Company v. Secretary of Labor,*—F.2d —, 2 MSHC 1065 (9th Cir. 1980)—substantial evidence supports the Commission's findings, and deference is required, because the Commission has specialized knowledge and is entitled to deference.

In addition to these review provisions, Section 108 of the Mine Act provides for obtaining injunctive relief from a federal district court if the Secretary of Labor is impeded in performing his inspection responsibilities. Section 108(b) also authorizes injunctive relief when it is believed that the mine operator is engaged in a pattern of violations which are deemed to constitute a continuing hazard to the health and safety of miners. Sections 110(j) and 106(b) authorize collection actions against operators who fail to pay final unappealed orders of the Commission requiring payment of civil penalties.

The Administrative Procedure Act, 5 U.S.C. Sections 551–559 and 701–706, is not generally applicable to proceedings under the Mine Act, unless it is expressly made applicable. Section 507 of the Mine Act.

J. STATE PLANS

There are no state plans, but state laws will remain in effect unless they come in conflict with MSHA. Section 506 of the Act.

K. RELATIONSHIP BETWEEN OSHA AND MSHA

The major area of possible conflict between OSHA and MSHA lies in surface construction activity related to mining. MSHA, for example, required that separate safety and health regulations be issued, as far as practicable, for mine construction activity on the surface. But currently, MSHA does not have such standards.

The OHSA Advisory Committee on Construction Safety and Health has opposed efforts to place construction related to mining under the jurisdiction of MSHA. MSHA argues that the Mine Act does not permit delegation of authority over surface construction at mine sites to OSHA; only delegation in mineral milling is permitted. Section 3(h)(1) of the Act. However, unions support OSHA jurisdiction. At the present time, borrow pits used for construction projects will not be subject to MSHA. An interagency agreement between MSHA and OSHA, 44 Fed. Reg. 22827 (April 17, 1979), has eliminated jurisdictional disputes, at least with respect to borrow pits and surface construction projects related to mining. *New York State Department of Transportation,* 1 MSHC 2481 (1980); *Southern Clay, Inc. v. Department of Labor,* 612 F.2d 574, 1 MSHC 2225 (3d Cir. 1979).

The net effect of the Mine Act upon the OSH Act has so far been relatively minor. In *Whirlpool Corp. v. Marshall,* 445 U.S. 1, 13, n.18 (1980), the Court cited in a footnote portions of the legislative history of the Mine Act recognizing a miner's right to refuse to work in conditions which are believed to be unsafe or unhealthful, in order to support the Secretary of Labor's interpretation of Section 11(c)(1) of the OSH Act (29 U.S.C. Section 660(c)(1)). In nearly all other instances, however, the Mine Act and the OSH Act have been contrasted, rather than treated similarly. See *Marshall v. Sun Petroleum Products Co.,* 622 F.2d 1176, 1184, 8 OSHC 1929 (3d Cir. 1980) *cert. denied,*—U.S.—(1980)—MSHRC but not OSHRC expressly granted power to review penalty settlement; *United Steelworkers of America, AFL-CIO CLC v. Marshall (lead standard case),*—F.2d—8 OSHC 1810, 1837, n.67. (D.C. Cir. 1980)—medical removal protection for miners, Mine Act dissimilar to OSH Act; *Republic Steel Corp.,* 1 MSHC 2002, 2005–2006 and n.12, (1979)—strict liability under predecessor mine statute, OSHA contrasted.

In *Pennsuco Cement and Aggregates, Inc.,* 8 OSHC 1378 (1980), OSHA's authority over a cement producer's kilns was preempted under Section 4(b)(1) of the OSH Act, since the Mining Enforcement and Safety Administration (precurser to MSHA) had exercised its authority to regulate the safety and health of employees in this area.

General Appendixes

APPENDIX A
The Occupational Safety and Health Act

To assure safe and healthful working conditions for working men and women; by authorizing enforcement of the standards developed under the Act; by assisting and encouraging the States in their efforts to assure safe and healthful working conditions; by providing for research, information, education, and training in the field of occupational safety and health; and for other purposes.

Be it enacted by the Senate and House of Representatives of the United States of America in Congress assembled, That this Act may be cited as the "Occupational Safety and Health Act of 1970".

Occupational Safety and Health Act of 1970.

CONGRESSIONAL FINDINGS AND PURPOSE

SEC. (2) The Congress finds that personal injuries and illnesses arising out of work situations impose a substantial burden upon, and are a hindrance to, interstate commerce in terms of lost production, wage loss, medical expenses, and disability compensation payments.

(b) The Congress declares it to be its purpose and policy, through the exercise of its powers to regulate commerce among the several States and with foreign nations and to provide for the general welfare, to assure so far as possible every working man and woman in the Nation safe and healthful working conditions and to preserve our human resources—

(1) by encouraging employers and employees in their efforts to reduce the number of occupational safety and health hazards at their places of employment, and to stimulate employers and employees to institute new and to perfect existing programs for providing safe and healthful working conditions;

(2) by providing that employers and employees have separate but dependent responsibilities and rights with respect to achieving safe and healthful working conditions;

(3) by authorizing the Secretary of Labor to set mandatory occupational safety and health standards applicable to businesses affecting interstate commerce, and by creating an Occupational Safety and Health Review Commission for carrying out adjudicatory functions under the Act;

(4) by building upon advances already made through employer and employee initiative for providing safe and healthful working conditions;

(5) by providing for research in the field of occupational safety and health, including the psychological factors involved, and by developing innovative methods, techniques, and approaches for dealing with occupational safety and health problems;

(6) by exploring ways to discover latent diseases, establishing causal connections between diseases and work in environmental conditions, and conducting other research relating to health problems, in recognition of the fact that occupational health standards present problems often different from those involved in occupational safety;

(7) by providing medical criteria which will assure insofar as practicable that no employee will suffer diminished health, functional capacity, or life expectancy as a result of his work experience;

(8) by providing for training programs to increase the number and competence of personnel engaged in the field of occupational safety and health;

(9) by providing for the development and promulgation of occupational safety and health standards;

(10) by providing an effective enforcement program which shall include a prohibition against giving advance notice of any inspection and sanctions for any individual violating this prohibition;

(11) by encouraging the States to assume the fullest responsibility for the administration and enforcement of their occupational safety and health laws by providing grants to the States to assist in identifying their needs and responsibilities in the area of occupational safety and health, to develop plans in accordance with the provisions of this Act, to improve the administration and enforcement of State occupational safety and health laws, and to conduct experimental and demonstration projects in connection therewith;

(12) by providing for appropriate reporting procedures with respect to occupational safety and health which procedures will help achieve the objectives of this Act and accurately describe the nature of the occupational safety and health problem;

(13) by encouraging joint labor-management efforts to reduce injuries and disease arising out of employment.

DEFINITIONS

SEC. 3. For the purposes of this Act—

(1) The term "Secretary" mean the Secretary of Labor.

(2) The term "Commission" means the Occupational Safety and Health Review Commission established under this Act.

(3) The term "commerce" means trade, traffic, commerce, transportation, or communication among the several States, or between a State and any place outside thereof, or within the District of Columbia, or a possession of the United States (other than the Trust Territory of the Pacific Islands), or between points in the same State but through a point outside thereof.

(4) The term "person" means one or more individuals, partnerships, associations, corporations, business trusts, legal representatives, or any organized group of persons.

(5) The term "employer" means a person engaged in a business affecting commerce who has employees, but does not include the United States or any State or political subdivision of a State.

(6) The term "employee" means an employee of an employer who is employed in a business of his employer which affects commerce.

(7) The term "State" includes a State of the United States, the District of Columbia, Puerto Rico, the Virgin Islands, American Samoa, Guam, and the Trust Territory of the Pacific Islands.

(8) The term "occupational safety and health standard" means a standard which requires conditions, or the adoption or use of one or more practices, means, methods, operations, or processes, reasonably necessary or appropriate to provide safe or healthful employment and places of employment.

(9) The term "national consensus standard" means any occupa-

tional safety and health standard or modification thereof which (1), has been adopted and promulgated by a nationally recognized standards-producing organization under procedures whereby it can be determined by the Secretary that persons interested and affected by the scope or provisions of the standard have reached substantial agreement on its adoption, (2) was formulated in a manner which afforded an opportunity for diverse views to be considered and (3) has been designated as such a standard by the Secretary, after consultation with other appropriate Federal agencies.

(10) The term "established Federal standard" means any operative occupational safety and health standard established by any agency of the United States and presently in effect, or contained in any Act of Congress in force on the date of enactment of this Act.

(11) The term "Committee" means the National Advisory Committee on Occupational Safety and Health established under this Act.

(12) The term "Director" means the Director of the National Institute for Occupational Safety and Health.

(13) The term "Institute" means the National Institute for Occupational Safety and Health established under this Act.

(14) The term "Workmen's Compensation Commission" means the National Commission on State Workmen's Compensation Laws established under this Act.

APPLICABILITY OF THIS ACT

SEC. 4. (a) This Act shall apply with respect to employment performed in a workplace in a State, the District of Columbia, the Commonwealth of Puerto Rico, the Virgin Islands, American Samoa, Guam, the Trust Territory of the Pacific Islands, Wake Island, Outer Continental Shelf lands defined in the Outer Continental Shelf Lands Act, Johnston Island, and the Canal Zone. The Secretary of the Interior shall, by regulation, provide for judicial enforcement of this Act by the courts established for areas in which there are no United States district courts having jurisdiction.

67 Stat. 462.
43 USC 1331 note.

(b) (1) Nothing in this Act shall apply to working conditions of employees with respect to which other Federal agencies, and State agencies acting under section 274 of the Atomic Energy Act of 1954, as amended (42 U.S.C. 2021), exercise statutory authority to prescribe or enforce standards or regulations affecting occupational safety or health.

73 Stat. 688.

(2) The safety and health standards promulgated under the Act of June 30, 1936, commonly known as the Walsh-Healey Act (41 U.S.C. 35 et seq.), the Service Contract Act of 1965 (41 U.S.C. 351 et seq.), Public Law 91-54, Act of August 9, 1969 (40 U.S.C. 333), Public Law 85-742, Act of August 23, 1958 (33 U.S.C. 941), and the National Foundation on Arts and Humanities Act (20 U.S.C. 951 et seq.) are superseded on the effective date of corresponding standards, promulgated under this Act, which are determined by the Secretary to be more effective. Standards issued under the laws listed in this paragraph and in effect on or after the effective date of this Act shall be deemed to be occupational safety and health standards issued under this Act, as well as under such other Acts.

49 Stat. 2036.
79 Stat. 1034.
83 Stat. 96.
72 Stat. 835.
79 Stat. 845;
Ante, p. 443.

(3) The Secretary shall, within three years after the effective date of this Act, report to the Congress his recommendations for legislation to avoid unnecessary duplication and to achieve coordination between this Act and other Federal laws.

Report to Congress.

(4) Nothing in this Act shall be construed to supersede or in any manner affect any workmen's compensation law or to enlarge or diminish or affect in any other manner the common law or statutory rights, duties, or liabilities of employers and employees under any law with respect to injuries, diseases, or death of employees arising out of, or in the course of, employment.

DUTIES

SEC. 5. (a) Each employer—

(1) shall furnish to each of his employees employment and a place of employment which are free from recognized hazards that are causing or are likely to cause death or serious physical harm to his employees;

(2) shall comply with occupational safety and health standards promulgated under this Act.

(b) Each employee shall comply with occupational safety and health standards and all rules, regulations, and orders issued pursuant to this Act which are applicable to his own actions and conduct.

OCCUPATIONAL SAFETY AND HEALTH STANDARDS

80 Stat. 381; 81 Stat. 195. 5 USC 500.

SEC. 6. (a) Without regard to chapter 5 of title 5, United States Code, or to the other subsections of this section, the Secretary shall, as soon as practicable during the period beginning with the effective date of this Act and ending two years after such date, by rule promulgate as an occupational safety or health standard any national consensus standard, and any established Federal standard, unless he determines that the promulgation of such a standard would not result in improved safety or health for specifically designated employees. In the event of conflict among any such standards, the Secretary shall promulgate the standard which assures the greatest protection of the safety or health of the affected employees.

(b) The Secretary may by rule promulgate, modify, or revoke any occupational safety or health standard in the following manner:

Advisory committee, recommendations.

(1) Whenever the Secretary, upon the basis of information submitted to him in writing by an interested person, a representative of any organization of employers or employees, a nationally recognized standards-producing organization, the Secretary of Health, Education, and Welfare, the National Institute for Occupational Safety and Health, or a State or political subdivision, or on the basis of information developed by the Secretary or otherwise available to him, determines that a rule should be promulgated in order to serve the objectives of this Act, the Secretary may request the recommendations of an advisory committee appointed under section 7 of this Act. The Secretary shall provide such an advisory committee with any proposals of his own or of the Secretary of Health, Education, and Welfare, together with all pertinent factual information developed by the Secretary or the Secretary of Health, Education, and Welfare, or otherwise available, including the results of research, demonstrations, and experiments. An advisory committee shall submit to the Secretary its recommendations regarding the rule to be promulgated within ninety days from the date of its appointment or within such longer or shorter period as may be prescribed by the Secretary, but in no event for a period which is longer than two hundred and seventy days.

(2) The Secretary shall publish a proposed rule promulgating, modifying, or revoking an occupational safety or health standard in the Federal Register and shall afford interested persons a period of thirty days after publication to submit written data or comments. Where an advisory committee is appointed and the Secretary determines that a rule should be issued, he shall publish the proposed rule within sixty days after the submission of the advisory committee's recommendations or the expiration of the period prescribed by the Secretary for such submission.

Publication in Federal Register.

(3) On or before the last day of the period provided for the submission of written data or comments under paragraph (2), any interested person may file with the Secretary written objections to the proposed rule, stating the grounds therefor and requesting a public hearing on such objections. Within thirty days after the last day for filing such objections, the Secretary shall publish in the Federal Register a notice specifying the occupational safety or health standard to which objections have been filed and a hearing requested, and specifying a time and place for such hearing.

Hearing, notice.

Publication in Federal Register.

(4) Within sixty days after the expiration of the period provided for the submission of written data or comments under paragraph (2), or within sixty days after the completion of any hearing held under paragraph (3), the Secretary shall issue a rule promulgating, modifying, or revoking an occupational safety or health standard or make a determination that a rule should not be issued. Such a rule may contain a provision delaying its effective date for such period (not in excess of ninety days) as the Secretary determines may be necessary to insure that affected employers and employees will be informed of the existence of the standard and of its terms and that employers affected are given an opportunity to familiarize themselves and their employees with the existence of the requirements of the standard.

(5) The Secretary, in promulgating standards dealing with toxic materials or harmful physical agents under this subsection, shall set the standard which most adequately assures, to the extent feasible, on the basis of the best available evidence, that no employee will suffer material impairment of health or functional capacity even if such employee has regular exposure to the hazard dealt with by such standard for the period of his working life. Development of standards under this subsection shall be based upon research, demonstrations, experiments, and such other information as may be appropriate. In addition to the attainment of the highest degree of health and safety protection for the employee, other considerations shall be the latest available scientific data in the field, the feasibility of the standards, and experience gained under this and other health and safety laws. Whenever practicable, the standard promulgated shall be expressed in terms of objective criteria and of the performance desired.

Toxic materials.

(6)(A) Any employer may apply to the Secretary for a temporary order granting a variance from a standard or any provision thereof promulgated under this section. Such temporary order shall be granted only if the employer files an application which meets the requirements of clause (B) and establishes that (i) he is unable to comply with a standard by its effective date because of unavailability of professional or technical personnel or of materials and equipment needed to come into compliance with the standard or because necessary construction or alteration of facilities cannot be completed by the effective date, (ii) he is taking all available steps to safeguard his employees against the hazards covered by the standard, and (iii) he has an effective program for coming into compliance with the standard as quickly as

Temporary variance order.

practicable. Any temporary order issued under this paragraph shall prescribe the practices, means, methods, operations, and processes which the employer must adopt and use while the order is in effect and state in detail his program for coming into compliance with the standard. Such a temporary order may be granted only after notice to employees and an opportunity for a hearing: *Provided*, That the Secretary may issue one interim order to be effective until a decision is made on the basis of the hearing. No temporary order may be in effect for longer than the period needed by the employer to achieve compliance with the standard or one year, whichever is shorter, except that such an order may be renewed not more than twice (I) so long as the requirements of this paragraph are met and (II) if an application for renewal is filed at least 90 days prior to the expiration date of the order. No interim renewal of an order may remain in effect for longer than 180 days.

Notice, hearing.

Renewal.

Time limitation.

(B) An application for a temporary order under this paragraph (6) shall contain:

(i) a specification of the standard or portion thereof from which the employer seeks a variance,

(ii) a representation by the employer, supported by representations from qualified persons having firsthand knowledge of the facts represented, that he is unable to comply with the standard or portion thereof and a detailed statement of the reasons therefor,

(iii) a statement of the steps he has taken and will take (with specific dates) to protect employees against the hazard covered by the standard,

(iv) a statement of when he expects to be able to comply with the standard and what steps he has taken and what steps he will take (with dates specified) to come into compliance with the standard, and

(v) a certification that he has informed his employees of the application by giving a copy thereof to their authorized representative, posting a statement giving a summary of the application and specifying where a copy may be examined at the place or places where notices to employees are normally posted, and by other appropriate means.

A description of how employees have been informed shall be contained in the certification. The information to employees shall also inform them of their right to petition the Secretary for a hearing.

(C) The Secretary is authorized to grant a variance from any standard or portion thereof whenever he determines, or the Secretary of Health, Education, and Welfare certifies, that such variance is necessary to permit an employer to participate in an experiment approved by him or the Secretary of Health, Education, and Welfare designed to demonstrate or validate new and improved techniques to safeguard the health or safety of workers.

Labels, etc.

(7) Any standard promulgated under this subsection shall prescribe the use of labels or other appropriate forms of warning as are necessary to insure that employees are apprised of all hazards to which they are exposed, relevant symptoms and appropriate emergency treatment, and proper conditions and precautions of safe use or exposure. Where appropriate, such standard shall also prescribe suitable protective equipment and control or technological procedures to be used in connection with such hazards and shall provide for monitoring or measuring employee exposure at such locations and intervals, and in such manner as may be necessary for the protection of employees. In

Protective equipment, etc.

addition, where appropriate, any such standard shall prescribe the type and frequency of medical examinations or other tests which shall be made available, by the employer or at his cost, to employees exposed to such hazards in order to most effectively determine whether the health of such employees is adversely affected by such exposure. In the event such medical examinations are in the nature of research, as determined by the Secretary of Health, Education, and Welfare, such examinations may be furnished at the expense of the Secretary of Health, Education, and Welfare. The results of such examinations or tests shall be furnished only to the Secretary or the Secretary of Health, Education, and Welfare, and, at the request of the employee, to his physician. The Secretary, in consultation with the Secretary of Health, Education, and Welfare, may by rule promulgated pursuant to section 553 of title 5, United States Code, make appropriate modifications in the foregoing requirements relating to the use of labels or other forms of warning, monitoring or measuring, and medical examinations, as may be warranted by experience, information, or medical or technological developments acquired subsequent to the promulgation of the relevant standard.

Medical examinations.

80 Stat. 383.

(8) Whenever a rule promulgated by the Secretary differs substantially from an existing national consensus standard, the Secretary shall, at the same time, publish in the Federal Register a statement of the reasons why the rule as adopted will better effectuate the purposes of this Act than the national consensus standard.

Publication in Federal Register.

(c)(1) The Secretary shall provide, without regard to the requirements of chapter 5, title 5, United States Code, for an emergency temporary standard to take immediate effect upon publication in the Federal Register if he determines (A) that employees are exposed to grave danger from exposure to substances or agents determined to be toxic or physically harmful or from new hazards, and (B) that such emergency standard is necessary to protect employees from such danger.

Temporary standard. Publication in Federal Register. 80 Stat. 381; 81 Stat. 195. 5 USC 500.

(2) Such standard shall be effective until superseded by a standard promulgated in accordance with the procedures prescribed in paragraph (3) of this subsection.

Time limitation.

(3) Upon publication of such standard in the Federal Register the Secretary shall commence a proceeding in accordance with section 6(b) of this Act, and the standard as published shall also serve as a proposed rule for the proceeding. The Secretary shall promulgate a standard under this paragraph no later than six months after publication of the emergency standard as provided in paragraph (2) of this subsection.

(d) Any affected employer may apply to the Secretary for a rule or order for a variance from a standard promulgated under this section. Affected employees shall be given notice of each such application and an opportunity to participate in a hearing. The Secretary shall issue such rule or order if he determines on the record, after opportunity for an inspection where appropriate and a hearing, that the proponent of the variance has demonstrated by a preponderance of the evidence that the conditions, practices, means, methods, operations, or processes used or proposed to be used by an employer will provide employment and places of employment to his employees which are as safe and healthful as those which would prevail if he complied with the standard. The rule or order so issued shall prescribe the conditions the employer must maintain, and the practices, means, methods, operations, and processes which he must adopt and utilize to the extent they differ from the standard in question. Such a rule or order may be modi-

Variance rule.

fied or revoked upon application by an employer, employees, or by the Secretary on his own motion, in the manner prescribed for its issuance under this subsection at any time after six months from its issuance.

Publication in Federal Register.

(e) Whenever the Secretary promulgates any standard, makes any rule, order, or decision, grants any exemption or extension of time, or compromises, mitigates, or settles any penalty assessed under this Act, he shall include a statement of the reasons for such action, which shall be published in the Federal Register.

Petition for judicial review.

(f) Any person who may be adversely affected by a standard issued under this section may at any time prior to the sixtieth day after such standard is promulgated file a petition challenging the validity of such standard with the United States court of appeals for the circuit wherein such person resides or has his principal place of business, for a judicial review of such standard. A copy of the petition shall be forthwith transmitted by the clerk of the court to the Secretary. The filing of such petition shall not, unless otherwise ordered by the court, operate as a stay of the standard. The determinations of the Secretary shall be conclusive if supported by substantial evidence in the record considered as a whole.

(g) In determining the priority for establishing standards under this section, the Secretary shall give due regard to the urgency of the need for mandatory safety and health standards for particular industries, trades, crafts, occupations, businesses, workplaces or work environments. The Secretary shall also give due regard to the recommendations of the Secretary of Health, Education, and Welfare regarding the need for mandatory standards in determining the priority for establishing such standards.

ADVISORY COMMITTEES; ADMINISTRATION

Establishment; membership.

SEC. 7. (a)(1) There is hereby established a National Advisory Committee on Occupational Safety and Health consisting of twelve members appointed by the Secretary, four of whom are to be designated by the Secretary of Health, Education, and Welfare, without regard to the provisions of title 5, United States Code, governing appointments in the competitive service, and composed of representatives of manangement, labor, occupational safety and occupational health professions, and of the public. The Secretary shall designate one of the public members as Chairman. The members shall be selected upon the basis of their experience and competence in the field of occupational safety and health.

80 Stat. 378. 5 USC 101.

(2) The Committee shall advise, consult with, and make recommendations to the Secretary and the Secretary of Health, Education, and Welfare on matters relating to the administration of the Act. The Committee shall hold no fewer than two meetings during each calendar year. All meetings of the Committee shall be open to the public and a transcript shall be kept and made available for public inspection.

Public transcript.

(3) The members of the Committee shall be compensated in accordance with the provisions of section 3109 of title 5, United States Code.

80 Stat. 416.

(4) The Secretary shall furnish to the Committee an executive secretary and such secretarial, clerical, and other services as are deemed necessary to the conduct of its business.

(b) An advisory committee may be appointed by the Secretary to assist him in his standard-setting functions under section 6 of this Act. Each such committee shall consist of not more than fifteen members and shall include as a member one or more designees of the Secretary of Health, Education, and Welfare, and shall include among its members an equal number of persons qualified by experience and affiliation to present the viewpoint of the employers involved, and of persons similarly qualified to present the viewpoint of the workers involved, as well as one or more representatives of health and safety agencies of the States. An advisory committee may also include such other persons as the Secretary may appoint who are qualified by knowledge and experience to make a useful contribution to the work of such committee, including one or more representatives of professional organizations of technicians or professionals specializing in occupational safety or health, and one or more representatives of nationally recognized standards-producing organizations, but the number of persons so appointed to any such advisory committee shall not exceed the number appointed to such committee as representatives of Federal and State agencies. Persons appointed to advisory committees from private life shall be compensated in the same manner as consultants or experts under section 3109 of title 5, United States Code. The Secretary shall pay to any State which is the employer of a member of such a committee who is a representative of the health or safety agency of that State, reimbursement sufficient to cover the actual cost to the State resulting from such representative's membership on such committee. Any meeting of such committee shall be open to the public and an accurate record shall be kept and made available to the public. No member of such committee (other than representatives of employers and employees) shall have an economic interest in any proposed rule.

80 Stat. 416.

Recordkeeping.

(c) In carrying out his responsibilities under this Act, the Secretary is authorized to—

(1) use, with the consent of any Federal agency, the services, facilities, and personnel of such agency, with or without reimbursement, and with the consent of any State or political subdivision thereof, accept and use the services, facilities, and personnel of any agency of such State or subdivision with reimbursement; and

(2) employ experts and consultants or organizations thereof as authorized by section 3109 of title 5, United States Code, except that contracts for such employment may be renewed annually; compensate individuals so employed at rates not in excess of the rate specified at the time of service for grade GS–18 under section 5332 of title 5, United States Code, including traveltime, and allow them while away from their homes or regular places of business, travel expenses (including per diem in lieu of subsistence) as authorized by section 5703 of title 5, United States Code, for persons in the Government service employed intermittently, while so employed.

Ante, p. 198-1.

80 Stat. 499;
83 Stat. 190.

INSPECTIONS, INVESTIGATIONS, AND RECORDKEEPING

SEC. 8. (a) In order to carry out the purposes of this Act, the Secretary, upon presenting appropriate credentials to the owner, operator, or agent in charge, is authorized—

(1) to enter without delay and at reasonable times any factory, plant, establishment, construction site, or other area, workplace or environment where work is performed by an employee of an employer; and

(2) to inspect and investigate during regular working hours and at other reasonable times, and within reasonable limits and in a reasonable manner, any such place of employment and all pertinent conditions, structures, machines, apparatus, devices, equipment, and materials therein, and to question privately any such employer, owner, operator, agent or employee.

Subpoena power.

(b) In making his inspections and investigations under this Act the Secretary may require the attendance and testimony of witnesses and the production of evidence under oath. Witnesses shall be paid the same fees and mileage that are paid witnesses in the courts of the United States. In case of a contumacy, failure, or refusal of any person to obey such an order, any district court of the United States or the United States courts of any territory or possession, within the jurisdiction of which such person is found, or resides or transacts business, upon the application by the Secretary, shall have jurisdiction to issue to such person an order requiring such person to appear to produce evidence if, as, and when so ordered, and to give testimony relating to the matter under investigation or in question, and any failure to obey such order of the court may be punished by said court as a contempt thereof.

Recordkeeping.

(c) (1) Each employer shall make, keep and preserve, and make available to the Secretary or the Secretary of Health, Education, and Welfare, such records regarding his activities relating to this Act as the Secretary, in cooperation with the Secretary of Health, Education, and Welfare, may prescribe by regulation as necessary or appropriate for the enforcement of this Act or for developing information regarding the causes and prevention of occupational accidents and illnesses. In order to carry out the provisions of this paragraph such regulations may include provisions requiring employers to conduct periodic inspections. The Secretary shall also issue regulations requiring that employers, through posting of notices or other appropriate means, keep their employees informed of their protections and obligations under this Act, including the provisions of applicable standards.

Work-related deaths, etc.; reports.

(2) The Secretary, in cooperation with the Secretary of Health, Education, and Welfare, shall prescribe regulations requiring employers to maintain accurate records of, and to make periodic reports on, work-related deaths, injuries and illnesses other than minor injuries requiring only first aid treatment and which do not involve medical treatment, loss of consciousness, restriction of work or motion, or transfer to another job.

(3) The Secretary, in cooperation with the Secretary of Health, Education, and Welfare, shall issue regulations requiring employers to maintain accurate records of employee exposures to potentially toxic materials or harmful physical agents which are required to be monitored or measured under section 6. Such regulations shall provide employees or their representatives with an opportunity to observe such monitoring or measuring, and to have access to the records thereof. Such regulations shall also make appropriate provision for each employee or former employee to have access to such records as will indicate his own exposure to toxic materials or harmful physical agents. Each employer shall promptly notify any employee who has been or is being exposed to toxic materials or harmful physical agents in concentrations or at levels which exceed those prescribed by an applicable occupational safety and health standard promulgated under section 6, and shall inform any employee who is being thus exposed of the corrective action being taken.

(d) Any information obtained by the Secretary, the Secretary of Health, Education, and Welfare, or a State agency under this Act shall be obtained with a minimum burden upon employers, especially those operating small businesses. Unnecessary duplication of efforts in obtaining information shall be reduced to the maximum extent feasible.

(e) Subject to regulations issued by the Secretary, a representative of the employer and a representative authorized by his employees shall be given an opportunity to accompany the Secretary or his authorized representative during the physical inspection of any workplace under subsection (a) for the purpose of aiding such inspection. Where there is no authorized employee representative, the Secretary or his authorized representative shall consult with a reasonable number of employees concerning matters of health and safety in the workplace.

(f)(1) Any employees or representative of employees who believe that a violation of a safety or health standard exists that threatens physical harm, or that an imminent danger exists, may request an inspection by giving notice to the Secretary or his authorized representative of such violation or danger. Any such notice shall be reduced to writing, shall set forth with reasonable particularity the grounds for the notice, and shall be signed by the employees or representative of employees, and a copy shall be provided the employer or his agent no later than at the time of inspection, except that, upon the request of the person giving such notice, his name and the names of individual employees referred to therein shall not appear in such copy or on any record published, released, or made available pursuant to subsection (g) of this section. If upon receipt of such notification the Secretary determines there are reasonable grounds to believe that such violation or danger exists, he shall make a special inspection in accordance with the provisions of this section as soon as practicable, to determine if such violation or danger exists. If the Secretary determines there are no reasonable grounds to believe that a violation or danger exists he shall notify the employees or representative of the employees in writing of such determination.

(2) Prior to or during any inspection of a workplace, any employees or representative of employees employed in such workplace may notify the Secretary or any representative of the Secretary responsible for conducting the inspection, in writing, of any violation of this Act which they have reason to believe exists in such workplace. The Secretary shall, by regulation, establish procedures for informal review of any refusal by a representative of the Secretary to issue a citation with respect to any such alleged violation and shall furnish the employees or representative of employees requesting such review a written statement of the reasons for the Secretary's final disposition of the case.

Reports, publication.

(g)(1) The Secretary and Secretary of Health, Education, and Welfare are authorized to compile, analyze, and publish, either in summary or detailed form, all reports or information obtained under this section.

Rules and regulations.

(2) The Secretary and the Secretary of Health, Education, and Welfare shall each prescribe such rules and regulations as he may deem necessary to carry out their responsibilities under this Act, including rules and regulations dealing with the inspection of an employer's establishment.

CITATIONS

SEC. 9. (a) If, upon inspection or investigation, the Secretary or his authorized representative believes that an employer has violated a requirement of section 5 of this Act, of any standard, rule or order promulgated pursuant to section 6 of this Act, or of any regulations prescribed pursuant to this Act, he shall with reasonable promptness issue a citation to the employer. Each citation shall be in writing and shall describe with particularity the nature of the violation, including a reference to the provision of the Act, standard, rule, regulation, or order alleged to have been violated. In addition, the citation shall fix a reasonable time for the abatement of the violation. The Secretary may prescribe procedures for the issuance of a notice in lieu of a citation with respect to de minimis violations which have no direct or immediate relationship to safety or health.

(b) Each citation issued under this section, or a copy or copies thereof, shall be prominently posted, as prescribed in regulations issued by the Secretary, at or near each place a violation referred to in the citation occurred.

Limitation.

(c) No citation may be issued under this section after the expiration of six months following the occurrence of any violation.

PROCEDURE FOR ENFORCEMENT

SEC. 10. (a) If, after an inspection or investigation, the Secretary issues a citation under section 9(a), he shall, within a reasonable time after the termination of such inspection or investigation, notify the employer by certified mail of the penalty, if any, proposed to be assessed under section 17 and that the employer has fifteen working days within which to notify the Secretary that he wishes to contest the citation or proposed assessment of penalty. If, within fifteen working days from the receipt of the notice issued by the Secretary the employer fails to notify the Secretary that he intends to contest the citation or proposed assessment of penalty, and no notice is filed by any employee or representative of employees under subsection (c) within such time, the citation and the assessment, as proposed, shall be deemed a final order of the Commission and not subject to review by any court or agency.

(b) If the Secretary has reason to believe that an employer has failed to correct a violation for which a citation has been issued within the period permitted for its correction (which period shall not begin to run until the entry of a final order by the Commission in the case of any review proceedings under this section initiated by the employer in good faith and not solely for delay or avoidance of penalties), the Secretary shall notify the employer by certified mail of such failure and of the penalty proposed to be assessed under section 17 by reason of such failure, and that the employer has fifteen working days within which to notify the Secretary that he wishes to contest the Secretary's notification or the proposed assessment of penalty. If, within fifteen working days from the receipt of notification issued by the Secretary, the employer fails to notify the Secretary that he intends to contest the notification or proposed assessment of penalty, the notification and assessment, as proposed, shall be deemed a final order of the Commission and not subject to review by any court or agency.

(c) If an employer notifies the Secretary that he intends to contest a citation issued under section 9(a) or notification issued under subsection (a) or (b) of this section, or if, within fifteen working days of the issuance of a citation under section 9(a), any employee or representative of employees files a notice with the Secretary alleging that the period of time fixed in the citation for the abatement of the violation is unreasonable, the Secretary shall immediately advise the Commission of such notification, and the Commission shall afford an opportunity for a hearing (in accordance with section 554 of title 5, United States Code, but without regard to subsection (a)(3) of such section). The Commission shall thereafter issue an order, based on findings of fact, affirming, modifying, or vacating the Secretary's citation or proposed penalty, or directing other appropriate relief, and such order shall become final thirty days after its issuance. Upon a showing by an employer of a good faith effort to comply with the abatement requirements of a citation, and that abatement has not been completed because of factors beyond his reasonable control, the Secretary, after an opportunity for a hearing as provided in this subsection, shall issue an order affirming or modifying the abatement requirements in such citation. The rules of procedure prescribed by the Commission shall provide affected employees or representatives of affected employees an opportunity to participate as parties to hearings under this subsection.

80 Stat. 384.

JUDICIAL REVIEW

SEC. 11. (a) Any person adversely affected or aggrieved by an order of the Commission issued under subsection (c) of section 10 may obtain a review of such order in any United States court of appeals for the circuit in which the violation is alleged to have occurred or where the employer has its principal office, or in the Court of Appeals for the District of Columbia Circuit, by filing in such court within sixty days following the issuance of such order a written petition praying that the order be modified or set aside. A copy of such petition shall be forthwith transmitted by the clerk of the court to the Commission and to the other parties, and thereupon the Commission shall file in the court the record in the proceeding as provided in section 2112 of title 28, United States Code. Upon such filing, the court shall have jurisdiction of the proceeding and of the question determined therein, and shall have power to grant such temporary relief or restraining order as it deems just and proper, and to make and enter upon the pleadings, testimony, and proceedings set forth in such record a decree affirming, modifying, or setting aside in whole or in part, the order of the Commission and enforcing the same to the extent that such order is affirmed or modified. The commencement of proceedings under this subsection shall not, unless ordered by the court, operate as a stay of the order of the Commission. No objection that has not been urged before the Commission shall be considered by the court, unless the failure or neglect to urge such objection shall be excused because of extraordinary circumstances. The findings of the Commission with respect to questions of fact, if supported by substantial evidence on the record considered as a whole, shall be conclusive. If any party shall apply to the court for leave to adduce additional evidence and shall show to the satisfaction of the court that such additional evidence is material and that there were reasonable grounds for the failure to adduce such evidence in the hearing before the Commission, the court may order such additional evidence to be taken before the Commission and to be made a part of the record. The Commission may modify its findings as to the facts, or make new findings, by reason of additional evidence so taken and filed, and it shall file such modified or new findings, which findings with respect to questions of fact, if supported by substantial evi-

72 Stat. 941;
80 Stat. 1323.

dence on the record considered as a whole, shall be conclusive, and its recommendations, if any, for the modification or setting aside of its original order. Upon the filing of the record with it, the jurisdiction of the court shall be exclusive and its judgment and decree shall be final, except that the same shall be subject to review by the Supreme Court of the United States, as provided in section 1254 of title 28, United States Code. Petitions filed under this subsection shall be heard expeditiously.

62 Stat. 928.

(b) The Secretary may also obtain review or enforcement of any final order of the Commission by filing a petition for such relief in the United States court of appeals for the circuit in which the alleged violation occurred or in which the employer has its principal office, and the provisions of subsection (a) shall govern such proceedings to the extent applicable. If no petition for review, as provided in subsection (a), is filed within sixty days after service of the Commission's order, the Commission's findings of fact and order shall be conclusive in connection with any petition for enforcement which is filed by the Secretary after the expiration of such sixty-day period. In any such case, as well as in the case of a noncontested citation or notification by the Secretary which has become a final order of the Commission under subsection (a) or (b) of section 10, the clerk of the court, unless otherwise ordered by the court, shall forthwith enter a decree enforcing the order and shall transmit a copy of such decree to the Secretary and the employer named in the petition. In any contempt proceeding brought to enforce a decree of a court of appeals entered pursuant to this subsection or subsection (a), the court of appeals may assess the penalties provided in section 17, in addition to invoking any other available remedies.

(c) (1) No person shall discharge or in any manner discriminate against any employee because such employee has filed any complaint or instituted or caused to be instituted any proceeding under or related to this Act or has testified or is about to testify in any such proceeding or because of the exercise by such employee on behalf of himself or others of any right afforded by this Act.

(2) Any employee who believes that he has been discharged or otherwise discriminated against by any person in violation of this subsection may, within thirty days after such violation occurs, file a complaint with the Secretary alleging such discrimination. Upon receipt of such complaint, the Secretary shall cause such investigation to be made as he deems appropriate. If upon such investigation, the Secretary determines that the provisions of this subsection have been violated, he shall bring an action in any appropriate United States district court against such person. In any such action the United States district courts shall have jurisdiction, for cause shown to restrain violations of paragraph (1) of this subsection and order all appropriate relief including rehiring or reinstatement of the employee to his former position with back pay.

(3) Within 90 days of the receipt of a complaint filed under the subsection the Secretary shall notify the complainant of his determination under paragraph 2 of this subsection.

THE OCCUPATIONAL SAFETY AND HEALTH REVIEW COMMISSION

Establishment; membership.

SEC. 12. (a) The Occupational Safety and Health Review Commission is hereby established. The Commission shall be composed of three members who shall be appointed by the President, by and with the advice and consent of the Senate, from among persons who by reason of training, education, or experience are qualified to carry out the functions of the Commission under this Act. The President shall designate one of the members of the Commission to serve as Chairman.

(b) The terms of members of the Commission shall be six years except that (1) the members of the Commission first taking office shall serve, as designated by the President at the time of appointment, one for a term of two years, one for a term of four years, and one for a term of six years, and (2) a vacancy caused by the death, resignation, or removal of a member prior to the expiration of the term for which he was appointed shall be filled only for the remainder of such unexpired term. A member of the Commission may be removed by the President for inefficiency, neglect of duty, or malfeasance in office.

Terms.

(c)(1) Section 5314 of title 5, United States Code, is amended by adding at the end thereof the following new paragraph:

80 Stat. 460.

"(57) Chairman, Occupational Safety and Health Review Commission."

(2) Section 5315 of title 5, United States Code, is amended by adding at the end thereof the following new paragraph:

Ante, p. 776.

"(94) Members, Occupational Safety and Health Review Commission."

(d) The principal office of the Commission shall be in the District of Columbia. Whenever the Commission deems that the convenience of the public or of the parties may be promoted, or delay or expense may be minimized, it may hold hearings or conduct other proceedings at any other place.

Location.

(e) The Chairman shall be responsible on behalf of the Commission for the administrative operations of the Commission and shall appoint such hearing examiners and other employees as he deems necessary to assist in the performance of the Commission's functions and to fix their compensation in accordance with the provisions of chapter 51 and subchapter III of chapter 53 of title 5, United States Code, relating to classification and General Schedule pay rates: *Provided*, That assignment, removal and compensation of hearing examiners shall be in accordance with sections 3105, 3344, 5362, and 7521 of title 5, United States Code.

5 USC 5101, 5331.
Ante, p. 198-1.

(f) For the purpose of carrying out its functions under this Act, two members of the Commission shall constitute a quorum and official action can be taken only on the affirmative vote of at least two members.

Quorum.

(g) Every official act of the Commission shall be entered of record, and its hearings and records shall be open to the public. The Commission is authorized to make such rules as are necessary for the orderly transaction of its proceedings. Unless the Commission has adopted a different rule, its proceedings shall be in accordance with the Federal Rules of Civil Procedure.

Public records.
28 USC app.

(h) The Commission may order testimony to be taken by deposition in any proceedings pending before it at any state of such proceeding. Any person may be compelled to appear and depose, and to produce books, papers, or documents, in the same manner as witnesses may be compelled to appear and testify and produce like documentary evidence before the Commission. Witnesses whose depositions are taken under this subsection, and the persons taking such depositions, shall be entitled to the same fees as are paid for like services in the courts of the United States.

(i) For the purpose of any proceeding before the Commission, the provisions of section 11 of the National Labor Relations Act (29 U.S.C. 161) are hereby made applicable to the jurisdiction and powers of the Commission.

61 Stat. 150;
Ante, p. 930.

Report.

(j) A hearing examiner appointed by the Commission shall hear, and make a determination upon, any proceeding instituted before the Commission and any motion in connection therewith, assigned to such hearing examiner by the Chairman of the Commission, and shall make a report of any such determination which constitutes his final disposition of the proceedings. The report of the hearing examiner shall become the final order of the Commission within thirty days after such report by the hearing examiner, unless within such period any Commission member has directed that such report shall be reviewed by the Commission.

(k) Except as otherwise provided in this Act, the hearing examiners shall be subject to the laws governing employees in the classified civil service, except that appointments shall be made without regard to section 5108 of title 5, United States Code. Each hearing examiner shall receive compensation at a rate not less than that prescribed for GS-16 under section 5332 of title 5, United States Code.

80 Stat. 453.

Ante, p. 198-1.

PROCEDURES TO COUNTERACT IMMINENT DANGERS

SEC. 13. (a) The United States district courts shall have jurisdiction, upon petition of the Secretary, to restrain any conditions or practices in any place of employment which are such that a danger exists which could reasonably be expected to cause death or serious physical harm immediately or before the imminence of such danger can be eliminated through the enforcement procedures otherwise provided by this Act. Any order issued under this section may require such steps to be taken as may be necessary to avoid, correct, or remove such imminent danger and prohibit the employment or presence of any individual in locations or under conditions where such imminent danger exists, except individuals whose presence is necessary to avoid, correct, or remove such imminent danger or to maintain the capacity of a continuous-process operation to resume normal operations without a complete cessation of operations, or where a cessation of operations is necessary, to permit such to be accomplished in a safe and orderly manner.

(b) Upon the filing of any such petition the district court shall have jurisdiction to grant such injunctive relief or temporary restraining order pending the outcome of an enforcement proceeding pursuant to this Act. The proceeding shall be as provided by Rule 65 of the Federal Rules, Civil Procedure, except that no temporary restraining order issued without notice shall be effective for a period longer than five days.

28 USC app.

(c) Whenever and as soon as an inspector concludes that conditions or practices described in subsection (a) exist in any place of employment, he shall inform the affected employees and employers of the danger and that he is recommending to the Secretary that relief be sought.

(d) If the Secretary arbitrarily or capriciously fails to seek relief under this section, any employee who may be injured by reason of such failure, or the representative of such employees, might bring an action against the Secretary in the United States district court for the district in which the imminent danger is alleged to exist or the employer has its principal office, or for the District of Columbia, for a writ of mandamus to compel the Secretary to seek such an order and for such further relief as may be appropriate.

REPRESENTATION IN CIVIL LITIGATION

SEC. 14. Except as provided in section 518(a) of title 28, United States Code, relating to litigation before the Supreme Court, the Solicitor of Labor may appear for and represent the Secretary in any civil litigation brought under this Act but all such litigation shall be subject to the direction and control of the Attorney General.

80 Stat. 613.

CONFIDENTIALITY OF TRADE SECRETS

SEC. 15. All information reported to or otherwise obtained by the Secretary or his representative in connection with any inspection or proceeding under this Act which contains or which might reveal a trade secret referred to in section 1905 of title 18 of the United States Code shall be considered confidential for the purpose of that section, except that such information may be disclosed to other officers or employees concerned with carrying out this Act or when relevant in any proceeding under this Act. In any such proceeding the Secretary, the Commission, or the court shall issue such orders as may be appropriate to protect the confidentiality of trade secrets.

62 Stat. 791.

VARIATIONS, TOLERANCES, AND EXEMPTIONS

SEC. 16. The Secretary, on the record, after notice and opportunity for a hearing may provide such reasonable limitations and may make such rules and regulations allowing reasonable variations, tolerances, and exemptions to and from any or all provisions of this Act as he may find necessary and proper to avoid serious impairment of the national defense. Such action shall not be in effect for more than six months without notification to affected employees and an opportunity being afforded for a hearing.

PENALTIES

SEC. 17. (a) Any employer who willfully or repeatedly violates the requirements of section 5 of this Act, any standard, rule, or order promulgated pursuant to section 6 of this Act, or regulations prescribed pursuant to this Act, may be assessed a civil penalty of not more than $10,000 for each violation.

(b) Any employer who has received a citation for a serious violation of the requirements of section 5 of this Act, of any standard, rule, or order promulgated pursuant to section 6 of this Act, or of any regulations prescribed pursuant to this Act, shall be assessed a civil penalty of up to $1,000 for each such violation.

(c) Any employer who has received a citation for a violation of the requirements of section 5 of this Act, of any standard, rule, or order promulgated pursuant to section 6 of this Act, or of regulations prescribed pursuant to this Act, and such violation is specifically determined not to be of a serious nature, may be assessed a civil penalty of up to $1,000 for each such violation.

(d) Any employer who fails to correct a violation for which a citation has been issued under section 9(a) within the period permitted for its correction (which period shall not begin to run until the date of the final order of the Commission in the case of any review proceeding under section 10 initiated by the employer in good faith and not solely for delay or avoidance of penalties), may be assessed a civil penalty of not more than $1,000 for each day during which such failure or violation continues.

(e) Any employer who willfully violates any standard, rule, or order promulgated pursuant to section 6 of this Act, or of any regulations prescribed pursuant to this Act, and that violation caused death to any employee, shall, upon conviction, be punished by a fine of not more than $10,000 or by imprisonment for not more than six months, or by both; except that if the conviction is for a violation committed after a first conviction of such person, punishment shall be by a fine of not more than $20,000 or by imprisonment for not more than one year, or by both.

(f) Any person who gives advance notice of any inspection to be conducted under this Act, without authority from the Secretary or his designees, shall, upon conviction, be punished by a fine of not more than $1,000 or by imprisonment for not more than six months, or by both.

(g) Whoever knowingly makes any false statement, representation, or certification in any application, record, report, plan, or other document filed or required to be maintained pursuant to this Act shall, upon conviction, be punished by a fine of not more than $10,000, or by imprisonment for not more than six months, or by both.

65 Stat. 721; 79 Stat. 234.

(h)(1) Section 1114 of title 18, United States Code, is hereby amended by striking out "designated by the Secretary of Health, Education, and Welfare to conduct investigations, or inspections under the Federal Food, Drug, and Cosmetic Act" and inserting in lieu thereof "or of the Department of Labor assigned to perform investigative, inspection, or law enforcement functions".

62 Stat. 756.

(2) Notwithstanding the provisions of sections 1111 and 1114 of title 18, United States Code, whoever, in violation of the provisions of section 1114 of such title, kills a person while engaged in or on account of the performance of investigative, inspection, or law enforcement functions added to such section 1114 by paragraph (1) of this subsection, and who would otherwise be subject to the penalty provisions of such section 1111, shall be punished by imprisonment for any term of years or for life.

(i) Any employer who violates any of the posting requirements, as prescribed under the provisions of this Act, shall be assessed a civil penalty of up to $1,000 for each violation.

(j) The Commission shall have authority to assess all civil penalties provided in this section, giving due consideration to the appropriateness of the penalty with respect to the size of the business of the employer being charged, the gravity of the violation, the good faith of the employer, and the history of previous violations.

(k) For purposes of this section, a serious violation shall be deemed to exist in a place of employment if there is a substantial probability that death or serious physical harm could result from a condition which exists, or from one or more practices, means, methods, operations, or processes which have been adopted or are in use, in such place of employment unless the employer did not, and could not with the exercise of reasonable diligence, know of the presence of the violation.

(l) Civil penalties owed under this Act shall be paid to the Secretary for deposit into the Treasury of the United States and shall accrue to the United States and may be recovered in a civil action in the name of the United States brought in the United States district court for the district where the violation is alleged to have occurred or where the employer has its principal office.

STATE JURISDICTION AND STATE PLANS

SEC. 18. (a) Nothing in this Act shall prevent any State agency or court from asserting jurisdiction under State law over any occupational safety or health issue with respect to which no standard is in effect under section 6.

(b) Any State which, at any time, desires to assume responsibility for development and enforcement therein of occupational safety and health standards relating to any occupational safety or health issue with respect to which a Federal standard has been promulgated under section 6 shall submit a State plan for the development of such standards and their enforcement.

(c) The Secretary shall approve the plan submitted by a State under subsection (b), or any modification thereof, if such plan in his judgment—

(1) designates a State agency or agencies as the agency or agencies responsible for administering the plan throughout the State,

(2) provides for the development and enforcement of safety and health standards relating to one or more safety or health issues, which standards (and the enforcement of which standards) are or will be at least as effective in providing safe and healthful employment and places of employment as the standards promulgated under section 6 which relate to the same issues, and which standards, when applicable to products which are distributed or used in interstate commerce, are required by compelling local conditions and do not unduly burden interstate commerce,

(3) provides for a right of entry and inspection of all workplaces subject to the Act which is at least as effective as that provided in section 8, and includes a prohibition on advance notice of inspections,

(4) contains satisfactory assurances that such agency or agencies have or will have the legal authority and qualified personnel necessary for the enforcement of such standards,

(5) gives satisfactory assurances that such State will devote adequate funds to the administration and enforcement of such standards,

(6) contains satisfactory assurances that such State will, to the extent permitted by its law, establish and maintain an effective and comprehensive occupational safety and health program applicable to all employees of public agencies of the State and its political subdivisions, which program is as effective as the standards contained in an approved plan,

(7) requires employers in the State to make reports to the Secretary in the same manner and to the same extent as if the plan were not in effect, and

(8) provides that the State agency will make such reports to the Secretary in such form and containing such information, as the Secretary shall from time to time require.

(d) If the Secretary rejects a plan submitted under subsection (b), he shall afford the State submitting the plan due notice and opportunity for a hearing before so doing.

Notice of hearing.

(e) After the Secretary approves a State plan submitted under subsection (b), he may, but shall not be required to, exercise his authority under sections 8, 9, 10, 13, and 17 with respect to comparable standards promulgated under section 6, for the period specified in the next sentence. The Secretary may exercise the authority referred to above until he determines, on the basis of actual operations under the

State plan, that the criteria set forth in subsection (c) are being applied, but he shall not make such determination for at least three years after the plan's approval under subsection (c). Upon making the determination referred to in the preceding sentence, the provisions of sections 5(a)(2), 8 (except for the purpose of carrying out subsection (f) of this section), 9, 10, 13, and 17, and standards promulgated under section 6 of this Act, shall not apply with respect to any occupational safety or health issues covered under the plan, but the Secretary may retain jurisdiction under the above provisions in any proceeding commenced under section 9 or 10 before the date of determination.

Continuing evaluation.

(f) The Secretary shall, on the basis of reports submitted by the State agency and his own inspections make a continuing evaluation of the manner in which each State having a plan approved under this section is carrying out such plan. Whenever the Secretary finds, after affording due notice and opportunity for a hearing, that in the administration of the State plan there is a failure to comply substantially with any provision of the State plan (or any assurance contained therein), he shall notify the State agency of his withdrawal of approval of such plan and upon receipt of such notice such plan shall cease to be in effect, but the State may retain jurisdiction in any case commenced before the withdrawal of the plan in order to enforce standards under the plan whenever the issues involved do not relate to the reasons for the withdrawal of the plan.

Plan rejection, review.

(g) The State may obtain a review of a decision of the Secretary withdrawing approval of or rejecting its plan by the United States court of appeals for the circuit in which the State is located by filing in such court within thirty days following receipt of notice of such decision a petition to modify or set aside in whole or in part the action of the Secretary. A copy of such petition shall forthwith be served upon the Secretary, and thereupon the Secretary shall certify and file in the court the record upon which the decision complained of was issued as provided in section 2112 of title 28, United States Code. Unless the court finds that the Secretary's decision in rejecting a proposed State plan or withdrawing his approval of such a plan is not supported by substantial evidence the court shall affirm the Secretary's decision. The judgment of the court shall be subject to review by the Supreme Court of the United States upon certiorari or certification as provided in section 1254 of title 28, United States Code.

72 Stat. 941; 80 Stat. 1323.

62 Stat. 928.

(h) The Secretary may enter into an agreement with a State under which the State will be permitted to continue to enforce one or more occupational health and safety standards in effect in such State until final action is taken by the Secretary with respect to a plan submitted by a State under subsection (b) of this section, or two years from the date of enactment of this Act, whichever is earlier.

FEDERAL AGENCY SAFETY PROGRAMS AND RESPONSIBILITIES

SEC. 19. (a) It shall be the responsibility of the head of each Federal agency to establish and maintain an effective and comprehensive occupational safety and health program which is consistent with the standards promulgated under section 6. The head of each agency shall (after consultation with representatives of the employees thereof)—

(1) provide safe and healthful places and conditions of employment, consistent with the standards set under section 6;

(2) acquire, maintain, and require the use of safety equipment, personal protective equipment, and devices reasonably necessary to protect employees;

(3) keep adequate records of all occupational accidents and illnesses for proper evaluation and necessary corrective action; Recordkeeping.

(4) consult with the Secretary with regard to the adequacy as to form and content of records kept pursuant to subsection (a) (3) of this section; and

(5) make an annual report to the Secretary with respect to occupational accidents and injuries and the agency's program under this section. Such report shall include any report submitted under section 7902(e)(2) of title 5, United States Code. Annual report. 80 Stat. 530.

(b) The Secretary shall report to the President a summary or digest of reports submitted to him under subsection (a)(5) of this section, together with his evaluations of and recommendations derived from such reports. The President shall transmit annually to the Senate and the House of Representatives a report of the activities of Federal agencies under this section. Report to President. Report to Congress.

(c) Section 7902(c)(1) of title 5, United States Code, is amended by inserting after "agencies" the following: "and of labor organizations representing employees".

(d) The Secretary shall have access to records and reports kept and filed by Federal agencies pursuant to subsections (a) (3) and (5) of this section unless those records and reports are specifically required by Executive order to be kept secret in the interest of the national defense or foreign policy, in which case the Secretary shall have access to such information as will not jeopardize national defense or foreign policy. Records, etc.; availability.

RESEARCH AND RELATED ACTIVITIES

SEC. 20. (a)(1) The Secretary of Health, Education, and Welfare, after consultation with the Secretary and with other appropriate Federal departments or agencies, shall conduct (directly or by grants or contracts) research, experiments, and demonstrations relating to occupational safety and health, including studies of psychological factors involved, and relating to innovative methods, techniques, and approaches for dealing with occupational safety and health problems.

(2) The Secretary of Health, Education, and Welfare shall from time to time consult with the Secretary in order to develop specific plans for such research, demonstrations, and experiments as are necessary to produce criteria, including criteria identifying toxic substances, enabling the Secretary to meet his responsibility for the formulation of safety and health standards under this Act; and the Secretary of Health, Education, and Welfare, on the basis of such research, demonstrations, and experiments and any other information available to him, shall develop and publish at least annually such criteria as will effectuate the purposes of this Act.

(3) The Secretary of Health, Education, and Welfare, on the basis of such research, demonstrations, and experiments, and any other information available to him, shall develop criteria dealing with toxic materials and harmful physical agents and substances which will describe exposure levels that are safe for various periods of employment, including but not limited to the exposure levels at which no employee will suffer impaired health or functional capacities or diminished life expectancy as a result of his work experience.

(4) The Secretary of Health, Education, and Welfare shall also conduct special research, experiments, and demonstrations relating to occupational safety and health as are necessary to explore new problems, including those created by new technology in occupational safety and health, which may require ameliorative action beyond that

which is otherwise provided for in the operating provisions of this Act. The Secretary of Health, Education, and Welfare shall also conduct research into the motivational and behavioral factors relating to the field of occupational safety and health.

Toxic substances, records.

(5) The Secretary of Health, Education, and Welfare, in order to comply with his responsibilities under paragraph (2), and in order to develop needed information regarding potentially toxic substances or harmful physical agents, may prescribe regulations requiring employers to measure, record, and make reports on the exposure of employees to substances or physical agents which the Secretary of Health, Education, and Welfare reasonably believes may endanger the health or safety of employees. The Secretary of Health, Education, and Welfare also is authorized to establish such programs of medical examinations and tests as may be necessary for determining the incidence of occupational illnesses and the susceptibility of employees to such illnesses. Nothing in this or any other provision of this Act shall be deemed to authorize or require medical examination, immunization, or treatment for those who object thereto on religious grounds, except where such is necessary for the protection of the health or safety of others. Upon the request of any employer who is required to measure and record exposure of employees to substances or physical agents as provided under this subsection, the Secretary of Health, Education, and Welfare shall furnish full financial or other assistance to such employer for the purpose of defraying any additional expense incurred by him in carrying out the measuring and recording as provided in this subsection.

Medical examinations.

Toxic substances, publication.

(6) The Secretary of Health, Education, and Welfare shall publish within six months of enactment of this Act and thereafter as needed but at least annually a list of all known toxic substances by generic family or other useful grouping, and the concentrations at which such toxicity is known to occur. He shall determine following a written request by any employer or authorized representative of employees, specifying with reasonable particularity the grounds on which the request is made, whether any substance normally found in the place of employment has potentially toxic effects in such concentrations as used or found; and shall submit such determination both to employers and affected employees as soon as possible. If the Secretary of Health, Education, and Welfare determines that any substance is potentially toxic at the concentrations in which it is used or found in a place of employment, and such substance is not covered by an occupational safety or health standard promulgated under section 6, the Secretary of Health, Education, and Welfare shall immediately submit such determination to the Secretary, together with all pertinent criteria.

Annual studies.

(7) Within two years of enactment of this Act, and annually thereafter the Secretary of Health, Education, and Welfare shall conduct and publish industrywide studies of the effect of chronic or low-level exposure to industrial materials, processes, and stresses on the potential for illness, disease, or loss of functional capacity in aging adults.

Inspections.

(b) The Secretary of Health, Education, and Welfare is authorized to make inspections and question employers and employees as provided in section 8 of this Act in order to carry out his functions and responsibilities under this section.

Contract authority.

(c) The Secretary is authorized to enter into contracts, agreements, or other arrangements with appropriate public agencies or private organizations for the purpose of conducting studies relating to his responsibilities under this Act. In carrying out his responsibilities under this subsection, the Secretary shall cooperate with the Secretary of Health, Education, and Welfare in order to avoid any duplication of efforts under this section.

(d) Information obtained by the Secretary and the Secretary of Health, Education, and Welfare under this section shall be disseminated by the Secretary to employers and employees and organizations thereof.

(e) The functions of the Secretary of Health, Education, and Welfare under this Act shall, to the extent feasible, be delegated to the Director of the National Institute for Occupational Safety and Health established by section 22 of this Act.

Delegation of functions.

TRAINING AND EMPLOYEE EDUCATION

SEC. 21. (a) The Secretary of Health, Education, and Welfare, after consultation with the Secretary and with other appropriate Federal departments and agencies, shall conduct, directly or by grants or contracts (1) education programs to provide an adequate supply of qualified personnel to carry out the purposes of this Act, and (2) informational programs on the importance of and proper use of adequate safety and health equipment.

(b) The Secretary is also authorized to conduct, directly or by grants or contracts, short-term training of personnel engaged in work related to his responsibilities under this Act.

(c) The Secretary, in consultation with the Secretary of Health, Education, and Welfare, shall (1) provide for the establishment and supervision of programs for the education and training of employers and employees in the recognition, avoidance, and prevention of unsafe or unhealthful working conditions in employments covered by this Act, and (2) consult with and advise employers and employees, and organizations representing employers and employees as to effective means of preventing occupational injuries and illnesses.

NATIONAL INSTITUTE FOR OCCUPATIONAL SAFETY AND HEALTH

SEC. 22. (a) It is the purpose of this section to establish a National Institute for Occupational Safety and Health in the Department of Health, Education, and Welfare in order to carry out the policy set forth in section 2 of this Act and to perform the functions of the Secretary of Health, Education, and Welfare under sections 20 and 21 of this Act.

Establishment.

(b) There is hereby established in the Department of Health, Education, and Welfare a National Institute for Occupational Safety and Health. The Institute shall be headed by a Director who shall be appointed by the Secretary of Health, Education, and Welfare, and who shall serve for a term of six years unless previously removed by the Secretary of Health, Education, and Welfare.

Director, appointment, term.

(c) The Institute is authorized to—

(1) develop and establish recommended occupational safety and health standards; and

(2) perform all functions of the Secretary of Health, Education, and Welfare under sections 20 and 21 of this Act.

(d) Upon his own initiative, or upon the request of the Secretary or the Secretary of Health, Education, and Welfare, the Director is authorized (1) to conduct such research and experimental programs as he determines are necessary for the development of criteria for new and improved occupational safety and health standards, and (2) after consideration of the results of such research and experimental programs make recommendations concerning new or improved occupational safety and health standards. Any occupational safety and health standard recommended pursuant to this section shall immediately be forwarded to the Secretary of Labor, and to the Secretary of Health, Education, and Welfare.

(e) In addition to any authority vested in the Institute by other provisions of this section, the Director, in carrying out the functions of the Institute, is authorized to—

(1) prescribe such regulations as he deems necessary governing the manner in which its functions shall be carried out;

(2) receive money and other property donated, bequeathed, or devised, without condition or restriction other than that it be used for the purposes of the Institute and to use, sell, or otherwise dispose of such property for the purpose of carrying out its functions;

(3) receive (and use, sell, or otherwise dispose of, in accordance with paragraph (2)), money and other property donated, bequeathed, or devised to the Institute with a condition or restriction, including a condition that the Institute use other funds of the Institute for the purposes of the gift;

(4) in accordance with the civil service laws, appoint and fix the compensation of such personnel as may be necessary to carry out the provisions of this section;

(5) obtain the services of experts and consultants in accordance with the provisions of section 3109 of title 5, United States Code;

80 Stat. 416.

(6) accept and utilize the services of voluntary and noncompensated personnel and reimburse them for travel expenses, including per diem, as authorized by section 5703 of title 5, United States Code;

83 Stat. 190.

(7) enter into contracts, grants or other arrangements, or modifications thereof to carry out the provisions of this section, and such contracts or modifications thereof may be entered into without performance or other bonds, and without regard to section 3709 of the Revised Statutes, as amended (41 U.S.C. 5), or any other provision of law relating to competitive bidding;

(8) make advance, progress, and other payments which the Director deems necessary under this title without regard to the provisions of section 3648 of the Revised Statutes, as amended (31 U.S.C. 529); and

(9) make other necessary expenditures.

Annual report to HEW, President, and Congress.

(f) The Director shall submit to the Secretary of Health, Education, and Welfare, to the President, and to the Congress an annual report of the operations of the Institute under this Act, which shall include a detailed statement of all private and public funds received and expended by it, and such recommendations as he deems appropriate.

GRANTS TO THE STATES

SEC. 23. (a) The Secretary is authorized, during the fiscal year ending June 30, 1971, and the two succeeding fiscal years, to make grants to the States which have designated a State agency under section 18 to assist them—

(1) in identifying their needs and responsibilities in the area of occupational safety and health,

(2) in developing State plans under section 18, or

(3) in developing plans for—

(A) establishing systems for the collection of information concerning the nature and frequency of occupational injuries and diseases;

(B) increasing the expertise and enforcement capabilities of their personnel engaged in occupational safety and health programs; or

(C) otherwise improving the administration and enforcement of State occupational safety and health laws, including standards thereunder, consistent with the objectives of this Act.

(b) The Secretary is authorized, during the fiscal year ending June 30, 1971, and the two succeeding fiscal years, to make grants to the States for experimental and demonstration projects consistent with the objectives set forth in subsection (a) of this section.

(c) The Governor of the State shall designate the appropriate State agency for receipt of any grant made by the Secretary under this section.

(d) Any State agency designated by the Governor of the State desiring a grant under this section shall submit an application therefor to the Secretary.

(e) The Secretary shall review the application, and shall, after consultation with the Secretary of Health, Education, and Welfare, approve or reject such application.

(f) The Federal share for each State grant under subsection (a) or (b) of this section may not exceed 90 per centum of the total cost of the application. In the event the Federal share for all States under either such subsection is not the same, the differences among the States shall be established on the basis of objective criteria.

(g) The Secretary is authorized to make grants to the States to assist them in administering and enforcing programs for occupational safety and health contained in State plans approved by the Secretary pursuant to section 18 of this Act. The Federal share for each State grant under this subsection may not exceed 50 per centum of the total cost to the State of such a program. The last sentence of subsection (f) shall be applicable in determining the Federal share under this subsection.

(h) Prior to June 30, 1973, the Secretary shall, after consultation with the Secretary of Health, Education, and Welfare, transmit a report to the President and to the Congress, describing the experience under the grant programs authorized by this section and making any recommendations he may deem appropriate.

Report to President and Congress.

STATISTICS

SEC. 24. (a) In order to further the purposes of this Act, the Secretary, in consultation with the Secretary of Health, Education, and Welfare, shall develop and maintain an effective program of collection, compilation, and analysis of occupational safety and health statistics. Such program may cover all employments whether or not subject to any other provisions of this Act but shall not cover employments excluded by section 4 of the Act. The Secretary shall compile accurate statistics on work injuries and illnesses which shall include all disabling, serious, or significant injuries and illnesses, whether or not involving loss of time from work, other than minor injuries requiring only first aid treatment and which do not involve medical treatment, loss of consciousness, restriction of work or motion, or transfer to another job.

(b) To carry out his duties under subsection (a) of this section, the Secretary may—

(1) promote, encourage, or directly engage in programs of studies, information and communication concerning occupational safety and health statistics;

(2) make grants to States or political subdivisions thereof in order to assist them in developing and administering programs dealing with occupational safety and health statistics; and

(3) arrange, through grants or contracts, for the conduct of such research and investigations as give promise of furthering the objectives of this section.

(c) The Federal share for each grant under subsection (b) of this section may be up to 50 per centum of the State's total cost.

(d) The Secretary may, with the consent of any State or political subdivision thereof, accept and use the services, facilities, and employees of the agencies of such State or political subdivision, with or without reimbursement, in order to assist him in carrying out his functions under this section.

Reports.

(e) On the basis of the records made and kept pursuant to section 8(c) of this Act, employers shall file such reports with the Secretary as he shall prescribe by regulation, as necessary to carry out his functions under this Act.

(f) Agreements between the Department of Labor and States pertaining to the collection of occupational safety and health statistics already in effect on the effective date of this Act shall remain in effect until superseded by grants or contracts made under this Act.

AUDITS

SEC. 25. (a) Each recipient of a grant under this Act shall keep such records as the Secretary or the Secretary of Health, Education, and Welfare shall prescribe, including records which fully disclose the amount and disposition by such recipient of the proceeds of such grant, the total cost of the project or undertaking in connection with which such grant is made or used, and the amount of that portion of the cost of the project or undertaking supplied by other sources, and such other records as will facilitate an effective audit.

(b) The Secretary or the Secretary of Health, Education, and Welfare, and the Comptroller General of the United States, or any of their duly authorized representatives, shall have access for the purpose of audit and examination to any books, documents, papers, and records of the recipients of any grant under this Act that are pertinent to any such grant.

ANNUAL REPORT

SEC. 26. Within one hundred and twenty days following the convening of each regular session of each Congress, the Secretary and the Secretary of Health, Education, and Welfare shall each prepare and submit to the President for transmittal to the Congress a report upon the subject matter of this Act, the progress toward achievement of the purpose of this Act, the needs and requirements in the field of occupational safety and health, and any other relevant information. Such reports shall include information regarding occupational safety and health standards, and criteria for such standards, developed during the preceding year; evaluation of standards and criteria previously developed under this Act, defining areas of emphasis for new criteria and standards; an evaluation of the degree of observance of applicable occupational safety and health standards, and a summary of inspection and enforcement activity undertaken; analysis and evaluation of research activities for which results have been obtained under governmental and nongovernmental sponsorship; an analysis of major occupational diseases; evaluation of available control and measurement technology for hazards for which standards or criteria have been developed during the preceding year; description of cooperative efforts undertaken between Government agencies and other interested parties in the implementation of this Act during the preceding year; a progress report on the development of an adequate supply of trained manpower in the field of occupational safety and health, including estimates of future needs and the efforts being made by Government and others to meet those needs; listing of all toxic substances in industrial usage for which labeling requirements, criteria, or standards have not yet been established; and such recommendations for additional legislation as are deemed necessary to protect the safety and health of the worker and improve the administration of this Act.

NATIONAL COMMISSION ON STATE WORKMEN'S COMPENSATION LAWS

SEC. 27. (a) (1) The Congress hereby finds and declares that—

(A) the vast majority of American workers, and their families, are dependent on workmen's compensation for their basic economic security in the event such workers suffer disabling injury or death in the course of their employment; and that the full protection of American workers from job-related injury or death requires an adequate, prompt, and equitable system of workmen's compensation as well as an effective program of occupational health and safety regulation; and

(B) in recent years serious questions have been raised concerning the fairness and adequacy of present workmen's compensation laws in the light of the growth of the economy, the changing nature of the labor force, increases in medical knowledge, changes in the hazards associated with various types of employment, new technology creating new risks to health and safety, and increases in the general level of wages and the cost of living.

(2) The purpose of this section is to authorize an effective study and objective evaluation of State workmen's compensation laws in order to determine if such laws provide an adequate, prompt, and equitable system of compensation for injury or death arising out of or in the course of employment.

(b) There is hereby established a National Commission on State Workmen's Compensation Laws.

Establishment.

(c) (1) The Workmen's Compensation Commission shall be composed of fifteen members to be appointed by the President from among members of State workmen's compensation boards, representatives of insurance carriers, business, labor, members of the medical profession having experience in industrial medicine or in workmen's compensation cases, educators having special expertise in the field of workmen's compensation, and representatives of the general public. The Secretary, the Secretary of Commerce, and the Secretary of Health, Education, and Welfare shall be ex officio members of the Workmen's Compensation Commission:

Membership.

(2) Any vacancy in the Workmen's Compensation Commission shall not affect its powers.

(3) The President shall designate one of the members to serve as Chairman and one to serve as Vice Chairman of the Workmen's Compensation Commission.

Quorum.

(4) Eight members of the Workmen's Compensation Commission shall constitute a quorum.

Study.

(d)(1) The Workmen's Compensation Commission shall undertake a comprehensive study and evaluation of State workmen's compensation laws in order to determine if such laws provide an adequate, prompt, and equitable system of compensation. Such study and evaluation shall include, without being limited to, the following subjects: (A) the amount and duration of permanent and temporary disability benefits and the criteria for determining the maximum limitations thereon, (B) the amount and duration of medical benefits and provisions insuring adequate medical care and free choice of physician, (C) the extent of coverage of workers, including exemptions based on numbers or type of employment, (D) standards for determining which injuries or diseases should be deemed compensable, (E) rehabilitation, (F) coverage under second or subsequent injury funds, (G) time limits on filing claims, (H) waiting periods, (I) compulsory or elective coverage, (J) administration, (K) legal expenses, (L) the feasibility and desirability of a uniform system of reporting information concerning job-related injuries and diseases and the operation of workmen's compensation laws, (M) the resolution of conflict of laws, extraterritoriality and similar problems arising from claims with multistate aspects, (N) the extent to which private insurance carriers are excluded from supplying workmen's compensation coverage and the desirability of such exclusionary practices, to the extent they are found to exist, (O) the relationship between workmen's compensation on the one hand, and old-age, disability, and survivors insurance and other types of insurance, public or private, on the other hand, (P) methods of implementing the recommendations of the Commission.

Report to President and Congress.

(2) The Workmen's Compensation Commission shall transmit to the President and to the Congress not later than July 31, 1972, a final report containing a detailed statement of the findings and conclusions of the Commission, together with such recommendations as it deems advisable.

Hearings.

(e)(1) The Workmen's Compensation Commission or, on the authorization of the Workmen's Compensation Commission, any subcommittee or members thereof, may, for the purpose of carrying out the provisions of this title, hold such hearings, take such testimony, and sit and act at such times and places as the Workmen's Compensation Commission deems advisable. Any member authorized by the Workmen's Compensation Commission may administer oaths or affirmations to witnesses appearing before the Workmen's Compensation Commission or any subcommittee or members thereof.

(2) Each department, agency, and instrumentality of the executive branch of the Government, including independent agencies, is authorized and directed to furnish to the Workmen's Compensation Commission, upon request made by the Chairman or Vice Chairman, such information as the Workmen's Compensation Commission deems necessary to carry out its functions under this section.

(f) Subject to such rules and regulations as may be adopted by the Workmen's Compensation Commission, the Chairman shall have the power to—

(1) appoint and fix the compensation of an executive director, and such additional staff personnel as he deems necessary, without regard to the provisions of title 5, United States Code, governing appointments in the competitive service, and without regard to the provisions of chapter 51 and subchapter III of chapter 53 of such title relating to classification and General Schedule

80 Stat. 378.
5 USC 101.

5 USC 5101, 5331.

pay rates, but at rates not in excess of the maximum rate for GS-18 of the General Schedule under section 5332 of such title, and

Ante, p. 198-1.

(2) procure temporary and intermittent services to the same extent as is authorized by section 3109 of title 5, United States Code.

80 Stat. 416.

Contract authorization.

(g) The Workmen's Compensation Commission is authorized to enter into contracts with Federal or State agencies, private firms, institutions, and individuals for the conduct of research or surveys, the preparation of reports, and other activities necessary to the discharge of its duties.

Compensation; travel expenses.

(h) Members of the Workmen's Compensation Commission shall receive compensation for each day they are engaged in the performance of their duties as members of the Workmen's Compensation Commission at the daily rate prescribed for GS-18 under section 5332 of title 5, United States Code, and shall be entitled to reimbursement for travel, subsistence, and other necessary expenses incurred by them in the performance of their duties as members of the Workmen's Compensation Commission.

Appropriation.

(i) There are hereby authorized to be appropriated such sums as may be necessary to carry out the provisions of this section.

Termination.

(j) On the ninetieth day after the date of submission of its final report to the President, the Workmen's Compensation Commission shall cease to exist.

ECONOMIC ASSISTANCE TO SMALL BUSINESSES

72 Stat. 387; 83 Stat. 802. 15 USC 636.

SEC. 28. (a) Section 7(b) of the Small Business Act, as amended, is amended—

(1) by striking out the period at the end of "paragraph (5)" and inserting in lieu thereof "; and"; and

(2) by adding after paragraph (5) a new paragraph as follows:

"(6) to make such loans (either directly or in cooperation with banks or other lending institutions through agreements to participate on an immediate or deferred basis) as the Administration may determine to be necessary or appropriate to assist any small business concern in effecting additions to or alterations in the equipment, facilities, or methods of operation of such business in order to comply with the applicable standards promulgated pursuant to section 6 of the Occupational Safety and Health Act of 1970 or standards adopted by a State pursuant to a plan approved under section 18 of the Occupational Safety and Health Act of 1970, if the Administration determines that such concern is likely to suffer substantial economic injury without assistance under this paragraph."

(b) The third sentence of section 7(b) of he Small Business Act, as amended, is amended by striking out "or (5)" after "paragraph (3)" and inserting a comma followed by "(5) or (6)".

80 Stat. 132. 15 USC 633.

(c) Section 4(c)(1) of the Small Business Act, as amended, is amended by inserting "7(b)(6)," after "7(b)(5),".

(d) Loans may also be made or guaranteed for the purposes set forth in section 7(b)(6) of the Small Business Act, as amended, pursuant to the provisions of section 202 of the Public Works and Economic Development Act of 1965, as amended.

79 Stat. 556. 42 USC 3142.

ADDITIONAL ASSISTANT SECRETARY OF LABOR

SEC. 29. (a) Section 2 of the Act of April 17, 1946 (60 Stat. 91) as amended (29 U.S.C. 553) is amended by—

75 Stat. 338.

(1) striking out "four" in the first sentence of such section and inserting in lieu thereof "five"; and

(2) adding at the end thereof the following new sentence, "One of such Assistant Secretaries shall be an Assistant Secretary of Labor for Occupational Safety and Health.".

80 Stat. 462.

(b) Paragraph (20) of section 5315 of title 5, United States Code, is amended by striking out "(4)" and inserting in lieu thereof "(5)".

ADDITIONAL POSITIONS

SEC. 30. Section 5108(c) of title 5, United States Code, is amended by—

(1) striking out the word "and" at the end of paragraph (8);

(2) striking out the period at the end of paragraph (9) and inserting in lieu thereof a semicolon and the word "and"; and

(3) by adding immediately after paragraph (9) the following new paragraph:

"(10) (A) the Secretary of Labor, subject to the standards and procedures prescribed by this chapter, may place an additional twenty-five positions in the Department of Labor in GS–16, 17, and 18 for the purposes of carrying out his responsibilities under the Occupational Safety and Health Act of 1970;

"(B) the Occupational Safety and Health Review Commission, subject to the standards and procedures prescribed by this chapter, may place ten positions in GS–16, 17, and 18 in carrying out its functions under the Occupational Safety and Health Act of 1970."

EMERGENCY LOCATOR BEACONS

72 Stat. 775.
49 USC 1421.

SEC. 31. Section 601 of the Federal Aviation Act of 1958 is amended by inserting at the end thereof a new subsection as follows:

"EMERGENCY LOCATOR BEACONS

"(d) (1) Except with respect to aircraft described in paragraph (2) of this subsection, minimum standards pursuant to this section shall include a requirement that emergency locator beacons shall be installed—

"(A) on any fixed-wing, powered aircraft for use in air commerce the manufacture of which is completed, or which is imported into the United States, after one year following the date of enactment of this subsection; and

"(B) on any fixed-wing, powered aircraft used in air commerce after three years following such date.

"(2) The provisions of this subsection shall not apply to jet-powered aircraft; aircraft used in air transportation (other than air taxis and charter aircraft); military aircraft; aircraft used solely for training purposes not involving flights more than twenty miles from its base; and aircraft used for the aerial application of chemicals."

SEPARABILITY

SEC. 32. If any provision of this Act, or the application of such provision to any person or circumstance, shall be held invalid, the remainder of this Act, or the application of such provision to persons or circumstances other than those as to which it is held invalid, shall not be affected thereby.

APPROPRIATIONS

SEC. 33. There are authorized to be appropriated to carry out this Act for each fiscal year such sums as the Congress shall deem necessary.

EFFECTIVE DATE

SEC. 34. This Act shall take effect one hundred and twenty days after the date of its enactment.

Approved December 29, 1970.

LEGISLATIVE HISTORY:

HOUSE REPORTS: No. 91-1291 accompanying H.R. 16785 (Comm. on Education and Labor) and No. 91-1765 (Comm. of Conference).
SENATE REPORT No. 91-1282 (Comm. on Labor and Public Welfare).
CONGRESSIONAL RECORD, Vol. 116 (1970):
Oct. 13, Nov. 16, 17, considered and passed Senate.
Nov. 23, 24, considered and passed House, amended, in lieu of H.R. 16785.
Dec. 16, Senate agreed to conference report.
Dec. 17, House agreed to conference report.

U.S. GOVERNMENT PRINTING OFFICE : 1979 O - 288-147

APPENDIX B
Sectional Conversion Table The Act to U.S.C.

Act: Section	29 U.S.C.: Section
1	—
2	651
2(b)(1)–(b)(13)	651(1)–(13)
3	652
4	653
5	654
6	655
7	656
8	657
9	658
10	659
11	660
12(a),(b)	661(a),(b)
12(c)	—
12(d)	661(c)
12(e)	661(d)
12(f)	661(e)
12(g)	661(f)
12(h)	661(g)
12(i)	661(h)
12(j)	661(i)
12(k)	661(j)
13	662
14	663
15	664

Act: Section	29 U.S.C.: Section
16	665
17(a)	666(a)
17(b)–(g)	666(b)–(g)
17(h)	—
17(i)	666(h)
17(j)	666(i)
17(k)	666(j)
17(l)	666(k)
18	667
19(a),(b)	668(a),(b)
19(c)	—
19(d)	668(c)
20	669
21	670
22	671
23	672
24	673
25	674
26	675
27	676
28–31	—
32	677
33	678
34	—

APPENDIX C
The Review Commission's Rules of Procedure

Subpart A—General Provisions

Rule 1 Definitions

As used herein:

(a) "Act" means the Occupational Safety and Health Act of 1970, 84 Stat. 1590, 29 U.S.C.A. 651, *et seq.*

(b) "Commission," "person," "employer," and "employee" have the meanings set forth in §3 of the Act.

(c) "Secretary" means the Secretary of Labor or his duly authorized representative.

(d) "Executive Secretary" means the Executive Secretary of the Commission.

(e) "Affected employee" means an employee of a cited employer who is exposed to the alleged hazard described in the citation, as a result of his assigned duties.

(f) "Judge" means a Hearing Examiner appointed by the Chairman of the Commission pursuant to §12(j) of the Act.

(g) "Authorized employee representative" means a labor organization which has a collective bargaining relationship with the cited employer and which represents affected employees.

(h) "Representative" means any person, including an authorized employee representative, authorized by a party or intervenor to represent him in a proceeding.

(i) "Citation" means a written communication issued by the Secretary to an employer pursuant to §9(a) of the Act.

(j) "Notification of proposed penalty" means a written communication issued by the Secretary to an employer pursuant to §10(a) or (b) of the Act.

(k) "Day" means a calendar day.

(l) "Working day" means all days except Saturdays, Sundays, or Federal Holidays.

(m) "Proceeding" means any proceeding before the Commission or before a Judge.

Rule 2 Scope of Rules; applicability of Federal Rules of Civil Procedure

(a) These rules shall govern all proceedings before the Commission and its Judges.

(b) In the absence of a specific provision, procedure shall be in accordance with the Federal Rules of Civil Procedure.

Rule 3 Use of gender and number

(a) Words importing the singular number may extend and be applied to the plural and vice versa.

(b) Words importing the masculine gender may be applied to the feminine gender.

Rule 4 Computation of time

(a) In computing any period of time prescribed or allowed in these rules, the day from which the designated period begins to run shall not be included. The last day of the period so computed shall be included unless it is a Saturday, Sunday, or Federal holiday, in which event the period runs until the end of the next day which is not a Saturday, Sunday, or federal holiday. When the period of time prescribed or allowed is less than 7 days, intermediate Saturdays, Sundays, and federal holidays shall be excluded in the computation.

(b) Where service of a pleading or document is by mail pursuant to §2200.7, 3 days shall be added to the time allowed by these rules for the filing of a responsive pleading.

Rule 5 Extensions of time

Requests for extensions of time for the filing of any pleading or document must be received in advance of the date on which the pleading or document is due to be filed.

Rule 6 Record address

The initial pleading filed by any person shall contain his name, address, and telephone number. Any change in such information must be communicated promptly in writing to the Judge or the Commission, as the case may be, and to all other parties and intervenors. A party or intervenor who fails to furnish such information shall be deemed to have waived his right to notice and service under these rules.

Rule 7 Service and notice

(a) At the time of filing pleadings or other documents, a copy thereof shall be served by the filing party or intervenor on every other party or intervenor.

(b) Service upon a party or intervenor who has appeared through a representative shall be made only upon such representative.

(c) Unless otherwise ordered, service may be accomplished by postage prepaid first-class mail or personal delivery. Service is deemed effected at the time of mailing (if by mail) or at the time of personal delivery (if by personal delivery).

(d) Proof of service shall be accomplished by a written statement of the same which sets forth the date and manner of service. Such statement shall be filed with the pleading or document.

(e) Where service is accomplished by posting, proof of such posting shall be filed not later than the first working day following the posting.

(f) Service and notice to employees represented by an authorized employee representative shall be deemed accomplished by serving the representative in the manner prescribed in paragraph (c) of this section.

(g) In the event that there are any affected employees who are not represented by an authorized employee representative, the employer shall, immediately upon receipt of notice of the docketing of the notice of contest or petition for modification of the abatement period, post, where the citation is required to be posted, a copy of the notice of contest and a notice informing such affected employees of their right to party status and of the availability of all pleadings for inspection and copying at reasonable times. A notice in the following form shall be deemed to comply with this paragraph:

[Name of employer]

Your employer has been cited by the Secretary of Labor for violation of the Occupational Safety and Health Act of 1970. The citation has been contested and will be the subject of a hearing before the OCCUPATIONAL SAFETY AND HEALTH REVIEW COMMISSION. Affected employees are entitled to participate in this hearing as parties under terms and conditions established by the OCCUPATIONAL SAFETY AND HEALTH REVIEW COMMISSION in its Rules of Procedure. Notice of intent to participate should be sent to:

Occupational Safety and Health
Review Commission
1825 K Street, N.W.
Washington, D.C. 20006

All papers relevant to this matter may be inspected at:

[Place reasonably convenient to employees, preferably at or near workplace].

Where appropriate, the second sentence of the above notice will be deleted and the following sentence will be substituted:

The reasonableness of the period prescribed by the Secretary of Labor for abatement of the violation has been contested and will be the subject of a hearing before the OCCUPATIONAL SAFETY AND HEALTH REVIEW COMMISSION.

(h) The authorized employee representative, if any, shall be served with the notice set forth in paragraph (g) of this rule and with a copy of the notice of contest.

(i) A copy of the notice of the hearing to be held before the Judge shall be served by the employer on affected employees who are not represented by an authorized employee representative by posting a copy of the notice of such hearing at or near the place where the citation is required to be posted.

(j) A copy of the notice of the hearing to be held before the Judge shall be served by the employer on the authorized employee representative of affected employees in the manner prescribed in paragraph (c) of this rule, if the employer has not been informed that the authorized employee representative has entered an appearance as of the date such notice is received by the employer.

(k) Where a notice of contest is filed by an affected employee who is not represented by an authorized employee representative and there are other affected employees who are represented by an authorized employee representative, the unrepresented employee shall, upon receipt of the statement filed in conformance with §2200.35, serve a copy thereof on such authorized employee representative in the manner prescribed in paragraph (c) of this rule and shall file proof of such service.

(l) Where a notice of contest is filed by an affected employee or an authorized employee representative, a copy of the notice of contest and response filed in support thereof shall be provided to the employer for posting in the manner prescribed in paragraph (g) of this rule.

(m) An authorized employee representative who files a notice of contest shall be responsible for serving any other authorized employee representative whose members are affected employees.

(n) Where posting is required by this section, such posting shall be maintained until the commencement of the hearing or until earlier disposition.

Rule 8 Filing

(a) Prior to the assignment of a case to a Judge, all papers shall be filed with the Executive Secretary at 1825 K Street, N.W., Washington, D.C. 20006. Subsequent to the assignment of the case to a Judge, and before the issuance of his decision, all papers shall be filed with the Judge at the address given in the notice informing of such assignment. Subsequent to the issuance of the decision of the Judge, all papers shall be filed with the Executive Secretary.

(b) Unless otherwise ordered, all filing may be accomplished by first class mail.

(c) Filing is deemed effected at the time of mailing.

Rule 9 Consolidation

Cases may be consolidated on the motion of any party, on the Judge's own motion, or on the Commission's own motion, where there exist common parties, common questions of law or fact, or both, or in such other circumstances as justice and the administration of the Act require.

Rule 10 Severance

Upon its own motion, or upon motion of any party or invervenor, the Commission or the Judge may, for good cause, order any proceeding severed with respect to some or all issues or parties.

Rule 11 Protection of trade secrets and other confidential information

Upon application by any person, in a proceeding where trade secrets or other matters may be divulged the confidentiality of which is protected by 18 U.S.C. 1905, the Judge shall issue such orders as may be appropriate to protect the confidentiality of such matters.

Subpart B—Parties and Representatives

Rule 20 Party status

(a) Affected employees may elect to participate as parties at any time before the commencement of the hearing before the Judge, unless, for good cause shown, the Commission or the Judge allows such election at a later time. See also §2200.21.

(b) Where a notice of contest is filed by an employee or by an authorized employee representative with respect to the reasonableness of the period for abatement of a violation, the employer charged with the responsibility of abating the violation may elect party status at any time before the commencement of the hearing before the Judge. See also §2200.21.

Rule 21 Intervention; appearance by non-parties

(a) A petition for leave to intervene may be filed at any stage of a proceeding before commencement of the hearing before the Judge.

(b) The petition shall set forth the interest of the petitioner in the proceeding and show that the participation of the petitioner will assist in the determination of the issues in question, and that the intervention will not unnecessarily delay the proceeding.

(c) The Commission or the Judge may grant a petition for intervention to such an extent and upon such terms as the Commission or the Judge shall determine.

Rule 22 Representatives of parties and intervenors

(a) Any party or intervenor may appear in person or through a representative.

(b) A representative of a party or intervenor shall be deemed to control all matters respecting the interest of such party or intervenor in the proceeding.

(c) Affected employees who are represented by an authorized employee representative may appear only through such authorized employee representative.

(d) Nothing contained herein shall be construed to require any representative to be an attorney at law.

(e) Withdrawal of appearance of any representative may be affected by filing a written notice of withdrawal and by serving a copy thereof on all parties and intervenors.

Subpart C—Pleadings and Motions

Rule 30 Form

(a) Except as provided herein, there are no specific requirements as to the form of any pleading. A pleading is simply required to contain a caption sufficient to identify the parties in accordance with §2200.31, which shall include the Commission's docket number, if assigned, and a clear and plain statement of the relief that is sought, together with the grounds therefor.

(b) Pleadings and other documents (other than exhibits) shall be typewritten, double spaced, on letter size opaque paper (approximately 8½ inches by 11 inches). The left margin shall be 1½ inches and the right margin 1 inch. Pleadings and other documents shall be fastened at the upper left corner.

(c) Pleadings shall be signed by the party filing or by his representative. Such signing constitutes a representation by the signer that he has read the document or pleading, that to the best of his knowledge, information, and belief the statements made therein are true, and that it is not interposed for delay.

(d) When a court decision is cited in which the first-listed parties on each side are the Secretary of Labor (or the name of a particular Secretary of Labor) and the Commission, the citation shall include in parenthesis the name of the respondent in the Commission proceeding. For example: *Brennan v. OSHRC (Vy Lactos Laboratories, Inc.)*, 494 F.2d 460 (8th Cir. 1974).

(e) The Commission may refuse for filing any pleading or document which does not comply with the requirements of paragraphs (a), (b), (c) and (d) of this section.

Rule 31 Caption; titles of cases

(a) Cases initiated by a notice of contest shall be titled:

Secretary of Labor,
Complainant
v.
(Name of Contestant),
Respondent.

(b) Cases initiated by a petition for modification of the abatement period shall be titled:

(Name of employer),
Petitioner
v.
Secretary of Labor,
Respondent.

(c) The titles listed in paragraphs (a) and (b) of this rule shall appear at the left upper portion of the initial page of any pleading or document (other than exhibits) filed.

(d) The initial page of any pleading or document (other than exhibits) shall show, at the upper right of the page, opposite the title, the docket number, if known, assigned by the Commission.

Rule 32 Notices of contest

The Secretary shall, within 7 days of receipt of a notice of contest, transmit the original to the Commission, together with copies of all relevant documents.

Rule 33 Employer contests

(a) Complaint

(1) The Secretary shall file a complaint with the Commission no later than 20 days after his receipt of the notice of contest.

(2) The complaint shall set forth all alleged violations and proposed penalties which are contested, stating with particularity:

(i) The basis for jurisdiction;

(ii) The time, location, place, and circumstances of each such alleged violation; and

(iii) The considerations upon which the period for abatement and the proposed penalty on each such alleged violation is based.

(3) Where the Secretary seeks in his complaint to amend his citation or proposed penalty, he shall set forth the reasons for amendment and shall state with particularity the change sought.

(b) Answer

(1) Within 15 days after service of the complaint, the party against whom the complaint was issued shall file an answer with the Commission.

(2) The answer shall contain a short and plain statement denying those allegations in the complaint which the party intends to contest. Any allegation not denied shall be deemed admitted.

Rule 34 Petitions for modification of abatement period

(a) An employer may file a petition for modification of abatement date when such employer has made a good faith effort to comply with the abatement requirements of a citation, but such abatement has not been completed because of factors beyond the employer's reasonable control.

(b) A petition for modification of abatement date shall be in writing and shall include the following information:

(1) All steps taken by the employer, and the dates of such action, in an effort to achieve compliance during the prescribed abatement period.

(2) The specific additional abatement time necessary in order to achieve compliance.

(3) The reasons such additional time is necessary including the unavailability of professional or technical personnel or of materials and equipment, or because necessary construction or alteration of facilities cannot be completed by the original abatement date.

(4) All available interim steps being taken to safeguard the employees against the cited hazard during the abatement period.

(c) A petition for modification of abatement date shall be filed with the Area Director of the United States Department of Labor who issued the citation no later than the close of the next working day following the date on which abatement was originally required. A later-filed petition shall be accompanied by the employer's statement of exceptional circumstances explaining the delay.

(1) A copy of such petition shall be posted in a conspicuous place where all affected employees will have notice thereof or near each location where the violation occurred. The petition shall remain posted for a period of ten (10) days.

(2) Affected employees or their representatives may file an objection in writing

to such petition with the aforesaid Area Director. Failure to file such objection within ten (10) working days of the date of posting of such petition shall constitute a waiver of any further right to object to said petition.

(3) The Secretary or his duly authorized agent shall have the authority to approve any petition for modification of abatement date filed pursuant to subparagraphs (b) and (c). Such uncontested petitions shall become final orders pursuant to sections 10(a) and (c) of the Act.

(4) The Secretary or his authorized representative shall not exercise his approval power until the expiration of fifteen (15) working days from the date the petition was posted pursuant to paragraphs (c)(1) and (2) by the employer.

(d) Where any petition is objected to by the Secretary or affected employees, such petition shall be processed as follows:

(1) The petition, citation, and any objections shall be forwarded to the Commission within three (3) working days after the expiration of the fifteen (15) day period set out in paragraph (c)(4).

(2) The Commission shall docket and process such petitions as expedited proceedings as provided for in §2200.101 of this Part.

(3) An employer petitioning for a modification of abatement period shall have the burden of proving in accordance with the requirements of 29 U.S.C. §659(c), that such employer has made a good faith effort to comply with the abatement requirements of the citation and that abatement has not been completed because of factors beyond the employer's control.

(4) Within ten (10) working days after the receipt of notice of the docketing by the Commission of any petition for modification of abatement date, each objecting party shall file a response setting forth the reasons for opposing the granting of a modification date different from that requested in the petition.

Rule 35 Employee contests

(a) Where an affected employee or authorized employee representative files a notice of contest with respect to the abatement period, the Secretary shall, within 10 days from his receipt of the notice of contest, file a clear and concise statement of the reasons the abatement period prescribed by him is not unreasonable.

(b) Not later than 10 days after receipt of the statement referred to in paragraph (a) of this rule, the contestant shall file a response.

(c) All contests under this section shall be handled as expedited proceedings as provided for in §2200.101 of this Part.

Rule 36 Statement of position

At any time prior to the commencement of the hearing before the Judge, any person entitled to appear as a party or any person who has been granted leave to intervene, may file a statement of position with respect to any or all issues to be heard.

Rule 37 Response to motions

Any party or intervenor upon whom a motion is served shall have 10 days from service of the motion to file a response.

Rule 38 Failure to file

Failure to file any pleading pursuant to these rules when due, may, in the discretion of the Commission or the Judge, constitute a waiver of the right to further participation in the proceedings.

Subpart D—Pre-hearing Procedures and Discovery

Rule 51 Pre-hearing conference

(a) At any time before a hearing, the Commission or the Judge, on their own motion or on the motion of a party, may direct the parties or their representatives to exchange information or to participate in a prehearing conference to consider settlement or matters which will tend to simplify issues or expedite the hearing.

(b) The Commission or the Judge may issue a pre-hearing order which includes the agreements reached by the parties. Such order shall be served on all parties and shall be a part of the record.

Rule 52 Requests for admissions

(a) At any time after the filing of responsive pleadings, any party may request of any other party admissions of facts to be made under oath. Each admission requested shall be set forth separately. The matter shall be deemed admitted unless, within 15 days after service of the request, or within such shorter or longer time as the Commission or the Judge may prescribe, the party to whom the request is directed serves upon the party requesting the admission a specific written response.

(b) Copies of all requests and responses shall be served on all parties in accordance with the provisions of §2200.7(a) and filed with the Commission within the time allotted and shall be a part of the record.

Rule 53 Discovery depositions and interrogatories

(a) Except by special order of the Commission or the Judge, discovery depositions of parties, intervenors, or witnesses, and interrogatories directed to parties, intervenors, or witnesses shall not be allowed.

(b) In the event the Commission or the Judge grants an application for the conduct of such discovery proceedings, the order granting the same shall set forth appropriate time limits governing the discovery.

Rule 54 Failure to comply with orders for discovery

If any party or intervenor fails to comply with an order of the Commission or the Judge to permit discovery in accordance with the provisions of these rules, the Commission or the Judge may issue appropriate orders.

Rule 55 Issuance of subpoenas; petitions to revoke or modify subpoenas; right to inspect or copy data

(a) Any member of the Commission shall, on the application of any party directed to the Commission, forthwith issue subpoenas requiring the attendance and testimony of witnesses and the production of any evidence, including relevant books, records, correspondence, or documents in his possession or under his control. Applications for subpoenas, if filed subsequent to the assignment of the case to a Judge, shall be filed with the Judge. A Judge shall grant the application on behalf of any member of the Commisison. Applications for subpoenas may be made *ex parte.* The subpoena shall show on its face the name and address of the party at whose request the subpoena was issued.

(b) Any person served with a subpoena, whether *ad testificandum* or *duces tecum,* shall, within 5 days after the date of service of the subpoena upon him, move in writing to revoke or modify the subpoena if he does not intend to comply. All motions to revoke or modify shall be served on the party at whose request the subpoena was issued. The Judge or the Commission, as the case may be, shall revoke or modify the subpoena if, in its opinion, the evidence whose production is required does not relate to any matter under investigation or in question in the proceedings, or the subpoena does not describe with sufficient particularity the evidence whose production is required, or if for any other reason sufficient in law the subpoena is otherwise invalid. The Judge or the Commission, as the case may be, shall make a simple statement of procedural or other grounds for the ruling on the motion to revoke or modify. The motion to revoke or modify any answer filed thereto, and any ruling thereon shall become a part of the record.

(c) Persons compelled to submit data or evidence at a public proceeding are entitled to retain, or on payment of lawfully prescribed costs, to procure copies of transcripts of the data or evidence submitted by them.

(d) Upon the failure of any person to comply with a subpoena issued upon the request of a party, the Commission by its counsel shall initiate proceedings in the appropriate district court for the enforcement thereof, if in its judgment the enforcement of such subpoena would be consistent with law and with policies of the Act. Neither the Commission nor its counsel shall be deemed thereby to have assumed responsibility for the effective prosecution of the same before the Court.

Subpart E—Hearings

Rule 60 Notice of hearing

Notice of the time, place, and nature of a hearing shall be given to the parties and intervenors at least 10 days in advance of such hearings, except as otherwise provided in §2200.101.

Rule 61 Postponement of hearing

(a) Postponement of a hearing ordinarily will not be allowed.

(b) Except in the case of an extreme emergency or in unusual circumstances, no such request will be considered unless received in writing at least 3 days in advance of the time set for the hearing.

(c) No postponement in excess of 30 days shall be allowed without Commission approval.

Rule 62 Failure to appear

(a) Subject to the provisions of paragraph (c) of this rule, the failure of a party to appear at a hearing shall be deemed to be a waiver of all rights except the rights to be served with a copy of the decision of the Judge and to request Commission review pursuant to §2200.91.

(b) Requests for reinstatement must be made, in the absence of extraordinary circumstances, within 5 days after the scheduled hearing date.

(c) The Commission or the Judge, upon a showing of good cause, may excuse such failure to appear. In such event, the hearing will be rescheduled.

Rule 63 Payment of witness fees and mileage; fees of persons taking depositions

Witnesses summoned before the Commission or the Judge shall be paid the same fees and mileage that are paid witnesses in the courts of the United States, and witnesses whose depositions are taken and the persons taking the same shall severally be entitled to the same fees as are paid for like services in the courts of the United States. Witness fees and mileage shall be paid by the party at whose instance the witness appears, and the person taking a deposition shall be paid by the party at whose instance the deposition is taken.

Rule 64 Reporter's fees

Reporter's fees shall be borne by the Commission, except as provided in §2200.63.

Rule 65 Transcript of testimony

Hearings shall be transcribed verbatim. A copy of the transcript of testimony taken at the hearing, duly certified by the reporter, shall be filed with the Judge before whom the matter was heard. The Judge shall promptly serve notice upon each of the parties and intervenors of such filing.

Rule 66 Duties and powers of Judges

It shall be the duty of the Judge to conduct a fair and impartial hearing, to assure that the facts are fully elicited, to adjudicate all issues and avoid delay. The Judge shall have authority with respect to cases assigned to him, between the time he is designated and the time he issues his decision, subject to the rules and regulations of the Commission, to:

(a) Administer oaths and affirmations;

(b) Issue authorized subpoenas;

(c) Rule upon petitions to revoke subpoenas;

(d) Rule upon offers of proof and receive relevant evidence;

(e) Take or cause depositions to be taken whenever the needs of justice would be served;

(f) Regulate the course of the hearing, and, if appropriate or necessary, exclude persons or counsel from the hearing for contemptuous conduct and strike all related testimony of witnesses refusing to answer any proper question;

(g) Hold conferences for the settlement or simplification of the issues;

(h) Dispose of procedural requests or similar matters, including motions referred to the Judge by the Commission and motions to amend pleadings; also to dismiss complaints or portions thereof, and to order hearings reopened or, upon motion, consolidated prior to issuance of his report;

(i) Make decisions in conformity with §557 of title 5, United States Code;

(j) Call and examine witnesses and to introduce into the record documentary or other evidence;

(k) Request the parties at any time during the hearing to state their respective positions concerning any issue in the case or theory in support thereof;

(l) Adjourn the hearing as the needs of justice and good administration require;

(m) Take any other action necessary under the foregoing and authorized by the published rules and regulations of the Commission.

Rule 67 Disqualification of Judge

(a) A Judge may withdraw from a proceeding whenever he deems himself disqualified.

(b) Any party may request the Judge, at any time following his designation and before the filing of his decision, to withdraw on ground of personal bias or disqualification, by filing with him promptly upon the discovery of the alleged facts an affidavit setting forth in detail the matters alleged to constitute grounds for disqualification.

(c) If, in the opinion of the Judge, the affidavit referred to in paragraph (b) of this rule is filed with due diligence and is sufficient on its face, the Judge shall forthwith disqualify himself and withdraw from the proceeding.

(d) If the Judge does not disqualify himself and withdraw from the proceeding, he shall so rule upon the record, stating the grounds for his ruling and shall proceed with the hearing, or, if the hearing has closed, he shall proceed with the issuance of his decision, and the provisions of §2200.90 shall thereupon apply.

Rule 68 Examination of witnesses

Witnesses shall be examined orally under oath. Opposing parties shall have the right to cross-examine any witness whose testimony is introduced by an adverse party.

Rule 69 Affidavits

An affidavit may be admitted as evidence in lieu of oral testimony if the matters therein contained are otherwise admissible and the parties agree to its admission.

Rule 70 Deposition in lieu of oral testimony; application; procedures, form; rulings

(a) An application to take the deposition of a witness in lieu of oral testimony shall be in writing and shall set forth the reasons such deposition should be taken, the name and address of the witness, the matters concerning which it is expected he will testify and the time and place proposed for the taking of the deposition, together with the name and address of the person before whom it is desired that the deposition be taken (for purposes of this section, hereinafter referred to as "the officer"). Such application shall be filed with the Commission or the Judge, as the case may be, and shall be served on all other parties and intervenors not less than 7 days (when the deposition is to be taken within the continental United States) and not less than 15 days (if the deposition is to be taken elsewhere) prior to the time when it is desired that the deposition be taken. Where good cause has been shown, the Commission or the Judge shall make and serve on the parties and intervenors an order which specifies the name of the witness whose deposition is to be taken and the time, place, and designation of the officer before whom the witness is to testify. Such officer may or may not be the officer specified in the application.

(b) Such deposition may be taken before any officer authorized to administer oaths by the laws of the United States or of the place where the examination is held. If the examination is held in a foreign country, it may be taken before any secretary of embassy or legation, consul general, consul, vice consul, or consular agent of the United States.

(c) At the time and place specified in the order, the officer designated to take such deposition shall permit the witness to be examined and cross-examined under oath by all parties appearing, and the testimony of the witness shall be reduced to typewriting by the officer or under his direction. All objections to questions or evidence shall be deemed waived unless made at the examination. The officer shall not have power to rule upon any objection, but he shall note them upon the deposition. The testimony shall be subscribed by the witness in the presence of the officer who shall attach his certificate stating that the witness was duly sworn by him, that the deposition is a true record of the testimony and exhibits given by the witness, and that the officer is not of counsel or attorney to any of the parties nor interested in the proceeding. If the deposition is not signed by the witness because he is ill, dead, cannot be found, or refuses to sign it, such fact shall be included in the certificate of the officer and the deposition may be used as fully as though signed. The officer shall immediately deliver an original and four copies of the transcript, together with his certificate, in person or by registered mail to the Executive Secretary at 1825 K Street, N.W., Washington, D.C. 20006.

(d) The Judge shall rule upon the admissibility of the deposition or any part thereof.

(e) All errors or irregularities in compliance with the provision of this section shall be deemed waived unless a motion to suppress the deposition or some part thereof is made with reasonable promptness after such defect is, or with due diligence might have been, discovered.

(f) If the parties so stipulate in writing, depositions may be taken before any person at any time or place, upon any notice and in any manner, and when so taken may be used as other depositions.

Rule 71 Exhibits

(a) All exhibits offered in evidence shall be numbered and marked with a designation identifying the party or intervenor by whom the exhibit is offered.

(b) In the absence of objection by another party or intervenor, exhibits shall be admitted into evidence as a part of the record, unless excluded by the Judge pursuant to §2200.72.

(c) Unless the Judge finds it impractical, a copy of each such exhibit shall be given to the other parties and intervenors.

(d) All exhibits offered, but denied admission into evidence, shall be identified as in paragraph (a) of this rule and shall be placed in a separate file designated for rejected exhibits.

Rule 72 Rules of evidence

Hearings before the Commission and its Judges shall be in accordance with §554 of Title 5 U.S.C. and insofar as practicable shall be governed by the rules of evidence applicable in the United States District Courts.

Rule 73 Burden of proof

(a) In all proceedings commenced by the filing of a notice of contest, the burden of proof shall rest with the Secretary.

(b) In proceedings commenced by a petition for modification of the abatement period, the burden of establishing the necessity for such modification shall rest with the petitioner.

Rule 74 Objections

(a) Any objection with respect to the conduct of the hearing, including any objection to the introduction of evidence or a ruling by the Judge may be stated orally or in writing, accompanied by a short statement of the grounds for the objection, and shall be included in the record. No such objection shall be deemed waived by further participation in the hearing.

(b) Whenever evidence is excluded from the record, the party offering such evidence may make an offer of proof, which shall be included in the record of the proceedings.

Rule 75 Interlocutory appeals

(a) Generally

A Judge's interlocutory ruling may be appealed to the Commission only in the manner prescribed by this rule.

(b) Certification

A party desiring to appeal from an interlocutory ruling shall file with the Judge a written request for certification of the appeal. The request and supporting documents shall be filed within 5 days after receipt of the Judge's ruling from which appeal is sought. Responses to the request, if any, shall be filed within 5 days after service of the request. The Judge shall certify an interlocutory appeal when the ruling involves an important question of law or policy about which there is substantial ground for difference of opinion and an immediate appeal of the ruling may materially expedite the proceedings.

(1) *Procedure after certification* Following certification, the Judge shall forward to the Executive Secretary the request for certification and supporting documents, responses filed by the other parties, the ruling from which appeal is taken, a copy of relevant portions of the record, and the Judge's order certifying the appeal.

(2) *Acceptance of certification-discretionary* The Commission at any time may decline to accept a certification.

(c) Petition for interlocutory appeal

Within 5 days following the receipt of a Judge's order denying certification, a party may file with the Commission a petition for interlocutory appeal. Responses to the petition, if any, shall be filed within 5 days following service of the petition. The Commission shall grant a petition for interlocutory appeal only in exceptional circumstances where it finds (1) that the appeal satisfies the criteria for certification of an appeal set forth in paragraph (b) of this section; and (2) that there is a substantial probability of reversal.

(d) Denial without prejudice

The Commission's action in declining to accept a certification or denying a petition for interlocutory appeal shall not preclude a party from raising an objection to the Judge's interlocutory ruling in a petition for discretionary review. A party whose request for certification of an interlocutory appeal is denied by a Judge and who elects not to file a petition for interlocutory appeal with the Commission shall not be precluded from raising in a petition for discretionary review an objection to the ruling from which interlocutory appeal was sought.

(e) Stay

(1) *Trade secret matters* The filing with a Judge of a request to certify an interlocutory appeal of a ruling concerning an alleged trade secret shall stay the effect of the ruling: (i) until the Judge denies the request; or (ii) if the request is granted, until the Commission rules on the appeal or declines to accept the certification. In the event such a request is denied, the Judge, upon motion of the requesting party, shall stay for a period of 5 days the effect of the ruling from which appeal was sought in order to allow the party to petition the commission for interlocutory appeal of the ruling. The filing with the Commission of a petition for interlocutory appeal of a ruling concerning an alleged trade secret shall stay the effect of the ruling until the Commission denies the petition or rules on the appeal.

(2) *Other cases* In all other cases, the filing or granting of a request to certify an interlocutory appeal or the filing or granting of a petition for interlocutory appeal, shall not stay a proceeding or the effect of a ruling unless otherwise ordered.

(f) Briefs

Should the Commission desire briefs on the issues raised by an interlocurory appeal, it shall give notice to the parties. See §2200.93 Briefs before the Commission.

Ruling 76 Filing of briefs and proposed findings with the Judge; oral argument at the hearing

Any party shall be entitled, upon request, to a reasonable period at the close of the hearing for oral argument, which shall be included in the stenographic report of the hearing. Any party shall be entitled, upon request made before the close of the hearing, to file a brief, proposed findings of fact and conclusions of law, or both, with the Judge. The Judge may fix a reasonable period of time for such filing, but such initial period may not exceed 20 days from the receipt by the party of the transcript of the hearing.

Subpart F—Post-hearing Procedures

Rule 90 Decisions and reports of Judges

(a) Upon completion of any proceeding, the Judge shall prepare a decision. When a hearing is held the decision shall comply with 5 U.S.C. 557. Copies of the decision shall be mailed to all parties. Thereafter, the Judge shall file with the Executive Secretary a report consisting of his decision, the record in support thereof, and any petitions for discretionary review of his decision, or statements in opposition to such petitions, that may be filed in accordance with §2200.91. The Judge shall file his report on the day following the close of the period for filing petitions for discretionary review, or statements in opposition to such petitions, but no later than the twenty-first day following the date of the mailing of the decision to the parties.

(b) (1) Promptly upon receipt of the Judge's report, the Executive Secretary shall docket the case and notify all parties of that fact. The date of docketing shall be the date that the Judge's report is made for purposes of section 12(j) of the Act (29 U.S.C. 661).

(2) On or after the date of docketing of the case, all pleadings or other documents that may be filed in the case shall be addressed to the Executive Secretary.

(3) In the event no Commission Member directs review of a decision on or before the thirtieth day following the date of docketing of the Judge's report, the decision of the Judge contained therein shall become a final order of the Commission.

Rule 91 Discretionary review; petitions for; statements in opposition

(a) A party aggrieved by the decision of a judge may submit a petition for discretionary review. An aggrieved party that fails to file a petition for such review by the Commission may be foreclosed from court review of any objection to the judge's decision. *Keystone Roofing Co., Inc. v. Dunlop*, 539 F.2d 960 (3rd Cir. 1976).

(b) (1) Except as provided in paragraphs (b)(2) and (3) of this section, any petition must be received by the Judge at his office on or before the twentieth day following his mailing of a copy of the decision to the parties.

(2) When there is no objection by any party, when an expedited proceeding has been directed pursuant to §2200.101, or for other good cause, the Judge is empowered to prescribe a shorter time for filing petitions for discretionary review following the mailing of his decision.

(3) Petitions for review of a Judge's decision may be filed directly with the Executive Secretary subsequent to the filing of the Judge's report. Such petitions will be considered to the extent that time and resources permit. Parties filing such petitions should be aware that any action by a Commission Member directing review must be taken within thirty (30) days following the filing of the Judge's report.

(4) In the case of proposed settlements or other proposed dispositions by consent of all parties, petitions for discretionary review shall not be allowed, except for good cause shown.

(c) A petition should contain a concise statement of each portion of the decision and order to which exception is taken and may be accompanied by a brief of points and authorities relied upon. The inclusion of precise citations to the record or legal authorities, as the case may be, will facilitate prompt review of the petition.

(d) Failure to act on such petition within the review period shall be deemed a denial thereof.

(e) Statements in opposition to petitions for discretionary review may be filed at the times and places specified in this section for the filing of petitions for discretionary review. Any Statement shall contain a concise statement on each portion of the petition to which it is addressed.

(f) An original and three copies of any petition or statement shall be filed with the Commission.

Rule 92 Review by the Commission

(a) Review is a matter of sound discretion of a member of the Commission.

(b) In exercising discretion, a Commission member will consider assertions of the following:

(1) A finding of material fact is not supported by a preponderance of the evidence.

(2) The decision is contrary to law or to the duly promulgated rules or decisions of the Commission.

(3) A substantial question of law, abuse of discretion, or policy is involved.

(4) A prejudicial error of procedure was committed.

(c) When a petition for discretionary review is granted, review shall be limited to the issues specified in the petition, unless the order for review expressly provides differently.

(d) At any time within thirty days after the filing of a decision of a judge, a case may also be directed for review by a member upon his own motion upon any ground that could be raised by a party, but the issues would normally be limited to novel questions of law or policy or questions involving conflict in Administrative Law Judge's decisions. Any direction for review shall state the issues with particularity. Except in extraordinary circumstances, the Commission's power to review is limited to issues of law or fact raised by the parties in the proceedings below.

Rule 93 Briefs before the Commission

(a) Requests for briefs

The Commission ordinarily will request the parties to file briefs on issues before the Commission. When briefs are requested, a party may file a letter setting forth its arguments instead of filing a brief. The provisions of this rule shall apply to such letters.

(b) Time for filing briefs

When briefs are requested under paragraph (a), a briefing notice shall be issued to the parties at a time reasonably in advance of the date when the case is scheduled for disposition at a Commission meeting. Unless the briefing notice provides otherwise, the time for filing of briefs shall be as follows:

(1) *Appeal by one party* A party whose petition for review or for interlocutory appeal is granted or whose interlocutory appeal is certified shall file a brief within 40 days after the date of the briefing notice. All other parties shall file briefs within 30 days after the brief of the petitioning or appealing party is served.

(2) *Appeals by two or more parties* When petitions of two or more parties are directed for review, each such party shall file an initial brief addressing the issues on which it appeals within 40 days after the date of the briefing notice and may file a brief responding to the initial brief of the other party within 30 days after the initial brief of the other party or parties is served. This sequence of briefing shall be followed in the event that two or more parties' interlocutory appeals are certified.

(3) *Direction for review on the motion of a Commission member* When no petition for discretionary review is granted and a member directs review of a Judge's decision on his own motion, all briefs shall be filed within 40 days after the date of the briefing notice.

(4) *Additional briefs* Additional briefs shall not be allowed except by leave of the Commission.

(c) Motion for extension of time for filing brief

Any extension of time to file a brief shall not be granted except in extraordinary circumstances. A motion for extension of time to file a brief shall be filed within the time limit prescribed in paragraph (b) of this section and shall include the following information: When the brief is due; the number and duration of extensions of time that have been granted to each party; the length of extension being requested; the specific reasons for the extension being requested; and an assurance that the brief will be filed within the time extension requested.

(d) Consequences of late filing of brief

The Commission may decline to accept a brief that is not timely filed.

(e) Length of brief

Except by permission of the Commission, a brief shall contain no more than 35 pages of text.

(f) Table of contents

A brief in excess of 15 pages shall include a table of contents.

(g) Failure to meet requirements

The Commission may return briefs that do not meet the requirements of paragraphs (e) and (f) of this section.

(h) Number of copies

Five copies of a brief shall be filed. See §2200.7(a).

Rule 94 Stay of final order

(a) Any party aggrieved by a final order of the Commission may, while the matter is within the jurisdiction of the Commission, file a motion for a stay.

(b) Such motion shall set forth the reasons a stay is sought and the length of the stay requested.

(c) The Commission may order such stay for the period requested or for such longer or shorter period as it deems appropriate.

Rule 95 Oral argument before the Commission

(a) Oral argument before the Commission ordinarily will not be allowed.

(b) In the event the Commission desires to hear oral argument with respect to any matter it will advise all parties to the proceeding of the date, hour, place, time allotted, and scope of such argument at least 10 days prior to the date set.

Rule 100 Settlement

(a) Policy

Settlement is permitted at any stage of the proceedings. Settlements submitted to consideration after the Judge's decision has been directed for review shall be filed with the Executive Secretary. A settlement proposal shall be approved when it is consistent with the provisions and objectives of the Act.

(b) Requirements

Every settlement proposal submitted to the Judge or Commission shall include, where applicable, the following:

(1) A motion to amend or withdraw a citation, notification of proposed penalty, notice of contest, or petition for modification of abatement;

(2) A statement that payment of the penalty has been tendered or a statement of a promise to pay; and

(3) A statement that the cited condition has been abated or a statement of the date by which abatement will be accomplished.

(c) Filing; service and notice.
When a settlement proposal is filed with the Judge or Commission, it shall also be served upon represented and unrepresented affected employees in the manner prescribed for notices of contest in §2200.7. Proof of service shall accompany the settlement proposal. A settlement proposal shall not be approved until at least 10 days following service of the settlement proposal on affected employees.

Rule 100a Withdrawal of notice of contest
At any stage of the proceedings, a party may move to withdraw its notice of contest or any portion of its notice of contest. The motion shall include a statement that a promise of another party has not led to the motion to withdraw the notice of contest. The rule on settlements, §2200.100, shall apply whenever a promise of another party has led to the party's motion to withdraw.

Rule 101 Expedited proceeding
(a) Upon application of any party or intervenor or upon his own motion, any Commissioner may order an expedited proceeding. Contests arising under §§2200.34 and 2200.35 shall be placed on a special docket and treated as expedited proceedings before Administrative Law Judges. Cases arising under these secitons which are directed for review before the Commission shall also be placed on a special docket for review, and shall be treated as expedited proceedings under this section.
(b) When such proceeding is ordered, the Executive Secretary shall notify all parties and intervenors.
(c) The Judge assigned in an expedited proceeding shall make necessary rulings with respect to time for filing of pleadings and with respect to all other matters, without reference to times set forth in these rules, shall order daily transcripts of the hearing, and shall do all other things necessary to complete the proceeding in the minimum time consistent with fairness.

Rule 102 Standards of conduct
All persons appearing in any proceeding shall conform to the standards of ethical conduct required in the courts of the United States.

Rule 103 Ex parte communication
(a) There shall be no *ex parte* communication, with respect to the merits of any case not concluded, between the Commission, including any member, officer, employee, or agent of the Commission who is employed in the decisional process, and any of the parties or intervenors.
(b) In the event such *ex parte* communication occurs, the Commission or the Judge may make such orders or take such action as fairness requires. Upon notice and hearing, the Commission may take such disciplinary action as is appropriate in the circumstances against any person who knowingly and willfully makes or solicits the making of a prohibited *ex parte* communication.

Rule 104 Restrictions as to participation by investigative or prosecuting officers
In any proceeding noticed pursuant to the rules, the Secretary shall not participate or advise with respect to the report of the Judge or the Commission decision.

Rule 105 Inspection and reproduction of documents
(a) Subject to the provisions of law restricting public disclosure of information, any person may, at the offices of the Commission, inspect and copy any document filed in any proceeding.
(b) Costs shall be borne by such person.

Rule 106 Restrictions with respect to former employees

(a) No former employee of the Commission or the Secretary (including a member of the Commission or the Secretary) shall appear before the Commission as an attorney or other representative for any party in any proceeding or other matter, formal or informal, in which he participated personally and substantially during the period of his employment.

(b) No former employee of the Commission or the Secretary (including a member of the Commission or the Secretary) shall appear before the Commission as an attorney or other representative for any party in any proceeding or other matter, formal or informal, for which he was personally responsible during the period of his employment, unless 1 year has elapsed since the termination of such employment.

Rule 107 Amendments to rules

The Commission may at any time upon its own motion or initiative, or upon written suggestion of any interested person setting forth reasonable grounds therefor, amend or revoke any of the rules contained herein. Such suggestions should be addressed to the Commission at 1825 K Street, N.W., Washington, D.C. 20006.

Rule 108 Special circumstances; waiver of rules

In special circumstances not contemplated by the provisions of these rules, or for good cause shown, the Commission may, upon application by any party or intervenor, or on its own motion, after 3 days notice to all parties and intervenors, waive any rule or make such orders as justice or the administration of the Act requires.

Rule 109 Penalties

(a) All penalties assessed by the Commission are Civil.

(b) The Commission has no jurisdiction under §17(e), (f), and (g) of the Act and will conduct no proceeding thereunder.

Rule 110 Official Seal Occupational Safety and Health Review Commission

The seal of the Commission shall consist of: A gold eagle outspread, head facing dexter, a shield with 13 vertical stripes superimposed on its breast, holding an olive branch in its claws, the whole superimposed over a plain solid white Greek cross with a green background, encircled by a white band edged in black and inscribed "Occupational Safety and Health Review Commission" in black letters.

Subpart M—Simplified Proceedings

Note: This subpart shall be instituted on an experimental basis for a period of 1 year from its effective date. The final status of this subpart will be determined by the Commission at the conclusion of the experimental period.

Rule 200 Purpose

(a) The purpose of this subpart is to provide simplified procedures for resolving contests under the Occupational Safety and Health Act of 1970, so that the parties before the Commission may save time and expense while preserving fundamental procedural fairness. The rules shall be construed and applied to accomplish these ends.

(b) Procedures under this subpart are simplified in a number of ways. The major differences between these procedures and those provided in subparts A through G of the Commission's rules of procedure are the following: (1) Pleadings generally are not permitted or required. Early discussions among the parties will inform the parties of the legal and factual matters in dispute and narrow the issues to the extent possible; (2) Discovery is generally not permitted; (3) The Federal Rules of Evidence do not apply; (4) Interlocutory appeals are not permitted.

Rule 201 Application

The rules in this subpart shall govern proceedings before an Administrative Law Judge when (a) the case is eligible for simplified proceedings under §2200.202, (b) any party requests simplified proceedings, and (c) no party files an objection to the request.

Rule 202 Eligibility for simplified proceedings

A case is eligible for simplified proceedings unless it concerns an alleged violation of section 5(a)(1) of the Act (29 U.S.C. 654(a)(1)) or an alleged failure to comply with a standard listed in table A.

Table A

All standards listed are found in title 29 of the Code of Federal Regulations.

§1910.94 §1910.96
§1910.95 §1910.97

§§1910.1000 to 1910.1045, and any occupational health standard that may be added to subpart Z of part 1910.

§1926.52 §1926.55
§1926.53 §1926.57
§1926.54 §1926.800(c)

Rule 203 Commencing simplified proceedings

(a) Requesting simplified proceedings

(1) *Who may request* Any party may request simplified proceedings.

(2) *When to request* After the Commission receives an employer's or employees' notice of contest or petition for modification of abatement, the Executive Secretary shall issue a notice indicating that the case has been docketed. A request for simplified proceedings, if any, shall be filed within 10 days after the notice of docketing is received, unless the notice of docketing states otherwise.

(3) *How to request* A simple statement is all that is necessary. For example, "I request simplified proceedings" will suffice. The request shall be filed with the Executive Secretary and served in the manner prescribed for notices of contest in §2200.7.

(4) *Effect of the request* For those cases eligible under §2200.202, simplified proceedings are in effect when any party requests simplified proceedings and no party files a timely objection to the request.

(b) Objecting to simplified proceedings

(1) *Who may object* Any party may object to a request for simplified proceedings.

(2) *When to object* An objection shall be filed within 15 days after the request for simplified proceedings is served.

(3) *How to object* A simple statement is all that is necessary. For example, "I object to simplified proceedings" will suffice. An objection shall be filed with the Executive Secretary and served in the manner prescribed for notices of contest in §2200.7.

(4) *Effect of the objection* The filing of a timely objection shall preclude the institution of simplified proceedings.

(c) Notice

(1) When the period for objecting to simplified proceedings expires and no

objection has been filed, the Commission shall notify all parties that simplified proceedings are in effect.

(2) When a party files a timely objection to a request for simplified proceedings, the Commission shall notify all parties that the case shall continue under conventional procedures (Subparts A through G).

Rule 204 Filing of pleadings

(a) Complaint and answer

There shall be no complaint or answer in simplified proceedings. If the Secretary has filed a complaint under §2200.33, a response to an employee contest under §2200.35, or a response to a petition under §2200.34, the complaint or response shall not be included in the record. No response to these documents shall be required.

(b) Motions

A primary purpose of simplified proceedings is to eliminate, as much as possible, motions and similar documents. A motion will not be viewed favorably if the subject of the motion has not been first discussed among the parties prior to the conference/hearing.

Rule 205 Discussion among parties

Within a reasonable time before the conference/hearing, the parties shall meet, or confer by telephone and discuss the following: Settlement of the case; the narrowing of issues; and agreed statement of issues and facts; defenses; witnesses and exhibits; motions; and any other pertinent matter.

Rule 206 Conference/Hearing

(a) The Judge shall schedule and preside over a conference/hearing, which shall be divided into two segments: a conference and a hearing.

(b) Conference

At the beginning of the conference, the Judge shall enter into the record all agreements reached by the parties as well as defenses raised during the discussion set forth in §2200.205. The parties and the Judge then shall attempt to resolve or narrow the remaining issues. At the conclusion of the conference, the Judge shall enter into the record any further agreements reached by the parties.

(c) Hearings

The Judge shall hold a hearing on any issue that remains in dispute at the conclusion of the conference. The hearing shall be in accordance with 5 U.S.C. 554.

(1) *Evidence* Oral or documentary evidence shall be received, but the Judge may exclude irrelevant or unduly repetitious evidence. Testimony shall be given under oath. The Federal Rules of Evidence shall not apply.

(2) *Oral and written argument* Each party may present oral argument at the close of the hearing. Parties wishing to present written argument shall notify the Judge at the conference/hearing so that the Judge may set a reasonable period for the prompt filing of written argument.

Rule 207 Reporter present; transcripts

A reporter shall be present at the conference/hearing. An official verbatim transcript of the hearing shall be prepared and filed with the Judge. Parties may purchase copies of the transcript from the reporter.

Rule 208 Decision of the Judge

(a) The Judge shall issue a written decision in accordance with §2200.90.

(b) After the issuance of the Judge's decision, the case shall proceed in the conventional manner (Subparts A through G).

Rule 209 Discovery

Discovery, including requests for admissions, shall not be allowed except by order of the Judge.

Rule 210 Interlocutory appeals not permitted

Appeals to the Commission of a ruling made by a Judge which is not the Judge's final disposition of the case are not permitted.

Rule 211 Applicability of Subparts A through G

Sections 2200.6, 2200.33, 2200.34(d)(4), 2200.35, 2200.36, 2200.38 and 2200.75 shall not apply to simplified proceedings. All other rules contained in subparts A through G of the Commission's rules of procedure shall apply when consistent with the rules in this subpart governing simplified proceedings.

Table of Cases

Table of Statutes

* cited for reference purposes.

Bibliography

Allen, Robert, *Industrial Hygiene* (Prentice Hall, 1976).

Anderson, C. Richard, *OSHA and Accident Control Through Training* (Industrial Press, 1975).

Anton, Thomas, J., *Occupational Safety and Health Management* (McGraw Hill, 1979).

Ashford, Nicholas, A., *Crisis in the Workplace* (MIT Press, 1976).

Bacow, Lawrence S., *Bargaining for Job Safety and Health* (MIT Press, 1980).

Berman, Daniel M., *Death on the Job: Occupational Health and Safety Struggles in the United States* (Monthly Review Press, 1978).

Binford, Charles M., *Loss Control in the OSHA Area* (McGraw-Hill, 1975).

Bishoff, Fred M., *Occupational Safety and Health Act* (1979).

Blake, Roland P., *Industrial Safety* (Hole, 1973).

Boley, Jack, *A Guide to Effective Industrial Safety* (Gulf Publications, 1977).

Brandt, Allen D., *Industrial Health Engineering* (John Wiley, 1947).

Chadd, Charles M., *Practice Under the Occupational Safety and Health Act* (BNA, 1978).

Chamber of Commerce of the United States of America, *What To Do About OSHA* (1978).

Chelius, James R., *Workplace Safety and Health* (1977).

Daubenspeck, G. Walker, *Occupational Health Hazards* (Exposition Press, 1974).

Davis, Keith B., *Health and Safety* (Van Nostrand Reinholdt, 1979).

DeReamer, Russell, *Modern Safety and Health Technology* (Wiely-Interscience, 1980).

Ffrench, Geoffrey E., *Occupational Health* (Lancaster Publication, 1973).

Heinrich, Herbert W., *Industrial Accident Prevention* (McGraw-Hill, 1980).

Hogan, Roscoe B., *Occupational Safety and Health Act* (Bender, 1977).

Interagency Taskforce on Workplace Safety and Health, *Making Prevention Pay, Final Report* (1978).

King, Ralph W., *Industrial Hazard and Safety Handbook* (Butterworths, 1979).

Kochan, Thomas A., *The Effectiveness of Union Management Safety and Health Committees* (Upjohn Institute, 1977).

Lowrance, William W., *Of Acceptable Risk: Science and the Determination of Safety* (W. Kaufmann, 1976).

Mayers, May R., *Occupational Health: Hazards of the Work Environment* (Williams and Wilkens & Co., 1969).

McElroy, Frank E., *Handbook of Occupational Safety and Health* (National Safety Council, 1975).

Mendeloff, John, *Regulating Safety: An Economic and Political Analysis of Occupational Safety and Health Policy* (MIT Press, 1979).

National Institute for Occupational Safety and Health, *Certified Equipment List, U.S. Department of Health and Human Services, Public Health Service Center for Disease Control* (1980).

National Institute for Occupational Safety and Health, *Occupational Diseases: A Guide to Their Recognition* (1977).

National Institute for Occupational Safety and Health, *Occupational Safety and Health Directory* (1980).

National Safety Council, *Handbook of Occupational Safety and Health* (1975).

NIOSH/OSHA Pocket Guide to Chemical Hazards, U.S. Department of Health, Education and Welfare, Public Health Service (1978).

Patty, Frank, *Industrial Hygiene and Toxicology* (Wiley, 1978).

Peterson, Dan, *The OSHA Compliance Manual* (McGraw-Hill, 1979).

Peterson, Jack, *Industrial Health* (Prentice Hall, 1977).

Proctor, N. and Hughes, J., *Chemical Hazards of the Workplace* (Lippincott, 1978).

Roberts, Joseph M., *OSHA Compliance Manual* (Reston Publishing Co., 1976).

Smith, Robert S., *The Occupational Safety and Health Act, Its Goals and Its Achievements* (American Enterprise Institute for Public Policy Research, 1976).

The President's Annual Report on Occupational Safety and Health, Including Reports by the United States Department of Labor, Occupational Safety and Health Administration, Occupational Safety and Health Review Commission, and the United States Department of Health Education and Welfare, National Institute for Occupational Safety and Health (1973–1980).

Trasko, Victoria M., *Occupational Health and Safety Legislation: A Compilation of State Laws and Regulations, United States Department of Health, Education and Welfare, Public Health Service, Bureau of State Services* (1964)

United States General Accounting Office, *How Effective Are OSHA's Complaint Procedures: Report to the Congress by the Comptroller General of the United States* (1979).

United States House of Representatives, Committee on Education and Labor, Select Subcommittee on Labor—Oversight Hearings on the Occupational Safety and Health Act (1973–1980)

United States Senate, Committee on Human Resources, Subcommittee on Labor—Oversight Hearings on the Administration of the Occupational Safety and Health Act (1973–1980)

United States, Senate Committee on Labor and Public Welfare, Subcommittee on Labor—Legislative History of the Occupational Safety and Health Act of 1970, P.L. 91–596 (1971)

Van Staaveren, Elizabeth K., Department of Labor, Office of the Assistant Secretary for Administration and Management, *Occupational Safety and Health: A Bibliography* (1978)

Index